Engineering
Network
Analysis

Engineering Network Analysis

GENE H. HOSTETTER
University of California, Irvine

1817

HARPER & ROW, PUBLISHERS, New York
Cambridge, Philadelphia, San Francisco,
London, Mexico City, São Paulo, Sydney

Sponsoring Editor: **Carl McNair**
Project Editor: **Robert Greiner**
Designer: **Michel Craig**
Production Manager: **Marion Palen**
Compositor: **Waldman Graphics, Inc.**
Printer and Binder: **Vail-Ballou Press, Inc.**
Art Studio: **Vantage Art, Inc.**

ENGINEERING NETWORK ANALYSIS

Copyright © 1984 by Gene H. Hostetter

Library of Congress Cataloging in Publication Data

Hostetter, G. H., 1939–
  Engineering network analysis.

  1. Electric network analysis. I. Title.
TK454.2.H668 1984   621.319′2   83-22765
ISBN 0-06-042907-0

To Donna, Kelly, and Kristen

Contents

Preface

There are few college or university courses where students are expected to assimilate so much so quickly as in the first engineering network analysis sequence. Extraordinary degrees of organization and perspective are required of both text and instructor if such a course is universally to encourage logical thinking and imagination. In this book, I have tried to capture the interest and excitement that can be felt by inquiring minds as they begin to explore a fascinating new subject. At the same time, I have tried to give a balance and depth that will leave the student not only excited, but equipped with solid analytical tools, understanding, and capability. The emphasis is definitely on topics that are highly relevant to the present and likely to be so far into the future.

I believe strongly in the need these days for textbooks to act as tutor for the student, to elaborate on and extend substantially beyond the material that can be emphasized in a lecture class. It is of the utmost importance that the horizons of a text be beyond those of the student and that it actively cultivate the thinking, visualizing, and reasoning that are crucial to modern engineering practice.

EDUCATIONAL OBJECTIVES

The text is intended for a two-term or year-long engineering course sequence at the sophomore or early junior level. It is designed to develop understanding and capability in these areas:

1. Source-resistor networks (Chapters 1–3).
2. Switched networks, especially those of first and second order (Chapters 4–6).
3. Sinusoidally driven networks, including topics relating to resonance and power (Chapters 7–9).
4. Series and transform methods and their application to networks (Chapters 10–12).
5. Introductory active network design (Chapter 13).

It is designed to give students:

1. Comfort with every aspect of source-resistor network analysis, including equivalent circuits and solutions of networks involving controlled sources.
2. The ability to design working voltage amplifier circuits using operational amplifier chips.
3. A solid background in the solution of switched first-order networks and extended acquaintance and facility with second- and higher-order switched networks.
4. A high degree of capability and confidence with steady state sinusoidal network response.
5. A sound foundation in series and transform methods and their application to network response calculation.
6. Experience with active network design and an appreciation for the design process.

Network analysis provides a nearly ideal vehicle for teaching series and transform methods. These subjects receive the same careful treatment as earlier topics because of their long-term importance and the undesirability of giving a highly superficial treatment now with the intention of making repairs later on.

Active networks give an invaluable, rich source of creative design applications.

PREREQUISITES

The book is as self-contained as possible, not assuming background or collateral study in Physics or Mathematics beyond that normal for a sophomore in a scientific/technical program. It is written to be understandable as the student progresses through it, without parts that are puzzle pieces to be assembled later. No attempt is made to interrelate network phenomena and Maxwell's equations.

Considerable mathematical review is included and placed in the text where the needs for these tools develop:

Cramer's rule (Section 2.2);
Solution of linear, time-invariant differential equations (Sections 4.7–4.9, 6.2, 7.3);
Exponential functions (Section 4.10);
Sinusoidal functions (Section 7.2); and
Complex algebra (Section 7.4).

Students who initially lack adequate preparation in the above will find discussion, examples, and drill material to guide them. For those who are well prepared, these sections emphasize the immediate concerns and provide a review and summary of the necessary background.

Review appendices seem to be much less effective than the integration of this material into the text.

ORGANIZATION AND PACE

The thirteen chapters are approximately equal in level of student difficulty and time requirement. Of course, at the start of their study most students need a great deal of reassurance and attention to detail. The development is thus gentle and deliberate in the early chapters, with explanations becoming more concise and thought-provoking as the reader's sophistication increases.

Color is an important element. Not only is it pleasing to the eye and helpful in holding visual interest, but the emphasis and discrimination it conveys appears to speed comprehension greatly.

The above-average number of pages here is attributable to the extraordinary number of exercises that are included. With this resource, it is feasible to rotate thoroughly the problem assignments and to draw quiz and examination problems routinely from the text.

Drill problems, with numerical answers given, are placed with each appropriate text section. Drill-problem numbering is prefaced with the letter "D" to help in distinguishing the drill problems from those at the end of each chapter. Answers are not printed for the end-of-chapter problems; a comprehensive solutions manual for these is available to instructors from the publisher.

ACKNOWLEDGMENTS

A great deal of student, teaching assistant, and instructor feedback has contributed in major and very positive ways to this book. I am greatly indebted to my colleagues, especially G. H. Cain, L. Ferguson, I. Foroutan, M. Hassul, T. Jordanides, H. John Lane, M. Santina, C. J. Savant, R. T. Stefani, A. R. Stubberud, and L. S. Yap, who class-tested the manuscript and helped greatly in incorporating improvements.

A distinguished group of reviewers from a representative cross-section of a dozen engineering schools across the United States contributed substantially to the final result and were important in achieving the best balance of topics. These are W. H. Boghosian, University of Pennsylvania; Arthur Broderson, Vanderbilt University; John D. Cowan, Ohio State University; E. D. Denman, University of Houston; L. P. Huelsman, University of Arizona; Ralph L. Knapp, The Cooper Union; William Middendorf, University of Cincinnati; Robert H. Miller, Virginia Polytechnic Institute; Arthur Siedman, Pratt Institute; Timothy Trick, University of Illinois; Jack Waintraub, Middlesex Community College; and C. H. Weaver, University of Tennessee.

Special thanks are due to Ms. Cindy Klepadlo, Mrs. Dorothy Shearer, and Mrs. Sarah Wilson, who supervised the typing of the manuscript.

GENE H. HOSTETTER

Introduction
to the Student

> Then said a teacher, Speak to us of Teaching.
> And he said:
> No man can reveal to you aught but that which
> already lies half asleep in the dawning of your knowledge.
>
> Kahil Gibran
> *The Prophet**

This book was conceived and written with a burning desire to provide a solid foundation of electrical engineering material, proceeding in a logical, organized, and interesting manner. It is a book of *concepts* and *analysis*, a way of thinking.

TEACHING AND LEARNING

There is a common idea to the effect that teaching consists of a transfer of knowledge from one mind to another by a hazy process resembling osmosis or the spread of germs. To the contrary, learning is a highly individual experience. A good teacher and a good textbook may allow you to proceed more rapidly than would otherwise be practical, by organizing and pacing the material. But *you* are the one who does the learning; *you* are the one who achieves the understanding.

One of the goals of a professional education is for you eventually to reach the point where you can make good progress in a subject on your own, without classes and teachers. Many of you will one day explore brand new subjects before there are teachers or textbooks to teach them.

EXPERIENCE AND CONFIDENCE

The understanding you are seeking does not involve much that can be mentally recorded and then played back for the next examination. What you will need beyond

*Quoted with the kind permission of the publisher, Alfred Knopf, New York (1961).

reasoning ability is experience and confidence, and this is most easily obtained by a steady effort, which includes solving a lot of problems.

Work toward *knowing*, by reason, that you have the correct answers to problems. The reinforcement of checking answers with the book can be very helpful, but do not waste your time misusing them by approaching problems in a trial and error fashion.

EFFECTIVE STUDYING

Too many students spend countless hours of agony imagining that they are studying when they are largely just punishing themselves. Here are some principles that may increase the effectiveness of your studying:

1. Study the subject regularly.
2. Give the subject your *full* attention and make every effort to eliminate interruptions.
3. Reinforce your learning by writing, in your own words, summaries of the material. Include brief examples of the application of new concepts.
4. Test your own understanding frequently by working problems. When you encounter difficulty, do additional problems of the same type.
5. Measure your level of understanding by how well you could explain the subject to a colleague for the first time, answering any questions that might arise. Explaining a subject to yourself builds your capability for conscious reasoning.

NONMAJORS

If your primary interest is in a different field, relax. We begin from the beginning. Here is a good opportunity to exercise your mind and to test your analytical skills. It is also a good chance to extend your horizons a bit: major innovation and invention generally come from minds that range far, minds that are able to apply knowledge in one area to problems in another area.

The previous remarks about studying are especially important to you if you wish to maximize the effectiveness of the time you invest.

ABOUT THE SUBJECT

Network analysis is a key subject upon which many, perhaps most, fields of electrical engineering build. It is the quantitative study of the behavior of voltage and current when electrical devices are interconnected. It is the electrical counterpart to the study of statics and dynamics of masses.

We begin now with basic ideas about voltage and current.

Engineering
Network
Analysis

Chapter One
Fundamental Concepts

It is this state of electricity in a series of electromotive and
conductive substances which I will briefly call electric
current; and as I will continuously be compelled to talk
about two opposed directions according to those in which
the two electricities move, I propose that each time the
subject comes up and in order to avoid repetition, I will
describe the direction of the electric current by referring to
that of the positive electricity.

Andre-Marie Ampère
From *Mémoires sur l'Electrodynamique*
French Academy of Sciences, Paris, 1820

1.1 PREVIEW

This first chapter contains a good deal of terminology, the two fundamental laws
governing the behavior of electrical networks, voltage and current relationships for
the resistor, and an introduction to electrical network problem solution. Each of the
concepts developed here is very important to the understanding and capability required for more advanced material in subsequent chapters.

All of the basic relationships for electrical networks could easily be printed on
half a page, but there is much more to know than this. To communicate with fellow
engineers and scientists, it is necessary to be familiar with many technical terms.
Systematic and efficient solution methods are necessary, methods that insure timely
answers to broad classes of problems. One needs to learn, without spending a life-
time rediscovering them, a number of important basic network properties. And a
great deal of experience in solving problems is helpful in gaining the insight that is
important for innovation and invention.

These first two chapters are concerned exclusively with networks consisting of
sources and resistors. In Chapter 1, the concepts of current and voltage are summarized and Kirchhoff's laws are given. Fixed (or independent) and controlled (or
dependent) voltage and current sources are introduced, and element power relations
are shown. The resistor model is discussed, and resistor power relations are derived.

Particular attention is given from the start to the routine incorporation of correct algebraic signs into the application of voltage-current relations and Kirchhoff's laws. Students who attempt to gloss over the matter of reference senses at this early stage will later typically have great difficulty writing network equations with correct algebraic signs. Most important, though, carelessness in signs detracts from a solid, assured, reasoning approach to problem solution.

General solution methods are developed for parallel and series source-resistor networks. The current divider and voltage divider rules are then derived and applied to a variety of situations, including various reference sense combinations.

The remainder of the chapter concerns equivalent circuits. A step-by-step approach is emphasized, keeping track of which signals are the same in the equivalent network at each stage. Equivalent circuits are of key importance to the study of networks at this beginning level for two reasons. First, they are preferred and are the usual solution method in much of electrical engineering practice. This is so because the method *applies* rather than *derives* results and because it retains a close and valuable identification with the original network. Second, equivalent circuits quickly supply a great deal of insight into network behavior. They foster a rapid visualization and understanding that would be difficult to obtain in any other way. It is this "feel" that is the beginning of an aptitude for design and innovation.

When you complete this chapter, you should know—

1. what electric current and voltage are;
2. Kirchhoff's current and voltage laws and how to apply them;
3. what a network diagram is and the meanings of the terms *element, conductor, node,* and *loop;*
4. how electrical power flow is related to an element's voltage and current;
5. what voltage sources, current sources, and resistors are;
6. how electrical power flow in a resistor is related to the resistor current and to the resistor voltage;
7. how to solve networks, quickly and easily, in situations where all elements are in parallel or where all elements are in series;
8. the current divider rule and the voltage divider rule and how to use them;
9. what it means to say that two elements are equivalent to one another;
10. how to construct equivalent elements for current sources in parallel, voltage sources in series, and other simple series and parallel equivalences;
11. how to find equivalent resistances for resistors in series and for resistors in parallel;
12. how to perform a Thévenin-Norton transformation;
13. how sources may always be substituted for known voltages and currents in a network;
14. how to use equivalent circuits to solve involved network problems.

The International System of units (Systéme International d'Unités) or "SI units" is used exclusively in network analysis. The appropriate, consistent unit will be stated when a quantity is introduced, but units will not be continually restated when it is cumbersome to do so. Table 1-1 is a summary of common SI units and their symbols. The most useful SI multipliers are given in Table 1-2.

Table 1-1   SELECTED SI UNITS AND ABBREVIATIONS

QUANTITY	EQUIVALENT UNITS	UNIT	ABBREVIATION
Length		meter	m
Mass		kilogram	kg
Time		second	s
Charge		coulomb	C
Force		newton	N
Energy	newton-meter	joule	J
Power	newton-meter/second	watt	W
Current	coulomb/second	ampere	A
Voltage	newton-meter/coulomb	volt	V
Magnetic Flux	volt-second	weber	Wb
Resistance	volt/ampere	ohm	Ω
Conductance	1/ohm	siemens	S
Capacitance	coulomb/volt	farad	F
Inductance	weber/ampere	henry	H
Frequency	1/second	hertz	Hz
Plane angle		radian	rad

1.2 ELECTRIC CURRENT

1.2.1 Current Flow in Conductors

An electric current is a flow of electric charge. In networks one is interested in charge flow in wires or *conductors*. Often, but not always, these conductors are made of a metal such as copper.

The outer electrons of each molecule of a metallic conductor are, in a sense, pooled and shared by neighboring molecules, giving rise to a "sea" of relatively free electrons throughout the material. The electron sea is uniformly distributed within the conductor; the attraction of the ionized molecules for nearby electrons and the repulsion between neighboring electrons tend to equalize the density of electrons in the sea.

Although the free electrons may mill about within the conductor, very few of them ever have sufficient energy to break free of the conductor surface unless the

Table 1-2   COMMON SI MULTIPLIERS

MULTIPLIER	PREFIX	ABBREVIATION
10^{12}	tera	T
10^{9}	giga	G
10^{6}	mega	M
10^{3}	kilo	k
10^{-1}	deci	d
10^{-2}	centi	c
10^{-3}	milli	m
10^{-6}	micro	μ
10^{-9}	nano	n
10^{-12}	pico	p

conductor temperature is very high or there is an extremely large electric force perpendicular to the conductor surface. The electron sea is thus confined to a conductor, much like water in a pipe. Conductors are also commonly surrounded by an *insulating* material, one in which charge motion is all but completely arrested, to keep conductors (and their free electrons) from coming into unwanted contact with one another.

The sea of electrons in a metallic conductor may be made to flow by the application of externally produced electric forces. In a network, where the propagation time of any disturbance is negligibly small, the forces between the particles within the material act to cause equal currents across each cross-section of the conductor. It makes sense then to speak of *the* current through a conductor in a network; the current is the same everywhere along a conductor.

Similar considerations apply to other conductors as to the metallic ones. An ionized solution such as salt water, for example, is a conductor in which the charge carriers are both Na^+ and Cl^- ions. The two types of ions may be made to flow in opposite directions, resulting in a net current.

1.2.2 Definition and Specification of Electric Current

Electric current is the equivalent rate of flow of positive charge in a given reference direction.

Network problems are only concerned with *equivalent* total charge flows, expressed in terms of either positive or negative charge. By convention, even though the mobile charges in a metallic conductor are the negative electrons, currents are described in terms of equivalent positive charge flow. The charge carriers in other devices (such as semiconductors) may be positive. And in others, both negative and positive charge motions contribute to the current.

The SI unit for current

$$\frac{\text{coulomb}}{\text{second}} = \text{ampere (abbreviation A)}$$

is named for Andre-Marie Ampère (1775–1836), who performed early experiments with magnetism and the magnetic effects of electric current. The usual symbol for a current is the function of time $i(t)$. If several currents are to be indicated, the symbol i is subscripted to distinguish different currents.

Because there are two possible directions of equivalent positive charge flow in a conductor, a careful definition of current must involve a directional sense, a reference direction for the current. The reference direction for a current may be indicated by an arrow beside the symbol for a conductor (a line) on a network diagram, as in Figure 1-1(a).

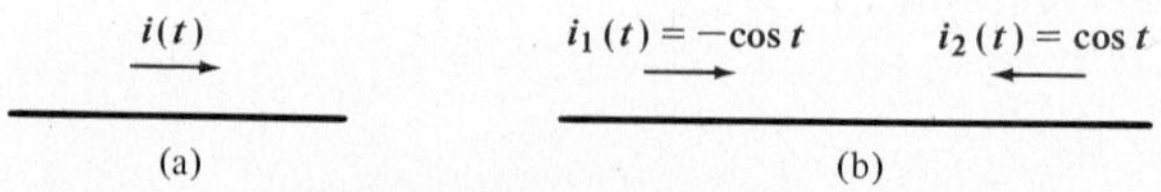

$i(t)$ →

(a)

$i_1(t) = -\cos t$ → $i_2(t) = \cos t$ ←

(b)

Figure 1-1   Current reference direction.
            (a) Indicating the reference direction of a current.
            (b) Reversing the sense of a current reference direction.

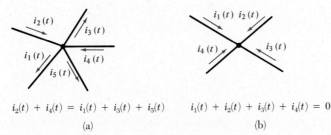

$$i_2(t) + i_4(t) = i_1(t) + i_3(t) + i_5(t) \qquad i_1(t) + i_2(t) + i_3(t) + i_4(t) = 0$$

(a) (b)

Figure 1-2   Examples of currents at junctions of conductors.

For constant currents it may be advantageous to choose reference directions, when it is convenient to do so, in the actual directions of equivalent positive charge flow. In more involved problems the actual directions of equivalent positive charge flow may not be obvious by inspection. Either direction may then be chosen (or assumed) as the reference direction. If the current with that reference direction turns out to be negative, then the actual direction of equivalent positive charge flow is counter to the reference direction.

For currents that are time varying and change actual direction from time to time, it is impractical to change the senses of the reference arrows whenever the current changes direction. One reference direction is selected to define each current of interest; the current with that reference direction may be negative some times and positive others, a negative current meaning that the equivalent rate of positive charge flow is counter to the direction of the arrow.

Reversing the reference direction reverses the algebraic sign of the current function, as in the example in Figure 1-1(b).

1.2.3 Kirchhoff's Current Law

In networks charge does not accumulate within a conductor. Thus the net rate of flow of charge into any conducting region is zero: The sum of the currents with reference directions entering a junction of conductors is equal to the sum of the currents with reference directions leaving the junction. This relationship is known as *Kirchhoff's current law*. [Gustav Kirchhoff (1824–1887) was Professor of Physics, Heidelberg.] The branching of a conductor may be represented by joined lines on a network diagram. Two examples of the application of Kirchhoff's current law to branching conductors are shown in Figure 1-2.

Given all but one of the junction currents, the remaining current is determined, as in Figure 1-3, where

$$i(t) = 6 \sin t - 10 + 5e^{-4t} - 3 \cos 2t \text{ A}$$

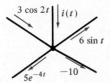

Figure 1-3   Determining one junction current from the others.

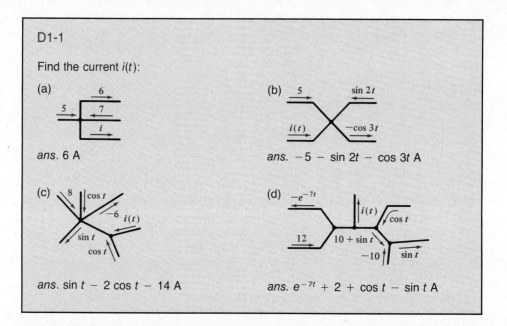

D1-1

Find the current $i(t)$:

(a)

ans. 6 A

(b)

ans. $-5 - \sin 2t - \cos 3t$ A

(c)

ans. $\sin t - 2 \cos t - 14$ A

(d)

ans. $e^{-7t} + 2 + \cos t - \sin t$ A

ELECTRIC CURRENT

Electric current is the equivalent rate of flow of positive charge in a given reference direction. The reference direction may be indicated by an arrow on a network diagram.

    The sum of the currents with reference directions entering a junction is equal to the sum of the currents with reference directions leaving the junction.

1.3 VOLTAGE

1.3.1 Electric Potential in Networks

The electric potential, ϕ, at a point in a network is the potential energy per unit charge of a charge located at that point. The electric potential is generally a function of position in the network. There are conditions in nature for which an electric potential function does not exist. This is to say that it is possible that the energy of a charge depends not only where it is but how it got there. But such is not the case in an electric network.

    Electric potential in a network is analogous to gravitational potential. As in the gravitational case, the zero level of electric potential is arbitrary; only differences in potential energy are of physical significance. Voltage is a difference in electric potential. The voltage between two points is the energy per unit charge necessary to move the charge through the network from one point to the other.

    The unit of voltage, the *volt*,

$$\frac{\text{newton-meter}}{\text{coulomb}} = \text{volt (abbreviation V)}$$

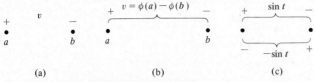

Figure 1-4 Voltage reference polarity.
(a) Plus and minus signs showing the sense of the potential difference.
(b) Addition of a bracket for clarity.
(c) Reversing the sense of a voltage reference polarity.

is named in honor of Alessandro Volta (1748–1827), who invented the battery and whose early electrical experiments stimulated the interest of many other great minds of his time.

The usual symbol for a voltage is the function of time $v(t)$. If several voltages are to be indicated, the symbol v is usually subscripted to distinguish the different voltages.

1.3.2 Specifying Voltages

Voltage is a difference in electrical potential between two points in a network. The points involved and the sense of the difference may be indicated on a diagram of the network with a plus and a minus sign such as that shown in Figure 1-4(a). The plus and minus signs indicate which potential is added and which is subtracted to form the voltage. This indication of the sense of the difference in potential is called the *voltage reference polarity*. The inclusion of a bracket, Figure 1-4(b), enhances the clarity of a drawing, especially when several voltages are shown on the same drawing. In this text, the bracket will be used in most cases.

It is occasionally advantageous to select one of the two possible reference polarities for a voltage in preference to the other. For a known constant voltage it is sensible to choose the reference so that the defined voltage is positive. If the actual polarity of a voltage is not known, either polarity may be chosen to define it. If that voltage turns out to be negative, so be it; the potential at the point by the plus sign is smaller than the potential at the other point.

The most interesting voltages are time varying and change in polarity from time to time. For these, one reference polarity is chosen and the voltage so defined is sometimes positive, sometimes negative. Reversing the reference polarity reverses the algebraic sign of the voltage function, as in Figure 1-4(c).

Plus signs are used for other purposes in connection with physical devices—for example, to identify the terminals of batteries, transformers, and meters. The plus sign on a voltage *reference polarity*, however, only indicates which potential is subtracted from which. It does *not* necessarily mean that the point nearest the plus sign is at a higher potential than the other point.

1.3.3 Kirchhoff's Voltage Law

The sum of the voltages around a loop, if the voltage reference polarities are chosen symmetrically, is zero. For example, for the four network voltages in Figure 1-5,

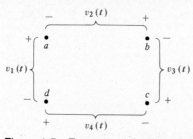

Figure 1-5   Example of voltages around a loop, with symmetric reference polarities.

$$v_1 + v_2 + v_3 + v_4 = \phi(a) - \phi(d) + \phi(b) - \phi(a) + \phi(c)$$

$$- \phi(b) + \phi(d) - \phi(c) = 0$$

If the voltage reference polarities do not happen to be symmetric, plus to minus, plus to minus, then the sum of the voltages with one reference polarity sense equals the sum of the voltages with the other reference polarity sense. This result is known as *Kirchhoff's voltage law*. Two examples of the application of Kirchhoff's voltage law are shown in Figure 1-6. The dots on the sketches represent points on conductors in an electrical network.

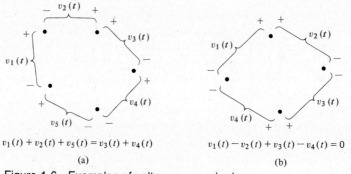

$$v_1(t) + v_2(t) + v_5(t) = v_3(t) + v_4(t) \qquad\qquad v_1(t) - v_2(t) + v_3(t) - v_4(t) = 0$$

(a)   (b)

Figure 1-6   Examples of voltages around a loop.

Given all but one voltage in a loop, the remaining voltage may be found, as in Figure 1-7, for which

$$v(t) = 3 \cos t + e^{-4t} + 5 - 8 \sin 2t \text{ V}$$

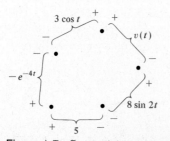

Figure 1-7   Determining one voltage from the others in a loop.

D1-2

Find the voltage $v(t)$:

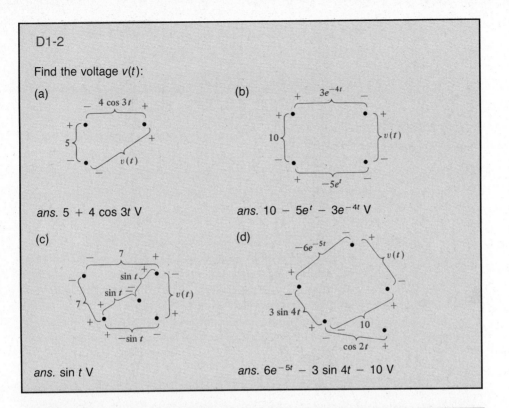

(a)

ans. 5 + 4 cos 3*t* V

(b)

ans. 10 − 5e^t − 3e^{-4t} V

(c)

ans. sin *t* V

(d)

ans. 6e^{-5t} − 3 sin 4*t* − 10 V

VOLTAGE

Voltage is a difference in electric potential between two points in a network. The sense of the potential difference, the reference polarity of a voltage, may be indicated by plus and minus signs on a network diagram.

Around a closed loop in an electrical network, the sum of voltages with one reference polarity sense equals the sum of the voltages around the loop with the other reference polarity sense.

1.4 NETWORK DIAGRAMS

1.4.1 Diagrams and Two-Terminal Elements

An electrical device terminating in two conductors within which net charge does not accumulate is called a *two-terminal element*. The current entering one terminal is the same as the current leaving the other terminal. There is then only one current (with either reference direction) and only one voltage (with either reference polarity) to speak of in regard to a two-terminal element. The general symbol for an element is shown in Figure 1-8, together with a defined current through the element $i(t)$ and voltage across it $v(t)$.

An element might represent a battery, a light bulb, an electric heater, and so on. Or the element might be more complicated, representing a combination of

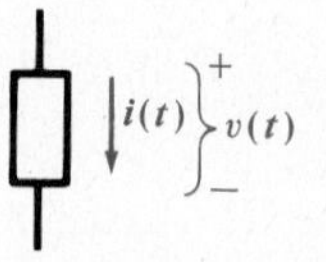

Figure 1-8 Element with defined current and voltage.

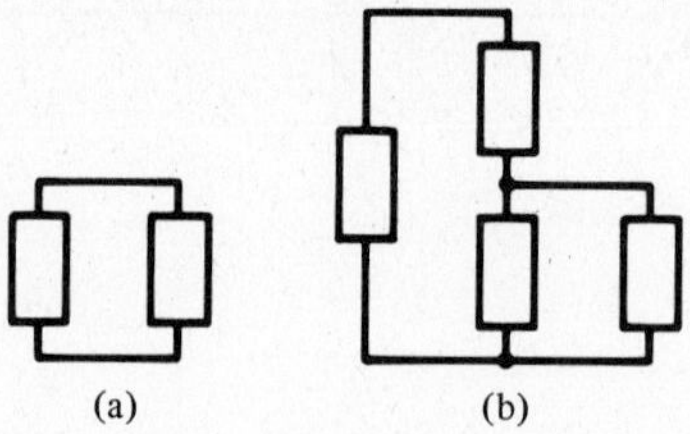

Figure 1-9 Element connections.

simpler devices such as all those connected to an automobile's battery. The important characteristic is that the device has two terminals: All the components could be placed within a box with just two conductors coming out of that box. Some physical devices, transistors, for example, are inherently three-or-more-terminal devices. Fortunately they may be represented by an equivalent set of two-terminal elements.

An electrical network is a connection of elements. The structure of the network may be indicated by a drawing, a network diagram, which indicates the interconnection of elements. Examples are given in Figure 1-9. Specific types of elements are given more specific symbols than the "box," which is used to indicate an element in general.

1.4.2 Ideal Conductors, Nodes, and Loops

An ideal *conductor* is indicated by a line on a network diagram, as in Figure 1-10(a). It is used to show the interconnection of elements. The same electric current flows at any position along a length of conductor in a network. Although a voltage, often small, may exist across a length of wire, the ideal conductor idealizes the wire to the point where there is zero voltage across its length (Figure 1-10(b)). To model a wire for which the voltage between the ends of the wire is not negligible, a more complicated element would be used.

A *node* is a complete set of conductor connections between elements, such that each node is separated from all other nodes by elements, and every two-terminal element connects between two nodes. In the example given in Figure 1-11(a), con-

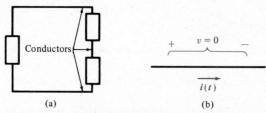

(a) (b)

Figure 1-10 Ideal conductors.
(a) Conductors joining elements in a network diagram.
(b) There is zero voltage across any length of conductor in a network diagram.

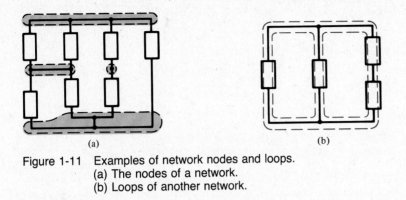

Figure 1-11   Examples of network nodes and loops.
(a) The nodes of a network.
(b) Loops of another network.

nections between conductors are indicated by dots, as is usual. The nodes are indicated by dashed lines.

A *loop* is a closed path through elements and conductors that does not pass through any node more than once. Figure 1-11(b) is an example in which the network loops are indicated by dashed lines. A network that contains one or more loops is called a *circuit*, so the terms *network* and *circuit* are usually used interchangeably.

1.4.3 Ideal Voltage and Current Sources

An element that maintains a specified voltage between its terminals, no matter what the current through the element, is called an ideal *voltage source*. The symbol for an ideal voltage source is a circle, a terminal-to-terminal voltage function for the element, and plus and minus signs indicating the polarity reference of the voltage. Figure 1-12(a) shows examples of ideal voltage sources with specific voltage functions.

If the voltage source function is stated in terms of some other network voltage or current, rather than explicitly as a function of time, the source is termed a *controlled* (or *dependent*) voltage source. The diamond-shaped symbol of Figure 1-12(b) is used then instead of the circle.

An element that maintains a specified current through its terminals, no matter what the voltage across the element, is called an ideal *current source*. The symbol for an ideal current source is a circle with an arrow indicating the source current reference direction. The current source function is indicated beside the symbol.

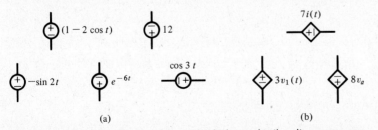

Figure 1-12   (a) Examples of fixed (or independent) voltage sources.
(b) Examples of controlled (or dependent) voltage sources.

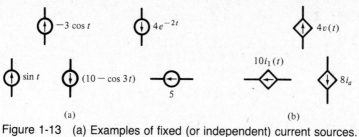

Figure 1-13 (a) Examples of fixed (or independent) current sources.
(b) Examples of controlled (or dependent) current sources.

Examples are given in Figure 1-13(a). Controlled (or dependent) current sources, for which the source function is expressed in terms of another network voltage or current, are denoted by the diamond symbol, as in Figure 1-13(b).

The ideal voltage and current source models used in network analysis are only approximations for physical sources such as batteries and generators. When the ideal elements are inadequate, more complicated models composed of more than one ideal element are used.

D1-3

How many nodes and how many loops do each of the following networks have?

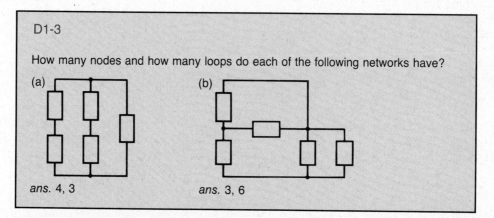

(a)

ans. 4, 3

(b)

ans. 3, 6

NETWORK DIAGRAMS

Network diagrams consist of elements, joined by conductors. A *node* is a complete set of conductor connections between elements. A *loop* is a closed path through elements and conductors that does not pass through any node more than once.

A voltage source maintains a specified voltage across its terminals. A current source maintains a specified current through its terminals.

1.5 ELECTRICAL POWER FLOW

1.5.1 Sink and Source Reference Relations

When the relationship between the voltage reference polarity and current reference direction for a two-terminal element is such that the current arrow has the sense entering the terminal nearest the plus sign, the voltage and current are said to be

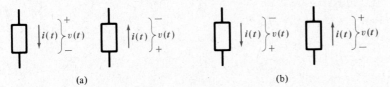

(a)   (b)

Figure 1-14   Sink and source reference relations for an element.
              (a) Sink reference relation.
              (b) Source reference relation.

defined with the *sink* relation in the element. The opposite situation is called the *source* relation. The four possible reference orientations are shown in Figure 1-14.

The names *sink* and *source* arose from the fact that if v and i have the source relation and if v and i are both positive, the element is a source of electrical energy; that is, electrical energy is flowing out of the element. Similarly, if v and i have the sink relation and both are positive, the electrical energy flow is into the element; the element is a sink of electrical energy. The most interesting elements, however, accept energy at times and supply energy at other times, so it is best to treat the terms *sink* and *source* just as labels.

1.5.2 Power Flow Relations

A charge Q that moves from potential $\phi(a)$ to potential $\phi(b)$ in passing through an element, as in Figure 1-15, loses the energy

$$W = Q[\phi(a) - \phi(b)] = Qv \qquad \text{newton-meter} = \text{joule (abbreviation J)}$$

That is, the electrical energy W is transferred from the charge to the element. If W is negative, which would be the case if the charge flow or v were negative, electrical energy is transferred from the element to the charge.

The rate of electrical energy flow, the electrical power flow, into an element is the rate of increase of W with time.

$$p_{\text{into}}(t) = \frac{dW}{dt} = \frac{dQ}{dt} v = iv \qquad \frac{\text{newton-meter}}{\text{second}} = \frac{\text{joule}}{\text{second}} = \text{watt} \quad \text{(abbreviation W)}$$

where i is the current in the sense from a to b, and v is the voltage $\phi(a) - \phi(b)$. The element voltage is not differentiated in this expression because, for all practical purposes, the effects of a time-varying current are propagated instantaneously in a network. The electrical power flow into a two-terminal element is then

$$p_{\text{into}}(t) = v(t)i(t)$$

where $v(t)$ and $i(t)$ have the sink reference relation.

$\phi(a)$ ╷a

$\phi(b)$ ╵b

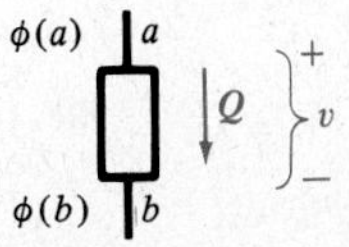

Figure 1-15   Charge flow through a potential difference.

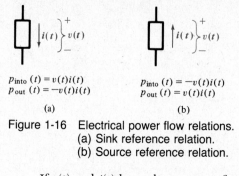

$p_{into}(t) = v(t)i(t)$
$p_{out}(t) = -v(t)i(t)$

(a)

$p_{into}(t) = -v(t)i(t)$
$p_{out}(t) = v(t)i(t)$

(b)

Figure 1-16 Electrical power flow relations.
(a) Sink reference relation.
(b) Source reference relation.

If $v(t)$ and $i(t)$ have the source reference relation, then

$$p_{into}(t) = -v(t)i(t)$$

as summarized in Figure 1-16. Negative power flow means that the actual flow of power is in the opposite direction.

The electrical power flow out of an element is the negative of the flow into the element:

$$p_{out}(t) = -p_{into}(t)$$

Several examples of electrical power flow calculations are shown in Figure 1-17.

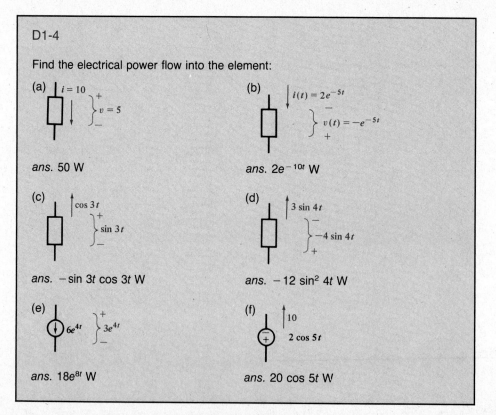

D1-4

Find the electrical power flow into the element:

(a) $i = 10$, $v = 5$

ans. 50 W

(b) $i(t) = 2e^{-5t}$, $v(t) = -e^{-5t}$

ans. $2e^{-10t}$ W

(c) $\cos 3t$, $\sin 3t$

ans. $-\sin 3t \cos 3t$ W

(d) $3 \sin 4t$, $-4 \sin 4t$

ans. $-12 \sin^2 4t$ W

(e) $6e^{4t}$, $3e^{4t}$

ans. $18e^{8t}$ W

(f) 10, $2 \cos 5t$

ans. $20 \cos 5t$ W

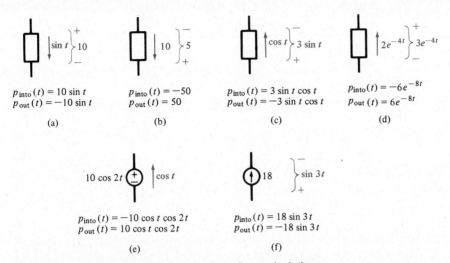

$p_{into}(t) = 10 \sin t$
$p_{out}(t) = -10 \sin t$

(a)

$p_{into}(t) = -50$
$p_{out}(t) = 50$

(b)

$p_{into}(t) = 3 \sin t \cos t$
$p_{out}(t) = -3 \sin t \cos t$

(c)

$p_{into}(t) = -6e^{-8t}$
$p_{out}(t) = 6e^{-8t}$

(d)

$p_{into}(t) = -10 \cos t \cos 2t$
$p_{out}(t) = 10 \cos t \cos 2t$

(e)

$p_{into}(t) = 18 \sin 3t$
$p_{out}(t) = -18 \sin 3t$

(f)

Figure 1-17 Examples of electrical power flow calculations.

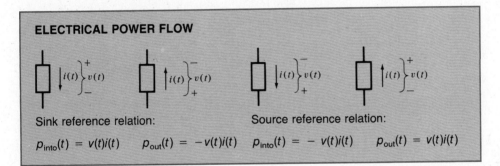

ELECTRICAL POWER FLOW

Sink reference relation:

$$p_{into}(t) = v(t)i(t) \qquad p_{out}(t) = -v(t)i(t)$$

Source reference relation:

$$p_{into}(t) = -v(t)i(t) \qquad p_{out}(t) = v(t)i(t)$$

1.6 THE RESISTOR

1.6.1 Voltage-Current Relations

An element for which

$$v(t) = Ri(t)$$

when $v(t)$ and $i(t)$ have the sink reference is called a *resistor*. The symbol is a zigzag line, and the constant of proportionality, the *resistance* of the element, is indicated beside the symbol, as in Figure 1-18(a). If $v(t)$ and $i(t)$, instead, have the source reference, Figure 1-18(b),

$$v(t) = -Ri(t)$$

The proportional voltage-current relationship for a resistor is called *Ohm's law*, in honor of Georg Ohm, a German physicist, who is believed to have first suggested the proportional voltage-current relationship. The SI unit of resistance is

$$R = \frac{\text{volt}}{\text{ampere}} = \text{ohm (abbreviation } \Omega)$$

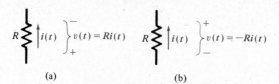

$$v(t) = Ri(t) \qquad v(t) = -Ri(t)$$

(a)   (b)

Figure 1-18 Sink and source reference voltage-current relations for the resistor.
(a) Sink reference relation.
(b) Source reference relation.

Several examples of the application of Ohm's law are shown in Figure 1-19. For a resistor of known resistance, if either the resistor voltage or the resistor current is known, the other is easily found via Ohm's law. The algebraic sign of the result is dependent on whether the defined element voltage and current have the sink or the source reference relation.

Physical devices called *resistors* are manufactured with a wide range of resistance, R, from a small fraction of an ohm to well over 100 million ohms. Resistors are marketed in many different sizes, the larger sizes capable of dissipating more heat power than the smaller sizes. The resistor is also a useful model of other devices, such as a length of wire, a light bulb, and an electric heater.

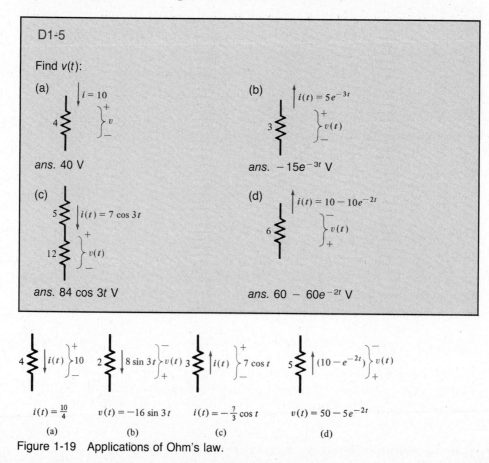

D1-5

Find $v(t)$:

(a) $i = 10$

4 v

ans. 40 V

(b) $i(t) = 5e^{-3t}$

3 $v(t)$

ans. $-15e^{-3t}$ V

(c) 5 $i(t) = 7\cos 3t$

12 $v(t)$

ans. 84 cos 3t V

(d) $i(t) = 10 - 10e^{-2t}$

6 $v(t)$

ans. $60 - 60e^{-2t}$ V

4 $i(t)$ 10 2 $8\sin 3t$ $v(t)$ 3 $i(t)$ $7\cos t$ 5 $(10 - e^{-2t})$ $v(t)$

$i(t) = \frac{10}{4}$ $v(t) = -16\sin 3t$ $i(t) = -\frac{7}{3}\cos t$ $v(t) = 50 - 5e^{-2t}$

(a)   (b)   (c)   (d)

Figure 1-19 Applications of Ohm's law.

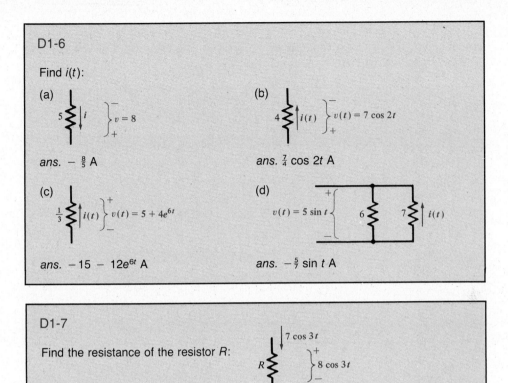

D1-6

Find $i(t)$:

(a)

5    $\downarrow i$ $\}\; v = 8$

ans. $-\frac{8}{5}$ A

(b)

4    $i(t)$ $\}\; v(t) = 7\cos 2t$

ans. $\frac{7}{4}\cos 2t$ A

(c)

$\frac{1}{3}$ $i(t)$ $\}\; v(t) = 5 + 4e^{6t}$

ans. $-15 - 12e^{6t}$ A

(d)

$v(t) = 5\sin t$ 6   7 $i(t)$

ans. $-\frac{5}{7}\sin t$ A

D1-7

Find the resistance of the resistor R:

$7\cos 3t$

R $\}\; 8\cos 3t$

ans. $\frac{8}{7}\ \Omega$

1.6.2 Resistor Power Relations

Using the voltage-current relation for the resistor, the power flow relations may be expressed in terms of the resistor voltage or the resistor current exclusively

$$p_{\text{into}}(t) = v(t)i(t) = i^2(t)R = \frac{v^2(t)}{R}$$

which hold for either the sink or the source reference since reversing the algebraic sign of v or i does not affect v^2 or i^2. Examples of resistor electrical power flow calculations are given in Figure 1-20.

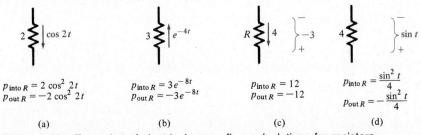

(a)

2    $\cos 2t$

$p_{\text{into }R} = 2\cos^2 2t$
$p_{\text{out }R} = -2\cos^2 2t$

(b)

3   e^{-4t}

$p_{\text{into }R} = 3e^{-8t}$
$p_{\text{out }R} = -3e^{-8t}$

(c)

R 4 -3

$p_{\text{into }R} = 12$
$p_{\text{out }R} = -12$

(d)

4    $\sin t$

$p_{\text{into }R} = \dfrac{\sin^2 t}{4}$
$p_{\text{out }R} = -\dfrac{\sin^2 t}{4}$

Figure 1-20   Examples of electrical power flow calculations for resistors.

For a resistor with positive resistance R, the power flow into R is always nonnegative. This is to say that electrical power flow is always into the resistor; the resistor always dissipates electrical energy.

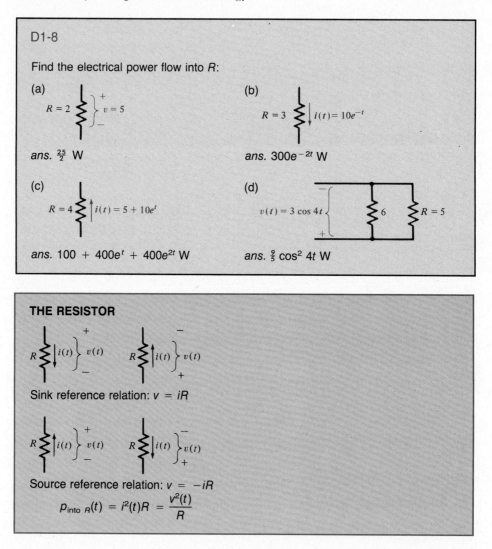

D1-8

Find the electrical power flow into R:

(a)

$R = 2$ $v = 5$

ans. $\frac{25}{2}$ W

(b)

$R = 3$ $i(t) = 10e^{-t}$

ans. $300e^{-2t}$ W

(c)

$R = 4$ $i(t) = 5 + 10e^{t}$

ans. $100 + 400e^{t} + 400e^{2t}$ W

(d)

$v(t) = 3 \cos 4t$ 6 $R = 5$

ans. $\frac{9}{5} \cos^2 4t$ W

THE RESISTOR

R $i(t)$ $v(t)$ R $i(t)$ $v(t)$

Sink reference relation: $v = iR$

R $i(t)$ $v(t)$ R $i(t)$ $v(t)$

Source reference relation: $v = -iR$

$$p_{\text{into } R}(t) = i^2(t)R = \frac{v^2(t)}{R}$$

1.7 SOLUTION OF PARALLEL NETWORKS

1.7.1 The Case of a Voltage Source between Nodes

Solving a network means finding any desired voltages and currents in the network. Consider first networks that have just two nodes, that is, networks where all of the elements are in *parallel*. If one of the parallel elements is a voltage source, then the only unknown element currents are resistor currents, which can be found by

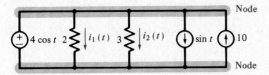

Figure 1-21   Parallel network with a voltage source between the nodes.

the application of the resistor voltage-current relations. For the network of Figure 1-21,

$$i_1(t) = \frac{4 \cos t}{2} \text{ A} \qquad i_2(t) = \frac{4 \cos t}{3} \text{ A}$$

There cannot be two different voltage sources in parallel because there can be only one specific node-to-node voltage.

1.7.2 Nontrivial Two-Node Networks

A more interesting problem involves the parallel connection of just current sources and resistors. An example is shown in Figure 1-22. If the node-to-node voltage $v(t)$ were known, it would be a simple matter to find any other unknown signal in the network. So the fundamental problem here is to find the node-to-node voltage with one of the two possible reference polarities.

To find $v(t)$, apply Kirchhoff's current law at one of the two nodes, say, the top node:

$$\sin t - i_1 + i_3 - 7 - i_4 = 0$$

(If Kirchhoff's current law is applied at the bottom node, the same equation results, multiplied through by -1.) Now substitute for the resistor currents in terms of the node-to-node voltage $v(t)$:

$$\sin t - \frac{v(t)}{2} + \left(-\frac{v(t)}{3}\right) - 7 - \frac{v(t)}{4} = 0$$

Rearranging,

$$(\tfrac{1}{2} + \tfrac{1}{3} + \tfrac{1}{4})v(t) = \sin t - 7$$

which is easily solved for $v(t)$:

$$v(t) = \tfrac{12}{13} \sin t - \tfrac{84}{13} \text{ V}$$

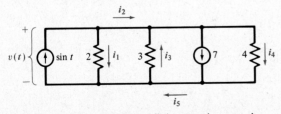

Figure 1-22   Nontrivial parallel network example.

The corresponding equation for any such parallel network has the form

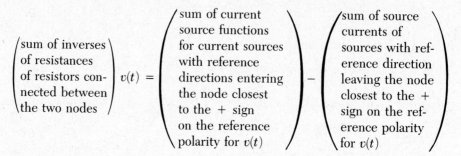

$$\left(\begin{array}{l}\text{sum of inverses}\\\text{of resistances}\\\text{of resistors con-}\\\text{nected between}\\\text{the two nodes}\end{array}\right) v(t) = \left(\begin{array}{l}\text{sum of current}\\\text{source functions}\\\text{for current sources}\\\text{with reference}\\\text{directions entering}\\\text{the node closest}\\\text{to the }+\text{ sign}\\\text{on the reference}\\\text{polarity for }v(t)\end{array}\right) - \left(\begin{array}{l}\text{sum of source}\\\text{currents of}\\\text{sources with ref-}\\\text{erence direction}\\\text{leaving the node}\\\text{closest to the }+\\\text{sign on the ref-}\\\text{erence polarity}\\\text{for }v(t)\end{array}\right)$$

The coefficient of $v(t)$ is the sum of the conductances (inverses of resistances) between the nodes.

Using $v(t)$, the various currents indicated in Figure 1-22 are found as follows:

$$i_1(t) = \frac{v(t)}{2} = \frac{6}{13}\sin t - \frac{42}{13}\text{ A}$$

$$i_2(t) = \sin t - i_1(t) = \frac{7}{13}\sin t + \frac{42}{13}\text{ A}$$

$$i_3(t) = -\frac{v(t)}{3} = \frac{28}{13} - \frac{4}{13}\sin t\text{ A}$$

$$i_4(t) = \frac{v(t)}{4} = \frac{3}{13}\sin t - \frac{21}{13}\text{ A}$$

$$i_5(t) = 7 + i_4(t) = 7 + \frac{v(t)}{4} = \frac{3}{13}\sin t + \frac{70}{13}\text{ A}$$

Alternatively, $i_2(t)$ and $i_5(t)$ can be expressed as

$$i_2(t) = 7 + i_4(t) - i_3(t) \qquad i_5(t) = \sin t + i_3(t) - i_1(t)$$

which, upon substitution, should give the same results.

Figure 1-23 shows another two-node network example. Its solution is as follows:

$$(\tfrac{1}{2} + \tfrac{1}{4} + \tfrac{1}{6})v(t) = 3\sin 5t + 10 - \cos t$$

$$v(t) = \tfrac{12}{11}(3\sin 5t + 10 - \cos t)$$

Given the node-to-node voltage, the other indicated signals are

$$i_1(t) = \frac{v(t)}{2} = \frac{6}{11}(3\sin 5t + 10 - \cos t)$$

$$i_2(t) = -\frac{v(t)}{4} = -\frac{3}{11}(3\sin 5t + 10 - \cos t)$$

$$i_3(t) = i_1(t) - 3\sin 5t - i_2(t) = -\frac{6}{11}\sin 5t - \frac{9}{11}\cos t + \frac{90}{11}$$

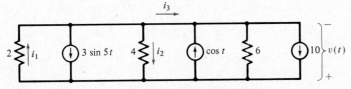

Figure 1-23   Another parallel network example.

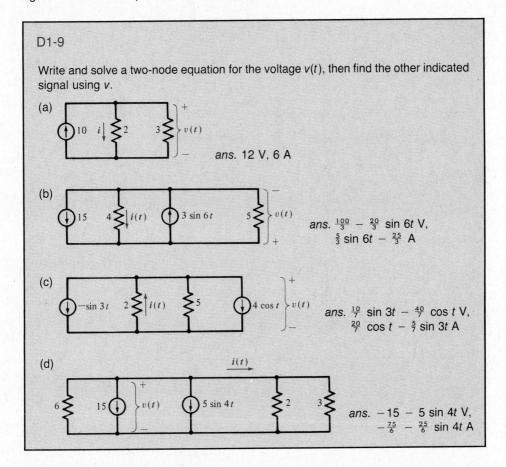

D1-9

Write and solve a two-node equation for the voltage $v(t)$, then find the other indicated signal using v.

(a)

ans. 12 V, 6 A

(b)

ans. $\frac{100}{3} - \frac{20}{3}$ sin 6t V,
$\frac{5}{3}$ sin 6t $- \frac{25}{3}$ A

(c)

ans. $\frac{10}{7}$ sin 3t $- \frac{40}{7}$ cos t V,
$\frac{20}{7}$ cos t $- \frac{5}{7}$ sin 3t A

(d)

ans. $-15 - 5$ sin 4t V,
$-\frac{75}{6} - \frac{25}{6}$ sin 4t A

1.7.3 The Current Divider Rule

Situations are often encountered in which a known current flows through the parallel connection of two resistors. Part of this net current $i(t)$ flows through one resistor, and the rest flows through the other resistor. Although this problem can be solved by writing a two-node equation every time it is encountered, it will save considerable time and effort to learn to write at a glance the fractions of $i(t)$ that flow through each of the two resistors. A general situation is pictured in Figure 1-24. It should be carefully noted that the voltage $v(t)$ is common to the two parallel resistors; if this is not the case in a problem under consideration, the following results do not apply.

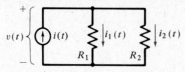

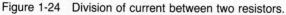

Figure 1-24 Division of current between two resistors.

The two-node equation for $v(t)$ in Figure 1-24 is

$$\left(\frac{1}{R_1} + \frac{1}{R_2}\right)v(t) = i(t)$$

and $v(t)$ and $i_1(t)$ are related by

$$v(t) = i_1(t)R_1$$

Solving the above two equations for $i_1(t)$,

$$i_1(t) = \frac{R_1 R_2}{R_1(R_1 + R_2)}\, i(t) = \frac{R_2}{R_1 + R_2}\, i(t) = \frac{\text{(opposite resistance)}}{\text{(sum of the two resistances)}}\, i(t)$$

Similarly,

$$i_2(t) = \frac{\text{(opposite resistance)}}{\text{(sum of the two resistances)}}\, i(t) = \frac{R_1}{R_1 + R_2}\, i(t)$$

This relationship, the *current divider rule*, applies directly only when the reference directions of the various currents have the relative senses in the drawing of Figure 1-24. The results for other relative senses of the current reference directions follow easily. An example is shown in Figure 1-25, for which

$$-i_1(t) = \frac{R_2}{R_1 + R_2}\, i(t) \qquad i_1(t) = \frac{-R_2}{R_1 + R_2}\, i(t)$$

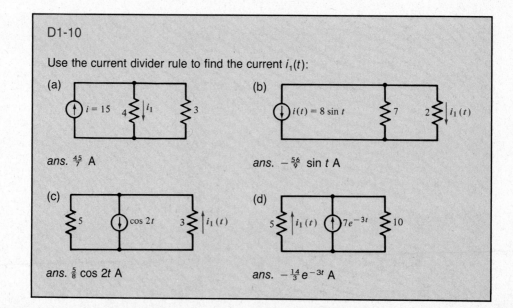

D1-10

Use the current divider rule to find the current $i_1(t)$:

(a)

$i = 15$ 4 i_1 3

ans. $\frac{45}{7}$ A

(b)

$i(t) = 8 \sin t$ 7 2 $i_1(t)$

ans. $-\frac{56}{9} \sin t$ A

(c)

5 $\cos 2t$ 3 $i_1(t)$

ans. $\frac{5}{8} \cos 2t$ A

(d)

5 $i_1(t)$ $7e^{-3t}$ 10

ans. $-\frac{14}{3} e^{-3t}$ A

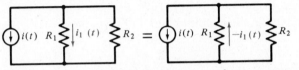

Figure 1-25   Current division with other reference senses.

SOLUTION OF PARALLEL NETWORKS

The systematic equation for the node-to-node voltage in a two-node network is of the form

$$\left(\frac{1}{R_1} + \frac{1}{R_2} + \cdots\right) v(t) = \left(\begin{array}{l}\text{sum of current}\\\text{source functions}\\\text{with references}\\\text{entering node}\\\text{nearest} + \text{sign on } v(t)\end{array}\right) - \left(\begin{array}{l}\text{sum of current}\\\text{source functions}\\\text{with references}\\\text{leaving node}\\\text{nearest} + \text{sign on } v(t)\end{array}\right)$$

The current divider rule is as follows:

$$i(t) \quad R_1 \qquad i_1(t) = \frac{R_2}{R_1 + R_2}\, i(t) \qquad R_2 \qquad i_2(t) = \frac{R_1}{R_1 + R_2}\, i(t)$$

$$i(t) \quad R_1 \qquad i_1(t) = -\frac{R_2}{R_1 + R_2}\, i(t) \qquad R_2 \qquad i_2(t) = -\frac{R_1}{R_1 + R_2}\, i(t)$$

1.8 SOLUTION OF SERIES NETWORKS

1.8.1 The Case of a Current Source in the Loop

A single-loop (or *single-mesh*) network is a *series* (end-to-end) connection of elements, as in the drawing of Figure 1-26. There cannot be two different current sources in the loop, otherwise one would be saying that the loop current is one function on the one hand and another function on the other.

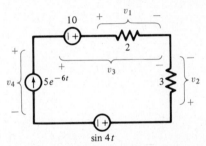

Figure 1-26   Series network with a current source in the loop.

If there is a current source in the loop, finding any voltage in the network is simple since the current through every element is known. For the example of Figure 1-26,

$$v_1(t) = 10e^{-6t} \text{ V}$$

$$v_2(t) = -15e^{-6t} \text{ V}$$

$$v_3(t) = v_1 - 10 = 10e^{-6t} - 10 \text{ V}$$

$$v_4(t) = -10 + v_1(t) - v_2(t) + \sin 4t$$

$$= -10 + 10e^{-6t} - (-15e^{-6t}) + \sin 4t$$

$$= \sin 4t - 10 + 25e^{-6t} \text{ V}$$

1.8.2 Nontrivial Single-Loop Networks

A more interesting problem is the single-loop network with only voltage sources and resistors in the loop, as the one in Figure 1-27. If the loop current $i(t)$ were known, it would be a simple matter to find any other signal in the network. Applying Kirchhoff's voltage law around the loop gives

$$7 \sin 5t - v_2(t) - v_1(t) - 12 + v_3(t) = 0$$

Substituting for the resistor voltages in terms of the loop current $i(t)$,

$$7 \sin 5t - 2i(t) - 3i(t) - 12 + (-4i(t)) = 0$$

$$(2 + 3 + 4)i(t) = 7 \sin 5t - 12$$

$$i(t) = \tfrac{1}{9}(7 \sin 5t - 12) \text{ A}$$

Every network with a single loop and no current source in the loop has a loop equation of the form

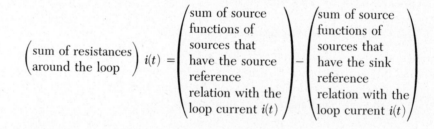

$$\begin{pmatrix} \text{sum of resistances} \\ \text{around the loop} \end{pmatrix} i(t) = \begin{pmatrix} \text{sum of source} \\ \text{functions of} \\ \text{sources that} \\ \text{have the source} \\ \text{reference} \\ \text{relation with the} \\ \text{loop current } i(t) \end{pmatrix} - \begin{pmatrix} \text{sum of source} \\ \text{functions of} \\ \text{sources that} \\ \text{have the sink} \\ \text{reference} \\ \text{relation with the} \\ \text{loop current } i(t) \end{pmatrix}$$

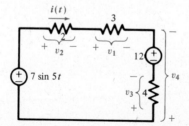

Figure 1-27   Nontrivial series network example.

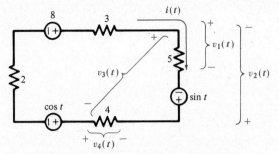

Figure 1-28   Another series network example.

For the example of Figure 1-27, the indicated voltages are as follows:

$v_1(t) = 3i(t) = \frac{1}{3}(7 \sin 5t - 12)$ V

$v_2(t) = 2i(t) = \frac{2}{9}(7 \sin 5t - 12)$ V

$v_3(t) = -4i(t) = -\frac{4}{9}(7 \sin 5t - 12)$ V

$v_4(t) = v_3(t) - 12 = -4i(t) - 12 = -\frac{4}{9}(7 \sin 5t - 12) - 12$

$\qquad = -\frac{28}{9} \sin 5t - \frac{60}{9}$ V

Figure 1-28 is another single-loop network example and is solved as follows:

$(2 + 3 + 5 + 4)i(t) = \sin t - \cos t + 8$

$\qquad\qquad i(t) = \frac{1}{14}(\sin t - \cos t + 8)$ A

Using $i(t)$, the other indicated network signals are as follows:

$v_1(t) = \frac{5}{14}(\sin t - \cos t + 8)$ V

$v_2(t) = \sin t - \frac{5}{14}(\sin t - \cos t + 8) = \frac{9}{14} \sin t + \frac{5}{14} \cos t - \frac{40}{14}$ V

$v_3(t) = v_1 - \sin t - v_4 = -\frac{5}{14} \sin t - \frac{9}{14} \cos t + \frac{72}{14}$ V

$v_4(t) = -\frac{4}{14}(\sin t - \cos t + 8)$ V

1.8.3 The Voltage Divider Rule

Often there are situations in which it is desired to determine how a known voltage, $v(t)$, is distributed across two resistors in series. Some fraction of $v(t)$ is $v_1(t)$, and the remaining fraction of $v(t)$ is $v_2(t)$, as indicated in Figure 1-29. It should be carefully noted that there is only the direct connection between the two resistors in this situation; the same current flows through each. This problem could be solved anew every time it is encountered but, better, it can be learned once and for all the fractions of $v(t)$ that are the voltages $v_1(t)$ and $v_2(t)$, thus saving considerable time and effort in the future.

The single-loop equation for the general network is

$(R_1 + R_2)i(t) = v(t)$

and v_1 is related to the loop current $i(t)$ by

$v_1(t) = R_1 i(t)$

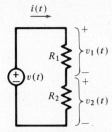

Figure 1-29 Division of voltage across two resistors.

Solving for v_1,

$$v_1(t) = \frac{R_1}{R_1 + R_2}\, v(t) = \frac{\text{same resistor}}{\text{sum of the two resistors}}\, v(t)$$

Similarly,

$$v_2(t) = \frac{\text{same resistor}}{\text{sum of the two resistors}}\, v(t) = \frac{R_2}{R_1 + R_2}\, v(t)$$

These results, similar to those for the current divider rule, may be easily applied to situations in which the references have different senses than those for the basic problem. An example is shown in Figure 1-30, for which

$$-v_1(t) = \frac{3}{3 + 4} \cdot 8e^{-7t} \qquad v_1(t) = -\frac{24}{7} e^{-7t}\ \text{V}$$

D1-11

Write and solve a single-loop equation for the current $i(t)$, then use $i(t)$ to find the other indicated signal:

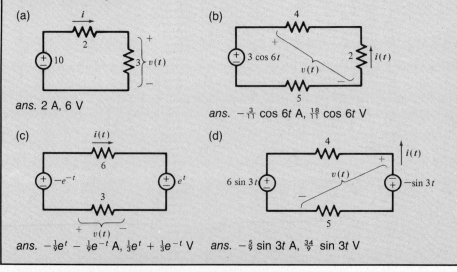

(a)

ans. 2 A, 6 V

(b)

ans. $-\frac{3}{11}$ cos $6t$ A, $\frac{18}{11}$ cos $6t$ V

(c)

ans. $-\frac{1}{9}e^t - \frac{1}{9}e^{-t}$ A, $\frac{1}{3}e^t + \frac{1}{3}e^{-t}$ V

(d)

ans. $-\frac{5}{9}$ sin $3t$ A, $\frac{34}{9}$ sin $3t$ V

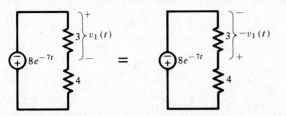

Figure 1-30 Voltage division with other reference polarities.

D1-12

Use the voltage divider rule to find the voltage $v_1(t)$:

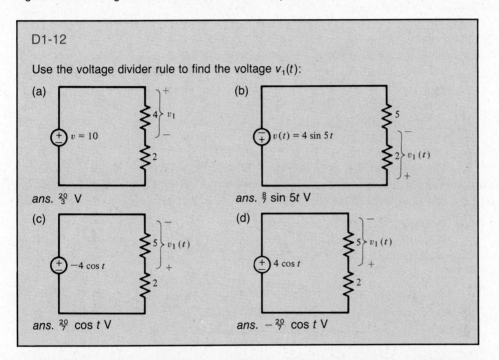

(a)

ans. $\frac{20}{3}$ V

(b)

ans. $\frac{8}{7}$ sin 5t V

(c)

ans. $\frac{20}{7}$ cos t V

(d)

ans. $-\frac{20}{7}$ cos t V

SOLUTION OF SERIES NETWORKS

The systematic equation for the loop current in a single-loop network is of the form

$$(R_1 + R_2 + R_3 + \cdots)i(t) = \left(\begin{array}{l}\text{sum of voltage source functions with} \\ \text{source reference relation with } i(t)\end{array}\right)$$

$$- \left(\begin{array}{l}\text{sum of voltage source functions with} \\ \text{sink reference relation with } i(t)\end{array}\right)$$

The voltage divider rule is as follows:

$$v_1(t) = \frac{R_1}{R_1 + R_2}\, v(t) \qquad\qquad v_1(t) = -\frac{R_1}{R_1 + R_2}\, v(t)$$

$$v_2(t) = \frac{R_2}{R_1 + R_2}\, v(t) \qquad\qquad v_2(t) = -\frac{R_2}{R_1 + R_2}\, v(t)$$

1.9 EQUIVALENT NETWORKS

1.9.1 Equivalence of Two-Terminal Elements

A very powerful technique in network solution is one of transformation. A complicated network is transformed to simpler, *equivalent* networks through one or more stages until its solution can be found easily. This transformation or equivalent circuit method is very commonly used for networks in all areas of electrical engineering because it is often the easiest solution method and because it retains an identification with the original network, which is lacking if the problem is immediately converted to equations.

A combination of elements that has two terminals is termed a *two-terminal network*. Two 2-terminal networks are said to be equivalent if both have the same voltage-current relations at their terminals. If a combination of elements is replaced with an equivalent in a network, all voltages and currents (signals) within the two networks are the same, *external* to the portion replaced. A network may be simplified in this way, often to the point where a two-node or single-loop equation may be written.

The following is a listing of some simple equivalences. A general relation is implied even though some examples show only a specific number of elements for simplicity. For clarity, the terminals of the equivalents are indicated now with small circles.

(a) *Current Sources in Parallel:* Figure 1-31(a) illustrates how current sources in parallel with one another may be combined into a single equivalent current source.

(b) *Voltage Sources in Series:* Voltage sources in series with one another may

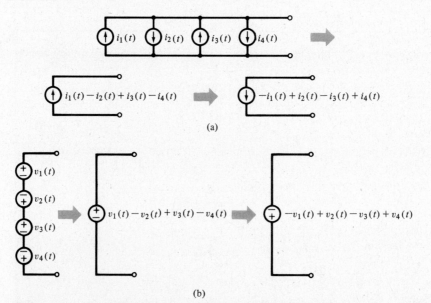

Figure 1-31 Equivalent current sources in parallel and voltage sources in series.

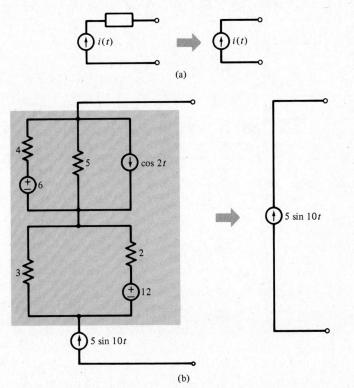

(a)

(b)

Figure 1-32   Equivalent of a current source in series with an element.

be combined into a single equivalent voltage source, as indicated in Figure 1-31(b).

(c) *Current Source in Series with Anything:* In the equivalence of Figure 1-32(a), the current $i(t)$ enters one terminal and leaves the other. Whether or not the series element is present makes a difference to the current source and, of course, to the series element; but external to the current source and series element, there may as well be only the current source present. The series element may be a two-terminal combination of more simple elements, as in the example of Figure 1-32(b).

(d) *Voltage Source in Parallel with Anything:* The terminal voltage is $v(t)$ in Figure 1-33(a), whether or not the parallel element is present. Of course, it makes a difference to the source whether or not the parallel element is present; the source must supply any additional current that flows through the element. It obviously makes a difference to the parallel element whether or not it is present. External to the combination, however, the presence or absence of the parallel element is of no consequence. The parallel element may be a two-terminal combination of simpler elements, as in Figure 1-33(b).

(e) *Series and Parallel Equivalences:* The order of any elements in a series or parallel connection may be changed without changing external voltages or currents. Figure 1-34(a) and (b) illustrate examples of this.

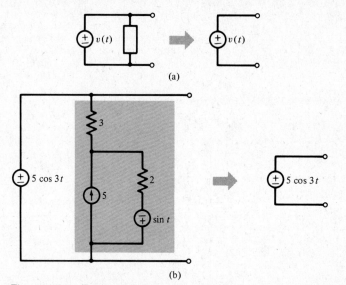

(a)

(b)

Figure 1-33 Equivalent of a voltage source in parallel with an element.

If a resistor is turned end for end, it still has the same voltage-current relation. An element such as a resistor is *bilateral*. Turning a source end for end in a network will change the network, however. Sources are *nonbilateral*. So it must be understood that in equivalents involving sources, the senses of these nonbilateral elements are to be preserved.

D1-13

Find the indicated signals by using equivalent circuits to reduce the problem to a single-loop or two-node problem, then solving:

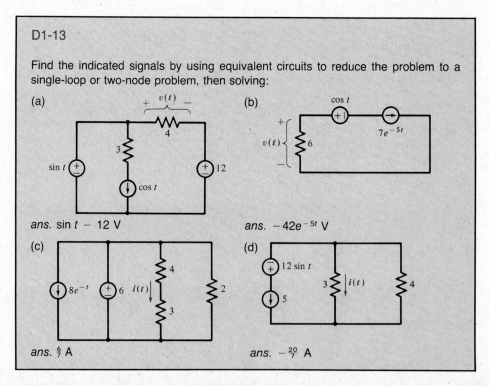

(a)

ans. $\sin t - 12$ V

(b)

ans. $-42e^{-5t}$ V

(c)

ans. $\frac{6}{7}$ A

(d)

ans. $-\frac{20}{7}$ A

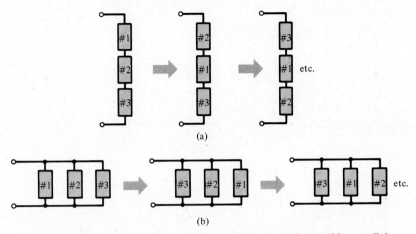

(a)

(b)

Figure 1-34   Equivalent interchange of elements in series and in parallel.

1.9.2 Resistors in Series

Any two-terminal combination of resistors is equivalent to a single resistor. For resistors in series, the equivalent single resistor has resistance that is just the sum of the individual resistances. In Figure 1-35 it is demonstrated that the series resistors and the appropriate single resistor have the same voltage-current relation; thus they are equivalent.

It should be noted that the equivalences presented here are between one two-terminal combination of more simple elements and another two-terminal element or combination of elements; and the resulting networks are equivalent only so far as voltages and currents external to the transformed portions are concerned. For example, resistors R_1 and R_2 in the network of Figure 1-36 cannot be replaced by an equivalent resistor of resistance $(R_1 + R_2)$ because of the connection between R_1 and R_2. The dashed box containing R_1 and R_2 has *three* terminals. R_3 and R_4, however, may be replaced by an equivalent resistor so far as any voltage or current external to that element is concerned.

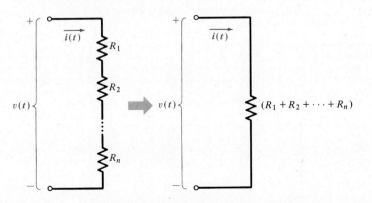

$$(R_1 + R_2 + \cdots + R_n)i(t) = v(t) \qquad v(t) = (R_1 + R_2 + \cdots + R_n)i(t)$$

Figure 1-35   Equivalent resistance of resistors in series.

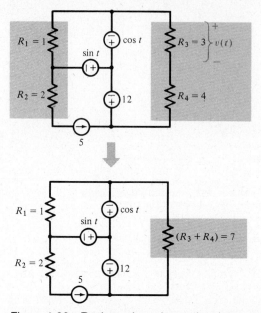

Figure 1-36   Resistors in series and resistors not in series.

The two networks in Figure 1-36 (which differ only so far as the dashed boxes to the right are concerned) are equivalent; they have the same voltages and currents external to the dotted boxes. The two networks are not equivalent so far as the voltage $v(t)$, within the box, is concerned, however. That voltage does not appear in the second network.

1.9.3 Resistors in Parallel

The equivalent resistance for a set of resistors in parallel is the inverse of the sum of the inverses of the individual resistances. In Figure 1-37, it is shown that the parallel resistors and the appropriate single resistor have the same voltage-current relation at their terminals.

The parallel resistor equivalence for just two resistors in parallel is

$$R_{\text{equiv}} = \frac{1}{(1/R_1) + (1/R_2)} = \frac{R_1 R_2}{R_1 + R_2}$$

This "product over the sum" relationship is well worth committing to memory be-

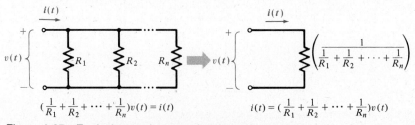

Figure 1-37   Equivalent resistance of resistors in parallel.

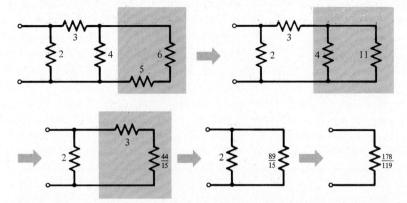

Figure 1-38   Finding the equivalent resistance of a two-terminal combination of resistors.

cause of the algebraic manipulation it saves each time it is applied. The corresponding relationships for three or more resistors in parallel are *not* "product over sum." They are more complicated, so the "inverse of the sum of the inverses" is probably as good a relationship to start with as any other for more than two resistors in parallel.

The equivalent resistance of a parallel combination of resistors is always less than any of the individual resistances in the combination.

The inverse of resistance is called *conductance* and has the symbol G. A resistor with resistance $R = 5 \ \Omega$ has conductance $G = \frac{1}{5} \ \Omega^{-1}$. The unit Ω^{-1} is named the siemens (abbreviation S) in honor of Karl Wilhelm Siemens (1823–1883), an early electrical experimenter and inventor.

The equivalent conductance of several resistors in parallel is the sum of the individual resistor conductances:

$$G_{equiv} = \frac{1}{R_{equiv}} = \frac{1}{R_1} + \frac{1}{R_2} + \cdots = G_1 + G_2 + \cdots$$

Dealing in terms of conductances instead of resistances is advantageous for networks where all of the elements are in parallel, but there is no advantage in more general (and more interesting) networks, since the equivalent conductance of resistors in series is analogous to resistances in parallel:

$$G_{equiv} = \frac{1}{R_{equiv}} = \frac{1}{R_1 + R_2 + \cdots} = \frac{1}{(1/G_1) + (1/G_2) + \cdots}$$

1.9.4 Two-Terminal Combinations of Resistors

The series and parallel resistor equivalences allow easy solution for equivalent resistances for more complicated two-terminal combinations of resistors. An example is given in Figure 1-38.

There are, however, two-terminal combinations of resistors that cannot be so reduced. In the two-terminal network of Figure 1-39 no resistor is in series or in parallel with any other resistor. Series and parallel combinations are not sufficient to find single equivalent resistors for *all* possible two-terminal resistor combinations, even though the equivalents exist.

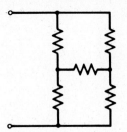

Figure 1-39 A two-terminal combination of resistors in which no resistors are in series or in parallel with one another.

D1-14

Find the equivalent resistances of the following two-terminal resistor networks:

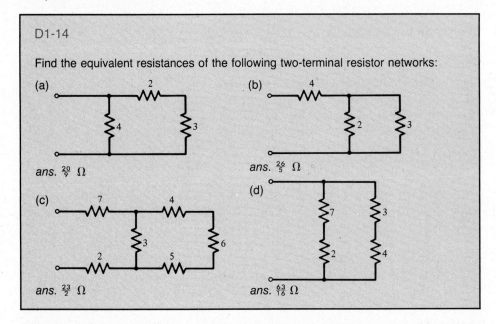

(a)

ans. $\frac{20}{9}$ Ω

(b)

ans. $\frac{26}{5}$ Ω

(c)

ans. $\frac{23}{2}$ Ω

(d)

ans. $\frac{63}{16}$ Ω

D1-15

Find the indicated signals using equivalent circuits. There are three signals to find in each network.

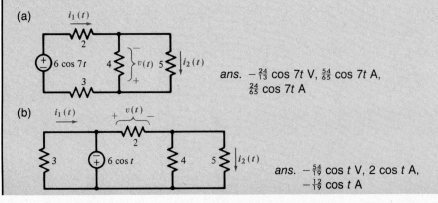

(a)

ans. $-\frac{24}{13}$ cos 7t V, $\frac{54}{65}$ cos 7t A, $\frac{24}{65}$ cos 7t A

(b)

ans. $-\frac{54}{19}$ cos t V, 2 cos t A, $-\frac{12}{19}$ cos t A

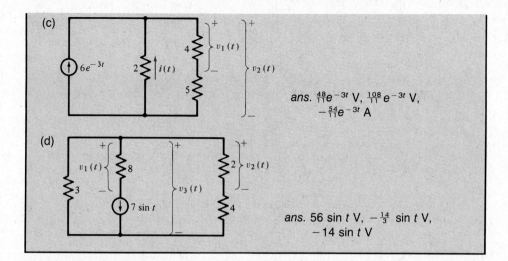

(c)

$ans.$ $\frac{48}{11}e^{-3t}$ V, $\frac{108}{11}e^{-3t}$ V,
$-\frac{54}{11}e^{-3t}$ A

(d)

$ans.$ 56 sin t V, $-\frac{14}{3}$ sin t V,
-14 sin t V

EQUIVALENT NETWORKS

Two elements are *equivalent* if both have the same voltage-current relations at their terminals. If an element is replaced by an equivalent element, all network voltages and currents external to that element are unchanged.

Equivalent resistances of resistors in series and in parallel are as follows:

Series: $R_{equiv} = R_1 + R_2 + \cdots$

Parallel: $R_{equiv} = \dfrac{1}{(1/R_1) + (1/R_2) + \cdots}$

For two resistors in parallel,

$$R_{equiv} = \frac{R_1 R_2}{R_1 + R_2}$$

1.10 MORE ABOUT EQUIVALENT NETWORKS

1.10.1 The Thévenin-Norton Equivalence

One of the most interesting equivalences is between a voltage source in series with a resistance and a current source in parallel with a resistance. The former is called a *Thévenin element* and the latter is called a *Norton element*. Each is illustrated in Figure 1-40.

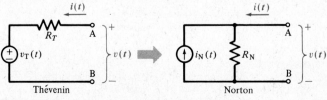

Figure 1-40   Thevenin and Norton elements.

The two-terminal networks of Figure 1-40 are equivalent, providing the relation between $v(t)$ and $i(t)$ is the same in both. For the Thévenin element,

$$R_T i(t) = v(t) - v_T(t)$$

$$v(t) = v_T(t) + R_T i(t)$$

For the Norton element,

$$\left(\frac{1}{R_N}\right) v(t) = i_N(t) + i(t)$$

$$v(t) = R_N i_N(t) + R_N i(t)$$

Thus the two are equivalent if

$$\begin{cases} v_T(t) = R_N i_N(t) \\ R_T = R_N \end{cases}$$

Since the Thévenin and Norton elements contain sources, the senses of which must be preserved, this equivalence holds only if terminal A (closest to the plus sign on v_T) is replaced by the other terminal A (toward which the reference for i_N points), and terminal B is replaced by the other terminal B.

Specific examples of conversions between Thévenin and Norton elements are given in Figure 1-41(a). A Thévenin-Norton or a Norton-Thévenin conversion, re-

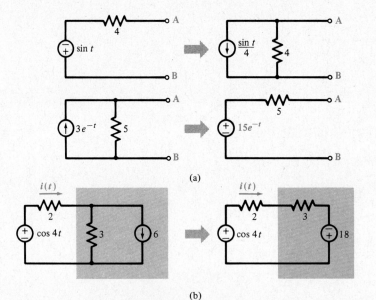

(a)

(b)

Figure 1-41   Thévenin-Norton equivalents.
(a) Examples of Thévenin-Norton and Norton-Thévenin conversions.
(b) Example of Norton-Thévenin conversion in network solution.

placing one element by its equivalent, is often very useful in finding network solutions. For the network of Figure 1-41(b), the indicated Thévenin-Norton transformation gives a resulting single-loop network, for which

$$(2 + 3)i(t) = \cos 4t + 18$$

$$i(t) = \tfrac{1}{5} \cos 4t + \tfrac{18}{5} \text{ A}$$

D1-16

Use the Thévenin-Norton element equivalence to find the indicated signals:

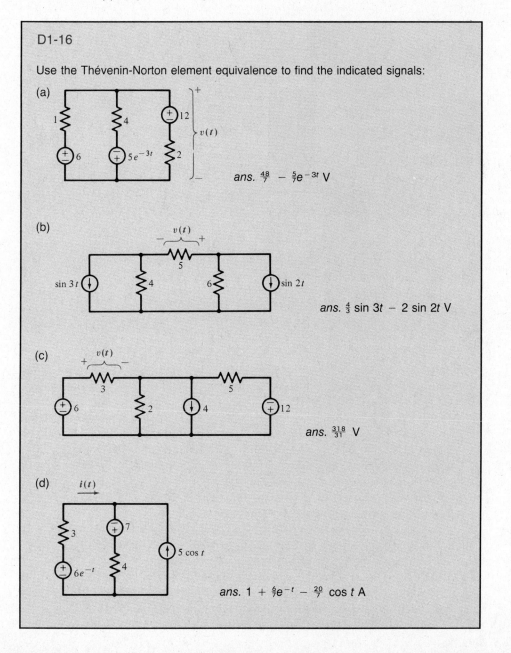

(a)

ans. $\tfrac{48}{7} - \tfrac{5}{7}e^{-3t}$ V

(b)

ans. $\tfrac{4}{3} \sin 3t - 2 \sin 2t$ V

(c)

ans. $\tfrac{318}{31}$ V

(d)

ans. $1 + \tfrac{6}{7}e^{-t} - \tfrac{20}{7} \cos t$ A

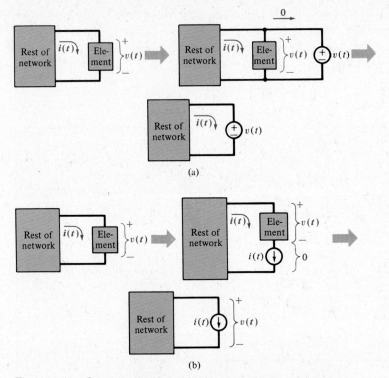

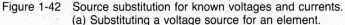

Figure 1-42   Source substitution for known voltages and currents.
  (a) Substituting a voltage source for an element.
  (b) Substituting a current source for an element.

1.10.2 Substitution of Sources for Known Voltages and Currents

Suppose that the voltage $v(t)$ across a certain element in a network is known. Then connecting a voltage source with source function $v(t)$ across the terminals of that element, as in Figure 1-42(a), will not affect the rest of the network. The original voltage was $v(t)$ and, with the source, it is still $v(t)$. Whatever the voltage-current relations of the element and of the original network, since the voltage is unchanged, the current $i(t)$ is unchanged. The current through the source is thus zero.

The voltage source in parallel with the element is equivalent, so far as the rest of the network is concerned, to just the voltage source. Thus any element in a network may be replaced by a voltage source with source function the element voltage, without changing any network voltage or current.

Similarly, any element in a network may be replaced by a current source with source function the element current, without changing any network voltage or current, as indicated in Figure 1-42(b).

These equivalences, known as the *substitution theorem*, are particularly useful in consolidating results found by using equivalent circuits and in picturing how an element affects a network. Whenever an element's voltage or current is known, so far as the rest of the network is concerned, the element may be considered to be a source.

D1-17

The voltage or current of the element that is represented by a box is given. Find the other indicated signal:

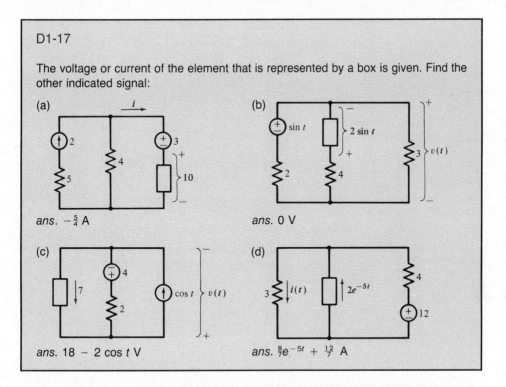

(a)

ans. $-\frac{5}{4}$ A

(b)

ans. 0 V

(c)

ans. $18 - 2\cos t$ V

(d)

ans. $\frac{8}{7}e^{-5t} + \frac{12}{7}$ A

1.10.3 Examples of Network Solution Using Equivalent Circuits

The equivalent circuits described so far may be used to find voltages and currents in almost any network, the exceptions being those networks that require three-terminal or, generally, multiterminal equivalences, such as two-terminal combinations of resistors where no resistor is in series or parallel with any other resistor. Further equivalences (called *delta-wye transformations*) may be developed for these special cases, and this will be done in Chapter 3 in connection with two-port network parameters.

The approach in solving a network using equivalent circuits is to reduce the problem, using equivalences, to a two-node or single-loop problem, solving for the node-to-node voltage or the loop current. Then, working from this solution, often using the current-divider and voltage-divider rules, we find the voltages and currents desired in the original network. With this method, there are numerous ways to proceed, depending on one's experience and skill in achieving a solution with minimal effort.

Figure 1-43 describes a network in which it is desired to find the voltage $v(t)$. With the indicated series of transformations, the problem is finally reduced to an equivalent single-loop network. For that network,

$$(6 + 4 + \tfrac{6}{5} + 5)i(t) = -7\sin 8t - 50$$

$$i(t) = \tfrac{5}{81}(-7\sin 8t - 50)\ \text{A}$$

$$v(t) = 6i(t) = -\tfrac{30}{81}(7\sin 8t + 50)\ \text{V}$$

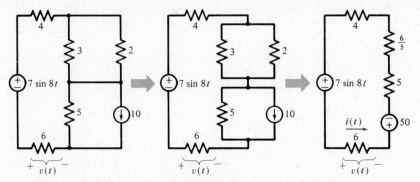

Figure 1-43 First example of network solution via equivalent circuits.

In Figure 1-44 it is desired to find $i(t)$. The transformations to an equivalent parallel network are such that $i(t)$ does not appear in the final network. Instead, the voltage $v(t)$ is first found:

$$\left(\frac{1}{3} + \frac{1}{4} + \frac{1}{5}\right)v(t) = \frac{3}{2}e^{-7t} - \frac{\sin t}{3} - \frac{12}{5}$$

$$v(t) = \frac{90}{47}e^{-7t} - \frac{20}{47}\sin t - \frac{144}{47}\ \text{V}$$

Returning to the original network, now knowing $v(t)$,

$$5i(t) = v(t) + 12$$

$$i(t) = \tfrac{18}{47}e^{-7t} - \tfrac{4}{47}\sin t + \tfrac{84}{47}\ \text{A}$$

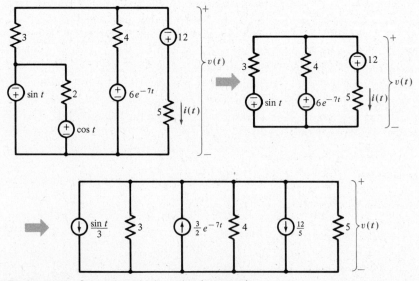

Figure 1-44 Second equivalent circuit example.

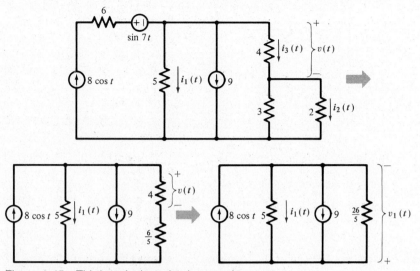

Figure 1-45   Third equivalent circuit example.

To find $v(t)$ and $i_2(t)$ in the network of Figure 1-45, the indicated transformations are chosen. A two-node equation is then solved for $v_1(t)$:

$$(\tfrac{1}{5} + \tfrac{5}{26})\, v_1(t) = 9 - 8 \cos t$$

$$v_1(t) = \tfrac{130}{51}\, (9 - 8 \cos t)\ \text{V}$$

Then

$$i_1(t) = -\frac{v_1(t)}{5} = \frac{26}{51}\,(8 \cos t - 9)\ \text{A}$$

Returning to the previous equivalent network, using the voltage-divider rule,

$$v(t) = \frac{-4}{4 + (\tfrac{6}{5})}\, v_1(t) = \frac{100}{51}\,(8 \cos t - 9)\ \text{V}$$

Returning to the original network,

$$i_3(t) = \frac{v(t)}{4} = \frac{25}{51}\,(8 \cos t - 9)\ \text{A}$$

Using the current-divider rule,

$$i_2(t) = \tfrac{3}{5}\, i_3(t) = \tfrac{15}{51}\,(8 \cos t - 9)\ \text{A}$$

The current in elements "hanging" from a network, as in the example of Figure 1-46, is zero because there is no closed path through them for current flow. Since voltage across a resistor is zero when its current is zero, the indicated voltage $v_1(t)$ is

$$v_1(t) = 12 \sin t - 8e^{-6t}\ \text{V}$$

Similarly,

$$v_2(t) = v_3(t) = 20 \sin t\ \text{V}$$

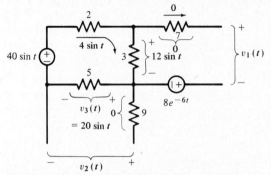

Figure 1-46   A network with hanging elements.

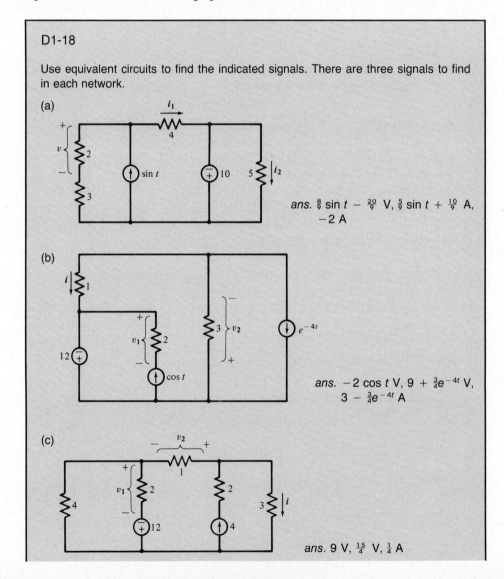

D1-18

Use equivalent circuits to find the indicated signals. There are three signals to find in each network.

(a)

ans. $\frac{8}{9}$ sin $t - \frac{20}{9}$ V, $\frac{5}{9}$ sin $t + \frac{10}{9}$ A, -2 A

(b)

ans. -2 cos t V, $9 + \frac{3}{4}e^{-4t}$ V, $3 - \frac{3}{4}e^{-4t}$ A

(c)

ans. 9 V, $\frac{15}{4}$ V, $\frac{1}{4}$ A

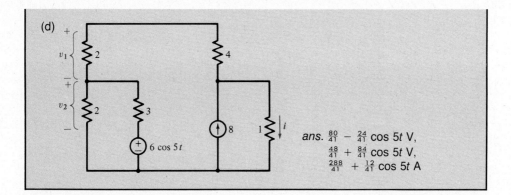

(d)

$ans.$ $\frac{80}{41} - \frac{24}{41}$ cos $5t$ V,

$\frac{48}{41} + \frac{84}{41}$ cos $5t$ V,

$\frac{288}{41} + \frac{12}{41}$ cos $5t$ A

MORE ABOUT EQUIVALENT NETWORKS

The Thévenin-Norton equivalence is as follows:

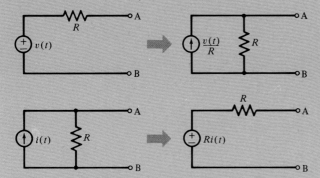

If any element in a network is replaced by a voltage source with source function equal to the element voltage, each network voltage and current is unchanged. If any element in a network is replaced by a current source with source function equal to the element current, each network voltage and current is unchanged.

Use equivalent circuits to simplify the network until a two-node or single-loop network results and is solved. Then, using signals found in the simplified network and previous equivalences as necessary, find other signals of interest in the original network.

CHAPTER ONE PROBLEMS

There are three types of problems at the end of each chapter. The *basic problems* consist of problems similar to the drill problems but which may require combining the techniques of several sections.

The *practical problems* involve practical numbers and SI units with multiplier prefixes and are generally of the type encountered in engineering practice. Some of these problems introduce common situations in electrical measurement and instrumentation.

The *advanced problems* introduce topics and problems of a more theoretical nature.

Basic Problems

Kirchhoff's Laws

1. Find the curent $i(t)$:

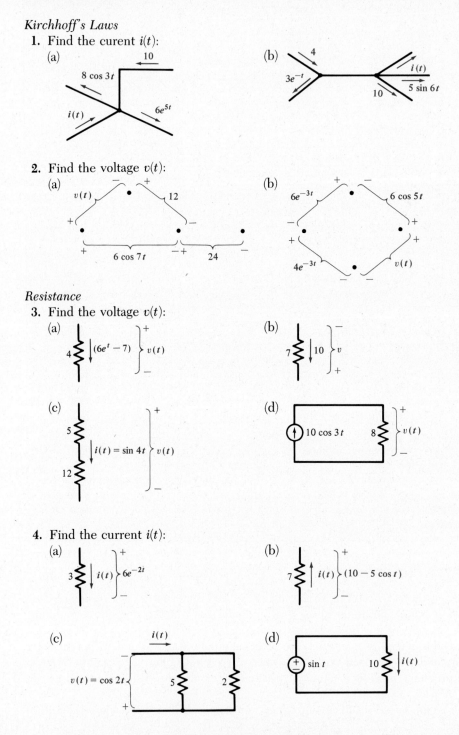

(a)

(b)

2. Find the voltage $v(t)$:

(a)

(b)

Resistance

3. Find the voltage $v(t)$:

(a)

(b)

(c)

(d)

4. Find the current $i(t)$:

(a)

(b)

(c)

(d)

Parallel Networks

5. Find the indicated voltages and currents:

(a)

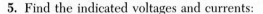

(b)

(c)

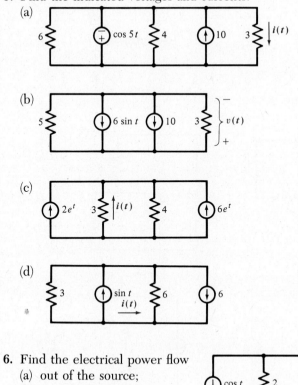

(d)

6. Find the electrical power flow
   (a) out of the source;
   (b) into the 10-Ω resistor;
   (c) into the 2-Ω resistor.

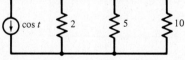

7. Find the indicated voltages and currents:

(a)   (b)

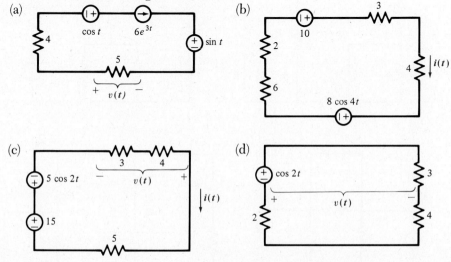

(c)   (d)

8. Find the electrical power flow
   (a) out of the sinusoidal source;
   (b) into the constant source;
   (c) into the 4-Ω resistor.

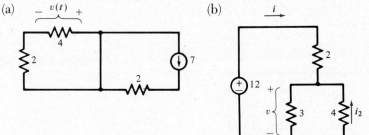

Equivalent Circuits
9. Find the indicated voltages and currents using equivalent circuits:

(a)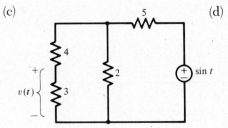

(b)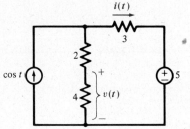

(c)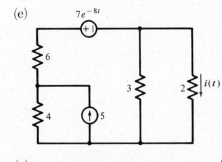

(d)

(e)

(f)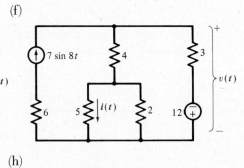

(g)

(h)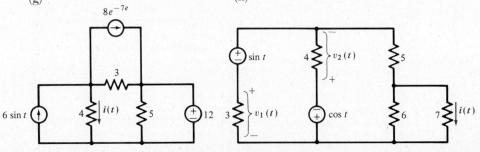

Equivalent Resistance

10. Find the equivalent resistance of each of the following two-terminal combinations of resistors:

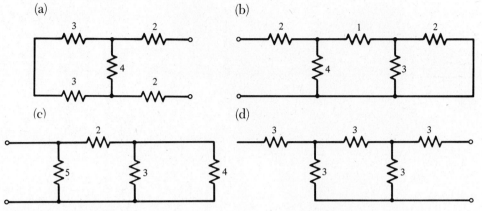

(a)

(b)

(c)

(d)

Practical Problems

Resistor Color Code

The color code of Table 1-3 is used for marking electronic component values. A color code for the percentage tolerance of components, Table 1-4, is also used.

Fixed composition-type resistors are constructed of a cylinder of resistive material, each end of which makes contact with a wire terminal. The resistive material is surrounded by an insulating case. This type of resistor is coded with four color bands as indicated in the sketch of (a). Bands A and B give the significant figures of the resistance, band C gives the decimal multiplier to those first two digits, and band D gives the percentage tolerance. For instance, the resistor depicted in (b) has value $25 \times 10^2 = 2500 \ \Omega \pm 5\%$.

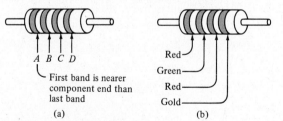

$A \ B \ C \ D$

First band is nearer component end than last band

(a)

Red
Green
Red
Gold

(b)

Table 1-3 COLOR CODE

COLOR	SIGNIFICANT FIGURE	DECIMAL MULTIPLIER
Black	0	$10^0 = 1$
Brown	1	$10^1 = 10$
Red	2	$10^2 = 100$
Orange	3	10^3
Yellow	4	10^4
Green	5	10^5
Blue	6	10^6
Violet	7	10^7
Gray	8	$10^{-2} = 0.01$
White	9	$10^{-1} = 0.1$

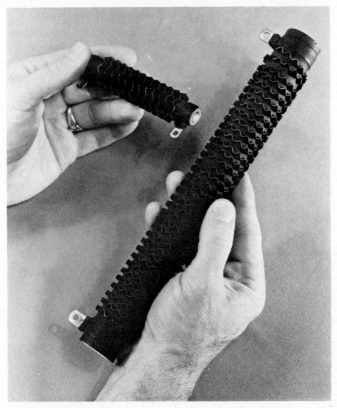

Power resistors, capable of dissipating relatively large amounts of electrical power as heat. (*Photo courtesy of Dale Electronics, Inc.*)

11. (a) For a resistor with color code as shown, what is the resistance and what is the tolerance? Specify the range of resistance within which the coded resistor must fall.

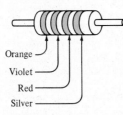

Orange ─
Violet ─
Red ─
Silver ─

(b) What is the color code for a 23-Ω ± 10% resistor?

Table 1-4 COMPONENT TOLERANCE

COLOR	PERCENT TOLERANCE
Gold	±5
Silver	±10
No color	±20

Table 1-5 SIGNIFICANT
FIGURES OF STANDARD
VALUES

±20%	±10%	±5%
10	10	10
		11
	12	12
		13
15	15	15
		16
	18	18
		20
22	22	22
		24
	27	27
		30
33	33	33
		36
	39	39
		43
47	47	47
		51
	56	56
		62
68	68	68
		75
	82	82
		91
100	100	100

Table 1-5 shows the significant figures of standard composition resistor values for each of the three common tolerances. The standard values are placed approximately 40% apart for ±20% tolerance resistors, 20% apart for ±10% tolerance, and 10% apart for ±5% tolerance.

12. (a) What is the closest ±10% standard value to 4075 Ω?

   (b) A certain application requires the use of a composition resistor with a resistance that lies between 11,450 and 15,200 Ω. Specify a standard value and tolerance for the resistor, using the largest possible resistor tolerance for greatest economy.

Resistor Power Rating
In addition to their resistance, physical resistors are also characterized by their *power rating*, the maximum average power they are designed to dissipate under normal conditions. Except possibly for very short intervals of time, $p_{\text{into } R}$ should not exceed the resistor power rating.

Standard power ratings for composition-type resistors are $\frac{1}{4}$, $\frac{1}{2}$, 1, and 2 W. Larger sized resistors, frequently made of a length of resistance wire wound on a tubular bobbin (called *wirewound* resistors), are available for applications requiring greater power dissipation.

13. (a) What is the maximum constant voltage that may be safely applied to a 100-Ω $\frac{1}{2}$-W resistor? What is the maximum safe constant current for the same resistor?

    (b) Two resistors, of resistance 3 and 10 Ω, are connected in series so that the same current will flow through each. If the 3-Ω resistor power rating should be 100 W, what is the smallest acceptable power rating for the 10-Ω resistor?

    (c) What is the smallest acceptable power rating for the 50-Ω resistor in the network shown?

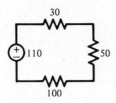

Rheostats and Adjustable Resistors

Adjustable resistors are given the descriptive symbol shown below. Often these resistors are constructed so that a metallic contact (represented by the arrow on the symbol) may be moved along a length of resistive material. If the device is intended to be adjusted frequently—for instance, if it is connected to a knob on an instrument panel or to a control motor shaft—it is called a *rheostat*. If it is intended to be adjusted infrequently—for instance, by loosening a screw and moving a contact plate—the device is called an *adjustable resistor*.

The specified resistance R of a rheostat or adjustable resistor is its maximum resistance. These devices are rated according to the maximum average power dissipation at the full resistance setting. At other than the full resistance setting, the device is capable of dissipating less power, the allowable power dissipation being proportional to the resistance setting. The power rating may be found by multiplying the full resistance by the square of the maximum current.

14. (a) It is desired to choose a resistor R_1 and a rheostat R_2 in the network so that the rheostat is capable of adjusting the current i from 1 A to 5 A. Find values for R_1 and R_2.

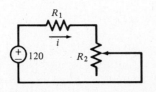

A printed circuit board is examined. The board consists of a sheet of insulating material upon which are bonded layers of copper conductors that have been shaped by a photographic etching process. (*Photo courtesy of Digital Equipment Corp.*)

(b) What is the smallest acceptable power rating for the rheostat in the network?

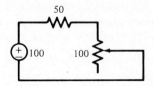

Resistor Models of Physical Devices

The resistor is also useful as a model for electrical devices, besides physical resistors, that convert electrical energy into other forms of energy. Lamps, lengths of wire, resistance heaters, even electric motors under the proper circumstances may be modeled by a resistor.

15. (a) A certain 110-V lamp draws 0.91 A when lit. What is the resistance of the lamp when lit? What is the electrical power flow into the lamp when it is connected to a constant 110-V source?

    (b) A certain electric heater consists of two identical 0.3-Ω heating elements that are simply lengths of nichrome wire strung on an insulating, heat-resistant holder. The two electrical elements are connected in series for low heat and in parallel for high heat. If the power source is a constant 120 V, compare the electrical power flows (into heat) at the low and at the high settings.

(c) A constant 220-V source is connected to eight 220-V, 500-W lamps connected in parallel. What current must the source provide?

Wire Gauge

The resistance of a conductor of uniform cross-sectional area A and length L is

$$R = \frac{\rho L}{A}$$

where ρ is the resistivity of the conducting material.

The American Wire Gauge (AWG) is a common way of specifying the diameter of wires of circular cross section. The wire diameter is often given in mils, which are 10^{-3} in. The cross-sectional area is then specified in *circular mils*, which is the square of the diameter in mils. Table 1-6 gives data for copper wire.

16. Specify the AWG gauge number of the smallest diameter of standard annealed solid copper wire for which 225 ft (67.58 m) has a resistance less than 0.25 Ω.

Resistor Temperature Coefficient

The resistance of a material generally depends somewhat on its temperature. A linear (or straight-line) approximation to a resistance, R as a function of temperature, T, is

$$R(T_2) = R(T_1)[1 + \alpha(T_2 - T_1)]$$

The number α is called the *temperature coefficient* of the resistor. It may be positive, indicating that the resistance increases with temperature, or for some materials negative, indicating a decrease in resistance with increasing temperature.

A *thermistor* is a resistive device that is especially constructed to have a large temperature coefficient. Thermistors are useful for temperature measurement.

17. A certain resistor has a temperature coefficient of $\alpha = 0.021/°C$ at 30°C. What is its resistance at 45°C if the resistance is 470 Ω at 30°C?

Ammeters

A permanent magnet type ammeter consists basically of a permanent magnet in proximity to a coil of wire through which the current to be measured flows. When current flows through the coil, it becomes an electromagnet, which is attracted to or repelled from the permanent magnet, depending on the direction of current flow. The coil is restrained by a spring, and the mechanism is usually designed so that the coil's deflection is proportional to the coil current.

Since the ammeter coil consists of a length of wire, a simple model for the ammeter is a resistance, called the *internal resistance* of the meter. In other words, to a first approximation at least, the meter "looks" like a resistor to the circuit in which it is connected.

Permanent magnet ammeters are often called *dc ammeters* because their most common use is for the measurement of constant or nearly constant currents. Such currents are called "direct current" (or dc).

18. Compare the current i in the circuit of (a) with the current i' of (b), which would

Table 1-6. DATA FOR STANDARD ANNEALED SOLID COPPER WIRE

AWG GAUGE NUMBER	DIAMETER IN mils	AREA IN CIRCULAR mils	RESISTANCE IN Ω/1000 ft	WEIGHT IN lb/1000 ft	TYPICAL MAX ALLOWABLE CURRENT IN AMPERES*
00	365	133,100	0.080	403	225
0	325	105,500	0.100	319	200
1	289	83,690	0.126	253	120
2	258	66,370	0.159	201	94.8
3	229	52,640	0.201	159	75.2
4	204	41,740	0.253	126	59.6
5	182	33,100	0.319	100	47.3
6	162	26,250	0.403	79.5	37.5
7	144	20,820	0.508	63.0	29.7
8	128	16,510	0.541	50.0	23.6
9	114	13,090	0.808	39.6	18.7
10	102	10,380	1.02	31.4	14.8
11	91	8,234	1.28	24.9	11.8
12	81	6,530	1.62	19.8	9.33
13	72	5,178	2.04	15.7	7.40
14	64	4,107	2.58	12.4	5.87
15	57	3,257	3.25	9.86	4.65
16	51	2,583	4.09	7.82	3.69
17	45	2,048	5.16	6.20	2.93
18	40	1,624	6.51	4.92	2.32
19	36	1,288	8.21	3.90	1.84
20	32	1,022	10.4	3.09	1.46
21	28.5	810	13.1	2.45	1.16
22	25.3	642	16.5	1.95	0.918
23	22.6	510	20.8	1.54	0.728
24	20.1	404	26.2	1.22	0.577
25	17.9	320	33.0	0.97	0.458

*Calculated on the basis of 700 circular mils per ampere.

be measured if an ammeter with 2-Ω internal resistance were inserted to measure the current. Find the percent of error due to the effect of the meter.

$$\left| \frac{i - i'}{i} \right| \times 100$$

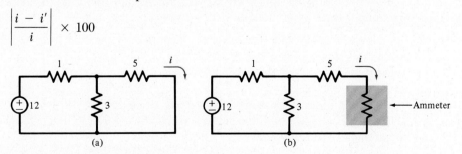

(a) (b)

19. An ammeter with internal resistance 0.5Ω is inserted in an electrical network as shown, and reads 6.25 A.

(a) What is the value of the resistance R?

(b) What would the percent error,

$$\left| \frac{R - R'}{R} \right| \times 100$$

in the calculated value of resistance R' be if the ammeter resistance were neglected?

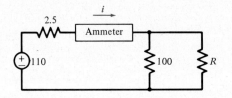

Ammeter Shunts

A convenient method of adjusting the scale of a permanent magnet ammeter so that the meter is less sensitive than it otherwise would be is to place a resistor, called a *meter shunt* resistor, in parallel with the meter coil. A fraction of the net current i through the shunt and coil combination flows through the ammeter coil itself:

$$i_{\text{coil}} = \left(\frac{R_{\text{shunt}}}{R_{\text{shunt}} + R_{\text{coil}}} \right) i$$

Thus if the shunt resistance is made equal to the coil resistance, an ammeter that would otherwise read 5 A full scale would, with such a shunt, read 10 A full scale.

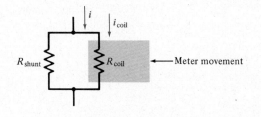

Although an ammeter shunt reduces the sensitivity of the ammeter, it also reduces the overall internal resistance of the instrument, thus reducing its effect on the network under test.

20. For an ammeter movement that reads 1 milliampere (1 mA) full scale and has an internal resistance of 250 Ω,
   (a) What should the shunt resistance be if the instrument, with the shunt, is to read 2.5 A full scale?
   (b) What is the internal resistance of this 2.5-A ammeter?
   (c) What electrical power will the shunt resistor be required to dissipate when the ammeter reads full scale?

Voltmeters

If the internal resistance of a permanent magnet ammeter is deliberately made very large, by winding the coil with many turns of fine wire, for instance, it may be used to measure voltage. The meter responds to its coil current, which is proportional to the voltage across the instrument's terminals.

As with the ammeter, a simple model for such a dc voltmeter is a resistance, the internal resistance of the meter.

21. Compare the voltage v in the circuit of (a) with the voltage v in (b), which would be measured if a voltmeter with 20 kilohms (kΩ) internal resistance were attached to measure the voltage. Find the percent of error due to the effect of the meter,

$$\left| \frac{v - v'}{v} \right| \times 100$$

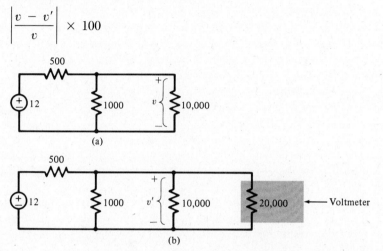

(a)

(b)

22. A voltmeter with internal resistance 5000 Ω is attached in an electrical network as shown and reads 32.5 V.
(a) What is the value of the resistance R?
(b) What would the percent error,

$$\left| \frac{R - R'}{R} \right| \times 100$$

in the calculated value of resistance R' be if the voltmeter resistance were neglected?

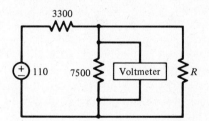

Voltmeter Scale Change
A convenient method of adjusting the scale of a permanent magnet voltmeter so that the meter is less sensitive than it otherwise would be is to place a resistor in

series with the meter coil. For a fixed voltage across the resistor and coil combination, less current will flow through the coil the higher this series resistance:

$$i = \frac{v}{R_{\text{series}} + R_{\text{coil}}}$$

Thus if the series resistance is made equal to the coil resistance, a voltmeter that would otherwise read 2 V full scale would, with such a series resistance, read 4 V full scale.

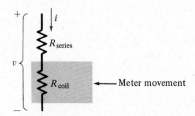

Although the series resistance reduces the sensitivity of the voltmeter, it also increases the overall resistance of the instrument, thus reducing its effect on the network under test.

It is common to construct permanent magnet voltmeters with standard internal resistance (including the series resistor) such as 20,000 ohms per volt of full scale reading. A 20,000 Ω/V voltmeter that reads 15 V full scale would have an internal resistance of 15(20,000) = 300 kΩ.

23. (a) For a 20,000 Ω/V voltmeter that reads 0.5 V full scale, what series resistance should be added to make the instrument read 25 V full scale?
    (b) Show that if a voltmeter internal resistance is R Ω/V of full-scale reading, the addition of a series resistor results in the same number of ohms per volt of the new full-scale reading.

Battery Internal Resistance
An adequate model for a battery under most circumstances consists of a constant voltage source in series with a resistor. The source voltage is called the *open-circuit voltage* or simply the battery voltage, and the resistor value for a given battery is called its *internal resistance*.

24. An ammeter with 0.7-Ω internal resistance is connected across a 110-V battery with a 0.65-Ω internal resistance. What current flows?
25. A 24-V battery is formed by connecting two 12-V batteries in series. If the 12-V batteries have internal resistances of 0.15 and 0.21 Ω, respectively, what is the internal resistance of the 24-V battery?
26. A 12.4-V battery with an internal resistance of 0.38 Ω is recharged by connecting it to a battery charger that may be modeled by a 14.2-V source in series with 10.5 Ω. Find the battery current. When recharging, the electrical power flow is into the battery, of course, so the battery current is in the opposite direction from that when the battery is supplying power.

Advanced Problems

Network Specification in a Computer Program

27. Carefully describe a method by which you could, without ambiguity, easily specify a network in words, not a drawing. When such a specification is given in a mutually understood format, it might be used to enter the description of a network into a digital computer.

Resistor Models

28. (a) Resistor models that have negative values of resistance are useful as models for some electronic devices. For the network of (a), which contains a negative resistor, find *i*.

(b) Resistor models for which the resistance varies with time are useful as models for some physical devices. For the network of (b), which contains a time varying resistor, find *v*(t).

(c) Resistor models for which the resistance depends on some other signal, such as a network voltage or current, are useful in modeling some electronic systems. For the network of (c), which contains a current-controlled resistor, find the two possible solutions for *v*.

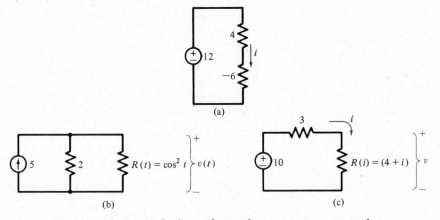

(a)

(b)   (c)

29. A short circuit is a length of (ideal) conductor between two terminals; an open circuit exists when there is no connection between the terminals. A short circuit may be thought of as a voltage source of 0 V or as a resistor of 0 Ω. What are analogous models for an open circuit?

Electrical Power Flow

30. The electrical power flow out of a certain two-terminal element is

$$p_{out}(t) = i^4(t)$$

where *i*(t) is the element current. What is the sink reference voltage-current relation for this element?

31. The power dissipated by a certain 10-Ω resistor as a function of time is

$$p_{into \ 10 \ \Omega}(t) = |\sin t|$$

Graph four of the infinite number of possible different resistor currents.

32. Two resistors are connected end-to-end (in series) so that the same current flows through each. The voltage across one resistor is twice that across the other. How do the power flows into the two resistors compare?

Element Equivalence
33. An element's voltage-current relation is unchanged if it is turned end for end, providing the voltage-current relation itself is unchanged if v is replaced by $-v$ and i is replaced by $-i$. This is to say that the element is equivalent to itself turned end for end. Such an element is said to be *bilateral*. For example, in the sink reference voltage-current relation for the resistor, $v = Ri$, replacing v by $-v$ and i by $-i$ still gives $v = Ri$.

   Apply this test to demonstrate that (nonzero) voltage sources and current sources are *not* bilateral elements. Would an element with voltage-current relation

   $$v = 10i^2$$

   be bilateral?

Equivalent Resistors
34. Show that the equivalent resistance of three resistors, R_1, R_2, and R_3, in parallel is

   $$R_{equiv} = \frac{R_1 R_2 R_3}{R_1 R_2 + R_2 R_3 + R_1 R_3}$$

35. Why is there no particular advantage to combining parallel resistors in a two-node network before writing the two-node equation?
36. Show that the sum of the electrical power flows into a parallel connection of resistors at any instant of time is equal to the electrical power flow into the equivalent resistor.

   It is true in general that the net electrical power flow into any two-terminal combination of resistors is the same as the electrical power flow into the equivalent resistor.

Thévenin-Norton Transformation
37. If a Thévenin element, a voltage source in series with a resistor, is replaced by this Norton equivalent in a network, show that
   (a) The electrical power flows into the Thévenin element and into the Norton element are identical.
   (b) The electrical power flows into the Thévenin resistor and into the Norton resistor are not necessarily the same.

   The Thévenin-Norton equivalence is only for voltages and currents external to the elements. The net power flows into the Thévenin and Norton elements may be expressed in terms of external voltages and currents, but the Thévenin and Norton resistor power flows involve a voltage or a current *internal* to the element.

Chapter Two
Source-Resistor Network Solutions

Yes! the apparatus of which I speak, and which will
doubtless astonish you, is only an assemblage of a number of
good conductors of different sorts arranged in a certain way.
Thirty, forty, sixty pieces or more of copper, or better of
silver, each in contact with a piece of tin, or what is much
better, of zinc and an equal number of layers of water or
some other liquid which is a better conductor than pure
water, such as salt-water or lye and so forth. . . .

Alessandro Volta
From a communication to the Royal Society,
Milan, 1800.

2.1 PREVIEW

Step-by-step methods of expressing network problems, no matter how complicated,
as a set of mathematical equations are important in several ways. Certain types of
problems, for instance those involving controlled sources, are very difficult to handle
in any other manner. When a large network problem is subjected to computer-aided
analysis, efficient programming requires a systematic procedure that is lacking in
equivalent circuit methods. Also, well-known properties of the equations describing
networks imply certain important properties of the networks themselves.

Networks consisting of just sources and resistors are considered here first because all of the techniques for the solution of source-resistor networks apply to the
solution of networks in general. Before proceeding to develop equation-writing methods
for source-resistor networks, linear algebraic equations are briefly reviewed and
Cramer's rule is summarized. Two systematic methods, nodal and mesh equations,
for writing sets of independent network equations in terms of a relatively small
number of variables are presented. These methods are extensions of the two-node
and single-loop approaches of Chapter 1.

After characterizing source-resistor networks by simultaneous linear algebraic
equations, some important properties of the networks that follow from corresponding
characteristics of their equations are investigated. These properties are fundamental
to a deeper understanding of networks and are most important in network design
and in the subsequent analysis of more advanced networks here.

Simultaneous network equations are applied to the determination of equivalent resistances, particularly in networks involving controlled sources, such as those commonly encountered in electronic design. Conditions for superimposing sources and source components are shown and demonstrated, then the Thévenin and Norton equivalents are derived, using Cramer's rule. The maximum power transfer theorem is developed, and network transfer ratios are discussed.

When you complete this chapter, you should know—

1. how to solve simultaneous linear algebraic equations, using Cramer's rule, evaluating determinants by Laplace expansion when necessary;
2. how to write systematic simultaneous nodal equations;
3. how to accommodate voltage sources in nodal equations;
4. how to write systematic simultaneous mesh equations;
5. how to accommodate current sources in mesh equations;
6. how to handle controlled sources in nodal equations and in mesh equations;
7. how to find the equivalent resistance of any two-terminal combination of resistors and controlled sources;
8. the superposition property and how to use it for sources, for source components, and in the presence of controlled sources;
9. how to find Thévenin and Norton equivalents, and how to use them in the solution of networks;
10. the maximum power transfer theorem and how to use it;
11. what network transfer ratios are and how to find and use them.

2.2 CRAMER'S RULE

The solution of a set of n independent linear algebraic equations in n variables,

$$a_{11}x_1 + a_{12}x_2 + \cdots + a_{1n}x_n = y_1$$
$$a_{21}x_1 + a_{22}x_2 + \cdots + a_{2n}x_n = y_2$$
$$\vdots$$
$$a_{n1}x_1 + a_{n2}x_2 + \cdots + a_{nn}x_n = y_n$$

is given by Cramer's rule,

$$x_i = \frac{\Delta_i}{\Delta}$$

where Δ is the determinant of the equations,

$$\Delta = \begin{vmatrix} a_{11} & a_{12} & \cdots & a_{1n} \\ a_{21} & a_{22} & \cdots & a_{2n} \\ \vdots & \vdots & & \vdots \\ a_{n1} & a_{n2} & \cdots & a_{nn} \end{vmatrix}$$

and Δ_i is the same as Δ except that the ith column is replaced by the column of "knowns":

ith column

$$\Delta_i = \begin{vmatrix} a_{11} & \cdots & y_1 & \cdots & a_{1n} \\ a_{21} & \cdots & y_2 & \cdots & a_{2n} \\ \vdots & & \vdots & & \vdots \\ a_{n1} & \cdots & y_n & \cdots & a_{nn} \end{vmatrix}$$

For example, in

$$\begin{cases} 2x_1 - x_2 - x_3 = 8\cos t \\ 3x_1 + 2x_2 \qquad\quad = 0 \\ -x_1 \qquad\quad + 4x_3 = 12 \end{cases}$$

$$x_1 = \frac{\begin{vmatrix} 8\cos t & -1 & -1 \\ 0 & 2 & 0 \\ 12 & 0 & 4 \end{vmatrix}}{\begin{vmatrix} 2 & -1 & -1 \\ 3 & 2 & 0 \\ -1 & 0 & 4 \end{vmatrix}}$$

$$x_2 = \frac{\begin{vmatrix} 2 & 8\cos t & -1 \\ 3 & 0 & 0 \\ -1 & 12 & 4 \end{vmatrix}}{\begin{vmatrix} 2 & -1 & -1 \\ 3 & 2 & 0 \\ -1 & 0 & 4 \end{vmatrix}}$$

$$x_3 = \frac{\begin{vmatrix} 2 & -1 & 8\cos t \\ 3 & 2 & 0 \\ -1 & 0 & 12 \end{vmatrix}}{\begin{vmatrix} 2 & -1 & -1 \\ 3 & 2 & 0 \\ -1 & 0 & 4 \end{vmatrix}}$$

The *cofactor* of any element of a square array is the determinant of the array formed by deleting the row and column of that element and multiplying by one or minus one, as given by the "checkerboard" of signs.

$$\begin{vmatrix} + & - & + & - & \cdots \\ - & + & - & + & \cdots \\ + & - & + & - & \cdots \\ \vdots & & & & \end{vmatrix}$$

For example, the cofactor of the element in row 2, column 3, in

$$\begin{vmatrix} 1 & -2 & 3 \\ 4 & 5 & ⑥ \\ 7 & 8 & -9 \end{vmatrix}$$

is

$$\Delta_{23} = - \begin{vmatrix} 1 & -2 \\ 7 & 8 \end{vmatrix} = -22$$

└─ from the "checkerboard" pattern

The value of a 2×2 (two rows by two columns) determinant is

$$\begin{vmatrix} c_1 & c_2 \\ c_3 & c_4 \end{vmatrix} = c_1 c_4 - c_2 c_3$$

Values of the determinants of larger square arrays may be found by *Laplace expansion*.

A determinant may be Laplace expanded along any row or any column. The value of the determinant is the sum of the products of each element in the row or column times their cofactors. The following is an example of Laplace expansion:

$$\begin{vmatrix} -1 & 2 & 3 & -4 \\ 5 & 6 & 0 & 8 \\ 0 & 0 & 9 & 1 \\ 4 & 0 & -3 & -2 \end{vmatrix} = -2 \begin{vmatrix} 5 & 0 & 8 \\ 0 & 9 & 1 \\ 4 & -3 & -2 \end{vmatrix} + 6 \begin{vmatrix} -1 & 3 & -4 \\ 0 & 9 & 1 \\ 4 & -3 & -2 \end{vmatrix}$$

$$= -2 \left[5 \begin{vmatrix} 9 & 1 \\ -3 & -2 \end{vmatrix} + 4 \begin{vmatrix} 0 & 8 \\ 9 & 1 \end{vmatrix} \right] + 6 \left[9 \begin{vmatrix} -1 & -4 \\ 4 & -2 \end{vmatrix} - \begin{vmatrix} -1 & 3 \\ 4 & -3 \end{vmatrix} \right]$$

$$= -2[5(-15) + 4(-72)] + 6[9(18) - (-9)]$$

An alternate method of evaluation for 3×3 determinants is to repeat the first two columns and sum three positive three-number products and three negative three-number products as in the example below:

$$\begin{vmatrix} 2 & 1 & -1 \\ 3 & 0 & 1 \\ -2 & 4 & 5 \end{vmatrix} = \begin{array}{l} (2)(0)(5) + (1)(1)(-2) + (-1)(3)(4) \\ -(-2)(0)(-1) - (4)(1)(2) - (5)(3)(1) \end{array}$$

$$= 0 - 2 - 12 + 0 - 8 - 15 = -37$$

Positive
products

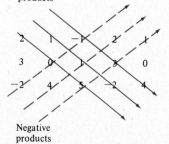

Negative
products

Consider the set of three linear algebraic equations in three variables

$$\begin{cases} x_1 - 2x_2 + x_3 = 4\cos 3t \\ -x_1 + 3x_2 \quad\quad = -12 \\ 2x_1 \quad\quad + 4x_3 = 0 \end{cases}$$

Using Cramer's rule, the solution for x_1, x_2, and x_3 are as follows:

$$x_1(t) = \frac{\Delta_1}{\Delta} = \frac{\begin{vmatrix} 4\cos 3t & -2 & 1 \\ -12 & 3 & 0 \\ 0 & 0 & 4 \end{vmatrix}}{\begin{vmatrix} 1 & -2 & 1 \\ -1 & 3 & 0 \\ 2 & 0 & 4 \end{vmatrix}} = \frac{48\cos 3t - 96}{2\begin{vmatrix} -2 & 1 \\ 3 & 0 \end{vmatrix} + 4\begin{vmatrix} 1 & -2 \\ -1 & 3 \end{vmatrix}} = -2$$

$$x_2(t) = \frac{\Delta_2}{\Delta} = \frac{\begin{vmatrix} 1 & 4\cos 3t & 1 \\ -1 & -12 & 0 \\ 2 & 0 & 4 \end{vmatrix}}{-2} = \frac{16\cos 3t - 24}{-2}$$

$$x_3(t) = \frac{\Delta_3}{\Delta} = \frac{\begin{vmatrix} 1 & -2 & 4\cos 3t \\ -1 & 3 & -12 \\ 2 & 0 & 0 \end{vmatrix}}{-2} = \frac{-24\cos 3t + 48}{-2}$$

For more than three equations in three variables, Cramer's rule with the Laplace expansion of determinants becomes very inefficient in comparison to other solution methods. The long-range interest in Cramer's rule here is twofold. First, it is simple and straightforward to apply in the case of a small number of equations, say two or three. It would be wise to seek digital computer aid in the solution of larger problems. Second, Cramer's rule is a very good theoretical tool for showing network properties.

D2-1

Solve the following simultaneous linear algebraic equations using Cramer's rule:

(a) $\begin{cases} -2x_1 + x_2 = 0 \\ 8x_1 - 3x_2 = -4 \end{cases}$ *ans.* $-2, -4$

(b) $\begin{cases} x_1 - 3x_2 = -\sin t \\ -x_1 + x_2 = 12 \end{cases}$ *ans.* $\frac{1}{2}\sin t - 18, \frac{1}{2}\sin t - 6$

(c) $\begin{cases} 3x_1 - x_2 - 3x_3 = -5 \\ -x_1 + 4x_2 \quad\quad = 0 \\ -3x_1 \quad\quad + 7x_3 = 6 \end{cases}$ *ans.* $-\frac{68}{41}, -\frac{17}{41}, \frac{6}{41}$

(d) $\begin{cases} 2x_1 + x_2 + x_3 = 0 \\ 7x_1 + 8x_2 - x_3 = 0 \\ -6x_1 + 9x_2 + x_3 = 0 \end{cases}$ *ans.* $0, 0, 0$

CRAMER'S RULE

The solution of a set of n independent linear algebraic equations in n variables,

$$
\begin{cases}
a_{11}x_1 + a_{12}x_2 + \cdots + a_{1n}x_n = y_1 \\
a_{21}x_1 + a_{22}x_2 + \cdots + a_{2n}x_n = y_2 \\
\quad \vdots \\
a_{n1}x_1 + a_{n2}x_2 + \cdots + a_{nn}x_n = y_n
\end{cases}
$$

is given by

$$
x_i = \frac{\Delta_i}{\Delta}
$$

2.3 SYSTEMATIC NODAL EQUATIONS

2.3.1 Node-to-Node Voltages and Notation

For a network with n nodes, the voltages between any one of the nodes and each of the remaining $(n - 1)$ nodes form a convenient set of independent variables. For example, in the network of Figure 2-1(a), which has five nodes, a set of four such node-to-node voltages is $v_1(t)$, $v_2(t)$, $v_3(t)$, and $v_4(t)$.

The network for this example has been drawn in a manner that makes the nodes especially evident, and this is generally helpful. It is convenient, too, to define the node-to-node voltages all with negative reference polarity signs at the same node, as shown in Figure 2-1(b). Instead of cluttering up the drawing with all those signs and brackets, the notation of numbering the nodes 0, 1, 2, . . . , shown in Figure 2-1(c), is adopted. The zero node (or *reference* node) is the node that is common to all the voltages. The voltage v_3, for example, is the voltage between node 3 and node 0, with the plus sign of the voltage reference polarity at node 3.

If these four voltages were known, any other voltage or current in the network would be quite easy to find. The indicated signals in Figure 2-2 are related to the node voltages and source functions as follows:

$$
i_1 = \frac{v_1}{4}
$$

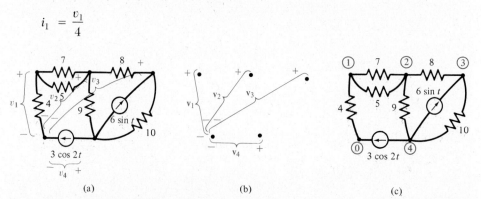

(a) (b) (c)

Figure 2-1   A network with five nodes.

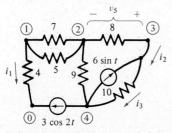

Figure 2-2   Finding other network signals from the node voltages.

$$v_5 = v_3 - v_2$$

$$i_2 = i_3 - 6 \sin t = \frac{v_3 - v_4}{10} - 6 \sin t$$

2.3.2 Equations for Each Independent Node

By applying Kirchhoff's current law, an independent equation may be written for each of the numbered nodes, 1, 2, ... For the first node in the example under discussion, Figure 2-3(a),

$$i_a + i_b + i_c = 0$$

Substituting for the resistor currents in terms of the node voltages,

$$\frac{v_1}{4} + \frac{v_1 - v_2}{5} + \frac{v_1 - v_2}{7} = 0$$

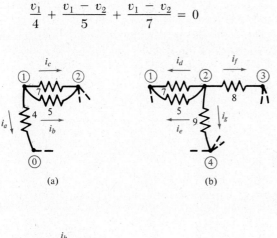

(a)   (b)

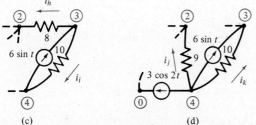

(c)   (d)

Figure 2-3   Node-by-node analysis of the example network.

For the second node, Figure 2-3(b),

$$i_d + i_e + i_f + i_g = 0$$

$$\frac{v_2 - v_1}{7} + \frac{v_2 - v_1}{5} + \frac{v_2 - v_3}{8} + \frac{v_2 - v_4}{9} = 0$$

For the third node, Figure 2-3(c),

$$i_h + i_i = 6 \sin t$$

$$\frac{v_3 - v_2}{8} + \frac{v_3 - v_4}{10} = 6 \sin t$$

For the fourth node, Figure 2-3(d),

$$i_j + i_k = -6 \sin t - 3 \cos 2t$$

$$\frac{v_4 - v_2}{9} + \frac{v_4 - v_3}{10} = -6 \sin t - 3 \cos 2t$$

Rearranging these four equations gives

$$\begin{cases} (\frac{1}{4} + \frac{1}{5} + \frac{1}{7})v_1 - (\frac{1}{5} + \frac{1}{7})v_2 - (0)v_3 - (0)v_4 = 0 \\ -(\frac{1}{5} + \frac{1}{7})v_1 + (\frac{1}{5} + \frac{1}{7} + \frac{1}{8} + \frac{1}{9})v_2 - (\frac{1}{8})v_3 - (\frac{1}{9})v_4 = 0 \\ -(0)v_1 - (\frac{1}{8})v_2 + (\frac{1}{8} + \frac{1}{10})v_3 - (\frac{1}{10})v_4 = 6 \sin t \\ -(0)v_1 - (\frac{1}{9})v_2 - (\frac{1}{10})v_3 + (\frac{1}{9} + \frac{1}{10})v_4 = -6 \sin t - 3 \cos 2t \end{cases}$$

which may be solved for the node-to-node voltages v_1, v_2, v_3, and v_4.

If an equation were to be written for the zero node, it would be a linear combination of the other equations and thus redundant.

2.3.3 General Form of the Equations

In general, systematic nodal equations have the form

$$g_{11}v_1 - g_{12}v_2 - g_{13}v_3 - \cdots = i_1$$

$$-g_{21}v_1 + g_{22}v_2 - g_{23}v_3 - \cdots = i_2$$

$$-g_{31}v_1 - g_{32}v_2 + g_{33}v_3 - \cdots = i_3$$

$$\vdots$$

where the conductances

g_{11} = sum of inverses of resistances connected directly to node 1

g_{22} = sum of inverses of resistances connected directly to node 2

g_{33} = sum of inverses of resistances connected directly to node 3

$$\vdots$$

and the other coefficients, which are conductances, are

g_{mn} = sum of inverses of resistances connected directly between nodes m and n

$\quad = g_{nm}$

For example,

g_{12} = sum of inverses of resistances connected directly between nodes 1 and 2

g_{13} = sum of inverses of resistances connected directly between nodes 1 and 3

g_{21} = sum of inverses of resistances connected directly between nodes 2 and 1

g_{23} = sum of inverses of resistances connected directly between nodes 2 and 3

The driving function terms involve the source currents as follows:

$$i_1 = \begin{pmatrix} \text{sum of currents from current} \\ \text{sources connected directly to} \\ \text{node 1, with reference} \\ \text{directions toward node 1} \end{pmatrix} - \begin{pmatrix} \text{sum of currents from current} \\ \text{sources connected directly to} \\ \text{node 1, with reference} \\ \text{directions away from node 1} \end{pmatrix}$$

$$i_2 = \begin{pmatrix} \text{sum of currents from current} \\ \text{sources connected directly to} \\ \text{node 2, with reference} \\ \text{directions toward node 2} \end{pmatrix} - \begin{pmatrix} \text{sum of currents from current} \\ \text{sources connected directly to} \\ \text{node 2 with reference} \\ \text{directions away from node 2} \end{pmatrix}$$

$$i_3 = \begin{pmatrix} \text{sum of currents from current} \\ \text{sources connected directly to} \\ \text{node 3, with reference} \\ \text{directions toward node 3} \end{pmatrix} - \begin{pmatrix} \text{sum of currents from current} \\ \text{sources connected directly to} \\ \text{node 3, with reference} \\ \text{directions away from node 3} \end{pmatrix}$$

$\vdots$

Each equation is the expression of Kirchhoff's current law at one node, with resistor currents written in terms of the node voltages.

Another example network is given in Figure 2-4. The systematic nodal equations, in terms of the indicated node voltages, are as follows:

$$\begin{cases} (\tfrac{1}{5} + \tfrac{1}{6} + \tfrac{1}{4})v_1 - (\tfrac{1}{6})v_2 - (\tfrac{1}{4})v_3 - (0)v_4 = 9\sin t - 12 + \cos 10t \\ -(\tfrac{1}{6})v_1 + (\tfrac{1}{2} + \tfrac{1}{3} + \tfrac{1}{6} + \tfrac{1}{7})v_2 - (\tfrac{1}{2} + \tfrac{1}{3})v_3 - (\tfrac{1}{7})v_4 = 12 \\ -(\tfrac{1}{4})v_1 - (\tfrac{1}{2} + \tfrac{1}{3})v_2 + (\tfrac{1}{2} + \tfrac{1}{3} + \tfrac{1}{4})v_3 - (0)v_4 = -\cos 10t \\ -(0)v_1 - (\tfrac{1}{7})v_2 - (0)v_3 + (\tfrac{1}{7} + \tfrac{1}{8})v_4 = 0 \end{cases}$$

Once these equations are solved for v_1, v_2, v_3, and v_4, then

$$i_1 = -\frac{v_4}{8} \qquad i_2 = \frac{v_3 - v_2}{2} \qquad v_5 = v_2 - v_4$$

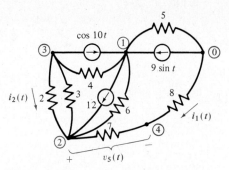

Figure 2-4 Another example network.

Note the symmetry of the coefficients about the diagonal elements. Any resistor connected between nodes a and b is also connected between nodes b and a. The negative *coupling* terms in each equation are also contained within the positive *self* terms. That is, the self terms in each equation are equal to or larger than the negative sum of the coupling terms.

If a current source is connected between two nodes for which equations are written, the source function appears with a plus sign in one equation and a minus sign in the other, since the reference direction of the source must be away from one node and toward the other. These properties provide an easy (although incomplete) check on the equations.

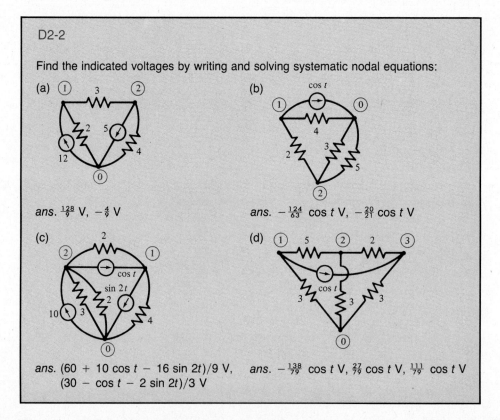

D2-2

Find the indicated voltages by writing and solving systematic nodal equations:

(a)

ans. $\frac{128}{9}$ V, $-\frac{4}{9}$ V

(b)

ans. $-\frac{124}{63}$ cos t V, $-\frac{20}{21}$ cos t V

(c)

ans. $(60 + 10 \cos t - 16 \sin 2t)/9$ V,
$(30 - \cos t - 2 \sin 2t)/3$ V

(d)

ans. $-\frac{138}{79}$ cos t V, $\frac{27}{79}$ cos t V, $\frac{111}{79}$ cos t V

D2-3

Write systematic nodal equations, solve for the node voltages, then find the two indicated signals:

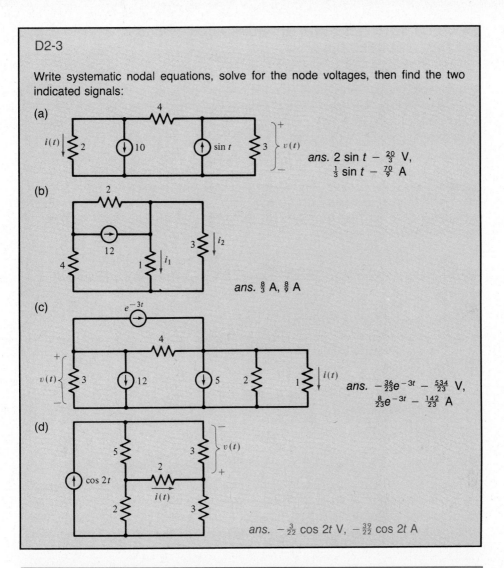

(a)

ans. $2 \sin t - \frac{20}{3}$ V, $\frac{1}{3} \sin t - \frac{70}{9}$ A

(b)

ans. $\frac{8}{3}$ A, $\frac{8}{9}$ A

(c)

ans. $-\frac{36}{23}e^{-3t} - \frac{534}{23}$ V, $\frac{8}{23}e^{-3t} - \frac{142}{23}$ A

(d)

ans. $-\frac{3}{22} \cos 2t$ V, $-\frac{39}{22} \cos 2t$ A

SYSTEMATIC SIMULTANEOUS NODAL EQUATIONS

Node voltages are indicated on the network diagram by numbering the nodes 0, 1, 2, ... An equation is written for each of the nodes except the zero node, in terms of the node voltages v_1, v_2, ...

The coefficient of v_n in the nth equation is the sum of the inverses of the resistances connected directly to node n. The coefficient of v_m, $m \neq n$, in the nth equation is the negative sum of the inverses of the resistances connected directly between node n and node m.

The driving function, on the opposite side of each equation from the variables, is, for the nth equation, the sum of the current source functions for sources connected directly to node n, with reference directions toward node n minus those with reference directions away from node n.

2.4 NODAL EQUATIONS FOR NETWORKS CONTAINING VOLTAGE SOURCES

If there is a voltage source between nodes, the current flowing through the source is unknown. On the other hand, the node-to-node voltage is known; it is fixed by the voltage-source function. Nodal equations for such a network may be written as in the usual case, where voltage sources are treated as current sources with unknown current functions. Then the known node-to-node voltages may be added to the set of equations.

An example network is shown in Figure 2-5. For this network the currents through each voltage source are defined, and the usual nodal equations are written in terms of these source currents. Then for each voltage source an additional equation is added to the set that expresses the known node-to-node voltage fixed by the voltage source:

$$\left\{\begin{array}{l} \tfrac{1}{3}v_1 - \tfrac{1}{3}v_2 = -i_1 \\ -\tfrac{1}{3}v_1 + (\tfrac{1}{3} + \tfrac{1}{2})v_2 = -i_2 \\ \tfrac{1}{4}v_3 = \sin t + i_2 \\ v_1 = 5 \cos t \\ -v_2 + v_3 = 12 \end{array}\right\} \begin{array}{l} \text{basic nodal equations with} \\ \text{unknown currents supplied} \\ \text{by voltage sources} \\ \text{node-to-node voltages fixed} \\ \text{by voltage sources} \end{array}$$

The mathematical problem is now to solve the five equations in five variables, v_1, v_2, v_3, i_1, and i_2.

$$\left\{\begin{array}{l} \tfrac{1}{3}v_1 - \tfrac{1}{3}v_2 + i_1 = 0 \\ -\tfrac{1}{3}v_1 + (\tfrac{1}{3} + \tfrac{1}{2})v_2 + i_2 = 0 \\ \tfrac{1}{4}v_3 - i_2 = \sin t \\ v_1 = 5 \cos t \\ -v_2 + v_3 = 12 \end{array}\right. \tag{2-1}$$

Although the voltage sources add to the number of variables and equations, it is always easy to eliminate the voltage-source current variables. In this example, adding the second and third equations to eliminate i_2 and ignoring the first equation (which is the only equation involving the variable i_1) reduces the problem to three simultaneous equations in v_1, v_2 and v_3:

$$\left\{\begin{array}{l} -\tfrac{1}{3}v_1 + (\tfrac{1}{3} + \tfrac{1}{2})v_2 + \tfrac{1}{4}v_3 = \sin t \\ v_1 = 5 \cos t \\ -v_2 + v_3 = 12 \end{array}\right. \tag{2-2}$$

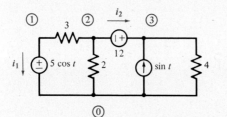

Figure 2-5   A network with voltage sources.

These have solution

$$v_1 = 5 \cos t \text{ V}$$

$$v_2 = -\tfrac{36}{13} + \tfrac{12}{13} \sin t + \tfrac{20}{13} \cos t \text{ V}$$

$$v_3 = \tfrac{120}{13} + \tfrac{12}{13} \sin t + \tfrac{20}{13} \cos t \text{ V}$$

The source currents are given by the first and third (or second) equations of the original set:

$$\tfrac{1}{3}v_1 - \tfrac{1}{3}v_2 + i_1 = 0 \tag{2-3}$$

$$i_1 = -\tfrac{12}{13} + \tfrac{4}{13} \sin t - \tfrac{15}{13} \cos t \text{ A}$$

and

$$\tfrac{1}{4}v_3 - i_2 = \sin t \tag{2-4}$$

$$i_2 = \tfrac{30}{13} - \tfrac{10}{13} \sin t + \tfrac{5}{13} \cos t \text{ A}$$

Rather than to develop additional procedures to form the decoupled equations (2-2), (2-3), (2-4) from the network, it is generally far easier (and there is less chance of error) to write the simple and straightforward coupled equations (2-1), then solve them, as was done here.

D2-4

Find the indicated signals by writing and solving nodal equations:

(a)

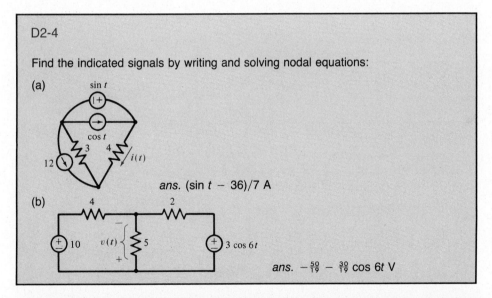

ans. $(\sin t - 36)/7$ A

(b)

ans. $-\tfrac{50}{19} - \tfrac{30}{19} \cos 6t$ V

VOLTAGE SOURCES IN NODAL EQUATIONS

Define the current through each voltage source and write the systematic nodal equations as if these source currents were known.

Augment the set of equations with one equation for each voltage source, expressing a known node voltage or difference between two node voltages.

2.5 CONTROLLED SOURCES AND NODAL EQUATIONS

In some applications, it is convenient to include voltage and current sources in which the source function is not given as a specific function of time but, rather, is expressed in terms of some other voltage or current in the network. These *controlled sources* are denoted by diamond-shaped symbols on a network diagram. To write nodal equations for networks containing controlled sources, first express the controlling signal in terms of the node voltages. Then write the ordinary nodal equations. The controlled source functions, which appear on the right side of the equations, will involve the node voltages. The equations may then be rearranged, placing all the unknowns on the left side of the equations.

An example is given in Figure 2-6(a). In Figure 2-6(b), the controlling signal for the controlled source, i_1, is expressed in terms of the node voltages as

$$i_1 = \frac{v_1}{5}$$

The ordinary nodal equations, with the source expressed this way, are

$$\begin{cases} (\frac{1}{3} + \frac{1}{5})v_1 & - \frac{1}{3}v_2 = 2v_1 \\ - \frac{1}{3}v_1 + (\frac{1}{3} + \frac{1}{4})v_2 = -6e^{-7t} \end{cases}$$

The equations may then be rearranged before solving as follows:

$$\begin{cases} (\frac{1}{3} + \frac{1}{5} - 2)v_1 & - \frac{1}{3}v_2 = 0 \\ - \frac{1}{3}v_1 + (\frac{1}{3} + \frac{1}{4})v_2 = -6e^{-7t} \end{cases}$$

Or,

$$\begin{cases} -22v_1 - 5v_2 = 0 \\ -4v_1 + 7v_2 = -72e^{-7t} \end{cases}$$

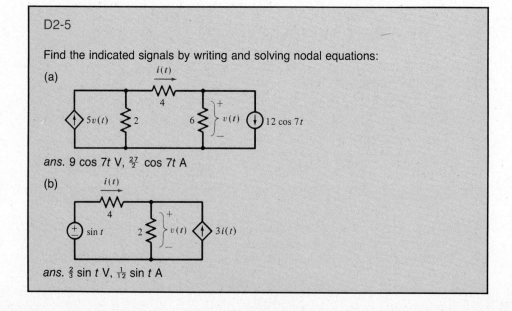

D2-5

Find the indicated signals by writing and solving nodal equations:

(a)

ans. $9 \cos 7t$ V, $\frac{27}{2} \cos 7t$ A

(b)

ans. $\frac{2}{3} \sin t$ V, $\frac{1}{12} \sin t$ A

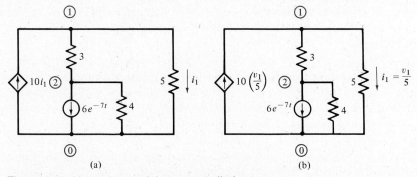

Figure 2-6 Network containing a controlled source.

CONTROLLED SOURCES IN NODAL EQUATIONS

Express the controlling signals in terms of the node voltages, then write the ordinary nodal equations as if the controlled source functions were known. Rearrange the equations, placing the unknowns on the left and the knowns on the right, and solve.

2.6 SYSTEMATIC MESH EQUATIONS

2.6.1 Mesh Currents for Planar Networks

A network is said to be *planar* if it can be drawn on a flat surface with no conductor crossovers, such as the network in Figure 2-7(a). A network with elements along the sides of a cube and the diagonals is an example of a nonplanar network. A planar network is composed of "boxes," or *meshes*, and current flows in such a network may be indicated by placing a current symbol in each of the "boxes" such as i_1 and i_2 in Figure 2-7(a). By this notation it is meant, for example, that the current i_a through the 2-Ω resistor is i_1. The current i_b is $i_1 - i_2$, and the current i_c is $i_2 - i_1$.

Mesh currents are a way of indicating Kirchhoff's current law schematically.

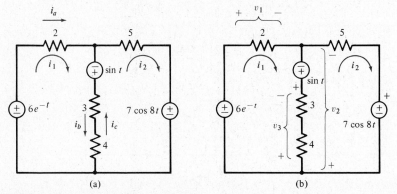

(a)   (b)

Figure 2-7 Example planar network.

Currents through elements that are in a single "box" are mesh currents. Currents through elements that are common to two "boxes" are the difference between two of the mesh currents.

The reference directions of the mesh currents are all chosen to have the same sense, all clockwise or all counterclockwise, so that currents in elements common to two "boxes" are always the difference between two mesh currents. It is not always possible to choose the reference directions so that these currents in elements common to two "boxes" are the sum of two mesh currents.

Knowing the mesh currents, it is simple to find any other current in the network and thus any network voltages as well. Referring to Figure 2-7(b), in which several signals in the example network have been indicated,

$$v_1 = 2i_1$$

$$v_2 = v_3 + \sin t = 7(i_2 - i_1) + \sin t$$

2.6.2 Equations for Each Mesh

For the example under discussion, defining each resistor voltage and applying Kirchhoff's voltage law around the first mesh, shown in Figure 2-8(a),

$$-6e^{-t} + v_a - \sin t + v_b + v_c = 0$$

Substituting for the resistor voltages in terms of the mesh currents,

$$-6e^{-t} + 2i_1 - \sin t + 3(i_1 - i_2) + 4(i_1 - i_2) = 0$$

$$(2 + 3 + 4)i_1 - (3 + 4)i_2 = 6e^{-t} + \sin t$$

For the second mesh, Figure 2-8(b),

$$v_d + 7 \cos 8t + v_f + v_e + \sin t = 0$$

$$5i_2 + 7 \cos 8t + 4(i_2 - i_1) + 3(i_2 - i_1) + \sin t = 0$$

$$-(3 + 4)i_1 + (3 + 4 + 5)i_2 = -7 \cos 8t - \sin t$$

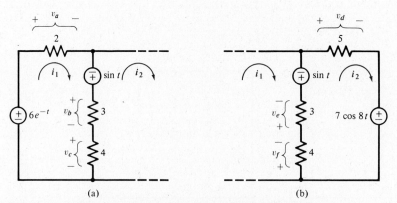

(a) (b)

Figure 2-8 Mesh-by-mesh analysis of the example network.

These two mesh equations may be solved simultaneously for i_1 and i_2.

This method is analogous to the process used to write systematic simultaneous nodal equations. For the nodal equations, the node numbering scheme expressed Kirchhoff's voltage-law relations. Equations were then written expressing Kirchhoff's current law, using the resistor voltage-current relations to put these equations in terms of the node voltages. An equation was not written for one of the nodes (the 0 node) because one application of Kirchhoff's current law is not independent of the others.

For the mesh equations, Kirchhoff's current-law relations are expressed by the mesh current scheme. Then Kirchhoff's voltage law is applied around each "box," or mesh, using the resistor voltage-current relations to express the equations in terms of the mesh currents. Although there are other loops besides the meshes for which one can write the voltage law, the resulting equations are not independent of the equations for the meshes; they are linear combinations of the mesh equations.

2.6.3 General Form of the Equations

The systematic simultaneous mesh equations have the following general form:

$$r_{11}i_1 - r_{12}i_2 - r_{13}i_3 - \cdots = v_1$$

$$-r_{21}i_1 + r_{22}i_2 - r_{23}i_3 - \cdots = v_2$$

$$-r_{31}i_1 - r_{32}i_2 + r_{33}i_3 - \cdots = v_3$$

$$\vdots$$

where

r_{11} = sum of resistances in the mesh through which i_1 flows

r_{22} = sum of resistances in the mesh through which i_2 flows

r_{33} = sum of resistances in the mesh through which i_3 flows

$$\vdots$$

and the other coefficients are of the form

r_{mn} = sum of resistances common to the i_m and i_n meshes = r_{nm}

For example,

r_{12} = sum of resistances common to the i_1 and i_2 meshes

r_{13} = sum of resistances common to the i_1 and i_3 meshes

r_{21} = sum of resistances common to the i_2 and i_1 meshes

r_{23} = sum of resistances common to the i_2 and i_3 meshes

The driving function terms involve the source voltages as follows:

$$v_1 = \left(\begin{array}{c}\text{sum of voltage source} \\ \text{functions for sources in} \\ \text{the } i_1 \text{ mesh that have the} \\ \text{source reference with } i_1\end{array}\right) - \left(\begin{array}{c}\text{sum of voltage source} \\ \text{functions for sources in} \\ \text{the } i_1 \text{ mesh that have the} \\ \text{sink reference with } i_1\end{array}\right)$$

$$v_2 = \left(\begin{array}{c}\text{sum of voltage source} \\ \text{functions for sources in} \\ \text{the } i_2 \text{ mesh that have the} \\ \text{source reference with } i_2\end{array}\right) - \left(\begin{array}{c}\text{sum of voltage source} \\ \text{functions for sources in} \\ \text{the } i_2 \text{ mesh that have the} \\ \text{sink reference with } i_2\end{array}\right)$$

$$v_3 = \left(\begin{array}{c}\text{sum of voltage source} \\ \text{functions for sources in} \\ \text{the } i_3 \text{ mesh that have the} \\ \text{source reference with } i_3\end{array}\right) - \left(\begin{array}{c}\text{sum of voltage source} \\ \text{functions for sources in} \\ \text{the } i_3 \text{ mesh that have the} \\ \text{sink reference with } i_3\end{array}\right)$$

$$\vdots$$

Each equation is the expression of Kirchhoff's voltage law around a mesh, where resistor voltages have been expressed in terms of the mesh currents.

The systematic loop equations have symmetries similar to those of the nodal equations: The "self" terms in each equation are positive, all the "coupling" terms are negative. The array of equation coefficients is symmetric about the diagonal, and the coupling terms are contained within the self term in each equation. If a voltage source is common to two meshes, the source function appears with a plus sign in one equation and with a minus sign in the other equation.

For the network of Figure 2-9 the systematic simultaneous mesh equations, in terms of the indicated mesh currents, are as follows:

$$\begin{cases} (2+5+6+3)i_1 - (5)i_2 - (3+6)i_3 - (0)i_4 = -\sin t - \cos t \\ -(5)i_1 + (5+4+7+8)i_2 - (0)i_3 - (8)i_4 = 0 \\ -(3+6)i_1 - (0)i_2 + (3+6+9)i_3 - (9)i_4 = 12 + \cos t \\ -(0)i_1 - (8)i_2 - (9)i_3 + (8+9)i_4 = -e^{-10t} \end{cases}$$

In terms of the mesh currents,

$$i_a = i_2, \qquad v_a = -2i_1, \qquad i_b = i_4 - i_2, \qquad v_b = 9(i_3 - i_4)$$

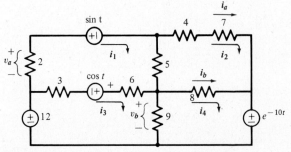

Figure 2-9 Another mesh equation example.

D2-6

Find the indicated mesh currents by writing and solving systematic equations:

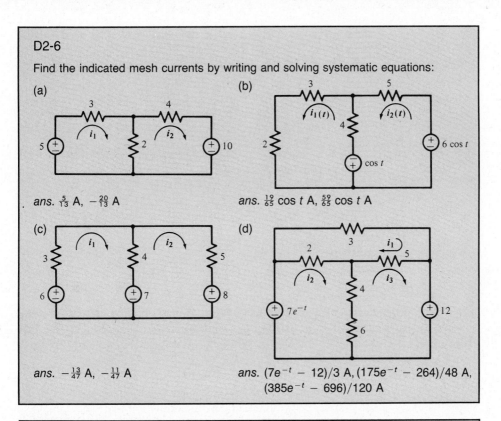

(a)

ans. $\frac{5}{13}$ A, $-\frac{20}{13}$ A

(b)

ans. $\frac{19}{65}$ cos t A, $\frac{59}{65}$ cos t A

(c)

ans. $-\frac{13}{47}$ A, $-\frac{11}{47}$ A

(d)

ans. $(7e^{-t} - 12)/3$ A, $(175e^{-t} - 264)/48$ A,
$(385e^{-t} - 696)/120$ A

D2-7

Write systematic mesh equations, solve for the mesh currents, then find the two indicated signals:

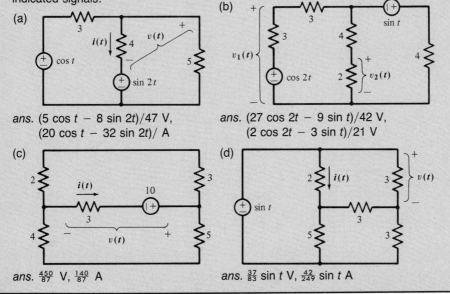

(a)

ans. $(5 \cos t - 8 \sin 2t)/47$ V,
$(20 \cos t - 32 \sin 2t)/$ A

(b)

ans. $(27 \cos 2t - 9 \sin t)/42$ V,
$(2 \cos 2t - 3 \sin t)/21$ V

(c)

ans. $\frac{450}{87}$ V, $\frac{140}{87}$ A

(d)

ans. $\frac{37}{83} \sin t$ V, $\frac{42}{249} \sin t$ A

SYSTEMATIC SIMULTANEOUS MESH EQUATIONS

Mesh currents are indicated, each with the same sense, inside each mesh (or "box") of the network. An equation is written for each of the meshes, in terms of the mesh currents i_1, i_2, ...

The coefficient of i_n in the nth equation is the sum of the resistances around the nth mesh. The coefficient of i_m, $m \neq n$ in the nth equation is the negative sum of the resistances that are common to mesh n and mesh m.

The driving function, on the opposite side of each equation from the variables, is, for the nth equation, the sum of the voltage-source functions for sources in mesh n with source reference relations with i_n minus those with sink reference relations with i_n.

2.7 MORE ABOUT MESH EQUATIONS

2.7.1 Incorporating Current Sources

Similar to the situation with nodal equations for a network that contains voltage sources, loop equations may be written for a network such as that in Figure 2-10 that contains current sources. Before the solution, the voltage across a current source is not generally known, but a mesh current or a difference between two mesh currents is fixed by the current source.

First, indicate by symbols each of the unknown voltages across the current sources and write the systematic mesh equations, including in them those unknown current source voltages, as if they were known.

$$(2 + 3)i_1 - \qquad 3i_2 = 12 - v$$

$$- 3i_1 + (3 + 4)i_2 = v$$

Then, augment the set by relating each current source function to the loop currents:

$$i_2 - i_1 = 5 \sin 6t$$

Finally, rearrange the equations before solving so that all unknowns are on the left:

$$\left\{ \begin{array}{l} 5i_1 - 3i_2 + v = 12 \\ -3i_1 + 7i_2 - v = 0 \\ - i_1 + i_2 \qquad = 5 \sin 6t \end{array} \right.$$

$\left. \begin{array}{l} \\ \end{array} \right\}$ basic mesh equations with unknown voltages supplied by current sources

current(s) fixed by current source(s) $\qquad$ **(2-5)**

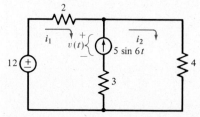

Figure 2-10 Network containing a current source.

When the equations are written in this manner, the unknown current source voltages always occur in an especially simple manner. Adding the first two equations to eliminate v gives

$$\begin{cases} 2i_1 + 4i_2 = 12 \\ -i_1 + i_2 = 5\sin 6t \end{cases} \tag{2-6}$$

so that

$$i_1 = \frac{\begin{vmatrix} 12 & 4 \\ 5\sin 6t & 1 \end{vmatrix}}{\begin{vmatrix} 2 & 4 \\ -1 & 1 \end{vmatrix}} = \frac{12 - 20\sin 6t}{6} = 2 - \frac{10}{3}\sin 6t \text{ A}$$

and

$$i_2 = \frac{\begin{vmatrix} 2 & 12 \\ -1 & 5\sin 6t \end{vmatrix}}{6} = \frac{10\sin 6t + 12}{6} = 2 + \frac{5}{3}\sin 6t \text{ A}$$

Then, substituting into the second equation of the original set,

$$-3i_1 + 7i_2 - v = 0$$

gives

$$v = 8 + \tfrac{65}{3}\sin 6t \text{ V} \tag{2-7}$$

For most people, writing the simpler coupled equations (2-5) initially, as was done here, is considerably easier than developing added procedures to write the decoupled equations (2-6) and the current source voltage equation (2-7) directly.

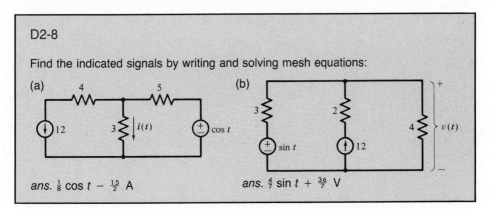

D2-8

Find the indicated signals by writing and solving mesh equations:

(a)

(b)

ans. $\tfrac{1}{8}\cos t - \tfrac{15}{2}$ A

ans. $\tfrac{4}{7}\sin t + \tfrac{36}{7}$ V

2.7.2 Mesh Equations with Controlled Sources

To accommodate controlled sources in mesh equations, the source controlling functions are expressed in terms of the mesh currents before writing the mesh equations.

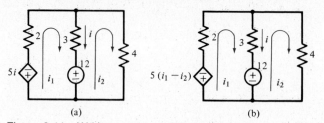

(a)   (b)

Figure 2-11   Writing mesh equations for a network with a controlled source.

For example, Figure 2-11 shows a network with a voltage source controlled by the current i. First, the controlling signal is expressed in terms of the mesh currents:

$$5i = 5(i_1 - i_2)$$

Then the ordinary mesh equations are written with the controlled source function expressed in terms of these variables:

$$\begin{cases} 5i_1 - 3i_2 = -5(i_1 - i_2) - 12 \\ -3i_1 + 7i_2 = 12 \end{cases}$$

Rearranging, placing the unknowns to the left,

$$\begin{cases} 10i_1 - 8i_2 = -12 \\ -3i_1 + 7i_2 = 12 \end{cases}$$

Figure 2-12 shows an example in which mesh equations are to be written for a network containing a controlled current source. Because of the presence of the current source, the unknown source voltage is defined as v_2 in Figure 2-12(b). The equations will be written in terms of v_2, then augmented by an equation involving the current source current, as usual. The source controlling signal is expressed in terms of the mesh currents,

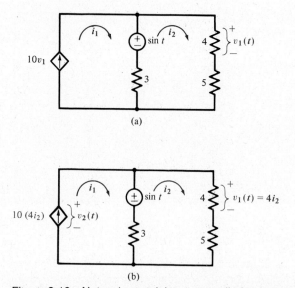

(a)

(b)

Figure 2-12   Network containing a controlled current source.

$$10v_1 = 10(4i_2) = 40i_2$$

before writing the equations.

The mesh equations are then as follows:

$$\begin{cases} 3i_1 - 3i_2 = v_2 - \sin t \\ -3i_1 + 12i_2 = \sin t \end{cases} \quad \left. \begin{array}{l} \text{basic mesh equations with unknown voltage} \\ \text{supplied by current source} \end{array} \right.$$

$$\left. \quad i_1 \qquad = 40i_2 \qquad\qquad \begin{array}{l} \text{loop current expressed in terms of current} \\ \text{source function} \end{array} \right.$$

Rearranging,

$$\begin{cases} 3i_1 - 3i_2 \quad - v_2 = -\sin t \\ -3i_1 + 12i_2 \qquad = \sin t \\ \quad i_1 - 40i_2 \qquad = 0 \end{cases}$$

2.7.3 Rearranging Meshes

In a problem such as that given in Figure 2-13(a), systematic simultaneous mesh equations for the network as it stands will yield two mesh equations plus a third equation relating the difference in mesh currents to the current source function, $6e^{-2t}$. If the equivalent circuit property is first used to interchange the two rightmost parallel branches, the network of Figure 2-13(b) results. Its meshes (the "boxes") are different from those of the original network diagram.

Since the current source appears in only one mesh in the new, equivalent drawing, the equations and their solution will be easier:

$$\begin{cases} 6i_1 - 2i_2 = -12 \\ -2i_1 + 5i_2 = 8 \sin 3t + v_1 \\ \quad i_2 = 6e^{-2t} \end{cases}$$

Substituting for i_2 from the third equation into the first equation,

$$6i_1 - 12e^{-2t} = -12$$

$$i_1(t) = 2e^{-2t} - 2$$

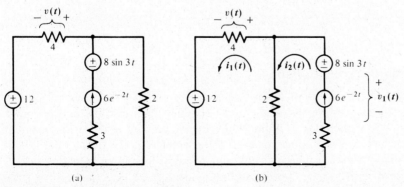

(a)   (b)

Figure 2-13   Rearranging meshes.

Then

$$v(t) = 4i_1 = 8e^{-2t} - 8$$

The most efficient network solutions usually involve a judicious combination of equivalent circuits and simultaneous equations. As in this example, interchanging parallel branches is a simple but important way of rearranging network meshes. Another view of this process, that of assigning loop current variables that are not necessarily mesh currents, is given in the next section. The technique, although involved, applies also to nonplanar networks.

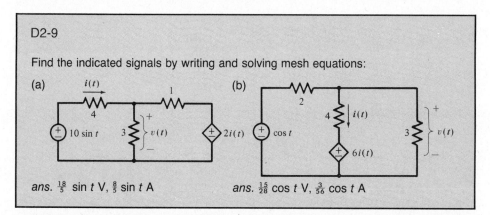

D2-9

Find the indicated signals by writing and solving mesh equations:

(a)

(b)

ans. $\frac{18}{5}$ sin t V, $\frac{8}{5}$ sin t A   ans. $\frac{15}{28}$ cos t V, $\frac{3}{56}$ cos t A

MORE ABOUT MESH EQUATIONS

To accommodate current sources in mesh equations, define the voltage across each current source and write the mesh equations as if these were known. Augment the set of equations with one equation for each current source, expressing a known mesh current or difference between two mesh currents.

    For controlled sources, express the controlling signal in terms of the mesh currents, then write the ordinary mesh equations as if the controlled source functions were known. Rearrange the equations, placing the unknowns on the left and the knowns on the right, and solve.

    Equivalent circuits involving interchanging the order of parallel branches may sometimes be used to advantage to rearrange meshes.

2.8 LOOP EQUATIONS FOR NONPLANAR NETWORKS

2.8.1 Selection of Loop Currents

The systematic mesh equations as presented previously are not a general method for network solution (as are systematic nodal equations) because the network must be planar for those methods to apply. One can always find a set of independent loops about which to write equations even if the network is not planar. For a non-planar network, however, a mesh current cannot simply be put in each of the "boxes" because, with crossovers, the "boxes" are not evident.

    To determine an independent set of loops, first form the *graph* of the network

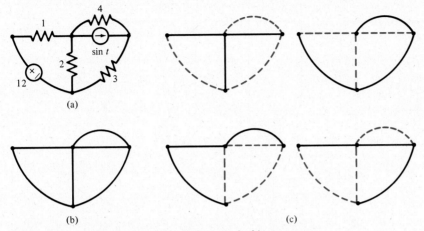

Figure 2-14   Network, its graph and several of its trees.

under consideration. The graph is merely a picture of the network's nodes, together with lines, called *branches*, indicating where elements connect from node to node. A planar network and its graph are shown in Figure 2-14(a) and (b).

A tree is any set of branches that connect all the nodes without forming any loops. For the example network, trees are indicated with solid lines in the drawings of Figure 2-14(c). The network also has several other trees. For a given tree, a *link* is any branch not in the tree. The links for each of the trees of Figure 2-14 are indicated with dashed lines. All of the trees for a network have the same number of links.

Given the network and a chosen tree, a set of loops about which Kirchhoff's voltage law may be applied independently consists of those loops, one for each link, that traverse a closed path through the tree and a single link each.

As with the systematic equations, a loop current may be indicated in each of these independent loops. Applying Kirchhoff's voltage law around the path of each loop current and expressing resistor voltages in terms of these currents yields a set of simultaneous algebraic equations.

2.8.2 Planar Network Example

Although the use of this general method is necessary only for nonplanar networks, it is sometimes used with planar networks instead of the systematic, one-current-in-each-"box" method. A simple example of the general method will now be applied to the planar network of Figure 2-15(a). In Figure 2-15(b) is shown the network graph. A tree of the graph, where the links have been numbered, is selected in Figure 2-15(c).

In the network itself, redrawn in Figure 2-15(d), the loop current i_1 is indicated in a closed path through the tree and link 1, while i_2 is in a closed path through the tree and link 2. Summing voltages around the outer loop and expressing them in terms of the loop currents i_1 and i_2 gives

$$v_1 + v_2 + 12 = 0$$

$$4(i_1 + i_2) + 2i_1 + 12 = 0$$

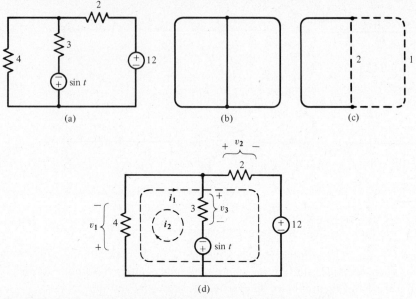

Figure 2-15 Example planar network with chosen loop currents.

Summing voltages around the smaller loop,

$$v_1 + v_3 - \sin t = 0$$

$$4(i_1 + i_2) + 3i_2 - \sin t = 0$$

Rearranging,

$$\begin{cases} 6i_1 + 4i_2 = -12 \\ 4i_1 + 7i_2 = \sin t \end{cases}$$

These equations do not have the same structure and symmetries as do the systematic equations.

2.8.3 Nonplanar Network Example

A more involved example, in which the general method is applied to a nonplanar network, is now given. For the network of Figure 2-16(a), a tree has been selected and the links numbered as in Figure 2-16(b).

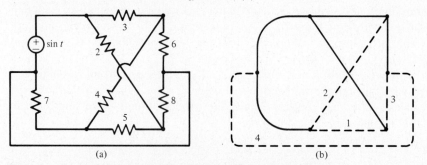

Figure 2-16 Nonplanar network and tree with numbered links.

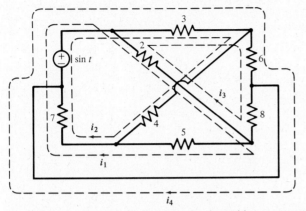

Figure 2-17   Example network and selected loop currents.

The loop currents corresponding to each link are placed in Figure 2-17 in such a manner that only the ith current flows through the ith link. Kirchhoff's voltage law is now written around each of the loops defined by the loop currents, expressing resistor voltages in terms of the loop currents:

$$\begin{cases} 7(i_1 + i_2) - \sin t + 2(i_1 - i_3) + 5i_1 = 0 \\ 7(i_1 + i_2) - \sin t + 3(i_2 + i_3 + i_4) + 4i_2 = 0 \\ 8i_3 + 2(i_3 - i_1) + 3(i_2 + i_3 + i_4) + 6(i_3 + i_4) = 0 \\ -\sin t + 3(i_2 + i_3 + i_4) + 6(i_3 + i_4) = 0 \end{cases}$$

Or,

$$\begin{cases} 14i_1 + 7i_2 - 2i_3 \qquad = \sin t \\ 7i_1 + 14i_2 + 3i_3 + 3i_4 = \sin t \\ -2i_1 + 3i_2 + 19i_3 + 9i_4 = 0 \\ \qquad 3i_2 + 9i_3 + 9i_4 = \sin t \end{cases}$$

D2-10

Write loop equations in terms of the indicated loop currents, then solve for the indicated voltage:

(a)   (b)

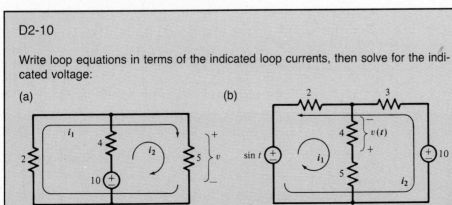

ans. $\frac{50}{19}$ V   *ans.* $(-80 - 12 \sin t)/51$ V

(c)

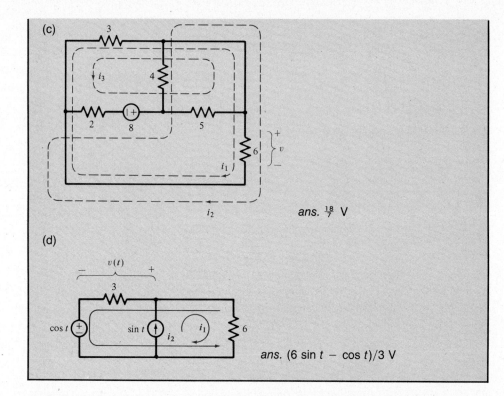

ans. $\frac{18}{7}$ V

(d)

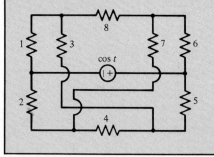

ans. $(6 \sin t - \cos t)/3$ V

D2-11

Write (but do not solve) loop equations for the following network. Show your choice of loop-current variables clearly:

LOOP EQUATIONS FOR NONPLANAR NETWORKS

For nonplanar networks, draw the network graph, form a tree, and define loop currents so that a different single loop current flows through each link. Then write Kirchhoff's voltage law around each of the loops defined by the loop currents, expressing resistor voltages in terms of the loop currents.

2.9 EQUIVALENT RESISTANCES

2.9.1 Equivalent Resistance Using Nodal Equations

Nodal equations offer a method of calculating equivalent resistances in those cases in which resistors are neither in series nor in parallel with one another, as in the two-terminal network of Figure 2-18(a).

Consider applying some current $i(t)$ through the element terminals, as in Figure 2-18(b), then finding the sink reference (for the element) voltage $v_1(t)$. The voltage $v_1(t)$ will be proportional to $i(t)$, and their ratio is the equivalent resistance of the element. A voltage source can be applied, measuring the current instead; but if nodal equations are to be written, they will be simpler with an applied source current.

For this example network, the systematic nodal equations are

$$\begin{cases} (\tfrac{1}{1} + \tfrac{1}{4})v_1 - & (\tfrac{1}{1})v_2 - & (\tfrac{1}{4})v_3 = i(t) \\ -(\tfrac{1}{1})v_1 + (\tfrac{1}{1} + \tfrac{1}{2} + \tfrac{1}{3})v_2 - & (\tfrac{1}{3})v_3 = 0 \\ -(\tfrac{1}{4})v_1 - & (\tfrac{1}{3})v_2 + (\tfrac{1}{3} + \tfrac{1}{4} + \tfrac{1}{5})v_3 = 0 \end{cases}$$

where the source current $i(t)$, although not specified, is treated as a known quantity. Every network voltage and current will be proportional to $i(t)$, whatever it is.

Simplifying the equations by eliminating the fractions,

$$\begin{cases} 5v_1 - 4v_2 - v_3 = 4i(t) \\ -6v_1 + 11v_2 - 2v_3 = 0 \\ -15v_1 - 20v_2 + 47v_3 = 0 \end{cases}$$

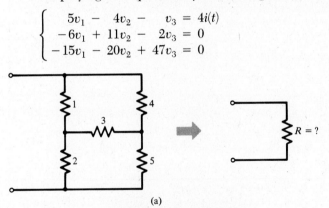

(a)

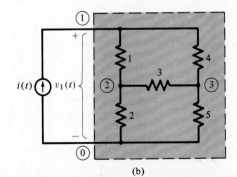

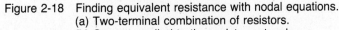

(b)

Figure 2-18 Finding equivalent resistance with nodal equations.
(a) Two-terminal combination of resistors.
(b) Current applied to the resistor network.

$$v_1(t) = \frac{\begin{vmatrix} 4i(t) & -4 & -1 \\ 0 & 11 & -2 \\ 0 & -20 & 47 \end{vmatrix}}{\begin{vmatrix} 5 & -4 & -1 \\ -6 & 11 & -2 \\ -15 & -20 & 47 \end{vmatrix}} = \frac{4i(t)\begin{vmatrix} 11 & -2 \\ -20 & 47 \end{vmatrix}}{5\begin{vmatrix} 11 & -2 \\ -20 & 47 \end{vmatrix} + 4\begin{vmatrix} -6 & -2 \\ -15 & 47 \end{vmatrix} - \begin{vmatrix} -6 & 11 \\ -15 & -20 \end{vmatrix}}$$

$$= \frac{1908i(t)}{5(477) + 4(-312) - (285)} = \frac{1908i(t)}{852}$$

The equivalent resistance of the two-terminal network is then the ratio

$$R_{\text{equiv}} = \frac{v_1(t)}{i(t)} = \frac{1908}{852} \ \Omega$$

If desired, a specific applied current $i(t)$ such as 1 or 10 or $\sin t$ could be used in the calculation.

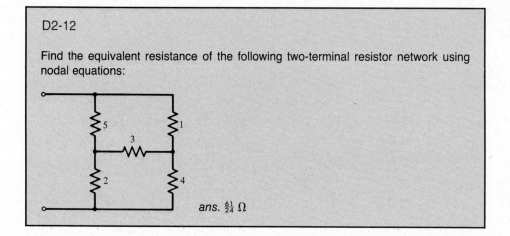

D2-12

Find the equivalent resistance of the following two-terminal resistor network using nodal equations:

ans. $\frac{61}{24} \ \Omega$

2.9.2 Equivalent Resistance Using Mesh Equations

Mesh equations may also be used to compute equivalent resistances of two-terminal combinations of resistors where the individual resistors are neither in series nor in parallel. The two-terminal network of Figure 2-19(a) is an example.

A voltage is applied to the network in Figure 2-19(b). The equivalent resistance is the ratio

$$R_{\text{equiv}} = \frac{v(t)}{i_1(t)}$$

Systematic mesh equations are solved for i_1 as follows:

$$\begin{cases} 2i_1 - i_2 - i_3 = v(t) \\ -i_1 + 6i_2 - 3i_3 = 0 \\ -i_1 - 3i_2 + 8i_3 = 0 \end{cases}$$

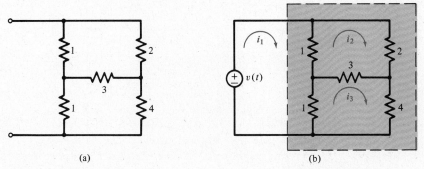

(a)   (b)

Figure 2-19   Two-terminal combination of resistors with applied voltage.

$$i_1 = \frac{\begin{vmatrix} v & -1 & -1 \\ 0 & 6 & -3 \\ 0 & -3 & 8 \end{vmatrix}}{\begin{vmatrix} 2 & -1 & -1 \\ -1 & 6 & -3 \\ -1 & -3 & 8 \end{vmatrix}} = \frac{v \begin{vmatrix} 6 & -3 \\ -3 & 8 \end{vmatrix}}{2\begin{vmatrix} 6 & -3 \\ -3 & 8 \end{vmatrix} + \begin{vmatrix} -1 & -3 \\ -1 & 8 \end{vmatrix} - \begin{vmatrix} -1 & 6 \\ -1 & -3 \end{vmatrix}}$$

$$= \frac{39v}{78 - 11 - 9}$$

Then

$$R_{\text{equiv}} = \frac{v(t)}{i_1(t)} = \frac{58}{39}\ \Omega$$

D2-13

Find the equivalent resistance of the following two-terminal resistor network using mesh equations:

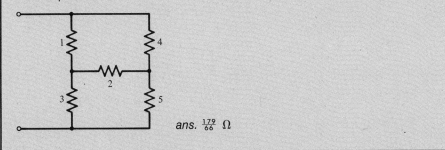

ans. $\frac{179}{66}$ Ω

2.9.3 Equivalent Resistances for Networks Involving Controlled Sources

A two-terminal network composed of only resistors and controlled sources is equivalent to a single resistor, providing that the controlling signals for the sources are voltages or currents within the two-terminal network. The equivalent resistance of

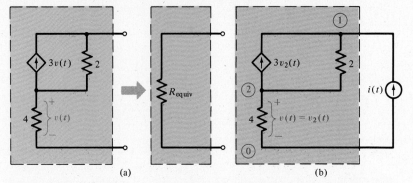

Figure 2-20 Two-terminal network involving resistors and a controlled source.

such a network may be found by methods similar to those of the previous sections in which a voltage or current source is connected to the element terminals and the ratio of the sink reference terminal voltage to terminal current is found.

An example network with resistors and a controlled source is given in Figure 2-20(a). In Figure 2-20(b) a current is applied to the network terminals. Systematic nodal equations are as follows:

$$\begin{cases} (\tfrac{1}{2})v_1 - (\tfrac{1}{2})v_2 = i(t) + 3v_2 \\ -(\tfrac{1}{2})v_1 + (\tfrac{1}{2} + \tfrac{1}{4})v_2 = -3v_2 \end{cases}$$

or

$$\begin{cases} v_1 - 7v_2 = 2i(t) \\ -2v_1 + 15v_2 = 0 \end{cases}$$

Then

$$v_1(t) = \frac{\begin{vmatrix} 2i(t) & -7 \\ 0 & 15 \end{vmatrix}}{\begin{vmatrix} 1 & -7 \\ -2 & 15 \end{vmatrix}} = 30i(t)$$

and

$$R_{\text{equiv}} = \frac{v_1(t)}{i(t)} = 30 \; \Omega$$

Since controlled sources may supply energy to a network, it is possible that within the element more energy is supplied than is dissipated, in which event the equivalent resistance will be negative. This situation is illustrated by the example of Figure 2-21, which is solved by applying a voltage to the terminals, in Figure 2-21(b), and finding the sink reference terminal current.

Systematic mesh equations for the network are

$$\begin{cases} 3i_1 - 3i_2 = v(t) \\ -3i_1 + 7i_2 = 5i_2 \end{cases}$$

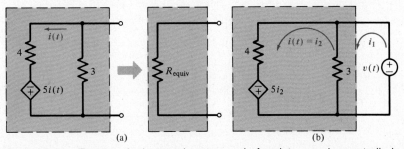

(a) (b)

Figure 2-21   Two-terminal network composed of resistors and a controlled source, with an applied voltage.

giving

$$i_1 = \frac{\begin{vmatrix} v(t) & -3 \\ 0 & 2 \end{vmatrix}}{\begin{vmatrix} 3 & -3 \\ -3 & 2 \end{vmatrix}} = \frac{2v(t)}{-3}$$

Then

$$R_{\text{equiv}} = \frac{v(t)}{i_1(t)} = -\frac{3}{2} \, \Omega$$

Because the 3-Ω resistor may be taken out of the "box" containing the rest of the elements and the controlling signal $i(t)$, the problem of Figure 2-21 could have been simplified by considering it to consist of the 3-Ω resistor in parallel with the rest of the element, the simpler element being that in Figure 2-22(a), for which

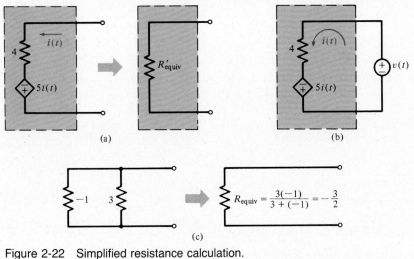

(a) (b)

(c)

Figure 2-22   Simplified resistance calculation.
(a) Component of the original network.
(b) Finding the equivalent resistance of the component.
(c) Original equivalent resistance found by parallel combination.

$$4i(t) = v(t) + 5i(t)$$

$$v(t) = -i(t)$$

$$R'_{equiv} = -1 \ \Omega$$

To obtain the net resistance of the entire network of the original problem, the parallel combination of 3 and $-1 \ \Omega$ is computed as shown in Figure 2-22(c).

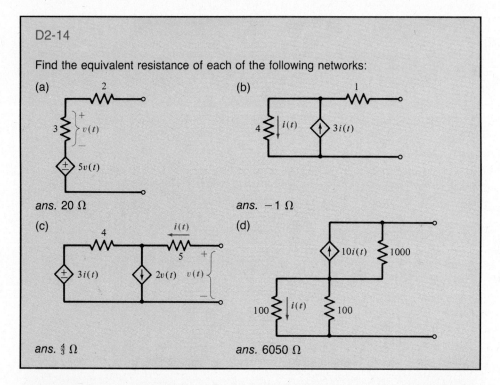

D2-14

Find the equivalent resistance of each of the following networks:

(a)

ans. 20 Ω

(b)

ans. -1 Ω

(c)

ans. $\frac{4}{3}$ Ω

(d)

ans. 6050 Ω

EQUIVALENT RESISTANCES

Any two-terminal network consisting of only resistors and controlled sources, which contains the controlling voltages and currents for the controlled sources, is equivalent to a single resistor.

The equivalent resistance may be found by solving the network problem in which a voltage or current source is placed across the element terminals and the sink reference ratio of terminal voltage to terminal current is found.

If controlled sources are present in the network, the equivalent resistance may be a negative number.

2.10 SUPERPOSITION OF SOURCES

2.10.1 Signal Components Due to Individual Sources

Linear equations have the property that if their driving functions (the "knowns," usually placed to the right of the equal signs) are decomposed into the sum of two or more component parts, the solution consists of a corresponding sum of component parts, each due to one driving function component acting separately.

For example, the solution to the equations

$$\begin{cases} x_1 - 2x_2 + x_3 = 4\cos 3t \\ -x_1 + 3x_2 \qquad\;\; = -12 \\ 2x_1 \qquad\;\; + 4x_3 = 0 \end{cases}$$

may be found by solving

$$\begin{cases} x_1{}' - 2x_2{}' + x_3{}' = 4\cos 3t \\ -x_1{}' + 3x_2{}' \qquad\;\; = 0 \\ 2x_1{}' \qquad\;\; + 4x_3{}' = 0 \end{cases}$$

then solving

$$\begin{cases} x_1{}'' - 2x_2{}'' + x_3{}'' = 0 \\ -x_1{}'' + 3x_2{}'' \qquad\;\; = -12 \\ 2x_1{}'' \qquad\;\; + 4x_3{}'' = 0 \end{cases}$$

The solutions for x_1, x_2, and x_3 in the original equations are then

$$x_1 = x_1{}' + x_1{}''$$

$$x_2 = x_2{}' + x_2{}''$$

$$x_3 = x_3{}' + x_3{}''$$

In the linear equations describing an electrical network if the driving function terms due to each individual, fixed network source are superimposed, the results are sets of equations, each set describing a single-source network, with all but one source replaced by zero. In other words, a network may be solved by solving a succession of single-source problems in which all but one fixed source at a time is set to zero. The response due to all of the sources acting together is then the sum of the single-source network solutions.

The solution for any network voltage or current then consists of a sum of terms, one term due to each fixed source acting separately. Each voltage and each current is said to consist of a sum of component parts, one component due to each separate fixed source.

2.10.2 Source Superposition

In an electrical network a voltage source is set to zero by replacing the source by a *short circuit* (a conductor, zero voltage). To set a current source to zero, replace it with an *open* circuit (remove it, leaving no connection, zero current). A network containing several fixed sources may be solved by solving it once for each source alone, setting the other sources to zero, then summing each of the single-source solutions to obtain the solution when all sources are present.

An example is shown in Figure 2-23. The component of $i(t)$ due to the voltage source is

$$i'(t) = \frac{3e^{-2t}}{9}\,\text{A}$$

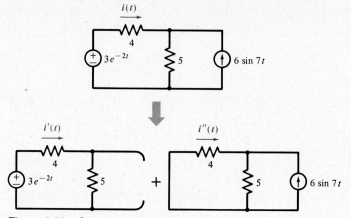

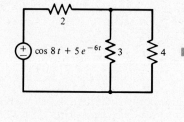

Figure 2-23   Superposition of sources.

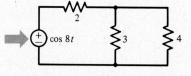

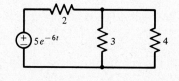

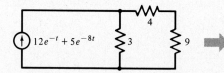

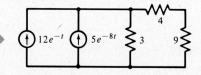

(a)

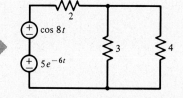

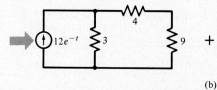

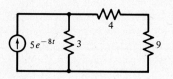

(b)

Figure 2-24   Source component superposition.
(a) Superposition of voltage source components.
(b) Superposition of current source components.

The component of $i(t)$ due to the current source is

$$i''(t) = -\tfrac{5}{9} \cdot 6 \sin 7t \text{ A}$$

so

$$i(t) = i'(t) + i''(t) = \tfrac{1}{3}e^{-2t} - \tfrac{10}{3} \sin 7t \text{ A}$$

A complicated source function may sometimes be simplified by expressing that source as the sum of simpler sources, as with the voltage and current source examples in Figure 2-24.

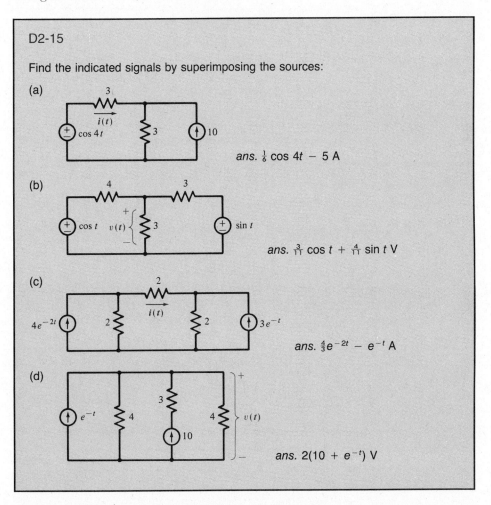

D2-15

Find the indicated signals by superimposing the sources:

(a)

ans. $\tfrac{1}{6} \cos 4t - 5$ A

(b)

ans. $\tfrac{3}{11} \cos t + \tfrac{4}{11} \sin t$ V

(c)

ans. $\tfrac{4}{3}e^{-2t} - e^{-t}$ A

(d)

ans. $2(10 + e^{-t})$ V

2.10.3 Source Superposition with Controlled Sources

Controlled sources cannot be superimposed because they appear in network equations in the same way as do resistors, not in the manner of fixed sources. All controlled sources should be left in all component solutions, as in the example of Figure 2-25.

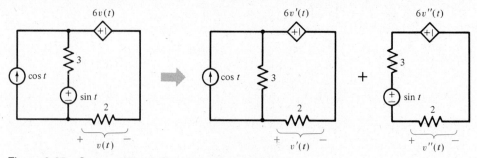

Figure 2-25 Superposition with a controlled source.

With source-resistor networks, superposition of sources and of source components has little to offer as a solution method. Solving several different networks for signal components is usually more difficult than solving the original network just once. The purpose here is to gain an understanding that will be fundamental in later work with more involved networks. Note that theoretically one could concentrate on networks with only a single fixed source, since all problems with multiple fixed sources may be reduced to a set of single-source problems.

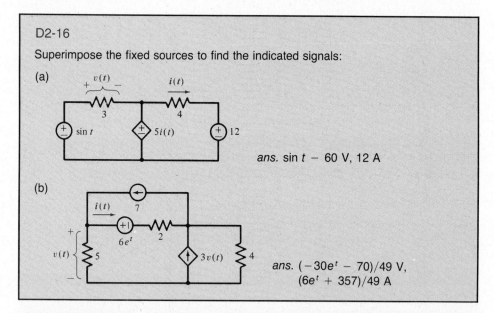

D2-16

Superimpose the fixed sources to find the indicated signals:

(a)

ans. sin t − 60 V, 12 A

(b)

ans. (−30e^t − 70)/49 V,
(6e^t + 357)/49 A

SUPERPOSITION OF SOURCES

A network containing several fixed sources may be solved by solving it once for each source alone, setting all other fixed sources to zero, then summing the solutions to obtain the entire solution. Controlled sources must be left intact in each of the component problems.

To set a voltage source to zero, replace it with a short circuit; a current source is set to zero by replacing it with an open circuit.

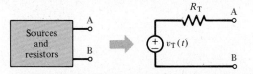

Figure 2-26   Network and its Thévenin equivalent.

2.11 THE THÉVENIN EQUIVALENT

2.11.1 Derivation of the Thévenin Equivalent

Any two-terminal network that contains only sources and resistors is equivalent (that is, has the same voltage-current relation at its terminals) to its *Thévenin equivalent*, consisting of a voltage source in series with a resistor, Figure 2-26. [Leon Thévenin (1857–1926) was a French telegraph engineer.] The voltage-source function in the Thévenin equivalent is called the *Thévenin voltage* and the resistance in the Thévenin equivalent is called the *Thévenin resistance*.

Suppose a current $i(t)$ is applied to the original two-terminal network. Imagine writing nodal equations for the network and the external current source, choosing the 0 node and the 1 node (in addition to others in the "box") for convenience as shown in Figure 2-27(a). The relation between the element voltage and current is then the relation between $v_1(t)$ and $i(t)$.

Nodal equations are chosen here because they are easily written for any network, not just the planar ones. Using Cramer's rule, the solution for v_1 has the form

$$v_1(t) = \frac{\Delta_1}{\Delta}$$

where Δ is just a number, the determinant of the equations; and the elements of the determinant Δ_1 are numbers except for the first-column elements, which are source functions. The determinant Δ_1 may be expanded as

$$\Delta_1 = \begin{vmatrix} i(t) + f_1(t) & \# & \cdots & \# \\ f_2(t) & \# & \cdots & \# \\ \vdots & \vdots & & \vdots \\ f_n(t) & \# & \cdots & \# \end{vmatrix} = i(t)\,\Delta_{11} + g(t)$$

where $g(t)$ is some function of time that depends on the sources within the "box."

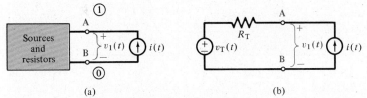

(a)   (b)

Figure 2-27   Deriving the Thévenin equivalent.
(a) Two-terminal network with current applied at the terminals.
(b) Thévenin network with current applied at the terminals.

Then

$$v_1(t) = \frac{\Delta_{11}}{\Delta} i(t) + \frac{g(t)}{\Delta}$$

The voltage-current relation of the Thévenin element, Figure 2-27(b), is

$$v_1(t) = R_T i(t) + v_T(t)$$

which is identical to that of the "box" for the correct choice of R_T and $v_T(t)$.

2.11.2 Calculations for Networks without Controlled Sources

Fortunately, in many cases it is not necessary to write and solve systematic equations to find the Thévenin equivalent. Consider measuring the voltage at the terminals of the Thévenin equivalent, with nothing connected externally to the terminals, as in Figure 2-28(a). There is no current through the Thévenin resistance, so there is no voltage across it, and the terminal voltage is the Thévenin voltage. The same measurement on the original network must give the same result. Thus the voltage across the original network terminals, with nothing connected externally to the terminals (called the *open-circuit* terminal voltage), is its Thévenin voltage, $v_T(t)$.

If all the fixed sources (not controlled sources) in the original network, Figure 2-28(a), are set to zero as in Figure 2-28(b), the solution for the terminal voltage $v_1(t)$, which in general is

$$v_1(t) = R_T i(t) + v_T(t)$$

becomes

$$v_1(t) = R_T i(t) \qquad \text{or} \qquad \frac{v_1(t)}{i(t)} = R_T$$

This is to say that the resistance looking into the terminals with all fixed sources set to zero is the Thévenin resistance.

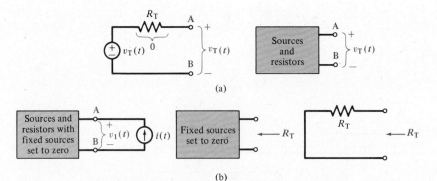

(a)

(b)

Figure 2-28 Finding Thévenin voltage and Thévenin resistance.
(a) Open-circuit voltage of the Thévenin equivalent and of the network.
(b) Setting fixed sources to zero to find Thévenin resistance.

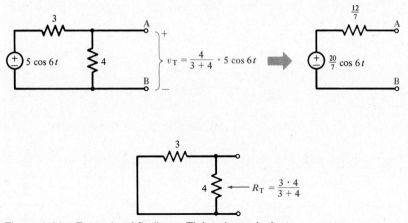

Figure 2-29   Example of finding a Thévenin equivalent.

An example of Thévenin equivalent calculation is shown ín Figure 2-29. The Thévenin voltage is the network's open-circuit voltage, and the Thévenin resistance is the resistance at the terminals when the fixed source is set to zero. Note that if the plus sign on the voltage reference polarity is closest to the A terminal when the calculation is made from the original network, the plus sign on the Thévenin voltage source in the equivalent is closest to the A terminal.

Another example Thévenin equivalent is found in Figure 2-30, where it is indicated that reversal of the Thévenin source polarity reverses the algebraic sign of the source function.

The only situation in which a Thévenin equivalent cannot be found is when the two-terminal source-resistor network is only a current source or is equivalent to only a current source. Then the open-circuit voltage cannot be found (one may think of it as being infinite) and the Thévenin resistance is infinite.

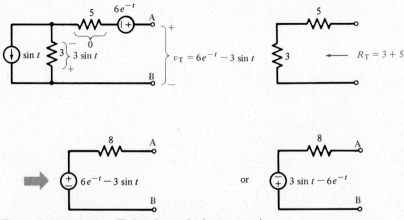

Figure 2-30   Another Thévenin equivalent example.

D2-17

Find the Thévenin equivalents of the following two-terminal networks. Indicate which is the A and which is the B terminal of the equivalent.

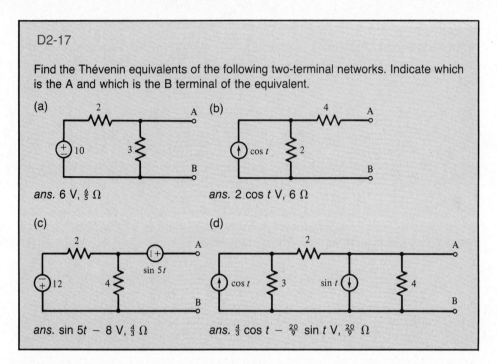

(a)

ans. 6 V, $\frac{6}{5}$ Ω

(b)

ans. 2 cos t V, 6 Ω

(c)

ans. sin 5t − 8 V, $\frac{4}{3}$ Ω

(d)

ans. $\frac{4}{3}$ cos t − $\frac{20}{9}$ sin t V, $\frac{20}{9}$ Ω

D2-18

Find the indicated signal using a Thévenin equivalent of the portion of the network in the dashed "box."

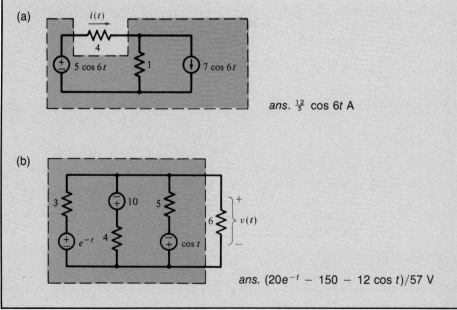

(a)

ans. $\frac{12}{5}$ cos 6t A

(b)

ans. (20e^{-t} − 150 − 12 cos t)/57 V

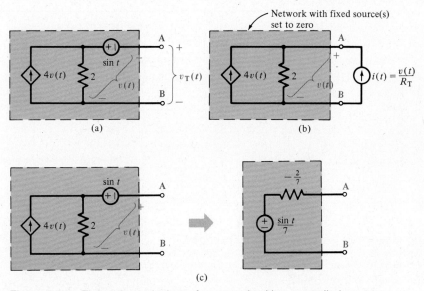

Figure 2-31   Thévenin equivalent of a network with a controlled source.

2.11.3 Networks with Controlled Sources

When it is desired to find the Thévenin equivalent of a two-terminal element containing controlled sources, the Thévenin resistance may be found by setting all fixed sources to zero, but leaving the controlled sources active. The resistance looking into the terminals then involves the equivalent resistance of a network with controlled sources in it, as in Section 2.9.3.

An example two-terminal network containing a controlled source is shown in Figure 2-31(a). Its open-circuit Thévenin voltage is given by

$$v(t) = 8v(t) - \sin t$$

$$7v(t) = \sin t$$

$$v_{\mathrm{T}}(t) = v(t) = \frac{\sin t}{7} \text{ V}$$

In Figure 2-31(b), the fixed source in the network is set to zero, and a current $i(t)$ is applied to the network terminals in order to calculate the Thévenin resistance. The terminal voltage in Figure 2-31(b) is given by

$$\tfrac{1}{2}v = 4v + i$$

$$-\tfrac{7}{2}v = i$$

$$R_{\mathrm{T}} = \frac{v}{i} = -\frac{2}{7} \,\Omega$$

a negative value, which is possible because of the presence of the controlled source. The complete Thévenin equivalent is shown in Figure 2-31(c).

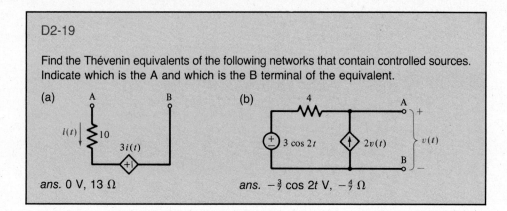

D2-19

Find the Thévenin equivalents of the following networks that contain controlled sources. Indicate which is the A and which is the B terminal of the equivalent.

(a)

(b)

ans. 0 V, 13 Ω

ans. $-\frac{3}{7}$ cos 2t V, $-\frac{4}{7}$ Ω

2.11.4 The Maximum Power Transfer Theorem

The Thévenin equivalence indicates that any complicated model of a two-terminal device that involves just sources and resistors can be reduced to a single-source, single-resistor model. In other words, the most complicated two-terminal source-resistor model may be replaced by its Thévenin equivalent.

A common problem is the following: Given a two-terminal source-resistor network, find the resistance that, when connected across the network terminals, results in the maximum electrical power transfer into that resistance. The two-terminal network might model an audio amplifier, for instance, and the resistance across its terminals might model a loudspeaker. One would be interested in getting the maximum amount of electrical power out of the amplifier into the loudspeaker.

So far as voltages and currents external to the two-terminal network are concerned, it may be replaced by its Thévenin equivalent, as in Figure 2-32(a). If the resistance connected across the network terminals, R_L (often called the *load re-*

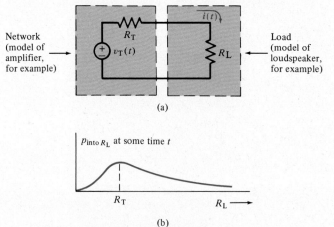

Figure 2-32   Maximum power transfer.
(a) Finding power transfer into the load resistor.
(b) Power into the load resistance as a function of load resistance.

sistance in such problems), is very large, the current through R_L is small and the electrical power flow into R_L,

$$p_{\text{into } R_L}(t) = i^2(t)R_L$$

will be small in comparison to what it could be were R_L smaller.

For very small R_L, $p_{\text{into } R_L}(t)$ will likewise be relatively small, because of the smallness of R_L itself. For some value of R_L between $R_L = 0$ and $R_L = \infty$ there will be a value of R_L for which $p_{\text{into } R_L}(t)$ will be maximum. Figure 2-32(b) is a sketch of power into the load resistance, as a function of the load resistance. In terms of the Thévenin voltage and resistance,

$$p_{\text{into } R_L} = i^2 R_L = \left[\frac{v_T(t)}{(R_T + R_L)} \right]^2 R_L$$

The maximum of $p_{\text{into } R_L}$ is given by the value of R_L for which

$$\frac{\partial p_{\text{into } R_L}}{\partial R_L} = \frac{(R_T + R_L)^2 v_T^2 - v_T^2 R_L \cdot 2(R_T + R_L)}{(R_T + R_L)^4}$$

$$= \frac{(R_T + R_L - 2R_L)v_T^2}{(R_T + R_L)^3} = 0$$

Solving for R_L,

$$R_L = R_T$$

To obtain maximum electrical power transfer into R_L, choose R_L equal to the Thévenin (or Norton) resistance of the rest of the network connected to the load resistance, as in the example of Figure 2-33, for which

$$R_L = \tfrac{13}{3} \ \Omega$$

for maximum power into R_L.

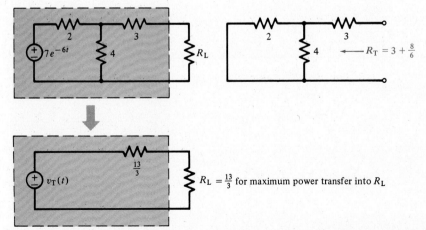

Figure 2-33   Example of Thévenin resistance calculation, for maximizing power transfer into load resistance.

One common misconception is that the maximum power transfer theorem applies in reverse: That is, one should choose the Thévenin resistance equal to the load resistance in order to maximize power transfer to the load. Not so. Differentiation of the power-flow equation with respect to R_T gives the quite obvious result that, if R_T is adjustable and R_L is not, R_T should be as small as possible for maximum electrical power transfer into R_L.

Another common misconception is that for maximum power transfer, the efficiency of the power transfer is 50 percent since equal powers are dissipated in R_T and in R_L. The Thévenin equivalent is equivalent so far as external voltages and currents are concerned, but it is *not* generally equivalent so far as *internal* electrical power is concerned. The efficiency may be less than 50 percent, as simple examples will demonstrate.

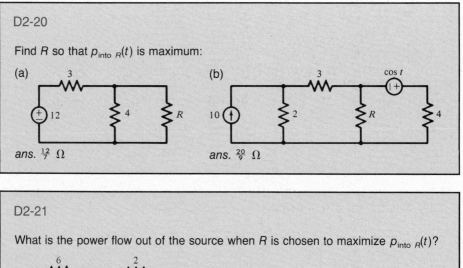

D2-20

Find R so that $p_{\text{into } R}(t)$ is maximum:

(a) [circuit: source +12, resistor 3, resistor 4, resistor R]

ans. $\frac{12}{7}$ Ω

(b) [circuit: source 10, resistor 3, resistor 2, resistor R, source $1+\cos t$, resistor 4]

ans. $\frac{20}{9}$ Ω

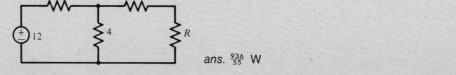

D2-21

What is the power flow out of the source when R is chosen to maximize $p_{\text{into } R}(t)$?

[circuit: source +12, resistor 6, resistor 4, resistor 2, resistor R]

ans. $\frac{936}{55}$ W

THE THÉVENIN EQUIVALENT

Any two-terminal combination of sources and resistors is equivalent to a single voltage source (the Thévenin voltage) in series with a single resistor (the Thévenin resistance).

The Thévenin voltage is the two-terminal network's open-circuit voltage and the Thévenin resistance is the resistance looking into the terminals when all fixed sources are set to zero.

Maximum electrical power is transferred into an adjustable resistor R when R is chosen to equal the Thévenin (or Norton) resistance of the two-terminal combination of sources and resistors that is connected to R.

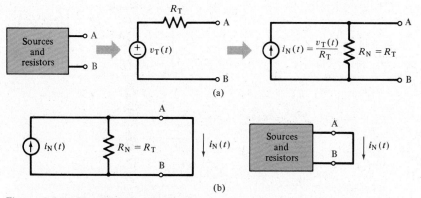

Figure 2-34 The Norton equivalent.
(a) Norton equivalent of a two-terminal source-resistor network.
(b) Norton current as the short-circuit current of the network.

2.12 THE NORTON EQUIVALENT

The equivalence of Thévenin and Norton networks (Section 1.10) shows that any two-terminal combination of sources and resistors is also equivalent to a current source in parallel with a resistor, Figure 2-34(a). An easy way to calculate the Norton equivalent is first to find the Thévenin equivalent and then convert to the Norton form. The Norton resistance is the same as the Thévenin resistance and may be calculated in the same way: Set all sources in the network, except controlled sources, to zero, then find the resistance looking into the terminals. The Norton source current is the Thévenin source voltage divided by the Thévenin resistance.

Alternatively, the Norton source current is the *short-circuit* network current, the current through the terminals when the network terminals are connected together, as shown in Figure 2-34(b). Note that the sense in which the reference direction for $i_N(t)$ is chosen for the measurement on the original network determines the sense of the Norton source in the Norton equivalent. The Norton/Thévenin resistance, being the ratio of the Thévenin voltage to the Norton current

$$R_N = R_T = \frac{v_T(t)}{i_N(t)}$$

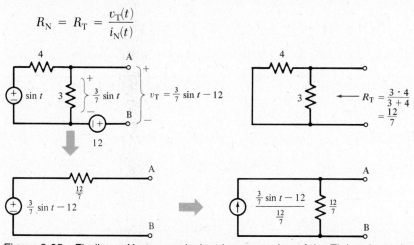

Figure 2-35 Finding a Norton equivalent by conversion of the Thévenin equivalent.

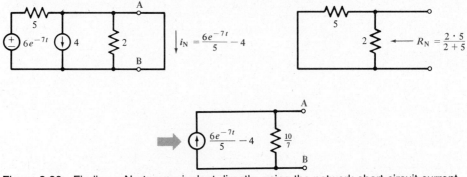

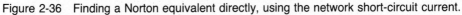

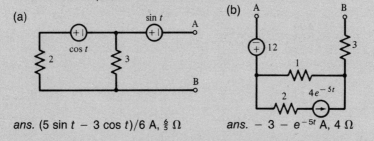

Figure 2-36 Finding a Norton equivalent directly, using the network short-circuit current.

is thus the ratio of network open-circuit voltage to short-circuit current. Occasionally, forming this ratio is easier than a direct calculation of $R_N = R_T$. The only situation for which a Norton equivalent does not exist is when the network is a voltage source or equivalent to just a voltage source.

Figure 2-35 is an example of finding the Norton equivalent by first finding the Thévenin equivalent. The example of Figure 2-36 shows how to find the Norton equivalent directly.

D2-22

Find the Norton equivalents of the following two-terminal networks by first finding the Thévenin equivalent then converting to the Norton form. Indicate the A and B terminals on the equivalent.

(a)

(b)

ans. (5 sin t − 3 cos t)/6 A, $\frac{6}{5}$ Ω   ans. − 3 − e^{-5t} A, 4 Ω

D2-23

Find the Norton equivalents of the following two-terminal networks directly, finding the short-circuit network current. Indicate the A and B terminals on the equivalent.

(a)

(b)

ans. −$\frac{9}{38}$ cos 3t A, $\frac{38}{9}$ Ω   ans. −3 sin 2t A, 4 Ω

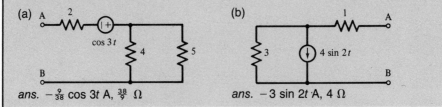

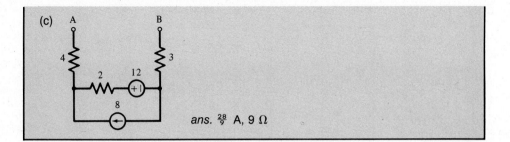

(c)

ans. $\frac{28}{9}$ A, 9 Ω

D2-24

Find the indicated signal using a Norton equivalent of the portion of the network in the dashed "box."

(a)

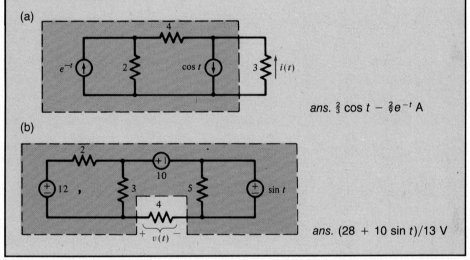

ans. $\frac{2}{3}$ cos $t - \frac{2}{5}e^{-t}$ A

(b)

ans. (28 + 10 sin t)/13 V

THE NORTON EQUIVALENT

Any two-terminal combination of sources and resistors is equivalent to a single current source (the Norton current) in parallel with a single resistor (the Norton resistance).

The Norton current is the network's short-circuit current, and the Norton resistance is identical to the Thévenin resistance, which is the resistance looking into the terminals when all fixed sources are set to zero.

The Norton equivalent may be found by first finding the Thévenin equivalent, then transforming the Thévenin element to an equivalent Norton element.

2.13 TRANSFER RATIOS FOR RESISTIVE NETWORKS

For a single-source source-resistor network, every voltage and every current in the network is proportional to the source function and thus to each other. The constants

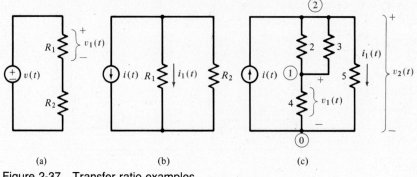

(a) (b) (c)

Figure 2-37 Transfer ratio examples.

of proportionality are known as the *transfer ratios* of the network. In general, a network transfer ratio is

$$T = \frac{\text{voltage or current of interest}}{\text{source function or another voltage or current of interest}} \tag{2-8}$$

For example, the transfer ratio that relates $v_1(t)$ to $v(t)$ in the network of Figure 2-37(a) is

$$T_1 = \frac{v_1(t)}{v(t)} = \frac{R_1}{R_1 + R_2}$$

For the network of Figure 2-37(b), the transfer ratio that relates $i_1(t)$ to $i(t)$ is

$$T_2 = \frac{i_1(t)}{i(t)} = \frac{-R_2}{R_1 + R_2}$$

Some writers distinguish between ratios (2-8) where the denominator is a source function and other ratios, calling the latter network *transmittance ratios*. We will call all quotients (2-8) *transfer ratios*, in keeping with the most common terminology.

The network of Figure 2-37(c) provides a more involved example of transfer ratio calculation. Systematic nodal equations for the network are

$$\begin{cases} (\tfrac{1}{4} + \tfrac{1}{2} + \tfrac{1}{3})v_1 - (\tfrac{1}{2} + \tfrac{1}{3})v_2 = 0 \\ -(\tfrac{1}{2} + \tfrac{1}{3})v_1 + (\tfrac{1}{2} + \tfrac{1}{3} + \tfrac{1}{5})v_2 = i(t) \end{cases}$$

or

$$\begin{cases} 13v_1 - 10v_2 = 0 \\ -25v_1 + 31v_2 = 30i(t) \end{cases}$$

The voltage v_1 is, in terms of the source, $i(t)$,

$$v_1(t) = \frac{\begin{vmatrix} 0 & -10 \\ 30i(t) & 31 \end{vmatrix}}{\begin{vmatrix} 13 & -10 \\ -25 & 31 \end{vmatrix}} = \frac{300}{153} i(t)$$

so the transfer ratio of the two is the constant

$$T_1 = \frac{v_1(t)}{i(t)} = \frac{300}{153} = \frac{100}{51}$$

The voltage v_2 is

$$v_2(t) = \frac{\begin{vmatrix} 13 & 0 \\ -25 & 30i(t) \end{vmatrix}}{153} = \frac{390}{153} i(t)$$

and the current $i_1(t)$ is, in terms of $i(t)$,

$$i_1(t) = \frac{v_2(t)}{5} = \frac{78}{153} i(t)$$

The transfer ratio of the two is the constant

$$T_2 = \frac{i_1(t)}{i(t)} = \frac{78}{153}$$

Since all network signals are proportional to $i(t)$, they are proportional to one another, so the ratio of any two signals is a constant. The ratio of i_1 to v_2, for instance, is

$$T_3 = \frac{i_1(t)}{v_2(t)} = \frac{i_1(t)/i(t)}{v_2(t)/i(t)} = \frac{T_2}{T_1} = \frac{78}{300} = \frac{13}{50}$$

D2-25

Find the transfer ratios:

(a) $T_1 = \dfrac{v_1}{v}$

(b) $T_2 = \dfrac{i_3}{v}$

(c) $T_3 = \dfrac{v_2}{v}$

(d) $T_4 = \dfrac{v_2}{i_2}$

(e) $T_5 = \dfrac{i_3}{i_1}$

ans. $\frac{15}{47}, \frac{3}{47}$ S, $\frac{6}{47}, -\frac{6}{5}$ Ω, $\frac{3}{8}$

TRANSFER RATIOS

In a single-source source-resistor network, every voltage and every current is proportional to the source function. Thus the ratio of any such network voltage or current to any other network voltage or current is a constant. Such ratios are called *transfer ratios*.

CHAPTER TWO PROBLEMS

Basic Problems

Simultaneous Linear Algebraic Equations

1. Solve, using Cramer's rule:

   (a) $\begin{cases} 12x_1 - 3x_2 - x_3 = 8\cos t \\ -3x_1 + 5x_2 = 6 - 8\cos t \\ -x_1 + 9x_3 = -6 \end{cases}$

   (b) $\begin{cases} 6x_1 - 3x_2 - x_3 = \cos 2t \\ -3x_1 + 5x_2 - 2x_3 = 8 - \cos 2t \\ -x_1 - 2x_2 + 8x_3 = 0 \end{cases}$

Systematic Nodal Equations

2. Write and solve systematic simultaneous nodal equations for the indicated signals:

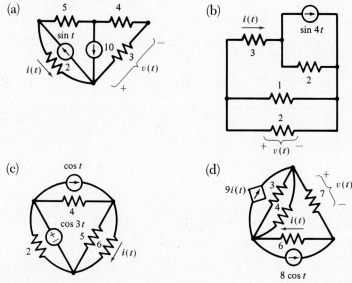

Systematic Mesh Equations

3. Write and solve systematic simultaneous mesh equations and determine the indicated signals:

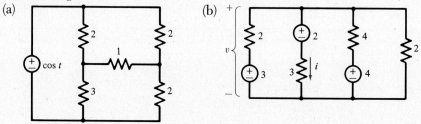

(c)

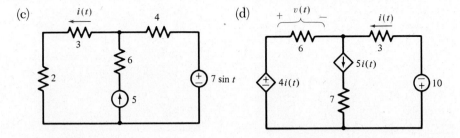

(d)

Equivalent Resistance

4. Find the equivalent resistances using nodal equations:

(a)

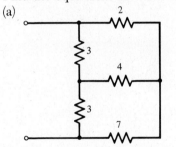

(b)

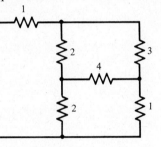

5. Find the equivalent resistances using mesh equations:

(a)

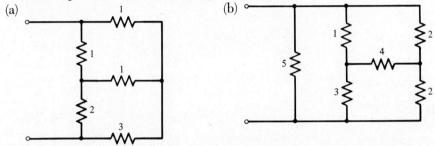

(b)

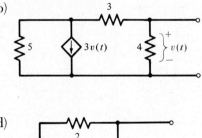

6. Find the equivalent resistances of the following elements:

(a)

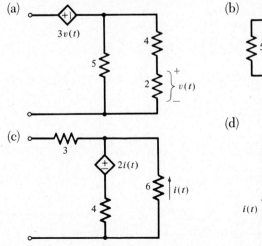

(b)

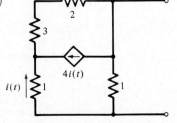

(c)

(d)

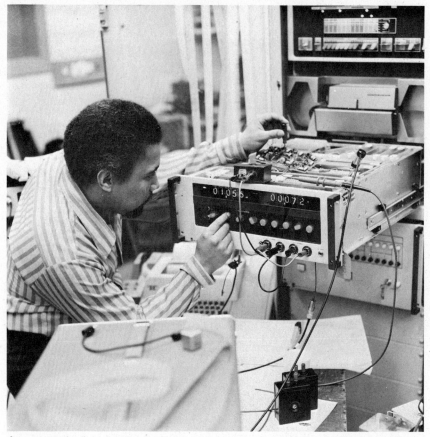

An automatically adjusting bridge is calibrated at the factory. (*Photo courtesy of GenRad, Inc.*)

Superposition of Sources

7. Find the indicated signal by superimposing the sources:

(a)

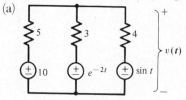

(b)

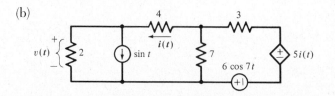

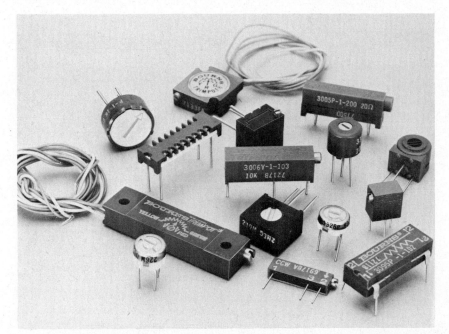

An assortment of screwdriver adjustable potentiometers. In most of these the contact arm is moved along a helical coil of resistance wire. (*Photo courtesy of Trimpot Products Division of Bourns, Inc.*)

Thévenin and Norton Equivalents

8. Find the Thévenin and Norton equivalents:

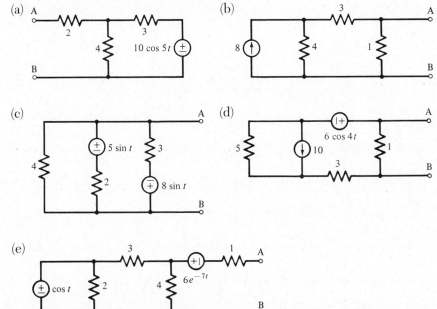

(f)

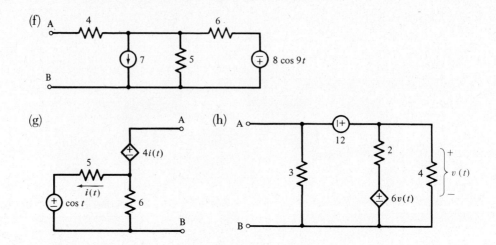

(g)

(h)

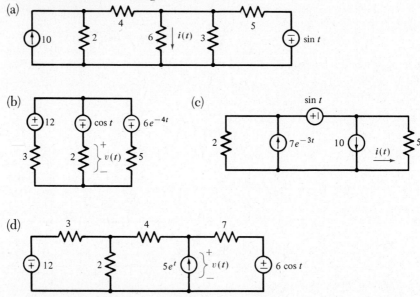

9. Use Thévenin and Norton equivalents (and other equivalents, too, as necessary) to find the indicated voltages and currents:

(a)

(b)

(c)

(d)

Maximum Power Transfer

10. Find the resistance R so that the electrical power flow into R is maximum. For that value of R, find $p_{\text{into } R}(t)$:

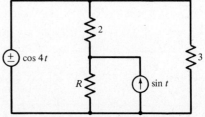

Transfer Ratios

11. Find the transfer ratios:

(a)
$$T_1 = \frac{i_1}{i}$$

(b)
$$T_2 = \frac{v_1}{i}$$

(c)
$$T_3 = \frac{i}{v_2}$$

(d)
$$T_4 = \frac{v_1}{v_2}$$

(e)
$$T_5 = \frac{i_2}{v_1}$$

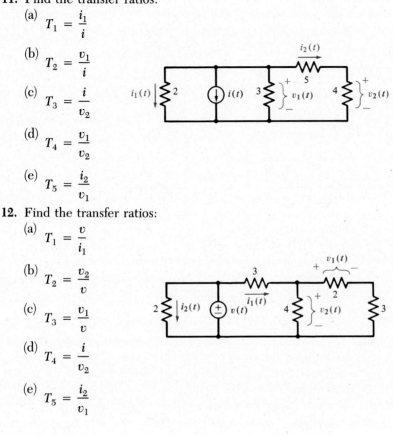

12. Find the transfer ratios:

(a)
$$T_1 = \frac{v}{i_1}$$

(b)
$$T_2 = \frac{v_2}{v}$$

(c)
$$T_3 = \frac{v_1}{v}$$

(d)
$$T_4 = \frac{i}{v_2}$$

(e)
$$T_5 = \frac{i_2}{v_1}$$

Practical Problems

Potentiometers

A potentiometer is an adjustable resistor similar to a rheostat, in which the movable contact and both ends of the resistive material are available for electrical connection. The symbol for a potentiometer is shown below. The indicated resistance R is the end-to-end resistance of the resistive material. Commonly, the device is designed so that the resistance between one end and the movable contact (or *arm*) is proportional to the angular position of a shaft on which the arm is mounted. The potentiometer is then said to have a *linear taper*. In other potentiometers, the end-to-arm resistance is made to vary in some other way with the arm position. For example, in volume controls the resistance taper is approximately logarithmic because perceived audio loudness varies logarithmically.

13. (a) A linear taper potentiometer is connected to a resistor as shown in (a). Find the resistance between the indicated terminals as a function of the potentiometer arm location and sketch the result.

(b) Potentiometers are often used for adjustment of electronic equipment. In the network of (b) a potentiometer controls the voltage v. What is the range over which v may be adjusted by the potentiometer? If the source voltage may vary as much as $\pm 10\%$, will it always be possible to adjust v to 18 V?

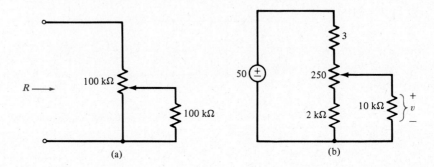

(a)

(b)

Potentiometric Voltage Measurement

Errors caused by the effects of voltmeter internal resistance on a network being measured can be eliminated by using a *potentiometric* measurement. In this method the unknown voltage is compared with a known voltage, often the voltage of a battery. Batteries especially constructed to supply stable, predictable voltages are called *standard cells*.

A potentiometer is used to develop an adjustable voltage v_a and is connected through an ammeter to the network to be measured as indicated in the drawing of (a). Zero current flows through the ammeter when the potentiometer voltage v_a equals the voltage to be measured, as indicated in (b). At that setting of the potentiometer slider, the potentiometer and the network may as well be disconnected from one another. The potentiometer voltage is then equal to the open circuit network voltage and is proportional to the slider position.

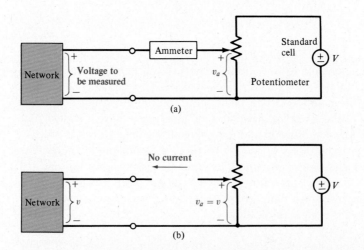

(a)

(b)

14. For the potentiometric voltage apparatus shown, which has a linear taper potentiometer, what is the voltage v if the potentiometer slider arm is one-third of the way from bottom to top of the resistance element? What fraction of the distance from bottom to top is the potentiometer slider when $v = 6.3$ V?

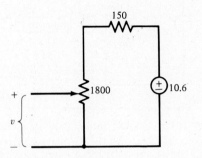

Resistance Measurement

One very fundamental way to measure the resistance of a resistor is to connect it in a circuit so that a current flows through it, then measure both the resistor voltage and the resistor current. The ratio of the sink reference voltage to current is the resistance, as diagrammed in (a).

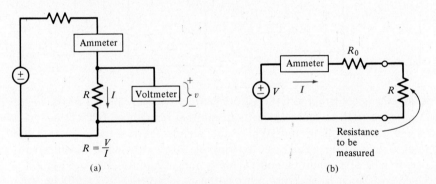

Because the measurement of resistance is so fundamental, special instruments called *ohmmeters* have been designed to give a direct reading of resistance on a single meter. Ohmmeters consist basically of a battery or other fixed, constant voltage source in series with a resistor and an ammeter, as shown in (b).

The ammeter current is

$$I = \frac{V}{R_0 + R}$$

and the scale of the ammeter may be calibrated in terms of R.

Different ranges of resistance R may be measured by changing the fixed resistance R_0.

15. (a) The ammeter has an internal resistance of 1 Ω and the voltmeter's internal resistance is 1000 Ω. In each of the two situations below, what is the ratio of the voltmeter reading to the ammeter reading? If the internal resistances of the two meters are each known, their effects may be taken into account.

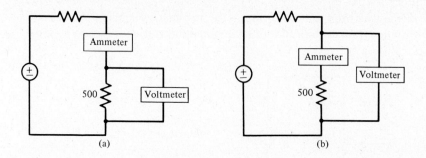

(b) In each of the two situations below, the ammeter internal resistance is 2.5 Ω, the voltmeter internal resistance is 5000 Ω, and the measured current and voltage are indicated. The voltage source V and the resistance R_0 are unknown. Find R_1 and R_2, taking the meter internal resistances into account.

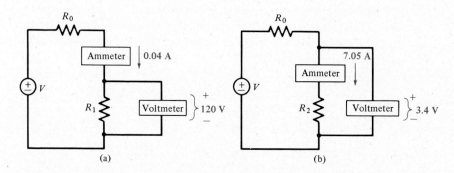

16. An ohmmeter is constructed from a 1.5-V battery with an internal resistance of 1.2 Ω in series with an ammeter with an internal resistance of 125 Ω in series with a resistor R_0 as indicated. What value should the resistor R_0 have if the ammeter is to register a full scale reading of 10 mA when the measured resistance R is 0 Ω (a short circuit)? This is the maximum current through a measured resistance R, and care must be taken in practice to avoid damaging delicate devices such as transistors with excessive ohmmeter currents.

   With this value for R_0, what measured resistances R will make the ammeter read three-quarters, one-half, and one-quarter of its full scale reading?

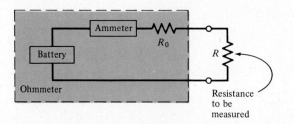

Resistance Bridge

A resistance bridge is a method of measuring an unknown resistance by comparing it to a known resistance. The circuit drawn in (a) is known as a *Wheatstone bridge*.

Zero current, I, flows through the ammeter when the voltage across the ammeter is zero, that is, when $v_1 = v_2$. Under that condition, the ammeter may as well be removed from the circuit since no current flows through it. This situation is shown in (b).

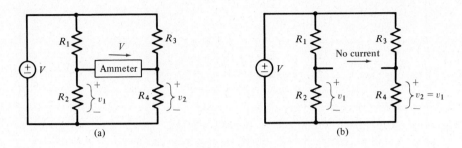

(a)   (b)

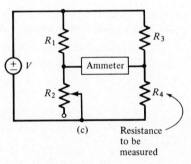

(c)   Resistance
to be
measured

The voltages $v_1 = v_2$ are then given by the voltage divider rule,

$$v_1 = \frac{R_2 V}{R_1 + R_2} = v_2 = \frac{R_4 V}{R_3 + R_4}$$

hence there is zero ammeter current when

$$\frac{R_2}{R_1 + R_2} = \frac{R_4}{R_3 + R_4} \qquad\qquad R_2 R_3 + R_2 R_4 = R_1 R_4 + R_2 R_4$$

$$R_2 R_3 = R_1 R_4 \qquad \text{or} \qquad \frac{R_1}{R_2} = \frac{R_3}{R_4}$$

Under this condition, the bridge is said to be *balanced*.

If $R_1 = R_3$, R_2 is adjustable and R_4 is the resistance to be measured, as in the arrangement of (c), the ammeter reads zero (the bridge is balanced) when R_2 is adjusted to be equal to R_4. In general, at balance,

$$R_2 = \frac{R_1}{R_3} R_4$$

so that various ratios (R_1/R_3) will give various proportional relationships between the measured resistance and the adjustable balancing resistance.

17. (a) The adjustable resistor in the Wheatstone bridge has a linear taper. What fraction of the full slider travel, from the end connected to the ammeter, is the slider setting when the bridge is balanced?

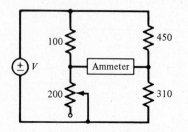

(b) Choose resistances R_1 and R_3 so that the bridge may measure resistances R in the range 0–100 Ω. Repeat for the range 0–1000 Ω.

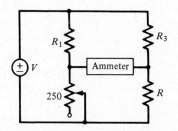

Thévenin Equivalent Measurement

A convenient method of finding the Thévenin voltage and the Thévenin resistance of a two-terminal network is the following. Find the Thévenin voltage by measuring the open-circuit terminal voltage with a voltmeter with very high internal resistance. Then connect an adjustable resistance across the terminals and adjust the resistance until the terminal voltage is exactly half of the open-circuit voltage. The adjustable resistance now equals the Thévenin resistance.

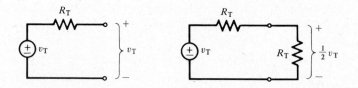

If the voltmeter internal resistance is not large compared to the Thévenin resistance, it must be taken into account in the calculations.

18. A voltmeter with internal resistance 5000 Ω is used in the following Thévenin equivalent measurements on a network with a constant Thévenin voltage. With the voltmeter across the otherwise open-circuited network terminals, the meter reads 80.6 V. When a 2000-Ω resistor is placed in parallel with the meter, across the network terminals, the meter reads 48.3 V. Find the Thévenin voltage and the Thévenin resistance of the network.

19. A voltmeter with negligibly large internal resistance measures an open-circuit, constant Thévenin voltage of 62.3 V. When a 1000-Ω resistor is placed in parallel with the meter, across the network terminals, the meter reads 58.1 V.

Suppose the actual voltage at the meter terminals is 1% higher than that indicated by the meter. Find the percentage error in the Thévenin voltage,

$$\frac{(v_T)_{actual} - (v_T)_{measured}}{(v_T)_{actual}} \times 100\%$$

and the similarly defined error in the Thévenin resistance.

Repeat if, instead, the 1000-Ω resistor is actually 1% higher in resistance value.

Thévenin Equivalent Model

20. The voltage-current characteristic for a certain silicon junction semiconductor diode is shown. The device has a nearly straight line relationship between voltage v and current i over a limited range of voltage and current. Find a Thévenin equivalent model that gives an accurate approximation to the diode characteristic for voltages and currents within the ranges indicated, where the curve is nearly a straight line.

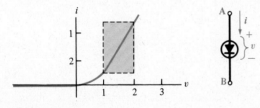

Advanced Problems

Linear Algebraic Equations

21. Show that if all driving functions (the "knowns") in a linearly independent set of n simultaneous linear algebraic equations in n variables are proportional to some function $f(t)$, then the solution of those equations (the "unknowns") is proportional to $f(t)$.

This is to say that if all the sources in a network are constant, all voltages and currents in the network are constant. If all sources vary as e^{3t}, all voltages and currents in the network vary as e^{3t}. If a network has a single source, all network voltages and currents are proportional to that source function.

22. Show that if all driving functions in a linearly independent set of n simultaneous linear algebraic equations in n variables are multiplied by a constant k, then the solution of those equations is multiplied by k.

This is to say that if all the fixed sources in a network are doubled, all voltages and currents in the network are doubled.

Network Equations

23. A set of nonsystematic equations for the network is as follows:

$$\begin{cases} v_1 + v_2 = 0 \\ v_1 - v_3 = 0 \\ v_1 + 12 - v_4 = 0 \\ v_2 + v_3 = 0 \\ v_2 - 12 + v_4 = 0 \\ v_3 + 12 - v_4 = 0 \\ \cos t + i_2 + i_3 - i_4 = 0 \\ i_1 - i_4 = 0 \\ -i_1 + i_3 + i_2 + \cos t = 0 \\ v_2 = 4i_2 \\ v_3 = -5i_3 \\ v_4 = 3i_4 \end{cases}$$

Solve these equations for i_3.

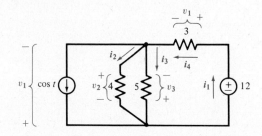

24. The voltage and the current in every element in the network have been defined. Write the equations for Kirchhoff's voltage law around *every* loop in the network, write the equations for Kirchhoff's current law at *every* node in the network, and write the equations for Ohm's law for *every* resistor in the network.

  Having converted the network problem to a mathematical problem, solve this set of simultaneous linear algebraic equations for v_3.

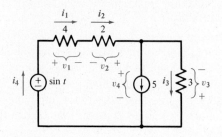

Controlled Sources

25. Show that if a network contains only resistors and controlled sources that have source functions proportional to other voltages and currents in the network, all network voltages and currents must be zero.

Network Topology

26. Construct a network with the graph shown and show that loop equations for the four loops indicated are not linearly independent of one another. Find another set of four loops, the equations for which are linearly independent.

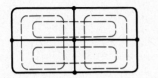

27. Show that for the graph of any network, the number of branches in a tree is one less than the number of nodes.

Ladder Networks

A network of the type sketched in (a) is called a *ladder network*. To solve such a network using systematic simultaneous equations generally involves quite a few mesh or nodal equations.

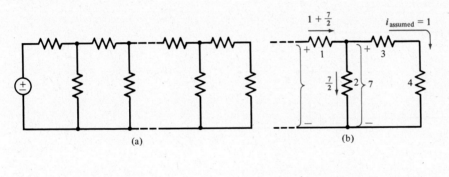

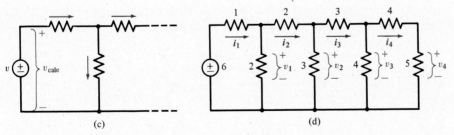

An easier method of solution is as follows. Start at the end of the network farthest from the source and *assume* some nonzero resistor current, $i_{assumed}$, say 1 A, as in (b). Using this current, the voltage across the next top-to-bottom resistor may be found. Using that voltage, the next current may be found, and so on. Eventually, the currents through the resistors nearest the source are found and a source voltage, v_{calc}, is found that would produce the originally assumed current and all the other voltages and currents found from the assumed current. This situation is illustrated in (c).

Unless a very lucky choice was made for the assumed current, v_{calc} will not be the same as the actual source voltage v. If v_{calc} would produce $i_{assumed}$, what current i, in the farthest resistor, is produced by the actual source voltage v? All of the actual voltages and currents in the network are (v/v_{calc}) times as large as the values calculated based upon $i_{assumed}$.

28. Use this method to find each of the indicated voltages and currents in the ladder of (d). Repeat for a source voltage function, instead, of $6 \sin t$.

Equivalent Resistance
29. Design a two-terminal network that has an equivalent resistance of $-100 \, \Omega$ that contains (in addition to other elements)
   (a) A voltage source controlled by a current
   (b) A voltage-controlled voltage source
   (c) A voltage-controlled current source
   (d) A current-controlled current source

Maximum Power Flow
30. Find the value of R that results in maximum electrical power flow into the 3-Ω resistor:

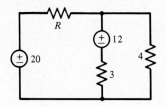

Transfer Ratios
31. Suppose that a network contains two sources, with source functions $f_1(t)$ and $f_2(t)$. Let the transfer ratio T_1 be the ratio of some specific network voltage or current to $f_1(t)$ when $f_2 = 0$. Similarly, let T_2 be the ratio of the same network voltage or current to $f_2(t)$ when $f_1 = 0$.

Show that the network voltage or current $g(t)$ is

$$g(t) = T_1 f_1(t) + T_2 f_2(t)$$

when both sources are nonzero.

Chapter Three
Operational Amplifiers and Two-Port Networks

The amount of current in a galvanic chain is directly proportional to the sum of all tensions and inversely proportional to the total reduced length of the chain.

Georg Simon Ohm
From *The Galvanic Chain, Mathematically Treated,*
Berlin, 1827

3.1 PREVIEW

The creation of high-performance amplifiers using inexpensive "op amp" integrated circuit chips is an important new capability that is useful in all branches of engineering. The analysis tools developed in the preceding chapters are now brought to bear on the design of amplifying circuits.

Then, the general concept of a two-port network is introduced. It is shown that a resistive network with two accessible pairs of terminal can be described in a variety of ways, each involving two equations linking the terminal voltages and currents. Two-port network models are used to derive the delta-wye equivalence between three-terminal connections of resistors and to describe amplifiers.

When you complete this chapter, you should know—

1. what operational amplifiers are, and how to construct and analyze network models for them;
2. how to design circuits to sum, difference, and amplify voltage signals using operational amplifiers;
3. what two-ports and two-port parameters are;
4. how to make measurements to determine two-port resistance, conductance, hybrid, and other parameters;
5. about reciprocity in resistive two-port networks;
6. how to apply delta-wye transformations;
7. how to use two-port parameters to simplify network models.

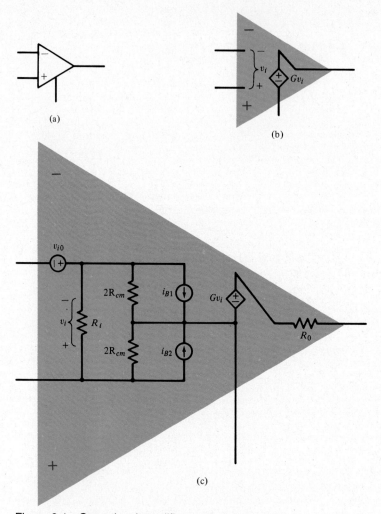

Figure 3-1 Operational amplifier symbol and models.
(a) Operational amplifier symbol.
(b) Simple model.
(c) More involved op amp model.

3.2 OPERATIONAL AMPLIFIERS AND NETWORK MODELS

The modern operational amplifier ("op amp" for short) is an integrated circuit that behaves approximately as a voltage-controlled voltage source. An operational amplifier symbol is given in Figure 3-1(a), and its simple controlled source model is shown in Figure 3-1(b). The controlling voltage is applied between the plus and minus *input* terminals of the op amp, and the controlled *output* voltage is produced between the other two terminals, as shown.

In order to function, an op amp generally must be powered by two constant voltage power supplies. These are considered to be part of the operational amplifier represented by the symbol shown, although several op amps may share the same power supply.

The coefficient G is termed the *gain* of the op amp and is typically on the order of 10^4 to 10^6. The value of G may vary by a factor of ten or more between op amps of the same type. The designer depends upon G being very large but not upon its precise value, trading large G for other desirable properties. In this section several important amplifying and signal summing circuits that use op amps as component parts are analyzed. In each design, simple signal relationships result for sufficiently large G, and important parameters of the overall circuit depend on externally connected resistors rather than on the op amp.

A more involved and more accurate op amp model is given in Figure 3-1(c). The output source becomes a Thévenin equivalent, tiny internal sources of current and voltage are included, and large resistances neglected in the simple model are shown here. Fortunately, in many circumstances, the simple controlled source op amp model is sufficiently accurate so that the additional complexity of the more involved model is not needed.

OPERATIONAL AMPLIFIERS AND NETWORK MODELS

Operational amplifiers are integrated circuits for which a simple model is a voltage-controlled voltage source. A symbol for an operational amplifier, which includes the required constant voltage power supplies that may be shared with other op amps, is shown below. Beside it is a simple network model for the op amp.

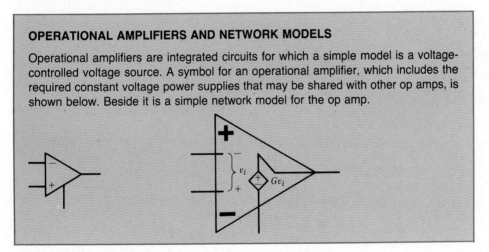

3.3 INVERTING AMPLIFIERS

3.3.1 Single-Input Inverters

When an operational amplifier is connected with external resistors R_A, R_F, and R_1 as shown in Figure 3-2(a), it forms an amplifier of the applied voltage v_{in}. The amplifier output voltage v_{out} is proportional to the negative of the input voltage v_{in}; hence this is termed an *inverting* amplifier. For a sufficiently large op amp gain G,

$$v_{out} = -\frac{R_F}{R_A} v_{in}$$

which is independent of the precise value of G and is fixed by the ratio of the external resistors R_F and R_A. The transfer ratio

$$\frac{v_{out}}{v_{in}} = -\frac{R_F}{R_A}$$

is the overall gain of the amplifier, which contains an operational amplifier as a component part.

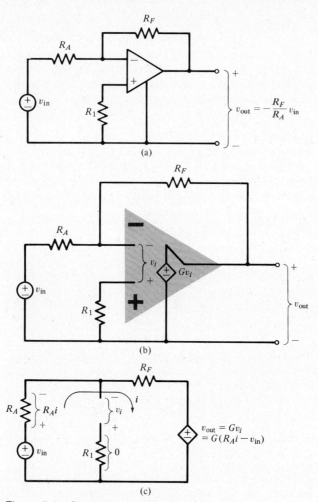

Figure 3-2 Single-input inverting configuration for the operational amplifier.
(a) Op amp diagram.
(b) Network model.
(c) Network model redrawn.

To derive this result, the operational amplifier is replaced by its network model in Figure 3-2(b). In Figure 3-2(c), the circuit has been redrawn to show clearly the single loop through which the current i has been defined. The controlled source controlling signal v_i is expressed in terms of the mesh current as

$$v_i = R_A i - v_{in}$$

and the loop equation for i is then

$$(R_A + R_F)i = v_{in} - G(R_A i - v_{in})$$

Solving for i in terms of v_{in},

$$[(1 + G)R_A + R_F]i = (1 + G)v_{in} \qquad i = \frac{1 + G}{(1 + G)R_A + R_F} v_{in}$$

The amplifier output voltage is given by

$$v_{out} = G(R_A i + v_{in})$$

and is, substituting for i,

$$v_{out} = G(R_A i + v_{in}) = G \frac{(1 + G)R_A}{(1 + G)R_A + R_F} v_{in} - v_{in}$$

$$= G \frac{(1 + G)R_A - (1 + G)R_A - R_F}{(1 + G)R_A + R_F} v_{in}$$

$$= \frac{-GR_F}{(1 + G)R_A + R_F} v_{in}$$

with typical op amp gains G on the order of 10^4 to 10^6 and resistor ratios

$$\frac{R_F}{R_A} << 10^4$$

this amplifier connection gives, for all practical purposes,

$$v_{out} = -\frac{R_F}{R_A} v_{in}$$

The design of an inverting amplifier is shown in Figure 3-3(a). An inexpensive type 741 operational amplifier, available in an 8-pin miniature dual in-line package (DIP), is used. For an overall amplifier gain of -5, resistors

$$R_F = 100,000 \ \Omega = 100 \ k\Omega \qquad R_A = 20,000 \ \Omega = 20 \ k\Omega$$

are chosen. Generally speaking, large resistor values such as these should be chosen so that the currents through them are relatively low. Otherwise, power is dissipated unnecessarily and for a fixed op amp output current capability, less current can be delivered to the load. On the other hand, for an inexpensive op amp such as the 741, when the feedback resistor R_F approaches about a megohm, the simple op amp model begins to fail. The op amp input resistance R_i in the more accurate model of Figure 3-1(c) is about a megohm and cannot be neglected when R_F is of comparable size.

So far as the simple model is concerned, the resistor R_1 does not affect the overall amplifier gain. The simple model applies when R_1 is not comparable to the megohm or so value of R_i in the more accurate model. There is another important consideration, however. Tiny constant *input bias currents*, due to the characteristics of the transistors comprising the op amp, flow through the external circuitry from the op amp input terminals. The sources of these currents are represented by the i_{B1} and i_{B2} current sources in the more accurate op amp model of Figure 3-1(c).

It is generally desirable to design an amplifier so that the average effects of the two input bias currents on v_{out} cancel one another. To do so, the external resistors should be chosen so that the net resistance connected to the plus op amp input terminal equals that connected to the minus input terminal. Provided that the Thévenin resistance input signal source is low, R_1 in this design should equal the parallel combination of R_F and R_A.

$$R_1 = \frac{R_F R_A}{R_F + R_A} = \frac{2000 \times 10^6}{120 \times 10^3} = 1.67 \times 10^4 = 16.7 \ k\Omega$$

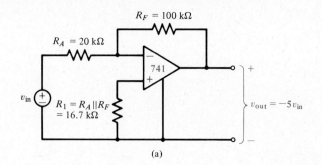

(a)

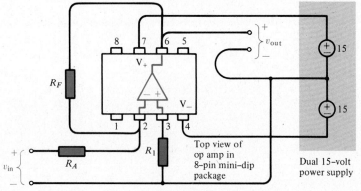

(b)

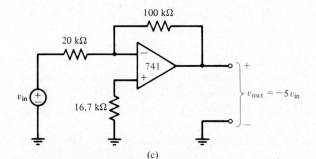

(c)

Figure 3-3   Designing an inverting amplifier.
(a) Op amp diagram.
(b) Pictorial diagram of amplifier connections.
(c) Alternative op amp diagram in which the power supply common connection is indicated by the ground symbol.

The pictorial diagram of Figure 3-3(b) shows how the designed inverting amplifier is connected in practice. A dual 15-V supply powers the amplifier. Inexpensive general purpose op amps such as the 741 require dual power supplies with voltages in the range from about 8–18 V. These op amps are capable of supplying output currents up to about ± 15 mA and output voltages to within about 2 V of the supply voltages. The design shown here will produce output voltages v_{out} within the range ± 13 V.

An alternative form is often used for op amp diagrams, particularly where the diagrams are complicated and involve several op amps. Connection to the common power supply terminal (between the two 15-V sources in the model above) is indicated by the "ground" symbol, as shown in Figure 3-3(c), which eliminates the need for drawing a number of lines on the diagram. Originally, the ground symbol meant a connection making contact with the earth, usually to pipes or plates buried under the ground. Such a connection insures a potential equal or nearly equal to that of the earth and prevents the connected equipment from acquiring a sizeable static charge. Here, as in many electronic applications, the ground symbol only indicates a common connection; it may or may not be at earth potential.

3.3.2 Inverting Input Resistance

To a single source, the input terminals of an amplifier appear to be a pure resistance that is termed the *input* resistance of the amplifier. After all, if there is a single input source, all network voltages and currents are proportional to that source function, so the ratio of input voltage to current is constant.

The resistance R_{in} in Figure 3-4(a), as viewed at the minus op amp terminal with the feedback resistor R_F connected, will now be calculated. At the amplifier input terminals, the resistance is the sum of R_A and R_{in}, as shown in Figure 3-4(b). If a voltage V is applied to the circuit as shown in Figure 3-4(c), the resistance R_{in} is the ratio of that voltage to the terminal current:

$$R_{in} = \frac{V}{I}$$

The network diagram is redrawn in Figure 3-4(d), where it is apparent that I satisfies the single mesh equation

$$R_F I = V + GV = (1 + G)V$$

giving

$$R_{in} = \frac{V}{I} = \frac{R_F}{1 + G}$$

For R_F and R_A in the usual few thousand to a few hundred thousand ohm range, and for an op amp gain G of 10^4 or 10^5, R_{in} is very small compared to R_A. Hence the inverting amplifier input resistance is very nearly R_A:

$$R_{input} = R_{in} + R_A \cong R_A$$

3.3.3 Multiple Inverting Inputs

Several input voltages can be combined and amplified with circuits having the structure shown in Figure 3-5(a). For simplicity, the specific case of three inputs is considered here, since it is easy to see then how fewer or more input signals may be combined. In Figure 3-5(b), the op amp has been replaced by its simple network model and in Figure 3-5(c) the fixed sources representing the input voltages are superimposed. Each of the individual superposition problems is identical in form; only the resistor and voltage subscripts differ.

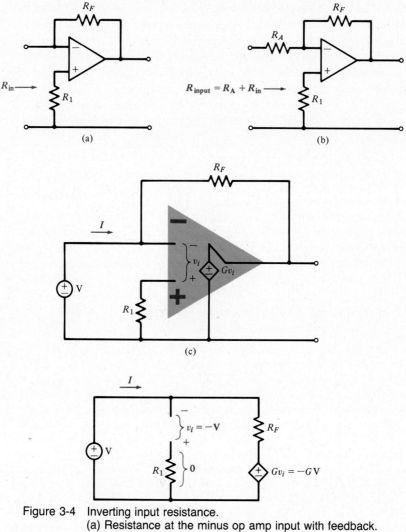

Figure 3-4   Inverting input resistance.
(a) Resistance at the minus op amp input with feedback.
(b) Resistance at the amplifier input terminals.
(c) Network model.
(d) Network model redrawn.

For the first superposition problem, involving the input voltage v_a, the parallel resistors R_b and R_c are combined and the loop currents i_1 and i_2 are defined in Figure 3-5(d). The controlled source controlling signal v_{i1} is, in terms of the loop currents,

$$v_{i1} = R(i_2 - i_1)$$

The systematic simultaneous loop equations for this network are then

$$(R + R_a)i_1 - Ri_2 = v_a$$

$$-Ri_1 + (R + R_F)i_2 = -Gv_{i1} = -GR(i_2 - i_1)$$

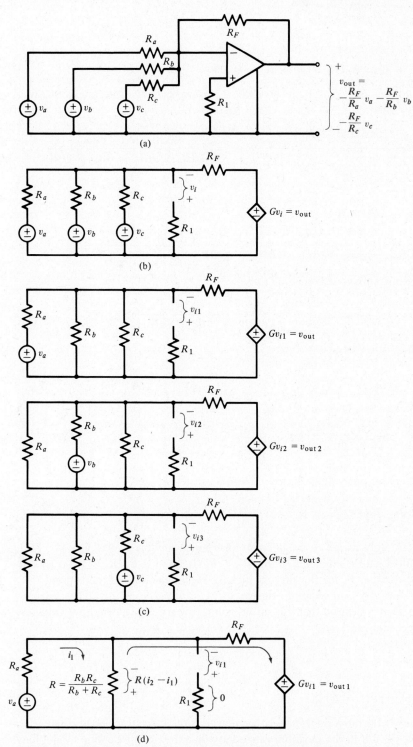

Figure 3-5   Multiple-input inverting operational amplifier configuration.

or,

$$(R + R_a)i_1 - \qquad\qquad Ri_2 = v_a$$

$$-(GR + R)i_1 + (GR + R + R_F)i_2 = 0$$

The solutions for i_1 and i_2, in terms of v_a, using Cramer's rule are as follows:

$$i_1 = \frac{\begin{vmatrix} v_a & -R \\ 0 & (GR + R + R_F) \end{vmatrix}}{\begin{vmatrix} (R + R_a) & -R \\ -(GR + R) & (GR + R + R_F) \end{vmatrix}} = \frac{(GR + R + R_F)v_a}{GRR_a + RR_a + RR_F + R_aR_F}$$

$$i_2 = \frac{\begin{vmatrix} (R + R_a) & v_a \\ -(GR + R) & 0 \end{vmatrix}}{GRR_a + RR_a + RR_F + R_aR_F} = \frac{(GR + R)v_a}{GRR_a + RR_a + RR_F + R_aR_F}$$

The output voltage due to v_a is thus

$$v_{\text{out }1} = Gv_{i1} = GR(i_2 - i_1) = \frac{-GRR_F}{GRR_a + RR_a + RR_F + R_aR_F} v_a$$

which for large G is virtually

$$v_{\text{out }1} = -\frac{R_F}{R_a} v_a$$

Similarly,

$$v_{\text{out }2} = -\frac{R_F}{R_b} v_b \qquad v_{\text{out }3} = -\frac{R_F}{R_c} v_c$$

so that the total output voltage in the circuit is

$$v_{\text{out}} = v_{\text{out }1} + v_{\text{out }2} + v_{\text{out }3} = -\frac{R_F}{R_a} v_a - \frac{R_F}{R_b} v_b - \frac{R_F}{R_c} v_c$$

This is an extremely useful result because for a given feedback resistance R_F, each of the individual gain coefficients, $-R_F/R_a$, $-R_F/R_b$, and $-R_F/R_c$, is fixed by one individual resistance, R_a for the first coefficient, R_b for the second, and so on.

Figure 3-6 gives an example design for a multiple-input inverting op amp amplifier where it is desired that the output voltage be related to two input voltages, v_1 and v_2, according to

$$v_{\text{out}} = -\tfrac{8}{5} v_1 - \tfrac{2}{3} v_2$$

The choice

$$R_F = 50,000 \ \Omega = 50 \ \text{k}\Omega$$

is made somewhat arbitrarily. It is relatively large, so that resistor power dissipation will be small, yet not in the megohm range for which the simple op amp model may

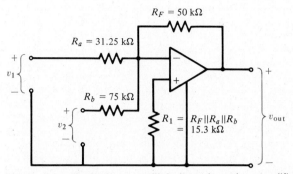

Figure 3-6 Designing a multiple-input inverting amplifier.

not be adequate. Then, the desired gain coefficients are obtained with

$$R_a = 31{,}250 \ \Omega = 31.25 \ \text{k}\Omega \qquad R_b = 75{,}000 \ \Omega = 75 \ \text{k}\Omega$$

The resistor R_1 is chosen to equal the parallel combination of R_F, R_a, and R_b to give cancellation of average input bias current effects at the output, assuming small source Thévenin resistances:

$$R_1 = \frac{1}{(1/R_F) + (1/R_a) + (1/R_b)} = \frac{1}{(1/50\text{k}) + (1/31.25\text{k}) + (1/75\text{k})}$$

$$= \frac{10^3}{0.02 + 0.032 + 0.0133} = \frac{10^3}{0.0653} = 15.3 \ \text{k}\Omega$$

Since the resistance at the minus op amp terminal with the feedback resistor R_F connected is small, each set of amplifier input terminals looks to its signal source as a resistance equal to the series resistor, R_a, R_b, . . . , for that input.

D3-1

Design single-amplifier op amp circuits which will, from input voltages v_1, v_2, . . ., produce the indicated output voltage v_{out}. Choose practical resistor values and arrange for approximate cancellation of the average input bias current effects assuming negligibly small signal source Thévenin resistances.

(a) $v_{\text{out}} = -3v_1$
    ans. One possibility: $R_F = 75 \ \text{k}\Omega$, $R_a = 25 \ \text{k}\Omega$, $R_1 = 18.75 \ \text{k}\Omega$

(b) $v_{\text{out}} = -\frac{1}{4}v_1$
    ans. One possibility: $R_F = 75 \ \text{k}\Omega$, $R_a = 300 \ \text{k}\Omega$, $R_1 = 60 \ \text{k}\Omega$

(c) $v_{\text{out}} = -2v_1 - \frac{1}{3}v_2$
    ans. One possibility: $R_F = 75 \ \text{k}\Omega$, $R_a = 37.5 \ \text{k}\Omega$, $R_b = 225 \ \text{k}\Omega$, $R_1 = 22.5 \ \text{k}\Omega$

(d) $v_{\text{out}} = -0.32v_1 - 0.44v_2 - 0.24v_3$
    ans. One possibility: $R_F = 75 \ \text{k}\Omega$, $R_a = 234 \ \text{k}\Omega$, $R_b = 170 \ \text{k}\Omega$, $R_c = 313 \ \text{k}\Omega$,
    $R_1 = 37.5 \ \text{k}\Omega$

INVERTING AMPLIFIERS

Inverting amplifier circuits with single and with multiple inputs are as follows:

Inverting Amplifier

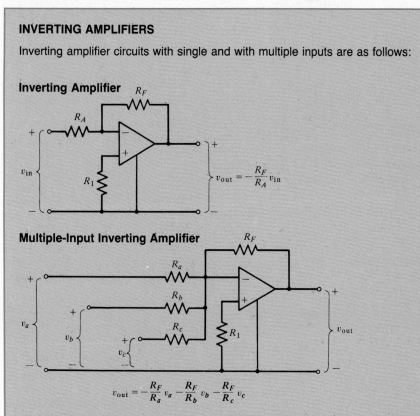

$$v_{\text{out}} = -\frac{R_F}{R_A} v_{\text{in}}$$

Multiple-Input Inverting Amplifier

$$v_{\text{out}} = -\frac{R_F}{R_a} v_a - \frac{R_F}{R_b} v_b - \frac{R_F}{R_c} v_c$$

Typical resistor values used range from a few thousand to a few hundred thousand ohms.

To cancel the effect on the output of the average input bias currents of the op amp,

$$R_1 = R_A \| R_F = \frac{R_A R_F}{R_A + R_F}$$

for the single-input amplifier, and

$$R_1 = R_F \| R_a \| R_b \| R_c \| \cdots$$

for the multiple-input amplifier.

With feedback, the voltage between the minus input terminal and the power common is very nearly zero. To input signal sources, the input terminals behave very nearly as the series input resistor:

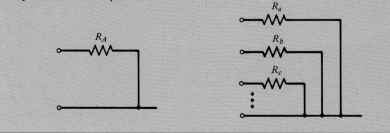

3.4 NONINVERTING AMPLIFIERS

3.4.1 Single-Input Noninverting Amplifiers

The operational amplifier connection shown in Figure 3-7(a) forms a noninverting amplifier. For this circuit, when the op amp gain G is sufficiently large,

$$v_{\text{out}} = \left(1 + \frac{R_F}{R_A}\right) v_{\text{in}}$$

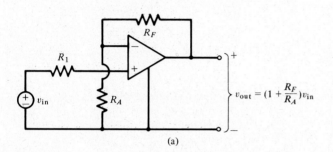

(a)

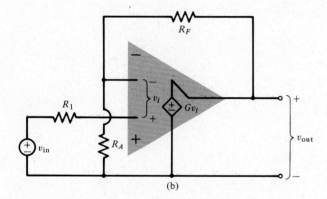

(b)

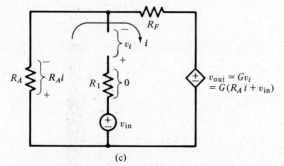

(c)

Figure 3-7   Single-input noninverting configuration for the operational amplifier.
(a) Op amp diagram.
(b) Network model.
(c) Network model redrawn.

which is fixed by the resistor ratio. The derivation of this result is as follows. In Figure 3-7(b), the operational amplifier is replaced by its network model and in Figure 3-7(c), the resulting network is redrawn and the mesh current i is defined. In terms of the mesh current, the controlled source controlling signal is

$$v_i = R_A i + v_{in}$$

The mesh equation for i is thus

$$(R_A + R_F)i = -Gv_i = -G(R_A i + v_{in})$$

or

$$[(1 + G)R_A + R_F]i = -Gv_{in}$$

$$i = \frac{-G}{(1 + G)R_A + R_F} v_{in}$$

The amplifier output voltage v_{out}, in terms of the input voltage v_{in}, is

$$v_{out} = G(R_A i + v_{in}) = G\left[\frac{-GR_A}{(1 + G)R_A + R_F}\right]v_{in} + v_{in}$$

$$= G\left[\frac{-GR_A + (1 + G)R_A + R_F}{(1 + G)R_A + R_F}\right]v_{in} = \frac{G(R_A + R_F)}{(1 + G)R_A + R_F} v_{in}$$

For op amp gains G on the order of 10^4 to 10^6 and resistor ratios

$$\frac{R_F}{R_A} << 10^4$$

the amplifier's input-output relation is virtually

$$v_{out} = \frac{R_A + R_F}{R_A} v_{in} = \left(1 + \frac{R_F}{R_A}\right)v_{in}$$

A noninverting amplifier design is shown in Figure 3-8(a). For an overall amplifier gain of 10, resistors

$$R_F = 100{,}000 \ \Omega = 100 \ \text{k}\Omega \qquad R_A = 11{,}100 \ \Omega = 11.1 \ \text{k}\Omega$$

have been chosen. R_A and R_F are large, yet R_F is considerably smaller than the value of a megohm, for which the simple op amp model might exhibit a sizeable error. The ratio

$$\frac{R_F}{R_A} = 9$$

so that

$$\left(1 + \frac{R_F}{R_A}\right) = 10$$

as desired. A pictorial diagram is given in Figure 3-8(b).

For cancellation of the average input bias current effects on v_{out}, the net re-

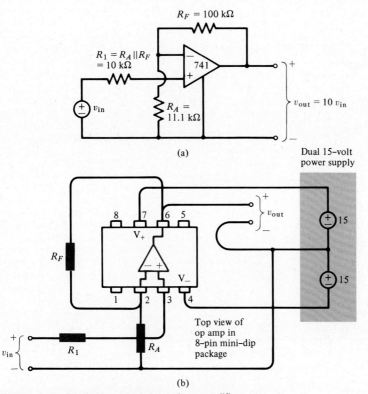

Figure 3-8 Designing a noninverting amplifier.
(a) Op amp diagram.
(b) Pictorial diagram of amplifier connections.

sistance connected to the minus op amp input terminal, the parallel combination of R_A and R_F, should equal that connected to the plus op amp input. The latter resistance is R_1 plus the Thévenin resistance of the network supplying the input voltage v_{in}. If the signal source Thévenin resistance is negligibly small, then

$$R_1 = \frac{R_A R_F}{R_A + R_F} = 10{,}000 \ \Omega = 10 \ \text{k}\Omega$$

as shown.

3.4.2 Noninverting Input Resistance

With the simple operational amplifier model, the resistance R_{IN} looking into the plus op amp input terminal is an open circuit, as is illustrated in Figure 3-9. For moderate overall amplifier gains, R_{IN} is actually equal to the *common-mode resistance* of the op amp, R_{cm}, which appears in the more involved op amp model of Figure 3-1(c).

The resistance seen by a signal source at the input terminals of the noninverting amplifier arrangement of Figure 3-9(a) is on the order of R_{cm}, which is approximately 200 megohms for a type 741 device. So long as R_1 is substantially less than

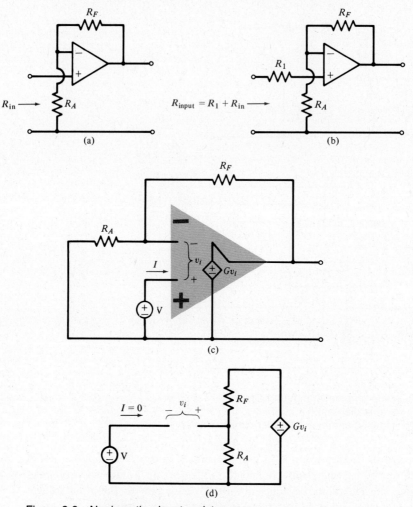

Figure 3-9 Noninverting input resistance.
(a) Resistance at the plus op amp terminal.
(b) Network model.
(c) Network model redrawn.

$R_{\text{IN}} = R_{cm}$, the resistor R_1 in the arrangement of Figure 3-9(b) does not affect the overall amplifier gain.

3.4.3 Gains Less Than Unity

The noninverting amplifier gain formula

$$v_{\text{out}} = \left(1 + \frac{R_F}{R_A}\right)v_{\text{in}}$$

does not allow gains less than unity. To obtain such gains when desired, a voltage divider is placed at the op amp input as shown in Figure 3-10(a). The input voltage is divided according to

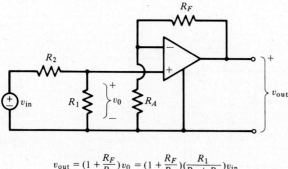

$$v_{out} = (1 + \frac{R_F}{R_A})v_0 = (1 + \frac{R_F}{R_A})(\frac{R_1}{R_1 + R_2})v_{in}$$

(a)

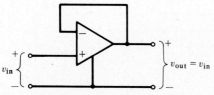

(b)

Figure 3-10 Noninverting amplifiers with gains of unity and less.
(a) Input voltage division.
(b) Unity gain follower.

$$v_o = \frac{R_1}{R_1 + R_2} v_{in}$$

since the op amp's noninverting input resistance is very nearly an open circuit. The output of the operational amplifier is then

$$v_{out} = \left(1 + \frac{R_F}{R_A}\right) v_o = \left(1 + \frac{R_F}{R_A}\right)\left(\frac{R_1}{R_1 + R_2}\right) v_{in}$$

Of course, for very large values of R_1 the 200 megohms or so common mode resistance should be taken into account.

The op amp connection of Figure 3-10(b) is a *unity-gain follower* amplifier. For it,

$$v_{out} = \left(1 + \frac{R_F}{R_A}\right)v_{in} = \left(1 + \frac{0}{\infty}\right)v_{in} = v_{in}$$

To input signal sources this amplifier's input terminals are nearly an open circuit, while its output, within the op amp's voltage and current limitations, is very nearly a perfect voltage source.

3.4.4 Multiple Noninverting Inputs

The arrangement of Figure 3-11(a) allows amplification and summation of several input signals, v_1, v_2, ... In Figure 3-11(b), using equivalent circuits, it is shown that the amplifier output voltage is related to the input voltages by

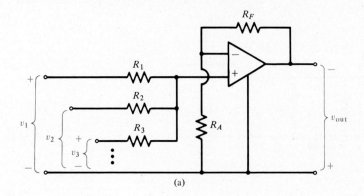

(a)

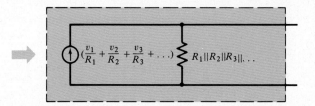

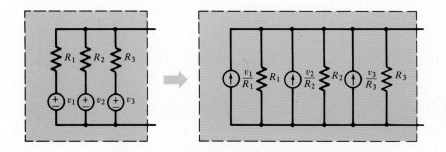

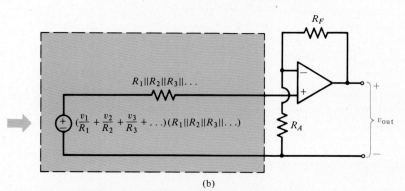

(b)

Figure 3-11   Multiple noninverting inputs.
(a) An amplifier with multiple noninverting inputs.
(b) Deriving the output voltage relation.

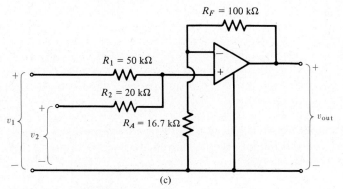

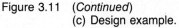

(c)

Figure 3.11   (*Continued*)
(c) Design example.

$$v_{\text{out}} = \left(1 + \frac{R_F}{R_A}\right)(R_1\|R_2\|R_3\| \cdots)\left(\frac{v_1}{R_1} + \frac{v_2}{R_2} + \frac{v_3}{R_3} + \cdots\right)$$

The notation

$$R_1\|R_2\|R_3\| \cdots = \frac{1}{(1/R_1) + (1/R_2) + (1/R_3) + \cdots}$$

is a shorthand for the parallel combination of the resistors. To balance the average input bias currents, the resistors should be chosen so that

$$R_1\|R_2\|R_3\| \cdots = R_F\|R_A$$

Suppose that it is desired to design an amplifier with

$$v_{\text{out}} = 2v_1 + 5v_2$$

In the arrangement of Figure 3-11(c), the two-input amplifier output voltage is given by

$$v_{\text{out}} = \left(1 + \frac{R_F}{R_A}\right)(R_1\|R_2)\left(\frac{v_1}{R_1} + \frac{v_2}{R_2}\right)$$

The resistors R_1 and R_2 must be in inverse proportion to the desired gains of 2 and 5. Choosing

$$R_1 = 50 \text{ k}\Omega \qquad R_2 = 20 \text{ k}\Omega$$

there results

$$v_{\text{out}} = \left(1 + \frac{R_F}{R_A}\right)(50\text{k}\|20\text{k})\left(\frac{v_1}{50\text{k}} + \frac{v_2}{20\text{k}}\right)$$

$$= \left(1 + \frac{R_F}{R_A}\right)(14.29\text{k})\left(\frac{v_1}{50\text{k}} + \frac{v_2}{20\text{k}}\right)$$

$$= \left(1 + \frac{R_F}{R_A}\right)(0.286v_1 + 0.715v_2) = 2v_1 + 5v_2$$

from which it is seen that

$$1 + \frac{R_F}{R_A} = \frac{2}{0.286} = 7 \qquad \frac{R_F}{R_A} = 6$$

The choice of

$$R_F = 60 \text{ k}\Omega \qquad R_A = 10 \text{ k}\Omega$$

will give the proper gain relations but will not result in a balance of the average input bias current effect on the output. If bias current balance is desired, then resistor values should be chosen so that

$$R_F \| R_A = R_1 \| R_2$$

Now either R_F and R_a or R_1 and R_2 can be scaled by an arbitrary factor without disturbing the gain relations, since the gains depend only on ratios of these resistances. Since

$$R_F \| R_A = 8.57 \text{ k}\Omega \qquad R_1 \| R_2 = 14.29 \text{ k}\Omega$$

multiplying R_F and R_A each by $14.29/8.57 = 1.67$ will give balance:

$$R_F = (1.67)(60 \text{ k}) = 100 \text{ k}\Omega \qquad R_A = (1.67)(10 \text{ k}) = 16.7 \text{ k}\Omega$$

When the desired noninverting amplifier gains are small, it is not possible to achieve them by just combining each signal through a resistor to the plus op amp input terminal. For the relation

$$v_{\text{out}} = 0.25v_1 + 0.4v_2$$

for example, using the circuit of Figure 3-12(a)

$$v_{\text{out}} = \left(1 + \frac{R_F}{R_A}\right)(R_1 \| R_2)\left(\frac{v_1}{R_1} + \frac{v_2}{R_2}\right)$$

and choosing

$$R_1 = 16 \text{ k}\Omega \qquad R_2 = 10 \text{ k}\Omega$$

gives

$$v_{\text{out}} = \left(1 + \frac{R_F}{R_A}\right)(6.15 \text{ k})\left(\frac{v_1}{16 \text{ k}} + \frac{v_2}{10 \text{ k}}\right)$$

$$= \left(1 + \frac{R_F}{R_A}\right)(0.38v_1 + 0.615v_2) = 0.25v_1 + 0.4v_2$$

which cannot be satisfied by any (positive) R_F and R_A. As R_1 and R_2 must always be in the same ratio, any other choices of R_1 and R_2 for which the gains are in the proper ratio will give the same result.

When an additional *dummy* input with zero input voltage is added as in Figure 3-12(b), the correct gains can be achieved, since the gain formula now involves the additional parallel resistance R_3:

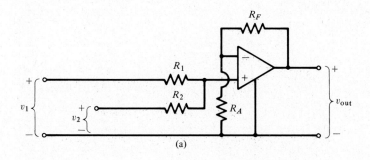

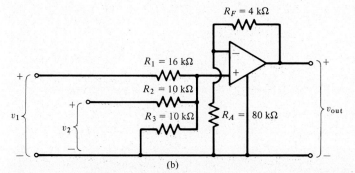

Figure 3-12   Design of another multiple-input noninverting amplifier.
(a) Attempted design.
(b) Design using a dummy input.

$$v_{out} = \left(1 + \frac{R_F}{R_A}\right)(R_1\|R_2\|R_3)\left(\frac{v_1}{R_1} + \frac{v_2}{R_2} + \frac{0}{R_3}\right)$$

with the choices

$$R_1 = 16 \text{ k}\Omega \qquad R_2 = 10 \text{ k}\Omega \qquad R_3 = 10 \text{ k}\Omega$$

then

$$v_{out} = \left(1 + \frac{R_F}{R_A}\right)(3.81 \text{ k})\left(\frac{v_1}{16 \text{ k}} + \frac{v_2}{10 \text{ k}}\right) = 0.25v_1 + 0.4v_2$$

from which

$$1 + \frac{R_F}{R_A} = \frac{0.25}{0.238} = 1.05 \qquad \frac{R_F}{R_A} = 0.05$$

The choices

$$R_F = 4 \text{ k}\Omega \qquad R_A = 80 \text{ k}\Omega$$

result in average bias current balance:

$$R_1\|R_2\|R_3 = 3.81 \text{ k}\Omega = R_F\|R_A$$

Many other solutions to this design problem are possible, of course, including those where R_A is an open circuit.

D3-2

Design single-amplifier op amp circuits which will, from the input voltages $v_1, v_2, \ldots,$ produce the indicated output voltage v_{out}. Choose practical resistor values and arrange for approximate cancellation of the average input bias current effects, assuming negligible small signal source Thévenin resistances.

(a) $v_{out} = 3v_1$
   ans. One possibility: $R_F = 100$ kΩ, $R_A = 50$ kΩ, $R_1 = 33.3$ kΩ

(b) $v_{out} = 4v_1 + 0.5v_2$
   ans. One possibility: $R_F = 40$ kΩ, $R_A = 11.4$ kΩ, $R_1 = 10$ kΩ, $R_2 = 80$ kΩ

(c) $v_{out} = 0.1v_1 + 0.25v_2$
   ans. One possibility: $R_F = 5$ kΩ, $R_1 = 50$ kΩ, $R_2 = 20$ kΩ, $R_3 = 7.69$ kΩ

(d) $v_{out} = 0.4v_1 + 3v_2 + 2.5v_3$
   ans. One possibility: $R_F = 152$ kΩ, $R_A = 5.25$ kΩ, $R_1 = 75$ kΩ, $R_2 = 10$ kΩ, $R_3 = 12$ kΩ

NONINVERTING AMPLIFIERS

Noninverting amplifier circuits with single and multiple inputs are the following:

Noninverting Amplifier

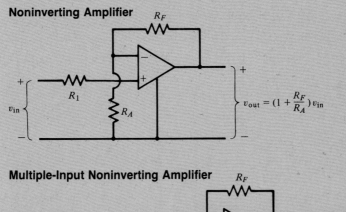

$$v_{out} = \left(1 + \frac{R_F}{R_A}\right) v_{in}$$

Multiple-Input Noninverting Amplifier

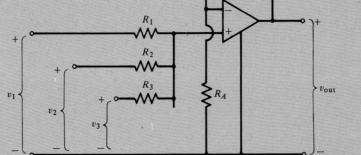

$$v_{out} = \left(1 + \frac{R_F}{R_A}\right)(R_1\|R_2\|R_3\| \ldots)\left(\frac{v_1}{R_1} + \frac{v_2}{R_2} + \frac{v_3}{R_3} + \cdots\right)$$

Typical resistor values used range from a few thousand ohms to a few hundred thousand ohms.

To cancel the effect on the output of the average input bias currents,

$$R_1 = R_A\|R_F$$

for the single-input amplifier, and

$$R_1\|R_2\|R_3\| \ldots = R_A\|R_F$$

for the multiple-input amplifier.

There is nearly an open circuit between the plus input terminal of the op amp and the power supply common. To input signal sources, single-input and multiple-input amplifiers appear as follows, for moderate overall amplifier gains:

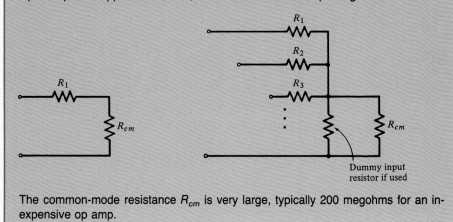

Dummy input
resistor if used

The common-mode resistance R_{cm} is very large, typically 200 megohms for an inexpensive op amp.

3.5 SUMMING AMPLIFIER DESIGN

3.5.1 Difference Amplifiers

The connections shown in Figure 3-13 are for an operational amplifier used for voltage differencing. For negligible or equal input source Thévenin resistances, the average input bias current effects cancel since the resistances connected to each op amp input terminal are equal. Superimposing the two input voltages, v_1 and v_2 in the figure, gives the result

$$v_{out} = \left(1 + \frac{R_F}{R_A}\right)\left(\frac{R_F}{R_F + R_A}\right)v_1 - \frac{R_F}{R_A}v_2$$

$$= \left(\frac{R_A + R_F}{R_A}\right)\left(\frac{R_F}{R_F + R_A}\right)v_1 - \frac{R_F}{R_A}v_2 = \frac{R_F}{R_A}(v_1 - v_2)$$

A voltage differencing amplifier design and a pictorial diagram of the amplifier connections are given in Figure 3-14.

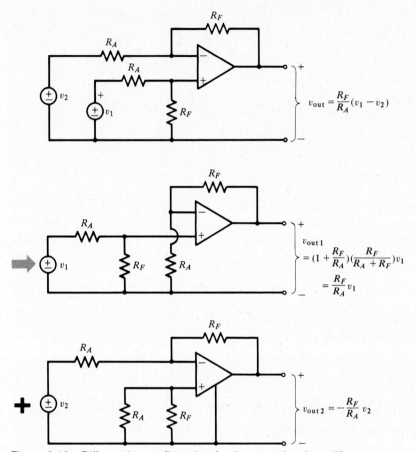

Figure 3-13 Differencing configuration for the operational amplifier.

D3-3

Design single-amplifier op amp circuits which will, from the input voltages $v_1, v_2, \ldots$, produce the indicated output voltage v_{out}. Choose practical resistor values and arrange for approximate cancellation of the average input bias current effects assuming negligibly small signal source Thévenin resistances.

(a) $v_{out} = 2v_1$
  ans. One possibility: $R_F = 100 \text{ k}\Omega$, $R_a = 50 \text{ k}\Omega$, $R_1 = 33.3 \text{ k}\Omega$

(b) $v_{out} = \frac{1}{4}v_1$
  ans. One possibility: $R_F = 100 \text{ k}\Omega$, $R_a = 100 \text{ k}\Omega$, $R_1 = 800 \text{ k}\Omega$, $R_2 = 114 \text{ k}\Omega$

(c) $v_{out} = 6v_1 - 6v_2$
  ans. One possibility: $R_F = 100 \text{ k}\Omega$, $R_a = 16.6 \text{ k}\Omega$, $R_1 = 16.6 \text{ k}\Omega$, $R_2 = 100 \text{ k}\Omega$

(d) $v_{out} = \frac{1}{3}(v_1 - v_2)$
  ans. One possibility: $R_F = 100 \text{ k}\Omega$, $R_a = 300 \text{ k}\Omega$, $R_1 = 300 \text{ k}\Omega$, $R_2 = 100 \text{ k}\Omega$

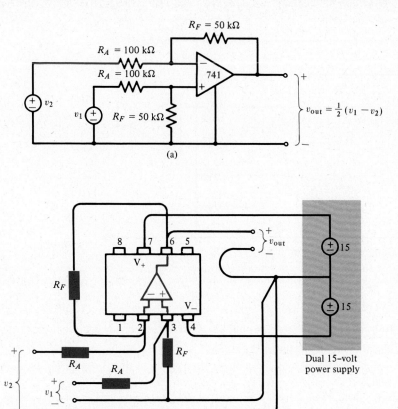

(a)

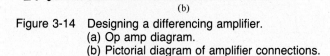

(b)

Figure 3-14 Designing a differencing amplifier.
(a) Op amp diagram.
(b) Pictorial diagram of amplifier connections.

3.5.2 General Single-Amplifier Circuits

A general summing amplifier, capable of combining input voltages with any algebraic signs, consists of a combination of the inverting and the noninverting amplifier circuits, as shown in Figure 3-15. For it,

$$v_{out} = -\frac{R_F}{R_a} v_a - \frac{R_F}{R_b} v_b - \frac{R_F}{R_c} v_c - \cdots$$

$$+ \left(1 + \frac{R_F}{R_A}\right)(R_1\|R_2\|R_3\| \ldots)\left(\frac{v_1}{R_1} + \frac{v_2}{R_2} + \frac{v_3}{R_3} + \cdots\right)$$

where

$$R_A = R_a\|R_b\|R_c\| \ldots$$

To cause cancellation of the effects on the output of the average input bias currents when it is important to do so, the resistors should be chosen so that

$$R_1\|R_2\|R_3\| \ldots = R_F\|R_a\|R_b\|R_c\| \ldots$$

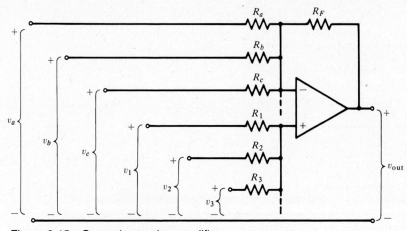

Figure 3-15 General summing amplifier.

Suppose that it is desired to obtain an amplifier input-output voltage relation

$$v_1 + 1.5v_2 - 4v_3$$

A step-by-step design procedure is as follows:

1. Design the inverting input(s) as would be done if the inputs were just the inverting ones. Figure 3-16(a) shows the results of this design step, where the choices

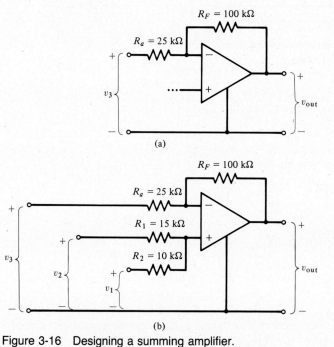

(a)

(b)

Figure 3-16 Designing a summing amplifier.
(a) Inverting input design.
(b) Attempted noninverting input design.

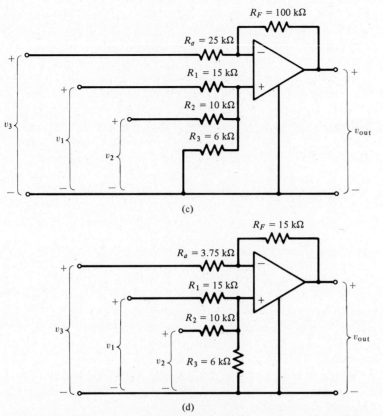

(c)

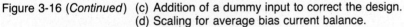

(d)

Figure 3-16 (*Continued*) (c) Addition of a dummy input to correct the design.
(d) Scaling for average bias current balance.

$$R_F = 100 \text{ k}\Omega \qquad R_a = 25 \text{ k}\Omega$$

have been made to obtain

$$v_{out} = -4v_3$$

provided that the noninverting input is properly connected.

2. Attempt to design the noninverting input(s) using the simplest noninverting input structure. Choosing

$$R_1 = 15 \text{ k}\Omega \qquad R_2 = 10 \text{ k}\Omega$$

in the structure of Figure 3-16(b) gives

$$v_{out} = \left(1 + \frac{R_F}{R_a}\right)(R_1\|R_2)\left(\frac{v_1}{R_1} + \frac{v_2}{R_2}\right) - \frac{R_F}{R_a}v_3$$

$$= (5)(6 \text{ k})\left(\frac{v_1}{15 \text{ k}} + \frac{v_2}{10 \text{ k}}\right) - 4v_3$$

$$= 2v_1 + 3v_2 - 4v_3 \neq v_1 + 1.5v_2 - 4v_3$$

The noninverting input gains need to be reduced in this case.

3. If necessary (and it almost always is), add a dummy input to correct the inverting or noninverting input gains. The dummy input in Figure 3-16(c) modifies the amplifier input-output voltage relation to

$$v_{\text{out}} = \left(1 + \frac{R_F}{R_a}\right)(R_1\|R_2\|R_3)\left(\frac{v_1}{R_1} + \frac{v_2}{R_2} + \frac{0}{R_3}\right) - \frac{R_F}{R_a}v_3$$

$$= (5)(6k\|R_3)\left(\frac{v_1}{15\ k} + \frac{v_2}{10\ k}\right) - 4v_3$$

Choosing R_3 so that

$$\frac{(5)(6\ k\|R_3)}{15\ k} = \frac{5(6000\ R_3)}{15000(6000 + R_3)} = 1 \qquad R_3 = 6\ k\Omega$$

gives

$$v_{\text{out}} = (5)(6\ k\|6\ k)\left(\frac{v_1}{15\ k} + \frac{v_2}{10\ k}\right) - 4v_3$$

$$= v_1 + 1.5v_2 - 4v_3$$

as desired.

4. Balance the average bias currents by scaling the resistances R_F, R_a, ... and/or the resistances R_1, R_2, ... Scale in such a way as to use reasonable resistance values, normally in the few thousand to the few hundred thousand ohm range. Scaling either or both sets of resistances will not affect the gain relations, which depend on ratios of resistances. Average bias current balance requires that

$$R_1\|R_2\|R_3 = R_F\|R_a$$

At present,

$$R_1\|R_2\|R_3 = 3\ k\Omega \qquad \text{and} \qquad R_F\|R_a = 20\ k\Omega$$

Multiplying R_F and R_a each by the factor $\frac{3}{20}$ gives

$$R_F = 15\ k\Omega \qquad R_a = 3.75\ k\Omega$$

for the final design, which is shown in Figure 3-16(d).

Another step-by-step design, this one for the amplifier input-output voltage relation

$$v_{\text{out}} = 3v_1 - 0.5v_2$$

is as follows:

1. Design the inverting input. The choices

$$R_F = 10\ k\Omega \qquad R_a = 20\ k\Omega$$

give

$$v_{\text{out}} = -0.5v_2$$

as in Figure 3-17(a).

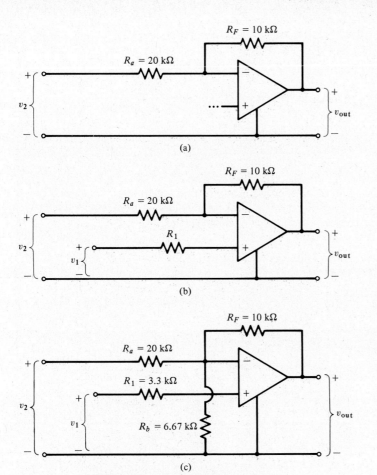

Figure 3-17   Another summing amplifier design.
(a) Inverting input design.
(b) Attempted noninverting input design.
(c) Addition of a dummy input to correct the design.

2. Attempt to design the noninverting input. Using the simplest noninverting structure, Figure 3-17(b), gives

$$v_{out} = \left(1 + \frac{R_F}{R_a}\right)(R_1)\left(\frac{v_1}{R_1}\right) - \frac{R_F}{R_a}v_2$$

$$= 1.5v_1 - 0.5v_2 \neq 3v_1 - 0.5v_2$$

3. Add a dummy input. In this example, the noninverting signal gain must be increased, so a dummy inverting input is added, as shown in Figure 3-17(c). The output voltage is then given by

$$v_{out} = \left(1 + \frac{R_F}{R_a\|R_b}\right)v_1 - \frac{R_F}{R_a}v_2$$

$$= \left(1 + \frac{10\ k}{20\ k\|R_b}\right)v_1 - 0.5v_2 = 3v_1 - 0.5v_2$$

It follows that

$$\frac{10 \text{ k}}{20 \text{ k} \| R_b} = 2 \qquad 20 \text{ k} \| R_b = 5 \text{ k} \qquad R_b = 6.67 \text{ k}\Omega$$

4. Balance the average bias currents. Balance is obtained with

$$R_1 = R_F \| R_a \| R_b = 3.3 \text{ k}\Omega$$

D3-4

Design single-amplifier op amp circuits that will, from the input voltages $v_1, v_2, \ldots,$ produce the indicated output voltage v_{out}. Choose practical resistor values and arrange for approximate cancellation of the average input bias current effects, assuming negligibly small Thévenin signal source resistances.

(a) $v_{out} = 6v_1 - v_2$
ans. One possibility: $R_F = 50 \text{ k}\Omega$, $R_a = 50 \text{ k}\Omega$, $R_b = 12.5 \text{ k}\Omega$, $R_1 = 8.3 \text{ k}\Omega$

(b) $v_{out} = v_1 + 2v_2 - 10v_3$
ans. One possibility: $R_F = 100 \text{ k}\Omega$, $R_a = 10 \text{ k}\Omega$, $R_1 = 99.5 \text{ k}\Omega$, $R_2 = 49.7 \text{ k}\Omega$, $R_3 = 12.4 \text{ k}\Omega$

(c) $v_{out} = 3v_1 - v_2 - 3v_3$
ans. One possibility: $R_F = 30 \text{ k}\Omega$, $R_a = 10 \text{ k}\Omega$, $R_b = 30 \text{ k}\Omega$, $R_1 = 10 \text{ k}\Omega$, $R_2 = 15 \text{ k}\Omega$

(d) $v_{out} = 0.5v_1 + 4v_2 - 0.5v_3$
ans. One possibility: $R_F = 10 \text{ k}\Omega$, $R_a = 20 \text{ k}\Omega$, $R_b = 3.32 \text{ k}\Omega$, $R_1 = 20 \text{ k}\Omega$, $R_2 = 2.5 \text{ k}\Omega$

3.5.3 Multiple-Amplifier Circuits

The interconnection of several individual operational amplifier stages, as in the example of Figure 3-18, will produce output voltages that are more involved linear combinations of several input voltages. In the diagram shown, "ground" symbols are used to signify connections to the common power supply terminal. In analyzing multiple amplifier diagrams such as these, one starts with the inputs and expresses each op amp output signal in terms of these, as has been done for the example. There are many ways to achieve a desired input-output relation with multiple amplifiers. Depending upon the design objectives, one may seek simplicity for servicing, a minimum number of components, a minimum number of operational amplifiers, or other properties.

D3-5

Design multiple op amp circuits that will, from the input voltages $v_1, v_2, \ldots,$ produce the indicated output voltage, v_{out}.

(a) $v_{out} = 3v_1 - \frac{1}{2}v_2$
(b) $v_{out} = 100v_1 + 30v_2$
(c) $v_{out} = 4v_1 - 3v_2 + \frac{1}{2}v_3 - \frac{1}{3}v_4$
(d) $v_{out} = 0.25v_1 + 0.7v_2 - 2.5v_3 - 50v_4$

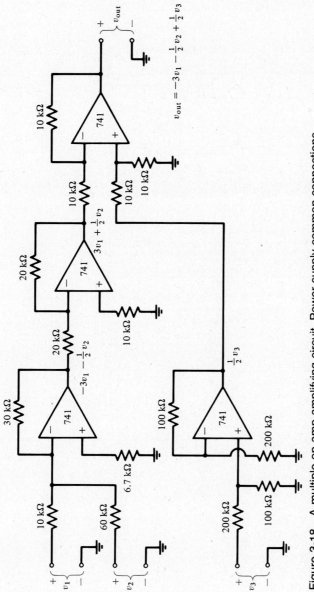

Figure 3-18 A multiple op amp amplifying circuit. Power supply common connections are indicated by the ground symbol.

SUMMING AMPLIFIER DESIGN

A differential amplifier develops an output voltage that is proportional to the difference between two input voltages:

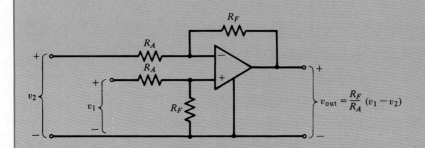

The most general multiple-input amplifier involving a single op amp is the following:

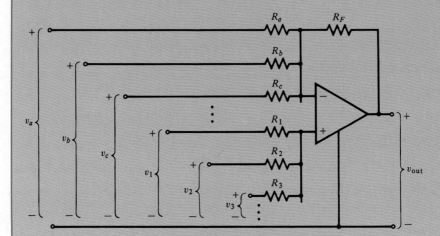

$$V_{out} = -\frac{R_F}{R_a} V_a - \frac{R_F}{R_b} V_b - \frac{R_F}{R_c} V_c - \cdots$$

$$+ \left(1 + \frac{R_F}{R_A} \right) (R_1 \| R_2 \| R_3 \| \ldots) \left(\frac{v_1}{R_1} + \frac{v_2}{R_2} + \frac{v_3}{R_3} + \cdots \right)$$

where

$$R_A = R_a \| R_b \| R_c \| \cdots$$

It is always possible to obtain a design solution, with balanced average bias current effects, for arbitrary gain coefficients, provided that the required span of resistor values is not impractical.

Amplifiers may be interconnected to form multiple op amp signal summation circuits, if desired. Simplified amplifier circuit diagrams result when the power supply common connections are indicated by the ground symbol.

3.6 TWO-PORT NETWORK PARAMETERS

A *two-port* is a network with two pairs of accessible terminals, as indicated in Figure 3-19(a). A port involves a pair of terminals, so another name for a two-terminal element is a *one*-port. In dealing with two-port networks, we will not consider external connections between ports; only connections of the type shown in Figure 3-19(b) will be used, so that at each of the ports, the current out of one of the pair of terminals is the current into the other terminal. If we did wish to consider connections between the ports, the connecting elements would be placed within the "box," forming a new two-port without external connections between ports, as in Figure 3-19(c).

Generally, each port of a two-port network has a Thévenin or Norton equivalent, as in the example model of Figure 3-20(a) that uses Thévenin equivalents. The source functions of the two Thévenin or Norton sources each have components proportional to the fixed sources within the two-port and a component proportional

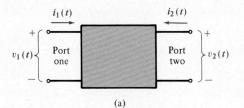

(a)

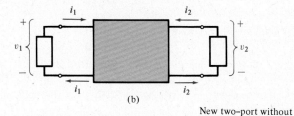

(b)

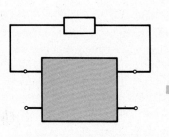
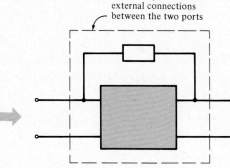

(c)

Figure 3-19　Two-port networks.
　　　　(a) Port voltages and currents.
　　　　(b) Equality of currents entering and leaving port terminals.
　　　　(c) A new two-port incorporating an external connection between the
　　　　　　original ports.

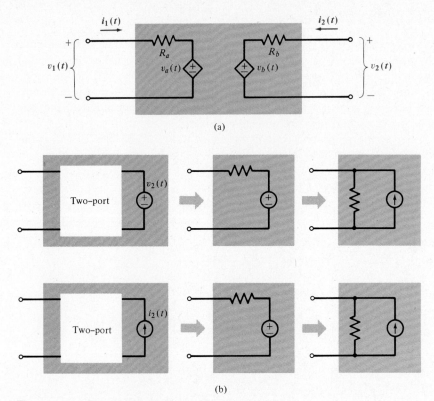

Figure 3-20   Two-port model.
            (a) Thévenin models of the ports.
            (b) Thévenin or Norton equivalence of each port.

to the voltage or current at the opposite port as indicated in Figure 3-20(b). From the standpoint of either port, the substitution theorem may be used to replace the voltage or current at the opposite port by a source. The resulting Thévenin or Norton equivalent's source function is seen to be a linear combination of the fixed source functions within the two-port and the voltage or current at the opposite port.

A *resistive* two-port consists of resistors and (possibly) controlled sources only. With no fixed sources present inside a two-port, each Thévenin or Norton source in the model has a source function that is just proportional to the voltage or current at the opposite port. It is characterized by two linear algebraic equations, one for each port, relating the port voltages and currents. The coefficients of these equations are termed the network's two-port parameters. As two equations in four variables can be arranged in many ways, there are many sets of equivalent two-port parameters.

TWO-PORT NETWORK PARAMETERS

A resistive two-port is a network of resistors and (possibly) controlled sources, with two accessible pairs of terminals. Its external behavior is described by two linear algebraic equations linking the port voltages and currents.

3.7 RESISTANCE PARAMETERS

3.7.1 Resistance Parameter Model

When a resistive two-port is expressed in terms of Thévenin port models controlled by opposite port currents, Figure 3-21, the two-port is described by equations of the form

$$\begin{cases} v_1 = r_{11}i_1 + r_{12}i_2 \\ v_2 = r_{21}i_1 + r_{22}i_2 \end{cases}$$

The numbers r_{11}, r_{12}, r_{21} and r_{22}, which have the dimensions of resistance, are called the *resistance parameters* of the two-port. Externally, a resistive two-port is completely described by its four resistance parameters, just as a two-terminal combination of resistors (a resistive one-port) is described externally by its single equivalent resistance. There may, of course, be special situations for which one or more of the parameters is infinite, in which event another description, possibly in terms of Norton equivalents and/or voltage-controlled sources, would be used.

One can solve directly for the resistance parameters, as will be done for the resistive two-port of Figure 3-22. Systematic simultaneous mesh equations in terms of the port voltages v_1 and v_2 are as follows:

$$\begin{cases} i_a - \quad i_b \qquad\quad = v_i \\ -i_a + \quad 10i_b - 4i_c = 0 \\ \qquad\quad -4i_b + 9i_c = -v_2 \end{cases}$$

In terms of the defined port currents

$$i_1 = i_a$$

$$i_2 = -i_c$$

these equations are

$$\begin{cases} i_1 - \quad i_b \qquad\quad = v_1 \\ -i_1 + \quad 10i_b + 4i_2 = 0 \\ \qquad\quad -4i_b - 9i_2 = -v_2 \end{cases}$$

Eliminating the current i_b by substituting from the first equation

$$i_b = i_1 - v_1$$

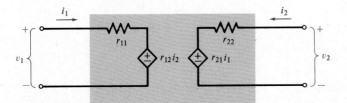

Figure 3-21   Resistance parameter model of a resistive two-port.

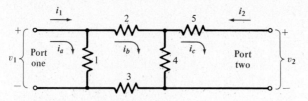

Figure 3-22 A resistive two-port.

into the second and third equations gives

$$\begin{cases} 9i_1 + 4i_2 = 10v_1 \\ -4i_1 - 9i_2 = -4v_1 - v_2 \end{cases} \tag{3-1}$$

These two equations relate the four quantities v_1, v_2, i_1, and i_2, and describe the external behavior of this two-port network.

Rearranging the equations (3-1) and solving for v_1 and v_2 in terms of i_1 and i_2,

$$\begin{cases} 10v_1 = 9i_1 + 4i_2 \\ -4v_1 - v_2 = -4i_1 - 9i_2 \end{cases}$$

$$v_1 = \tfrac{9}{10}i_1 + \tfrac{4}{10}i_2 = r_{11}i_1 + r_{12}i_2$$

$$v_2 = \dfrac{\begin{vmatrix} 10 & (9i_1 + 4i_2) \\ -4 & (-4i_1 - 9i_2) \end{vmatrix}}{\begin{vmatrix} 10 & 0 \\ -4 & -1 \end{vmatrix}} = \tfrac{4}{10}i_1 + \tfrac{74}{10}i_2 = r_{21}i_1 + r_{22}i_2$$

shows the two-port resistance parameters to be

$$r_{11} = \tfrac{9}{10}\ \Omega$$

$$r_{12} = \tfrac{2}{5}\ \Omega$$

$$r_{21} = \tfrac{2}{5}\ \Omega$$

$$r_{22} = \tfrac{37}{5}\ \Omega$$

3.7.2 Parameter Measurements

A generally easier method of finding resistance parameters is to make simple tests or measurements. In the equations

$$\begin{cases} v_1 = r_{11}i_1 + r_{12}i_2 \\ v_2 = r_{21}i_1 + r_{22}i_2 \end{cases}$$

if $i_2 = 0$, then

$$v_1 = r_{11}i_1$$

Setting $i_2 = 0$ simply means to leave port two open-circuited. Thus

$$r_{11} = \dfrac{v_1}{i_1}\bigg|_{i_2=0}$$

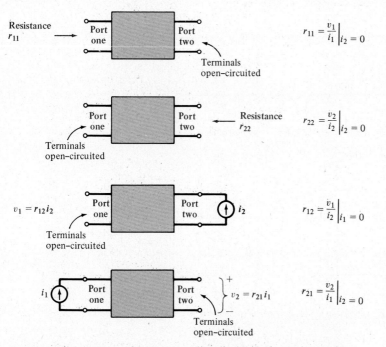

$$r_{11} = \frac{v_1}{i_1}\bigg|_{i_2 = 0}$$

$$r_{22} = \frac{v_2}{i_2}\bigg|_{i_2 = 0}$$

$$r_{12} = \frac{v_1}{i_2}\bigg|_{i_1 = 0}$$

$$r_{21} = \frac{v_2}{i_1}\bigg|_{i_2 = 0}$$

Figure 3-23   Measurements to determine resistance parameters.

which is the resistance looking into port one when port two is open circuited, as shown in Figure 3-23. Similarly

$$r_{22} = \frac{v_2}{i_2}\bigg|_{i_1 = 0}$$

$$r_{12} = \frac{v_1}{i_2}\bigg|_{i_1 = 0}$$

$$r_{21} = \frac{v_2}{i_1}\bigg|_{i_2 = 0}$$

For example, these calculations applied to the previous resistive two-port give

$$r_{11} = \tfrac{9}{10} \ \Omega \qquad \text{and} \qquad r_{22} = \tfrac{37}{5} \ \Omega$$

as shown in Figure 3-24. In the calculations of r_{12} and r_{21}, equivalent circuits are used to obtain

$$r_{12} = \frac{v_1}{i_2}\bigg|_{i_1 = 0} = \tfrac{2}{5} \ \Omega$$

$$r_{21} = \frac{v_2}{i_1}\bigg|_{i_2 = 0} = \tfrac{2}{5} \ \Omega$$

Resistance parameters are sometimes termed *open-circuit parameters* to emphasize the nature of the simplest measurements of these quantities.

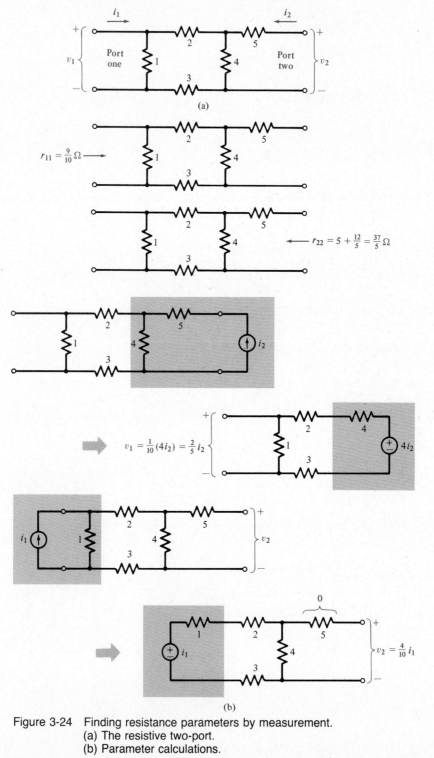

Figure 3-24 Finding resistance parameters by measurement.
(a) The resistive two-port.
(b) Parameter calculations.

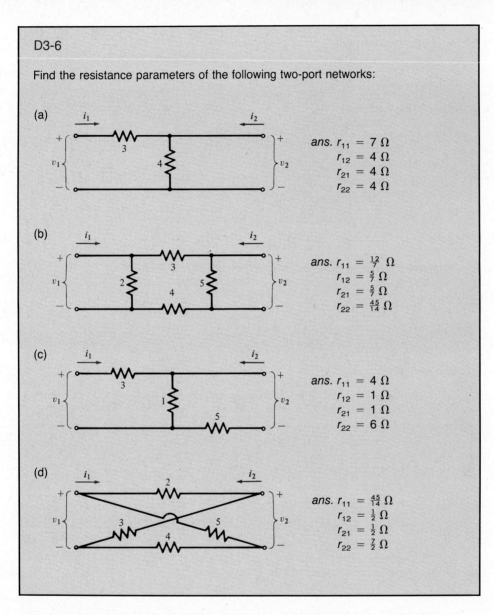

D3-6

Find the resistance parameters of the following two-port networks:

(a)

ans. $r_{11} = 7\ \Omega$
$r_{12} = 4\ \Omega$
$r_{21} = 4\ \Omega$
$r_{22} = 4\ \Omega$

(b)

ans. $r_{11} = \frac{12}{7}\ \Omega$
$r_{12} = \frac{5}{7}\ \Omega$
$r_{21} = \frac{5}{7}\ \Omega$
$r_{22} = \frac{45}{14}\ \Omega$

(c)

ans. $r_{11} = 4\ \Omega$
$r_{12} = 1\ \Omega$
$r_{21} = 1\ \Omega$
$r_{22} = 6\ \Omega$

(d)

ans. $r_{11} = \frac{45}{14}\ \Omega$
$r_{12} = \frac{1}{2}\ \Omega$
$r_{21} = \frac{1}{2}\ \Omega$
$r_{22} = \frac{7}{2}\ \Omega$

3.7.3 Using Two-Port Models

Using two-port parameters, one may replace arbitrarily complicated resistive two-ports by a two-port model. In the example of Figure 3-25(a) a two-port network that might represent, for example, a very complicated electric power distribution system has been modeled by its resistance parameters. Using the resistance parameter model, Figure 3-25(b), the equivalent network is solved for an external signal of interest, $v_2(t)$.

In terms of the clockwise mesh currents, systematic simultaneous equations for the network are

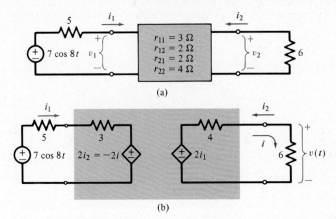

(a)

(b)

Figure 3-25 Using resistance parameters to represent a two-port.

$$\begin{cases} 8i_1 = 7 \cos 8t - (-2i) \\ 10i = 2i_1 \end{cases} \quad \text{or} \quad \begin{cases} 8i_1 - 2i = 7 \cos 8t \\ -2i_1 + 10i = 0 \end{cases}$$

$$i = \frac{\begin{vmatrix} 8 & 7 \cos 8t \\ -2 & 0 \end{vmatrix}}{\begin{vmatrix} 8 & -2 \\ -2 & 10 \end{vmatrix}} = \frac{14 \cos 8t}{76}$$

Then

$$v_2(t) = 6i = \tfrac{21}{19} \cos 8t$$

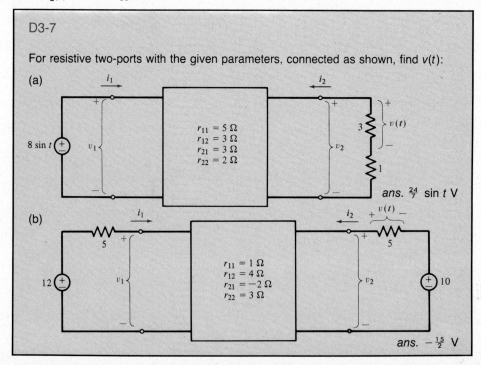

D3-7

For resistive two-ports with the given parameters, connected as shown, find $v(t)$:

(a)

$$r_{11} = 5\ \Omega$$
$$r_{12} = 3\ \Omega$$
$$r_{21} = 3\ \Omega$$
$$r_{22} = 2\ \Omega$$

ans. $\tfrac{24}{7} \sin t$ V

(b)

$$r_{11} = 1\ \Omega$$
$$r_{12} = 4\ \Omega$$
$$r_{21} = -2\ \Omega$$
$$r_{22} = 3\ \Omega$$

ans. $-\tfrac{15}{2}$ V

RESISTANCE PARAMETERS

When a resistive two-port is modelled by Thévenin equivalents at each port,

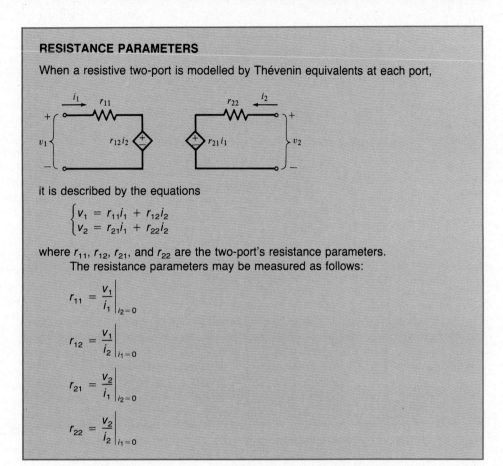

it is described by the equations

$$\begin{cases} v_1 = r_{11}i_1 + r_{12}i_2 \\ v_2 = r_{21}i_1 + r_{22}i_2 \end{cases}$$

where r_{11}, r_{12}, r_{21}, and r_{22} are the two-port's resistance parameters.
The resistance parameters may be measured as follows:

$$r_{11} = \left.\frac{v_1}{i_1}\right|_{i_2=0}$$

$$r_{12} = \left.\frac{v_1}{i_2}\right|_{i_1=0}$$

$$r_{21} = \left.\frac{v_2}{i_1}\right|_{i_2=0}$$

$$r_{22} = \left.\frac{v_2}{i_2}\right|_{i_1=0}$$

3.8 CONDUCTANCE, HYBRID AND OTHER PARAMETERS

3.8.1 Conductance Parameters

The four variables v_1, v_2, i_1, and i_2 that describe the external behavior of a resistive two-port are related by two equations. Expressing the port voltages in terms of the port currents describes the network in terms of its resistance parameters:

$$\begin{cases} v_1 = r_{11}i_1 + r_{12}i_2 \\ v_2 = r_{21}i_1 + r_{22}i_2 \end{cases}$$

Solving instead for the port currents in terms of the voltages

$$\begin{cases} i_1 = g_{11}v_1 + g_{12}v_2 \\ i_2 = g_{21}v_1 + g_{22}v_2 \end{cases}$$

defines the two-port *conductance parameters*, g_{11}, g_{12}, g_{21}, g_{22}. The conductance parameter model of a resistive two-port is given in Figure 3-26. Conductance and other parameters are useful in network analysis when alternate two-port models are more convenient, similar to the situation where either a Thévenin or Norton model may be used for a one-port.

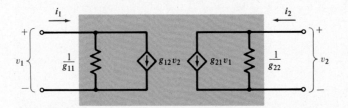

Figure 3-26 Conductance parameter model of a resistive two-port.

Measurements to determine two-port conductance parameters are summarized in Figure 3-27, and calculation of the conductance parameters of an example network is done in Figure 3-28. The parameters found are

$$g_{11} = \tfrac{1}{5} \text{ S}$$

$$g_{22} = \tfrac{9}{20} \text{ S}$$

$$g_{12} = -\tfrac{1}{5} \text{ S}$$

$$g_{21} = -\tfrac{1}{5} \text{ S}$$

Negative values for g_{12} and g_{21} are typical. Conductance parameters are sometimes called *short-circuit parameters* to emphasize the nature of these measurements.

3.8.2 Relation Between Conductance and Resistance Parameters

The relations between the conductance and the resistance parameters are easily found by applying Cramer's rule to solve for v_1 and v_2 in terms of i_1 and i_2:

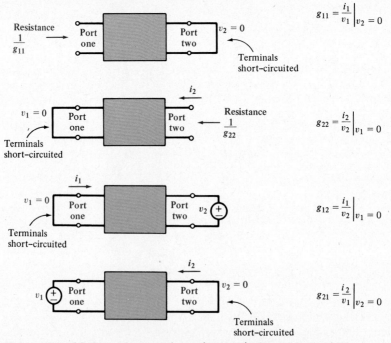

$$g_{11} = \frac{i_1}{v_1}\bigg|_{v_2 = 0}$$

$$g_{22} = \frac{i_2}{v_2}\bigg|_{v_1 = 0}$$

$$g_{12} = \frac{i_1}{v_2}\bigg|_{v_1 = 0}$$

$$g_{21} = \frac{i_2}{v_1}\bigg|_{v_2 = 0}$$

Figure 3-27 Measurements to determine conductance parameters.

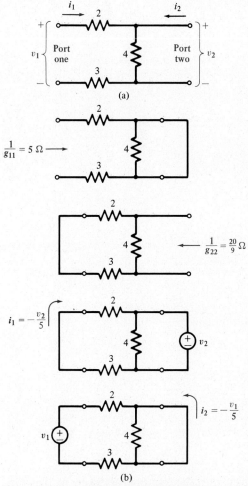

Figure 3-28   A resistive two-port and its conductance parameters.
(a) The resistive two-port.
(b) Parameter calculations.

$$\begin{cases} g_{11}v_1 + g_{12}v_2 = i_1 \\ g_{21}v_1 + g_{22}v_2 = i_2 \end{cases}$$

$$v_1 = \frac{\begin{vmatrix} i_1 & g_{12} \\ i_2 & g_{22} \end{vmatrix}}{\begin{vmatrix} g_{11} & g_{12} \\ g_{21} & g_{22} \end{vmatrix}} = \frac{g_{22}i_1 - g_{12}i_2}{g_{11}g_{22} - g_{21}g_{12}} = r_{11}i_1 + r_{12}i_2$$

$$v_2 = \frac{\begin{vmatrix} g_{11} & i_1 \\ g_{21} & i_2 \end{vmatrix}}{g_{11}g_{22} - g_{21}g_{12}} = \frac{g_{11}i_2 - g_{21}i_1}{g_{11}g_{22} - g_{21}g_{12}} = r_{21}i_1 + r_{22}i_2$$

Thus,

$$r_{11} = \frac{g_{22}}{g_{11}g_{22} - g_{21}g_{12}}$$

$$r_{12} = \frac{-g_{12}}{g_{11}g_{22} - g_{21}g_{12}}$$

$$r_{21} = \frac{-g_{21}}{g_{11}g_{22} - g_{21}g_{12}}$$

$$r_{22} = \frac{g_{11}}{g_{11}g_{22} - g_{21}g_{12}}$$

(3-2)

Similarly, the g's can be found in terms of the r's:

$$g_{11} = \frac{r_{22}}{r_{11}r_{22} - r_{21}r_{12}}$$

$$g_{12} = \frac{-r_{12}}{r_{11}r_{22} - r_{21}r_{12}}$$

$$g_{21} = \frac{-r_{21}}{r_{11}r_{22} - r_{21}r_{12}}$$

$$g_{22} = \frac{r_{11}}{r_{11}r_{22} - r_{21}r_{12}}$$

(3-3)

3.8.3 Reciprocity

For a two-port network consisting entirely of resistors, it always happens that

$$r_{21} = r_{12}$$

and that

$$g_{21} = g_{12}$$

as in the previous examples. This result is known as the *reciprocity theorem.* Reciprocity follows from the fact that *symmetric* equations, either mesh or nodal, can always be written for such a network. The elimination of variables from these equations to obtain the two-port resistance equations may always be done through a series of steps that preserves the symmetry of the equations. The final set of equations, of the form

$$\begin{cases} v_1 = r_{11}i_1 + r_{12}i_2 \\ v_2 = r_{21}i_1 + r_{22}i_2 \end{cases} \quad \text{or} \quad \begin{cases} i_1 = g_{11}v_1 + g_{12}v_2 \\ i_2 = g_{21}v_1 + g_{22}v_2 \end{cases}$$

are thus symmetric, which is to say that $r_{21} = r_{12}$ and $g_{21} = g_{12}$.

In any network of resistors, no matter how complicated,

$$r_{12} = \left.\frac{v_1}{i_2}\right|_{i_1=0} = r_{21} = \left.\frac{v_2}{i_1}\right|_{i_2=0} = r$$

Suppose that a current $i(t)$ is applied at one port, producing the voltage

$$v(t) = ri(t)$$

at a second port. Then if the current is, instead, applied at the second port, the same voltage $v(t)$ will be produced at the first port. This idea is illustrated in Figure 3-29(a). Similarly, since

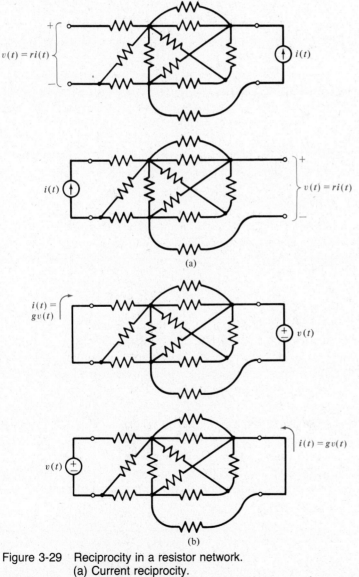

(a)

(b)

Figure 3-29 Reciprocity in a resistor network.
(a) Current reciprocity.
(b) Voltage reciprocity.

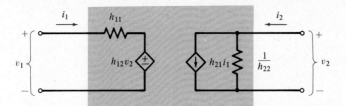

Figure 3-30 Hybrid parameter model of a resistive two-port.

$$g_{12} = \left.\frac{i_1}{v_2}\right|_{v_1=0} = g_{21} = \left.\frac{i_2}{v_1}\right|_{v_2=0} = g$$

an applied voltage produces the same short-circuit port current at opposite ports, as illustrated in Figure 3-29(b).

When a two-port network contains controlled sources in addition to resistors, with the source controlling signals within the two-port, the two-port network is still governed by two linear algebraic equations relating the port voltages and currents, but the symmetry of the network equations is no longer necessarily present. Except in special cases,

$$r_{21} \neq r_{12} \quad \text{and} \quad g_{21} \neq g_{12}$$

3.8.4 Hybrid and Other Parameters

The external behavior of a resistive two-port is characterized by two equations linking the four port signals. Hybrid parameters, used extensively in transistor circuit analysis and design, are defined by

$$\begin{cases} v_1 = h_{11}i_1 + h_{12}v_2 \\ i_2 = h_{21}i_1 + h_{22}v_2 \end{cases}$$

and have the network model given in Figure 3-30. For these and most other possible sets of parameters, the conditions for reciprocity do *not* make, say, $h_{21} = h_{12}$; this simple result occurs only when the two-port equations used can be obtained from an original symmetric set of symmetric operations. In fact, for the hybrid parameters, reciprocity gives $h_{21} = -h_{12}$.

Another set of two-port parameters of special usefulness are the transmission parameters

$$\begin{cases} v_1 = Av_2 - Bi_2 \\ i_1 = Cv_2 - Di_2 \end{cases}$$

which are especially convenient for the description of the port-to-port joining of two or more two-ports. Reciprocity gives $AD - BC = 1$.

D3-8

Find the conductance parameters of the following two-port networks:

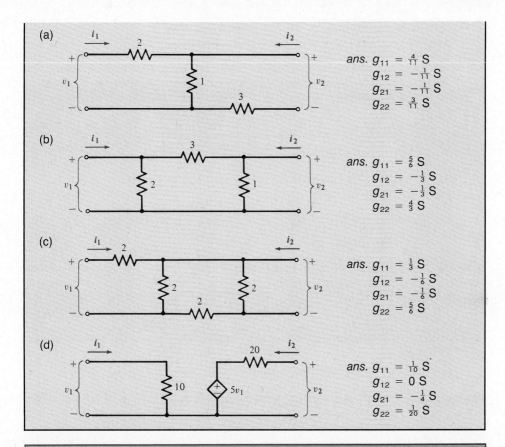

(a)

ans. $g_{11} = \frac{4}{11}$ S
$g_{12} = -\frac{1}{11}$ S
$g_{21} = -\frac{1}{11}$ S
$g_{22} = \frac{3}{11}$ S

(b)

ans. $g_{11} = \frac{5}{6}$ S
$g_{12} = -\frac{1}{3}$ S
$g_{21} = -\frac{1}{3}$ S
$g_{22} = \frac{4}{3}$ S

(c)

ans. $g_{11} = \frac{1}{3}$ S
$g_{12} = -\frac{1}{6}$ S
$g_{21} = -\frac{1}{6}$ S
$g_{22} = \frac{5}{6}$ S

(d)

ans. $g_{11} = \frac{1}{10}$ S
$g_{12} = 0$ S
$g_{21} = -\frac{1}{4}$ S
$g_{22} = \frac{1}{20}$ S

CONDUCTANCE, HYBRID AND OTHER PARAMETERS

Two-port conductance parameters correspond to a model with Norton equivalents at each port:

$$\begin{cases} i_1 = g_{11}v_1 + g_{12}v_2 \\ i_2 = g_{21}v_1 + g_{22}v_2 \end{cases}$$

The conductance parameters may be measured as follows:

$$g_{11} = \left. \frac{i_1}{v_1} \right|_{v_2 = 0}$$

$$g_{12} = \left. \frac{i_1}{v_2} \right|_{v_1 = 0}$$

$$g_{21} = \left. \frac{i_2}{v_1} \right|_{v_2 = 0}$$

$$g_{22} = \left. \frac{i_2}{v_2} \right|_{v_1 = 0}$$

They are related to the resistance parameters by equations (3-2) and (3-3). Hybrid parameters correspond to the following model

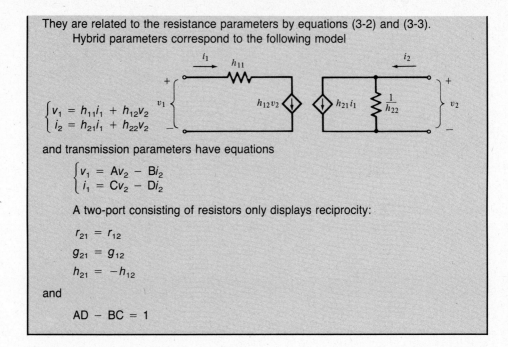

$$\begin{cases} v_1 = h_{11}i_1 + h_{12}v_2 \\ i_2 = h_{21}i_1 + h_{22}v_2 \end{cases}$$

and transmission parameters have equations

$$\begin{cases} v_1 = Av_2 - Bi_2 \\ i_1 = Cv_2 - Di_2 \end{cases}$$

A two-port consisting of resistors only displays reciprocity:

$$r_{21} = r_{12}$$

$$g_{21} = g_{12}$$

$$h_{21} = -h_{12}$$

and

$$AD - BC = 1$$

3.9 DELTA-WYE TRANSFORMATION

The "tee" network of Figure 3-31(a) has very simple resistance parameters. As derived in Figure 3-31(b), the resistance parameters are, in terms of the tee-network resistors

$$r_{11} = R_A + R_C$$

$$r_{12} = r_{21} = R_C$$

$$r_{22} = R_B + R_C$$

or

$$R_A = r_{11} - r_{12}$$

$$R_B = r_{22} - r_{12}$$

$$R_C = r_{12} = r_{21}$$

as in Figure 3-31(c). The tee network is often used as an equivalent for a resistive two-port.

Similarly, the "pi" network of Figure 3-32 has especially simple conductance parameters and so is also used as an equivalent for a more complicated resistive two-port. Conductance parameters of the pi network, derived in Figure 3-32(b), are

$$g_{11} = \frac{R_1 + R_3}{R_1 R_3} = \frac{1}{R_1} + \frac{1}{R_3} = G_1 + G_3$$

$$g_{12} = g_{21} = -\frac{1}{R_3} = -G_3$$

$$g_{22} = \frac{R_2 + R_3}{R_2 R_3} = \frac{1}{R_2} + \frac{1}{R_3} = G_2 + G_3$$

where the G's denote resistor conductances. The pi network resistances and con-

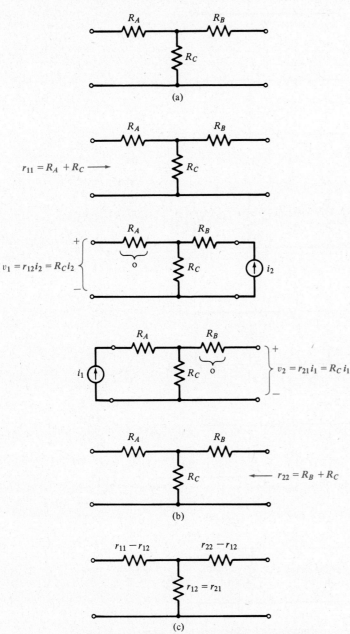

(a)

$r_{11} = R_A + R_C \longrightarrow$

$v_1 = r_{12}i_2 = R_C i_2$

i_1

$v_2 = r_{21} i_1 = R_C i_1$

$\longleftarrow r_{22} = R_B + R_C$

(b)

$r_{11} - r_{12}$

$r_{22} - r_{12}$

$r_{12} = r_{21}$

(c)

Figure 3-31   Resistance parameters of a resistive tee network.
(a) Tee network.
(b) Finding resistance parameters of the tee network.
(c) Tee network in terms of resistance parameters.

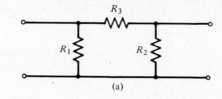

(a)

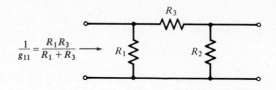

$$\frac{1}{g_{11}} = \frac{R_1 R_3}{R_1 + R_3} \longrightarrow$$

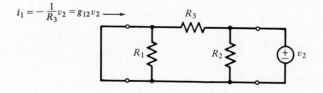

$$i_1 = -\frac{1}{R_3} v_2 = g_{12} v_2 \longrightarrow$$

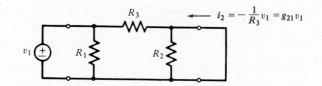

$$\longleftarrow i_2 = -\frac{1}{R_3} v_1 = g_{21} v_1$$

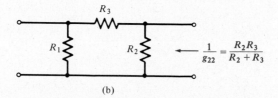

$$\longleftarrow \frac{1}{g_{22}} = \frac{R_2 R_3}{R_2 + R_3}$$

(b)

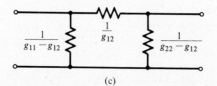

(c)

Figure 3-32   Conductance parameters of a resistive pi network.
(a) Pi network.
(b) Finding conductance parameters of the pi network.
(c) Pi network in terms of conductance parameters.

ductances are, in terms of the conductance parameters,

$$R_1 = \frac{1}{g_{11} + g_{12}} \qquad G_1 = g_{11} + g_{12}$$

$$R_2 = \frac{1}{g_{22} + g_{12}} \qquad G_2 = g_{22} + g_{12}$$

$$R_3 = -\frac{1}{g_{12}} = -\frac{1}{g_{21}} \qquad G_3 = -g_{12} = -g_{21}$$

Viewed in another way, the tee and the pi networks are *three-terminal* combinations of resistors, as shown in Figure 3-33. The three-terminal resistor connections are called *wye* and *delta* connections, respectively, in practice. As one can find an equivalent two-port pi network for the tee network and vice versa, wye connections of resistors have equivalent delta connections, and vice versa. This equivalence between three terminal connections of resistors is termed the delta-wye transformation. It is an additional equivalent circuit that may be used to simplify networks, particularly those in which there are connections of resistors that are neither in series nor in parallel.

To derive the relations between equivalent delta and wye connections of resistors, the two-port resistance parameters of the delta (or pi) will be equated to the resistance parameters of the wye (or tee) network. The resistance parameters of the delta network are found in Figure 3-34. Equating these to the resistance parameters of the wye network gives

$$r_{11} = \frac{R_1 R_2 + R_2 R_3}{R_1 + R_2 + R_3} = R_A + R_C$$

$$r_{12} = r_{21} = \frac{R_1 R_2}{R_1 + R_2 + R_3} = R_C$$

$$r_{22} = \frac{R_1 R_2 + R_1 R_3}{R_1 + R_2 + R_3} = R_B + R_C$$

from which it is evident that the delta and the wye resistor connections are indistinguishable at their terminals, provided that

$$R_A = \frac{R_2 R_3}{R_1 + R_2 + R_3}$$

$$R_B = \frac{R_1 R_3}{R_1 + R_2 + R_3} \qquad \text{(3-4)}$$

$$R_C = \frac{R_1 R_2}{R_1 + R_2 + R_3}$$

Each of these equations is of the form

$$R_i = \frac{\text{product of the two resistances connected to terminal } i}{\text{sum of the three resistances}}$$

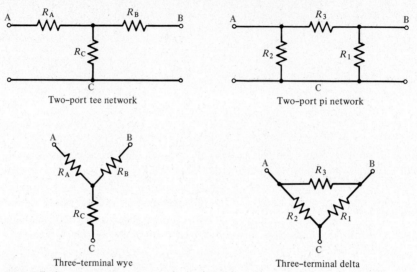

Figure 3-33   Tee and pi networks as delta and wye connections of resistors.

And each wye resistance is positive if the delta resistances are positive.

By forming ratios of the equations (3-4) two at a time, the delta resistances may be found in terms of the wye ones:

$$\frac{R_A}{R_B} = \frac{R_2 R_3/(R_1 + R_2 + R_3)}{R_1 R_3/(R_1 + R_2 + R_3)} = \frac{R_2}{R_1}$$

$$\frac{R_A}{R_C} = \frac{R_3}{R_1}$$

$$\frac{R_B}{R_C} = \frac{R_3}{R_2}$$

Substituting into

$$R_C = \frac{R_1 R_2}{R_1 + R_2 + R_3} = \frac{R_1}{(R_1/R_2) + 1 + (R_3/R_2)}$$

gives

$$R_C = \frac{R_1}{(R_B/R_A) + 1 + (R_B/R_C)}$$

$$R_1 = \frac{R_A R_B + R_A R_C + R_B R_C}{R_A}$$

Similarly,

$$R_2 = \frac{R_A R_B + R_A R_C + R_B R_C}{R_B}$$

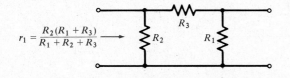

$$r_1 = \frac{R_2(R_1 + R_3)}{R_1 + R_2 + R_3} \longrightarrow$$

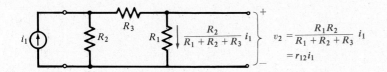

$$v_2 = \frac{R_1 R_2}{R_1 + R_2 + R_3} i_1$$
$$= r_{12} i_1$$

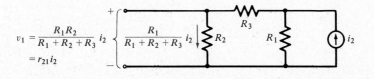

$$v_1 = \frac{R_1 R_2}{R_1 + R_2 + R_3} i_2$$
$$= r_{21} i_2$$

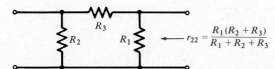

$$r_{22} = \frac{R_1(R_2 + R_3)}{R_1 + R_2 + R_3}$$

Figure 3-34   Resistance parameters of a pi network.

$$R_3 = \frac{R_A R_B + R_A R_C + R_B R_C}{R_C}$$

Each of these equations is of the form

$$R_i = \frac{\text{sum of products of resistances taken two at a time}}{\text{resistance connected to terminal opposite to } R_i}$$

The equivalent delta resistances are positive if the wye resistances are all positive.

An example of a specific delta-to-wye conversion is shown in Figure 3-35(a). If a delta connection is part of a more complicated network such as the one in Figure 3-35(b), a delta-wye transformation may simplify the network considerably, as it does in this example.

Figure 3-36(a) is an example of wye-to-delta conversion. As with delta-to-wye transformation, a wye-to-delta transformation may be used to simplify a network such as one in which resistors are neither in series nor in parallel with one another. An example is shown in Figure 3-36(b).

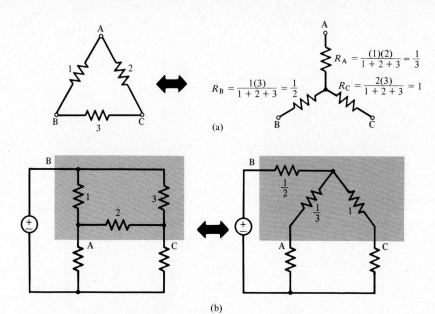

$$R_A = \frac{(1)(2)}{1+2+3} = \frac{1}{3}$$

$$R_B = \frac{1(3)}{1+2+3} = \frac{1}{2} \qquad R_C = \frac{2(3)}{1+2+3} = 1$$

(a)

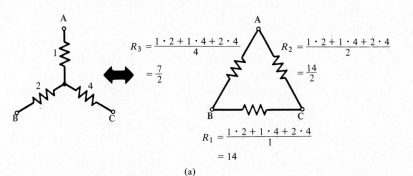

(b)

Figure 3-35   Delta-to-wye conversion.
(a) Numerical example.
(b) Using delta-wye equivalence to simplify a network.

$$R_3 = \frac{1\cdot2+1\cdot4+2\cdot4}{4} \qquad R_2 = \frac{1\cdot2+1\cdot4+2\cdot4}{2}$$

$$= \frac{7}{2} \qquad\qquad = \frac{14}{2}$$

$$R_1 = \frac{1\cdot2+1\cdot4+2\cdot4}{1}$$

$$= 14$$

(a)

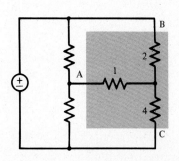 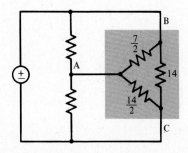

(b)

Figure 3-36   Wye-to-delta conversion.
(a) Numerical example.
(b) Using wye-delta equivalence to simplify a network.

D3-9

Convert the delta networks to equivalent wyes and convert the wye networks to equivalent deltas. Label the terminals A, B, and C in each equivalent network:

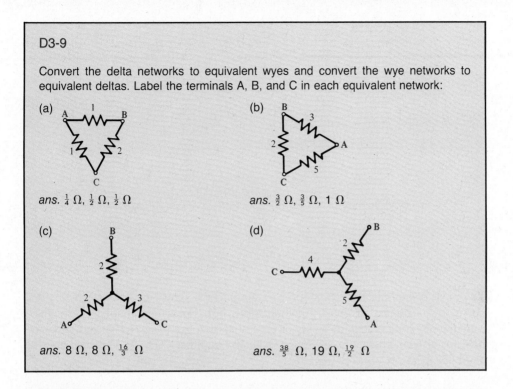

(a)

ans. $\frac{1}{4}$ Ω, $\frac{1}{2}$ Ω, $\frac{1}{2}$ Ω

(b)

ans. $\frac{3}{2}$ Ω, $\frac{3}{5}$ Ω, 1 Ω

(c)

ans. 8 Ω, 8 Ω, $\frac{16}{3}$ Ω

(d)

ans. $\frac{38}{5}$ Ω, 19 Ω, $\frac{19}{2}$ Ω

D3-10

Use delta-wye transformation and series and parallel resistance combinations to find the equivalent resistance of the following network. Repeat, using a wye-delta transformation.

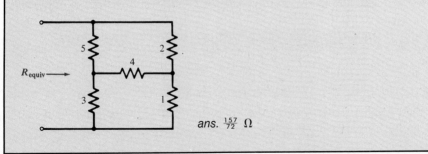

ans. $\frac{157}{72}$ Ω

DELTA-WYE TRANSFORMATION

The three-terminal delta and wye resistor networks are equivalent to one another, with the following relations between the resistor values:

$$R_1 = \frac{R_A R_B + R_A R_C + R_B R_C}{R_A} \qquad R_A = \frac{R_2 R_3}{R_1 + R_2 + R_3}$$

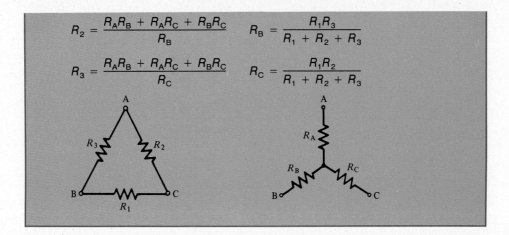

$$R_2 = \frac{R_A R_B + R_A R_C + R_B R_C}{R_B} \qquad R_B = \frac{R_1 R_3}{R_1 + R_2 + R_3}$$

$$R_3 = \frac{R_A R_B + R_A R_C + R_B R_C}{R_C} \qquad R_C = \frac{R_1 R_2}{R_1 + R_2 + R_3}$$

3.10 AMPLIFIER MODELS

The most commonly used model for an amplifier is shown in Figure 3-37(a). It is not in the standard two-port equation form, but this arrangement is particularly convenient because it is in terms of quantities that are easily visualized and have been used from the earliest times to specify amplifiers. The model describes an amplifier in terms of the equations

$$v_{in} = R_{in} i_{in} + \rho i_{out}$$

$$v_{out} = \mu v_{in} - R_{out} i_{out}$$

(3-5)

Measurements to determine the model parameters are shown in Figure 3-37(b). The parameter R_{in}, the amplifier *input resistance*, is the resistance looking into the amplifier input terminals when no amplifier output current is flowing. The parameter R_{out}, the amplifier *output resistance*, is the resistance of the output port when the input voltage is zero. The amplifier's *open circuit voltage gain*, μ, is the ratio of output voltage to input voltage when no output current is flowing:

$$\mu = \left. \frac{v_{out}}{v_{in}} \right|_{i_{out}=0}$$

And ρ, the amplifier's *reverse gain*, describes the effect of nonzero amplifier output current on the amplifier input port:

$$\rho = \left. \frac{v_{in}}{i_{out}} \right|_{i_{in}=0}$$

Often the effect is negligible so that $\rho = 0$, and the input port is described solely by its input resistance.

The single-input op amp inverting amplifier, Figure 3-38(a), provides a good example of calculating amplifier parameters. Approximate expressions were found earlier for this amplifier's input resistance and voltage gain. Now all four parameters will be found, using an improved op amp model.

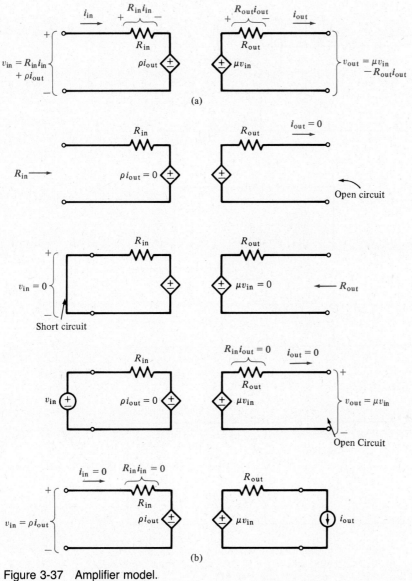

Figure 3-37   Amplifier model.
(a) The model.
(b) Measurements to determine the model parameters.

In Figure 3-38(b) and (c), the operational amplifier is replaced by a simple network model that now includes the Thévenin resistance R_o of the op amp output. R_o is on the order of 100 Ω for an inexpensive device such as the type 741. The systematic simultaneous mesh equations for the resulting network are

$$\begin{cases} (R_A + R_F + R_o)\,i_{in} - R_o i_{out} = v_{in} - G(R_A i_{in} - v_{in}) \\ \qquad\qquad - R_o i_{in} + R_o i_{out} = G(R_A i_{in} - v_{in}) - v_{out} \end{cases}$$

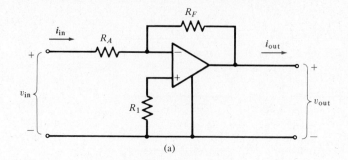

(a)

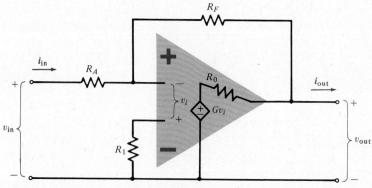

(b)

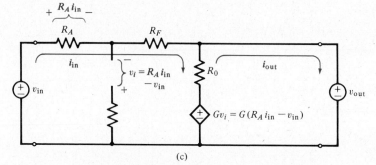

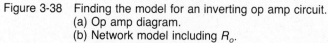

(c)

Figure 3-38   Finding the model for an inverting op amp circuit.
(a) Op amp diagram.
(b) Network model including R_o.
(c) Network model redrawn.

or,

$$\begin{cases} [(1 + G)R_A + R_F + R_o]i_{in} - R_o i_{out} = (1 + G)v_{in} \\ \qquad - (R_o + GR_A)i_{in} + R_o i_{out} = -Gv_{in} - v_{out} \end{cases}$$

(3-6)

To place the first of these equations in the amplifier model form (3-5), it is only necessary to divide each side of the equation by $(1 + G)$:

$$v_{in} = \left(R_A + \frac{R_F + R_o}{1 + G}\right)i_{in} - \frac{R_o}{1 + G}i_{out} = R_{in}i_{in} + \rho i_{out}$$

Here it is seen that

$$R_{in} = R_A + \frac{R_F + R_o}{1 + G} \qquad \rho = -\frac{R_o}{1 + G}$$

which are very nearly

$$R_{in} \cong R_A \qquad \rho \cong 0$$

for large G and resistances of the usual sizes.

To place the second equation of (3-6) in the form

$$v_{out} = \mu v_{in} - R_{out}i_{out}$$

the variable i_{in} is eliminated by substitution of the first equation into the second:

$$-(R_o + GR_A)\left[\frac{R_o i_{out} + (1 + G)v_{in}}{(1 + G)R_A + R_F + R_o}\right] + R_o i_{out} = -Gv_{in} - v_{out}$$

Or,

$$v_{out} = \left[\frac{(R_o + GR_A)(1 + G)}{(1 + G)R_A + R_F + R_o} - G\right]v_{in}$$

$$+ \left[\frac{(R_o + GR_A)R_o}{(1 + G)R_A + R_F + R_o} - R_o\right]i_{out}$$

$$= \left[\frac{-GR_F + R_o}{(1 + G)R_A + R_F + R_o}\right]v_{in} + \left[\frac{-R_oR_F - R_oR_A}{(1 + G)R_A + R_F + R_o}\right]i_{out}$$

$$= \mu v_{in} - R_{out}i_{out}$$

from which it is seen that

$$\mu = \frac{R_o - GR_F}{(1 + G)R_A + R_F + R_o} \qquad R_{out} = \frac{R_o(R_F + R_A)}{(1 + G)R_A + R_F + R_o}$$

For large G and resistances of the usual sizes,

$$\mu \cong -\frac{R_F}{R_A} \qquad R_{out} \cong \frac{R_o(R_F + R_A)}{(1 + G)R_A} \cong 0$$

D3-11

Find R_{in}, R_{out}, μ, and ρ for an amplifier described by the following network:

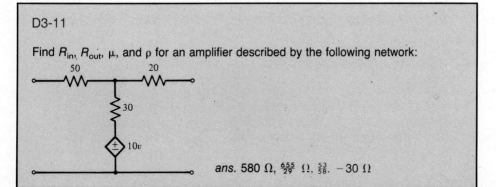

ans. 580 Ω, $\frac{655}{29}$ Ω, $\frac{53}{58}$. -30 Ω

AMPLIFIER MODELS

Amplifiers are often represented by the following two-port:

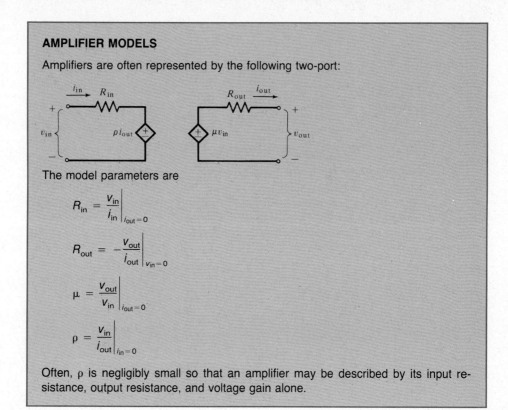

The model parameters are

$$R_{in} = \left. \frac{v_{in}}{i_{in}} \right|_{i_{out}=0}$$

$$R_{out} = \left. -\frac{v_{out}}{i_{out}} \right|_{v_{in}=0}$$

$$\mu = \left. \frac{v_{out}}{v_{in}} \right|_{i_{out}=0}$$

$$\rho = \left. \frac{v_{in}}{i_{out}} \right|_{i_{in}=0}$$

Often, ρ is negligibly small so that an amplifier may be described by its input resistance, output resistance, and voltage gain alone.

CHAPTER THREE PROBLEMS

Basic Problems

Operational Amplifier Design

1. Design single-amplifier op amp circuits that will, from the input voltages v_1, v_2, . . ., produce the indicated output voltage v_{out}. Choose practical resistor values and arrange for approximate cancellation of the average input bias current effects, assuming negligibly small signal source Thévenin resistances.

    (a) $v_{out} = -3.5v_1$

    (b) $v_{out} = -0.3v_1$

    (c) $v_{out} = -0.4v_1 - 2.3v_2$

    (d) $v_{out} = -v_1 - 2v_2 - 5v_3$

    (e) $v_{out} = 7v_1$

    (f) $v_{out} = 0.5v_1$

    (g) $v_{out} = (\frac{1}{3})(v_1 - v_2)$

    (h) $v_{out} = 10(v_2 - v_1)$

    (i) $v_{out} = 3v_1 - 0.5v_2$

    (j) $v_{out} = 0.5v_1 + 0.8v_2 - 2.5v_3$

    (k) $v_{out} = v_1 + 1.5v_2 - 4v_3$

    (l) $v_{out} = 3v_1 + 4v_2 - 2v_3$

Dual in-line packages (DIPs) are used for a variety of electronic components, including operational amplifiers and sets of resistors. (*Photo courtesy of Beckman Instruments, Inc., Electronic Technologies Group.*)

2. For each of the following op amp diagrams, find the relation between the output voltage(s) v_{out} and the input voltages v_1, v_2, ...

(a)

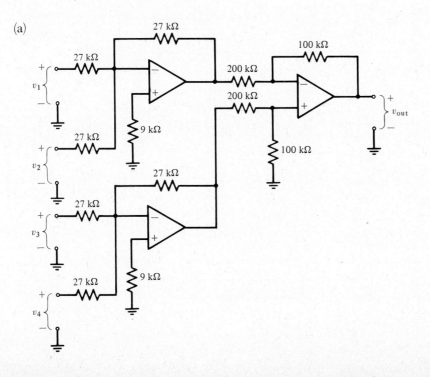

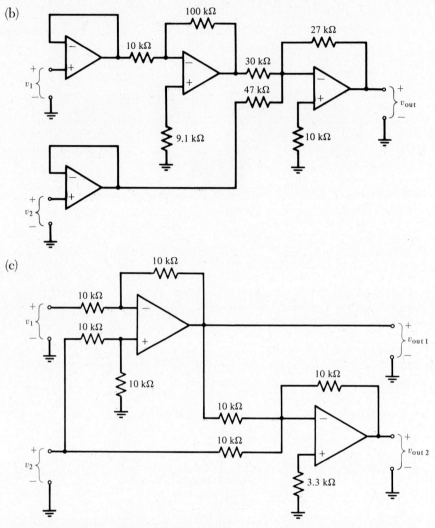

3. Design multiple-amplifier op amp circuits that will, from input voltages v_1, v_2, ..., produce the indicated voltage v_{out}:

(a) $v_{out} = 3v_1 - 2v_2$

(b) $v_{out} = v_1 + v_2 + v_3$

(c) $v_{out} = 10v_1 - 5v_2 + 0.5v_3$

(d) $v_{out} = 2v_1 + 0.3v_2 - 4v_3 - 0.5v_4$

Two-Port Models

4. Find the resistance parameters of each of the following two-ports:

(a)

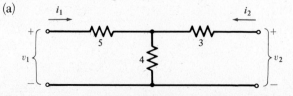

A radio station studio. The large knobs on the control board adjust the volume of the various audio sources via potentiometers with logarithmic tapers. (*Photo courtesy of the Radio Corporation of America.*)

(b)

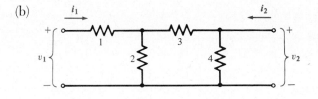

(c)

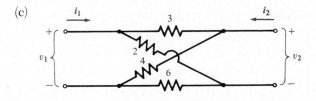

(d)

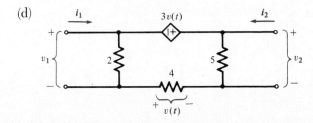

5. Find the conductance parameters of each of the following two-ports:

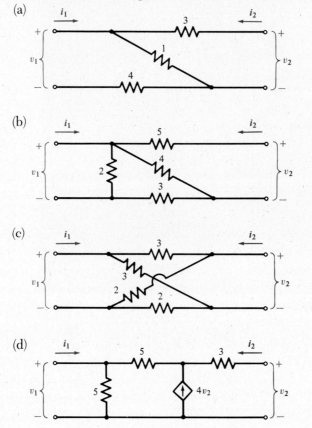

(a)

(b)

(c)

(d)

6. For resistive two-ports with the given parameters, connected as shown, find the voltages $v(t)$:

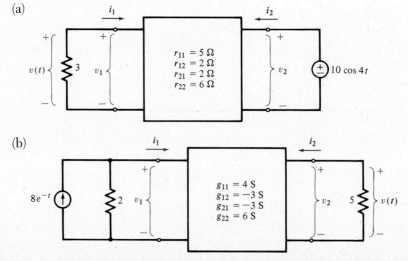

(a)

$$r_{11} = 5\ \Omega$$
$$r_{12} = 2\ \Omega$$
$$r_{21} = 2\ \Omega$$
$$r_{22} = 6\ \Omega$$

$10 \cos 4t$

(b)

$8e^{-t}$

$$g_{11} = 4\ \text{S}$$
$$g_{12} = -3\ \text{S}$$
$$g_{21} = -3\ \text{S}$$
$$g_{22} = 6\ \text{S}$$

$v(t)$

(c)

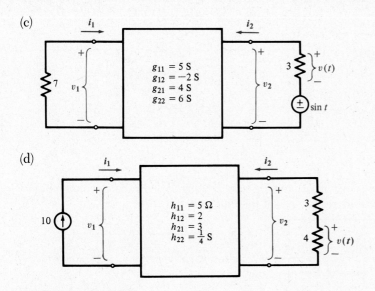

$$g_{11} = 5 \text{ S}$$
$$g_{12} = -2 \text{ S}$$
$$g_{21} = 4 \text{ S}$$
$$g_{22} = 6 \text{ S}$$

(d)

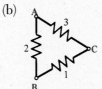

$$h_{11} = 5 \text{ }\Omega$$
$$h_{12} = 2$$
$$h_{21} = 3$$
$$h_{22} = \frac{1}{4} \text{ S}$$

Delta-Wye Transformation

7. Convert the following to equivalent wye networks. Label the terminals A, B, and C in the equivalent networks.

(a)

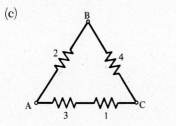

(b)

(c)

(d)

8. Convert the following to equivalent delta networks. Label the terminals A, B, and C in the equivalent networks.

(a)

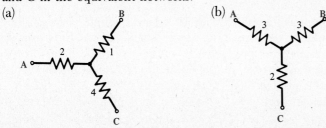

(b)

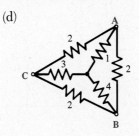

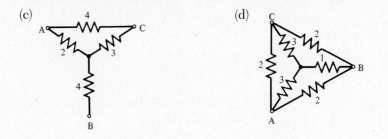

(c)

(d)

9. Use delta-wye transformation and series and parallel resistance combinations to find the equivalent resistances of the following networks:

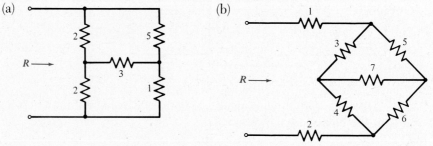

(a)

(b)

Two-Port Amplifier Models
10. Find R_{in}, R_{out}, μ, and ρ for amplifiers described by the following networks:

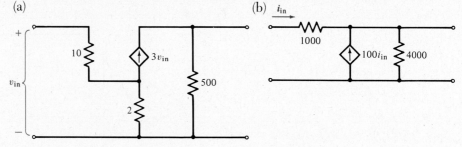

(a)

(b)

Practical Problems

Operational Amplifiers
11. The type 741 operational amplifier with 15 V power supplies is typically capable of providing output voltages in the range ± 13 V and output currents up to ± 15 mA.
    (a) For the circuit shown, what are the maximum and minimum input voltages v_{in} such that the output voltage is within ± 13 V?
    (b) Neglecting current flow in R_F, what is the minimum load resistance R_L for

which output voltages within the range ± 13 V do not cause excessive output current to flow?

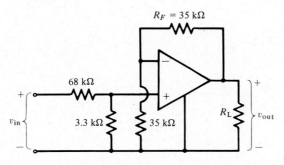

$R_F = 35$ kΩ

68 kΩ

v_{in}

3.3 kΩ 35 kΩ

R_L

v_{out}

12. Small asymmetries in manufactured operational amplifiers result in a nonzero output voltage when the voltage across the input terminals is zero. This effect is expressed as the *input offset voltage* of the op amp and is represented by the additional voltage source in the simple model of (a). The value and polarity of v_{io} varies from device to device, so manufacturers specify the maximum magnitude of v_{io} under various conditions.

In many operational amplifier types, provision is made to connect an external *offset null* potentiometer with which the offset voltage may be cancelled. For the type 741 op amp, a 10-kΩ offset null potentiometer is connected as shown in (b).

The manufacturer of a type 741 op amp guarantees that the magnitude of v_{io} will not be greater than 4 mV at 25°C and that when the integrated circuit is at a higher temperature, the input offset voltage will not increase by more than 15 μV per °C. If no offset nulling is used, what is the maximum magnitude of the component of v_{out} due to v_{io} in the circuit of (c) when the temperature is 50°C?

(a)   (b)

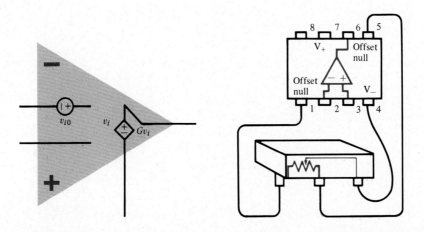

(c)

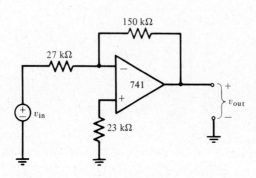

Common-Mode Rejection

The physical operational amplifier has slight asymmetries that make the output voltage depend slightly upon the average of the two input terminal voltages as well as the terminal voltage difference. A network model that accounts for this effect is shown in (a). G is the ordinary *differential-mode* gain, and H is the *common-mode* gain.

When the common-mode gain of the op amp is taken into account, a differential amplifier will have an output voltage with both differential and common-mode components, although the common-mode term is generally very small:

$$v_{\text{out}} = A(v_1 - v_2) + B\left(\frac{v_1 + v_2}{2}\right)$$

The common-mode rejection ratio (CMRR) of such an amplifier is defined as the magnitude of the ratio of the two gains:

$$\text{CMRR} = \left|\frac{A}{B}\right|$$

13. Using the amplifier model of (a) with

$$G = 10^5 \qquad H = 10^3$$

for the differential amplifier of (b), find the common-mode rejection ratio.

14. Using the amplifier model of (a) with

$$G = 10^5 \qquad H = 0$$

for the amplifier of (c) with slightly incorrect resistor values, find the common-mode rejection ratio.

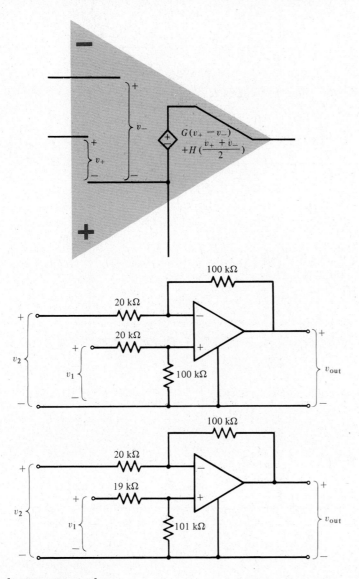

$$G(v_+ - v_-)$$
$$+H\left(\frac{v_+ + v_-}{2}\right)$$

Delta-Wye Equivalence

15. Convert the following network to an equivalent three-terminal element consisting of just three resistors in the wye configuration:

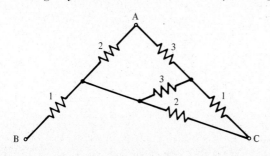

Advanced Problems

Operational Amplifiers

16. Below is shown an operational amplifier model and network in which all fixed sources except those representing the input bias currents have been set to zero. Show that for equal currents i_{B1} and i_{B2}, the component of v_{out} due to the bias currents is zero if R_1 equals the parallel combination of R_F and R_A.

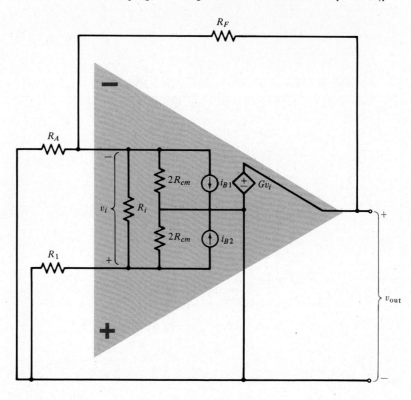

Two-Port Parameters

17. Show that resistance parameters r_{11}, r_{12}, r_{21}, r_{22}, and the hybrid parameters h_{11}, h_{12}, h_{21}, h_{22} are related by

$$r_{11} = \frac{h_{11}h_{22} - h_{12}h_{22}}{h_{22}}$$

$$r_{12} = \frac{h_{12}}{h_{22}}$$

$$r_{21} = \frac{-h_{21}}{h_{22}}$$

$$r_{22} = \frac{1}{h_{22}}$$

and by

$$h_{11} = \frac{r_{11}r_{22} - r_{12}r_{21}}{r_{22}}$$

$$h_{12} = \frac{r_{12}}{r_{22}}$$

$$h_{21} = \frac{-r_{21}}{r_{22}}$$

$$h_{22} = \frac{1}{r_{22}}$$

18. Find the resistance parameters of the "tee" network. Then find a tee network (with R_4 present) with resistance parameters

$$r_{11} = 10 \ \Omega$$

$$r_{12} = r_{21} = 3 \ \Omega$$

$$r_{22} = 6 \ \Omega$$

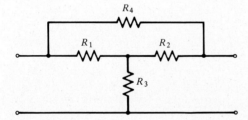

Delta-Wye Transformation

19. Derive the relations for converting from a delta to a wye connection of resistors from the three equations equating the resistances looking into each pair of terminals when the remaining terminal is not connected to anything.

Amplifier Models

20. Find the resistance parameters of the two-port amplifier model with parameters R_{in}, R_{out}, μ, and ρ.

Chapter Four
Network Differential Equations

The particles of an insulating dielectric whilst under
induction may be compared to a series of small magnetic
needles, or more correctly still to a series of small insulated
conductors. If the space round a charged globe were filled
with a mixture of an insulating dielectric, as oil of turpentine
or air, and small globular conductors, as shot, the latter
being at a little distance from each other so as to be
insulated, then these would in their condition and action
exactly resemble what I consider to be the condition and
action of the particles of the insulating dielectric itself.

Michael Faraday
From *Experimental Researches in Electricity*,
The Royal Institution, 1847

In the minds of many, the electro-magnetic telegraph is
associated with the many chimerical projects constantly
brought before the public and particularly with schemes so
popular a year or two ago for the application of electricity as
a moving power in the arts. All schemes for this purpose, I
have from the first asserted, are premature and formed
without proper scientific knowledge. The case, however, is
entirely different in regard to the electro-magnetic
telegraph. The science is now fully ripe for such an
application of its principles, and I have not the least doubt,
if proper means be afforded, of the perfect success of this
invention.

Joseph Henry
From a letter to Samuel F. B. Morse,
Washington, D.C., 1842

4.1 PREVIEW

In this chapter the remaining basic network elements, the capacitor, the inductor,
and coupled inductors, are introduced. The voltage-current relationships for these
elements involve time derivatives and integrals, and so networks containing the
elements are described by integrodifferential equations.

After discussion of these new elements, methods for writing systematic simultaneous nodal and mesh equations are developed. All of the effort and care invested in writing source-resistor network equations will now pay added dividends. Writing equations with inductors and capacitors simply involves substituting appropriate derivative or integral relations for the new elements.

The incorporation of inductive coupling is considerably simplified by the introduction of a controlled source equivalent circuit. In writing network equations, it is then unnecessary to learn a new set of procedures for inductive coupling; the coupling is replaced by controlled sources that are routinely incorporated into the equations just as they were in the source-resistor case.

In preparation for solving networks described by differential equations, pertinent aspects of the subject, including solution methods, are discussed in the concluding sections of this chapter. Properties of the exponential functions that are involved in differential equation solutions are reviewed.

When you complete this chapter, you should know—

1. the voltage, current, power, and energy relations for capacitors and how to use them;
2. the voltage, current, power, and energy relations for inductors and how to use them;
3. voltage, current, winding sense, power, and energy relations for mutually coupled inductors and controlled source equivalents;
4. how to write systematic simultaneous nodal equations for networks containing resistors, capacitors, inductors, mutually coupled inductors, and fixed and controlled sources;
5. how to write systematic simultaneous mesh equations for networks containing resistors, capacitors, inductors, mutually coupled inductors, and fixed and controlled sources;
6. how to find the general solution of linear, constant-coefficient homogeneous differential equations of any order, including those with repeated and complex characteristic roots;
7. how to find the general solution of linear, constant-coefficient differential equations with constant driving functions;
8. how to apply boundary conditions to the general solution of a differential equation to obtain a specific solution;
9. the shape of the exponential function, how to graph it, and how to find and use time constants.

4.2 VOLTAGE-CURRENT, POWER, AND ENERGY RELATIONS FOR THE CAPACITOR

4.2.1 Defining Relations

The physical capacitor consists of two conductors in close proximity, separated by an insulator. The two conductors are called *plates*, and under the network conditions of confined current flow and negligible propagation times for disturbances, the current flowing onto one plate is identical to the current leaving the other plate. It makes sense, then, to speak of the current flow *through* the capacitor; even though charge does not really flow between the plates, it appears at the capacitor terminals as though it does.

The symbol for a capacitor model is shown in Figure 4-1. The *capacitance C* of the element is indicated beside the symbol for the capacitor. The unit of capacitance is the *farad* (abbreviated F), which is named in honor of Michael Faraday (1791–1867), a brilliant experimenter who made great contributions to electromagnetic theory.

The voltage across the capacitor, $v(t)$, is the charge, $q(t)$, on the plate nearest the plus sign on the voltage polarity reference $[-q(t)$ is on the other plate] divided by the capacitance:

$$v(t) = \frac{1}{C} q(t)$$

In terms of the sink reference current $i(t)$, the capacitor charge is

$$q(t) = \int_{-\infty}^{t} i(t) \, dt$$

In other words, $q(t)$ reflects the whole past history—from "way back," which is $t \to -\infty$ mathematically, to now, time t—of the rate of charge flow $i(t)$. This is known as a "running integral"; it is a function of the present time t.

It is sometimes preferable to write integrals such as the above with a different variable of integration, as

$$q(t) = \int_{-\infty}^{t} i(x) \, dx$$

for example. This will not be done here, since there is little chance of confusing the variable of integration with the integral limits.

Sometimes the whole past history of a current is not known. What is known is the net charge on the capacitor (or the capacitor voltage, which is the charge divided by the capacitance) at some specific time t_0, and the history of the current from time t_0 on. Then the running integral may be written as

$$q(t) = \int_{-\infty}^{t_0} i(t) \, dt + \int_{t_0}^{t} i(t) \, dt = q(t_0) + \int_{t_0}^{t} i(t) \, dt$$

which holds, of course, only for times after t_0.

Figure 4-1   Capacitor symbol.

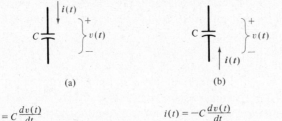

(a)   (b)

$$i(t) = C\frac{dv(t)}{dt}$$

$$v(t) = \frac{1}{C}\int_{-\infty}^{t} i(t)dt$$

$$= v(t_0) + \frac{1}{C}\int_{t_0}^{t} i(t)dt, \quad t > t_0$$

$$i(t) = -C\frac{dv(t)}{dt}$$

$$v(t) = -\frac{1}{C}\int_{-\infty}^{t} i(t)dt$$

$$= v(t_0) - \frac{1}{C}\int_{t_0}^{t} i(t)dt, \quad t > t_0$$

Figure 4-2   Capacitor voltage-current relations.
(a) Sink reference relation.
(b) Source reference relation.

It is most convenient to choose the origin of the time scale to make $t_0 = 0$, giving

$$q(t) = q(0) + \int_0^t i(t)\, dt$$

after $t = 0$. There is then no information about the detailed behavior of $q(t)$ before $t = 0$.

Sink and source reference relation capacitor voltage-current relations are summarized in Figure 4-2. Given a capacitor voltage, the capacitor current may be found by differentiation. As with the resistor, there is a minus sign in the voltage-current relationship if the voltage and current have the source reference relationship.

Examples of finding a capacitor current from its voltage are shown in Figure 4-3. In Figure 4-3(b), the capacitor voltage is specified by a source, and the voltage and current have the source reference relation in the capacitor.

Given the entire past history of a capacitor current, the capacitor charge, and thus the capacitor voltage, may be found by integration, as in the example of Figure 4-4(a), for which

$$v(t) = -\frac{1}{2}\int_{-\infty}^t e^{3t}\, dt = -\frac{1}{2}\frac{e^{3t}}{3}\Big|_{-\infty}^t = -\frac{1}{6}e^{3t} \text{ V}$$

In the network of Figure 4-4(b), the given current is described piecewise, as is the solution, $v(t)$. In general,

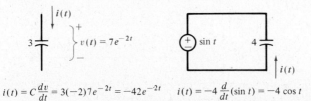

$$i(t) = C\frac{dv}{dt} = 3(-2)7e^{-2t} = -42e^{-2t} \qquad i(t) = -4\frac{d}{dt}(\sin t) = -4\cos t$$

Figure 4-3   Determining capacitor current from the capacitor voltage.

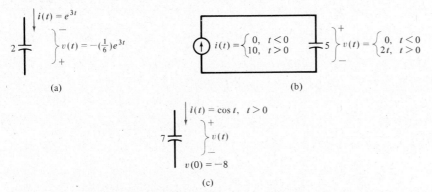

(a)   (b)

(c)

Figure 4-4   Determining capacitor voltage from the capacitor current. In (c), the initial capacitor voltage summarizes the effect of the current up to time $t = 0$.

$$v(t) = \frac{1}{C} \int_{-\infty}^{t} i(t)\, dt$$

Before $t = 0$, $i(t) = 0$, so

$$v(t) = \tfrac{1}{5} \int_{-\infty}^{t} 0\, dt = 0, \qquad t < 0$$

After time $t = 0$, $i(t) = 10$, and

$$v(t) = \tfrac{1}{5} \int_{0}^{t} 10\, dt = 2t \text{ V}, \qquad t > 0$$

If the capacitor current is known after some time, for convenience time $t = 0$, and the capacitor voltage is known at $t = 0$, the capacitor voltage may be found after $t = 0$, as in the example of Figure 4-4(c). The net effect of the whole past history of the current before $t = 0$ is the capacitor voltage at $t = 0$. After $t = 0$,

$$v(t) = \frac{1}{C} \int_{-\infty}^{0} i(t)\, dt + \frac{1}{C} \int_{0}^{t} i(t)\, dt$$

$$= v(0) + \frac{1}{C} \int_{0}^{t} i(t)\, dt = -8 + \frac{1}{7} \int_{0}^{t} \cos t\, dt$$

$$= -8 + \frac{1}{7} \sin t \Big|_{0}^{t} = -8 + \frac{1}{7} \sin t \text{ V}$$

D4-1

Find the indicated currents and voltages:

(a)

$i(t)$

sin $3t$ $\underset{\hphantom{=}}{=}\, 2$

ans. 6 cos 3t A

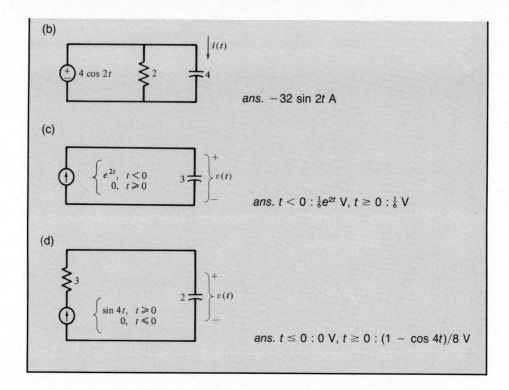

(b)

ans. −32 sin 2t A

(c)

ans. $t < 0 : \frac{1}{6}e^{2t}$ V, $t \geq 0 : \frac{1}{6}$ V

(d)

ans. $t \leq 0 : 0$ V, $t \geq 0 : (1 - \cos 4t)/8$ V

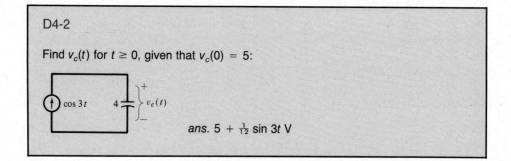

D4-2

Find $v_c(t)$ for $t \geq 0$, given that $v_c(0) = 5$:

ans. $5 + \frac{1}{12} \sin 3t$ V

4.2.2 Capacitor Power and Energy

Capacitors are energy-storing elements. With the sink reference, the electrical power flow into a capacitor is

$$p_{\text{into}}(t) = v(t)i(t) = Cv(t)\frac{dv(t)}{dt}$$

The electrical energy stored in the capacitor is proportional to the square of the capacitor voltage, since

$$W_C(t) = \int_{-\infty}^{t} p_{\text{into}}(t) \, dt = \tfrac{1}{2} \, Cv^2(t)$$

The SI unit of energy is the newton-meter = joule (abbreviation J).

4.2.3 Two-Terminal Combinations of Capacitors

Any two-terminal combination of capacitors is equivalent to a single capacitor. No matter how the terminal voltage is distributed from capacitor to capacitor, the terminal current is proportional to the time rate of change of that voltage.

The equivalent capacitance of several capacitors in *parallel* is just the sum of the individual capacitances, as shown in Figure 4-5. This equivalence is similar to the relation for resistors in *series*. For capacitors in series, the equivalent capacitance is the inverse of the sum of the inverses of the individual capacitances. This equivalence is developed in Figure 4-6.

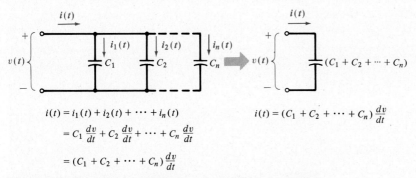

$$i(t) = i_1(t) + i_2(t) + \cdots + i_n(t)$$
$$= C_1 \frac{dv}{dt} + C_2 \frac{dv}{dt} + \cdots + C_n \frac{dv}{dt}$$
$$= (C_1 + C_2 + \cdots + C_n) \frac{dv}{dt}$$

$$i(t) = (C_1 + C_2 + \cdots + C_n) \frac{dv}{dt}$$

Figure 4-5 Equivalent capacitance of capacitors in parallel.

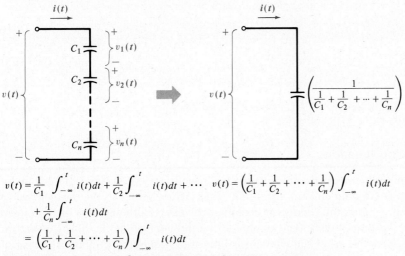

$$v(t) = \frac{1}{C_1} \int_{-\infty}^{t} i(t)dt + \frac{1}{C_2} \int_{-\infty}^{t} i(t)dt + \cdots$$
$$+ \frac{1}{C_n} \int_{-\infty}^{t} i(t)dt$$
$$= \left(\frac{1}{C_1} + \frac{1}{C_2} + \cdots + \frac{1}{C_n} \right) \int_{-\infty}^{t} i(t)dt$$

$$v(t) = \left(\frac{1}{C_1} + \frac{1}{C_2} + \cdots + \frac{1}{C_n} \right) \int_{-\infty}^{t} i(t)dt$$

Figure 4-6 Equivalent capacitance of capacitors in series.

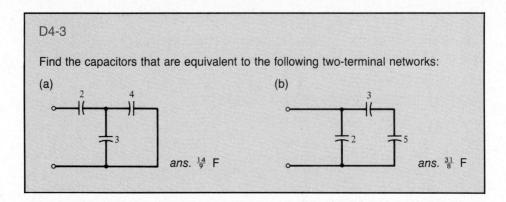

D4-3

Find the capacitors that are equivalent to the following two-terminal networks:

(a)　　　　　　　　　　　　　　　　　(b)

ans. $\frac{14}{9}$ F　　　　　　　　　　　*ans.* $\frac{31}{8}$ F

CAPACITOR RELATIONS

A capacitor is an energy-storing element that has the sink reference voltage-current relationships

$$i(t) = C \frac{dv(t)}{dt}$$

$$v(t) = \frac{1}{C} \int_{-\infty}^{t} i(t)\, dt$$

There are minus signs in these equations if v and i have the source reference relation.

The stored energy in a capacitor is

$$W_C(t) = \tfrac{1}{2} C v^2(t)$$

where v is the capacitor voltage.

Any two-terminal combination of capacitors is equivalent to a single capacitor. The single-capacitor equivalent of capacitors in parallel is the sum of the individual capacitances. The single-capacitor equivalent of capacitors in series is the inverse of the sum of the inverses of the individual capacitances.

4.3 VOLTAGE-CURRENT, POWER, AND ENERGY RELATIONS FOR THE INDUCTOR

4.3.1 Defining Relations

The physical inductor consists of a conductor, usually coiled, in which the magnetic field of the conductor current exerts an appreciable influence on the conductor current itself. Coiling the conductor magnifies the effect, although it is present even in a straight section of wire. The symbol for an inductor model is shown in Figure 4-7. The *inductance*, L, of the element is indicated beside the symbol. The unit of inductance is the *henry* (abbreviated H), which is named in honor of Joseph Henry (1797–1878), an early electrical experimenter.

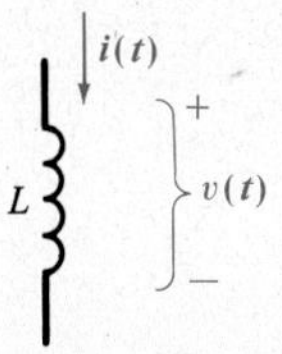

Figure 4-7 Inductor symbol.

Applying Faraday's law, the voltage induced in the inductor is

$$v(t) = \frac{d\psi}{dt}$$

where ψ is the magnetic flux cutting the inductor. The SI unit of magnetic flux is the volt-second = weber (abbreviation Wb). If there are no magnetic materials (such as iron) present in the vicinity of the conductor, ψ is in turn proportional to the conductor current. With the sink reference,

$$\psi(t) = Li(t)$$

giving

$$v(t) = L\frac{di}{dt}$$

If, instead, $v(t)$ and $i(t)$ have the source reference relation,

$$v(t) = -L\frac{di}{dt}$$

The reverse voltage-current relation for the inductor is analogous to the capacitor relations. With the sink reference,

$$i(t) = \frac{1}{L}\int_{-\infty}^{t} v(t)\, dt = i(t_0) + \frac{1}{L}\int_{t_0}^{t} v(t)\, dt$$

A summary of inductor voltage-current relations is given in Figure 4-8.

$$v(t) = L\frac{di(t)}{dt}$$

$$i(t) = \frac{1}{L}\int_{-\infty}^{t} v(t)dt$$

$$= i(t_0) + \frac{1}{L}\int_{t_0}^{t} v(t)dt, \quad t > t_0$$

$$v(t) = -L\frac{di(t)}{dt}$$

$$i(t) = -\frac{1}{L}\int_{-\infty}^{t} v(t)dt$$

$$= i(t_0) - \frac{1}{L}\int_{t_0}^{t} v(t)dt, \quad t > t_0$$

Figure 4-8 Inductor voltage-current relations.
(a) Sink reference relation.
(b) Source reference relation.

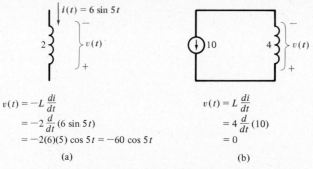

$$v(t) = -L\frac{di}{dt}$$
$$= -2\frac{d}{dt}(6\sin 5t)$$
$$= -2(6)(5)\cos 5t = -60\cos 5t$$

(a)

$$v(t) = L\frac{di}{dt}$$
$$= 4\frac{d}{dt}(10)$$
$$= 0$$

(b)

Figure 4-9   Determining inductor voltage from the inductor current.

Given an inductor current, the inductor voltage may be found by differentiation. Two examples of finding an inductor voltage from the inductor current are given in Figure 4-9.

From the entire past history of an inductor voltage, the inductor current may be found by integration. An example is given in Figure 4-10(a). Before time $t = 0$,

$$i(t) = \frac{1}{3}\int_{-\infty}^{t} 0\, dt = 0$$

After $t = 0$,

$$i(t) = \frac{1}{3}\int_{0}^{t} 6\sin 4t\, dt = \frac{1}{3}\frac{-6\cos 4t}{4}\bigg|_{0}^{t} = \frac{-\cos 4t + 1}{2}\; A$$

Another example of finding an inductor current from its voltage is shown in Figure 4-10(b), where the inductor voltage is specified by a source. For this network,

$$i(t) = -\frac{1}{L}\int_{-\infty}^{t} v(t)\, dt = -\frac{1}{5}\int_{-\infty}^{t} 4e^{5t}\, dt = -\frac{1}{5}\frac{4e^{5t}}{5}\bigg|_{-\infty}^{t}$$

$$= -\frac{4}{25}e^{5t}\; A$$

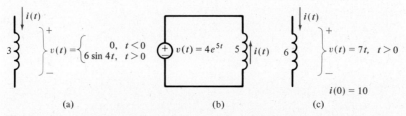

(a)

$$v(t) = \begin{cases} 0, & t<0 \\ 6\sin 4t, & t>0 \end{cases}$$

(b)

$$v(t) = 4e^{5t}$$

(c)

$$v(t) = 7t, \quad t>0$$
$$i(0) = 10$$

Figure 4-10   Determining inductor current from the inductor voltage. In (c), the initial inductor current summarizes the effect of the voltage up to time $t = 0$.

If the inductor voltage is known after some time, say $t = 0$, and the inductor current is known at $t = 0$, the inductor current can be found after $t = 0$. For the example of Figure 4-10(c), after $t = 0$,

$$i(t) = \frac{1}{L} \int_{-\infty}^{0} v(t)\, dt + \frac{1}{L} \int_{0}^{t} v(t)\, dt = i(0) + \frac{1}{L} \int_{0}^{t} v(t)\, dt$$

$$= 10 + \frac{1}{6} \int_{0}^{t} 7t\, dt = 10 + \frac{7}{12} t^2 \text{ A}$$

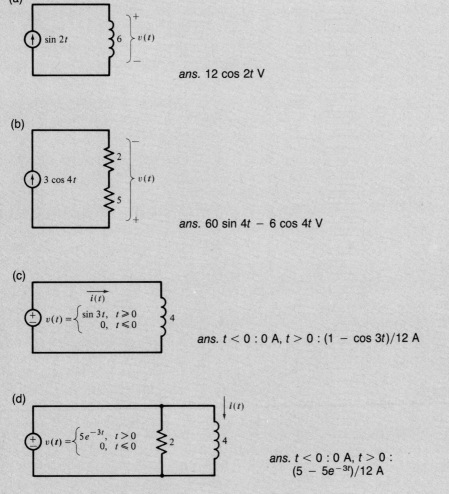

D4-4

Find the indicated voltages and currents:

(a)

sin 2t   6 $v(t)$

ans. 12 cos 2t V

(b)

3 cos 4t   2   5 $v(t)$

ans. 60 sin 4t − 6 cos 4t V

(c)

$i(t)$

$v(t) = \begin{cases} \sin 3t, & t \geqslant 0 \\ 0, & t \leqslant 0 \end{cases}$ 4

ans. $t < 0$: 0 A, $t > 0$: (1 − cos 3t)/12 A

(d)

$i(t)$

$v(t) = \begin{cases} 5e^{-3t}, & t > 0 \\ 0, & t \leqslant 0 \end{cases}$ 2   4

ans. $t < 0$: 0 A, $t > 0$:
(5 − 5e^{-3t})/12 A

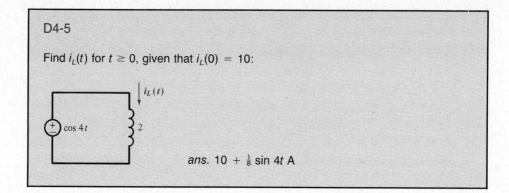

D4-5

Find $i_L(t)$ for $t \geq 0$, given that $i_L(0) = 10$:

ans. $10 + \frac{1}{8} \sin 4t$ A

4.3.2 Inductor Power and Energy

In the inductor, energy is stored in the magnetic field in an amount proportional to the square of the current. With the sink reference, the electrical power flow into the inductor is

$$p_{\text{into}}(t) = v(t)i(t) = Li(t) \frac{di(t)}{dt}$$

The electrical energy stored in the inductor is

$$W_L(t) = \int_{-\infty}^{t} p_{\text{into}}(t) \, dt = \frac{1}{2} Li^2(t)$$

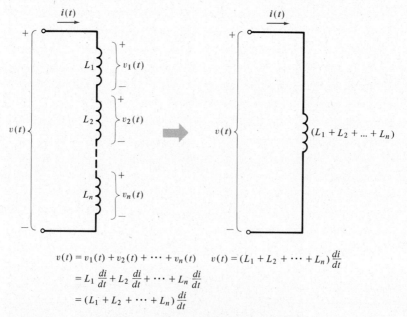

$$v(t) = v_1(t) + v_2(t) + \cdots + v_n(t) \qquad v(t) = (L_1 + L_2 + \cdots + L_n) \frac{di}{dt}$$

$$= L_1 \frac{di}{dt} + L_2 \frac{di}{dt} + \cdots + L_n \frac{di}{dt}$$

$$= (L_1 + L_2 + \cdots + L_n) \frac{di}{dt}$$

Figure 4-11   Equivalent inductance of inductors in series.

4.3.3 Two-Terminal Combinations of Inductors

Any two-terminal combination of inductors is equivalent to a single inductor. In such a combination, however the terminal current branches through each of the individual inductors, the terminal voltage is proportional to the time rate of change of that current.

The equivalent inductance of several inductors in series (with no magnetic coupling between inductors) is the sum of the individual inductances. This equivalence is developed in Figure 4-11. For inductors in parallel, the equivalent inductance is the inverse of the sum of the inverses of the individual inductances. This equivalence is developed in Figure 4-12. These relations are analogous to those for resistors in series and in parallel.

For two inductors in parallel,

$$L_{\text{equiv}} = \frac{L_1 L_2}{L_1 + L_2}$$

D4-6

Find the inductors that are equivalent to the following two-terminal networks:

(a)

(b)

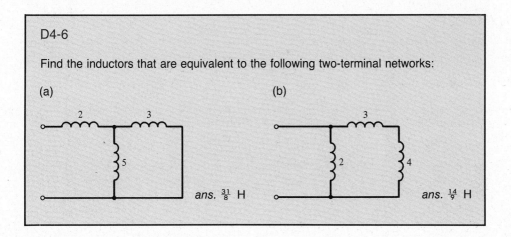

ans. $\frac{31}{8}$ H

ans. $\frac{14}{9}$ H

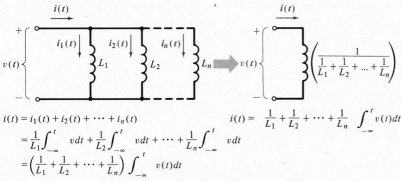

$i(t) = i_1(t) + i_2(t) + \cdots + i_n(t)$

$= \frac{1}{L_1}\int_{-\infty}^{t} v\,dt + \frac{1}{L_2}\int_{-\infty}^{t} v\,dt + \cdots + \frac{1}{L_n}\int_{-\infty}^{t} v\,dt$

$= \left(\frac{1}{L_1} + \frac{1}{L_2} + \cdots + \frac{1}{L_n}\right)\int_{-\infty}^{t} v(t)\,dt$

$i(t) = \frac{1}{L_1} + \frac{1}{L_2} + \cdots + \frac{1}{L_n} \int_{-\infty}^{t} v(t)\,dt$

Figure 4-12   Equivalent inductance of inductors in parallel.

INDUCTOR RELATIONS

An inductor is an energy-storing element that has the sink reference voltage-current relationships

$$v(t) = L\frac{di(t)}{dt}$$

$$i(t) = \frac{1}{L}\int_{-\infty}^{t} v(t)\, dt$$

There are minus signs in these equations if v and i have the source reference relation. The stored energy in an inductor is

$$W_L(t) = \tfrac{1}{2}Li^2(t)$$

where i is the inductor current.

Any two-terminal combination of inductors is equivalent to a single inductor. The single-inductor equivalent of inductors in series is the sum of the individual inductances. The single-inductor equivalent of inductors in parallel is the inverse of the sum of the inverses of the individual inductances.

4.4 INDUCTIVE COUPLING

4.4.1 Voltage-Current Relations and Winding Senses

When a time-varying magnetic field from one inductor passes through another inductor, a voltage is induced, just as when the time-varying magnetic field is self-produced. For two magnetically coupled coils, in the absence of magnetic materials such as iron that make the relations nonlinear, the magnetic flux cutting coil number one consists of a term proportional to coil one's current plus a term proportional to coil two's current:

$$\psi_1(t) = L_1 i_1(t) \pm M i_2(t)$$

Similarly, the flux cutting coil number two involves a term proportional to coil two's current plus a term proportional to coil one's current:

$$\psi_2(t) = \pm M i_1(t) + L_2 i_2(t)$$

L_1 and L_2 are called the *self-inductances* of coils one and two, respectively. The *mutual inductance*, M, is the same in the two relations. It is taken to be positive, with the algebraic signs used in the relations being dependent on the relative senses of the windings of the two coils.

The sink reference voltages in the coils are

$$v_1 = \frac{d\psi_1(t)}{dt} = L_1\frac{di_1}{dt} \pm M\frac{di_2}{dt}$$

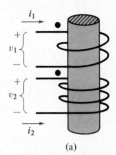

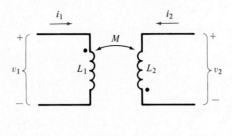

(a)   (b)

Figure 4-13   Mutually coupled inductors.
(a) Physical arrangement.
(b) Network model.

$$v_2 = \frac{d\psi_2(t)}{dt} = \pm M \frac{di_1}{dt} + L_2 \frac{di_2}{dt}$$

as indicated in Figure 4-13. The algebraic signs to be used with the mutual coupling terms depend on the relative winding senses. If a positive i_2 produces flux in coil one in the same sense as does a positive i_1, the plus sign applies; if a positive i_2 produces coil-one flux that opposes the flux produced by a positive i_1, the negative sign applies. The relative senses of the coil fluxes are indicated by large dots, as in the drawings of Figure 4-13.

It is desirable to be able to write down easily the voltage-current relations for coupled inductors for any of the many possible combinations of voltage and current references and flux senses. There are, however, $2^6 = 64$ possible different combinations of the two voltage references, the two current references, and the two flux sense dots, so a moderate degree of complexity is understandable. First, the correct algebraic signs for the self-inductance terms are found by imagining temporarily that there is no mutual coupling. Then, the algebraic signs of the mutual coupling terms are determined as follows. Consider the mutually induced voltage in one of the coils, with polarity in the sense with the plus sign at the same terminal as the dot for that coil. That voltage is the mutual inductance M, times the time derivative of the opposite coil's current, with reference direction in the sense into the dot at the opposite coil. The mutually induced voltage in the other coil may be found in the same manner.

The mutual inductance relationships are most easily and routinely accounted for by representing the mutually induced voltage terms with controlled sources. A systematic way of doing this is illustrated in Figure 4-14. Replace the mutual inductive coupling with controlled sources in series with each inductor, as shown, each source oriented in the sense of the flux reference dot for that inductor. The controlled source function is the mutual coupling M, times the time derivative of the current in the opposite inductor, in the sense into the dot in that inductor. In the example, i_2 is in the sense into the dot for the right inductor, so the left inductor's equivalent source controlling function is $M(di_2/dt)$. Similarly, i_1 is in the sense into the dot at the left inductor so the right inductor's equivalent source-controlling

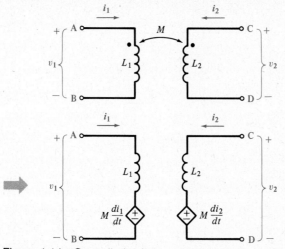

Figure 4-14 Controlled voltage-source equivalent of mutually coupled inductors.

function is $M(di_1/dt)$. The voltage-current relations at the coupled inductor terminals are thus

$$\begin{cases} v_1 = L_1 \dfrac{di_1}{dt} + M \dfrac{di_2}{dt} \\[2mm] v_2 = M \dfrac{di_1}{dt} + L_2 \dfrac{di_2}{dt} \end{cases}$$

In the example of Figure 4-15, equivalent controlled sources replace the mutual inductive coupling. Each controlled source has been given polarity in the sense of the dot at its series inductor. The left source-controlling function is the mutual coupling M, times the time derivative of the current into the dot at the right inductor. Since i_2 is defined with sense out of the dot in the right inductor, the current into the dot is $-i_2$. The left source-controlling function is then $-2(di_2/dt)$. The right controlled source has controlling function M, times the time derivative of the current into the dot in the left inductor: $2(di_1/dt)$.

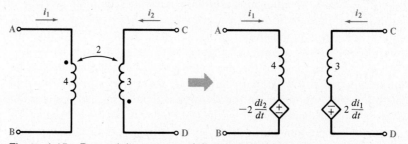

Figure 4-15 Determining senses of the controlled sources.

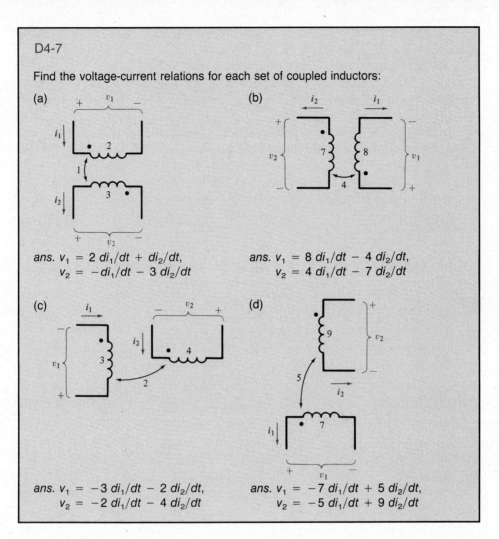

D4-7

Find the voltage-current relations for each set of coupled inductors:

(a)

ans. $v_1 = 2\, di_1/dt + di_2/dt,$
$v_2 = -di_1/dt - 3\, di_2/dt$

(b)

ans. $v_1 = 8\, di_1/dt - 4\, di_2/dt,$
$v_2 = 4\, di_1/dt - 7\, di_2/dt$

(c)

ans. $v_1 = -3\, di_1/dt - 2\, di_2/dt,$
$v_2 = -2\, di_1/dt - 4\, di_2/dt$

(d)

ans. $v_1 = -7\, di_1/dt + 5\, di_2/dt,$
$v_2 = -5\, di_1/dt + 9\, di_2/dt$

4.4.2 Power and Energy

The electrical power flow into a set of two coupled inductors is the sum of the power flows into each of the coils. With sink reference relations for the voltage and current at each coil,

$$p_{\text{into}}(t) = v_1(t)i_1(t) + v_2(t)i_2(t)$$

$$= L_1 i_1 \frac{di_1}{dt} \pm M i_1 \frac{di_2}{dt} + L_2 i_2 \frac{di_2}{dt} \pm M i_2 \frac{di_1}{dt}$$

$$= \frac{d}{dt}\left(\frac{1}{2} L_1 i_1{}^2 + \frac{1}{2} L_2 i_2{}^2 \pm M i_1 i_2\right)$$

where the algebraic sign associated with the mutual coupling term depends on the relative orientation of the two coils.

The energy stored in the two coupled inductors is then

$$W_M = \int_{-\infty}^{t} p_{into}(t) \; dt = \tfrac{1}{2}L_1 i_1{}^2 + \tfrac{1}{2}L_2 i_2{}^2 \pm M i_1 i_2$$

By considering various possible currents, i_1 and i_2, and using the fact that the stored energy can never be negative, it can be shown that

$$M \leq \sqrt{L_1 L_2}$$

The *coupling coefficient* of two inductors is defined by

$$k = \frac{M}{\sqrt{L_1 L_2}}$$

where k can range from zero, no coupling, to unity, the maximum possible coupling.

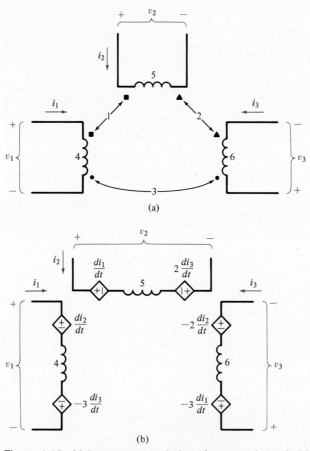

(a)

(b)

Figure 4-16   Voltage-current relations for several coupled inductors.
(a) The coupled inductors.
(b) Controlled voltage source equivalent.

4.4.3 Several Coupled Inductors

When more than two coils are mutually coupled, each to one another, a great number of different sets of coupling senses are possible because the arrangement of coils can be three-dimensional. The flux senses for the couplings between each pair of coils are indicated by separate symbols such as dots, squares, and triangles. The algebraic signs of the mutual coupling terms may be determined by considering the coils two at a time.

As an example, the voltage-current relations for the three mutually coupled inductors of Figure 4-16 are as follows:

$$v_1 = 4\frac{di_1}{dt} + \frac{di_2}{dt} + 3\frac{di_3}{dt}$$

$$v_2 = \frac{di_1}{dt} + 5\frac{di_2}{dt} - 2\frac{di_3}{dt}$$

$$v_3 = -3\frac{di_1}{dt} + 2\frac{di_2}{dt} - 6\frac{di_3}{dt}$$

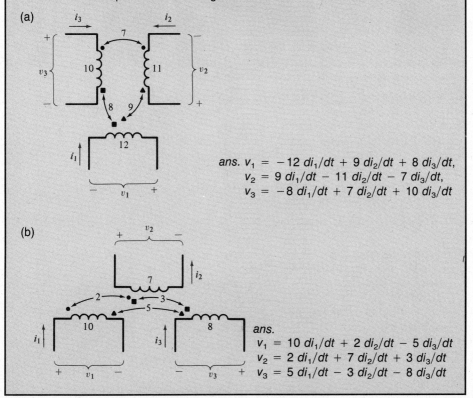

D4-8

Find the three coupled inductor voltages in terms of the inductor currents:

(a)

ans. $v_1 = -12\,di_1/dt + 9\,di_2/dt + 8\,di_3/dt$,
$v_2 = 9\,di_1/dt - 11\,di_2/dt - 7\,di_3/dt$,
$v_3 = -8\,di_1/dt + 7\,di_2/dt + 10\,di_3/dt$

(b)

ans.
$v_1 = 10\,di_1/dt + 2\,di_2/dt - 5\,di_3/dt$
$v_2 = 2\,di_1/dt + 7\,di_2/dt + 3\,di_3/dt$
$v_3 = 5\,di_1/dt - 3\,di_2/dt - 8\,di_3/dt$

COUPLED INDUCTORS

When two inductors are magnetically coupled, the inductor voltages and currents, with the sink reference relation for each inductor, are of the form

$$
\begin{cases}
v_1 = L_1 \dfrac{di_1}{dt} \pm M \dfrac{di_2}{dt} \\[3mm]
v_2 = \pm M \dfrac{di_1}{dt} + L_2 \dfrac{di_2}{dt}
\end{cases}
$$

The correct algebraic sign of each mutual coupling term, $\pm M$, depends on the relative orientations of the two coils, which is indicated by dots at one side of the symbol for each inductor.

A very useful equivalent circuit replaces the mutual inductive coupling by controlled voltage sources in series with the inductors. Each source polarity is chosen with the same sense as the flux polarity dot for its series inductor. The source-controlling signals are the mutual coupling M, times the time derivative of the current with sense into the dot in the opposite inductor.

The energy stored in a pair of mutually coupled inductors is given by

$$
W_M = \tfrac{1}{2} L_1 i_1{}^2 + \tfrac{1}{2} L_2 i_2{}^2 \pm M i_1 i_2
$$

The value of the mutual inductance, M, cannot exceed the geometric mean of the two self-inductances, L_1 and L_2:

$$
M \le \sqrt{L_1 L_2}
$$

When several inductors are mutually coupled, the mutual coupling terms may be determined by taking the inductors two at a time.

4.5 SYSTEMATIC SIMULTANEOUS MESH EQUATIONS

To write systematic simultaneous equations for a network containing inductors and capacitors as well as sources and resistors, one need only write the inductor and capacitor voltages in terms of their sink reference currents, in place of the resistor voltage-current relation that would be written if the element were a resistor. The equations are most easily written in *operator notation* in which, for example,

$$
\left(3 + 4 \frac{d}{dt} + 5 \int_{-\infty}^{t} dt \right) i
$$

stands for

$$
3i + 4 \frac{di}{dt} + 5 \int_{-\infty}^{4} i \, dt
$$

In summing the voltages around each mesh, simply write R for each resistor, $L\,d/dt$ for each inductor, and $(1/C)\int_{-\infty}^{t} dt$ for each capacitor.

Consider the network of Figure 4-17. Systematic simultaneous mesh equations for this network are as follows:

$$
\left(1 + 2 \frac{d}{dt} + \frac{1}{3} \int_{-\infty}^{t} dt + 6 + 5 \frac{d}{dt} + 4 \right) i_1 - \left(4 + 5 \frac{d}{dt} \right) i_2 - (6) i_3 = 12 \sin t + 18
$$

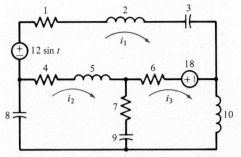

Figure 4-17   Planar *RLC* network without mutually coupled inductors.

$$-\left(4 + 5\frac{d}{dt}\right)i_1 + \left(4 + 5\frac{d}{dt} + \frac{1}{8}\int_{-\infty}^{t} dt + 7 + \frac{1}{9}\int_{-\infty}^{t} dt\right)i_2 - \left(7 + \frac{1}{9}\int_{-\infty}^{t} dt\right)i_3 = 0$$

$$-(6)i_1 - \left(7 + \frac{1}{9}\int_{-\infty}^{t} dt\right)i_2 + \left(7 + \frac{1}{9}\int_{-\infty}^{t} dt + 6 + 10\frac{d}{dt}\right)i_3 = -18$$

Mutually coupled inductors may be accommodated in mesh equations by first replacing them with the controlled voltage-source equivalent circuit. Incorporating controlled sources into the mesh equations is done in the same manner as for source-resistor networks.

Consider the network containing coupled inductors shown in Figure 4-18. The coupling has been replaced by its controlled voltage-source equivalent. Systematic simultaneous mesh equations are then

$$\begin{cases} \left(3 + 10\frac{d}{dt}\right)i_1 - \left(10\frac{d}{dt}\right)i_2 = 8\cos 9t - 4\frac{di_2}{dt} \\ -\left(10\frac{d}{dt}\right)i_1 + \left(17\frac{d}{dt} + 2 + \frac{1}{5}\int_{-\infty}^{t} dt\right)i_2 = 4\frac{di_2}{dt} + 4\frac{d}{dt}(i_2 - i_1) \end{cases}$$

or

$$\begin{cases} \left(3 + 10\frac{d}{dt}\right)i_1 - \left(6\frac{d}{dt}\right)i_2 = 8\cos 9t \\ -\left(6\frac{d}{dt}\right)i_1 + \left(9\frac{d}{dt} + 2 + \frac{1}{5}\int_{-\infty}^{t} dt\right)i_2 = 0 \end{cases}$$

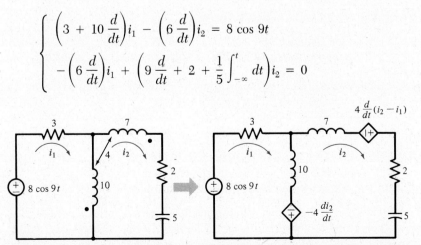

Figure 4-18   Network where mutually coupled inductors are replaced by their controlled voltage-source equivalent.

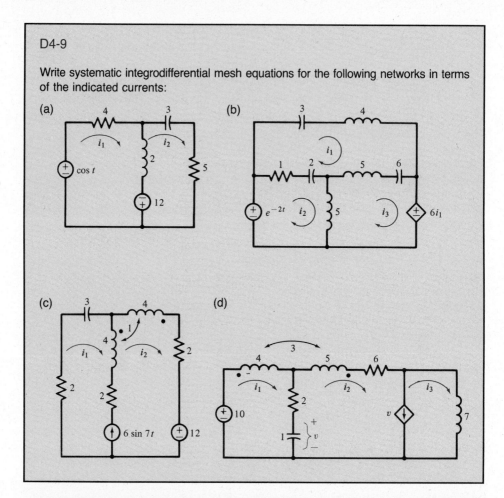

D4-9

Write systematic integrodifferential mesh equations for the following networks in terms of the indicated currents:

SYSTEMATIC SIMULTANEOUS MESH EQUATIONS

Mesh equations may be written in operator notation for *RLC* networks in the same manner as for source-resistor networks. For each inductor, instead of *R*, write *Ld/dt*. For each capacitor, write $(1/C) \int_{-\infty}^{t} dt$.

When coupled inductors are involved, it is helpful to replace the mutual coupling by a controlled voltage-source equivalent circuit.

4.6 SYSTEMATIC SIMULTANEOUS NODAL EQUATIONS

4.6.1 Equations for Networks without Inductive Coupling

For nodal equations, write the inductor and capacitor currents in terms of the element voltages. For a resistor, write $1/R$; for an inductor write, instead, $(1/L) \int_{-\infty}^{t} dt$. For a capacitor, write Cd/dt. An example network is given in Figure 4-19, for which the systematic nodal equations are as follows:

$$
\begin{cases}
\left(\dfrac{d}{dt} + 2\dfrac{d}{dt} + \dfrac{1}{3}\displaystyle\int_{-\infty}^{t} dt + \dfrac{1}{4}\right)v_1 - \left(2\dfrac{d}{dt} + \dfrac{1}{3}\displaystyle\int_{-\infty}^{t} dt + \dfrac{1}{4}\right)v_2 - (0)v_3 = 10\cos t \\[4mm]
-\left(2\dfrac{d}{dt} + \dfrac{1}{3}\displaystyle\int_{-\infty}^{t} dt + \dfrac{1}{4}\right)v_1 + \left(2\dfrac{d}{dt} + \dfrac{1}{3}\displaystyle\int_{-\infty}^{t} dt + \dfrac{1}{4} + \dfrac{1}{7}\displaystyle\int_{-\infty}^{t} dt\right. \\[4mm]
\qquad\qquad\qquad \left. + \dfrac{1}{6} + \dfrac{1}{5}\displaystyle\int_{-\infty}^{t} dt\right)v_2 - \left(\dfrac{1}{6} + \dfrac{1}{5}\displaystyle\int_{-\infty}^{t} dt\right)v_3 = -12e^{-t} \\[4mm]
\qquad\quad -(0)v_1 - \left(\dfrac{1}{6} + \dfrac{1}{5}\displaystyle\int_{-\infty}^{t} dt\right)v_2 + \left(\dfrac{1}{6} + \dfrac{1}{5}\displaystyle\int_{-\infty}^{t} dt + \dfrac{1}{9} + 8\dfrac{d}{dt}\right)v_3 = 12e^{-t}
\end{cases}
$$

The same sorts of symmetries present in the systematic algebraic source-resistor network nodal equations are present in the *operations* for each term in these integrodifferential equations.

4.6.2 Controlled Current-Source Equivalent for Coupled Inductors

The controlled voltage-source inductive coupling equivalent is well suited to mesh equations, but is very poorly suited for nodal equations. A better equivalent for nodal equations is derived as follows. The pair of coupled inductors given in Figure 4-20(a) are described by the equations

$$
v_1 = L_1\frac{di_1}{dt} + M\frac{di_2}{dt} \qquad v_2 = M\frac{di_1}{dt} + L_2\frac{di_2}{dt}
$$

Solving these two simultaneous equations for the current derivatives gives

$$
\frac{di_1}{dt} = \frac{\begin{vmatrix} v_1 & M \\ v_2 & L_2 \end{vmatrix}}{\begin{vmatrix} L_1 & M \\ M & L_2 \end{vmatrix}} = \frac{L_2}{L_1L_2 - M^2}v_1 - \frac{M}{L_1L_2 - M^2}v_2
$$

$$
\frac{di_2}{dt} = \frac{\begin{vmatrix} L_1 & v_1 \\ M & v_2 \end{vmatrix}}{L_1L_2 - M^2} = -\frac{M}{L_1L_2 - M^2}v_1 + \frac{L_1}{L_1L_2 - M^2}v_2
$$

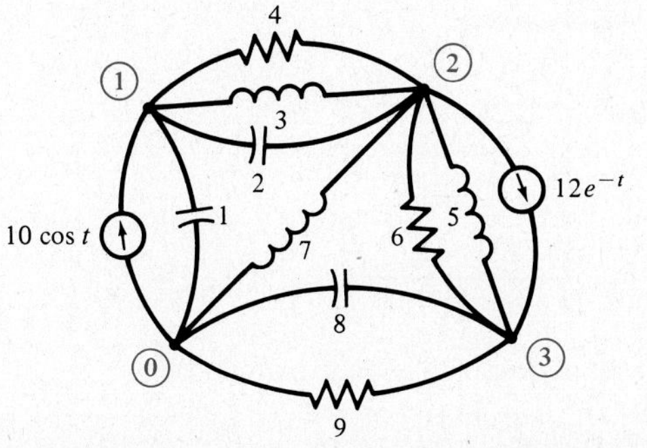

Figure 4-19   *RLC network without mutually coupled inductors.*

Or,

$$i_1 = \frac{L_2}{L_1 L_2 - M^2} \int_{-\infty}^{t} v_1 \, dt - \frac{M}{L_1 L_2 - M^2} \int_{-\infty}^{t} v_2 \, dt$$

$$i_2 = -\frac{M}{L_1 L_2 - M^2} \int_{-\infty}^{t} v_1 \, dt + \frac{L_1}{L_1 L_2 - M^2} \int_{-\infty}^{t} v_2 \, dt$$

when

$$M^2 \neq L_1 L_2$$

The voltage-controlled, current-source equivalent network represented by these equations is shown in Figure 4-20(b), and a numerical example of conversion to the equivalent is given in Figure 4-20(c). For the example equivalent,

$$L_1 L_2 - M^2 = (8)(10) - (3)^2 = 71$$

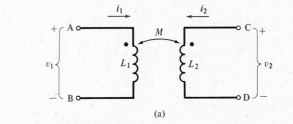

(a)

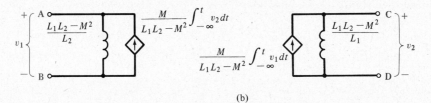

(b)

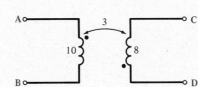

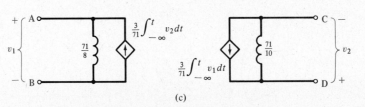

(c)

Figure 4-20 Controlled current source equivalent of mutually coupled inductors.
(a) Mutually coupled inductors.
(b) Controlled current source equivalent.
(c) A numerical example.

The controlling voltages are those with the senses of the flux dots, and the controlled current sources have reference directions toward the dots. When

$$M^2 = L_1 L_2$$

an equivalent circuit of this form does not exist.

Consider the network of Figure 4-21, which contains coupled inductors. In the equivalent network, the magnetically coupled circuits have been replaced by a controlled current-source equivalent. For the equivalent,

$$L_1 L_2 - M^2 = (3)(8) - (2)^2 = 20$$

and the controlling voltages are expressed in terms of the defined node voltages. The systematic simultaneous nodal equations for the network are then as follows:

$$\begin{cases} \left(4\dfrac{d}{dt} + \dfrac{2}{5}\int_{-\infty}^{t} dt + \dfrac{3}{20}\int_{-\infty}^{t} dt\right)v_1 - \left(\dfrac{3}{20}\int_{-\infty}^{t} dt\right)v_2 = -\dfrac{1}{10}\int_{-\infty}^{t} v_1\, dt - \dfrac{1}{10}\int_{-\infty}^{t} (v_1 - v_2)\, dt \\[2mm] -\left(\dfrac{3}{20}\int_{-\infty}^{t} dt\right)v_1 + \left(\dfrac{1}{5} + \dfrac{3}{20}\int_{-\infty}^{t} dt\right)v_2 = 6\sin 7t + \dfrac{1}{10}\int_{-\infty}^{t} v_1\, dt \end{cases}$$

Placing the unknowns, v_1 and v_2, on the left,

$$\begin{cases} \left(4\dfrac{d}{dt} + \dfrac{2}{5}\int_{-\infty}^{t} dt + \dfrac{3}{20}\int_{-\infty}^{t} dt + \dfrac{1}{5}\int_{-\infty}^{t} dt\right)v_1 - \left(\dfrac{20}{3}\int_{-\infty}^{t} dt + \dfrac{1}{10}\int_{-\infty}^{t} dt\right)v_2 = 0 \\[2mm] -\left(\dfrac{3}{20}\int_{-\infty}^{t} dt + \dfrac{1}{10}\int_{-\infty}^{t} dt\right)v_1 + \left(\dfrac{1}{5} + \dfrac{3}{20}\int_{-\infty}^{t} dt\right)v_2 = 6\sin 7t \end{cases}$$

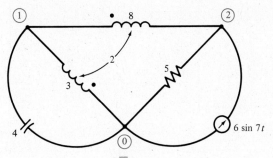

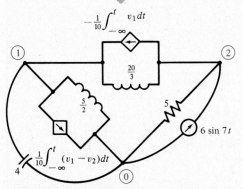

Figure 4-21   Network where mutually coupled inductors are replaced by their controlled current-source equivalent.

or,

$$\begin{cases} \left(4\dfrac{d}{dt} + \dfrac{3}{4}\displaystyle\int_{-\infty}^{t} dt\right)v_1 - \left(\dfrac{203}{30}\displaystyle\int_{-\infty}^{t} dt\right)v_2 = 0 \\[3mm] -\left(\dfrac{1}{4}\displaystyle\int_{-\infty}^{t} dt\right)v_1 + \left(\dfrac{1}{5} + \dfrac{3}{20}\displaystyle\int_{-\infty}^{t} dt\right)v_2 = 6\sin 7t \end{cases}$$

D4-10

Write systematic integrodifferential nodal equations for the following networks in terms of the indicated node voltages:

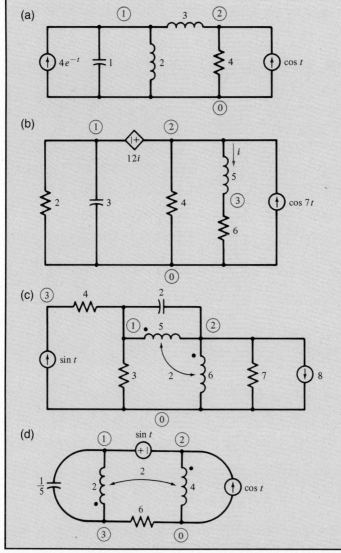

> **SYSTEMATIC SIMULTANEOUS NODAL EQUATIONS**
>
> Nodal equations may be written in operator notation for *RLC* networks in the same manner as for source-resistor networks. For each capacitor, instead of $1/R$, write $C\,d/dt$. For each inductor, write $(1/L)\int_{-\infty}^{t} dt$.
>
> When coupled inductors are involved, it is helpful to replace the coupling by a controlled current-source equivalent circuit.

4.7 LINEAR, TIME-INVARIANT DIFFERENTIAL EQUATIONS

4.7.1 Equation Form and Classification

An equation of the form

$$a_n \frac{d^n y}{dt^n} + a_{n-1} \frac{d^{n-1}y}{dt^{n-1}} + \cdots + a_1 \frac{dy}{dt} + a_0 y = f(t)$$

where $a_n, a_{n-1}, \ldots, a_1, a_0$ are constants, is called a linear, constant-coefficient differential equation. This type of differential equation is termed "linear" because the function $y(t)$ and derivatives of $y(t)$ appear linearly in the equation. It is constant coefficient (or *time-invariant* or *fixed*) because $a_n, a_{n-1}, \ldots, a_1, a_0$ are constants, not functions of time. The function $f(t)$ is called the *driving function* of the equation.

If the highest derivative to appear in the equation is the nth derivative, the equation is said to be of nth order. For example,

$$3 \frac{dy}{dt} + 2y = \sin t$$

is a first-order linear, time-invariant differential equation and

$$-\frac{d^3 y}{dt^3} + 2 \frac{dy}{dt} = 7e^{-4t} + 10$$

is third order.

A function $y(t)$ that satisfies a differential equation is termed a *solution* to that equation. For example, in the equation

$$\frac{dy}{dt} - y = 0$$

$y(t) = e^t$ is a solution.

A linear differential equation has many solutions. Not only is $y(t) = e^t$ a solution to the above equation, but $y = 0$, $y = 2e^t$, and $y = 10e^t$ are also solutions. In fact, $y = Ke^t$, where K is any constant, is a solution of the equation.

The *general solution* of a differential equation is the collection of all possible solutions. It is characteristic of linear differential equations that the general solutions have *arbitrary constants* such as K, above. Arbitrary constants in the general solution are certainly to be expected. After all, the problem of integration is the solution of the differential equation

$$\frac{dy}{dt} = f(t)$$

which is

$$y(t) = \int f(t)dt + K$$

The arbitrary constant K appears in the solution because a whole family of functions $y(t)$ have derivative $f(t)$. So it is with all first-order linear differential equations.

The general solution to an nth-order linear differential equation contains exactly n independent arbitrary constants. This is to say that for an nth-order equation, once a set of solutions with n independent arbitrary constants in it is found, one can stop looking for further solutions. There are no other solutions.

4.7.2 Solution of First-Order Homogeneous Equations

A *homogeneous* differential equation is an equation in which the driving function is zero. The first-order homogeneous equation is of the form

$$a_1 \frac{dy}{dt} + a_0 y = 0$$

Rearranging,

$$\frac{dy}{dt} = -\frac{a_0}{a_1} y$$

which is to ask, within the constant $(-a_0/a_1)$, what function is its own derivative? The exponential function is the one that is proportional to its derivative.

Trying the function

$$y = Ke^{st}$$

where K and s are constants, as a possible solution to the first-order homogeneous equation gives

$$a_1 \frac{dy}{dt} + a_0 y = 0$$

$$a_1 sKe^{st} + a_0 Ke^{st} = 0$$

$$(a_1 s + a_0)Ke^{st} = 0$$

Dividing both sides of the equation by Ke^{st},

$$a_1 s + a_0 = 0; \quad s = -\frac{a_0}{a_1}$$

That is,

$$y(t) = Ke^{(-a_0/a_1)t}$$

satisfies the equation, as is easily checked by substitution. This solution contains one arbitrary constant and so is the general solution to the equation.

As a numerical example, consider

$$\frac{dy}{dt} + 3y = 0$$

Substituting $y = Ke^{st}$ into the equation,

$$sKe^{st} + 3Ke^{st} = 0$$

$$s + 3 = 0$$

$$s = -3$$

Thus

$$y(t) = Ke^{-3t}$$

is the general solution of this equation.

4.7.3 Solution of Higher-Order Homogeneous Equations

Exponential functions also are solutions to the higher-order homogeneous equations. For the general homogeneous equation,

$$a_n \frac{d^n y}{dt^n} + a_{n-1} \frac{d^{n-1} y}{dt^{n-1}} + \cdots + a_1 \frac{dy}{dt} + a_0 y = 0$$

substituting $y(t)$ of the form Ke^{st} gives

$$a_n s^n K e^{st} + a_{n-1} s^{n-1} K e^{st} + \cdots + a_1 s K e^{st} + a_0 K e^{st} = 0$$

Dividing both sides by Ke^{st}, there results the *characteristic equation* (or *auxiliary equation*),

$$a_n s^n + a_{n-1} s^{n-1} + \cdots + a_1 s + a_0 = 0$$

It is easy to write the characteristic equation directly from the original differential equation.

The characteristic equation is an nth-order polynomial in s, which has n solutions or *roots*, $s_1, s_2, \ldots, s_n$. In other words, there are n values of s for which Ke^{st} is a solution to an nth-order homogeneous equation.

Since the equation is linear, not only are $Ke^{s_1 t}$ and $Ke^{s_2 t}$, and so on, solutions to the equation, but sums of solutions are also solutions. In fact, the arbitrary constants involved with each exponential term can be different numbers, so that any function of the form

$$y(t) = K_1 e^{s_1 t} + K_2 e^{s_2 t} + \cdots + K_n e^{s_n t}$$

for any numbers $K_1, K_2, \ldots, K_n$, satisfies the differential equation. So long as the roots of the characteristic equation, $s_1, s_2, \ldots, s_n$, are distinct, the above solution contains n independent arbitrary constants, $K_1, K_2, \ldots, K_n$, and so is the general solution to the nth-order equation.

Consider the second-order equation

$$\frac{d^2y}{dt^2} + 5\frac{dy}{dt} + 6y = 0$$

Substituting $y(t) = Ke^{st}$ gives

$$s^2Ke^{st} + 5sKe^{st} + 6Ke^{st} = 0$$

$$s^2 + 5s + 6 = 0$$

which is the characteristic equation. The characteristic equation factors as

$$(s + 2)(s + 3) = 0$$

and so has solutions

$$s_1 = -2 \quad \text{and} \quad s_2 = -3$$

The general solution to the original second-order differential equation is then

$$y(t) = K_1e^{-2t} + K_2e^{-3t}$$

which has two independent arbitrary constants.

For the equation

$$2\frac{d^2y}{dt^2} + 4\frac{dy}{dt} + y = 0$$

the characteristic equation is

$$2s^2 + 4s + 1 = 0$$

which does not factor easily. Using the quadratic formula,

$$s_1, s_2 = \frac{-4 \pm \sqrt{4^2 - 4 \cdot 2 \cdot 1}}{2 \cdot 2} = \frac{-4 \pm \sqrt{8}}{4}$$

$$= -1 + \frac{\sqrt{2}}{2}, -1 - \frac{\sqrt{2}}{2}$$

The general solution is then

$$y(t) = K_1e^{-1.71t} + K_2e^{-0.293t}$$

The equation

$$\frac{d^2y}{dt^2} + 5\frac{dy}{dt} = 0$$

has characteristic equation

$$s^2 + 5s = 0, \quad s(s + 5) = 0$$

and general solution

$$y(t) = K_1e^{0t} + K_2e^{-5t} = K_1 + K_2e^{-5t}$$

4.7.4 Repeated Characteristic Roots

A special situation occurs when two or more of the roots of the characteristic equation are the same number. For example,

$$\frac{d^2y}{dt^2} + 2\frac{dy}{dt} + y = 0$$

has characteristic equation

$$s^2 + 2s + 1 = (s + 1)(s + 1) = 0$$

giving

$$s_1 = -1, \qquad s_2 = -1$$

Now

$$K_1 e^{-t} + K_2 e^{-t} = (K_1 + K_2)e^{-t} = Ke^{-t}$$

cannot be the general solution to this second-order differential equation since the arbitrary constants K_1 and K_2 are not independent.

It can be shown that the general solution in this case is

$$y(t) = K_1 e^{-t} + K_2 t e^{-t}$$

If a root s_i is repeated three times, the corresponding terms in the solution of the homogeneous equation are

$$K_1 e^{s_i t} + K_2 t e^{s_i t} + K_3 t^2 e^{s_i t}$$

and so on.

The case of repeated roots is of limited practical importance, since the numbers in the equation have to be "just right" for repeated roots to occur.

4.7.5 Complex Roots

It is possible that some of the roots of the characteristic equation will be complex numbers. The solution to the homogeneous equation is still of the form

$$y(t) = K_1 e^{s_1 t} + K_2 e^{s_2 t} + \cdots + K_n e^{s_n t}$$

but some algebraic manipulation is then expedient so that the solution is in a more easily visualized form.

For differential equations with real coefficients, complex roots always occur in conjugate pairs. Each set of complex conjugate roots, for example,

$$s_1, s_2 = \alpha \pm j\beta$$

where $j = \sqrt{-1}$ is the imaginary unit, gives rise to terms in the homogeneous solution

$$y(t) = K_1 e^{(\alpha + j\beta)t} + K_2 e^{(\alpha - j\beta)t} = e^{\alpha t}(K_1 e^{j\beta t} + K_2 e^{-j\beta t})$$

Algebraic manipulations using Euler's relation (discussed in detail in Chapter 6) may be used to convert the above to the equivalent form

$$y(t) = e^{\alpha t}(D \cos \beta t + E \sin \beta t)$$

where D and E are the arbitrary constants. This trigonometric form is generally easier to deal with and to visualize for equations of relatively low order.

For example, the equation

$$\frac{d^2y}{dt^2} + 4\frac{dy}{dt} + 13y = 0$$

has characteristic equation

$$s^2 + 4s + 13 = 0$$

Using the quadratic formula, the characteristic roots are

$$s_1, s_2 = \frac{-4 \pm \sqrt{16 - 52}}{2} = -2 \pm j3$$

One form of the general solution to this homogeneous equation is

$$y(t) = K_1 e^{s_1 t} + K_2 e^{s_2 t} = K_1 e^{(-2+j3)t} + K_2 e^{(-2-j3)t}$$

$$= e^{-2t}(K_1 e^{j3t} + K_2 e^{-j3t})$$

The alternate form is

$$y(t) = e^{-2t}(D \cos 3t + E \sin 3t)$$

where D and E are arbitrary constants.

The homogeneous equation

$$4\frac{d^3y}{dt^3} + 4\frac{d^2y}{dt^2} + \frac{dy}{dt} = 0$$

has characteristic equation

$$4s^3 + 4s^2 + 5s = 4(s)(s + 0.5 + j)(s + 0.5 - j) = 0$$

$$s_1, s_2, s_3 = 0, -0.5 + j, -0.5 - j$$

The alternate solution form is

$$y(t) = K_1 + e^{-0.5t}(D \cos t + E \sin t)$$

D4-11

Find the general solutions to the following homogeneous differential equations:

(a) $2\dfrac{dy}{dt} + y = 0$ ans. $Ke^{-(1/2)t}$

(b) $-\dfrac{dy}{dt} - 3y = 0$ ans. Ke^{-3t}

(c) $\dfrac{d^2y}{dt^2} + 6\dfrac{dy}{dt} + 8y = 0$ ans. $K_1 e^{-2t} + K_2 e^{-4t}$

(d) $2\dfrac{d^2y}{dt^2} + \dfrac{dy}{dt} = 0$ ans. $K_1 + K_2e^{-(1/2)t}$

(e) $3\dfrac{d^2y}{dt^2} + 5\dfrac{dy}{dt} + y = 0$ ans. $K_1e^{s_1t} + K_2e^{s_2t}$,
where $s_1, s_2 = (-5 \pm \sqrt{13})/6$

(f) $\dfrac{d^3y}{dt^3} + 2\dfrac{d^2y}{dt^2} + \dfrac{dy}{dt} = 0$ ans. $K_1 + K_2e^{-t} + K_3te^{-t}$

(g) $\dfrac{d^4y}{dt^4} = 0$ ans. $K_1 + K_2t + K_3t^2 + K_4t^3$

(h) $2\dfrac{d^3y}{dt^3} + 8\dfrac{d^2y}{dt^2} + \dfrac{dy}{dt} = 0$ ans. $K_1 + K_2e^{s_1t} + K_3e^{s_2t}$,
where $s_1, s_2 = (-4 \pm \sqrt{14})/2$

(i) $9\dfrac{d^2y}{dt^2} + 6\dfrac{dy}{dt} + 5y = 0$ ans. $e^{-(1/3)t}(D \cos \tfrac{2}{3}t + E \sin \tfrac{2}{3}t)$

(j) $9\dfrac{d^2y}{dt^2} - 6\dfrac{dy}{dt} + 5y = 0$ ans. $e^{(1/3)t}(D \cos \tfrac{2}{3}t + E \sin \tfrac{2}{3}t)$

(k) $\dfrac{d^3y}{dt^3} + 18\dfrac{d^2y}{dt^2} + 82\dfrac{dy}{dt} = 0$ ans. $K_1 + e^{-9t}(D \cos t + E \sin t)$

(l) $\dfrac{d^4y}{dt^4} - 16y = 0$ ans. $K_1e^{2t} + K_2e^{-2t} + D \cos 2t$
$+ E \sin 2t$

GENERAL SOLUTIONS TO HOMOGENEOUS DIFFERENTIAL EQUATIONS

An nth-order homogeneous linear, constant-coefficient differential equation is of the form

$$a_n\frac{d^ny}{dt^n} + a_{n-1}\frac{d^{n-1}y}{dt^{n-1}} + \cdots + a_1\frac{dy}{dt} + a_0y = 0$$

The general solution to this equation is

$$y(t) = K_1e^{s_1t} + K_2e^{s_2t} + \cdots + K_ne^{s_nt}$$

where $s_1, s_2, \ldots, s_n$ are the roots of the characteristic equation

$$a_ns^n + a_{n-1}s^{n-1} + \cdots + a_1s + a_0 = 0$$

providing the roots are distinct.

If there are repeated roots, s_1, the terms corresponding to these roots appear in the general solution as

$$K_1e^{s_1t} + K_2te^{s_1t} + K_3t^2e^{s_1t} + \cdots$$

For complex conjugate pairs of roots

$$s_1, s_2 = \alpha \pm j\beta$$

the corresponding solution terms are

$$K_1e^{(\alpha+j\beta)t} + K_2e^{(\alpha-j\beta)t} = e^{\alpha t}(D \cos \beta t + E \sin \beta t)$$

4.8 GENERAL SOLUTIONS TO DRIVEN EQUATIONS

4.8.1 Forced and Natural Components of Solutions

For an equation with a nonzero driving function, the corresponding homogeneous equation is the original equation with the driving function replaced by zero. The general solution to the entire equation is the *general* solution to the homogeneous equation plus any *one* solution to the entire equation.

For the equation

$$a_n \frac{d^n y}{dt^n} + a_{n-1} \frac{d^{n-1} y}{dt^{n-1}} + \cdots + a_1 \frac{dy}{dt} + a_0 y = f(t)$$

let the general solution to the homogeneous equation be

$$y_n(t) = K_1 e^{s_1 t} + K_2 e^{s_2 t} + \cdots$$

so that

$$\left[a_n \frac{d^n}{dt^n} + a_{n-1} \frac{d^{n-1}}{dt^{n-1}} + \cdots + a_1 \frac{d}{dt} + a_0 \right] y_n(t) = 0$$

Let $y_f(t)$ be any one solution to the original equation:

$$\left[a_n \frac{d^n}{dt^n} + a_{n-1} \frac{d^{n-1}}{dt^{n-1}} + \cdots + a_1 \frac{d}{dt} + a_0 \right] y_f(t) = f(t)$$

Then

$$y(t) = y_n(t) + y_f(t)$$

satisfies the driven equation,

$$\left[a_n \frac{d^n}{dt^n} + a_{n-1} \frac{d^{n-1}}{dt^{n-1}} + \cdots + a_1 \frac{d}{dt} + a_0 \right] [y_n(t) + y_f(t)] = f(t)$$

Since the function

$$y(t) = y_n(t) + y_f(t)$$

satisfies the driven nth order equation and contains n independent, arbitrary constants [in $y_n(t)$], it is the general solution.

The solution to the homogeneous equation is called the *natural* (or *homogeneous* or *transient*) component of the solution. Physically, this is the solution of the network with the fixed sources set to zero. It is how the network would respond "naturally." Usually this component of the solution dies out in time; energy that is initially stored in inductors and capacitors is eventually dissipated by the network's resistors. The single solution to the whole equation is called the *forced* (or *particular* or *steady state*) component of the solution.

4.8.2 Forced Constant Response

Finding a single solution to the entire equation, a forced solution, is rather complicated in general. There are some types of driving functions, however, for which the solution is easy.

Except in a special case, if the driving function of the equation is a constant, the forced solution is also a constant. For example, for

$$\frac{d^3y}{dt^3} + 8\frac{d^2y}{dt^2} + 7\frac{dy}{dt} + 6y = 10 \qquad\qquad y_f = \frac{10}{6}$$

a constant, satisfies the equation.

To find the forced solution due to a constant driving function, simply substitute a constant $y_f = A$ into the equation:

$$a_n\frac{d^nA}{dt^n} + a_{n-1}\frac{d^{n-1}A}{dt^{n-1}} + \cdots + a_1\frac{dA}{dt} + a_0A = f$$

Since all derivatives of a constant are zero, this gives

$$a_0A = f; \qquad y_f = A = \frac{f}{a_0}$$

For the equation

$$\frac{d^2y}{dt^2} + 5\frac{dy}{dt} + 6y = -7$$

the homogeneous equation is

$$\frac{d^2y_n}{dt^2} + 5\frac{dy_n}{dt} + 6y_n = 0$$

and the characteristic equation is

$$s^2 + 5s + 6 = (s + 2)(s + 3) = 0$$

$$s_1 = -2, \qquad s_2 = -3$$

The natural part of the solution is then

$$y_n(t) = K_1e^{-2t} + K_2e^{-3t}.$$

Substituting a trial forced solution,

$$y_f = A$$

a constant, into the entire equation gives

$$\frac{d^2A}{dt^2} + 5\frac{dA}{dt} + 6A = -7$$

$$6A = -7$$

$$y_f = A = -\tfrac{7}{6}$$

The general solution to the equation is thus

$$y(t) = y_n(t) + y_f = K_1e^{-2t} + K_2e^{-3t} - \tfrac{7}{6}$$

A special case occurs when the differential equation has no term proportional to the function itself, that is, when $a_0 = 0$. The equation

$$\frac{d^2y}{dt^2} + 8\frac{dy}{dt} = 10$$

is of this type. Substitution of a constant $y_f = A$ into the equation gives

$$0 = 10$$

so this cannot be the correct forced solution. This complication occurs because the natural component of the equation solution contains an arbitrary constant term:

$$\frac{d^2y_n}{dt^2} + 8\frac{dy_n}{dt} = 0$$

$$s^2 + 8s = s(s + 8) = 0$$

$$y_n(t) = K_1 + K_2e^{-8t}$$

Since the operation

$$\left(\frac{d^2}{dt^2} + 8\frac{d}{dt}\right)K_1 = 0$$

for any constant K_1, a constant cannot be the equation's forced solution. The forced solution to the equation is, instead, a constant times t:

$$y_f(t) = At$$

$$\frac{d^2(At)}{dt^2} + 8\frac{d(At)}{dt} = 0 + 8A = 10$$

$$A = \tfrac{5}{4}$$

The equation's general solution is then

$$y(t) = K_1 + K_2e^{-8t} + \tfrac{5}{4}t$$

Should there be a term of the form Kt in the natural component of the solution of an equation with a constant driving function, as in the equation

$$\frac{d^4y}{dt^4} + 5\frac{d^3y}{dt^3} + 6\frac{d^2y}{dt^2} = 12$$

one would then try a forced solution

$$y_f(t) = At^2$$

Except for this special case where there is a constant term in the natural component of the solution, if the driving function differential equation is a constant, the forced component of the solution is a constant.

4.8.3 Other Forced Responses and Superposition

There are several other driving functions for which determination of a forced solution to a linear, constant-coefficient differential equation is rather easy. Exponential and sinusoidal driving functions will be of great importance in later chapters. Except in a special case, if the driving function is exponential, the forced solution is exponential with the same exponential constant as the driving function. Except in a special case, if the driving function is sinusoidal, the forced solution is sinusoidal with the same frequency as the driving function.

As with linear algebraic equations, driving functions of linear differential equations may be superimposed. The solution due to a sum of two driving function component terms is the sum of the solutions for the individual terms.

D4-12

Find the general solutions to the following constant driving function differential equations:

(a) $3\dfrac{dy}{dt} + 4y = -2$ *ans.* $-\frac{1}{2} + Ke^{-(4/3)t}$

(b) $\dfrac{dy}{dt} - 3y = 4$ *ans.* $-\frac{4}{3} + Ke^{3t}$

(c) $\dfrac{d^2y}{dt^2} + 5\dfrac{dy}{dt} + 6y = 10$ *ans.* $\frac{5}{3} + K_1e^{-2t} + K_2e^{-3t}$

(d) $\dfrac{d^2y}{dt^2} + 2\dfrac{dy}{dt} + y = 3$ *ans.* $3 + K_1e^{-t} + K_2te^{-t}$

(e) $\dfrac{d^2y}{dt^2} - 9y = 8$ *ans.* $-\frac{8}{9} + K_1e^{3t} + K_2e^{-3t}$

(f) $-\dfrac{dy}{dt} = 3$ *ans.* $-3t + K$

(g) $\dfrac{d^2y}{dt^2} + 8\dfrac{dy}{dt} + 3y = 2$ *ans.* $\frac{2}{3} + K_1e^{s_1t} + K_2e^{s_2t}$,
where $s_1, s_2 = -4 \pm \sqrt{13}$

(h) $\dfrac{d^3y}{dt^3} + 8\dfrac{d^2y}{dt^2} = 10$ *ans.* $\frac{5}{8}t^2 + K_1 + K_2t + K_3e^{-8t}$

GENERAL SOLUTIONS TO DRIVEN DIFFERENTIAL EQUATIONS

The general solution to a linear, constant-coefficient differential equation consists of the sum of the general solution to the homogeneous equation plus a single solution to the entire equation.

The general solution of the homogeneous equation is called the *natural* component of the solution. It contains the arbitrary constants. The single solution to the entire equation is called the *forced* component of the solution.

Except in the special case in which one of the terms in the natural component of the solution is a constant, if the driving function of an equation is constant, the forced solution of the equation is constant.

Driving function components may be superimposed, as in linear algebraic equations.

4.9 SPECIFIC SOLUTIONS AND BOUNDARY CONDITIONS

The general solutions to the differential equations describing networks indicate that there is a whole family of possible network solutions, different possibilities for each different set of arbitrary constants. Of course, only one of these possibilities applies to any specific situation. Which of all the possible solutions applies depends on when and how the elements of the network were connected together.

For example, the solution for $v(t)$ in the network of Figure 4-22 depends on the time at which the network is connected together and on the amount of charge on the capacitor plates (or, equivalently, the capacitor voltage) when the network was first connected. After the network is connected, $v(t)$ satisfies the same differential equation, no matter what the circumstances of the network connection; but different initial capacitor charges will result in different possible solutions for $v(t)$ applying.

Which of all of the possible solutions of a differential equation is the specific one that applies to a given situation may be described with boundary conditions. Suppose it is known that of all the solutions of the form

$$y(t) = Ke^{-3t} + 7$$

the one for which the boundary condition

$$y(0) = 5$$

applies in a certain situation. Then the arbitrary constant K may be evaluated.

$$y(0) = Ke^0 + 7 = K + 7 = 5; \qquad K = -2.$$

Of all the possible solutions,

$$y(t) = -2e^{-3t} + 7$$

is the one that satisfies the boundary condition. It should be carefully noted that boundary conditions apply to the general solution of a differential equation, not just to its natural component.

Mathematically, the most convenient time t for a boundary condition is $t = 0$, since at $t = 0$ an exponential function Ke^{st} has value

$$Ke^0 = K$$

For this reason, the origin of the time scale in a problem is usually chosen so that boundary conditions occur at $t = 0$.

A second-order differential equation has two arbitrary constants in the general solution and requires two independent boundary conditions to select one specific solution from the general solution.

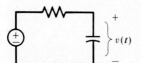

Figure 4-22   Simple network involving a capacitor.

Consider the equation

$$\frac{d^2y}{dt^2} + 5\frac{dy}{dt} + 4y = 12$$

with the boundary conditions

$$y(0) = 2 \quad \text{and} \quad \left.\frac{dy}{dt}\right|_{t=0} = 7$$

The homogeneous equation is

$$\frac{d^2y_n}{dt^2} + 5\frac{dy_n}{dt} + 4y_n = 0$$

the characteristic equation is

$$s^2 + 5s + 4 = (s + 1)(s + 4) = 0$$

and the natural component of the general solution is

$$y_n(t) = K_1e^{-t} + K_2e^{-4t}$$

The forced solution is given by

$$y_f = A$$

$$\frac{d^2A}{dt^2} + 5\frac{dA}{dt} + 4A = 12$$

$$4A = 12$$

$$y_f = A = 3$$

The general solution is thus

$$y(t) = y_n(t) + y_f = K_1e^{-t} + K_2e^{-4t} + 3$$

Applying the first boundary condition,

$$y(0) = K_1 + K_2 + 3 = 2$$

$$K_1 + K_2 = -1$$

The second boundary condition in this example involves the derivative of the function.

$$\frac{dy}{dt} = -K_1e^{-t} - 4K_2e^{-4t}$$

$$\left.\frac{dy}{dt}\right|_{t=0} = -K_1 - 4K_2 = 7$$

Collecting the conditions on K_1 and K_2,

$$\begin{cases} K_1 + K_2 = -1 \\ -K_1 - 4K_2 = 7 \end{cases}$$

$$K_1 = \frac{\begin{vmatrix} -1 & 1 \\ 7 & -4 \end{vmatrix}}{\begin{vmatrix} 1 & 1 \\ -1 & -4 \end{vmatrix}} = \frac{-3}{-3} = 1$$

$$K_2 = \frac{\begin{vmatrix} 1 & -1 \\ -1 & 7 \end{vmatrix}}{-3} = \frac{6}{-3} = -2$$

So of all the possible solutions of the form

$$y(t) = K_1 e^{-t} + K_2 e^{-4t} + 3$$

the specific solution

$$y(t) = e^{-t} - 2e^{-4t} + 3$$

is the one that applies in this situation.

Consider the equation

$$\frac{d^2 y}{dt^2} + 4 \frac{dy}{dt} = 12$$

with boundary conditions

$$y(0) = 0$$

$$\left. \frac{dy}{dt} \right|_{t=0} = 11$$

The homogeneous equation is

$$\frac{d^2 y_n}{dt^2} + 4 \frac{dy_n}{dt} = 0$$

and has characteristic equation

$$s^2 + 4s = s(s + 4) = 0$$

so the natural component of the solution is

$$y_n(t) = K_1 + K_2 e^{-4t}$$

In view of the constant term in $y_n(t)$, the forced solution of this equation cannot be a constant. Substituting

$$y_f(t) = At$$

there results

$$0 + 4A = 12; \qquad A = 3$$

giving the general equation solution

$$y(t) = K_1 + K_2 e^{-4t} + 3t$$

Applying the two boundary conditions,

$$y(0) = K_1 + K_2 = 0$$

$$\left.\frac{dy}{dt}\right|_{t=0} = -4K_2 + 3 = 11$$

gives

$$K_1 = 2, \qquad K_2 = -2$$

so the specific solution to this problem, satisfying the differential equation and the boundary conditions is

$$y(t) = 2 - 2e^{-4t} + 3t$$

The equation

$$\frac{d^2y}{dt^2} + 4\frac{dy}{dt} + 29y = 12$$

with boundary conditions

$$y(0) = -3$$

$$\left.\frac{dy}{dt}\right|_{t=0} = 2$$

has natural solution component as follows:

$$\frac{d^2y_n}{dt^2} + 4\frac{dy_n}{dt} + 29y_n = 0$$

$$s^2 + 4s + 29 = 0$$

$$s_1, s_2 = \frac{-4 \pm \sqrt{16 - 116}}{2} = -2 \pm j5$$

$$y_n(t) = e^{-2t}(D \cos 5t + E \sin 5t)$$

The constant forced solution component is

$$y_f = A = \tfrac{12}{29}$$

so the general solution to the equation is

$$y(t) = e^{-2t}(D \cos 5t + E \sin 5t) + \tfrac{12}{29}$$

Application of the boundary conditions gives the following specific solution:

$$y(0) = D + \tfrac{12}{29} = -3$$

$$D = -\tfrac{99}{29}$$

$$\frac{dy}{dt} = e^{-2t}(-5D \sin 5t + 5E \cos 5t) - 2e^{-2t}(D \cos 5t + E \sin 5t)$$

$$\frac{dy}{dt}\bigg|_{t=0} = 5E - 2D = 2$$

$$E = -\tfrac{28}{29}$$

$$y(t) = e^{-2t}(-\tfrac{99}{29}\cos 5t - \tfrac{28}{29}\sin 5t) + \tfrac{12}{29}$$

D4-13

Find the specific solutions to the following differential equations with the indicated boundary conditions:

(a) $\dfrac{dy}{dt} + 3y = 6$

    $y(0) = 0$ ans. $2 - 2e^{-3t}$

(b) $\dfrac{dy}{dt} + y = 7$

    $y(0) = 5$ ans. $7 - 2e^{-t}$

(c) $2\dfrac{dy}{dt} + 5y = 12$

    $y(0) = -2$ ans. $\tfrac{12}{5} - \tfrac{22}{5}e^{-(5/2)t}$

(d) $4\dfrac{dy}{dt} + 3y = 12$

    $y(1) = 1$ ans. $4 - [3e^{(3/4)}]e^{-(3/4)t}$

(e) $2\dfrac{dy}{dt} + 5y = 0$

    $\dfrac{dy}{dt}\bigg|_{t=0} = 3$ ans. $-\tfrac{6}{5}e^{-(5/2)t}$

(f) $\dfrac{dy}{dt} + 6y = 10$

    $\dfrac{dy}{dt}\bigg|_{t=0} = 0$ ans. $\tfrac{5}{3}$

(g) $\dfrac{d^2y}{dt^2} + 3\dfrac{dy}{dt} + 2y = 10$

    $y(0) = 5$

    $\dfrac{dy}{dt}\bigg|_{t=0} = 0$ ans. 5

(h) $\dfrac{d^2y}{dt^2} + \dfrac{dy}{dt} = 4$

    $y(0) = 0$

$$\frac{dy}{dt}\bigg|_{t=0} = 2$$

ans. $4t - 2 + 2e^{-t}$

(i) $\dfrac{d^2y}{dt^2} + 4\dfrac{dy}{dt} + 4y = 12$

$y(0) = 6$

$$\frac{dy}{dt}\bigg|_{t=0} = 4$$

ans. $3e^{-2t} + 10te^{-2t} + 3$

(j) $\dfrac{d^2y}{dt^2} + 8\dfrac{dy}{dt} + 20y = 5$

$y(0) = -2$

$$\frac{dy}{dt}\bigg|_{t=0} = 11$$

ans. $e^{-4t}(-\frac{9}{4}\cos 2t + \sin 2t) + \frac{1}{4}$

(k) $4\dfrac{d^2y}{dt^2} + y = 10$

$y(0) = 5$

$$\frac{dy}{dt}\bigg|_{t=0} = 0$$

ans. $10 - 5\cos\frac{1}{2}t$

(l) $\dfrac{d^2y}{dt^2} + 2\dfrac{dy}{dt} + 5y = -15$

$y(0) = 4$

$$\frac{dy}{dt}\bigg|_{t=0} = -3$$

ans. $e^{-t}(7\cos 2t + 2\sin 2t) - 3$

BOUNDARY CONDITIONS

One specific solution to a differential equation, of all the possibilities given by the general solution, may be selected through the application of boundary conditions.

4.10 THE EXPONENTIAL FUNCTION

An exponential function is a function of the form

$$f(t) = Ke^{st}$$

where K and s are constants. The constant K is called the *amplitude* of the exponential function, and the constant s is called the *exponential constant* of the function.

If the exponential constant is positive (and a real number), the exponential function is said to be expanding, since e^{st} gets larger and larger with time, t. The larger the value of s, the faster e^{st} "blows up." Of the two expanding exponential functions in Figure 4-23(a), e^{3t} expands more quickly than e^{2t}. If the exponential

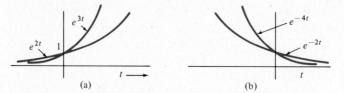

Figure 4-23 Expanding and decaying exponential functions.

constant is negative (and a real number), the exponential function is said to be decaying, since e^{st} approaches zero with time, t. The more negative the value of s, the faster e^{st} "dies out." Examples of rapid and not-so-rapidly decaying exponential functions are given in Figure 4-23(b).

The effect of the amplitude K is illustrated in the examples of Figure 4-24.

The *time constant* of an exponential function (with a real value of s) is the inverse of the magnitude of the exponential constant. For

$$f(t) = 7e^{-2t}$$

the time constant is

$$\tau = \tfrac{1}{2}$$

In one time constant, a decaying exponential function decays by a factor of $(1/e)$, or to about 37 percent of its initial value. This function has value 7 at time $t = 0$. In a time of one time constant, it has value $7/e$. After another time-constant time interval has passed, the value of the function has decayed by another factor of $1/e$, or to $7/e^2$, as illustrated in Figure 4-25(a).

For

$$f(t) = 8e^{3t}$$

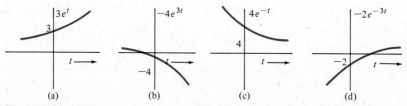

Figure 4-24 Examples of exponential functions with various amplitudes.

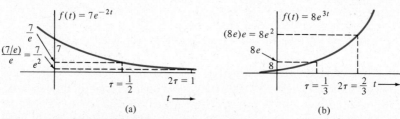

Figure 4-25 Time constants.
(a) Decaying exponential function.
(b) Expanding exponential function.

the time constant is

$$\tau = \tfrac{1}{3}$$

An expanding exponential expands by a factor of e in each time interval of one time constant, as illustrated for this function in Figure 4-25(b).

D4-14

Sketch the following exponential functions. Find the amplitude and time constant of each.

(a) $f(t) = -e^{3t}$ *ans.* $-1, \tfrac{1}{3}$

(b) $f(t) = 7e^{-5t}$ *ans.* $7, \tfrac{1}{5}$

(c) $f(t) = -10e^{-(4/3)t}$ *ans.* $-10, \tfrac{3}{4}$

(d) $f(t) = 10^4e^{-10^6 t}$ *ans.* $10^4, 10^{-6}$

D4-15

Find the exponential functions (approximately) from the plots:

(a)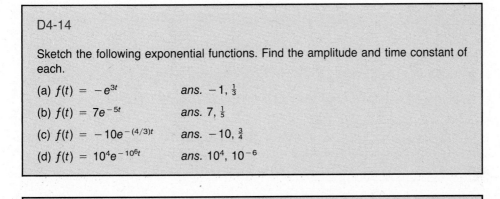

(b)

ans. $15e^{-0.2t}$ *ans.* $-150e^{-54t}$

THE EXPONENTIAL FUNCTION

Exponential functions have the form

$$f(t) = Ke^{st}$$

where the amplitude K and the exponential constant s are constants.
    For s, a real number, the time constant of an exponential function is

$$\tau = \frac{1}{|s|}$$

It is the time interval over which a decaying exponential decays by a factor of $e^{-1} = 1/e$ and over which an expanding exponential function grows by a factor of $e^1 = e$.

CHAPTER FOUR PROBLEMS

Basic Problems

Capacitor Relations

1. Find $i(t)$, $p_{\text{into C}}(t)$, and $W_C(t)$:

(a)

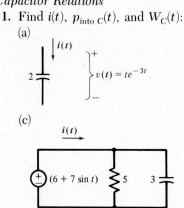

$v(t) = te^{-3t}$

(b)

$$v(t) = \begin{cases} 10, & t \leq 0 \\ 10e^{-2t}, & t \geq 0 \end{cases}$$

(c)

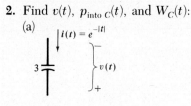

$(6 + 7 \sin t)$

(d)

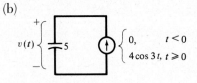

$\sin 4t$

2. Find $v(t)$, $p_{\text{into C}}(t)$, and $W_C(t)$:

(a)

$i(t) = e^{-|t|}$

$v(t)$

(b)

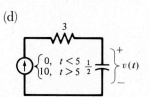

$$v(t) \quad \begin{cases} 0, & t < 0 \\ 4 \cos 3t, & t \geq 0 \end{cases}$$

(c)

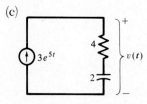

$3e^{5t}$

$v(t)$

(d)

$$\begin{cases} 0, & t < 5\tfrac{1}{2} \\ 10, & t > 5\tfrac{1}{2} \end{cases} \quad v(t)$$

3. The given current flows for $t \geq 0$. Find $v(t)$ for $t \geq 0$, given $v(0)$.

(a)

$8 \sin t$ $\tfrac{1}{4}$ $v(t)$

$v(0) = 10$

(b)

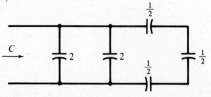

$v(0) = 12$

4. Find the equivalent capacitance:

(a)

(b)

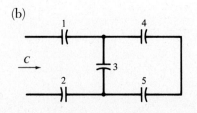

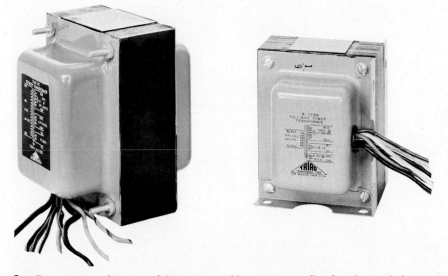

Small power transformers of the type used in power supplies for electronic instruments. These consist of coupled inductors and are used to develop several voltages larger or smaller than the power line voltage. (*Photo courtesy of Triad-Utrad.*)

Inductor Relations

5. Find $v(t)$, $p_{\text{into } L}(t)$, and $W_L(t)$:

(a)

$i(t) = 8 \cos 7t$

$\frac{1}{2}$ $v(t)$

(b)

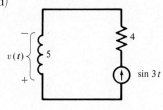

$$\begin{cases} 4, & t \le 0 \\ 4e^{-2t}, & t \ge 0 \end{cases} \quad 3 \quad v(t)$$

(c)

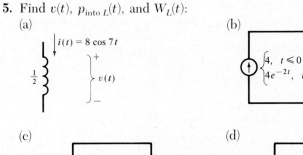

$v(t)$ $(3 + 4e^{-6t})$ 5 2

(d)

$v(t)$ 5 4 $\sin 3t$

6. Find $i(t)$, $p_{\text{into } L}(t)$, and $W_L(t)$:

(a)

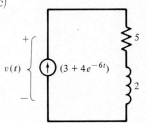

$i(t)$

6 $v(t) = \begin{cases} 0, & t < 0 \\ \cos 4t, & t \ge 0 \end{cases}$

(b)

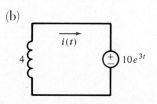

$i(t)$

4 $10e^{3t}$

A special-purpose computer is used to extensively test a complicated electronic circuit. Test results are summarized on the viewing screen, and the user is able to control and modify the test sequence as it progresses. (*Photo courtesy of Gould, Inc.*)

(c)   (d)

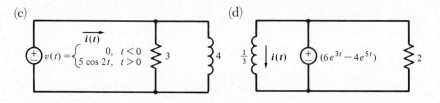

7. The given voltage is applied for $t \geq 0$. Find $i(t)$ for $t \geq 0$, given $i(0)$.

(a)   (b)

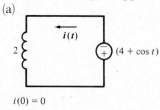

$i(0) = 0$

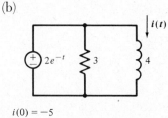

$i(0) = -5$

8. Find the equivalent inductance:

(a)   (b)

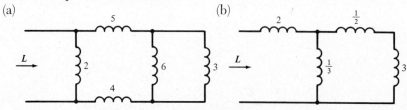

Mutually Coupled Inductor Relations

9. Find the voltage-current relations:

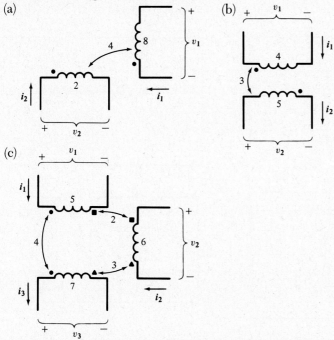

(a) (b) (c)

10. Find the equivalent controlled voltage-source models for the following sets of coupled inductors in terms of the indicated terminal voltages and currents:

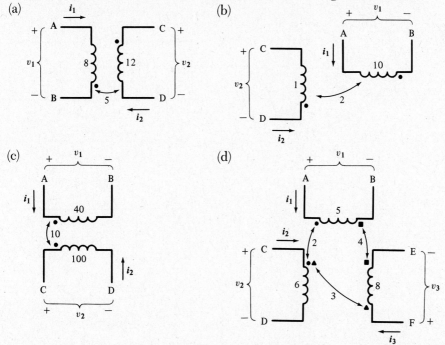

(a) (b) (c) (d)

Systematic Simultaneous Equations

11. Write (but do not attempt to solve) systematic integrodifferential mesh equations for the following networks, in terms of the indicated currents:

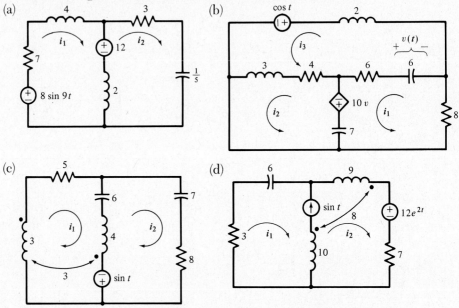

(a)   (b)

12. Write (but do not attempt to solve) systematic integrodifferential nodal equations for the following networks, in terms of the indicated variables:

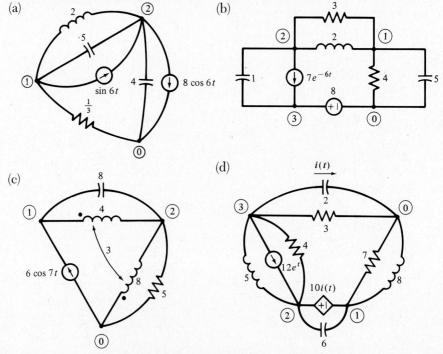

(a)   (b)

(c)   (d)

Differential Equations

13. Find the general solutions to the following differential equations:

(a) $3\dfrac{dy}{dt} + y = 0$

(b) $4\dfrac{dy}{dt} + 3y = 5$

(c) $\dfrac{d^2y}{dt^2} + 6\dfrac{dy}{dt} = 0$

(d) $3\dfrac{d^2y}{dt^2} + 7\dfrac{dy}{dt} + 4y = -2$

(e) $\dfrac{d^2y}{dt^2} + 6\dfrac{dy}{dt} + 9y = 18$

(f) $\dfrac{d^2y}{dt^2} + 4\dfrac{dy}{dt} + 29y = -6$

(g) $\dfrac{d^3y}{dt^3} + 4\dfrac{d^2y}{dt^2} + 20\dfrac{dy}{dt} = 0$

(h) $\dfrac{d^2y}{dt^2} + 8\dfrac{dy}{dt} = 10$

14. Find the specific solutions to the following differential equations with the indicated boundary conditions:

(a) $4\dfrac{dy}{dt} + 2y = 3$

$y(0) = 5$

(b) $\dfrac{dy}{dt} + 4y = -2$

$y(0) = -3$

(c) $\dfrac{d^2y}{dt^2} + 7\dfrac{dy}{dt} + 6y = 10$

$y(0) = 5$

$\left.\dfrac{dy}{dt}\right|_{t=0} = 0$

(d) $3\dfrac{d^2y}{dt^2} + 8\dfrac{dy}{dt} + 2y = 6$

$y(0) = 4$

$\left.\dfrac{dy}{dt}\right|_{t=0} = -3$

(e) $\dfrac{d^2y}{dt^2} + 9y = 0$

$y(0) = -4$

$\left.\dfrac{dy}{dt}\right|_{t=0} = 6$

(f) $\dfrac{d^2y}{dt^2} - 2\dfrac{dy}{dt} + 5y = 10$

$y(0) = -2$

$\dfrac{dy}{dt}\bigg|_{t=0} = -3$

(g) $\dfrac{d^2y}{dt^2} = 3$

$y(0) = -2$

$\dfrac{dy}{dt}\bigg|_{t=0} = 5$

(h) $3\dfrac{d^2y}{dt^2} + \dfrac{dy}{dt} = 5$

$y(0) = 0$

$\dfrac{dy}{dt}\bigg|_{t=0} = 4$

Practical Problems

Capacitors

15. The current through a certain $\frac{1}{2}$-F capacitor is graphed for a 7-s interval of time. Draw graphs of three of the possible capacitor voltages during this time interval.

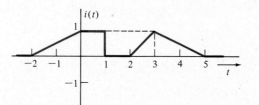

16. Physical capacitors will withstand only limited voltages and currents. The voltage limit is that beyond which the dielectric material between the capacitor plates will "break down" and become a conductor. Manufacturers generally rate capacitors in terms of a guaranteed maximum constant (or dc) *working voltage.*

　　Capacitors with air, vacuum, oil, and similar dielectrics have the advantage of recovery of their properties if excessive voltage has caused conduction through the dielectric. Dielectric breakdown in other materials, such as paper and plastic, generally destroys the capacitor.

　　Exceeding the current limit in a physical capacitor results in excessive heating, due to small resistances of the plates and the dielectric. The maximum safe current for a capacitor is usually only of concern in applications such as power distribution and radio transmission, where relatively large power flows are involved.

In an *electrolytic* capacitor a large capacitance is formed by depositing an extremely thin layer of insulating material on one plate of an electrolytic cell. It is important that the voltage applied to such a capacitor (and other *polarized* types of capacitors) be only of a single, specified polarity. The wrong polarity of applied voltage will cause destruction of the deposited insulating material.

If a 10-μF capacitor has the applied voltage shown, what is the largest capacitor current? Could an electrolytic capacitor be used in this situation?

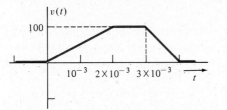

17. The leakage of current through the dielectric material between the plates of a physical capacitor may be modeled by the capacitance in parallel with a leakage resistance, as indicated. Capacitors designed for small leakage can have leakage resistances that cause a charged-capacitor voltage to decay only a few percent in a year's time. For such devices, the current leakage through contaminants on the surface of their package may be much greater than the leakage through the dielectric. More typical low-leakage capacitors have voltage decay rates of perhaps a few percent per minute.

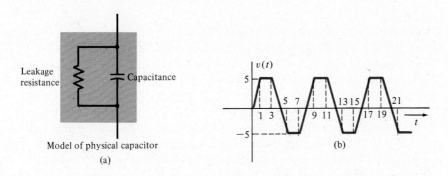

Model of physical capacitor
(a)

(b)

In many applications, resistors are deliberately connected across capacitors to increase the rate of decay of their voltages when equipment power is turned off, insuring that dangerous amounts of charge do not remain long on the plates. Capacitors present a danger if charged to lethal voltages and by their capability of delivering extremely high currents if accidentally short-circuited. When large amounts of energy are stored, a capacitor's short-circuit current may be capable of vaporizing wires, screwdrivers, and the like.

Suppose a 0.1-μF capacitor has a 150-kΩ leakage resistance. For the capacitor voltage shown, carefully sketch the current through this physical capacitor.

Inductors

18. The current through a certain 3-H inductor is sketched. Draw a sketch of the inductor voltage for the same time interval.

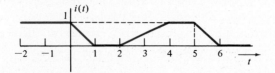

19. Physical inductors are composed of a coil of wire. At normal temperatures, the resistance of that wire, called the *winding resistance*, may be significant. A model for a physical inductor that takes its winding resistance into account is shown in (a).

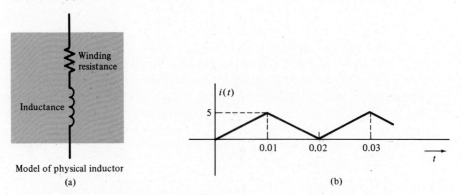

Model of physical inductor
(a)

(b)

Suppose a 2-H inductor has a 100-Ω winding resistance. For the current given in (b), carefully sketch the voltage across this inductor.

20. The inductance of a coil of wire may be greatly increased by channeling the magnetic flux through a core of ferromagnetic material. Most of the magnetic flux due to any loop of the coil may then be made to pass through all of the other loops.

Unfortunately, ferromagnetic materials have a nonlinear relation between current and flux, exhibiting permanent magnetism and saturation effects. Thus air gaps are commonly built into inductor cores to improve the linearity of the current–magnetic flux relation. "Soft" ferromagnetic core material such as iron may respond rather sluggishly to a changed current with a changed flux, making the inductor effective only for relatively slowly varying currents.

To reduce the effects of induced currents in the core material itself, an iron inductor core may be *laminated*—that is, cut into thin strips that are electrically insulated from one another and placed side by side; or a conductive ferromagnetic material may be finely powdered and pressed or glued so that the conductive particles are insulated from one another.

Suppose that for a certain coil and core, the relation between coil current i and core flux ψ is

$$\psi = 20i + i^3$$

Neglecting any resistance of the coil, what is the voltage-current relation of this nonlinear inductor?

21. The current limitation for a physical inductor is determined by the maximum heating of the conductor and core that may be sustained. An inductor's maximum voltage is the voltage at which breakdown of the insulation occurs. Safe voltage and current limits are termed "working voltages" and "working currents."

   The current $i(t)$ through a 100-mH inductor with a 16-Ω winding resistance R_W is sketched.

   (a) What is the maximum inductor voltage?
   (b) What is the maximum amount of energy stored in the inductor? At what time is the maximum energy stored?
   (c) Approximating the power loss in the inductor by

   $$p_{loss}(t) = i^2(t)R_W$$

   what is the total energy dissipated in the inductor in the time interval shown?

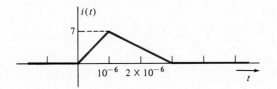

Exponential Functions

22. Radioactive decay is described by a decaying exponential function, and the half-life of a radioactive material is the time interval over which the decay is to one-half its initial value. Relate the half-life of an exponential to its time constant.

23. When exponential data are available only for a span of time that is small compared to a time constant, it is helpful to be able to identify the exponential function from its value and slope at a given time. Carefully describe how this may be done for the time $t = 0$. Repeat, assuming that the value and slope at some time $t = a$ are known, instead.

24. It is commonly asserted that for all practical purposes an exponential function has decayed to zero in a time span of five time constants. By what factor has such a function actually decayed in that time?

Advanced Problems

Inductor and Capacitor Voltage-Current Relations

25. (a) Develop voltage-divider and current-divider rules for inductors.
    (b) Develop voltage-divider and current-divider rules for capacitors.

Power and Stored Energy

26. Show that the electrical power flow into a series connection of capacitors at any instant of time is equal to the power flow into the equivalent capacitor.

It is generally true that the power flow into any two-terminal combination of capacitors (or a two-terminal combination of inductors) is the same as the power flow into the equivalent capacitor (or inductor).

27. Suppose a certain electrical element stores energy according to

$$W(t) = v^4(t)$$

Using $p_{into}(t) = dW/dt$, find the sink reference voltage-current relation for the element.

Nonlinear and Time-Varying Capacitance and Inductance

28. If the capacitance of a capacitor is time-varying, its voltage-current relation is

$$i(t) = \frac{d}{dt} [C(t)v(t)]$$

For a parallel plate capacitor, the capacitance varies nearly inversely with plate spacing d,

$$C = \frac{\epsilon A}{d}$$

where ϵ is the permittivity of the dielectric between the plates and A is the area of each plate.

Suppose that a capacitor with variable plate spacing is connected in parallel with a constant voltage source so that the capacitor voltage is a constant, v.

(a) In terms of ϵ, A, and v, find the nonlinear differential equation relating $i(t)$ and $d(t)$. A device such as this can function as a velocity transducer. It is the basis of the capacitive microphone.

(b) If the plate spacing is

$$d(t) = 100 + \cos t$$

find the current $i(t)$ in terms of ϵ, A, and v.

29. The behavior of inductors with ferromagnetic cores is occasionally modeled by a nonlinear inductance L, dependent upon the inductor current:

$$v(t) = \frac{d}{dt} [L(i)i(t)]$$

Let

$$i(t) = 10 \cos 30t$$

and let

$$L(i) = 6e^{-0.1i}$$

At sufficiently large currents, the inductance decreases, reflecting the decreasing number of magnetic domains available for alignment with increasing magnetization.

(a) Find $v(t)$.

(b) What time-varying inductance $L(t)$ would, with this current $i(t)$, produce the same inductor voltage?

(c) What current $i(t)$ would, with a constant inductance $L = 6$, produce the same inductor voltage $v(t)$?

Mutual Inductance

30. Develop the indicated equivalent circuit for two coupled inductors. This equivalent holds even when $M^2 = L_1L_2$ but is more difficult to use for nodal equations because the controlling signals are not currents in single elements.

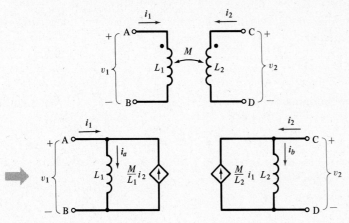

31. Find the equivalent inductances of the two series-connected coupled inductor arrangements shown. Any two-terminal combination of inductors, including those that are mutually coupled, is equivalent to a single inductor.

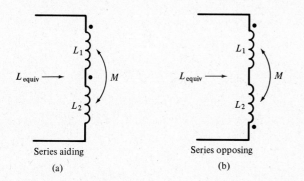

Series aiding
(a)

Series opposing
(b)

32. Find the equivalent inductances of the two parallel-connected coupled inductors shown.

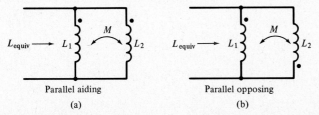

Parallel aiding
(a)

Parallel opposing
(b)

Simultaneous Equations

33. Find the equivalent capacitance of the two-terminal combination of capacitors:

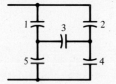

Differential Equations

34. One method of differential equation solution is to find an *integrating factor*. For a first-order equation,

$$a_1 \frac{dy}{dt} + a_0 y = f(t)$$

multiplying by the integrating factor

$$p(t) = e^{(a_0/a_1)t}$$

gives

$$a_1 e^{(a_0/a_1)t} \frac{dy}{dt} + a_0 e^{(a_0/a_1)t} y = f(t)$$

$$a_1 \frac{d}{dt}(e^{(a_0/a_1)t}y) = f(t)$$

$$\frac{d}{dt}(e^{(a_0/a_1)t}y) = \frac{f(t)}{a_1}$$

which may be solved by integration.

Find the general solution to

$$-2 \frac{dy}{dt} + 3y = e^{-4t}$$

using the integrating factor method.

Unfortunately, finding an integrating factor for a second-order equation involves solving another second-order differential equation, and the situation is similar for equations of higher order.

35. Show that if

$$y(t) = K_1 e^{(\alpha + j\beta)t} + K_2 e^{(\alpha - j\beta)t}$$

is the general solution of a second-order differential equation, for arbitrary K_1 and K_2, then

$$y(t) = e^{\alpha t}(A \cos \beta t + B \sin \beta t)$$

is also the general solution, for arbitrary A and B.

36. Find a differential equation for which

$$y(t) = K_1 e^{4t} + K_2 e^{-2t} + K_3 t e^{-2t} + 6$$

is the general solution, where K_1, K_2, and K_3 are arbitrary constants.

Chapter Five
Switched Networks

The electric oscillations of open induction coils have a period
of vibration which is measured by a ten-thousandth of a
second. The vibrations in the oscillatory discharges of
Leyden jars [capacitors] . . . follow each other about a
hundred times as rapidly. Theory admits the possibility of
oscillations even more rapid than these in open wire circuits
of good conductivity, provided that the ends are not loaded
with large capacities; but at the same time theory does not
enable us to decide whether such oscillations can be actually
excited on such a scale as to admit of their being observed.

Heinrich Hertz
From *Electric Waves*,
Berlin, 1893

5.1 PREVIEW

The complete solution of first-order networks, those containing a single energy-
storing element, is now developed, using many examples and paralleling classical
solution of the underlying differential equations. Switched constant-source, first-
order inductive and capacitive networks are covered in detail, proceeding step by
step in each case. Both switches and step functions are used.

Series and parallel *RLC* switched networks are then discussed, and general
solution methods are given. Examples with distinct, repeated, and complex characteristic roots are given.

When you complete this chapter, you should know—

1. how to solve first-order switched inductive networks and to graph any
   signal in the network;
2. what the unit step function is and how it may be used to describe source
   functions in switched networks;
3. how to solve first-order switched capacitive networks and to graph any
   signal in the network;
4. how to solve switched series and parallel *RLC* networks and how second-
   order response is classified.

5.2 SWITCHED FIRST-ORDER INDUCTIVE NETWORKS

5.2.1 Differential Equation for the Inductor Current

A first-order network is a network that contains sources, resistors, and one energy-storing element, either a capacitor or an inductor. If the energy-storing element is an inductor, the network is said to be an inductive first-order network.

So far as the inductor is concerned, the rest of a first-order inductive network consists of sources and resistors, and so may be represented by a Thévenin equivalent, as in the diagram of Figure 5-1. In terms of the Thévenin voltage and current, the mesh equation for the inductor current $i(t)$ is

$$L\frac{di}{dt} + R_T i = v_T(t)$$

The solution of the differential equation for the inductor current in a first-order inductive network consists of the natural component plus the forced component.

The natural component of $i(t)$ is the solution to the homogeneous equation

$$L\frac{di_n}{dt} + R_T i_n = 0$$

Substituting

$$i_n = Ke^{st}$$

$$sL + R_T = 0; \qquad s = -\frac{R_T}{L}$$

$$i_n(t) = Ke^{-(R_T/L)t}$$

For the usual case of positive R_T and L, this part of the solution dies out in time, with time constant L/R_T.

This general result applies to any first-order inductive network: The natural component of the inductor current is as given above, where R_T is the Thévenin resistance of the rest of the network that is connected to the inductor.

The forced component of the inductor current depends on the driving function $v_T(t)$. If $R_T \neq 0$ and if all of the network sources are constant, v_T will be constant, and finding the forced component of the solution is particularly easy. For a constant v_T,

$$i_f = A$$

is constant. Substituting into the differential equation,

$$L\frac{dA}{dt} + R_T A = v_T$$

$$R_T A = v_T$$

$$i_f = A = \frac{v_T}{R_T}$$

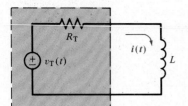

Figure 5-1   Thévenin equivalent used for first-order inductive network.

So far as the *forced* response to *constant* sources is concerned, the inductor voltage,

$$L \frac{di_f}{dt} = L \frac{dA}{dt} = 0$$

is zero; that is, the inductor behaves as a short circuit. To find the forced component of the inductor current due to *constant* sources, one can replace the inductor by a short circuit and solve for the corresponding current.

Consider the network of Figure 5-2(a). The natural component of the inductor current is given by

$$i_n(t) = Ke^{-(R_T/L)t} = Ke^{-(6/35)t}$$

where the Thévenin resistance is found by setting the sources to zero as in Figure 5-2(b). Since the network involves only constant sources, the forced component of the inductor current may be found by replacing the inductor by a short circuit, which is done in Figure 5-2(c).

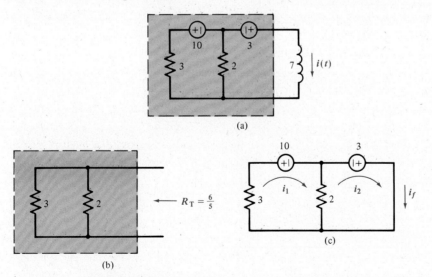

Figure 5-2   (a) Original network.
(b) Finding Thévenin resistance of sources and resistors connected to the inductor.
(c) Finding forced behavior due to constant sources by replacing the inductor with a short circuit.

Solving,

$$\begin{cases} 5i_1 - 2i_2 = -10 \\ -2i_1 + 2i_2 = 3 \end{cases}$$

$$i_f = i_2 = \frac{\begin{vmatrix} 5 & -10 \\ -2 & 3 \end{vmatrix}}{\begin{vmatrix} 5 & -2 \\ -2 & 2 \end{vmatrix}} = \frac{-5}{6}$$

The general solution for the inductor current is then

$$i(t) = i_n(t) + i_f = Ke^{-(6/35)t} - \tfrac{5}{6}$$

D5-1

Find the Thévenin equivalents of the two-terminal networks connected to an inductor. Then write the single-loop differential equation in terms of the inductor current i.

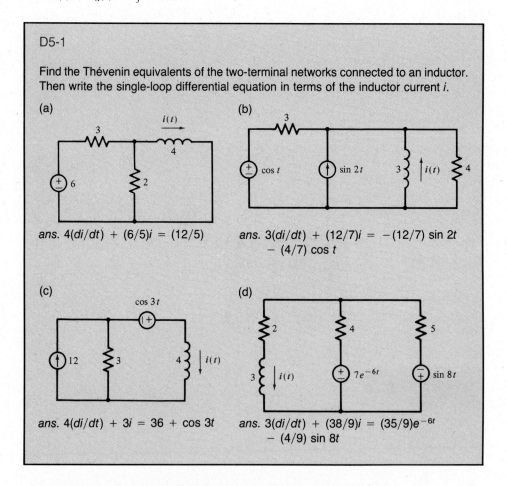

(a)

ans. $4(di/dt) + (6/5)i = (12/5)$

(b)

ans. $3(di/dt) + (12/7)i = -(12/7) \sin 2t - (4/7) \cos t$

(c)

ans. $4(di/dt) + 3i = 36 + \cos 3t$

(d)

ans. $3(di/dt) + (38/9)i = (35/9)e^{-6t} - (4/9) \sin 8t$

5.2.2 Switched Networks and Continuity of Inductor Current

Changes in a network may be represented with switches, Figure 5-3, beside which are indicated the times of openings and closings. In most cases the origin of the

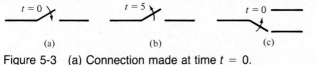

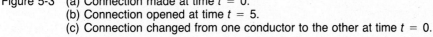

(a)   (b)   (c)

Figure 5-3   (a) Connection made at time $t = 0$.
(b) Connection opened at time $t = 5$.
(c) Connection changed from one conductor to the other at time $t = 0$.

time scale is best chosen so that a switching is made at time $t = 0$ for numerical convenience. Switched networks with a single switching are a type of problem in which the forced network response before the switching is sought, and both forced and natural components of the response are desired after the switching. It is implied that the network has been connected a long time before the switching and that any natural response arising from the original connection has long since decayed, leaving the forced response.

The network, then, is changed by the switching. The response changes to a new forced component and a new natural component, governed by the equations for the new network after the switching.

Only one of all the possible solutions given by the general solution after the switching applies in a particular situation. It is the one for which the inductor current is continuous. If the inductor current were discontinuous (if it "jumped" from one value to another), the inductor voltage,

$$v = L \frac{di}{dt}$$

(and the inductor power flow) would be infinite. Just the right amount of the natural component of inductor current is present after the switching so that the inductor current is continuous.

Consider the network in Figure 5-4(a). Since no information is given as to when or in what manner the network was connected before $t = 0$, it is implied that it has been connected, with the switch open, for a long time prior to $t = 0$. If there was any natural response when the network was first wired up, long ago, it has died out and is not of interest. So up to time $t = 0$, only the forced parts of the network signals are to be found. Because the network source is disconnected from the network before $t = 0$, the Thévenin voltage and thus the forced component of $i(t)$ are zero before $t = 0$, as shown in Figure 5-4(b).

At time $t = 0$, the network is suddenly changed as the switch is closed. Now both the forced and the natural components of the inductor current are of interest. Because the source is constant, the forced component of the current may be found, as it was before $t = 0$, by replacing the inductor by a short circuit. The network is different now that the switch is closed, so i_f will differ from what it was before time $t = 0$. As indicated in Figure 5-4(c),

$$i_f = \tfrac{5}{2} \text{ A}$$

after $t = 0$.

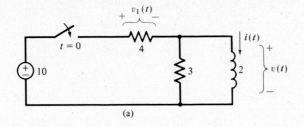

(a)

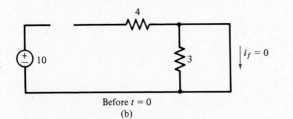

Before $t = 0$

(b)

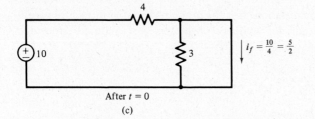

After $t = 0$

(c)

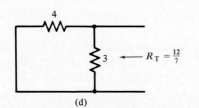

(d)

Figure 5-4 (a) Complete problem.
(b) Forced inductor current before $t = 0$.
(c) Forced inductor current after $t = 0$.
(d) Thévenin resistance after $t = 0$.

The natural component of $i(t)$ after $t = 0$ is

$$i_n(t) = Ke^{-(R_T/L)t} = Ke^{-(6/7)t}$$

R_T is the Thévenin resistance of the source-resistor network connected to the inductor after $t = 0$, after the switch has closed, Figure 5-4(d). The general solution for the inductor current after $t = 0$ is thus

$$i(t) = i_n(t) + i_f = Ke^{-(6/7)t} + \tfrac{5}{2}$$

Of all the possible solutions, for each possible value of K in the general solution,

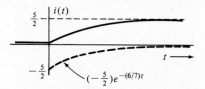

Figure 5-5 Sketch of the inductor current.

only one is the solution to this specific problem. It is the solution for which the inductor current is continuous through the switching time, $t = 0$. Before $t = 0$, the inductor current was $i_f = 0$. Just after $t = 0$, when the general solution above applies, the current must be zero also:

$$i(0) = Ke^0 + \tfrac{5}{2} = K + \tfrac{5}{2} = 0$$

$$K = -\tfrac{5}{2}$$

So the solution for the inductor current in this problem is

$$i(t) = \begin{cases} 0 \text{ A}, & t \le 0 \\ \tfrac{5}{2} - \tfrac{5}{2}e^{-(6/7)t} \text{ A}, & t \ge 0 \end{cases}$$

A sketch of $i(t)$ is shown in Figure 5-5. The inductor current changes from one forced value ($i_f = 0$) to another ($i_f = \tfrac{5}{2}$). It cannot change instantaneously however, because the inductor current must be continuous. The current changes from one forced value to the other along the characteristic exponential curve for the network. Just the right amount of natural current added to the forced current after $t = 0$ makes the inductor current continuous.

5.2.3 Finding Other Network Signals

Once the inductor current is found, any other voltage or current in the network may be easily found. Suppose, in the previous network, it is also desired to find $v(t)$ and $v_1(t)$ in Figure 5-4(a). The voltage $v(t)$ is

$$v(t) = 2\frac{di(t)}{dt} = 2\frac{d}{dt}\begin{cases} 0, & t \le 0 \\ \dfrac{5}{2} - \dfrac{5}{2}e^{-(6/7)t}, & t \ge 0 \end{cases}$$

$$= 2\begin{cases} 0, & t < 0 \\ -\left(\dfrac{6}{7}\right)\left(-\dfrac{5}{2}\right)e^{-(6/7)t}, & t > 0 \end{cases} = \begin{cases} 0 \text{ V}, & t < 0 \\ \dfrac{30}{7}e^{-(6/7)t} \text{ V}, & t > 0 \end{cases}$$

A sketch of $v(t)$ is shown in Figure 5-6(a). The inductor current must be continuous, but other network voltages and currents, such as this one, need not be continuous.

Before $t = 0$, $v_1 = 0$ since, with the switch open, there is no current through the 4-Ω resistor. After $t = 0$,

$$v_1(t) = 10 - v(t) = 10 - \tfrac{30}{7}e^{-(6/7)t} \text{ V}$$

as sketched in Figure 5-6(b).

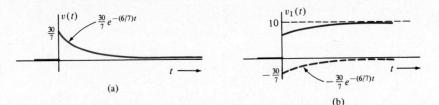

Figure 5-6 Other switched network signals.

An alternative solution procedure to using voltage-current relations, as above, is to use the substitution theorem: Replace the inductor with a current source with source function the known inductor current. All network voltages and currents are unchanged, and the resulting source-resistor network may be solved for any other network voltage or current.

5.2.4 Systematic Solutions

Consider the switched first-order inductive network of Figure 5-7(a). A step-by-step, systematic solution for the indicated signals v_1 and v_2 follows.

1. Before $t = 0$, find the forced inductor current. Since the network source is constant, the forced inductor current will be constant and may be found by replacing the inductor by a short circuit, Figure 5-7(b).

   $i_f = 3$ A, $t < 0$

2. After $t = 0$, find the forced inductor current. The same procedure is used to find the forced inductor current after $t = 0$, Figure 5-7(c):

   $i_f = \frac{12}{5}$ A, $t > 0$

3. After $t = 0$, find the Thévenin resistance looking back from the inductor terminals. See Figure 5-7(d).

   $R_T = \frac{10}{3}$ Ω

4. Form the general solution after $t = 0$ and apply the boundary condition of continuous inductor current. After $t = 0$, the general solution for the inductor current is

   $i(t) = \frac{12}{5} + Ke^{-(R_T/L)t} = \frac{12}{5} + Ke^{-(10/9)t}$

   Before $t = 0$, $i = 3$ so that at time $t = 0$,

   $i(0) = \frac{12}{5} + K = 3; \qquad K = \frac{3}{5}$

   and

   $i(t) = \begin{cases} 3 \text{ A,} & t \leq 0 \\ \frac{12}{5} + \frac{3}{5}e^{-(10/9)t} \text{ A,} & t \geq 0 \end{cases}$

   A sketch of $i(t)$ is given in Figure 5-8(a).

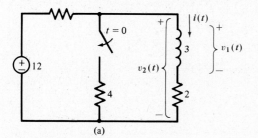

(a)

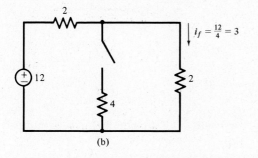

(b)

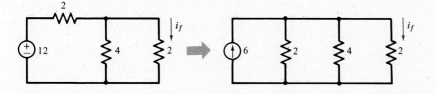

(c)

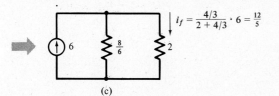

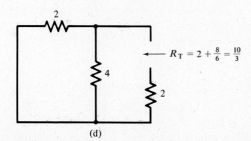

(d)

Figure 5-7 Systematic solution of a switched first-order inductive network.
(a) Original network.
(b) Forced inductor current before t = 0.
(c) Forced inductor current after t = 0.
(d) Thévenin resistance after t = 0.

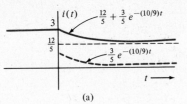

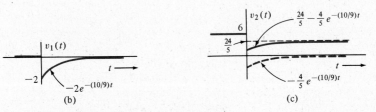

Figure 5-8   Signals in the example network.

5. Using the inductor current, find any other signals of interest in the network.

$$v_1(t) = L\frac{di}{dt} = \begin{cases} 0\text{ V}, & t < 0 \\ 3\left(-\frac{10}{9}\right)\left(\frac{3}{5}\right)e^{-(10/9)t} = -2e^{-(10/9)t}\text{ V}, & t > 0 \end{cases}$$

This signal is sketched in Figure 5-8(b).

$$v_2(t) = 3\frac{di}{dt} + 2i$$

$$= \begin{cases} 6\text{ V}, & t < 0 \\ -2e^{-(10/9)t} + 2\left(\frac{12}{5} + \frac{3}{5}e^{-(10/9)t}\right) = \frac{24}{5} - \frac{4}{5}e^{-(10/9)t}\text{ V}, & t > 0 \end{cases}$$

A sketch is shown in Figure 5-8(c).

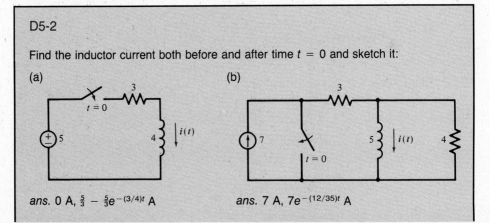

D5-2

Find the inductor current both before and after time $t = 0$ and sketch it:

(a)   (b)

ans. 0 A, $\frac{5}{3} - \frac{5}{3}e^{-(3/4)t}$ A   ans. 7 A, $7e^{-(12/35)t}$ A

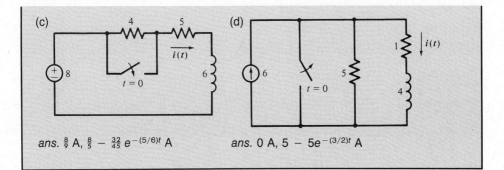

(c)

ans. $\frac{8}{9}$ A, $\frac{8}{5} - \frac{32}{45} e^{-(5/6)t}$ A

(d)

ans. 0 A, $5 - 5e^{-(3/2)t}$ A

D5-3

Find the indicated signals before and after time $t = 0$. Sketch both the inductor current and the signal of interest:

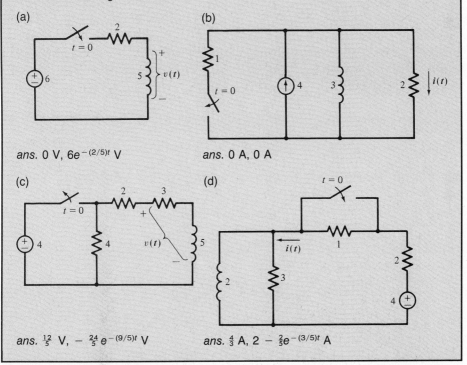

(a)

ans. 0 V, $6e^{-(2/5)t}$ V

(b)

ans. 0 A, 0 A

(c)

ans. $\frac{12}{5}$ V, $-\frac{24}{5} e^{-(9/5)t}$ V

(d)

ans. $\frac{4}{3}$ A, $2 - \frac{2}{5}e^{-(3/5)t}$ A

SWITCHED NETWORKS CONTAINING A SINGLE INDUCTOR

The natural component of the inductor current in a first-order network is of the form

$$i_n(t) = Ke^{-(R_T/L)t}$$

where R_T is the Thévenin resistance of all of the source-resistor network connected to the inductor.

The forced component of the inductor current due to *constant* sources may be found by replacing the inductor by a short circuit and solving for the corresponding current.

In switched networks with a single switching time, it is desired to find the forced network response prior to the switching and both forced and natural components of the response after the switching. The specific solution for the inductor current after the switching is found from the general solution by applying the boundary condition of continuous inductor current.

5.3 THE UNIT STEP FUNCTION

The unit step function is defined as follows:

$$u(t) = \begin{cases} 0, & t < 0 \\ 1, & t \geq 0 \end{cases}$$

Its graph is shown in Figure 5-9(a).

The unit step function offers a convenient alternative to switches in switched network problems, particularly in complicated situations. A switched-on voltage source is easily represented with this notation, Figure 5-9(b), as is a switched-on current source, Figure 5-9(c).

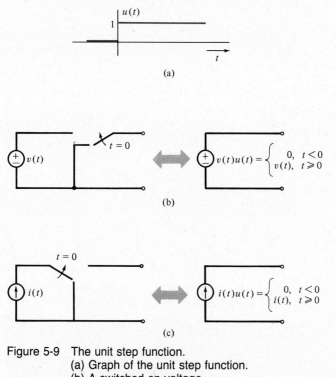

(a)

(b)

(c)

Figure 5-9   The unit step function.
  (a) Graph of the unit step function.
  (b) A switched-on voltage.
  (c) A switched-on current.

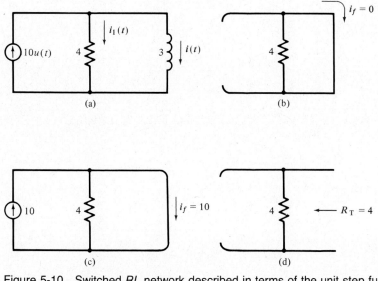

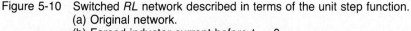

Figure 5-10   Switched *RL* network described in terms of the unit step function.
(a) Original network.
(b) Forced inductor current before $t = 0$.
(c) Forced inductor current after $t = 0$.
(d) Thévenin resistance after $t = 0$.

The following is a step-by-step example of the solution of a switched network where the switching is described by the step function. The network is given in Figure 5-10(a).

1. Before $t = 0$, find the forced inductor current. The current source function is zero prior to $t = 0$, Figure 5-10(b), giving $i_f = 0$.

2. After $t = 0$, find the forced inductor current. Since the network source is constant, the forced inductor current may be found by replacing the inductor by a short circuit, as in Figure 5-10(c), where it is seen that $i_f = 10$.

3. After $t = 0$, find the Thévenin resistance looking back from the inductor terminals. From Figure 5-10(d), $R_T = 4$.

4. Form the general solution after $t = 0$ and apply the boundary condition of continuous inductor current.

$$i(t) = 10 + Ke^{-(4/3)t}, \qquad t \geq 0$$

$$i(0) = 10 + K = 0; \qquad K = -10$$

so

$$i(t) = \begin{cases} 0 \text{ A}, & t \leq 0 \\ 10 - 10e^{-(4/3)t} \text{ A}, & t \geq 0 \end{cases}$$

A sketch of $i(t)$ is given in Figure 5-11(a).

5. Using the inductor current, find any other signals of interest in the net-

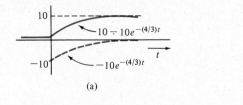

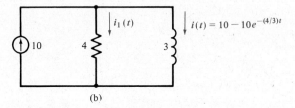

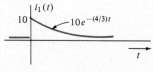

(a)

(b)

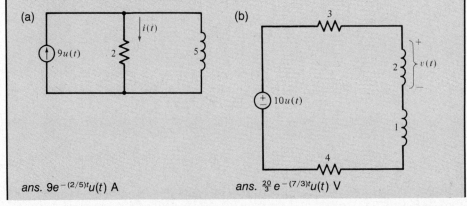

(c)

Figure 5-11 Signals in the switched *RL* network.
  (a) Inductor current.
  (b) Network after *t* = 0.
  (c) Another network signal.

work. Before $t = 0$, $i_1 = 0$. After $t = 0$, the network is as in Figure 5-11(b) and

$$i_1(t) = 10 - i(t) = 10e^{-(4/3)t} \text{ A}$$

This current is sketched in Figure 5-11(c).

D5-4

Find and sketch the indicated signals before and after time $t = 0$. For comparison, also sketch the inductor current.

(a)

(b)

ans. $9e^{-(2/5)t}u(t)$ A

ans. $\frac{20}{3}e^{-(7/3)t}u(t)$ V

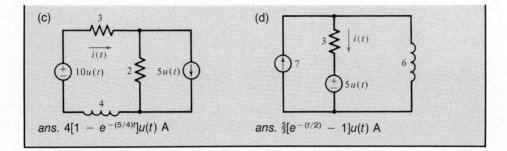

ans. $4[1 - e^{-(5/4)t}]u(t)$ A

ans. $\frac{5}{3}[e^{-(t/2)} - 1]u(t)$ A

THE UNIT STEP FUNCTION

The unit step function is defined as

$$u(t) = \begin{cases} 0, & t < 0 \\ 1, & t \geq 0 \end{cases}$$

It is a convenient alternative to switches in describing switched networks.

5.4 SWITCHED FIRST-ORDER CAPACITIVE NETWORKS

5.4.1 Differential Equation for the Capacitor Voltage

If a first-order network contains a capacitor, the solution may be obtained in a manner similar to that for first-order inductive networks, except that the Norton equivalent should be used to give an equivalent two-node network and a differential equation in terms of the capacitor voltage.

For a network containing sources, resistors, and a single capacitor, construct the Norton equivalent of the source-resistor network connected to the capacitor, as indicated in Figure 5-12. The two networks are equivalents so far as the capacitor voltage is concerned, and that voltage satisfies the two-node differential equation

$$\left(C\frac{d}{dt} + \frac{1}{R_\mathrm{T}}\right)v(t) = i_\mathrm{N}(t)$$

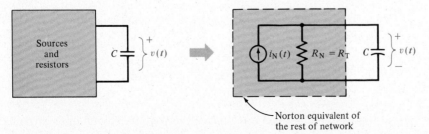

Norton equivalent of the rest of network

Figure 5-12   Norton equivalent used for first-order capacitive network.

Had a Thévenin equivalent been used to convert to an equivalent single-loop problem, the loop-current equation would have involved an integral. The equation should then be differentiated or a change of variables made to convert it to a differential equation. The problem is better formulated in terms of the capacitor voltage because a differential equation results without further manipulation and because the boundary condition applies directly to the capacitor voltage.

The solution of the differential equation for the capacitor voltage in a first-order capacitive network consists of the natural component plus the forced component. The natural component of $v(t)$ is the solution to the homogeneous equation

$$C \frac{dv_n}{dt} + \frac{1}{R_T} v_n = 0$$

Substituting

$$v_n = Ke^{st}$$

$$sC + \frac{1}{R_T} = 0; \qquad s = -\frac{1}{R_T C}$$

$$v_n(t) = Ke^{-(1/R_T C)t}$$

If, as usual, R_T and C are positive, this part of the solution dies out in time, with time constant $R_T C$. This is a general result that applies to any first-order capacitive network. The natural component of the capacitor voltage is as given above, where R_T is the Thévenin (or Norton) resistance of the rest of the network that is connected to the capacitor.

The forced component of the capacitor voltage depends on the driving function $i_N(t)$. If all of the network sources are constant, i_N will be a constant, and

$$v_f = A$$

is constant. Substituting into the differential equation,

$$C \frac{dA}{dt} + \frac{1}{R_T} A = i_N$$

$$\frac{A}{R_T} = i_N$$

$$v_f = A = i_N R_T$$

So far as the *forced* response to *constant* sources is concerned, the capacitor current is zero:

$$C \frac{dv_f}{dt} = C \frac{dA}{dt} = 0$$

The capacitor behaves as an open circuit. To find the forced component of the capacitor voltage due to *constant* sources, replace the capacitor by an open circuit and solve for the corresponding voltage.

For a switched capacitive network, it is the capacitor voltage that must remain continuous since if it were discontinuous, the capacitor current,

$$i = C\frac{dv}{dt}$$

would be infinite.

D5-5

Find the Norton equivalents of the two-terminal networks connected to the capacitor. Then write the two-node differential equation in terms of the capacitor voltage $v(t)$.

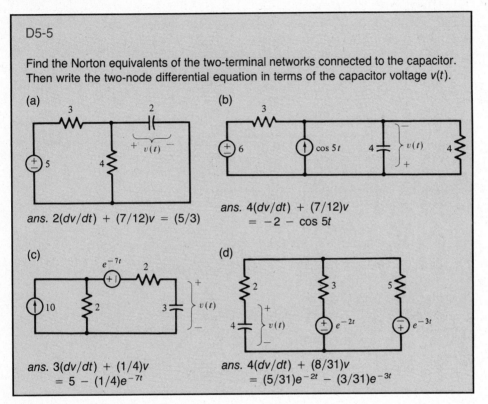

(a)

ans. $2(dv/dt) + (7/12)v = (5/3)$

(b)

ans. $4(dv/dt) + (7/12)v$
$= -2 - \cos 5t$

(c)

ans. $3(dv/dt) + (1/4)v$
$= 5 - (1/4)e^{-7t}$

(d)

ans. $4(dv/dt) + (8/31)v$
$= (5/31)e^{-2t} - (3/31)e^{-3t}$

5.4.2 Systematic Solutions

Consider the first-order switched capacitive network of Figure 5-13(a). A step-by-step solution is as follows.

1. Before $t = 0$, find the forced capacitor voltage. Since the network source is constant, the forced capacitor voltage will be constant and may be found by replacing the capacitor by an open circuit, Figure 5-13(b):

    $v_f = \frac{15}{7}$ V

    before $t = 0$.

2. After $t = 0$, find the forced capacitor voltage. As shown in Figure 5-13(c),

    $v_f = 5$ V

    after $t = 0$.

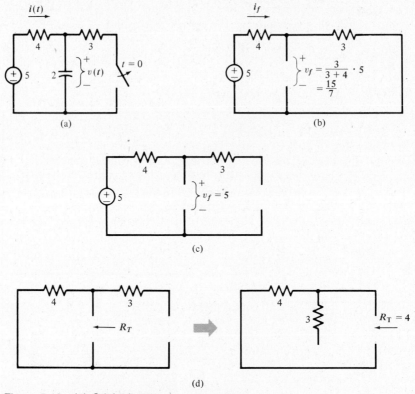

Figure 5-13 (a) Original network.
(b) Forced capacitor voltage before $t = 0$.
(c) Forced capacitor voltage after $t = 0$.
(d) Thévenin resistance after $t = 0$.

3. After $t = 0$, find the Thévenin resistance looking back from the capacitor terminals.

$$R_T = 4\ \Omega$$

as in Figure 5-13(d).

4. Form the general solution after $t = 0$ and apply the boundary condition of continuous capacitor voltage. After $t = 0$, the general solution for the capacitor voltage is

$$v(t) = 5 + Ke^{-(1/R_T C)t} = 5 + Ke^{-(1/8)t}$$

Before $t = 0$, $v = 15/7$, so that at time $t = 0$,

$$v(0) = 5 + K = \tfrac{15}{7}; \qquad K = -\tfrac{20}{7}$$

and

$$v(t) = \begin{cases} \tfrac{15}{7}\ \text{V}, & t \le 0 \\ 5 - (\tfrac{20}{7})e^{-(1/8)t}\ \text{V}, & t \ge 0 \end{cases}$$

A sketch of $v(t)$ is shown in Figure 5-14(a).

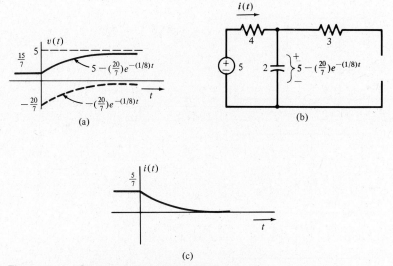

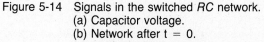

(c)

Figure 5-14   Signals in the switched *RC* network.
(a) Capacitor voltage.
(b) Network after t = 0.
(c) Another network signal.

5. Using the capacitor voltage, find any other signals of interest in the network. Before $t = 0$,

$$i_f = \tfrac{5}{7} \text{ A}$$

from Figure 5-14(b). After $t = 0$, the network is as shown in Figure
5-14(c), and

$$i(t) = \frac{5 - 5 + (20/7)e^{-(1/8)t}}{4} = \left(\frac{5}{7}\right)e^{-(1/8)t} \text{ A}$$

A sketch of $i(t)$ is shown in Figure 5-14(c).

Consider the first-order switched capacitive network of Figure 5-15(a), which
is described in terms of the unit step function.

1. Before $t = 0$, find the forced capacitor voltage. This calculation is done in
Figure 5-15(b), where it is seen that

$$v_f = 6 \text{ V}$$

2. After $t = 0$, find the forced capacitor voltage. Using equivalent circuits,
Figure 5-15(c),

$$(\tfrac{1}{2} + \tfrac{1}{3})v_f = 5 - 6$$

$$v_f = -\tfrac{6}{5} \text{ V}$$

3. After $t = 0$, find the Thévenin resistance looking back from the capacitor
terminals. From Figure 5-15(d),

$$R_T = \tfrac{6}{5} \ \Omega$$

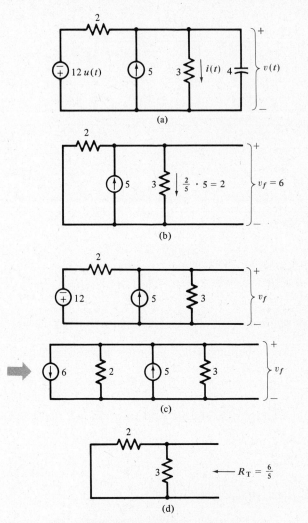

Figure 5-15 (a) Original network.
(b) Forced capacitor voltage before $t = 0$.
(c) Forced capacitor voltage after $t = 0$.
(d) Thévenin resistance after $t = 0$.

4. Form the general solution after $t = 0$ and apply the boundary condition of continuous capacitor voltage.

$$v(t) = -\tfrac{6}{5} + Ke^{-(1/R_TC)t} = -\tfrac{6}{5} + Ke^{-(5/24)t}$$

$$v(0) = -\tfrac{6}{5} + K = 6; \qquad K = \tfrac{36}{5}$$

$$v(t) = \begin{cases} 6\text{ V}, & t \le 0 \\ (-\tfrac{6}{5}) + (\tfrac{36}{5})e^{-(5/24)t}\text{ V}, & t \ge 0 \end{cases}$$

A sketch of the capacitor voltage is shown in Figure 5-16.

5. Using the capacitor voltage, find any other signals of interest in the network.

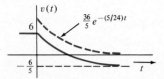

Figure 5-16 Sketch of the capacitor voltage.

$$i(t) = \frac{v(t)}{3} = \begin{cases} 2 \text{ A}, & t \leq 0 \\ \left(-\dfrac{2}{5}\right) + \left(\dfrac{12}{5}\right)e^{-(5/24)t} \text{ A}, & t \geq 0 \end{cases}$$

D5-6

Find the capacitor voltage both before and after time $t = 0$ and sketch it:

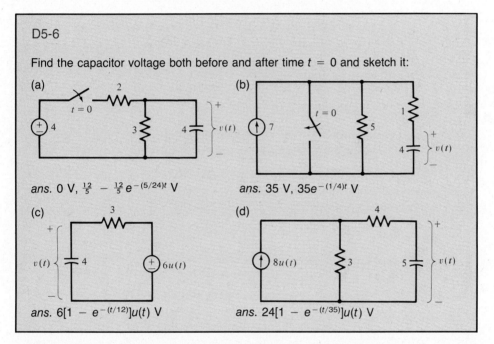

(a)

ans. 0 V, $\frac{12}{5} - \frac{12}{5}e^{-(5/24)t}$ V

(b)

ans. 35 V, $35e^{-(1/4)t}$ V

(c)

ans. $6[1 - e^{-(t/12)}]u(t)$ V

(d)

ans. $24[1 - e^{-(t/35)}]u(t)$ V

D5-7

Find and sketch the indicated signals before and after time $t = 0$. For comparison, sketch the capacitor voltage also:

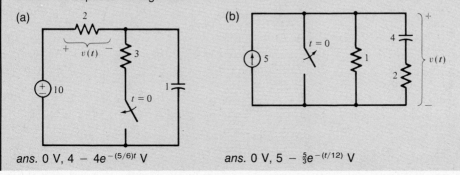

(a)

ans. 0 V, $4 - 4e^{-(5/6)t}$ V

(b)

ans. 0 V, $5 - \frac{5}{3}e^{-(t/12)}$ V

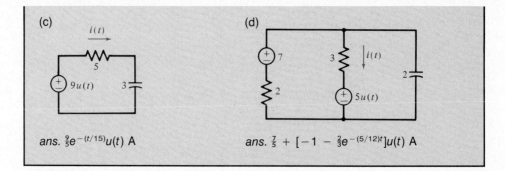

(c)

ans. $\frac{9}{5}e^{-(t/15)}u(t)$ A

(d)

ans. $\frac{7}{5} + [-1 - \frac{2}{3}e^{-(5/12)t}]u(t)$ A

SWITCHED NETWORKS CONTAINING A SINGLE CAPACITOR

The natural component of the capacitor voltage in a first-order network is of the form

$$v_n(t) = Ke^{-(1/R_TC)t}$$

where R_T is the Thévenin resistance of all of the source-resistor network connected to the capacitor.

The forced component of the capacitor voltage due to *constant* sources may be found by replacing the capacitor by an open circuit and solving for the corresponding voltage.

In switched networks, the specific solution for the capacitor voltage after the switching may be found from the general solution by applying the boundary condition of continuous capacitor voltage.

5.5 SERIES *RLC* NETWORKS

5.5.1 Differential Equation for the Capacitor Voltage

The series *RLC* circuit is diagrammed in Figure 5-17. It contains two energy-storing elements, the inductor and the capacitor, and so is described by a second-order integrodifferential equation. The loop equation for $i(t)$ is

$$L\frac{di}{dt} + Ri + \frac{1}{C}\int_{-\infty}^{t} i\, dt = v(t)$$

If this equation is rewritten in terms of the capacitor voltage,

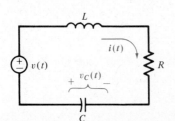

Figure 5-17 Series *RLC* network.

$$v_C(t) = \frac{1}{C} \int_{-\infty}^{t} i \, dt$$

a differential equation, without integrals, results. The other network voltages and currents are easily expressed in terms of the capacitor voltage, since

$$i(t) = C \frac{dv_C}{dt}$$

In terms of $v_C(t)$, the equation is

$$LC \frac{d^2v_C}{dt^2} + RC \frac{dv_C}{dt} + v_C = v(t)$$

The boundary conditions for this network are that the capacitor voltage, v_C, and the inductor current,

$$i(t) = C \frac{dv_C}{dt}$$

thus dv_C/dt must be continuous.

The natural components of the voltages and current in a series *RLC* network are governed by the characteristic equation

$$LCs^2 + RCs + 1 = 0$$

which has characteristic roots

$$s_1, s_2 = \frac{-RC \pm \sqrt{R^2C^2 - 4LC}}{2}$$

If

$$4L < R^2C$$

the roots of the network natural behavior are real and distinct. The network's natural response consists of two real exponential terms,

$$v_{C_n}(t) = K_1 e^{s_1 t} + K_2 e^{s_2 t}$$

and is then said to be *overdamped*.

When

$$4L > R^2C$$

the characteristic roots are complex numbers, and the network response is termed *underdamped*. The corresponding natural behavior consists of a decaying exponential times a sinusoid, a damped oscillation:

$$v_{C_n}(t) = K_1 e^{s_1 t} + K_2 e^{s_2 t} = K_1 e^{(\alpha + j\beta)t} + K_2 e^{(\alpha - j\beta)t}$$

$$= e^{\alpha t}(D \cos \beta t + E \sin \beta t)$$

The case

$$4L = R^2C$$

is termed *critical damping* and is the borderline between exponential and oscillatory behavior, where both characteristic roots are identical:

$$s_1 = s_2 = \frac{-RC}{2}$$

The natural response of the network then consists of the sum of an exponential and a *t*-times-exponential term:

$$v_{C_n}(t) = K_1 e^{s_1 t} + K_2 t e^{s_1 t}$$

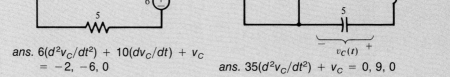

D5-8

Write (but do not solve) differential equations for the capacitor voltage, $v_C(t)$, after time $t = 0$. Also find the boundary conditions $v_C(0)$ and $(dv_C/dt)_{t=0}$:

(a)

ans. $6(d^2v_C/dt^2) + 8(dv_C/dt) + v_C$
= 5, 0, 0

(b)

ans. $6(d^2v_C/dt^2) + 15(dv_C/dt) + v_C$
= 10, 10, 0

(c)

ans. $6(d^2v_C/dt^2) + 10(dv_C/dt) + v_C$
= −2, −6, 0

(d)

ans. $35(d^2v_C/dt^2) + v_C = 0, 9, 0$

5.5.2 Real Roots and Overdamped Response

Consider the specific network of Figure 5-18. Before time $t = 0$, the source voltage is zero, and all voltages and currents are zero. After $t = 0$, the network is governed by

$$4\frac{di}{dt} + 3i + \frac{1}{2}\int_{-\infty}^{t} i\, dt = 5$$

Eliminating the running integral by dealing with the capacitor voltage,

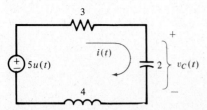

Figure 5-18 Example series *RLC* network.

$$8\frac{d^2v_C}{dt^2} + 6\frac{dv_C}{dt} + v_C = 5$$

The solution of this network problem is thus formulated mathematically as the solution of the above differential equation with the boundary conditions

$$v_C(0) = 0$$

$$\left.\frac{dv_C}{dt}\right|_{t=0} = 0$$

For this network, the homogeneous equation is

$$8\frac{d^2v_{C_n}}{dt^2} + 6\frac{dv_{C_n}}{dt} + v_{C_n} = 0$$

for which the characteristic equation is

$$8s^2 + 6s + 1 = 0$$

$$s_1, s_2 = \frac{-6 \pm \sqrt{36 - 32}}{16} = -\frac{1}{4}, \ -\frac{1}{2}$$

The natural component of the capacitor voltage is therefore of the form

$$v_{C_n}(t) = K_1 e^{-(1/4)t} + K_2 e^{-(1/2)t}$$

Substituting a constant forced solution

$$v_{C_f} = A$$

gives

$$8\frac{d^2}{dt^2}(A) + 6\frac{d}{dt}(A) + (A) = 5$$

$$A = 5 = v_{C_f}$$

so that the general solution for the capacitor voltage is

$$v_C(t) = v_{C_f} + v_{C_n}(t)$$

$$= 5 + K_1 e^{-(1/4)t} + K_2 e^{-(1/2)t}$$

Applying the boundary conditions gives

$$v_C(0) = 5 + K_1 + K_2 = 0$$

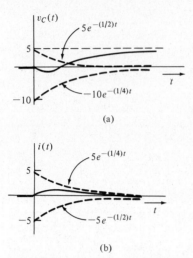

(a)

(b)

Figure 5-19   Sketches of signals in the series *RLC* network.

and

$$\frac{dv_C}{dt}\bigg|_{t=0} = -\frac{1}{4}K_1 - \frac{1}{2}K_2 = 0$$

which may be solved for K_1 and K_2:

$$\begin{cases} K_1 + K_2 = -5 \\ -\frac{1}{4}K_1 - \frac{1}{2}K_2 = 0 \end{cases}$$

$$K_1 = \frac{\begin{vmatrix} -5 & 1 \\ 0 & -\frac{1}{2} \end{vmatrix}}{\begin{vmatrix} 1 & 1 \\ -\frac{1}{4} & -\frac{1}{2} \end{vmatrix}} = \frac{5/2}{-1/4} = -10$$

$$K_2 = \frac{\begin{vmatrix} 1 & -5 \\ -\frac{1}{4} & 0 \end{vmatrix}}{-\frac{1}{4}} = 5$$

Hence the solution for the capacitor voltage after $t = 0$ in this network is

$$v_C(t) = 5 - 10e^{-(1/4)t} + 5e^{-(1/2)t} \text{ V}, \qquad t \geq 0$$

which is sketched in Figure 5-19(a).

Other network signals are easily found from $v_C(t)$. For example,

$$i(t) = C\frac{dv_C}{dt} = 2\frac{dv_C}{dt}$$

$$= 5e^{-(1/4)t} - 5e^{-(1/2)t} \text{ A}$$

after $t = 0$. This signal is sketched in Figure 5-19(b).

D5-9

Find and sketch the indicated signals before and after time $t = 0$. There are two signals of interest in each network:

(a)

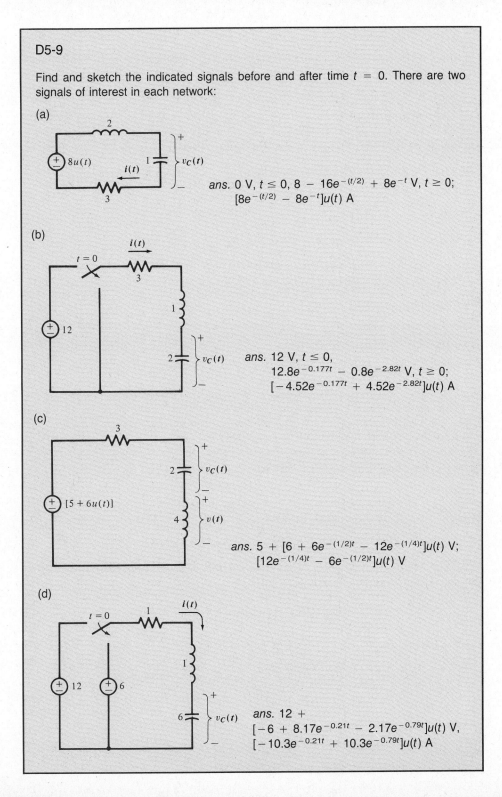

ans. 0 V, $t \leq 0$, $8 - 16e^{-(t/2)} + 8e^{-t}$ V, $t \geq 0$;
$[8e^{-(t/2)} - 8e^{-t}]u(t)$ A

(b)

ans. 12 V, $t \leq 0$,
$12.8e^{-0.177t} - 0.8e^{-2.82t}$ V, $t \geq 0$;
$[-4.52e^{-0.177t} + 4.52e^{-2.82t}]u(t)$ A

(c)

ans. $5 + [6 + 6e^{-(1/2)t} - 12e^{-(1/4)t}]u(t)$ V;
$[12e^{-(1/4)t} - 6e^{-(1/2)t}]u(t)$ V

(d)

ans. $12 +$
$[-6 + 8.17e^{-0.21t} - 2.17e^{-0.79t}]u(t)$ V,
$[-10.3e^{-0.21t} + 10.3e^{-0.79t}]u(t)$ A

5.5.3 Repeated Roots and Critically Damped Response

Consider the series RLC network of Figure 5-20. Before time $t = 0$, the inductor current and capacitor voltage are zero, so

$$i(0) = \frac{1}{16}\frac{dv_C}{dt}\bigg|_{t=0} = 0$$

$$v_C(0) = 0$$

After $t = 0$, the network loop equation is

$$\frac{di}{dt} + 8i + 16\int_{-\infty}^{t} i \, dt = 4$$

In terms of the capacitor voltage,

$$i = \frac{1}{16}\frac{dv_C}{dt}, \qquad \frac{di}{dt} = \frac{1}{16}\frac{d^2v_C}{dt^2}$$

this equation becomes

$$\frac{1}{16}\frac{d^2v_C}{dt^2} + \frac{1}{2}\frac{dv_C}{dt} + v_C = 4$$

The homogeneous equation for $v_C(t)$ is

$$\frac{1}{16}\frac{d^2v_{C_n}}{dt^2} + \frac{1}{2}\frac{dv_{C_n}}{dt} + v_{C_n} = 0$$

and the corresponding characteristic equation is

$$\frac{1}{16}(s^2 + 8s + 16) = \frac{1}{16}(s + 4)^2 = 0$$

which has the characteristic root $s = -4$ repeated. The natural component of the solution for $v_C(t)$ is then of the form

$$v_{C_n}(t) = K_1 e^{-4t} + K_2 t e^{-4t}$$

The forced component of v_C is found by substituting a constant solution (since the driving function is a constant)

$$v_{C_f} = A$$

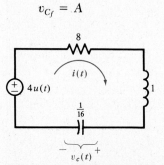

Figure 5-20   Critically damped series RLC network.

into the entire equation:

$$\frac{1}{16}(0) + \frac{1}{2}(0) + A = 4$$

$$v_{C_f} = A = 4$$

The general solution for the capacitor voltage is then

$$v_C(t) = v_{C_f} + v_{C_n}(t) = 4 + K_1 e^{-4t} + K_2 t e^{-4t}$$

Applying the boundary conditions gives

$$v_C(0) = 4 + K_1 = 0; \qquad K_1 = -4$$

$$\frac{dv_C}{dt} = -4K_1 e^{-4t} + K_2 e^{-4t} - 4K_2 t e^{-4t}$$

$$\left.\frac{dv_C}{dt}\right|_{t=0} = -4K_1 + K_2 = 0; \qquad K_2 = -16$$

so that

$$v_C(t) = 4 - 4e^{-4t} - 16te^{-4t} \text{ V}$$

The capacitor voltage and the mesh current

$$i(t) = \frac{1}{16}\frac{dv_C}{dt} = 4te^{-4t} \text{ A}$$

are sketched after $t = 0$ in Figure 5-21.

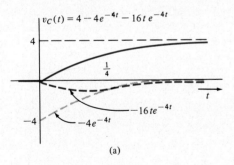

(a)

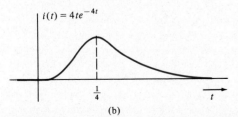

(b)

Figure 5-21   Capacitor voltage and mesh current in the critically damped series *RLC* network.

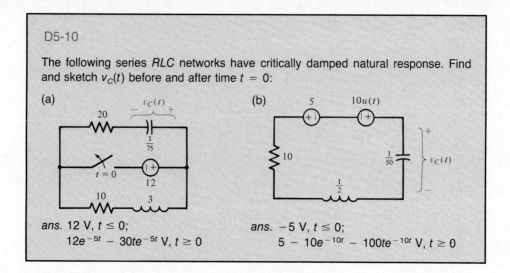

5.5.4 Complex Roots and Underdamped Response

The switched series *RLC* network of Figure 5-22(a) is described by a second-order differential equation with complex characteristic roots. Its natural response is thus underdamped (or oscillatory). After time $t = 0$, the mesh equation for $i(t)$ is

$$\frac{di}{dt} + 2i + 5 \int_{-\infty}^{t} i\, dt = 10$$

In terms of the capacitor voltage, $v_C(t)$, this equation is

$$\frac{1}{5}\frac{d^2 v_C}{dt^2} + \frac{2}{5}\frac{dv_C}{dt} + v_C = 10$$

Continuity of the capacitor voltage requires that

$$v_C(0) = 0$$

Before time $t = 0$, the network is as sketched in Figure 5-22(b). In view of the constant source, the forced inductor current before $t = 0$ may be found by replacing

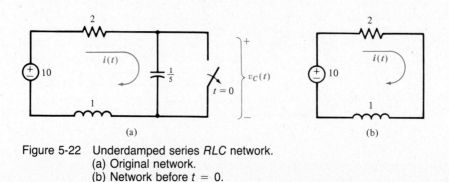

(a)   (b)

Figure 5-22 Underdamped series *RLC* network.
(a) Original network.
(b) Network before $t = 0$.

the inductor by a short circuit. It is

$$i = \frac{10}{2} = 5$$

Inductor current continuity then requires that the solution after $t = 0$ have

$$i(0) = 5$$

In terms of the capacitor voltage, this means that

$$i = C \frac{dv_C}{dt}$$

$$\frac{1}{5} \frac{dv_C}{dt}\bigg|_{t=0} = 5 \quad \text{or} \quad \frac{dv_C}{dt}\bigg|_{t=0} = 25$$

Proceeding now to the solution of this network's differential equation with the stated boundary conditions, the homogeneous equation is

$$\frac{1}{5} \frac{d^2 v_{C_n}}{dt^2} + \frac{2}{5} \frac{dv_{C_n}}{dt} + v_{C_n} = 0$$

which has characteristic equation

$$\tfrac{1}{5}s^2 + \tfrac{2}{5}s + 1 = 0$$

$$s_1, s_2 = \frac{-(2/5) \pm \sqrt{(4/25) - (4/5)}}{2/5} = -1 \pm \sqrt{-4}$$

$$= -1 \pm j2$$

The natural component of the capacitor voltage is of the form

$$v_{C_n} = K_1 e^{s_1 t} + K_2 e^{s_2 t}$$

$$= K_1 e^{(-1+j2)t} + K_2 e^{(-1-j2)t}$$

$$= e^{-t}(K_1 e^{j2t} + K_2 e^{-j2t})$$

$$= e^{-t}(D \cos 2t + E \sin 2t)$$

The forced component of the capacitor voltage, due to the constant source, is found by substituting a constant trial solution into the differential equation, obtaining

$$v_{C_f} = 10$$

The general solution of the equation is thus

$$v_C(t) = v_{C_f} + v_{C_n}(t) = 10 + e^{-t}(D \cos 2t + E \sin 2t)$$

after time $t = 0$.

Applying the boundary conditions,

$$v_C(0) = 10 + D = 0$$

$$D = -10$$

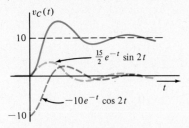

$\frac{15}{2} e^{-t} \sin 2t$

$-10 e^{-t} \cos 2t$

Figure 5-23 Oscillatory *RLC* network capacitor voltage.

and

$$\frac{dv_C}{dt} = -e^{-t}(D \cos 2t + E \sin 2t) + e^{-t}(-2D \sin 2t + 2E \cos 2t)$$

$$\left.\frac{dv_C}{dt}\right|_{t=0} = -D + 2E = 25$$

$$E = \frac{25 - 10}{2} = \frac{15}{2}$$

The solution to this problem,

$$v_C(t) = \begin{cases} 0 \text{ V}, & t \le 0 \\ 10 + e^{-t}(-10 \cos 2t + \frac{15}{2} \sin 2t) \text{ V}, & t \ge 0 \end{cases}$$

$$= [10 + e^{-t}(-10 \cos 2t + \frac{15}{2} \sin 2t)]u(t) \text{ V}$$

is sketched in Figure 5-23.

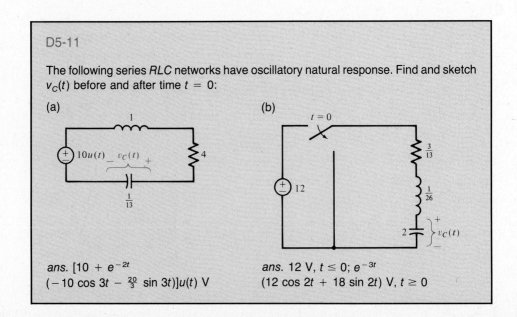

D5-11

The following series *RLC* networks have oscillatory natural response. Find and sketch $v_C(t)$ before and after time $t = 0$:

(a)

(b)

ans. $[10 + e^{-2t}$
$(-10 \cos 3t - \frac{20}{3} \sin 3t)]u(t)$ V

ans. 12 V, $t \le 0$; e^{-3t}
$(12 \cos 2t + 18 \sin 2t)$ V, $t \ge 0$

THE SERIES *RLC* NETWORK

The series *RLC* network has loop equation

$$L \frac{di}{dt} + Ri + \frac{1}{C} \int_{-\infty}^{t} i \, dt = v(t)$$

Rewriting the loop equation in terms of the capacitor voltage,

$$v_C(t) = \frac{1}{C} \int_{-\infty}^{t} i \, dt, \qquad i = C \frac{dv_C}{dt}$$

gives

$$LC \frac{d^2 v_C}{dt^2} + RC \frac{dv_C}{dt} + v_C = v(t)$$

which does not involve an integral.

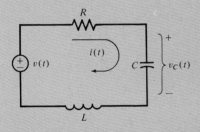

The two boundary conditions are that the capacitor voltage v_C must be continuous and that the inductor current,

$$i(t) = C \frac{dv_C}{dt}$$

must be continuous. In terms of the capacitor voltage, continuous inductor current means dv_C/dt must be continuous.

Second-order network response is classified as overdamped, critically damped, or underdamped, according to whether the characteristic roots of its differential equation are real and distinct, equal, or complex. Critical damping occurs in the series network when

$$R = 2 \sqrt{\frac{L}{C}}$$

The network is overdamped when

$$R > 2 \sqrt{\frac{L}{C}}$$

and it is underdamped (or oscillatory) when

$$R < 2 \sqrt{\frac{L}{C}}$$

5.6 PARALLEL *RLC* NETWORKS

5.6.1 Differential Equation for the Inductor Current

The two-node equation for a parallel *RLC* network. Figure 5-24, is

$$C\frac{dv}{dt} + \frac{1}{R}v + \frac{1}{L}\int_{-\infty}^{t} v\,dt = i(t)$$

If this equation is rewritten in terms of the inductor current,

$$i_L(t) = \frac{1}{L}\int_{-\infty}^{t} v\,dt, \qquad v = L\frac{di_L}{dt}$$

a differential equation results:

$$LC\frac{d^2 i_L}{dt^2} + \frac{L}{R}\frac{di_L}{dt} + i_L = i(t)$$

The boundary conditions for this network are that the inductor current i_L must be continuous and that the capacitor voltage,

$$v = L\frac{di_L}{dt}$$

must be continuous. In terms of the i_L, continuous capacitor voltage means that the slope of i_L, di_L/dt, must be continuous.

The natural response of the parallel *RLC* network is given by the characteristic equation

$$LCs^2 + \frac{L}{R}s + 1 = 0$$

and the characteristic roots

$$s_1, s_2 = \frac{-\dfrac{L}{R} \pm \sqrt{\left(\dfrac{L}{R}\right)^2 - 4LC}}{2LC}$$

If

$$4C < \frac{L}{R^2}$$

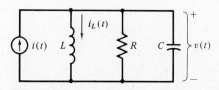

Figure 5-24   Parallel *RLC* network.

the characteristic roots are real and distinct and the network's response is over-damped. If

$$4C = \frac{L}{R^2}$$

the characteristic roots are equal and the network has critically damped response. If

$$4C > \frac{L}{R^2}$$

the characteristic roots are complex and network response is underdamped (or oscillatory).

D5-12

Write (but do not solve) differential equations for the inductor current $i_L(t)$ after time $t = 0$. Also find the boundary conditions $i_L(0)$ and $(di_L/dt)|_{t=0}$:

(a)

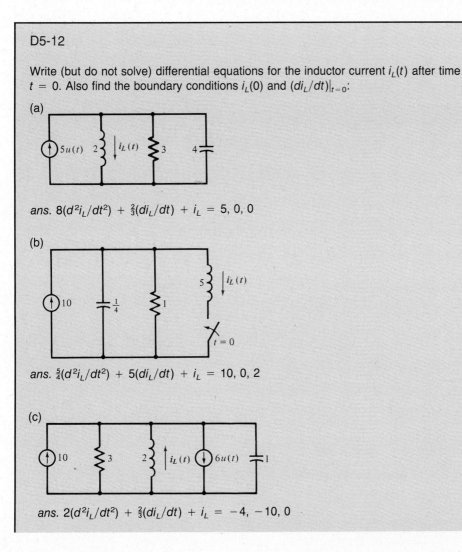

ans. $8(d^2i_L/dt^2) + \frac{2}{3}(di_L/dt) + i_L = 5, 0, 0$

(b)

ans. $\frac{5}{4}(d^2i_L/dt^2) + 5(di_L/dt) + i_L = 10, 0, 2$

(c)

ans. $2(d^2i_L/dt^2) + \frac{2}{3}(di_L/dt) + i_L = -4, -10, 0$

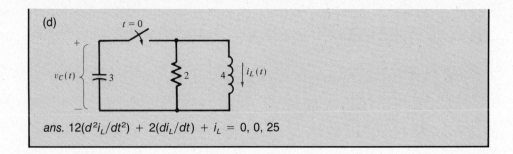

ans. $12(d^2i_L/dt^2) + 2(di_L/dt) + i_L = 0, 0, 25$

5.6.2 Example with Overdamped Response

Consider the specific network of Figure 5-25. Before time $t = 0$, the network is governed by

$$9\frac{d^2i_L}{dt^2} + 48\frac{di_L}{dt} + i_L = 6$$

The forced solution to this equation is

$$i_L = 6$$

The forced capacitor voltage before $t = 0$ is

$$v = L\frac{di_L}{dt} = 0$$

After $t = 0$, the equation becomes

$$9\frac{d^2i_L}{dt^2} + 48\frac{di_L}{dt} + i_L = 13$$

with boundary conditions

$$i_L(0) = 6$$

and

$$v(0) = L\frac{di_L}{dt}\bigg|_{t=0} = 0 \quad \text{or} \quad \frac{di_L}{dt}\bigg|_{t=0} = 0$$

The homogeneous equation is

$$9\frac{d^2i_{L_n}}{dt^2} + 48\frac{di_{L_n}}{dt} + i_{L_n} = 0$$

and has the characteristic equation

$$9s^2 + 48s + 1 = 0$$

$$s_1, s_2 = \frac{-48 \pm \sqrt{2304 - 36}}{18} = -0.021, -5.31$$

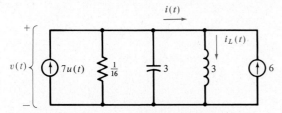

Figure 5-25   Example parallel *RLC* network.

So,

$$i_{L_n}(t) = K_1 e^{-0.021t} + K_2 e^{-5.31t}$$

The forced component of i_L after $t = 0$, due to the constant source, is easily found by substituting

$$i_{L_f} = A$$

into the entire equation, giving

$$i_{L_f} = A = 13$$

so that the general solution for $i_L(t)$ is

$$i_L(t) = 13 + K_1 e^{-0.021t} + K_2 e^{-5.31t}, \qquad t \geq 0$$

Applying the boundary conditions,

$$i_L(0) = 13 + K_1 + K_2 = 6$$

$$\left. \frac{di_L}{dt} \right|_{t=0} = -0.021 K_1 - 5.31 K_2 = 0$$

gives

$$K_1 = \frac{\begin{vmatrix} -7 & 1 \\ 0 & -5.31 \end{vmatrix}}{\begin{vmatrix} 1 & 1 \\ -0.021 & -5.31 \end{vmatrix}} = \frac{37.17}{-5.29} = -7.03$$

$$K_2 = \frac{\begin{vmatrix} 1 & -7 \\ -0.021 & 0 \end{vmatrix}}{-5.29} = 0.03$$

The solution is thus

$$i_L(t) = \begin{cases} 6 \text{ A,} & t \geq 0 \\ 13 - 7.03 e^{-0.021t} + 0.03 e^{-5.31t} \text{ A,} & t \geq 0. \end{cases}$$

A sketch of $i_L(t)$ is shown in Figure 5-26. Other signals of interest are easily found from $i_L(t)$. For example, the signal $i(t)$ in the network diagram is

$$i(t) = i_L(t) - 6 = \begin{cases} 0 \text{ A,} & t \leq 0 \\ 7 - 7.03 e^{-0.021t} + 0.03 e^{-5.31t} \text{ A,} & t \geq 0 \end{cases}$$

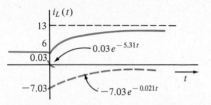

Figure 5-26 Sketch of the parallel *RLC* network inductor current.

D5-13

Find and sketch the indicated signals before and after time $t = 0$. There are two signals of interest in each network:

(a)

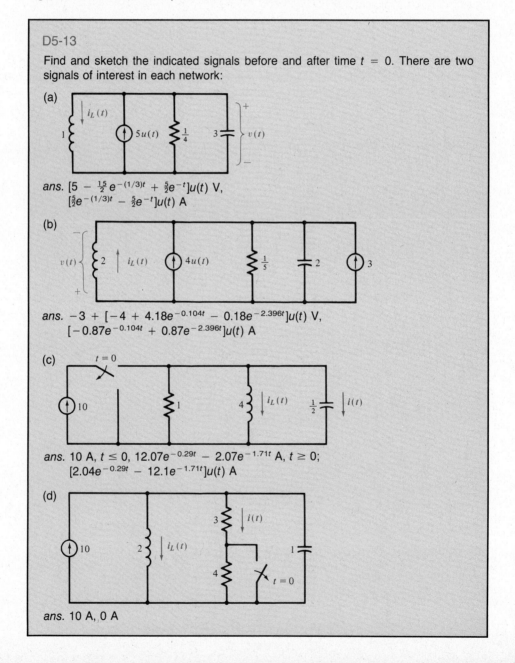

ans. $[5 - \frac{15}{2}e^{-(1/3)t} + \frac{5}{2}e^{-t}]u(t)$ V,
$[\frac{5}{2}e^{-(1/3)t} - \frac{5}{2}e^{-t}]u(t)$ A

(b)

ans. $-3 + [-4 + 4.18e^{-0.104t} - 0.18e^{-2.396t}]u(t)$ V,
$[-0.87e^{-0.104t} + 0.87e^{-2.396t}]u(t)$ A

(c)

ans. 10 A, $t \leq 0$, $12.07e^{-0.29t} - 2.07e^{-1.71t}$ A, $t \geq 0$;
$[2.04e^{-0.29t} - 12.1e^{-1.71t}]u(t)$ A

(d)

ans. 10 A, 0 A

5.6.3 Example with Critical Damping

The parallel *RLC* network of Figure 5-27 is critically damped. Before time $t = 0$, the capacitor voltage equals 5, the source voltage, and the inductor current is zero. After $t = 0$, the nodal equation for the network is

$$\frac{1}{18}\frac{dv}{dt} + \frac{1}{3}v + \frac{1}{2}\int_{-\infty}^{t} v \, dt = 0$$

In terms of the inductor current,

$$v = 2\frac{di_L}{dt}, \qquad \frac{dv}{dt} = 2\frac{d^2i_L}{dt^2}$$

this equation becomes

$$\frac{1}{9}\frac{d^2i_L}{dt^2} + \frac{2}{3}\frac{di_L}{dt} + i_L = 0$$

with the boundary conditions

$$i_L(0) = 0$$

$$2\frac{di_L}{dt}\bigg|_{t=0} = v(0) = 5$$

The differential equation, which is homogeneous, has characteristic equation

$$\tfrac{1}{9}s^2 + \tfrac{2}{3}s + 1 = \tfrac{1}{9}(s + 3)^2 = 0$$

The repeated characteristic root is $s = -3$, so the general solution to the differential equation is

$$i_L(t) = K_1 e^{-3t} + K_2 t e^{-3t}$$

Applying the boundary conditions gives the specific solution for the inductor current after $t = 0$:

$$i_L(0) = K_1 = 0$$

$$\frac{di_L}{dt} = -3K_1 e^{-3t} + K_2 e^{-3t} - 3K_2 t e^{-3t}$$

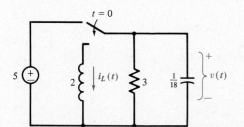

Figure 5-27   Critically damped parallel *RLC* network.

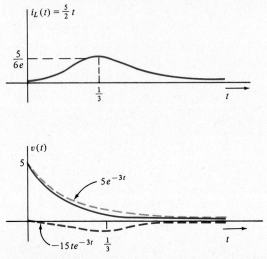

Figure 5-28   Inductor current and node-to-node voltage in the critically damped parallel *RLC* network.

$$\frac{di_L}{dt}\bigg|_{t=0} = -3K_1 + K_2 = \frac{5}{2}; \qquad K_2 = \frac{5}{2}$$

$$i_L(t) = \frac{5}{2} te^{-3t} \text{ A}$$

This function and the node-to-node voltage

$$v(t) = 2\frac{di_L}{dt} = 5e^{-3t} - 15te^{-3t} \text{ V}$$

are sketched after $t = 0$ in Figure 5-28.

D5-14

The parallel *RLC* networks below have critically damped natural response. Find and sketch $i_L(t)$ before and after time $t = 0$:

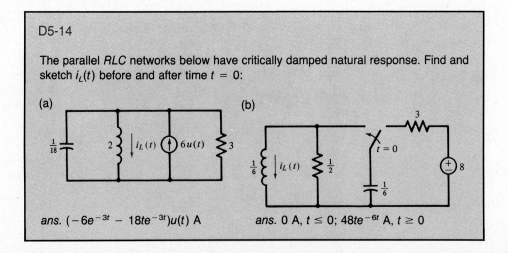

(a)

ans. $(-6e^{-3t} - 18te^{-3t})u(t)$ A

(b)

ans. 0 A, $t \le 0$; $48te^{-6t}$ A, $t \ge 0$

5.6.4 Underdamped Response Example

Whereas the series *RLC* network has oscillatory response for sufficiently small *R*; in the parallel *RLC* network, large *R* gives oscillatory response. The example parallel network of Figure 5-29 has nodal equation

$$\frac{1}{20}\frac{dv}{dt} + \frac{1}{10}v + \frac{1}{2}\int_{-\infty}^{t} v\ dt = 5$$

after time $t = 0$. In terms of the inductor current,

$$v = 2\frac{di_L}{dt}, \qquad \frac{dv}{dt} = 2\frac{d^2i_L}{dt^2}$$

this equation becomes

$$\frac{1}{10}\frac{d^2i_L}{dt^2} + \frac{1}{5}\frac{di_L}{dt} + i_L = 5$$

The boundary conditions are

$$i_L(0) = 0$$

$$2\frac{di_L}{dt}\bigg|_{t=0} = v(0) = 0$$

The natural response component $i_{L_n}(t)$ satisfies the homogeneous equation for which the characteristic equation is

$$\tfrac{1}{10}s^2 + \tfrac{1}{5}s + 1 = \tfrac{1}{10}(s^2 + 2s + 10) = 0$$

The characteristic roots are then

$$s_1, s_2 = \frac{-2 \pm \sqrt{4 - 40}}{2} = -1 \pm j3$$

and the form of the natural response is

$$i_{L_n}(t) = K_1 e^{(-1+j3)t} + K_2 e^{(-1-j3)t}$$

$$= e^{-t}(D \cos 3t + E \sin 3t)$$

Since the driving function is constant, the forced response component is constant:

$$i_{L_f} = A$$

$$\tfrac{1}{10}(0) + \tfrac{1}{5}(0) + A = 5$$

$$i_{L_f} = A = 5$$

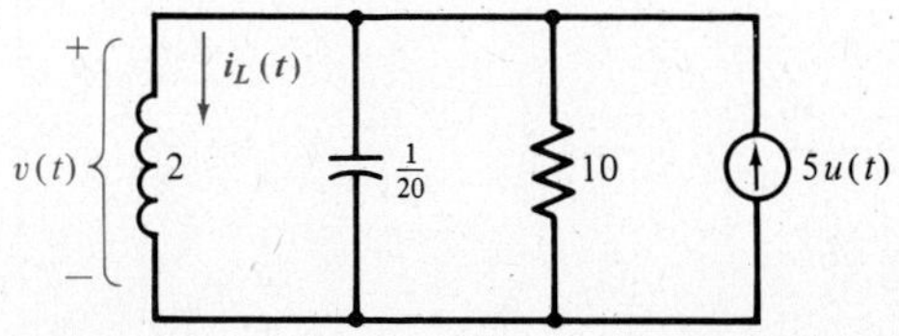

Figure 5-29 Underdamped parallel *RLC* network.

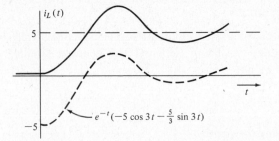

Figure 5-30   Inductor current in the underdamped parallel *RLC* network.

The general solution for the inductor current is thus

$$i_L(t) = 5 + e^{-t}(D \cos 3t + E \sin 3t)$$

Applying the boundary conditions gives

$$i_L(0) = 5 + D = 0; \qquad D = -5$$

$$\frac{di_L}{dt} = -e^{-t}(D \cos 3t + E \sin 3t) + e^{-t}(-3D \sin 3t + 3E \cos 3t)$$

$$\left.\frac{di_L}{dt}\right|_{t=0} = -D + 3E = 0; \qquad E = -\frac{5}{3}$$

so

$$i_L(t) = 5 + e^{-t}\left(-5 \cos 3t - \frac{5}{3} \sin 3t\right) \text{ A}$$

A sketch of $i_L(t)$ is shown in Figure 5-30.

D5-15

The parallel *RLC* networks below have oscillatory natural response. Find and sketch $i_L(t)$ before and after time $t = 0$:

(a)

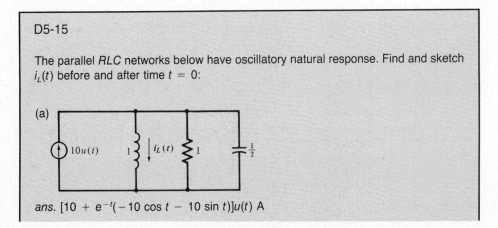

ans. $[10 + e^{-t}(-10 \cos t - 10 \sin t)]u(t)$ A

(b)

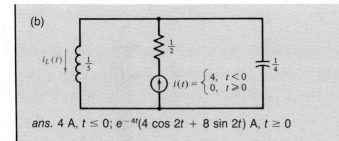

ans. 4 A, $t \leq 0$; $e^{-4t}(4 \cos 2t + 8 \sin 2t)$ A, $t \geq 0$

THE PARALLEL *RLC* NETWORK

The two-node equation for $v(t)$ is

$$C\frac{dv}{dt} + \frac{1}{R}v + \frac{1}{L}\int_{-\infty}^{t} v\, dt = i(t)$$

Rewriting the nodal equation in terms of the inductor current,

$$i_L(t) = \frac{1}{L}\int_{-\infty}^{t} v(t)\, dt, \qquad v = L\frac{di_L}{dt}$$

gives

$$LC\frac{d^2 i_L}{dt^2} + \frac{L}{R}\frac{di_L}{dt} + i_L = i(t)$$

which does not involve an integral.

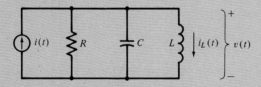

   The two boundary conditions are that the inductor current i_L must be continuous and that the capacitor voltage,

$$v(t) = L\frac{di_L}{dt}$$

must be continuous. In terms of the inductor current, continuous capacitor voltage means di_L/dt must be continuous.
   The network's response is critically damped when

$$R = \frac{1}{2}\sqrt{\frac{L}{C}}$$

It is oscillatory for

$$R > \frac{1}{2}\sqrt{\frac{L}{C}}$$

CHAPTER FIVE PROBLEMS

Basic Problems

Switched First-Order Inductive Networks
1. Find the indicated inductor currents before and after time $t = 0$ and sketch them:

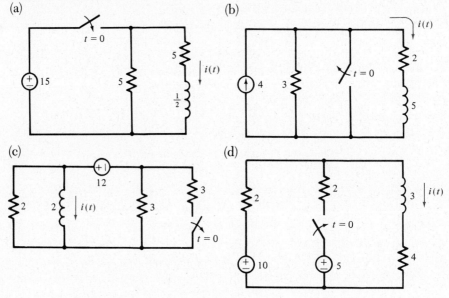

2. Find the indicated signals before and after time $t = 0$ and sketch them:

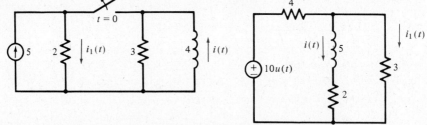

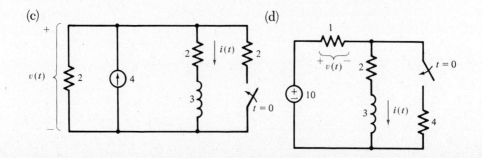

Assortment of switches. A wide variety of types and contact arrangements are shown. (*Photo courtesy of Cutler-Hammer Co.*)

3. Find the indicated voltages and currents before and after $t = 0$ and sketch them:

(a)

(b)

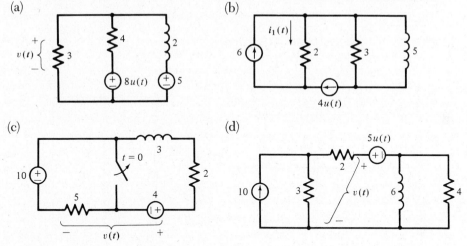

Switched First-Order Capacitive Networks

4. Find the indicated capacitor voltages before and after time $t = 0$ and sketch them:

(a)

(b)

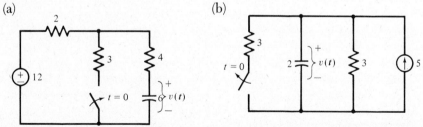

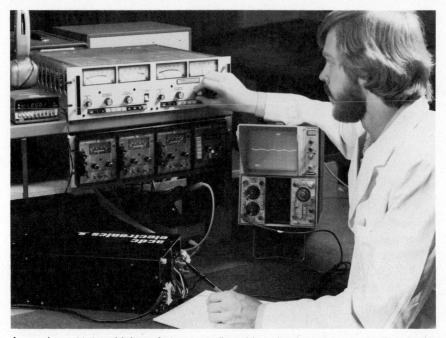

An engineer tests a high-performance adjustable voltage power supply. This instrument maintains a nearly constant voltage at its terminals, for currents up to at least 20 amperes. (*Photo courtesy of ACDC Electronics.*)

(c)

(d)

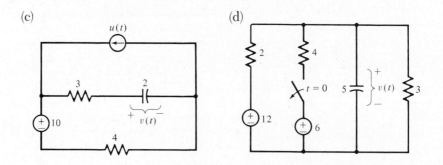

5. Find the indicated signals before and after time $t = 0$ and sketch them:

(a)

(b)

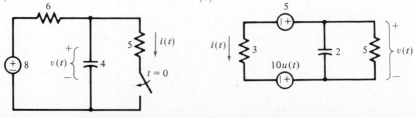

(c)   (d)

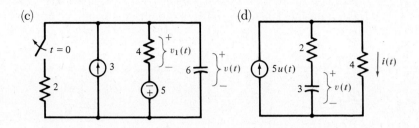

6. Find the indicated voltages and currents before and after $t = 0$ and sketch them:

(a)   (b)

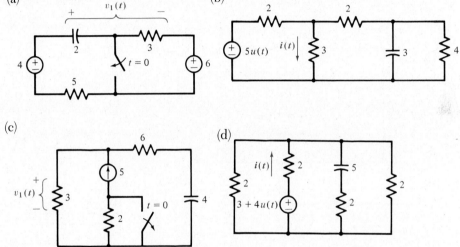

(c)   (d)

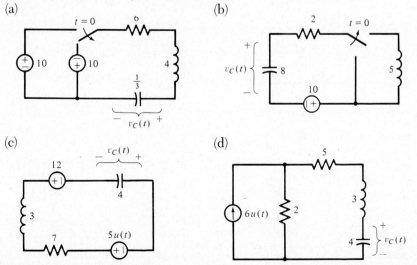

Series RLC Networks

7. Find the differential equation for the capacitor voltage $v_C(t)$ after $t = 0$, and the boundary conditions $v_C(0)$ and $(dv_C/dt)|_{t=0}$:

(a)   (b)

(c)   (d)

8. Find and sketch the indicated capacitor voltages and mesh currents before and after time $t = 0$:

(a)

(b)

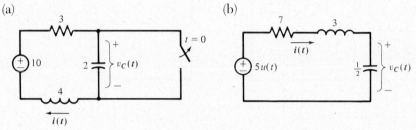

(c)

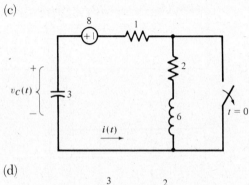

(d)

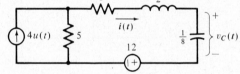

9. Find and sketch the indicated signals before and after time $t = 0$:

(a)

(b)

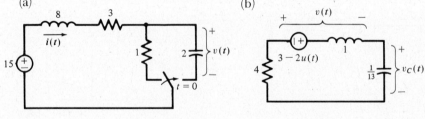

(c)

(d)

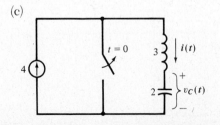

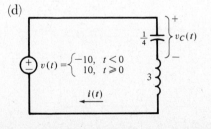

Parallel RLC *Networks*

10. Find the differential equation for the inductor current $i_L(t)$ after $t = 0$, and the boundary conditions $i_L(0)$ and $(di_L/dt)|_{t=0}$:

(a)

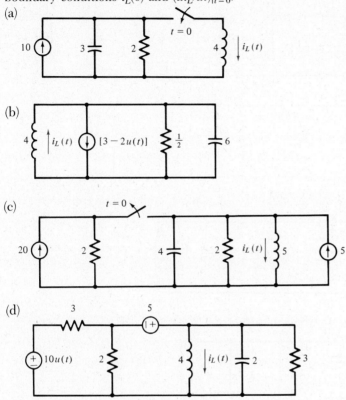

(b)

(c)

(d)

11. Find and sketch the indicated inductor currents and node voltages before and after time $t = 0$:

(a)

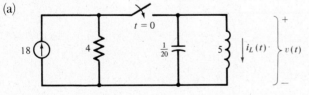

(b)

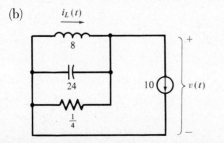

(c)

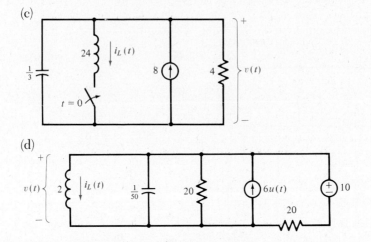

(d)

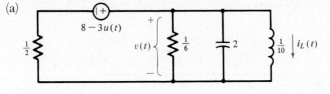

12. Find and sketch the indicated signals before and after time $t = 0$:

(a)

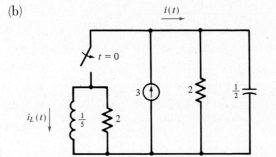

(b)

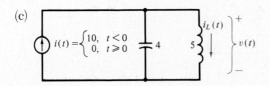

(c)

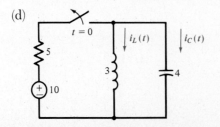

(d)

Practical Problems

Inductors

13. Inductors may be used to produce very high voltages by rapidly changing the current through them. This method is used in most television receivers to develop the 15- to 30-kV accelerating voltage for the cathode-ray tube.

    If the current through a 90-mH inductor is "triangular" shaped as in the sketch, what must the amplitude A be in order to generate an inductor voltage with pulses that have an amplitude 10 kV?

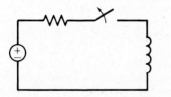

14. Suppose one attempts to interrupt an inductor current with a switch. If the inductor current were changed instantly, an infinite voltage would be produced across the switch terminals. In practice, the voltage developed at the switch will rise rapidly, as the switch disconnects, until the air surrounding the switch contacts becomes ionized. An electric arc at the switch will then develop.

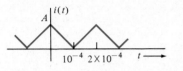

    If the switch contacts are moved far enough apart sufficiently rapidly, the voltage developed across the inductor will cause the insulation around the inductor wires to break down, causing arcing between turns in the inductor winding itself. The inductor current *will* be continuous, even if it must complete its flow outside of the conductors.

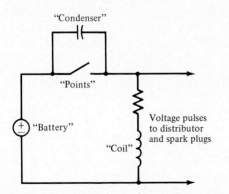

    Automobile ignition circuits, which are capable of generating 5- to 30-kV voltage pulses, interrupt an inductor current produced by the storage battery.

The "points" are a set of cam-driven switch contacts that open each time a voltage pulse is desired. A simplified ignition circuit is shown below.

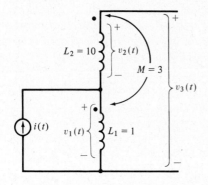

Rather than to allow an arc to occur at the points, a capacitor (often called a "condenser" in this application) is placed across the points to slow slightly the rise and fall of the point voltage so that the arcing will, instead, take place at the spark plug. A faulty "condenser" will allow arcing at the points and cause their rapid destruction. In electronic ignition systems, the switching action is done by a transistor rather than the mechanical "points." Performance of the system is improved by using coupled coils to magnify the voltage still further.

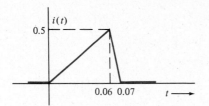

If the current through the coupled inductor L_1 is the given waveform, sketch the open-circuit voltages $v_1(t)$, $v_2(t)$, and $v_3(t)$.

Capacitors

15. One method of measuring the capacitance of large-valued capacitors is to charge them to some convenient voltage, then allow them to discharge through a voltmeter, as in the drawing. With a known voltmeter internal resistance, the time constant of the capacitor discharge $R_{meter}C$, which may be obtained by timing the voltage decay, is used to calculate C.

Resistor R to
limit current
during charging

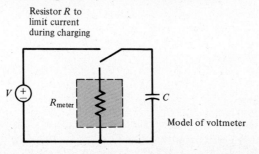

Model of voltmeter

For a voltmeter with 20,000-Ω internal resistance, a certain capacitor voltage decays from 100 V to 84.5 V in 10 s. What is the value of the capacitor?

16. When a capacitor is discharged by short-circuiting it with a switch, the resistance of the connecting wires and of the switch contacts must be taken into account

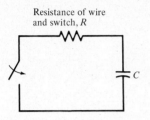

for accurate results. For an initial capacitor voltage V, the current that flows at the instant the switch is closed is V/R, since the capacitor voltage is continuous. This initial current may be very large if R is sufficiently small.

If 1 J of energy is stored in a 10-μF capacitor, what will the magnitude of the initial current be if the capacitor is discharged through 0.3 Ω?

17. Low voltages may be momentarily converted to higher voltages by charging several capacitors in parallel, then connecting the capacitors in series for their discharge. This technique is often used in battery-powered geiger counters and photoflash units, where high voltage pulses, which would be inconvenient to obtain directly from a battery, are needed.

Five initially uncharged 1-μF capacitors are charged for 1 s as shown in (a). What is the voltage across the capacitors? The capacitors are then connected as in (b). What is the voltage v after $\frac{1}{10}$ s of discharge?

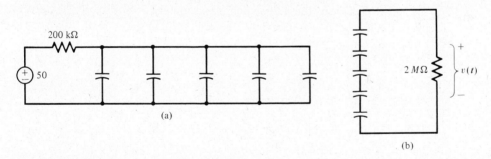

(a)

(b)

Periodic Switching

18. The repetitive square-wave voltage sketched is the source voltage for the *RL* network. Sketch i(t). [It is not required to find i(t) analytically.]

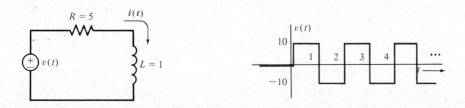

When, as in this case, the network time constant is much smaller than the time between switchings of the square wave, the response becomes essentially a switched RL network response repeated over and over. Sketch $i(t)$ also for $R = 5$, $L = 10$.

19. Sketch the voltage $v_C(t)$.

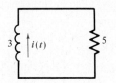

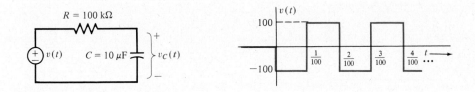

Advanced Problems

Switched Inductive Networks

20. At time $t = 0$, 20 J of energy are stored in the inductor. Find the two possible inductor currents after $t = 0$.

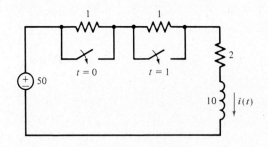

21. For the network shown, the first switching occurs at $t = 0$. Before the natural component of the response dies out, a second switching occurs at time $t = 1$. Find the inductor current $i(t)$.

Unit Step Function

22. Sketch a graph of the function

$$f(t) = 3u(-t) + 4u(t) - 3u(t - 2) - u(t - 3)$$

23. Sketch the function

$$f(t) = u(t^2 - t)$$

Switched Capacitive Networks

24. The voltage across an ideal capacitor, when nothing is connected across it, can be anything. Thus in the network, it is necessary to specify the capacitor voltage prior to $t = 0$ to be able to solve the network. Find $i(t)$ if $v_C(0) = -5$.

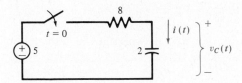

25. A capacitor is to be used to store energy supplied by a voltage generator that is modeled by a Thévenin equivalent. The energy storage might be used in place of storage batteries for an electric vehicle. The capacitor voltage v will eventually reach 100 V.

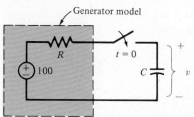

Show that the energy eventually stored in the capacitor equals the net energy that is dissipated in (that is, flows into) the resistance. This result holds regardless of the (positive) values of R and C.

26. The problem of finding how a constant voltage divides across two ideal capacitors is indeterminant. A more accurate model of the physical capacitor includes a large parallel resistance representing the leakage of current between the capac-

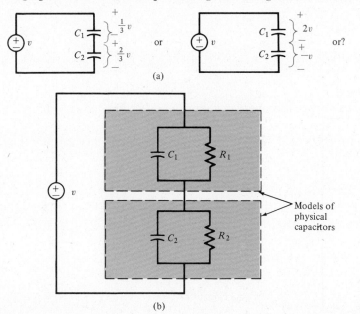

itor plates. Show that with this model, a net constant voltage is *eventually* distributed as

$$v_1 = \frac{R_1}{R_1 + R_2} v$$

RLC *Networks*

27. Find two different switched first-order inductive networks in which the inductor current after $t = 0$ satisfies

$$3\frac{di}{dt} + 4i = 5$$

with the boundary condition

$$i(0) = -3$$

Repeat, finding two different switched capacitive networks for which the capacitor voltage satisfies

$$3\frac{dv}{dt} + 4v = 5$$

after $t = 0$, with the boundary condition

$$v(0) = -3$$

28. The network shown is first-order even though it contains two inductors, because the two inductors in series are equivalent to a single inductor. Find $v(t)$.

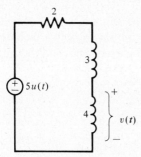

29. Find the indicated signals before and after time $t = 0$ and sketch them:
(a)

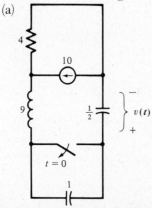

(b)

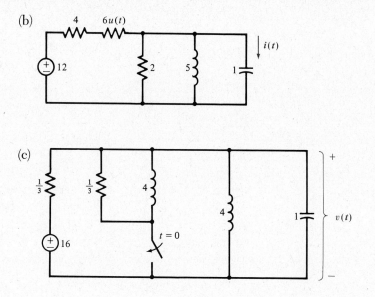

(c)

(d)

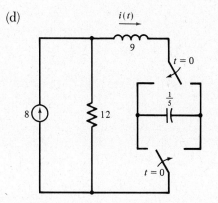

Chapter Six
The Impedance Concept

But of utmost and supreme importance is the proper discussion of and approach to the impedance concept. In this connection we cannot regard transient analysis as an advanced topic to be dealt with later on. Transient analysis *must* precede the discussion of a-c steady state response in order that the true character of the impedance function may be recognized.

Ernst A. Guillemin [1898–1970]
Professor of Electrical Communication,
Massachusetts Institute of Technology
From the Preface to *Introductory
Circuit Theory* (John Wiley & Sons), 1953

6.1 PREVIEW

The key property of exponential signals is now developed: If all network voltages and currents are exponential, varying as e^{st}, the voltage-current ratio for every network element is a constant, and each element is equivalent to a resistor.

A foundation is constructively developed for understanding impedance, phasor methods, frequency response, and, ultimately, the key nature of the Fourier and Laplace transforms. The generalized impedance $Z(s)$ is considered from the beginning, to be specialized in later chapters to $Z(s = j\omega)$ for sinusoidal response.

The relationship between network functions and a network's natural response modes is explored in the final section of the chapter.

When you complete this chapter, you should know—

1. how to find the general solution to linear, time-invariant differential equations with exponential driving functions;
2. what impedance is and the impedances of capacitors, inductors, and coupled inductors;
3. how to find forced exponential network response, using the impedance method;
4. how to superimpose sources to accommodate exponential functions with different time constants;
5. how to solve simple switched networks with exponential sources;
6. what network transfer functions are;
7. the relationship between impedance and network natural behavior.

6.2 SOLUTIONS TO DIFFERENTIAL EQUATIONS WITH EXPONENTIAL DRIVING FUNCTIONS

Except in a very special case, if the driving function to a differential equation is an exponential function, varying as e^{st}, the forced component of the solution of the differential equation is also exponential, with the same value of the exponential constant, s.

For example, consider

$$\frac{d^2y}{dt^2} + 5\frac{dy}{dt} + 4y = 10e^{-2t}$$

Substituting,

$$y(t) = Ae^{-2t}$$

$$(-2)^2Ae^{-2t} + 5(-2)Ae^{-2t} + 4Ae^{-2t} = 10e^{-2t}$$

$$-2Ae^{-2t} = 10e^{-2t}$$

$$A = -5$$

and

$$y_f(t) = -5e^{-2t}$$

satisfies this equation.

The characteristic equation is

$$s^2 + 5s + 4 = (s + 1)(s + 4) = 0$$

so that the natural component of the solution is

$$y_n(t) = K_1e^{-t} + K_2e^{-4t}$$

The general solution of the equation is thus

$$y(t) = y_n(t) + y_f(t) = K_1e^{-t} + K_2e^{-4t} - 5e^{-2t}$$

A second example is as follows. The equation

$$\frac{d^3y}{dt^3} + 5\frac{d^2y}{dt^2} + 6\frac{dy}{dt} = 7e^{4t}$$

has characteristic equation

$$s^3 + 5s^2 + 6s = s(s - 2)(s + 3) = 0$$

so

$$y_n(t) = K_1 + K_2e^{-2t} + K_3e^{-3t}$$

Substituting a trial forced solution

$$y(t) = Ae^{4t}$$

into the entire equation gives

$$4^3Ae^{4t} + 5 \cdot 4^2Ae^{4t} + 6 \cdot 4Ae^{4t} = 7e^{4t}$$

$$168Ae^{4t} = 7e^{4t}$$

$$A = \tfrac{7}{168} = \tfrac{1}{24}$$

so

$$y_f(t) = Ae^{4t} = \tfrac{1}{24}e^{4t}$$

and the general solution is

$$y(t) = y_n(t) + y_f(t) = K_1 + K_2e^{-2t} + K_3e^{-3t} + \tfrac{1}{24}e^{4t}$$

A special case occurs when the driving function has the same time constant as one of the terms in the natural component of the solution. For example, for the equation

$$\frac{d^2y}{dt^2} + 3\frac{dy}{dt} + 2y = 4e^{-t}$$

the characteristic equation is

$$s^2 + 3s + 2 = (s + 1)(s + 2) = 0$$

giving

$$y_n(t) = K_1e^{-t} + K_2e^{-2t}$$

Substituting a trial forced solution

$$y_f(t) = Ae^{-t}$$

into the original equation in an attempt to find a forced solution gives

$$0 = 4e^{-t}$$

A forced solution of the form Ae^{-t} cannot be found in this case.

The forced solution is, instead, of the form

$$y_f(t) = Ate^{-t}$$

Substituting,

$$\frac{d}{dt}(Ate^{-t}) = Ae^{-t} - Ate^{-t}$$

$$\frac{d^2}{dt^2}(Ate^{-t}) = -2Ae^{-t} + Ate^{-t}$$

there results

$$\frac{d^2y_f}{dt^2} + 3\frac{dy_f}{dt} + 2y_f = -2Ae^{-t} + Ate^{-t} + 3(Ae^{-t} - Ate^{-t}) + 2Ate^{-t}$$

$$= 4e^{-t}$$

$$A = 4$$

$$y_f(t) = 4te^{-t}$$

Except in this special case, then, if the driving function of a differential equation varies as e^{st}, the forced solution is of the form Ae^{st}.

D6-1

Find the general solutions of the following differential equations:

(a) $\dfrac{dy}{dt} + 3y = 4e^{5t}$ ans. $\frac{1}{2}e^{5t} + Ke^{-3t}$

(b) $3\dfrac{dy}{dt} + 2y = e^{-4t}$ ans. $-\frac{1}{10}e^{-4t} + Ke^{-(2/3)t}$

(c) $\dfrac{d^2y}{dt^2} + 3\dfrac{dy}{dt} + 2y = 5e^{4t}$ ans. $\frac{1}{6}e^{4t} + K_1e^{-t} + K_2e^{-2t}$

(d) $\dfrac{d^2y}{dt^2} + 5\dfrac{dy}{dt} + 6y = -2e^{-3t}$ ans. $2te^{-3t} + K_1e^{-2t} + K_2e^{-3t}$

DIFFERENTIAL EQUATIONS WITH EXPONENTIAL DRIVING EQUATIONS

Except when the driving function is of the same form as one of the natural behavior terms, the forced solution to a differential equation with an exponential driving function, varying as e^{st}, is exponential, with the same exponential constant, s.

The forced solution, Ae^{st}, may be found by substituting into the equation and solving for A.

6.3 IMPEDANCE

When the network sources are exponential, all with the same exponential constant, the differential equations for each network voltage and current have exponential driving functions. The forced components of every voltage and every current are then exponential, with the same exponential constant as the sources. Since the forced component of every voltage and every current varies as e^{st}, the ratio of forced voltage to forced current in any element is a constant, just as it is for a resistor. Generally the forced exponential voltage-current ratio in an element depends on the exponential constant, s.

The impedance of a two-terminal element is defined as follows:

$$Z(s) = \frac{\text{voltage across the element, } v(t)}{\text{current through the element, } i(t)} \bigg|_{\substack{\text{when } v(t) \text{ and } i(t) \text{ have the sink reference relation} \\ \text{and when } v(t) \text{ and } i(t) \text{ each vary as } e^{st}}}$$

The unit of impedance is the same as that of resistance, the ohm. The inverse of impedance, $1/Z(s)$, is called *admittance* and is given the symbol $Y(s)$.

The impedance of a resistor is just its resistance, R:

$$Z_R = \frac{v(t)}{i(t)} = R$$

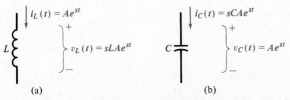

Figure 6-1   Exponential voltages and currents for the inductor and the capacitor.

With the inductor and capacitor, the ratio of sink reference voltage to current is *not* a constant unless both the voltage and current are exponential.

For the inductor, Figure 6-1(a), if the inductor current is

$$i_L(t) = Ae^{st}$$

the sink reference inductor voltage is also exponential and is

$$v_L(t) = L\frac{di_L}{dt} = sLAe^{st}$$

so the impedance of an inductance is

$$Z_L = \frac{v_L(t)}{i_L(t)} = \frac{sLAe^{st}}{Ae^{st}} = sL$$

For the capacitor, Figure 6-1(b), if the capacitor voltage is exponential,

$$v_C(t) = Ae^{st}$$

the sink reference capacitor current is also exponential,

$$i_C(t) = C\frac{dv_C(t)}{dt} = sCAe^{st}$$

The impedance of a capacitor is thus

$$Z_C = \frac{v_C(t)}{i_C(t)} = \frac{Ae^{st}}{sCAe^{st}} = \frac{1}{sC}$$

D6-2

The impedance of a certain element, as a function of the exponential constant s is

$$Z(s) = \frac{s}{2s + 3}$$

If the element voltage and current are both exponential and the current through the element is

$$i(t) = 4e^{-5t} \text{ A}$$

what is the sink reference element voltage?   *ans.* $\frac{20}{7}e^{-5t}$ V

IMPEDANCE

When the sources in a network are all exponential, with the same exponential constant s, the forced components of each voltage and current vary as e^{st}.

The ratio of forced exponential voltage to current in any element is a constant. Thus for forced exponential voltages and currents, every network element has the same voltage-current relation as, and so is equivalent to, a resistor.

This equivalent resistance, the sink reference voltage-current ratio when both the voltage and current vary as e^{st}, is called the *impedance* of the element. The impedance of an inductor is

$$Z_L = sL$$

and the impedance of a capacitor is

$$Z_C = \frac{1}{sC}$$

The inverse of an impedance is termed *admittance*.

6.4 SOLUTIONS FOR FORCED EXPONENTIAL RESPONSE

6.4.1 The Impedance Equivalent

To find the forced component of any voltage or current in a network driven by one or more exponential sources with the same exponential constant s, simply convert the original network to an equivalent network in which the inductors and capacitors are replaced by equivalent resistances that are those element's impedances. The exponential constants of the sources must all be the same, and each of the impedances is evaluated at the value of s of the sources to obtain the equivalent resistance of the element.

Consider the network of Figure 6-2. The impedance of the inductor is $sL = s \cdot 3$ which, for exponential signals of the form of the source e^{2t}, is $sL = 2 \cdot 3$. For signals that vary as e^{2t}, the inductor is equivalent to a 6-Ω resistor. Similarly, the capacitor is equivalent to a resistor of $1/sC = 1/(2 \cdot 4) = \frac{1}{8}$ Ω. The inductor and the capacitor are replaced by their impedances in the equivalent network. All of the

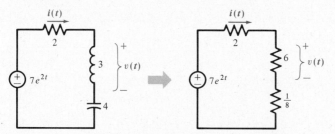

Figure 6-2   Impedance equivalent for network solution.

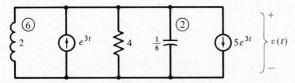

Figure 6-3   Example of using impedance equivalent.

solution methods for source-resistor networks then apply to this equivalent source-resistor network problem.

$$i(t) = \frac{7e^{2t}}{2 + 6 + \frac{1}{8}} = \frac{56e^{2t}}{65} \text{ A}$$

$$v(t) = \frac{42}{2 + 6 + \frac{1}{8}} e^{2t} = \frac{336}{65} e^{2t} \text{ V}$$

Instead of redrawing the network, these equivalent resistances, the element impedances, may be indicated on the original diagram. It is helpful to circle the impedances to distinguish them from the element values, as in the network of Figure 6-3, for which

$$(\tfrac{1}{6} + \tfrac{1}{4} + \tfrac{1}{2})v(t) = e^{3t} - 5e^{3t}$$

$$v(t) = -\tfrac{48}{11}e^{3t} \text{ V}$$

D6-3

Using impedance methods, find the indicated signals:

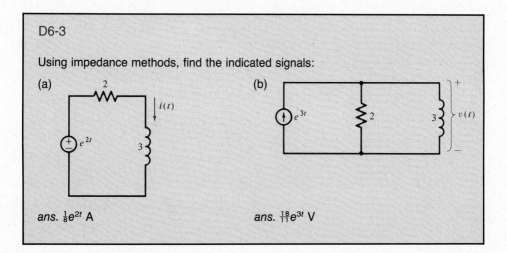

(a)

(b)

ans. $\frac{1}{8}e^{2t}$ A

ans. $\frac{18}{11}e^{3t}$ V

6.4.2 Constant Sources

Constant sources are of the form

$$Ae^{0t} = A$$

for which the exponential constant s is zero, so forced exponential response includes forced constant response as a special case.

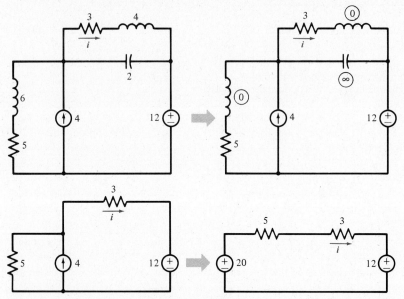

Figure 6-4 Impedance methods for a network with constant sources.

For constant sources, $s = 0$, the impedance of an inductor is $sL = 0$, and the impedance of a capacitor is $1/sC = \infty$. This is to say that the forced response of a network with constant sources may be found by replacing all inductors by short circuits (zero impedance) and all capacitors by open circuits (infinite impedance). This result, which is a special case of the impedance equivalent, was introduced in Chapter 5 in connection with switched networks.

The example network of Figure 6-4 has constant sources. For it,

$$8i = 20 - 12 \qquad i = 1 \text{ A}$$

D6-4

Using impedance methods, find the indicated signals:

(a)

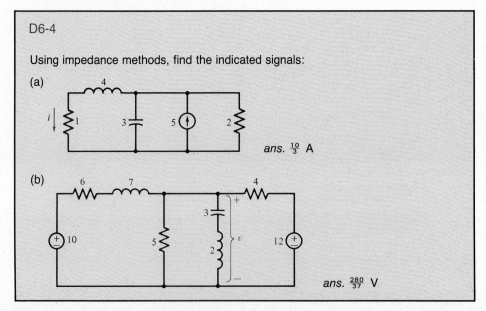

ans. $\frac{10}{3}$ A

(b)

ans. $\frac{280}{37}$ V

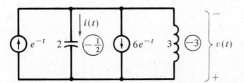

Figure 6-5   Network with negative impedances.

6.4.3 Negative Element Impedances

For sources with negative exponential constants, s, the impedances of inductors and capacitors are negative, but they are handled just as are positive resistors so far as the mathematics is concerned. For the network of Figure 6-5,

$$(-2 - \tfrac{1}{3})v(t) = 6e^{-t} - e^{-t}$$

$$v(t) = -\tfrac{15}{7} e^{-t} \text{ V}$$

$$i(t) = \frac{-v(t)}{-\tfrac{1}{2}} = -\frac{30}{7}e^{-t} \text{ A}$$

Figure 6-6 is an example of impedance solution in which equivalent circuits are used. When different types of elements, each of which is equivalent to a resistance are combined to form an overall impedance, the combination element is denoted by the general symbol for an element, a "box." For this network, using a Thévenin equivalent as shown,

$$(-4 - \tfrac{15}{2})i(t) = 3e^{-2t} - 4e^{-2t}$$

$$i(t) = \tfrac{2}{23}e^{-2t} \text{ A}$$

$$v(t) = -\tfrac{15}{2} i(t) = -\tfrac{15}{23}e^{-2t} \text{ V}$$

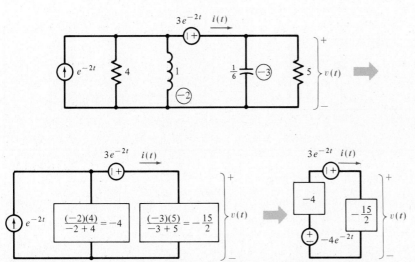

Figure 6-6   Equivalent circuit solution resulting in combined impedances.

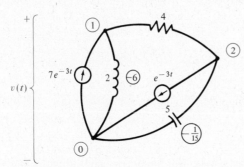
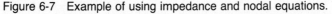
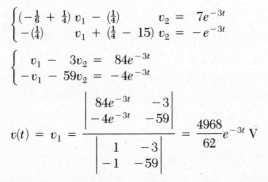

Figure 6-7 Example of using impedance and nodal equations.

An example in which nodal equations are used in the solution is given in Figure 6-7. For this network,

$$\begin{cases} (-\frac{1}{6} + \frac{1}{4})\,v_1 - (\frac{1}{4})\quad\quad v_2 = 7e^{-3t} \\ -(\frac{1}{4})\quad\quad v_1 + (\frac{1}{4} - 15)\,v_2 = -e^{-3t} \end{cases}$$

$$\begin{cases} v_1 - 3v_2 = 84e^{-3t} \\ -v_1 - 59v_2 = -4e^{-3t} \end{cases}$$

$$v(t) = v_1 = \frac{\begin{vmatrix} 84e^{-3t} & -3 \\ -4e^{-3t} & -59 \end{vmatrix}}{\begin{vmatrix} 1 & -3 \\ -1 & -59 \end{vmatrix}} = \frac{4968}{62}e^{-3t}\ \text{V}$$

D6-5

Find the indicated signals:

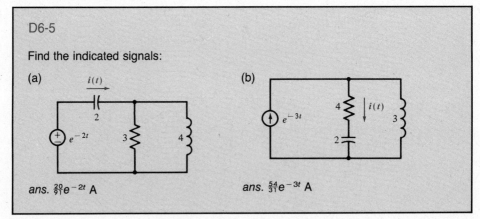

(a)

(b)

ans. $\frac{20}{91}e^{-2t}$ A

ans. $\frac{54}{31}e^{-3t}$ A

6.4.4 Mutual Inductive Coupling

Figure 6-8 is an example in which the network has mutual inductive coupling. Replacing the coupling by the controlled voltage-source equivalent, the equivalent network results. For exponential currents i_1 and i_2, each varying as e^{2t},

$$\frac{di_1}{dt} = 2i_1 \quad\quad \text{and} \quad\quad \frac{di_2}{dt} = 2i_2$$

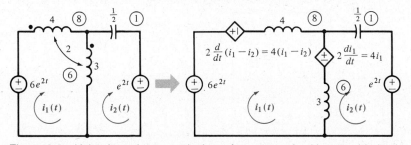

Figure 6-8 Using impedance equivalents for a network with mutual inductive coupling.

so the mutual coupling term derivatives may also be eliminated. The network equations are as follows:

$$\begin{cases} (6 + 8)i_1 - \qquad 6i_2 = 6e^{2t} - 4i_1 + 4i_2 - 4i_1 \\ \qquad -6i_1 + (6 + 1)i_2 = -e^{2t} + 4i_1 \end{cases}$$

$$\begin{cases} 22i_1 - 10i_2 = 6e^{2t} \\ -10i_1 + 7i_2 = -e^{2t} \end{cases}$$

$$i_1 = \frac{\begin{vmatrix} 6e^{2t} & -8 \\ -e^{2t} & 7 \end{vmatrix}}{\begin{vmatrix} 22 & -10 \\ -10 & 7 \end{vmatrix}} = \frac{34e^{2t}}{54}\ \text{A}$$

$$i_2 = \frac{\begin{vmatrix} 18 & 6e^{2t} \\ -8 & -e^{2t} \end{vmatrix}}{54} = \frac{30e^{2t}}{54}\ \text{A}$$

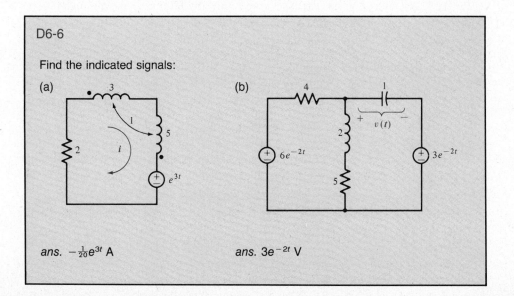

D6-6

Find the indicated signals:

(a)

(b)

ans. $-\frac{1}{20}e^{3t}$ A

ans. $3e^{-2t}$ V

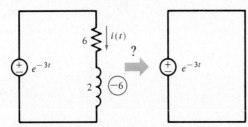

Figure 6-9 Special case in which the forced network response is not exponential.

6.4.5 Singular Cases

In the network of Figure 6-9, a difficulty occurs. This situation is indicative of the special case of a network differential equation in which the forced network response is not exponential. The current $i(t)$ satisfies

$$2\frac{di}{dt} + 6i = e^{-3t}$$

The homogeneous equation is

$$2\frac{di_n}{dt} + 6i_n = 0$$

and the characteristic equation is

$$2s + 6 = 0 \qquad s = -3$$

The natural component of $i(t)$ is

$$i_n(t) = Ke^{-3t}$$

which is of the same form as the driving function.

An inconsistency such as this in the equivalent impedance network will always occur in these special cases because under these conditions, the forced voltage-current ratio is not a constant. For the example, a forced current of the form

$$i(t) = Ate^{-3t}$$

will satisfy the differential equation:

$$2(Ae^{-3t} - 3Ate^{-3t}) + 6Ate^{-3t} = e^{-3t}$$

$$A = \tfrac{1}{2}$$

$$i(t) = \tfrac{1}{2}te^{-3t}$$

These special cases are of limited practical importance because of the difficulty of obtaining exactly the right element values in a network for the situation to occur.

FINDING FORCED EXPONENTIAL RESPONSE

When all sources in a network are exponential, Ae^{st}, with the same exponential constant s, all network forced voltages and currents vary as e^{st}. The ratio of forced voltage to current in an inductor or capacitor under these circumstances is a constant, the impedance of the element.

> In terms of the exponential constant s, the impedance of an inductance L is
>
> $$Z_L = sL$$
>
> and the impedance of a capacitance C is
>
> $$Z_C = \frac{1}{sC}$$
>
> To solve for any forced voltage or current in such a network, replace the inductors and capacitors by their impedances and solve the corresponding source-resistor network.

6.5 SOURCE SUPERPOSITION

If a network has two or more sources that are exponential, but with different exponential constants s, the sources may be superimposed so that in each subproblem all voltages and currents are exponential with a single value of s. An example is shown in Figure 6-10. For this network,

$$(-5 - 2 + 4)i_1 = 3e^{-t}$$

$$i_1(t) = -e^{-t}$$

and

$$i_2(t) = \frac{6}{6 + 5}e^t = \frac{6}{11}e^t$$

Then

$$i(t) = i_1(t) + i_2(t) = \frac{6}{11}e^t - e^{-t} \text{ A}$$

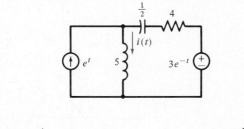

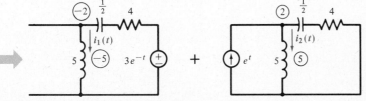

Figure 6-10   Superposition of sources.

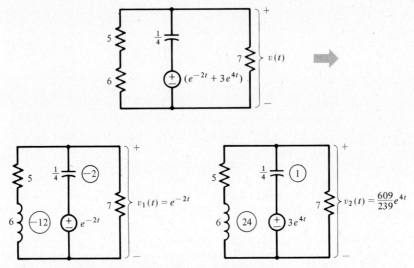

Figure 6-11 Superposition of source components.

Figure 6-11 shows an example of superposition for which the source has more than one exponential component. In this example,

$$v(t) = v_1(t) + v_2(t) = 6e^{-2t} + \tfrac{609}{239}e^{4t} \text{ V}$$

D6-7

Find the indicated signals using impedance methods:

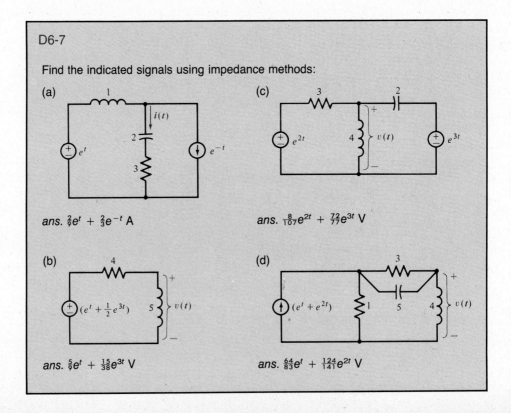

(a)

ans. $\tfrac{2}{9}e^t + \tfrac{2}{3}e^{-t}$ A

(b)

ans. $\tfrac{5}{9}e^t + \tfrac{15}{38}e^{3t}$ V

(c)

ans. $\tfrac{8}{107}e^{2t} + \tfrac{72}{77}e^{3t}$ V

(d)

ans. $\tfrac{64}{83}e^t + \tfrac{124}{141}e^{2t}$ V

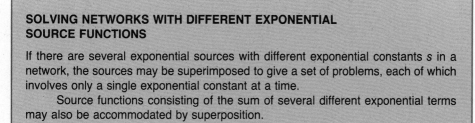

SOLVING NETWORKS WITH DIFFERENT EXPONENTIAL SOURCE FUNCTIONS

If there are several exponential sources with different exponential constants s in a network, the sources may be superimposed to give a set of problems, each of which involves only a single exponential constant at a time.

   Source functions consisting of the sum of several different exponential terms may also be accommodated by superposition.

6.6 SWITCHED NETWORKS WITH EXPONENTIAL SOURCES

The same development as was done previously for the switched networks with constant sources could now be carried through for switched networks with exponential sources. The same methods of solution apply, of course, except that the calculation of the forced response differs.

   An example is given in Figure 6-12. Since the network source is disconnected before $t = 0$, the forced inductor current is zero prior to $t = 0$. After time $t = 0$, the source is exponential, so the impedance method with $s = 3$ applies:

$$(2 + \tfrac{15}{8})i'(t) = 4e^{3t}$$

$$i_f(t) = \tfrac{5}{8}i'(t) = \tfrac{20}{31}e^{3t}$$

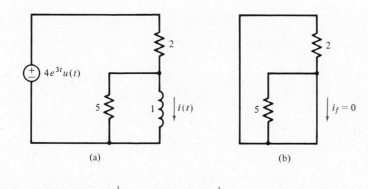

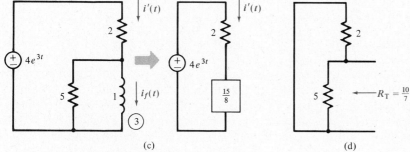

Figure 6-12 (a) Original network.
           (b) Forced inductor current before $t = 0$.
           (c) Forced inductor current after $t = 0$.
           (d) Thévenin resistance after $t = 0$.

After $t = 0$, the general solution for the inductor current is thus

$$i(t) = \tfrac{20}{31}e^{3t} + Ke^{-(R_T/L)t} = \tfrac{20}{31}e^{3t} + Ke^{-(10/7)t}$$

Before $t = 0$, $i = 0$, so

$$i(0) = \tfrac{20}{31} + K = 0$$

$$K = -\tfrac{20}{31}$$

$$i(t) = \begin{cases} 0\ \text{A} & t \le 0 \\ \tfrac{20}{31}e^{3t} - \tfrac{20}{31}e^{-(10/7)t}\ \text{A} & t \ge 0 \end{cases}$$

D6-8

Find the indicated signal and sketch it. Show the signal slightly before $t = 0$ on the sketch:

(a)   (b)

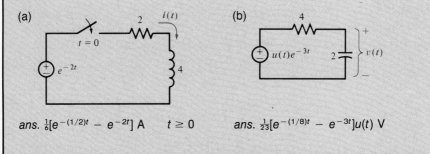

ans. $\tfrac{1}{6}[e^{-(1/2)t} - e^{-2t}]$ A $t \ge 0$ ans. $\tfrac{1}{23}[e^{-(1/8)t} - e^{-3t}]u(t)$ V

SWITCHED NETWORKS WITH EXPONENTIAL SOURCES

When exponential sources are involved in switched networks, the forced exponential response may be found by using impedance methods.

6.7 TRANSFER FUNCTIONS

Every forced voltage and current due to a single exponential source in a network varies as does the source, as e^{st}. The ratio of *any* forced voltages or currents in a single-source network is thus constant. Impedance is an exponential signal ratio of this kind, that of the sink reference exponential voltage to current in one element. Other ratios of one forced exponential voltage or current to another are called *transfer functions* (or *network functions*) and are generally given the symbol $T(s)$.

To calculate network transfer functions, as functions of the exponential constant s, let the source be the general exponential function Ae^{st}. All elements then behave as impedances so far as the forced network signals are concerned. Solve for the exponential signals involved, in terms of s, and form the desired ratio.

For the network of Figure 6-13, the source $v(t)$ has been taken to be a general

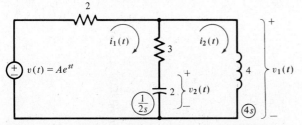

Figure 6-13   Network with a single exponential source.

exponential function, and the network elements have been represented by their impedances, as functions of s. Suppose that it is desired to find the transfer function

$$T_1(s) = \left. \frac{v_1(t)}{v(t)} \right|_{\text{when all signals vary as } e^{st}}$$

The systematic mesh equations for the network are, in terms of the exponential constant s,

$$\begin{cases} \left(5 + \dfrac{1}{2s}\right)i_1 - \left(3 + \dfrac{1}{2s}\right)i_2 \quad = v \\[3mm] -\left(3 + \dfrac{1}{2s}\right)i_1 + \left(3 + \dfrac{1}{2s} + 4s\right)i_2 = 0 \end{cases}$$

Solving for i_2 in terms of v,

$$i_2 = \frac{\begin{vmatrix} \left(5 + \dfrac{1}{2s}\right) & v \\[3mm] -\left(3 + \dfrac{1}{2s}\right) & 0 \end{vmatrix}}{\begin{vmatrix} \left(5 + \dfrac{1}{2s}\right) & -\left(3 + \dfrac{1}{2s}\right) \\[3mm] -\left(3 + \dfrac{1}{2s}\right) & \left(3 + \dfrac{1}{2s} + 4s\right) \end{vmatrix}}$$

$$= \frac{\left(3 + \dfrac{1}{2s}\right)v}{15 + \dfrac{5}{2s} + 20s + \dfrac{3}{2s} + \dfrac{1}{4s^2} + 2 - 9 - \dfrac{3}{s} - \dfrac{1}{4s^2}}$$

$$= \frac{(6s + 1)v}{40s^2 + 16s + 2}$$

The voltage v_1 is then, for exponential signals,

$$v_1 = (4s)i_2 = \frac{(24s^2 + 4s)v}{40s^2 + 16s + 2}$$

and the desired transfer function is

$$T_1(s) = \left. \frac{v_1(t)}{v(t)} \right|_{\substack{\text{when all signals} \\ \text{vary as } e^{st}}} = \frac{12s^2 + 2s}{20s^2 + 8s + 1}$$

To find the transfer function

$$T_2(s) = \left. \frac{v_2(t)}{v_1(t)} \right|_{\substack{\text{when all signals} \\ \text{vary as } e^{st}}}$$

the voltage divider rule may be used. For exponential signals,

$$\frac{v_2(t)}{v_1(t)} = \frac{\frac{1}{2}s}{3 + \frac{1}{2}s} = \frac{1}{6s + 1} = T_2(s)$$

Other network transfer functions may be similarly found.

As network transfer functions may be voltage-current, current-voltage, voltage-voltage, or current-current ratios, they may have the units of resistance or conductance, or they may be dimensionless.

Some authors distinguish those ratios of voltages and currents to the source function from other ratios of network signals, calling the former *transfer functions* and the latter *transmittances*.

D6-9

For the given network, find the following transfer functions, as a function of the exponential constant *s*:

(a) $T_a(s) = \left. \dfrac{i_1(t)}{v(t)} \right|_{\substack{\text{when all signals} \\ \text{vary as } e^{st}}}$

(b) $T_b(s) = \left. \dfrac{v_1(t)}{v(t)} \right|_{\substack{\text{when all signals} \\ \text{vary as } e^{st}}}$

(c) $T_c(s) = \left. \dfrac{i_2(t)}{v(t)} \right|_{\substack{\text{when all signals} \\ \text{vary as } e^{st}}}$

(d) $T_d(s) = \left. \dfrac{i_2(t)}{i_1(t)} \right|_{\substack{\text{when all signals} \\ \text{vary as } e^{st}}}$

(e) $T_e(s) = \left. \dfrac{i_2(t)}{v_1(t)} \right|_{\substack{\text{when all signals} \\ \text{vary as } e^{st}}}$

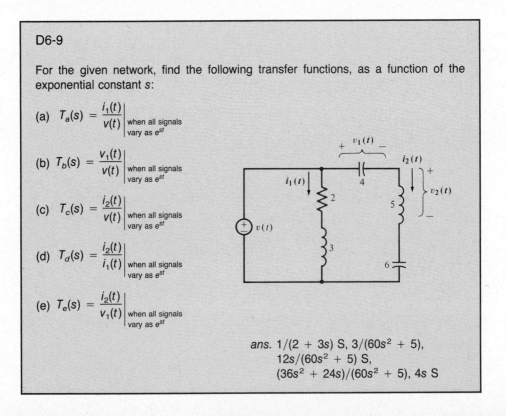

ans. 1/(2 + 3s) S, 3/(60s² + 5),
12s/(60s² + 5) S,
(36s² + 24s)/(60s² + 5), 4s S

TRANSFER FUNCTIONS

In a network with a single exponential source, with source function Ae^{st}, the forced component of every voltage and every current is also exponential, varying as e^{st}.

A network transfer function is the ratio of one forced exponential voltage or current to another, as a function of the exponential constant s.

6.8 POLES AND ZEROS

6.8.1 Factors of Rational Functions

A *rational* function is the ratio of two polynomials. Network transfer functions, including impedances, are rational functions. The *zeros* of a rational function $T(s)$ are the values of the variable s for which

$$T(s) = 0$$

Its *poles* are the values of s for which

$$\frac{1}{T(s)} = 0$$

that is, those values of s that make $T(s)$ infinite.

When a rational function is expressed as the ratio of two polynomials, its zeros are the roots of the numerator polynomial and its poles are the roots of the denominator polynomial. For example, the impedance indicated in Figure 6-14(a) is

$$Z_1(s) = 2s + \frac{(4 + \frac{1}{3}s)(1)}{4 + \frac{1}{3}s + 1} = 2s + \frac{12s + 1}{15s + 1}$$

$$= \frac{30s^2 + 14s + 1}{15s + 1}$$

The zeros of this impedance are given by

$$30s^2 + 14s + 1 = 0$$

$$s_1, s_2 = \frac{-14 \pm \sqrt{196 - 120}}{60} = -0.38, -0.089$$

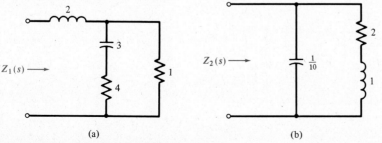

(a)   (b)

Figure 6-14   Impedance poles and zeros.

Its pole is given by

$$15s + 1 = 0 \qquad s = -\tfrac{1}{15}$$

This impedance has the following numerator and denominator factors:

$$Z_1(s) = \frac{30(s + 0.38)(s + 0.089)}{15(s + \tfrac{1}{15})}$$

The impedance of Figure 6-14(b) is

$$Z_2(s) = \frac{(10/s)(2 + s)}{(10/s) + 2 + s} = \frac{10s + 20}{s^2 + 2s + 10}$$

$$= \frac{10(s + 2)}{(s + 1 + j3)(s + 1 - j3)}$$

has a zero at $s = -2$ and poles at $s = -1 - j3$ and at $s = -1 + j3$.

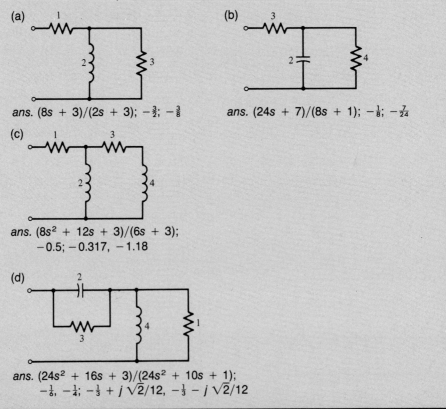

D6-10

Find the impedance of each of the following elements, as a function of the exponential constant s, and express as the ratio of two polynomials in s. Then find the poles and the zeros of each impedance:

(a)

ans. $(8s + 3)/(2s + 3);\ -\tfrac{3}{2};\ -\tfrac{3}{8}$

(b)

ans. $(24s + 7)/(8s + 1);\ -\tfrac{1}{8};\ -\tfrac{7}{24}$

(c)

ans. $(8s^2 + 12s + 3)/(6s + 3);$
$\quad -0.5;\ -0.317,\ -1.18$

(d)

ans. $(24s^2 + 16s + 3)/(24s^2 + 10s + 1);$
$\quad -\tfrac{1}{6},\ -\tfrac{1}{4};\ -\tfrac{1}{3} + j\,\sqrt{2}/12,\ -\tfrac{1}{3} - j\,\sqrt{2}/12$

6.8.2 Pole-Zero Plots

When the poles and zeros of a rational function $T(s)$ are plotted on the complex plane, the result is termed a *pole-zero plot*. The zeros of $T(s)$ are plotted with O's and the poles of $T(s)$ are plotted with X's, as in the example plots in Figure 6-15 for

$$T_1(s) = \frac{-4(s + 6)(s - 1)}{3s(s + 2 + j5)(s + 2 - j5)}$$

and

$$T_2(s) = \frac{(s + 2j)(s - 2j)}{8(s + 3)^2}$$

As network transfer functions always involve real coefficients, their poles and zeros, when complex, always occur in complex conjugate pairs. Therefore, their pole-zero plots are always symmetric about the real axis.

Complete description of a rational function

$$T(s) = \frac{b_m s^m + b_{m-1} s^{m-1} + \cdots + b_0}{a_n s^n + a_{n-1} s^{n-1} + \cdots + a_0}$$

$$= \left(\frac{b_m}{a_n}\right) \frac{(s - s_a)(s - s_b) \cdots (s - s_m)}{(s - s_1)(s - s_2) \cdots (s - s_n)}$$

includes not only its zeros $s_a, s_b, \ldots, s_m$ and its poles, $s_1, s_2, \ldots, s_n$, but also the multiplying constant (b_m/a_n). It is convenient to place the multiplying constant in a box to the right of the plot, as in the examples of Figure 6-15, so that the function is given entirely by the plot. The function represented by the pole-zero plot of Figure 6-16 is, for example,

$$T(s) = \frac{10(s + 5)(s - 2 - j3)(s - 2 + j3)}{(s + 3 - j6)^2(s + 3 + j6)^2}$$

$$= \frac{10(s + 5)(s^2 - 4s + 13)}{(s^2 + 6s + 45)^2}$$

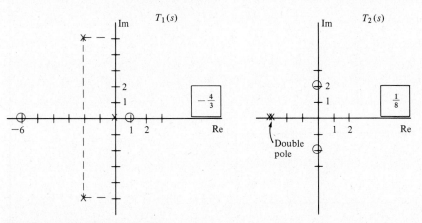

Figure 6-15   Pole-zero plots.

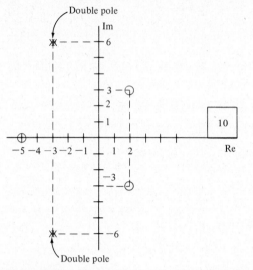

Figure 6-16 Finding the function from its pole-zero plot.

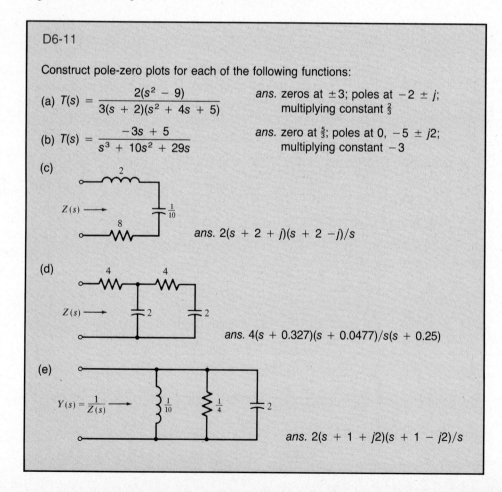

D6-11

Construct pole-zero plots for each of the following functions:

(a) $T(s) = \dfrac{2(s^2 - 9)}{3(s + 2)(s^2 + 4s + 5)}$ *ans.* zeros at ± 3; poles at $-2 \pm j$; multiplying constant $\frac{2}{3}$

(b) $T(s) = \dfrac{-3s + 5}{s^3 + 10s^2 + 29s}$ *ans.* zero at $\frac{5}{3}$; poles at 0, $-5 \pm j2$; multiplying constant -3

(c)

$Z(s) \longrightarrow$

ans. $2(s + 2 + j)(s + 2 - j)/s$

(d)

$Z(s) \longrightarrow$

ans. $4(s + 0.327)(s + 0.0477)/s(s + 0.25)$

(e)

$Y(s) = \dfrac{1}{Z(s)} \longrightarrow$

ans. $2(s + 1 + j2)(s + 1 - j2)/s$

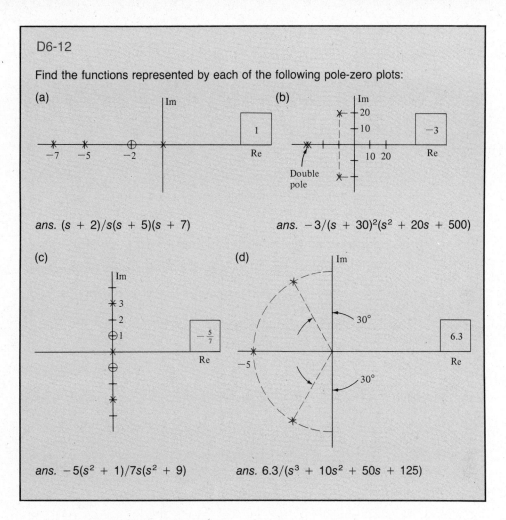

D6-12

Find the functions represented by each of the following pole-zero plots:

(a)

ans. $(s + 2)/s(s + 5)(s + 7)$

(b)

ans. $-3/(s + 30)^2(s^2 + 20s + 500)$

(c)

ans. $-5(s^2 + 1)/7s(s^2 + 9)$

(d)

ans. $6.3/(s^3 + 10s^2 + 50s + 125)$

6.8.3 Graphical Evaluation

When it is desired to evaluate a rational function at a specific value of the variable $s = p$,

$$T(s) = \left(\frac{b_m}{a_n}\right)\frac{(s - s_a)(s - s_b) \cdots (s - s_m)}{(s - s_1)(s - s_2) \cdots (s - s_n)}\bigg|_{s=p}$$

$$= \left(\frac{b_m}{a_n}\right)\frac{(p - s_a)(p - s_b) \cdots (p - s_m)}{(p - s_1)(p - s_2) \cdots (p - s_n)}$$

calculation may be performed graphically, using the function's pole-zero plot. Each numerator and denominator factor of the form $(p - s_i)$ is a complex number with magnitude the distance between the point s_i and the point p, as illustrated in Figure 6-17. The angle of the factor is the angle of the directed line segment from s_i to p. Thus,

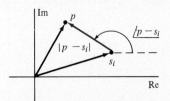

Figure 6-17   Magnitude and angle of a pole or zero factor.

$$|T(s = p)| = \left(\frac{b_m}{a_n}\right)\frac{\text{product of distances of zeros to } p}{\text{product of distances of poles to } p}$$

and

$$\underline{/T(s = p)} = (\text{sum of zero angles to } p) - (\text{sum of pole angles to } p)$$

$$(+ 180° \text{ if the multiplying constant is negative})$$

For the example of Figure 6-18(a)

$$|T_1(s = 2 + j2)| = \frac{(5)(2)}{(4)(4\sqrt{2})(2\sqrt{2})} = \frac{10}{64}$$

and

$$\underline{/T_1(s = 2 + j2)} = 90° - 0° - 45° - 45° = 0°$$

so that

$$T_1(s = 2 + j2) = \tfrac{5}{32}e^{j0°} = \tfrac{5}{32}$$

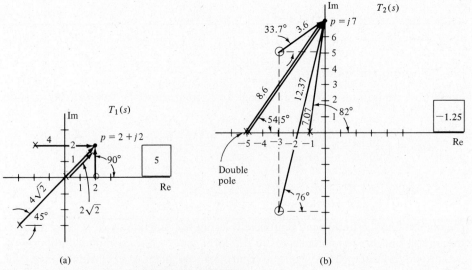

(a)   (b)

Figure 6-18   Examples of graphical evaluation.

In the example of Figure 6-18(b),

$$|T_2(s = j7)| = \frac{(1.25)(3.6)(12.37)}{(8.6)^2(7.07)} = 0.1065$$

$$\underline{/T_2(s = j7)} = 33.7° + 76° - 54.5° - 54.5° - 82° + 180°$$

$$= 98.7°$$

giving

$$T_2(s = j7) = 0.1065e^{j98.7°}$$

The major advantages of graphical evaluation methods are the abilities to obtain approximate results rapidly and to visualize the effects of changing pole and zero locations.

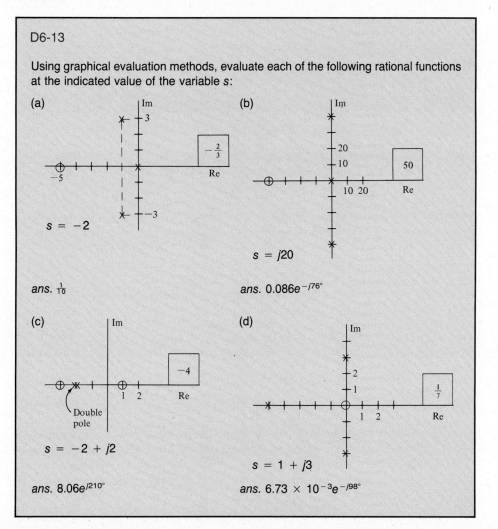

D6-13

Using graphical evaluation methods, evaluate each of the following rational functions at the indicated value of the variable s:

(a)

$s = -2$

ans. $\frac{1}{10}$

(b)

$s = j20$

ans. $0.086e^{-j76°}$

(c)

$s = -2 + j2$

ans. $8.06e^{j210°}$

(d)

$s = 1 + j3$

ans. $6.73 \times 10^{-3}e^{-j98°}$

POLES AND ZEROS

The zeros of a rational function

$$T(s) = \left(\frac{b_m}{a_n}\right)\frac{(s - s_a)(s - s_b) \cdots (s - s_m)}{(s - s_1)(s - s_2) \cdots (s - s_n)}$$

are the values

$$s = s_a, s_b, \ldots, s_m$$

for which

$$T(s) = 0$$

while the function's poles are the values

$$s = s_1, s_2, \ldots, s_n$$

for which $T(s)$ is infinite or

$$\frac{1}{T(s)} = 0$$

A pole-zero plot of $T(s)$ shows the zeros of $T(s)$ marked with O's and the poles of $T(s)$ marked with X's on the complex plane. For a rational function with real coefficients, complex zeros and complex poles occur in conjugate pairs. When the multiplying constant (b_m/a_n) is included with the plot, the pole-zero plot completely describes the function.

At a specific value of the variable $s = p$, the value $T(s = p)$ has magnitude equal to the product of the magnitude of the multiplying constant times the distances between each of the zero locations and the point p, divided by the product of the distances between each of the pole locations and p. The angle of $T(s = p)$ is the sum of the angles of each of the zeros to p, minus the angles of each of the poles to p, plus 180° if the multiplying constant is negative.

6.9 USING IMPEDANCE TO FIND NATURAL BEHAVIOR

6.9.1 Loop Impedance Method

The natural component of a voltage or current in any network is of the form

$$y_n(t) = K_1 e^{s_1 t} + K_2 e^{s_2 t} + \cdots + K_n e^{s_n t}$$

where $K_1, K_2, \ldots, K_n$ are arbitrary constants that are dependent on the network initial conditions. The exponential constants $s_1, s_2, \ldots, s_n$ are called the *characteristic roots* of the natural behavior.

The natural component of any network voltage or current is the general solution of the homogeneous differential equations describing the signal. In terms of the network, to form the homogeneous equation is to set all fixed sources to zero. Thus the natural components of network voltages and currents are those signals that can exist in the network without fixed sources.

Since each natural behavior term, $K_i e^{s_i t}$, is exponential, element impedances may be used in determining which exponential signals may exist in the source-free

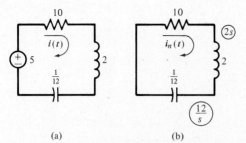

(a)   (b)

Figure 6-19   Exponential signals in a source-free network.

network. For example, for the single-loop network in Figure 6-19(a), setting the source to zero results in the source-free network of Figure 6-19(b). For an exponential natural current

$$i_n(t) = Ke^{st}$$

the element impedances may be used, and the loop equation is

$$\left(2s + 10 + \frac{12}{s}\right)i_n(t) = 0$$

The current $i_n(t)$ can be nonzero only if

$$\left(2s + 10 + \frac{12}{s}\right) = \frac{2s^2 + 10s + 12}{s} = 0$$

$$2(s + 2)(s + 3) = 0$$

$$s = -2, -3$$

That is, only exponential loop currents of the shape e^{-2t} and e^{-3t} can exist in the source-free network. The most general such current is

$$i_n(t) = K_1 e^{-2t} + K_2 e^{-3t}$$

and this is the form of the natural part of the loop current.

Equating the loop impedance, through which a current flows, to zero to find the roots of the natural component of that current is to ask if a source-free loop with impedance $Z(s)$ can have a nonzero current in it. Generally it can, for exponential currents with those values of s for which $Z(s) = 0$.

The drawings of Figure 6-20 illustrate this concept for a more complicated network. The original network in Figure 6-20(a) has all fixed sources set to zero in Figure 6-20(b). There being no sources present, there are only natural voltages and currents in Figure 6-20(b), which are the natural components of the corresponding signals in the original network. The conductor through which the natural current of interest, $i_n(t)$, flows is imagined pulled out from the rest of the network; and the entire impedance looking back into the rest of the network from this loop, $Z(s)$ in Figure 6-20(c), is calculated, as a function of the exponential constant s.

In Figure 6-20(d) the current $i_n(t)$ is seen to be nonzero only if

$$Z(s)i_n(t) = 0$$

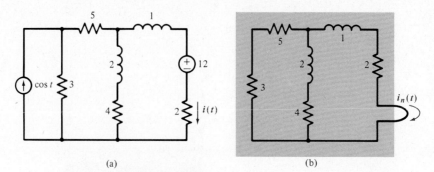

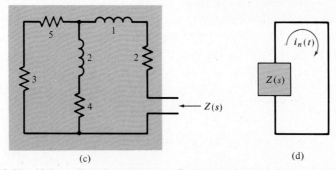

Figure 6-20 Using a loop impedance to find network natural response.

that is, for those values of s for which

$$Z(s) = 0$$

The loop impedance $Z(s)$ in the example is

$$Z(s) = s + 2 + \frac{(2s + 4)(8)}{2s + 4 + 8} = \frac{2s^2 + 32s - 56}{2s + 12}$$

$$= \frac{s^2 + 16s + 28}{s + 6}$$

Equating $Z(s)$ to zero, the zeros of the function are

$$Z(s) = \frac{(s + 2)(s + 14)}{s + 6} = 0$$

$$s_1, s_2 = -2, -14$$

So the natural component of this current $i(t)$ is of the form

$$i_n(t) = K_1 e^{-2t} - K_2 e^{-14t}$$

6.9.2 Characteristic Roots of the Natural Behavior

The natural component of each network voltage and current consists of a sum of exponential terms of the form

$$y_n(t) = K_1 e^{s_1 t} + K_2 e^{s_2 t} + \cdots + K_n e^{s_n t}$$

The operations involved in expressing one source-free network voltage or current in terms of another are differentiation, integration, multiplication by a constant, and addition, all of which only modify the amplitudes of each exponential term.

Hence the natural component of every network signal consists of the *same* exponential terms, in varying amounts. Every voltage and current is of the above form, but with generally different amplitudes K_1, K_2, . . ., K_n. The characteristic roots of the natural behavior of a network, s_1, s_2, . . ., s_n, are characteristic of the network; the same roots are involved in the natural component of every voltage and every current in the network. Except in special cases, every network loop impedance should yield the same characteristic roots. For example, for the network of Figure 6-21, in which all sources have been set to zero, all of the indicated loop impedances have the same zeros, which are the characteristic roots of the network:

$$Z_1(s) = 2s + 3 - \frac{(4/s)}{4 + (1/s)} = \frac{8s^2 + 14s + 7}{4s + 1}$$

$$Z_2(s) = \frac{1}{s} + \frac{(2s + 3)4}{2s + 3 + 4} = \frac{8s^2 + 14s + 7}{2s^2 + 7s}$$

$$Z_3(s) = 4 + \frac{(2s + 3)/s}{2s + 3 + (1/s)} = \frac{8s^2 + 14s + 7}{2s^2 + 3s + 1}$$

6.9.3 Number of Natural Behavior Roots

The number of characteristic natural behavior roots n for a network is equal to the number of independent energy-storing elements in it; n is the number of independent initial energy storages that may be made. Since each inductor current and each capacitor voltage must be continuous, n is also the number of boundary conditions that apply to the network. Mutual inductances are not counted in addition to self-inductances as independent energy-storing elements because the amount of energy stored in a mutual inductance is fixed once the initial currents in the associated self-inductances are given.

Ordinarily one needs only to total the number of inductors and capacitors in the network to determine the number of characteristic roots expected. However, there are some special circumstances for which some inductors and capacitors may not contribute to the number of characteristic roots. After setting all fixed network

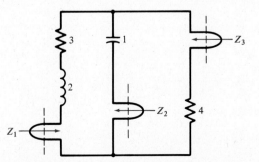

Figure 6-21 Finding the same characteristic roots using any loop impedance.

sources to zero, if there are several inductors that may be replaced by a single equivalent inductor, as in Figure 6-22(a), only the single equivalent inductor should be counted as an independent energy-storing element. A similar consideration holds for equivalent capacitors.

If an inductor current is fixed by a current source or a capacitor voltage is fixed by a voltage source, the initial energy storage in that element is not arbitrary, and the element does not contribute to the number of characteristic roots. When the network fixed sources are set to zero, an inductor in series with a current source will be left hanging and a capacitor in parallel with a voltage source will be left short-circuited, so these situations are quite obvious, as in the example network of Figure 6-22(b).

Less obvious situations where the energies stored in the network energy-storing elements are not independent of one another involve a node joining only inductors or a loop involving only capacitors, as shown in Figure 6-22(c). In the case of the inductors, one inductor current is expressible in terms of the others, so the number of independent energy storages in inductors joined exclusively at a node is one less than the number of inductors. Similarly, one of the capacitor voltages in the loop of capacitors is dependent on the others, so the energy stored in one of these capacitors is not independent of that stored in the others.

6.9.4 Terminal-Pair Impedance Method

The characteristic roots of the natural behavior of a network may also be found by considering the nonzero exponential network *voltages* that may exist with the fixed sources set to zero. Consider a network such as the one in Figure 6-23(a) for which the fixed sources have been set to zero. The natural voltage $v_n(t)$, for example, may be considered to be the voltage across an impedance $Z'(s)$, as in Figure 6-23(b). A terminal pair has been drawn, connected to the network, to aid in visualizing $Z'(s)$.

In Figure 6-23(c), the natural voltage $v_n(t)$ is seen to be nonzero only if it satisfies the source-free two-node equation

$$\left(\frac{1}{Z'(s)}\right)v_n(t) = 0$$

or

$$\frac{1}{Z'(s)} = 0$$

For the example network

$$Z'(s) = \frac{4\left[s + \dfrac{(4/s)}{2 + (2/s)}\right]}{4 + s + \dfrac{(4/s)}{2 + (2/s)}} = \frac{4\left(s + \dfrac{2}{s + 1}\right)}{4 + s + \dfrac{2}{s + 1}}$$

$$= \frac{4s^2 + 4s + 8}{s^2 + 5s + 6}$$

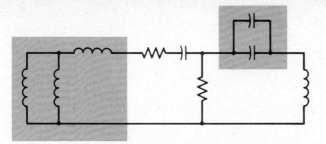

Four independent energy–storing elements

(a)

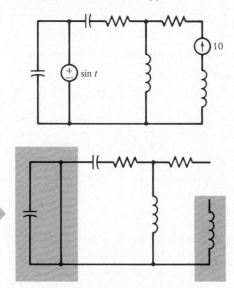

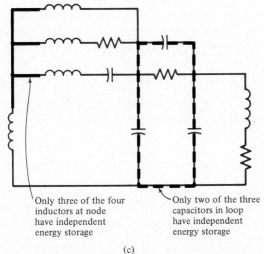

Two independent energy–storing elements

(b)

Only three of the four
inductors at node
have independent
energy storage

Only two of the three
capacitors in loop
have independent
energy storage

(c)

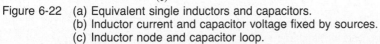

Figure 6-22   (a) Equivalent single inductors and capacitors.
(b) Inductor current and capacitor voltage fixed by sources.
(c) Inductor node and capacitor loop.

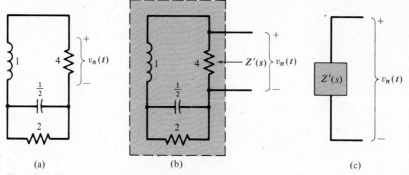

(a)   (b)   (c)

Figure 6-23   Using a terminal-pair impedance to find network natural response.

The inverse of this terminal-pair impedance is

$$\frac{1}{Z'(s)} = \frac{s^2 + 5s + 6}{4s^2 + 4s + 8} = \frac{(s + 2)(s + 3)}{4s^2 + 4s + 8}$$

and equating to zero gives its poles, which are the natural behavior roots, -2 and -3. Then

$$v_n(t) = K_1 e^{-2t} + K_2 e^{-3t}$$

and the natural components of all other network voltages and currents also consist of an e^{-2t} term plus an e^{-3t} term.

Thus an alternate method of finding the characteristic roots of a network is to set the fixed sources to zero, connect a terminal pair to the network, and calculate the impedance looking into the terminal pair. The *poles* of a terminal-pair impedance are the network characteristic roots, since they describe the exponential terminal-pair voltages that may exist in the source-free network.

Except in special cases, the poles of other terminal-pair impedances will yield the same characteristic roots, as will the loop impedance zeros. For the example network, redrawn in Figure 6-24, the terminal-pair impedance Z_2 is

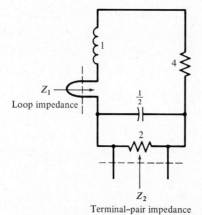

Terminal–pair impedance

Figure 6-24   Loop and terminal-pair impedances of the example network.

$$Z_2(s) = \cfrac{1}{\dfrac{s}{2} + \dfrac{1}{2} + \dfrac{1}{s+4}}$$

Its inverse is

$$\frac{1}{Z_2(s)} = \frac{s}{2} + \frac{1}{2} + \frac{1}{s+4} = \frac{s^2 + 5s + 6}{2s + 8}$$

which has the same roots as found previously, $s = -2$ and $s = -3$.

The loop impedance zeros

$$Z_1(s) = 4 + s + \frac{(4/s)}{2 + (2/s)} = 4 + s + \frac{1}{s+1}$$

$$= \frac{s^2 + 5s + 6}{s+1}$$

also gives the same roots, as do the other terminal-pair and loop impedances.

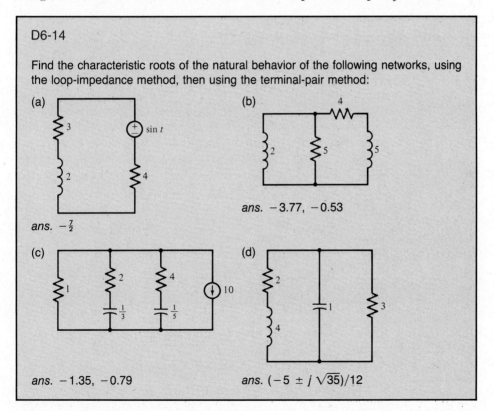

D6-14

Find the characteristic roots of the natural behavior of the following networks, using the loop-impedance method, then using the terminal-pair method:

(a)

ans. $-\frac{7}{2}$

(b)

ans. $-3.77, -0.53$

(c)

ans. $-1.35, -0.79$

(d)

ans. $(-5 \pm j\sqrt{35})/12$

6.9.5 Roots Unobservable from a Loop

It could happen that all of the network characteristic roots cannot be obtained from one of the loops or one of the terminal-pair impedances. That obviously will be the

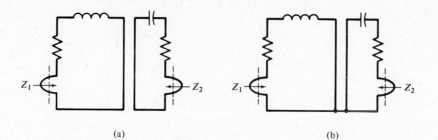

(a)   (b)

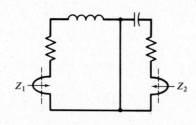

(c)

Figure 6-25   Networks with two uncoupled parts.

case if the network consists of two or more parts that are not coupled to one another, as in the examples of Figure 6-25.

In very special cases, quite ordinary-looking networks might also have loop or terminal-pair impedances from which not all of the characteristic roots are obtainable. The network of Figure 6-26 is an example. The loop impedance Z_1 is

$$Z_1(s) = 2s + 6 + \frac{4(s + 3)}{s + 7} = \frac{2s^2 + 24s + 54}{s + 7}$$

$$= \frac{2(s + 3)(s + 9)}{s - 7}$$

which has roots $s = -3$ and $s = -9$. The loop impedance Z_2 is

$$Z_2(s) = 4 + \frac{2s + 6}{3} = \frac{2s - 18}{3}$$

which displays only the root $s = -9$.

This property of this special network is understandable when its behavior is analyzed for exponential signals of the form Ke^{-3t}. Evaluating the impedances at $s = -3$, it is seen in Figure 6-26(d) that the Z_2 loop is decoupled from the other loop for $s = -3$.

Occasionally networks are deliberately designed in this way so that some of the characteristic roots are not observable in parts of the network. Thus one may see in practice, for example, networks involving three or more energy-storing elements, designed for second-order behavior so far as a certain part of the network is concerned.

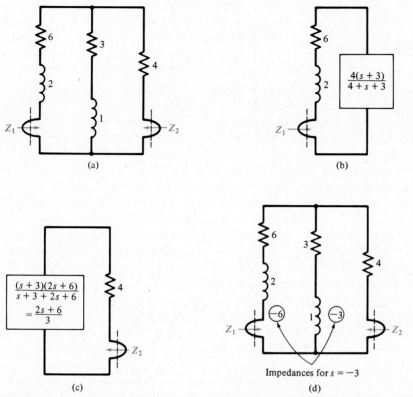

Figure 6-26 Network where a natural behavior root is not observable from one of the loops.

If a network loop or terminal-pair impedance yields a lesser number of roots than the network has independent energy-storing elements, other impedances must be examined to obtain the remaining roots.

USING IMPEDANCE TO FIND NATURAL RESPONSE

The natural component of each voltage and current in a network consists of a sum of exponential terms.

$$y_n(t) = K_1 e^{s_1 t} + K_2 e^{s_2 t} + \cdots + K_n e^{s_n t}$$

where n is the number of independent energy-storing elements in the network.

The set of exponential constants $s_1, s_2, \ldots, s_n$ (the *characteristic roots* of the natural behavior) are the same for every network voltage and current, although the constants $K_1, K_2, \ldots, K_n$ are generally different for each signal.

The characteristic roots of a network may be found by setting a loop impedance to zero or by setting the inverse of a network terminal-pair impedance to zero. These characteristic natural behavior roots are the zeros of the loop impedances, and they are the poles of the terminal-pair impedances.

CHAPTER 6 PROBLEMS

Basic Problems

Differential Equations
1. Find the solutions to each of the following equations with the given boundary conditions:

(a) $4\dfrac{dy}{dt} + 3y = 10e^{-2t}$

$y(0) = 5$

(b) $\dfrac{d^2y}{dt^2} + 7\dfrac{dy}{dt} + 12y = 5e^{-5t}$

$y(0) = -3$

$\dfrac{dy}{dt}\bigg|_{t=0} = 2$

(c) $\dfrac{d^2y}{dt^2} + 2\dfrac{dy}{dt} + y = 3e^{4t}$

$y(0) = 0$

$\dfrac{dy}{dt}\bigg|_{t=0} = 0$

(d) $\dfrac{dy}{dt} - 3y = 5e^{-3t}$

$y(0) = -2$

(e) $\dfrac{d^3y}{dt^3} + 4\dfrac{dy}{dt} = 3e^{2t}$

$y(0) = 8$

$\dfrac{dy}{dt}\bigg|_{t=0} = -4$

$\dfrac{d^2y}{dt^2}\bigg|_{t=0} = 3$

(f) $\dfrac{dy}{dt} = 10e^{-t}$

$y(2) = 7$

(g) $\dfrac{d^2y}{dt^2} + 2\dfrac{dy}{dt} + 10y = 4e^{2t}$

$y(0) = 0$

$\dfrac{dy}{dt}\bigg|_{t=0} = 0$

A television station master control room. The large-screen display devices in the top two rows are television monitors, which are similar to ordinary television sets. Below the monitors are oscilloscopes that display the television signals as functions of time. (*Photo courtesy of Radio Corporation of America.*)

(h) $\dfrac{d^2y}{dt^2} + 4\dfrac{dy}{dt} + 5y = 2e^{-3t}$

$y(0) = 10$

$\dfrac{dy}{dt}\bigg|_{t=0} = 6$

2. Find the forced solutions to each of the following differential equations:

(a) $\dfrac{dy}{dt} + 6y = 10e^{2t}$

(b) $3\dfrac{d^2y}{dt^2} + 4\dfrac{dy}{dt} + 2y = 3e^{-3t}$

(c) $9\dfrac{d^3y}{dt^3} - 3\dfrac{d^2y}{dt^2} + 6\dfrac{dy}{dt} - 7y = -4e^{-2t}$

(d) $6\dfrac{d^2y}{dt^2} + \dfrac{dy}{dt} + 4y = 7e^{2t}$

(e) $3\dfrac{d^4y}{dt^4} + 7\dfrac{d^3y}{dt^3} + 3y = -2e^{t}$

(f) $\dfrac{d^3y}{dt^3} + 2\dfrac{d^2y}{dt^2} + 2\dfrac{dy}{dt} + 3y = 6e^{3t}$

A telephone switching center. Long-distance audio, video and data communications signals are routed worldwide. Electronic systems monitor the flow of communications "traffic", routing connections quickly and efficiently. Extensive periodic testing and maintenance of the equipment is required to achieve the necessary high levels of reliability. (*Photo courtesy of AT&T—Long Lines, San Francisco, CA.*)

Impedance Concept

3. The impedance of a certain element, as a function of the exponential constant s, is

$$Z(s) = \frac{s - 2}{3s + 1}$$

If the element voltage and current are both exponential and the element voltage is

$$v(t) = -5e^{-4t}$$

what is the sink reference element current?

4. Find the impedances of elements with the following sink reference relation voltage-current relations:

(a) $v = 4\dfrac{di}{dt} - 3i$

(b) $i = 5\dfrac{dv}{dt} + \displaystyle\int_{-\infty}^{t} v\,dt$

(c) $i = \dfrac{d^3v}{dt^3}$

(d) $v(t) = i(t - 3)$

Networks with Real Exponential Sources

5. Using impedance methods with equivalent circuits, find the indicated signals.

There are two signals to find in each network.

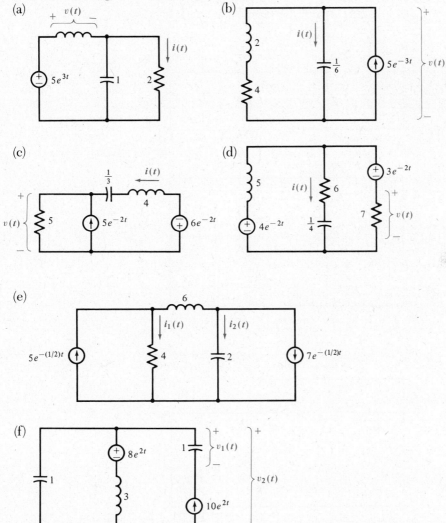

(a)

(b)

(c)

(d)

(e)

(f)

6. Using impedance methods and systematic nodal equations, find the indicated signals:

(a)

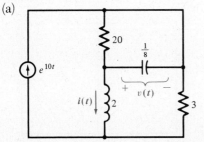

(b)

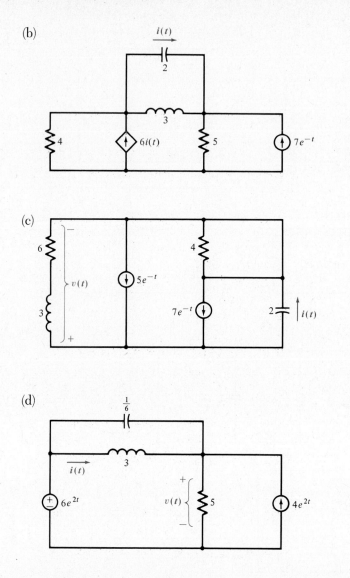

(c)

(d)

7. Using impedance methods and systematic mesh equations, find the indicated signals:

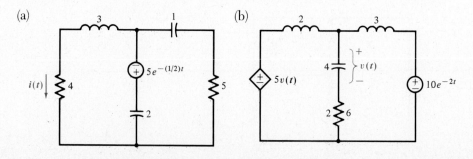

(c)

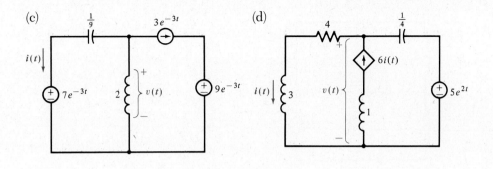

(d)

(e)

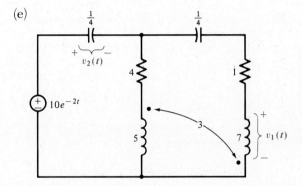

(f)

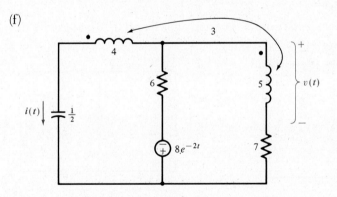

Superposition of Exponential Sources

8. Using impedance methods, find the indicated signals:

(a)

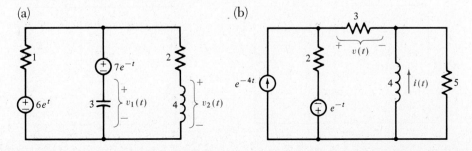

(b)

(c)

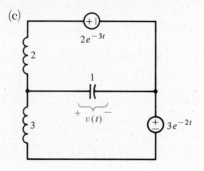

(d)

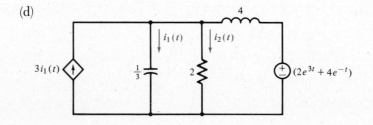

Switched Networks

9. Find the indicated signals both before and after time $t = 0$ and sketch them.

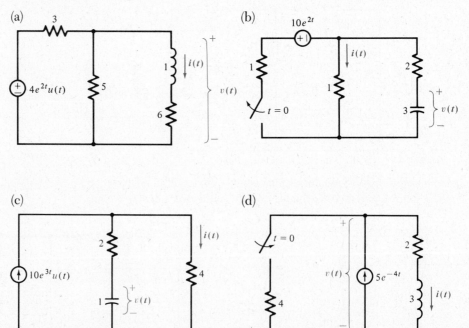

Transfer Functions

10. Find the indicated transfer functions for each of the following networks:

(a) $T(s) = \dfrac{v(t)}{i(t)}\bigg|_{\text{when all signals vary as } e^{st}}$

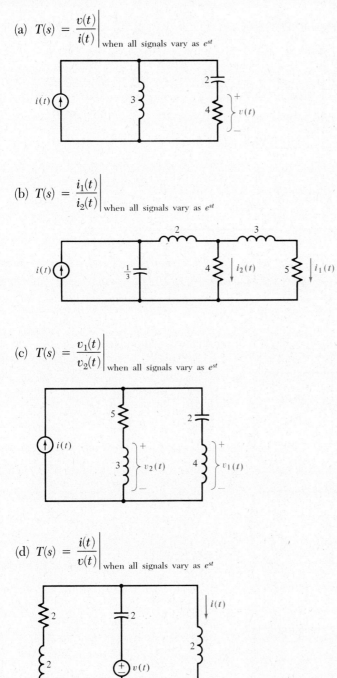

(b) $T(s) = \dfrac{i_1(t)}{i_2(t)}\bigg|_{\text{when all signals vary as } e^{st}}$

(c) $T(s) = \dfrac{v_1(t)}{v_2(t)}\bigg|_{\text{when all signals vary as } e^{st}}$

(d) $T(s) = \dfrac{i(t)}{v(t)}\bigg|_{\text{when all signals vary as } e^{st}}$

(e) $T(s) = \dfrac{i_2(t)}{i_1(t)}\bigg|_{\text{when all signals vary as } e^{st}}$

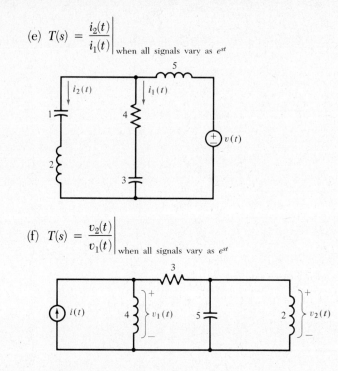

(f) $T(s) = \dfrac{v_2(t)}{v_1(t)}\bigg|_{\text{when all signals vary as } e^{st}}$

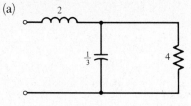

Poles and Zeros

11. Find the poles and zeros of each of the following impedances:

(a)

(b)

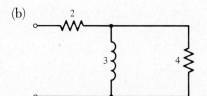

(c)

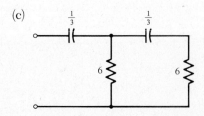

(d)

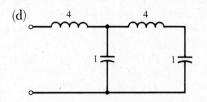

12. Construct pole-zero plots for each of the following functions:

(a) $T(s) = \dfrac{-3(s^2 + 2s + 1)}{9s^2 + 1}$

(b) $T(s) = \dfrac{s^2 + 3s + 1}{(s^2 + 6s + 10)^2}$

(c)

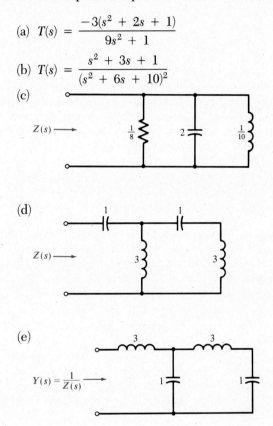

(d)

(e)

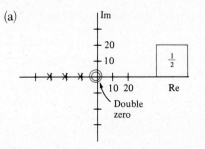

13. Find the functions represented by each of the following pole-zero plots:

(a)

(b)

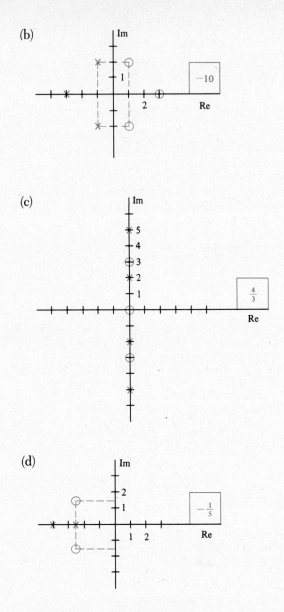

(c)

(d)

14. Using graphical methods, evaluate each of the following rational functions at the indicated value of the variable s:

(a)

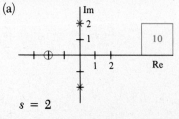

$s = 2$

(b)

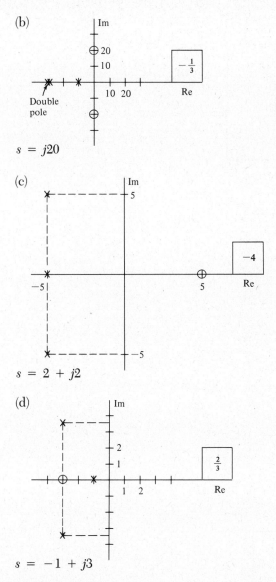

$s = j20$

(c)

$s = 2 + j2$

(d)

$s = -1 + j3$

Natural Behavior

15. Find the equations satisfied by the roots of the natural behavior of the following networks, using the loop-impedance method, then using the terminal-pair method:

(a)

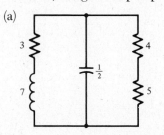

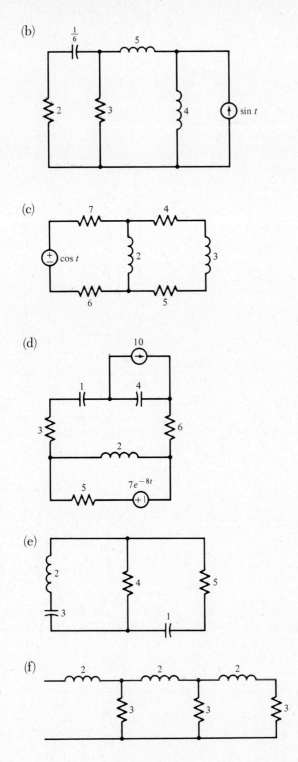

Practical Problems

Adjustable Inductors

16. Adjustable inductors may be constructed by using an adjustable slider contact that selects the number of coil turns used, or by using a movable magnetic-core material, the position of which controls the degree of magnetic-flux coupling from turn to turn of the coil. The inductor symbol with an arrow through it denotes an adjustable inductor.

To what value of inductance should the inductor L in the network be adjusted to produce a voltage $v_2(t)$ that is five times the voltage $v_1(t)$? What value of L will produce v_2 that is ten times v_1?

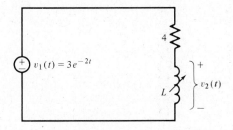

17. When an inductance changes with time, the sink reference voltage v is related to the inductor current i by

$$v = \frac{d}{dt}(Li)$$

For a constant inductor current $i = 5$, find an inductance as a function of time $L(t)$ to produce the constant inductor voltage, $v = 100$.

Adjustable Capacitors

18. Adjustable capacitors may be constructed with movable sets of plates for which the plate separation or degree of overlap of the plates is changeable, as indicated below.

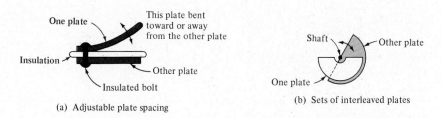

(a) Adjustable plate spacing

(b) Sets of interleaved plates

What value of capacitance C will produce precisely zero voltage v_2 for an exponential voltage v_1 with exponential constant $a = -3$ in the network? What value of C will produce zero voltage v_2 when $a = -4$?

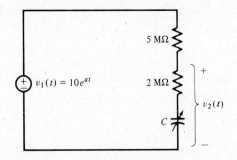

The capacitor symbol with an arrow through it denotes an adjustable capacitor.

19. When a capacitance is changed with time, the sink reference capacitor voltage v and current i are related by

$$i = \frac{d}{dt}(Cv)$$

Suppose that over an interval of time an open-circuited capacitance is increased according to

$$C(t) = 10 + 2t$$

If the capacitor voltage is 100 V at time $t = 0$, what is the capacitor voltage as a function of time thereafter?

Advanced Problems

Exponentially Driven Differential Equations
20. Consider the system described by

$$\frac{dy}{dt} + 2y = e^{at}$$

with $y(0) = 0$. Find the solution $y(t)$ in terms of a for $a \neq -2$.

Impedance
21. Find the impedance of an element for which the sink reference voltage $v(t)$ and current $i(t)$ are related by

$$2\frac{dv}{dt} = 3i - 4\int_{-\infty}^{t} i \, dt$$

What is this element's admittance?
22. The impedance of a certain element, as a function of the exponential constant s, is

$$Z(s) = \frac{3s - 10}{s + 4}$$

Find a differential equation relating the sink reference voltage and current in this element.
23. The impedance of a certain combination of elements is known to be of the form

$$Z(s) = \frac{a_1 s}{a_2 s + 3}$$

where a_1 and a_2 are unknown constants. Find $Z(s)$ if it is known that a current $i(t) = 3e^{-t}$ produces the voltage $v(t) = 3e^{-t}$ and that a current $i(t) = -4e^t$ produces the voltage $v(t) = -2e^t$.

24. Using R's, L's, and C's as needed, design two-terminal networks with the following impedances:

(a) $Z(s) = \dfrac{1}{s + 2}$

(b) $Z(s) = 2s + \dfrac{3}{s + 4} = \dfrac{2s^2 + 8s + 3}{s + 4}$

(c) $Z(s) = \dfrac{4s}{s + 3}$

(d) $Z(s) = \dfrac{s + 3}{4s}$

Switched Exponential-Source Networks

25. For the network shown, write and solve the single-loop differential equation for the current $i(t)$ after $t = 0$. Repeat, using impedance methods and compare the two solution methods.

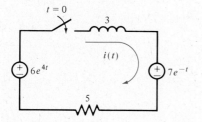

26. Find $v(t)$:

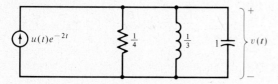

27. By visualizing graphical evaluation from the pole-zero plot for $T(s)$, roughly sketch $|T(s = j\omega)|$ and $\underline{/T(s = j\omega)}$ as functions of ω:

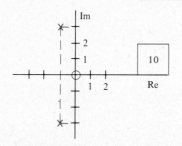

Natural Behavior

28. Design a two-terminal network that has natural behavior of the form $K_1 e^{-2t}$ when the terminals are open-circuited and $K_2 e^{-3t}$ when the terminals are short-circuited.

29. By considering the form of its open-circuit and short-circuited natural response, show that it is not possible to construct a two-terminal network involving positive L's, R's, and C's with an impedance that has poles or zeros that are positive numbers.

30. A certain two-terminal network has impedance

$$Z(s) = \frac{s^2 + 3s + 2}{s^2 + 5s + 6}$$

Find the characteristic roots of the natural behavior if
(a) the terminals are left unconnected
(b) the terminals are short-circuited
(c) the terminals are connected to a voltage source
(d) the terminals are connected to a current source
(e) the terminals are connected to a 4-Ω resistor

Chapter Seven
Response to
Sinusoidal Sources

While alternating waves can be, and frequently are, represented graphically in rectangular coordinates, with the time as abscissa and the instantaneous values of the wave as ordinates, the best insight with regard to the mutual relation of different alternating waves is given by their representation in polar coordinates, with the time as the angle; the amplitude one complete period being represented by one revolution and the instantaneous values as radius vectors.

Charles Proteus Steinmetz
Theory and Calculation of Alternating Current Phenomena, Fourth Edition, 1908

7.1 PREVIEW

This chapter begins with a discussion of sinusoidal functions and the relations between the various ways in which they are commonly represented. Then differential equations with sinusoidal driving functions are examined. Because complex algebra plays a key role in all of the analysis to follow, considerable effort is devoted initially to a careful review of this topic.

Three impedance methods for solving for forced sinusoidal network response are then developed. Each of the three methods is closely related, and each is commonly used in one or another branch of electrical engineering. Expanding the sinusoid into Euler components, superposition of the two components and solution using impedance is the first, obvious method. This *method of components* is very straightforward and is fundamental to later work with the exponential form of the Fourier series.

The redundancy of the two Euler components leads quickly to the *method of sinors* (or rotating phasors), in which it is recognized that only one of the two component problems must be solved.

The *method of phasors* simply involves a "snapshot" of the sinor solution at time $t = 0$. The phasor method is emphasized by drill that includes networks with inductive coupling and careful acquaintance with graphical solutions. The phasors used in this chapter and the next have magnitudes equal to the amplitudes of the sinusoids they represent; rms phasors are introduced and used in Chapter 9, in connection with electrical power.

When you complete this chapter, you should know—

1. how sinusoidal functions are represented and the meaning of the terms *amplitude*, *frequency*, and *phase*;
2. how to solve differential equations that have sinusoidal driving functions;
3. how to perform algebraic operations with complex numbers, including Euler's relation and conversions between rectangular and polar forms;
4. how to use the method of components to solve networks with sinusoidal sources;
5. how to solve sinusoidally driven networks, using the sinor method;
6. how to streamline sinusoidal network solution further with the phasor method;
7. graphical methods for networks with a single sinusoidal source and how to use them.

7.2 SINUSOIDAL FUNCTIONS

7.2.1 Amplitude, Frequency, and Phase

A sinusoidal function (or sinusoid) is a function that may be expressed in the form

$$f(t) = A \cos (\omega t + \theta)$$

where A is a positive constant called the *amplitude* of the sinusoidal function, ω is a positive constant called its *radian frequency* (in radians per second), and θ is a constant called the *phase angle*. A sketch of this function is given in Figure 7-1. Since the maximum and minimum values of the cosine function are $+1$ and -1, respectively, $f(t)$ has maximum and minimum values of $+A$ and $-A$.

The position of the wave with respect to the origin is fixed by the phase angle θ, which may be expressed within a 2π rad (or 360°) range. When

$$\omega t + \theta = 0$$

the argument of the cosine function is zero and the value of the function $f(t)$ is maximum. This occurs at the time

$$t = -\frac{\theta}{\omega}$$

and periodically, at intervals of $(2\pi/\omega)$ s before and after that time.

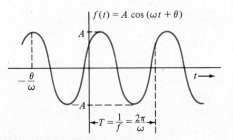

Figure 7-1   Sinusoidal function.

Although, strictly speaking, the argument of the cosine function is in radians, there is no reason why θ cannot be described in terms of degrees, for example,

$$f(t) = 10 \cos (3t + 45°)$$

so long as the use of degrees is clearly indicated. Then the angle in degrees can be converted to radians according to

$$(\text{angle in radians}) = \frac{2\pi}{360} (\text{angle in degrees}) \cong 0.0175 \, (\text{angle in degrees})$$

if and when it is necessary to do so.

The radian frequency ω of a sinusoid is always expressed as a positive number. The trigonometric identity that cosine is an even function,

$$\cos (-x) = \cos x$$

allows conversion of any argument so that the multiplier of t is positive. For example,

$$8 \cos (-3t + 39°) = 8 \cos (3t - 39°)$$

The amplitude A of a sinusoid may also always be expressed as a nonnegative number. Since

$$\cos (x + \pi) = -\cos x$$

a negative amplitude may always be converted to a positive one by adding π rad, 180°, to the phase angle. For example,

$$-6 \cos (4t + 73°) = 6 \cos (4t + 253°)$$

Of course, any integer multiple of 2π rad, 360°, may be added to, or subtracted from, the argument of a sinusoid without changing the function. So

$$6 \cos (4t + 253°) = 6 \cos (4t + 253° - 360°)$$
$$= 6 \cos (4t - 107°)$$

for example. The phase angle of a sinusoid may thus be expressed within a 360° range, although common practice is to express it within ±360° rather than to bother with deciding which 360° range (±180°, 0–360°, or some other) should be the standard.

D7-1

Find, approximately, the sinusoidal functions from the plots. Express in the form $A \cos (\omega t + \theta)$ with $A \geq 0$, $\omega \geq 0$, and $-360° < \theta < 360°$:

(a) $f(t)$

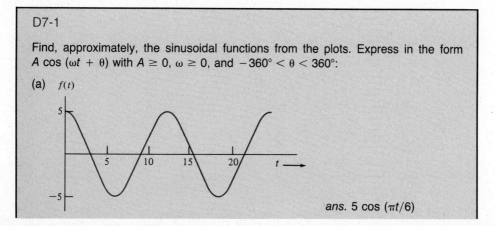

ans. 5 cos (πt/6)

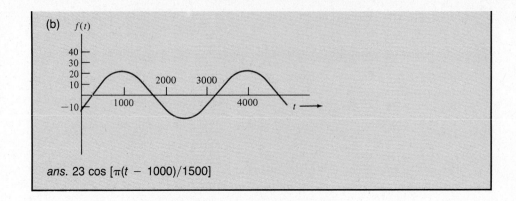

(b) $f(t)$

ans. $23 \cos [\pi(t - 1000)/1500]$

7.2.2 Representation in Terms of the Sine Function

Sinusoidal functions may also be expressed in terms of the sine function. As indicated by the trigonometric identities of Figure 7-2,

$$\sin x = \cos (x - 90°) \qquad \cos x = \sin (x + 90°)$$

the only distinction between sine and cosine is a 90° difference in phase angle.

To convert a sinusoidal function expressed in terms of sine to an expression in terms of cosine, subtract 90° from the argument. For example,

$$7 \sin (8t - 105°) = 7 \cos (8t - 195°)$$

To convert a sinusoidal representation in terms of cosine to one in terms of sine, add 90° to the argument. For example,

$$10 \cos (20t - 140°) = 10 \sin (20t - 50°)$$

Sine is an odd function, so

$$\sin (-x) = -\sin x$$

As with cosine,

$$\sin (x + \pi) = -\sin x$$

And, adding or subtracting any integer multiple of 2π, 360°, to the phase angle leaves the sine unchanged.

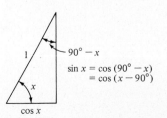

$90° - x$

$\sin x = \cos (90° - x)$
$\quad\quad = \cos (x - 90°)$

1

x

$\cos x$

Figure 7-2   Relation between sine and cosine.

D7-2

Find the amplitude, radian frequency, and phase angle (in terms of the cosine function) for each of the following sinusoids:

(a) $-10 \cos \left(377t - \dfrac{5\pi}{6} \right)$ *ans.* $10 \cos (377t + \pi/6)$

(b) $10^5 \cos (-2\pi t + 700°)$ *ans.* $10^5 \cos (2\pi t + 20°)$

(c) $50 \sin \left(\dfrac{3\pi}{5} - 377t \right)$ *ans.* $50 \cos (377t - \pi/10)$

(d) $-100 \sin (3t + 40°)$ *ans.* $100 \cos (3t + 130°)$

D7-3

Sketch plots of the following sinusoidal functions, indicating amplitude, period, and time to the first maximum after $t = 0$:

(a) $f(t) = 3 \cos \left(8t - \dfrac{\pi}{4} \right)$ *ans.* $3, 2\pi/8, \pi/32$ s

(b) $f(t) = -10 \cos (10^6 t + 30°)$ *ans.* $10, 2\pi \times 10^{-6}, 2.6 \times 10^{-6}$ s

(c) $f(t) = 7 \sin (0.3t + 120°)$ *ans.* $7, 20\pi/3, 19.2$ s

(d) $f(t) = -10^4 \sin \left(t - \dfrac{3\pi}{5} \right)$ *ans.* $10^4, 2\pi, \pi/10$ s

---

7.2.3 Quadrature Representation

Another useful way to represent a sinusoidal function is as the sum of a cosine and a sine, as

$$A \cos (\omega t + \theta) = D \cos \omega t + E \sin \omega t$$

This is known as the *quadrature* form for a sinusoid because it is in terms of cosine and sine, which differ in phase by 90°, or are in quadrature to one another.

The constants D and E are related to A and θ by

$$D = A \cos \theta \qquad E = -A \sin \theta$$

as is evident from the trigonometric identity

$$\cos (x + y) = \cos x \cos y - \sin x \sin y$$

That is,

$$A \cos (\omega t + \theta) = (A \cos \theta) \cos \omega t + (-A \sin \theta) \sin \omega t$$

An example of the conversion of a cosine and angle form to the quadrature form is the following:

$$10 \cos (3t + 60°) = (10 \cos 60°) \cos 3t - (10 \sin 60°) \sin 3t$$

$$= \left(\frac{10}{2}\right) \cos 3t + \left(\frac{-10 \sqrt{3}}{2}\right) \sin 3t$$

In terms of D and E, the amplitude and phase angle of the quadrature sinusoid are

$$A = \sqrt{D^2 + E^2}$$

$$\theta = -\arctan \left(\frac{E}{D}\right)$$

To derive this identity, multiply and divide by $\sqrt{D^2 + E^2}$, giving

$$D \cos \omega t + E \sin \omega t$$

$$= \sqrt{D^2 + E^2} \left[\left(\frac{D}{\sqrt{D^2 + E^2}}\right) \cos \omega t + \left(\frac{E}{\sqrt{D^2 + E^2}}\right) \sin \omega t\right]$$

Defining the angle ϕ as in the sketch of Figure 7-3(a), and using the identity

$$\cos (x - y) = \cos x \cos y + \sin x \sin y$$

there results

$$D \cos \omega t + E \sin \omega t = \sqrt{D^2 + E^2} \, [\cos \phi \cos \omega t + \sin \phi \sin \omega t]$$

$$= \sqrt{D^2 + E^2} \cos (\omega t - \phi)$$

where

$$\phi = -\theta = \arctan \left(\frac{E}{D}\right)$$

It should be noted that the individual algebraic signs of the coefficients D and E must be examined to fix the quadrant of ϕ.

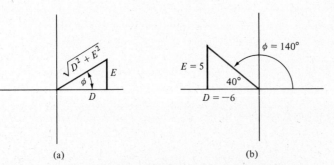

(a)   (b)

Figure 7-3   Phase angle of a sinusoid in quadrature form.

The following is an example of conversion from the quadrature to the cosine and angle form:

$$-6 \cos 10t + 5 \sin 10t = \sqrt{6^2 + 5^2} \cos [10t - \arctan (5/-6)]$$

$$= 7.8 \cos (10t - 140°)$$

The quadrant of the angle $\arctan (5/-6)$ is indicated in Figure 7-3(b).

By converting to the quadrature form and back again, it is straightforward to convert sums of sinusoidal functions with the same frequency to a single sinusoidal function. For example,

$$4 \cos (2t - 60°) + 3 \cos (2t + 45°)$$

$$= \left(\frac{4}{2}\right) \cos 2t + \left(\frac{4\sqrt{3}}{2}\right) \sin 2t + \left(\frac{3}{\sqrt{2}}\right) \cos 2t - \left(\frac{3}{\sqrt{2}}\right) \sin 2t$$

$$= \left(2 + \frac{3}{\sqrt{2}}\right) \cos 2t + \left(2\sqrt{3} - \frac{3}{\sqrt{2}}\right) \sin 2t$$

$$= (4.12) \cos 2t + (1.63) \sin 2t$$

$$= \sqrt{(4.12)^2 + (1.63)^2} \cos (2t - \arctan \tfrac{1.63}{4.12})$$

$$= (4.43) \cos (2t - 23°)$$

D7-4

Convert to the form $A \cos (\omega t + \theta)$ with $A \geq 0$, $\omega > 0$, and $-360° < \theta < 360°$:

(a) $f(t) = -2 \cos 5t + \sin 5t$

ans. $\sqrt{5} \cos (5t + 206.5°)$

(b) $f(t) = -6 \cos 10^7t + 3 \sin (-10^7t)$

ans. $6.7 \cos (10^7t + 153.5°)$

(c) $f(t) = -7 \cos 10t + 6 \cos (10t - 45°)$

ans. $5.06 \cos (10t + 237°)$

(d) $f(t) = -3 \cos (2t - 90°) - 4 \cos (2t + 120°)$

ans. $2.05 \cos (2t - 13°)$

SINUSOIDAL FUNCTIONS

A sinusoidal function may be expressed in the form

$$f(t) = A \cos (\omega t + \theta)$$

with amplitude A nonnegative, radian frequency ω positive, and phase angle θ in a 360° range.

The following identities allow conversion of sinusoidal functions in other forms to this standard form:

$$\cos(-x) = \cos x \qquad \sin(-x) = -\sin x$$

$$\cos(x + 180°) = -\cos x \qquad \sin(x + 180°) = -\sin x$$

$$\sin x = \cos(x - 90°) \qquad \cos x = \sin(x + 90°)$$

and

$$D\cos \omega t + E\sin \omega t = A\cos(\omega t + \theta)$$

where

$$A = \sqrt{D^2 + E^2}, \qquad \theta = -\arctan\frac{E}{D}$$

$$D = A\cos\theta, \qquad E = -A\sin\theta$$

7.3 SOLUTIONS TO DIFFERENTIAL EQUATIONS WITH SINUSOIDAL DRIVING FUNCTIONS

For a linear, time-invariant differential equation with sinusoidal driving function,

$$a_n\frac{d^n y}{dt^n} + a_{n-1}\frac{d^{n-1} y}{dt^{n-1}} + \cdots + a_1\frac{dy}{dt} + a_0 y = A\cos(\omega t + \theta)$$

the forced component of the solution is, except in a special case, sinusoidal with the same radian frequency, ω, as the driving function.

After all, the time derivative of a sinusoid,

$$\frac{d}{dt}[A\cos(\omega t + \theta)] = -A\omega\sin(\omega t + \theta)$$

is another sinusoid with the same frequency. Sums of sinusoids of the same frequency are sinusoidal with that frequency. So if the operations

$$a_n\frac{d^n}{dt^n} + a_{n-1}\frac{d^{n-1}}{dt^{n-1}} + \cdots + a_1\frac{d}{dt} + a_0$$

are performed on a sinusoid with frequency ω, the result is a sinusoid with frequency ω. The problem is then to find the amplitude and phase angle of the solution that yields the correct driving function amplitude A and phase angle θ. As the following example illustrates, it is easiest, in directly solving differential equations with sinusoidal driving functions, to deal with the sinusoids in quadrature form.

Consider the equation,

$$\frac{dy}{dt} + 2y = 4\cos 3t + 2\sin 3t$$

where the driving function has been expressed in quadrature form. Substituting a trial forced solution with the same frequency as the driving function,

$$y_f(t) = D\cos 3t + E\sin 3t$$

gives

$$\frac{dy_f}{dt} + 2y_f = -3D \sin 3t + 3E \cos 3t + 2D \cos 3t + 2E \sin 3t$$

$$= (2D + 3E) \cos 3t + (-3D + 2E) \sin 3t$$

$$= 4 \cos 3t + 2 \sin 3t$$

Equating coefficients,

$$\begin{cases} 2D + 3E = 4 \\ -3D + 2E = 2 \end{cases}$$

$$D = \frac{\begin{vmatrix} 4 & 3 \\ 2 & 2 \end{vmatrix}}{\begin{vmatrix} 2 & 3 \\ -3 & 2 \end{vmatrix}} = \frac{2}{13}, \qquad E = \frac{\begin{vmatrix} 2 & 4 \\ -3 & 2 \end{vmatrix}}{13} = \frac{16}{13}$$

So

$$y_f(t) = \tfrac{2}{13} \cos 3t + \tfrac{16}{13} \sin 3t$$

which could, if desired, be expressed in the cosine and angle form:

$$y_f(t) = \sqrt{(\tfrac{2}{13})^2 + (\tfrac{16}{13})^2} \cos \left(3t - \arctan \frac{\tfrac{16}{13}}{\tfrac{2}{13}} \right)$$

$$= 1.24 \cos (3t - 82.9°)$$

The form of the natural component of the solution is found, as usual, from the homogeneous equation,

$$\frac{dy_n}{dt} + 2y_n = 0$$

Substituting

$$y_n(t) = Ke^{st}$$

$$(s + 2)Ke^{st} = 0$$

$$s = -2$$

and

$$y_n(t) = Ke^{-2t}$$

The general solution to the equation is then

$$y(t) = y_f(t) + y_n(t)$$

$$= \tfrac{2}{13} \cos 3t + \tfrac{16}{13} \sin 3t + Ke^{-2t}$$

which involves the one arbitrary constant.

The singular case, similar to that with exponential driving functions, occurs if the equation is such that a natural component of its solution is sinusoidal with the same frequency as the driving function. Then it will not be possible to equate coefficients for the forced component of the solution. Similar to the exponential case, the forced solution in this special circumstance then involves t times the sinusoid.

D7-5

Find the general solutions of each of the following differential equations:

(a) $\dfrac{dy}{dt} = 3 \cos 2t$ *ans.* $K + \frac{3}{2} \sin 2t$

(b) $\dfrac{dy}{dt} + 2y = 6 \sin 3t$ *ans.* $Ke^{-2t} + (-\frac{18}{13}) \cos 3t + \frac{12}{13} \sin 3t$

(c) $2\dfrac{dy}{dt} + 3y = 4 \cos (2t - 45°)$ *ans.* $Ke^{-(3/2)t} - 0.113 \cos 2t + 0.8 \sin 2t$

(d) $\dfrac{d^2y}{dt^2} + 3\dfrac{dy}{dt} + 2y = 4 \cos 5t$ *ans.* $K_1e^{-t} + K_2e^{-2t} + (-\frac{46}{377}) \cos 5t$
                                     $+ \frac{30}{377} \sin 5t$

SOLUTIONS TO SINUSOIDALLY DRIVEN DIFFERENTIAL EQUATIONS

The sum of two sinusoids of the same frequency is another sinusoid of that frequency. The derivative of a sinusoid is a sinusoid of the same frequency.

The forced component of the solution of a linear, time-invariant differential equation with a sinusoidal driving function is (except in the singular case) another sinusoid with the same frequency as the driving function.

7.4 THE ALGEBRA OF COMPLEX NUMBERS

7.4.1 Rectangular Form

The rather large number of trigonometric identities involved in the solution of the differential equations describing networks may be avoided, or at least simplified, through the use of complex numbers. The principles of complex algebra are now summarized, as preparation for the use of complex numbers throughout the remainder of the text.

A complex number $m = a + jb$ consists of a real part.

Re $[m] = a$

plus $j = \sqrt{-1}$ times an imaginary part.

Im $[m] = b$

Mathematicians often use the symbol i for the imaginary unit, but because of the conflict with the symbol i for electric current (which was chosen before the usefulness of complex numbers in network analysis was generally recognized), engineers and most scientists prefer to use the symbol j.

The algebra of complex numbers is the same as real number algebra, where j is just treated as the constant it is. Powers of j are

$$j^2 = (\sqrt{-1})^2 = -1$$

$$j^3 = (\sqrt{-1})^3 = -1\sqrt{-1} = -j$$

$$j^4 = (\sqrt{-1})^4 = (-1)^2 = 1$$

$$j^5 = j$$

and so on.

A real number k times a complex number m gives

$$km = (ka) + j(kb)$$

The *complex conjugate* of a complex number m is denoted by m^* and is defined to be

$$m^* = (a + jb)^* = a - jb = a + j(-b)$$

The complex conjugate has the same real part as the original number, but the negative of its imaginary part. The solutions given by the quadratic formula, when they are complex, are complex conjugates.

If

$$m_1 = a_1 + jb_1 \qquad \text{and} \qquad m_2 = a_2 + jb_2$$

the sum of the complex numbers is given by

$$m_1 + m_2 = a_1 + jb_1 + a_2 + jb_2 = (a_1 + a_2) + j(b_1 + b_2)$$

Their difference is

$$m_1 - m_2 = a_1 + jb_1 - a_2 - jb_2 = (a_1 - a_2) + j(b_1 - b_2)$$

The sum of a complex number and its complex conjugate is twice the real part of either number:

$$m + m^* = a + jb + a - jb = 2a = 2 \operatorname{Re}[m] = 2 \operatorname{Re}[m^*]$$

The product and quotient of two complex numbers have real and imaginary parts which also may be found using the ordinary rules of algebra. For the product,

$$m_1 m_2 = (a_1 + jb_1)(a_2 + jb_2) = a_1 a_2 + j^2 b_1 b_2 + jb_1 a_2 + jb_2 a_1$$

$$= (a_1 a_2 - b_1 b_2) + j(b_1 a_2 + a_1 b_2)$$

The product of a complex number with its complex conjugate is a real number:

$$mm^* = (a + jb)(a - jb) = a^2 + jab - jab + b^2$$

$$= a^2 + b^2$$

The quotient of two complex numbers requires a little more manipulation. Multiply numerator and denominator by the complex conjugate of the denominator, leaving j's only in the numerator:

$$\frac{m_1}{m_2} = \frac{a_1 + jb_1}{a_2 + jb_2} = \frac{(a_1 + jb_1)(a_2 - jb_2)}{(a_2 + jb_2)(a_2 - jb_2)} = \frac{(a_1a_2 + b_1b_2) + j(b_1a_2 - a_1b_2)}{a_2^2 + b_2^2}$$

For example,

$$\frac{3 - j2}{-4 + j5} = \frac{(3 - j2)(-4 - j5)}{(-4 + j5)(-4 - j5)}$$

$$= \frac{-12 - j15 + j8 - 10}{16 + 25}$$

$$= -\frac{22}{41} - j\frac{7}{41}$$

D7-6

Express the following complex numbers in the form $a + jb$:

(a) $(2 - j3)(j4)$ ans. $12 + j8$

(b) $\dfrac{(1 - j2)}{(5 + j6)}$ ans. $-\frac{7}{61} - j\frac{16}{61}$

(c) $\dfrac{(-2 - j8)(4 + j1)}{(3 - j5)}$ ans. $5 - j3$

(d) $\dfrac{(-4 + j3)}{(-3 + j2)(5 + j9)}$ ans. $\frac{81}{1378} + j - \frac{167}{1378}$

7.4.2 Polar Coordinates

It is often convenient to plot the real and imaginary parts of complex numbers on a two-axis plot, which is called the *complex plane*, Figure 7-4. An arrow is drawn from the origin to the point representing a complex number. Every complex number is

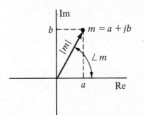

Figure 7-4   Complex number plotted on the complex plane.

represented by a unique point on the complex plane, and every point on the plane represents a complex number.

The distance from the origin of the complex plane to the point on the plane representing a complex number is called the *magnitude* of a complex number m and is denoted by $|m|$. In terms of the number's real and imaginary parts,

$$|m| = |a + jb| = \sqrt{a^2 + b^2}$$

The *angle* of a complex number, denoted by $\angle m$, is the angle a line from the origin to the point representing the complex number makes with the positive real axis:

$$\angle m = \arctan \frac{b}{a}$$

The algebraic signs of both b and a must be examined to determine the quadrant of $\angle m$; the ratio alone, (b/a), is not sufficient to determine the angle. For example, if

$$m = -3 + j3$$

$$|m| = \sqrt{3^2 + 3^2} = 3\sqrt{2} \quad \text{and} \quad \angle m = \arctan \frac{3}{-3} = 135°$$

as shown in Figure 7-5.

In terms of its magnitude and angle, the real and imaginary parts of a complex number are

$$\text{Re } [m] = |m| \cos (\angle m) \qquad \text{Im } [m] = |m| \sin (\angle m)$$

If

$$|m| = 4 \quad \text{and} \quad \angle m = 60°$$

$$m = a + jb = |m| \cos (\angle m) + j|m| \sin (\angle m)$$

$$= 4\left(\frac{1}{2}\right) + j4\left(\frac{\sqrt{3}}{2}\right) = 2 + j2\sqrt{3}$$

Multiplication of a complex number m by a positive real number k multiplies the magnitude by k and leaves the angle unchanged:

$$\begin{cases} |km| = k|m| \\ \angle km = \angle m, \quad \text{if } k \text{ is positive} \end{cases}$$

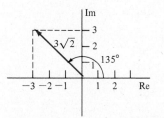

Figure 7-5 Example of magnitude and angle of a complex number.

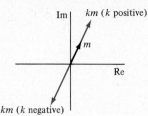

Figure 7-6 Graphical multiplication of a complex number by a real number.

If the real number is negative, then the magnitude of the product is changed by the factor $|k|$, and the direction of the product is reversed; that is, the angle is changed by $180°$:

$$\begin{cases} |km| = |k|\,|m| \\ \angle km = \angle m + 180°, \qquad \text{if } k \text{ is negative} \end{cases}$$

Figure 7-6 shows these relationships.

The addition of two complex numbers is easy to perform graphically on the complex plane. If a parallelogram is constructed from the sides m_1 and m_2, the diagonal of the parallelogram is the arrow for the sum $(m_1 + m_2)$. Subtraction of one complex number from another,

$$m_1 - m_2$$

may be easily done graphically on the complex plane by first constructing $(-m_2)$ from m_2, by reversing the direction of the arrow for m_2, and then adding m_1 and $(-m_2)$ by constructing the addition parallelogram.

If

$$m_1 = 2 - j3 \qquad \text{and} \qquad m_2 = -1 + j4$$

the sum

$$m_1 + m_2 = 1 + j$$

constructed graphically is shown in Figure 7-7(a). The difference,

$$m_1 - m_2 = 3 - j7$$

is constructed graphically in Figure 7-7(b).

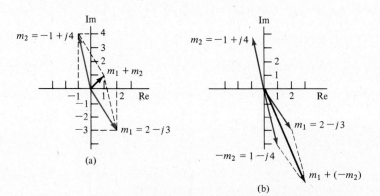

Figure 7-7 Graphical addition and subtraction of complex numbers.

D7-7

Sketch plots of the following complex numbers on the complex plane. Show the axes and scales carefully:

(a) $2 + j2$

(b) $2 - j2$

(c) $-2 + j2$

(d) $-2 - j2$

D7-8

Calculate the magnitudes and angles of each of the following complex numbers:

(a) $1 + j2$ *ans.* $\sqrt{5}$, 63.4°

(b) $3 - j4$ *ans.* 5, $-53.1°$

(c) $-4 + j3$ *ans.* 5, 143.1°

(d) $-2 - j1$ *ans.* $\sqrt{5}$, 206.6°

D7-9

Find approximate graphical solutions for the magnitudes and angles of the following complex numbers, using a sketch of the number on the complex plane:

(a) $7 - j8$ *ans.* 10.6, $-48.7°$

(b) $-10^4 + j10^5$ *ans.* 1.005×10^5, 95.7°

(c) $-0.13 - j0.52$ *ans.* 0.54, $-104°$

(d) $1.7 + j4.6$ *ans.* 4.9, 69.7°

7.4.3 Euler's Relation

Specifying the magnitude and angle of a complex number is equivalent to specifying its real and imaginary parts. There is a difference, though, in that up to now $m = a + jb$ is the complex number, and $(|m|, \angle m)$ are only a set of coordinates for the number. Leonard Euler (1707–1783), a Swiss mathematical genius, changed this viewpoint considerably. He showed that any complex number $m = a + jb$ could also be expressed in the polar form

$$m = a + jb = Ke^{j\theta}$$

where

$$K = |m| \quad \text{and} \quad \theta = \angle m$$

Consider the power series for the following functions:

$$e^x = 1 + \frac{x}{1!} + \frac{x^2}{2!} + \frac{x^3}{3!} + \cdots$$

$$\cos x = 1 - \frac{x^2}{2!} + \frac{x^4}{4!} - \frac{x^6}{6!} + \cdots$$

$$\sin x = \frac{x}{1!} - \frac{x^3}{3!} + \frac{x^5}{5!} - \frac{x^7}{7!} + \cdots$$

The angle x must be expressed in *radians* for the series for $\cos x$ and $\sin x$ to be in this simple form.

If in the series for e^x, x is replaced in the series by $j\alpha$, there results

$$e^{j\alpha} = 1 + \frac{j\alpha}{1!} + \frac{j^2\alpha^2}{2!} + \frac{j^3\alpha^3}{3!} + \frac{j^4\alpha^4}{4!} + \frac{j^5\alpha^5}{5!} + \frac{j^6\alpha^6}{6!} + \frac{j^7\alpha^7}{7!} + \cdots$$

Now $j^2 = (\sqrt{-1})^2 = -1$, $j^3 = j(j^2) = -j$, $j^4 = j(j^3) = 1$, $j^5 = j(j^4) = j$, and so on, so

$$e^{j\alpha} = \left(1 - \frac{\alpha^2}{2!} + \frac{\alpha^4}{4!} - \frac{\alpha^6}{6!} + \cdots\right) + j\left(\frac{\alpha}{1!} - \frac{\alpha^3}{3!} + \frac{\alpha^5}{5!} - \frac{\alpha^7}{7!} + \cdots\right)$$

$$= \cos \alpha + j \sin \alpha$$

This result is known as Euler's relation.

A complex number may thus be expressed in terms of its magnitude

$$K = |m| = \sqrt{a^2 + b^2}$$

and angle

$$\theta = \angle m = \arctan \frac{b}{a}$$

as

$$m = a + jb = K \cos \theta + jK \sin \theta = K(\cos \theta + j \sin \theta)$$

$$= Ke^{j\theta}$$

For example,

$$-3 - j2 = 3.61e^{j214°} = 3.61e^{-j146°}$$

as shown in Figure 7-8. An alternate magnitude and angle notation

$$3.61e^{-j146°} = 3.61 \underline{/-146°}$$

is also used, but we will not do so here until later because it is important to emphasize the exponential nature of the polar form.

Some identities that will be very useful later in this chapter may be obtained as follows. Substituting $-\alpha$ for α in Euler's relation,

$$e^{-j\alpha} = \cos(-\alpha) + j \sin(-\alpha) = \cos \alpha - j \sin \alpha$$

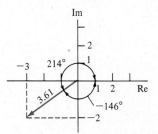

Figure 7-8 Rectangular to polar conversion.

Adding $e^{j\alpha}$ and $e^{-j\alpha}$ gives

$$e^{j\alpha} + e^{-j\alpha} = 2\cos\alpha$$

or

$$\cos\alpha = \frac{e^{j\alpha} + e^{-j\alpha}}{2}$$

Similarly,

$$\sin\alpha = \frac{e^{j\alpha} - e^{-j\alpha}}{2j}$$

D7-10

Express the following complex numbers in polar form, using the Euler identity. Do not use graphical methods:

(a) $-4 + j4$ *ans.* $4\sqrt{2}e^{j135°}$

(b) $3 - j4$ *ans.* $5e^{-j53.1°}$

(c) $-4 - j3$ *ans.* $5e^{j216.8°}$

(d) $j4(2 - j2)$ *ans.* $8\sqrt{2}e^{j45°}$

D7-11

Express the following complex numbers in rectangular form. Do not use graphical methods:

(a) $6e^{-j45°}$ *ans.* $(6/\sqrt{2}) - j(6/\sqrt{2})$

(b) $3e^{j30°}$ *ans.* $2.6 + j1.5$

(c) $10^6 e^{-j60°}$ *ans.* $5 \times 10^5 - j8.7 \times 10^5$

(d) $-5e^{j120°}$ *ans.* $2.5 - j4.33$

7.4.4 Multiplication and Division in Polar Form

Using the exponential properties

$$e^x e^y = e^{(x+y)} \quad \text{and} \quad \frac{e^x}{e^y} = e^{(x-y)}$$

multiplication and division of complex numbers is seen to be very easy when performed in the polar form.

The product of two complex numbers,

$$(K_1 e^{j\theta_1})(K_2 e^{j\theta_2}) = K_1 K_2 e^{j(\theta_1 + \theta_2)}$$

has magnitude the product of the individual magnitudes and angle that is the sum of the individual angles. For example,

$$(3e^{j36°})(4e^{-j120°}) = 12e^{-j84°}$$

The quotient of two complex numbers,

$$\frac{(K_1 e^{j\theta_1})}{(K_2 e^{j\theta_2})} = \frac{K_1}{K_2} e^{j(\theta_1 - \theta_2)}$$

has magnitude the quotient of the individual magnitudes and angle equal to the numerator angle minus the denominator angle. For example,

$$\frac{7e^{-j100°}}{10e^{j70°}} = \frac{7}{10} e^{-j170°}$$

More complicated products and quotients are similarly formed:

$$\frac{(11e^{j30°})(5e^{-j50°})}{(4e^{-j120°})(6e^{j70°})} = \frac{(11)(5)}{(4)(6)} e^{j(30° - 50° + 120° - 70°)} = \frac{55}{24} e^{j30°}$$

D7-12

Find the polar form, $Ke^{j\theta}$ with $K \geq 0$ and $-360° < \theta < 360°$, of each of the following complex numbers. Use graphical aid where appropriate:

(a) $(6e^{j40°})(-2e^{j610°})$ ans. $12e^{j110°}$

(b) $\dfrac{5e^{j30°}(2 - j2)}{-2e^{j60°}}$ ans. $7.08e^{j105°}$

(c) $\dfrac{7e^{-j50°}}{(-j4)(3 - j4)}$ ans. $0.35e^{j93°}$

(d) $\dfrac{3e^{j70°}(-4 + j3)}{-3e^{j150°}}$ ans. $5e^{j243°}$

COMPLEX ALGEBRA

A complex number in rectangular form

$$m = a + jb$$

consists of real part a and imaginary part b. The complex conjugate of a complex number m is denoted by m^* and is

$$m^* = a - jb$$

Complex algebra is identical to real number algebra, where $j = \sqrt{-1}$ is treated as a constant. Addition and subtraction of complex numbers are easiest to do in rectangular form and may be done graphically, using the parallelogram rule.

A complex number may alternatively be expressed in polar form,

$$m = Ke^{j\theta}$$

where the magnitude is

$$K = |m| = \sqrt{a^2 + b^2}$$

and the angle of the complex number is

$$\theta = \angle m = \arctan\left(\frac{b}{a}\right)$$

Multiplication and division of complex numbers is easiest to perform in polar form. Euler's relation

$$e^{\pm j\alpha} = \cos \alpha \pm j \sin \alpha$$

leads to the identities

$$\cos \alpha = \frac{e^{j\alpha} + e^{-j\alpha}}{2}$$

$$\sin \alpha = \frac{e^{j\alpha} - e^{-j\alpha}}{2j}$$

7.5 THE METHOD OF COMPONENTS

7.5.1 Superposition of Euler Components

The Euler's relation identity

$$\cos \alpha = \tfrac{1}{2}e^{j\alpha} + \tfrac{1}{2}e^{-j\alpha}$$

allows expression of a sinusoidal source as the sum of two exponential components, which may be superimposed. Each component is exponential, so that impedance methods rather than involved trigonometric identities may be used for solution. For example, in the network of Figure 7-9(a), the sinusoidal source has been expressed as the sum of two complex exponential components and the two terms, which involve different values of s, superimposed.

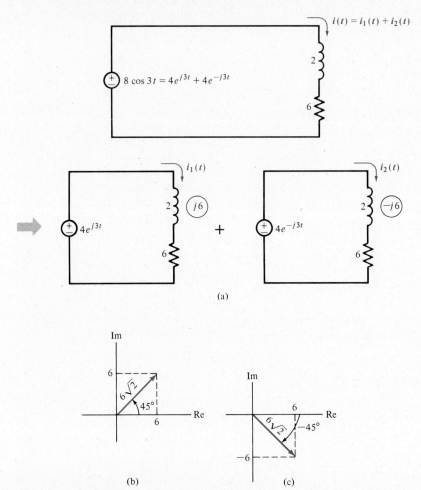

Figure 7-9   Finding forced sinusoidal network response using the component method.
(a) Superposition of complex source components.
(b) and (c) Rectangular to polar complex number conversions.

For the first component problem, $s = j3$ and the current is

$$i_1(t) = \frac{4e^{j3t}}{6 + j6}$$

To perform the indicated division, convert the denominator to polar form, as indicated in Figure 7-9(b),

$$6 + j6 = 6\sqrt{2}e^{j45°}$$

Then

$$i_1(t) = \frac{4e^{j3t}}{6 + j6} = \frac{4e^{j3t}}{6\sqrt{2}e^{j45°}} = \frac{2}{3\sqrt{2}}e^{j(3t - 45°)}$$

Similarly, the second component problem has $s = -j3$ and the second current component is

$$i_2(t) = \frac{4e^{-j3t}}{6 - j6}$$

Converting to polar form, Figure 7-9(c),

$$6 - j6 = 6\sqrt{2}e^{-j45°}$$

and

$$i_2(t) = \frac{4e^{-j3t}}{6 - j6} = \frac{4e^{-j3t}}{6\sqrt{2}e^{-j45°}} = \frac{2}{3\sqrt{2}}e^{j(-3t+45°)}$$

$$= \frac{2}{3\sqrt{2}}e^{-j(3t-45°)}$$

The component currents, being complex functions of time, could not be measured in the laboratory; physical currents are real functions of time. Their sum, however, is the physical current of interest:

$$i(t) = i_1(t) + i_2(t) = \frac{2}{3\sqrt{2}}e^{j(3t-45°)} + \frac{2}{3\sqrt{2}}e^{-j(3t-45°)}$$

$$= \frac{2}{3\sqrt{2}}[e^{j(3t-45°)} + e^{-j(3t-45°)}] = \frac{2}{3\sqrt{2}}[2\cos(3t - 45°)]$$

$$= \frac{4}{3\sqrt{2}}\cos(3t - 45°) \text{ A}$$

D7-13

Expand the following sinusoidal functions into Euler components:

(a) 8 cos 7t   ans. $4e^{j7t} + 4e^{-j7t}$

(b) 8 cos (6t + 58°)   ans. $4e^{j(6t+58°)} + 4e^{-j(6t+58°)}$

(c) $\frac{1}{20}$ cos (100t − 80°)   ans. $\frac{1}{40}(e^{-j80°})e^{j100t} + \frac{1}{40}(e^{j80°})e^{-j100t}$

(d) 10 sin 5t   ans. $(5e^{-j90°})e^{j5t} + (5e^{j90°})e^{-j5t}$

7.5.2 Voltage-Source Example

In the more general problem, where the sinusoidal source function has a nonzero phase angle,

$$f(t) = A\cos(\omega t + \theta)$$

the Euler components are

$$f(t) = A \cos (\omega t + \theta) = \frac{A}{2} e^{j(\omega t + \theta)} + \frac{A}{2} e^{-j(\omega t + \theta)}$$

$$= \left(\frac{A}{2} e^{j\theta} \right) e^{j\omega t} + \left(\frac{A}{2} e^{-j\theta} \right) e^{-j\omega t}$$

The first component,

$$\left(\frac{A}{2} e^{j\theta} \right) e^{j\omega t}$$

is of the form Ke^{st} with

$$K = \frac{A}{2} e^{j\theta}$$

and $s = j\omega$. For the second component,

$$K = \frac{A}{2} e^{-j\theta}$$

and $s = -j\omega$.

A numerical network solution example is shown in Figure 7-10(a). The first current component is

$$i_1(t) = \frac{\frac{7}{2} e^{j(4t + 30°)}}{2 - j\frac{5}{4}}$$

To place $i_1(t)$ in polar form, the denominator is converted to the polar form

$$2 - j\frac{5}{4} = 2.36 e^{-j32°}$$

as indicated graphically in Figure 7-10(b). Then

$$i_1(t) = \frac{3.5 e^{j(4t + 30°)}}{2.36 e^{j(-32°)}} = 1.48 e^{j(4t + 62°)}$$

For the second component,

$$i_2(t) = \frac{\frac{7}{2} e^{-j(4t + 30°)}}{2 + j\frac{5}{4}} = \frac{3.5 e^{-j(4t + 30°)}}{2.36 e^{j32°}}$$

$$= 1.48 e^{-j(4t + 62°)}$$

which gives

$$i(t) = i_1(t) + i_2(t) = 1.48[e^{j(4t + 62°)} + e^{-j(4t + 62°)}]$$

$$= 1.48[2 \cos (4t + 62°)]$$

$$= 2.96 \cos (4t + 62°) \text{ A}$$

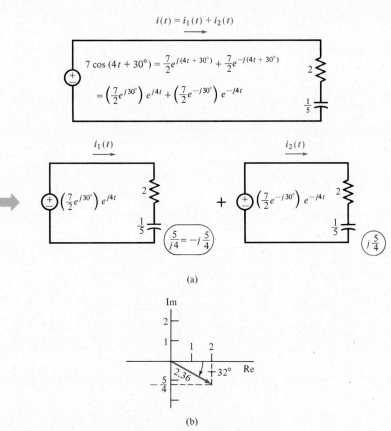

(a)

(b)

Figure 7-10   Example network with voltage source.
(a) Superposition of Euler components.
(b) Rectangular to polar conversion.

7.5.3 Current-Source Example

The network of Figure 7-11(a) involves a current source. In this example, the original problem is decomposed into the superposition of the two Euler component problems, to which impedance methods apply. Then equivalent circuits are used to solve each of the component problems by placing them in the form of current dividers.
The solution to the first component problem is

$$i_1(t) = \frac{(-j\frac{1}{6})[\frac{15}{2} e^{j(3t-100°)}]}{-j\frac{1}{6} + 1 + j\frac{3}{4}} = \frac{(\frac{1}{6}e^{-j90°})[\frac{15}{2} e^{j(3t-100°)}]}{1 + j\frac{7}{12}}$$

The $-j\frac{1}{6}$ term in the numerator is easily converted to polar form. The denominator term is converted to polar form as indicated in Figure 7-11(b). Then

$$i_1(t) = \frac{(\frac{1}{6}e^{-j90°})[\frac{15}{2} e^{j(3t-100°)}]}{1.16e^{j30.2°}}$$

$$= 1.08e^{j(3t-220.2°)}$$

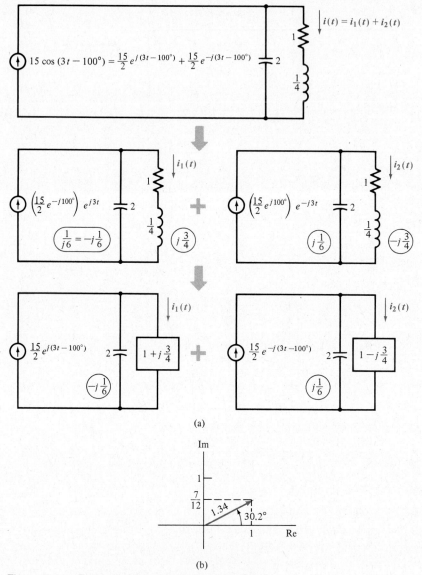

(a)

(b)

Figure 7-11   Example network with current source.
(a) Superposition of Euler components.
(b) Rectangular to polar conversion.

Similarly, using the current divider rule for the second component problem,

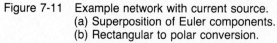

$$i_2(t) = \frac{(j\frac{1}{6})[\frac{15}{2} e^{-j(3t-100°)}]}{j\frac{1}{6} + 1 - j\frac{7}{12}} = \frac{(\frac{1}{6}e^{j90°})[\frac{15}{2} e^{-j(3t-100°)}]}{1 - j\frac{7}{12}}$$

$$= \frac{(\frac{1}{6}e^{j90°})[\frac{15}{2} e^{-j(3t-100°)}]}{1.16e^{-j30.2°}} = 1.08e^{j(-3t+220.2°)}$$

$$= 1.08e^{-j(3t-220.2°)}$$

The current $i(t)$ is thus

$$i(t) = i_1(t) + i_2(t) = 1.08[e^{j(3t - 220.2°)} + e^{-j(3t - 220.2°)}]$$

$$= 1.08[2 \cos (3t - 220.2°)]$$

$$= 2.16 \cos (3t - 220.2°) \text{ A}$$

Once a sinusoidally driven network problem is converted to equivalent component problems with impedances, any source-resistor network solution technique is applicable.

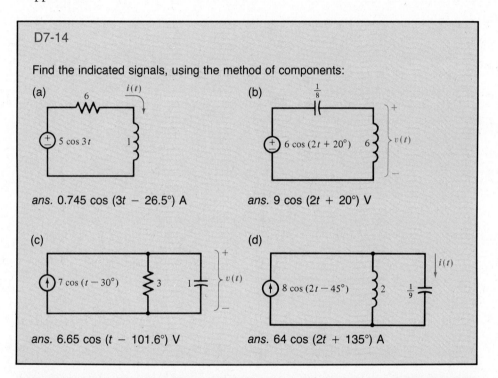

D7-14

Find the indicated signals, using the method of components:

(a)

ans. 0.745 cos (3t − 26.5°) A

(b)

ans. 9 cos (2t + 20°) V

(c)

ans. 6.65 cos (t − 101.6°) V

(d)

ans. 64 cos (2t + 135°) A

THE METHOD OF COMPONENTS

A sinusoidal source may be decomposed into the sum of two complex exponential sources via Euler's relation. The two complex exponential sources may be superimposed, and impedance methods may be used separately for each component.

7.6 THE METHOD OF SINORS

7.6.1 Redundancy of the Two Component Solutions

Considerable savings in the effort needed to find forced network response to a sinusoidal input may be made. The two Euler component responses, like the input components, are always complex conjugates of one another. That is, their real parts

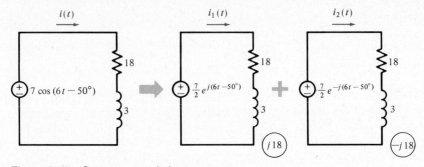

Figure 7-12 Component solution.

are identical, and their imaginary parts are negatives of one another. Thus it is only necessary to *solve* for one of the two components; the second component may be easily found from the first.

As an example, the component solution of a network is outlined in Figure 7-12. The second component problem is related to the first by

$$i_2(t) = i_1^*(t)$$

The original problem solution is thus

$$i(t) = i_1(t) + i_2(t) = i_1(t) + i_1^*(t)$$

$$= 2 \, \text{Re} \, [i_1(t)]$$

the sum of a complex number and its complex conjugate being twice the real part of either number.

7.6.2 The Sinor Solution

In the sinor solution method, the division by two to form the first component and the later multiplication by two (taking twice the real part) is eliminated. The response due to twice the first component is found, then one times the real part of that result is the desired signal. The sinusoidal source,

(amplitude) cos (argument)

is replaced by the *sinor* source,

(amplitude)$e^{j(\text{argument})}$

converting the original problem into the equivalent sinor problem, as in Figure 7-13. Since inductor and capacitor impedances are imaginary numbers, circling impedances to distinguish them from element values on the network diagram here is unnecessary.

Impedance methods apply to the sinor problem, which, for the example, gives

$$\underline{i}(t) = \frac{7e^{j(6t-50°)}}{18 + j18} = \frac{7e^{j(6t-50°)}}{18\sqrt{2}e^{j45°}}$$

$$= \frac{7}{18\sqrt{2}} e^{j(6t-95°)}$$

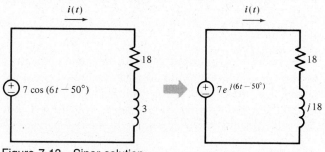

Figure 7-13   Sinor solution.

The sinor current is distinguished from the original current by the underbar. The original current is then

$$i(t) = \text{Re}\,[\underline{i}(t)] = \text{Re}\left[\frac{7}{18\sqrt{2}}\,e^{j(6t-95°)}\right]$$

$$= \frac{7}{18\sqrt{2}}\cos(6t - 95°)\text{ A}$$

Figure 7-14 shows another example of the sinor solution method. The original, sinusoidal problem is replaced by the corresponding sinor network. Combining the individual impedances into a single equivalent impedance gives

$$\underline{v}(t) = \frac{[10e^{j((1/2)t+80°)}](j2)(3 - j)}{(3 + j)}$$

Next, $\underline{v}(t)$ is expressed in polar form:

$$\underline{v}(t) = \frac{[10e^{j((1/2)t+80°)}](2e^{j90°})(3.16e^{-j18.3°})}{(3.16e^{j18.3°})}$$

$$= 20e^{j((1/2)t+133.4°)}$$

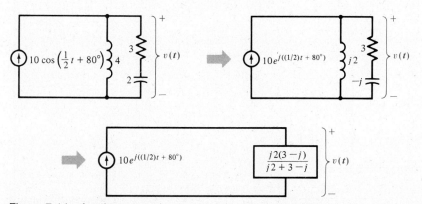

Figure 7-14   Another network solution using the sinor method.

The desired voltage $v(t)$ is then

$$v(t) = \text{Re} \, [\underline{v}(t)] = 20 \cos \left(\tfrac{1}{2}t + 133.4°\right) \text{ V}$$

The real part of each voltage and each current sinor in the sinor network is the corresponding voltage of current of interest in the original, sinusoidal network.

The sinor may alternatively be interpreted as the addition of the imaginary part $jA \sin (\omega t + \theta)$ to the actual sinusoidal source, $A \cos (\omega t + \theta)$. The real part of every sinor network voltage and current is the solution due to the real part of the source.

D7-15

Find the sinors that represent the following sinusoidal functions:

(a) $6 \cos (3t + 47°)$ *ans.* $6e^{j(3t + 47°)}$

(b) $8 \cos \left(8t - \dfrac{3\pi}{4} \right)$ *ans.* $8e^{j[8t - (3\pi/4)]}$

(c) $10^5 \cos 10^6 t$ *ans.* $10^5 e^{j10^6 t}$

(d) $2.6 \cos (377t - 120°)$ *ans.* $2.6e^{j(377t - 120°)}$

D7-16

Find the indicated signals, using the method of sinors:

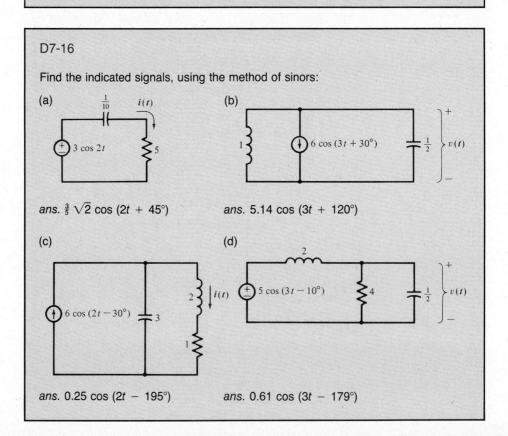

(a)

ans. $\tfrac{3}{5} \sqrt{2} \cos (2t + 45°)$

(b)

ans. $5.14 \cos (3t + 120°)$

(c)

ans. $0.25 \cos (2t - 195°)$

(d)

ans. $0.61 \cos (3t - 179°)$

7.6.3 Multiple Sources

To find forced response in a network with more than a single sinusoidal source, the sources must be superimposed if they are of different frequency. After all, the source sinors will have different values of s. An example of such a network is given in Figure 7-15. The sources are superimposed, and each of the two single-source problems is solved, using sinors. For one source, $s = j1$ and for the other, $s = j2$.

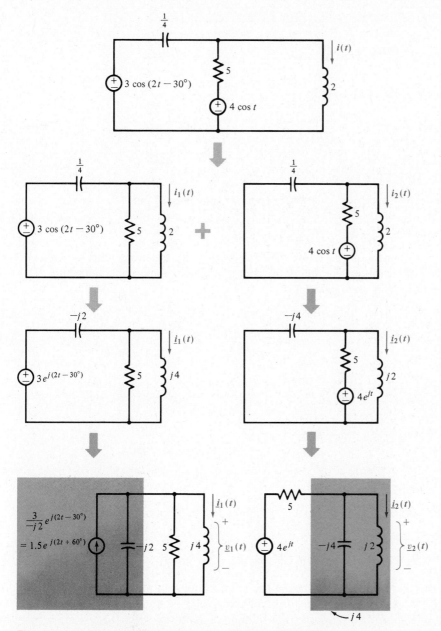

Figure 7-15 Superimposing sources of different frequency.

For the solution of the first superposition problem, where $s = j2$, the sinor voltage source in series with the impedance $-j2$ has been converted into an equivalent current source in parallel with the impedance. The node-to-node voltage $\underline{v}_1$ of the resulting network is given by

$$\left(\frac{1}{-j2} + \frac{1}{5} + \frac{1}{j4}\right)\underline{v}_1(t) = 1.5e^{j(2t+60°)}$$

$$(-10 + j4 + 5)\underline{v}_1(t) = (j20)1.5e^{j(2t+60°)}$$

$$\underline{v}_1(t) = \frac{(j20)1.5e^{j(2t+60°)}}{-5 + j4}$$

The sinor current of interest is

$$\underline{i}_1 = \frac{\underline{v}_1(t)}{j4} = \frac{7.5e^{j(2t+60°)}}{-5 + j4}$$

$$= \frac{7.5e^{j(2t+60°)}}{6.4e^{j141.3°}} = 1.17e^{j(2t-81.3°)}$$

Thus

$$i_1(t) = 1.17 \cos (2t - 81.3°) \text{ A}$$

In the second superposition problem, where $s = j$, the two parallel impedances of $-j4$ and $j2$ have been combined to give a single equivalent impedance of $j4$. The voltage divider rule then gives

$$\underline{v}_2(t) = \frac{j4}{5 + j4} 4e^{jt}$$

Then,

$$\underline{i}_2(t) = \frac{\underline{v}_2(t)}{j2} = \frac{(2)4e^{jt}}{5 + j4}$$

$$= \frac{8e^{jt}}{6.4e^{j39°}} = 1.25e^{j(t-39°)}$$

so that

$$i_2(t) = 1.25 \cos (t - 39°) \text{ A}$$

The net forced current in the original multiple-frequency source problem is

$$i(t) = i_1(t) + i_2(t)$$

$$= 1.17 \cos (2t - 81.3°) + 1.25 \cos (t - 39°) \text{ A}$$

If there are multiple sources of the same frequency, they need not be superimposed, because they involve the same value of s in the sinor equivalent. The example of Figure 7-16 illustrates the solution of a multiple-source network, where the sources all have the same frequency. Systematic simultaneous loop equations

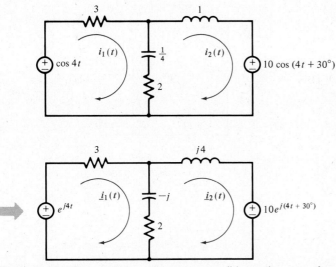

Figure 7-16   Sinor solution when sources all have the same frequency.

for this sinor network are as follows:

$$\begin{cases} (5 - j)\underline{i}_1(t) - (2 - j)\underline{i}_2(t) = e^{j4t} \\ -(2 - j)\underline{i}_1(t) + (2 + j3)\underline{i}_2(t) = -10e^{j(4t+30°)} \end{cases}$$

These may be solved for $\underline{i}_1(t)$ and $\underline{i}_2(t)$. Other sinor voltages and currents are easily found from $\underline{i}_1(t)$ and $\underline{i}_2(t)$, and the real part of each sinor signal is the corresponding sinusoidal signal of interest.

7.6.4 Sources Expressed in Terms of the Sine Function

If a network source is expressed in terms of the sine function instead of cosine, it may be converted into a cosine expression by subtracting 90° from the argument:

$$A \sin (\omega t + \theta) = A \cos (\omega t + \theta - 90°)$$

Or, some savings in computational effort may be made by recognizing that if a source with phase angle θ produces a response with phase angle ϕ, then a source phase, instead, of $(\theta - 90°)$ will give response with phase angle $(\phi - 90°)$.

Figure 7-17 shows an example of a phase-shifted sinor technique. The source $7 \sin (4t - 20°)$ is treated as if it were $7 \cos (4t - 20°)$. Using the voltage divider rule,

$$\underline{v}(t) = \frac{[-4j/(2 - 2j)]7e^{j(4t-20°)}}{-4j/(2 - 2j) + 12j} = \frac{(-4j)[7e^{j(4t-20°)}]}{24 + 20j}$$

$$= \frac{(4e^{-j90°})[7e^{j(4t-20°)}]}{31.2e^{j39.8°}} = 0.9e^{j(4t-149.8°)}$$

The solution for the desired voltage is then, replacing cosine by sine for the result,

$$v(t) = 0.9 \sin (4t - 149.8°) \text{ V}$$

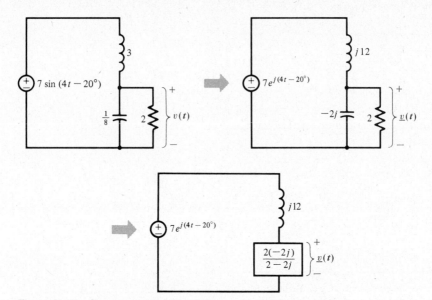

Figure 7-17   Source and solution in terms of the sine function.

Of course, this technique will not work if the network has several sources, some expressed in terms of cosine and some in terms of sine. All of the sources must be placed in terms of one or the other, all cosine or all sine.

Alternatively, it may be interpreted that the *imaginary* parts of the sinor quantities are the actual voltages and currents of interest in this situation.

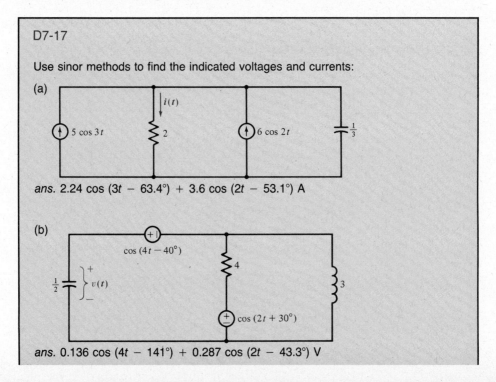

D7-17

Use sinor methods to find the indicated voltages and currents:

(a)

ans. 2.24 cos (3t − 63.4°) + 3.6 cos (2t − 53.1°) A

(b)

ans. 0.136 cos (4t − 141°) + 0.287 cos (2t − 43.3°) V

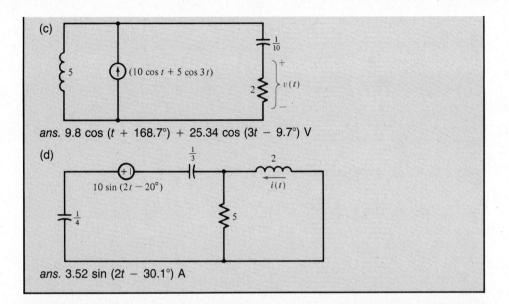

(c)

ans. $9.8 \cos (t + 168.7°) + 25.34 \cos (3t - 9.7°)$ V

(d)

ans. $3.52 \sin (2t - 30.1°)$ A

THE METHOD OF SINORS

The responses of the two Euler components of a sinusoidal source are always complex conjugates of one another. Thus it is only necessary to solve one of the component problems; the solution to the second component problem is the complex conjugate of the first.

In the sinor method, a sinusoidal source

$$A \cos (\omega t + \theta)$$

is replaced by the sinor source

$$A e^{j(\omega t + \theta)}$$

The real part of every voltage and current in the sinor network is the solution for the corresponding signal in the sinusoidal network.

7.7 THE METHOD OF PHASORS

7.7.1 "Snapshots" of Sinors

When plotted on the complex plane, the sinor

$$\underline{y}(t) = A e^{j(\omega t + \theta)}$$

circles counterclockwise about the origin at the angular rate of ω rad/s, as indicated in the drawing of Figure 7-18(a). The projection of the sinor on the real axis, its real part, is the desired sinusoidal function, Figure 7-18(b). All of the sinors for the various voltages and currents in a sinusoidally driven network involve the same radian frequency ω and so rotate together on the complex plane at the same rate of ω rad/s. These sinors maintain the same angular relationships to one another as they rotate.

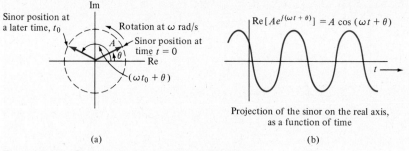

(a)   (b)

Figure 7-18   Sinor rotating on the complex plane.

The complex value of a sinor at time $t = 0$ is called the *phasor* for the corresponding network signal:

$$\mathbf{Y} = \underline{y}(0) = Ae^{j\theta}$$

Phasor quantities are identified by boldface type in printed text. Hand-written characters representing phasors are underlined with a wavy line (like a cycle of a sinusoid) to distinguish them. The phasor magnitude is the amplitude of the sinusoidal signal it represents,

$$|\mathbf{Y}| = A,$$

and its angle is the sinusoidal phase angle:

$$\angle\mathbf{Y} = \theta$$

D7-18

Find the phasors that represent the following sinusoidal functions:

(a) $7 \cos (10t - 80°)$ *ans.* $7e^{-j80°}$

(b) $10^4 \cos (t + \pi/3)$ *ans.* $10^4 e^{j\pi/3}$

(c) $1.2 \cos (0.1t - 60°)$ *ans.* $1.2e^{-j60°}$

(d) $50 \cos 10^4 t$ *ans.* 50

7.7.2 The Phasor Solution

The phasor solution of a sinusoidally driven network is the solution of the sinor network at time $t = 0$. The sinors at time $t = 0$ convey all of the needed information about the solutions for the various sinusoidal network signals, their amplitudes, and their phase angles. Their frequencies are all the same as that of the source or sources. In the phasor solution method, the original problem is replaced by the phasor problem, which is the sinor network at time $t = 0$, as in the example of Figure 7-19. The net effect is to drop the factors of $e^{j\omega t}$ from each of the sinors to form the phasors.

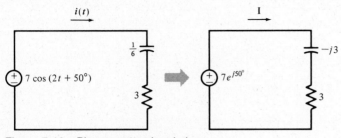

Figure 7-19   Phasor network solution.

The sinusoidal source, $7 \cos (2t + 50°)$, is replaced by the phasor source, $7e^{j50°}$, with magnitude equal to the source amplitude and angle the source's phase angle. The impedances are evaluated at $s = j\omega = j2$, as with sinors, resulting in a complex number source-resistor network.

The phasor **I** is the value of the sinor $\underline{i}(t)$ at time $t = 0$. It is a complex number with magnitude equal to the amplitude of the corresponding current of interest $i(t)$ and angle equal to the phase angle of $i(t)$. The single-loop equation for **I** is

$$(3 - j3)\mathbf{I} = 7e^{j50°}$$

$$\mathbf{I} = \frac{7e^{j50°}}{3 - j3} = \frac{7e^{j50°}}{3\sqrt{2}e^{-j45°}} = \frac{7}{3\sqrt{2}}\,e^{j95°}$$

The current $i(t)$ thus has amplitude $\frac{7}{3}\sqrt{2}$ and phase angle 95°. Its radian frequency is the same as the network source, $\omega = 2$:

$$i(t) = \frac{7}{3\sqrt{2}}\cos(2t + 95°)\ \text{A}$$

Another example is shown in Figure 7-20. The original network, in which it is desired to find $i(t)$ and $v(t)$, is replaced by the phasor network. Equivalent circuits will be used to solve for **I** and **V** in the phasor network. Combining the impedances 3 and $j4$ in series and using the current divider rule gives

$$\mathbf{I} = \frac{j8(5e^{-j70°})}{3 + j4 + j8} = \frac{(8e^{j90°})(5e^{-j70°})}{(12.37e^{j76°})}$$

$$= 3.23e^{-j56°}$$

In the original network, then

$$i(t) = 3.23\cos(2t - 56°)\ \text{A}$$

The phasor **V** is now found:

$$\mathbf{V} = -(j4)\mathbf{I}$$

The minus sign in the relation between **V** and **I** is due to the source relationship between the references for **V** and **I**.

$$\mathbf{V} = -j4\mathbf{I} = (4e^{-j90°})(3.23e^{-j56°})$$

$$= 12.92e^{-j146°}$$

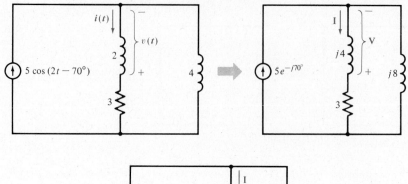

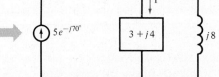

Figure 7-20   Phasor solution using equivalent circuits.

So

$$v(t) = 12.92 \cos (2t - 146°) \text{ V}$$

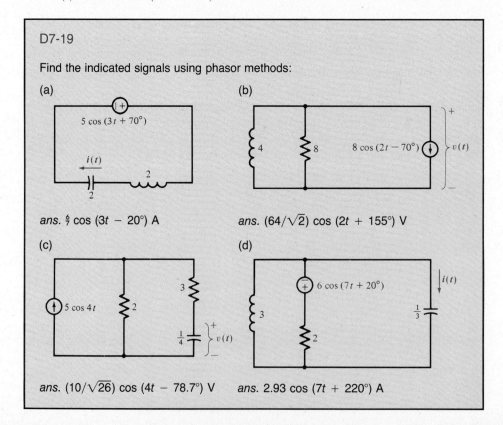

D7-19

Find the indicated signals using phasor methods:

(a)

5 cos (3t + 70°)

i(t)

2

2

ans. $\frac{6}{7}$ cos (3t − 20°) A

(b)

8 cos (2t − 70°)   v(t)

4   8

ans. $(64/\sqrt{2})$ cos (2t + 155°) V

(c)

5 cos 4t

2

3

$\frac{1}{4}$ v(t)

ans. $(10/\sqrt{26})$ cos (4t − 78.7°) V

(d)

6 cos (7t + 20°)

i(t)

3

2

$\frac{1}{3}$

ans. 2.93 cos (7t + 220°) A

D7-20

Find the two indicated signals in each network using phasor methods with equivalent circuits:

(a)

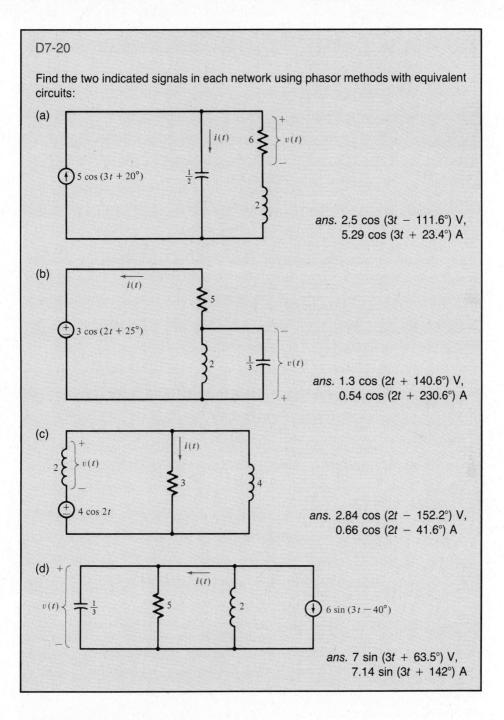

ans. 2.5 cos (3*t* − 111.6°) V,
5.29 cos (3*t* + 23.4°) A

(b)

ans. 1.3 cos (2*t* + 140.6°) V,
0.54 cos (2*t* + 230.6°) A

(c)

ans. 2.84 cos (2*t* − 152.2°) V,
0.66 cos (2*t* − 41.6°) A

(d)

ans. 7 sin (3*t* + 63.5°) V,
7.14 sin (3*t* + 142°) A

7.7.3 Simultaneous Equations

Mesh and nodal equations are written for phasor networks exactly as for source-resistor networks. The phasor sources and impedances are generally complex num-

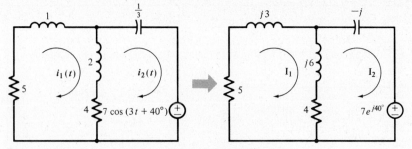

Figure 7-21   Phasor solution using mesh equations.

bers, so that complex algebra is involved, but the equations and their solutions proceed just as they did for source-resistor networks.

Suppose it is desired to find the current $i_1(t)$ in the network of Figure 7-21. The systematic, simultaneous mesh equations for the phasor network are as follows:

$$\begin{cases} (9 + j9)\mathbf{I}_1 - (4 + j6)\mathbf{I}_2 = 0 \\ -(4 + j6)\mathbf{I}_1 + (4 + j5)\mathbf{I}_2 = -7e^{j40°} \end{cases}$$

The phasor $\mathbf{I}_1$ may be found via Cramer's rule. The algebra involved is rather lengthy, but straightforward:

$$\mathbf{I}_1 = \frac{\begin{vmatrix} 0 & -(4 + j6) \\ -7e^{j40°} & (4 + j5) \end{vmatrix}}{\begin{vmatrix} (9 + j9) & -(4 + j6) \\ -(4 + j6) & (4 + j5) \end{vmatrix}} = \frac{(7e^{j40°})(-4 - j6)}{(9 + j9)(4 + j5) - (4 + j6)(4 + j6)}$$

$$= \frac{(7e^{j40°})(-4 - j6)}{36 - 45 + j36 + j45 - 16 + 36 - j24 - j24} = \frac{(7e^{j40°})(-4 - j6)}{11 + j33}$$

Converting to polar form,

$$\mathbf{I}_1 = \frac{(7e^{j40°})(7.21e^{j236.5°})}{(34.8e^{j71.7°})} = 1.45e^{j204.8°}$$

Then

$$i_1(t) = 1.45 \cos(3t + 204.8°) \text{ A}$$

A more complicated example involving coupled inductors is given in Figure 7-22. The coupled inductors are first replaced by their controlled source equivalent, then the phasor network is drawn. In the language of phasors, $d/dt(\ )$ becomes $j\omega(\ )$, so the controlled sources' controlling signals may be expressed in terms of the phasors.

Systematic simultaneous mesh equations are as follows:

$$\begin{cases} (5 + j)\mathbf{I}_1 - j\mathbf{I}_2 = -1 - j2\mathbf{I}_2 \\ -j\mathbf{I}_1 + (j5)\mathbf{I}_2 = -8 + j2\mathbf{I}_2 - j2(\mathbf{I}_1 - \mathbf{I}_2) \end{cases}$$

Rearranging,

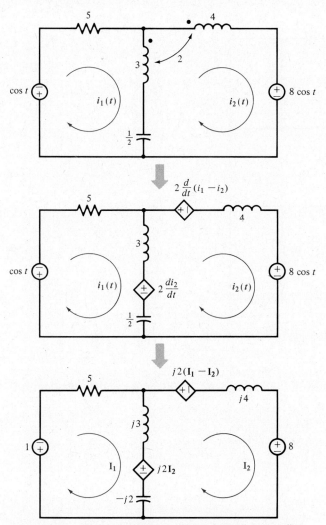

Figure 7-22   Phasor solution involving inductive coupling.

$$\begin{cases} (5 + j)\mathbf{I}_1 + j\mathbf{I}_2 = -1 \\ \quad j\mathbf{I}_1 + j\mathbf{I}_2 = -8 \end{cases}$$

If it is desired to find the current $i_2(t)$,

$$\mathbf{I}_2 = \frac{\begin{vmatrix} (5 + j) & -1 \\ j & -8 \end{vmatrix}}{\begin{vmatrix} (5 + j) & j \\ j & j \end{vmatrix}} = \frac{-40 - j7}{j5} = \frac{40.6e^{j190°}}{5e^{j90°}}$$

$$= 8.12e^{j100°}$$

Then

$$i_2 = 8.12 \cos (t + 100°) \text{ A}$$

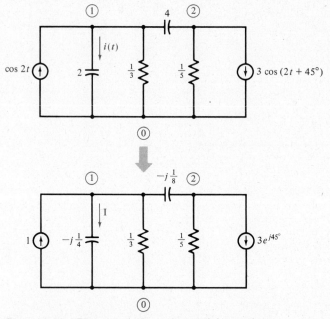

Figure 7-23 Phasor solution using nodal equations.

A nodal equation example is given in Figure 7-23. The original network is replaced by the phasor network, and systematic simultaneous nodal equations are as follows:

$$\begin{cases} (3 + j12)\mathbf{V}_1 - j8\mathbf{V}_2 = 1 \\ -j8\mathbf{V}_1 + (5 + j8)\mathbf{V}_2 = -3e^{j45°} \end{cases}$$

If it is desired to find the current $i(t)$, the node voltage phasor $\mathbf{V}_1$ is first found:

$$\mathbf{V}_1 = \frac{\begin{vmatrix} 1 & -j8 \\ -3e^{j45°} & (5 + j8) \end{vmatrix}}{\begin{vmatrix} (3 + j12) & -j8 \\ -j8 & (5 + j8) \end{vmatrix}} = \frac{(5 + j8) + (-j8)(3e^{j45°})}{15 + j24 + j60 - 96 + 64}$$

$$= \frac{5 + j8 + (-j8)(2.12 + j2.12)}{-17 + j84} = \frac{22 - j9}{-17 + j84}$$

$$= \frac{23.8e^{-j22.2°}}{85.7e^{j101.4°}} = 0.28e^{-j123.6°}$$

The phasor $\mathbf{I}$ is then

$$\mathbf{I} = \frac{\mathbf{V}_1}{-j(1/4)} = j4\mathbf{V}_1 = (4e^{j90°})(0.28e^{-j123.6°})$$

$$= 1.12e^{-j33.6°}$$

So

$$i(t) = 1.12 \cos(2t - 33.6°) \text{ A}$$

7.7.4 Sources Expressed in Terms of the Sine Function

For networks in which all sources are expressed in terms of the sine function, the same phasors may be used, as if the sines were cosines. Then phasor magnitudes represent the sinusoidal amplitudes and phasor angles represent the phase angles within the sine function rather than within the cosine. Of course, if there is a mixture of cosine- and sine-function sources, one or the other function will have to be used exclusively as the basis for all of the phase angles.

Consider the network of Figure 7-24. The source $6 \sin (3t + 110°)$ is represented by the phasor $6e^{j110°}$. To solve the phasor network, a Thévenin-Norton transformation has been made to convert it to an equivalent two-node problem. The two-node equation for $\mathbf{V}$ is

$$\left(\frac{1}{2} - \frac{1}{j3} - \frac{1}{j4}\right)\mathbf{V} = 2e^{j200°}$$

Multiplying each side of this equation by $j12$ gives

$$(j6 - 7)\mathbf{V} = (j12)2e^{j200°}$$

$$\mathbf{V} = \frac{(j12)(2e^{j200°})}{7 - j6} = \frac{(12e^{j90°})(2e^{j200°})}{9.22e^{-j40.6°}}$$

$$= 2.6e^{j330.6°}$$

Since the original source was described in terms of the sine function, the solution involves sine:

$$v(t) = 2.6 \sin (3t + 330.6°) \text{ V}$$

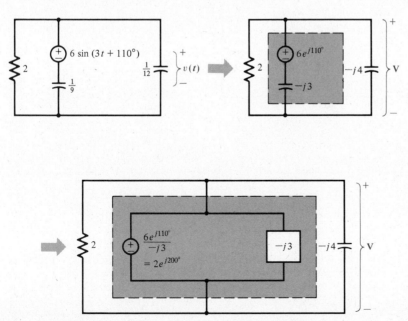

Figure 7-24 Phasor solution in terms of the sine function.

D7-21

Find the indicated signal, using phasor methods and mesh equations:

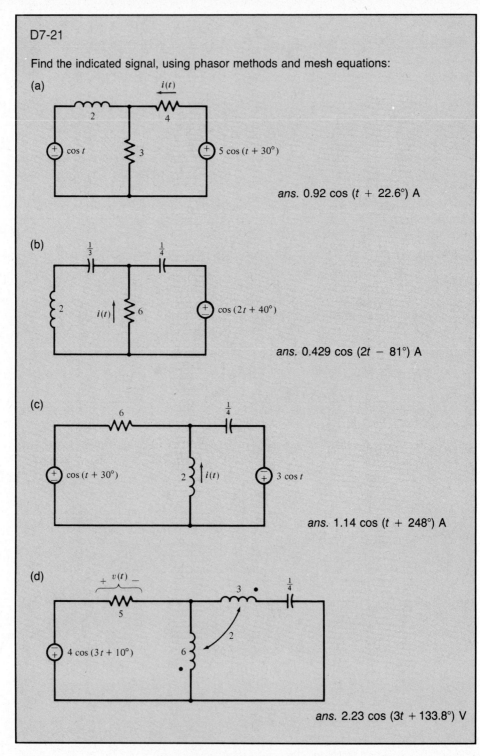

(a)

$i(t)$

2

4

$\cos t$

3

$5 \cos (t + 30°)$

ans. 0.92 cos (t + 22.6°) A

(b)

$\frac{1}{3}$

$\frac{1}{4}$

2

$i(t)$

6

$\cos (2t + 40°)$

ans. 0.429 cos (2t − 81°) A

(c)

6

$\frac{1}{4}$

$\cos (t + 30°)$

2

$i(t)$

$3 \cos t$

ans. 1.14 cos (t + 248°) A

(d)

$+ \ v(t) \ -$

5

3

$\frac{1}{4}$

$4 \cos (3t + 10°)$

6

2

ans. 2.23 cos (3t + 133.8°) V

D7-22

Find the indicated signal, using phasor methods and nodal equations:

(a)

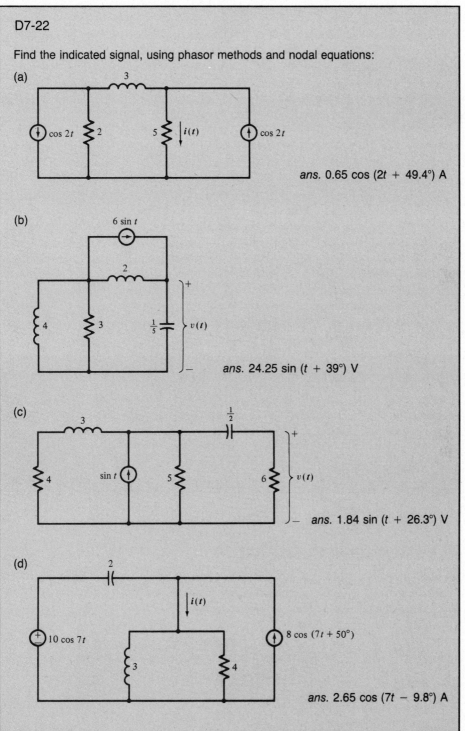

ans. 0.65 cos (2t + 49.4°) A

(b)

ans. 24.25 sin (t + 39°) V

(c)

ans. 1.84 sin (t + 26.3°) V

(d)

ans. 2.65 cos (7t − 9.8°) A

7.7.5 Perspective on the Three Solution Methods

In the method of components, a sinusoidal source is simply decomposed into its two complex exponential components, which are then superimposed. The components being exponential, impedance methods apply, and the corresponding exponential responses may be found by source-resistor network methods. The added complexity of complex numbers is generally far less complicated than the alternative of dealing directly with the sinusoidal functions and trigonometric identities.

In our later work, where more general source functions are expanded into an infinite series of sinusoids of different frequencies, components are particularly attractive for solution because the sum of components is the function itself; no conversions from sinor or phasor representations are required.

The method of sinors is perhaps the most widely used solution method for sinusoidally driven linear systems of all kinds. The method is particularly popular in such fields as quantum mechanics, communications, control systems, and electromagnetics.

The phasor method requires slightly less writing than sinors, $e^{j\omega t}$ factors being eliminated. Network analysts, particularly those concerned with electrical power, deal almost exclusively in terms of phasors. Phasors have a substantial advantage in computer-aided network analysis in that only complex constants, not functions of time, occur as equation driving functions. And phasors are easily plotted and manipulated graphically on the complex plane.

THE METHOD OF PHASORS

In the phasor method, a sinusoidal source $A \cos (\omega t + \theta)$ is replaced by the phasor source $Ae^{j\theta}$, which is the source sinor at time $t = 0$. Network impedances are evaluated at $s = j\omega$, as with sinors.

The magnitude of each phasor signal is the amplitude of the corresponding sinusoidal signal in the original network, and each phasor angle is the corresponding signal's phase angle.

7.8 GRAPHICAL PHASOR METHODS

7.8.1 Graphical Addition of Phasors

Phasors offer an easy graphical method of calculating sums of same-frequency sinusoidal signals with various amplitudes and phase angles. For example, the phasors $\mathbf{Y}_1$ and $\mathbf{Y}_2$, which represent the sinusoidal signals

$$y_1(t) = 8 \cos (100t - 40°)$$

and

$$y_2(t) = 5 \cos (100t + 120°)$$

are drawn on the complex plane in Figure 7-25(a). The sum of the two phasors, which may be found by graphical addition, is

$$\mathbf{Y}_1 + \mathbf{Y}_2 = 3.7e^{-j12.6°}$$

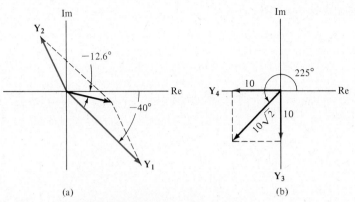

Figure 7-25   Graphical addition of phasors.
(a) Finding the sum of two sinusoidal signals of the same frequency.
(b) Sine and cosine sum.

so

$$y_1(t) + y_2(t) = 3.7 \cos (100t - 12.6°)$$

Consider the sum

$$y_3(t) + y_4(t) = 10 \sin (10^4 t) - 10 \cos (10^4 t)$$

The sine function is related to the cosine by

$$A \sin (\omega t + \phi) = A \cos (\omega t + \phi - 90°)$$

so

$$y_3(t) = 10 \sin (10^4 t)$$

has the phasor along the negative imaginary axis, as in Figure 7-25(b). From that diagram,

$$y_3(t) + y_4(t) = 10 \sin (10^4 t) + 10 \cos (10^4 t + 180°)$$

$$= 10 \sqrt{2} \cos (10^4 t + 225°)$$

D7-23

Find the sums of the following sinusoidal signals using graphical addition of phasors on the complex plane. Express in the form $A \cos (\omega t + \theta)$, with $A \geq 0$, $\omega > 0$, and $-360° < \theta < 360°$:

(a) $3 \cos 4t + 4 \cos (4t + 30°)$ *ans.* 6.8 cos (4t + 17°)

(b) $\cos (10^5 t + 60°) + \cos (10^5 t - 60°)$ *ans.* cos $10^5 t$

(c) $3 \cos (377t + 10°) - \sin (377t + 45°)$ *ans.* 2.6 cos (377t + 28°)

(d) $2 \sin (7t - 30°) + 3 \sin 7t$ *ans.* 4.8 cos (7t - 102°)

7.8.2 Graphical Relations for *R, L,* and *C*

The sink reference voltage and current phasors for resistors, inductors, and capacitors have the simple relations shown in Figure 7-26, where only the relative positions of the phasors are indicated. For the resistor, the voltage and current phasors have the same angle (the voltage and current are said to be *in phase*), and the length of the voltage phasor is *R* times the length (or magnitude) of the current phasor.

For the inductor,

$$\frac{\mathbf{V}_L}{\mathbf{I}_L} = j\omega L$$

or

$$\mathbf{V}_L = j\omega L\mathbf{I}_L = (\omega L)e^{j90°}\mathbf{I}_L$$

The length of the phasor $\mathbf{V}_L$ is (ωL) times the length of $\mathbf{I}_L$, and the angle of $\mathbf{V}_L$ is 90° larger than that of $\mathbf{I}_L$. The current is thus said to *lag* the voltage by 90°.

For the capacitor,

$$\frac{\mathbf{V}_C}{\mathbf{I}_C} = \frac{1}{j\omega C}$$

$$\mathbf{V}_C = \left(\frac{1}{\omega C}\right)e^{-j90°}\mathbf{I}_C$$

The capacitor voltage phasor has length $(1/\omega C)$ times that of the capacitor current phasor. The capacitor current leads the capacitor voltage by 90°.

7.8.3 Solution of Trivial Networks

The *R, L,* and *C* phasor relationships may be used to solve simple sinusoidal network problems graphically. For example, for the network in Figure 7-27(a), the source

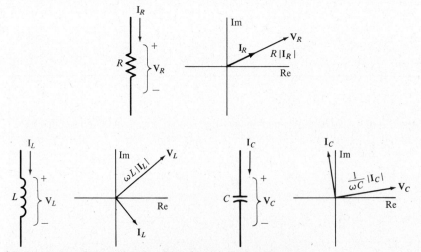

Figure 7-26   Phasor relationships for the basic elements.

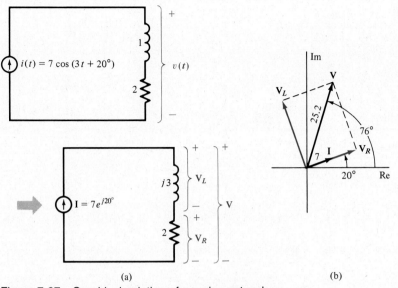

Figure 7-27 Graphical solution of a series network.

phasor is plotted on the phasor diagram, Figure 7-27(b). **I** is the current through the inductor, so the inductor voltage magnitude is ωL times the magnitude of the current. The angle of $\mathbf{V}_L$ is 90° greater than that of the current. The resistor voltage phasor $\mathbf{V}_R$ has the same angle as **I** and twice its magnitude.

The voltage **V** is

$$\mathbf{V} = \mathbf{V}_L + \mathbf{V}_R$$

which is found by graphically summing $\mathbf{V}_L$ and $\mathbf{V}_R$ with a parallelogram. Using a ruler and a protractor, it is found to be approximately

$$\mathbf{V} = 25.2e^{j76°}$$

so

$$v(t) = 25.2 \cos (3t + 76°) \text{ V}$$

For the network of Figure 7-28(a), the phasor **V** is plotted on the complex plane in Figure 7-28(b). $\mathbf{I}_R$ is in the same direction as **V** and has one-third of its magnitude. The phasor $\mathbf{I}_L$ is of magnitude

$$|\mathbf{I}_L| = \left(\frac{1}{\omega L}\right)|\mathbf{V}| = |\mathbf{V}|$$

and lags **V** by 90°. The phasor $\mathbf{I}_C$ has magnitude

$$|\mathbf{I}_C| = \omega C|\mathbf{V}| = \tfrac{1}{2}|\mathbf{V}|$$

and angle 90° greater than that of **V**.

The net current

$$\mathbf{I} = \mathbf{I}_R + \mathbf{I}_L + \mathbf{I}_C$$

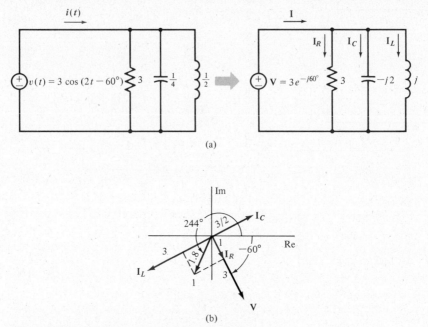

(a)

(b)

Figure 7-28   Graphical solution of a parallel network.

is constructed graphically on the phasor diagram and is found to be approximately

$$\mathbf{I} = 1.8e^{j244°}$$

So

$$i(t) = 1.8 \cos (2t + 244°) \text{ A}$$

7.8.4 The Trial Source Method

It is only in rather trivial network problems that the phasors of interest are directly related to the source function as in the previous examples. To solve more involved networks graphically, use is made of the fact that for a single-source network the solution for any phasor signal is proportional to the source phasor. Thus if the magnitude of the source is changed by some factor, the magnitudes of all network phasors are changed by the same factor. Doubling the source magnitude, for instance, doubles all the other phasor magnitudes. If the source phase angle is changed, all phasor angles are changed by the same amount.

For the network of Figure 7-29(a), the voltages $\mathbf{V}_R$ and $\mathbf{V}_L$ are not simply related to the source $\mathbf{V}$. If the current $\mathbf{I}$ were known, finding $\mathbf{V}$ would be easy, but it is $\mathbf{V}$ that is known. Although the impedances 2 and $j3$ could be combined and dealt with as a single element, to do so would be complicated in comparison to using the simple in-phase and right-angle phasor relationships of the basic elements.

Instead, the graphical method is to try a convenient current $\mathbf{I}$, perhaps a guess at what it might be. Using the simple phasor relations, the corresponding $\mathbf{V}$ may be found. The $\mathbf{V}$ found, which is unlikely to be the correct value, is the source voltage

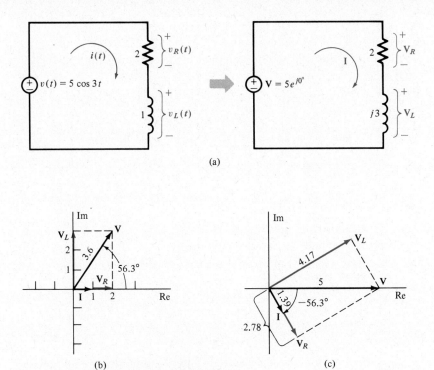

(a)

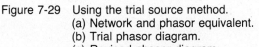

(b)

(c)

Figure 7-29   Using the trial source method.
(a) Network and phasor equivalent.
(b) Trial phasor diagram.
(c) Revised phasor diagram.

that would produce the assumed current. Then the correct current and the correct values of the other phasors are found by considering how the magnitude and angle of the trial **I** must be changed to give the correct **V**.

For the example, a convenient trial current

$$\mathbf{I} = 1e^{j0°}$$

is first chosen. The resistor and inductor phasor voltages are, with this current,

$$\mathbf{V}_R = 2e^{j0°} \qquad \mathbf{V}_L = 3e^{j90°}$$

as plotted on the trial phasor diagram of Figure 7-29(b). The sum of the resistor voltage and the inductor voltage is the source voltage **V**, which if the current were the trial current, would have to be

$$\mathbf{V} = 3.6e^{j56.3°}$$

Of course, the source voltage is not the calculated value based on **I**; it would be a great coincidence if it were.

It is now easy to determine the correct value for **I**, however. If **I** were at the angle $-56.3°$, all other phasors would be rotated clockwise on the phasor diagram by 56.3°, and **V** would be at 0°, as it should be. If **I** were $\frac{5}{3.6}$ units long instead of 1 unit long, all the other phasors on the diagram would be $\frac{5}{3.6}$ times as long and, in particular, the magnitude of $\mathbf{V}$ would be the correct 5 units instead of 3.6 units. A corrected phasor diagram for the network is given in Figure 7-29(c).

Another example of the trial solution method involves the network of Figure 7-30(a). Trying

$$\mathbf{I}_1 = 1e^{j0°}$$

gives

$$\mathbf{V}_R = 3\mathbf{I}_1 = 3e^{j0°} \qquad \mathbf{V}_L = j4\mathbf{I}_1 = 4e^{j90°}$$

and

$$\mathbf{V} = \mathbf{V}_R + \mathbf{V}_L = 5e^{j53°}$$

as shown on the trial phasor diagram, Figure 7-30(b).

From the capacitor voltage $\mathbf{V}$, the capacitor current is

$$\mathbf{I}_2 = \frac{\mathbf{V}}{-j2} = \left(\frac{1}{2}e^{j90°}\right)\mathbf{V}$$

which is half the magnitude of $\mathbf{V}$ and leads it by 90°. The source current is then

$$\mathbf{I} = \mathbf{I}_1 + \mathbf{I}_2$$

as shown, which with the assumed value for $\mathbf{I}_1$ is found graphically to be approximately

$$\mathbf{I} = 1.8e^{j124°}$$

To achieve

$$\mathbf{I} = 5e^{-j60°}$$

all phasors on the trial phasor diagram must be rotated clockwise 184° and expanded in length by the factor $\frac{5}{1.8}$, as shown in Figure 7-30(c).

The correct current $\mathbf{I}_1$ is thus approximately

$$\mathbf{I}_1 = 2.78e^{-j184°}$$

The trial source method will generally not work for networks with more than a single source because it depends on the proportionality of phasor network signals to the source. For the graphical solution of multiple-source networks, the sources should be superimposed.

The choice of starting signal, $\mathbf{I}_1$ in the previous example, is made by thinking through the problem: From $\mathbf{I}_1$, $\mathbf{V}_R$ and $\mathbf{V}_L$ may be constructed; $\mathbf{V}$ may be constructed from $\mathbf{V}_R$ and $\mathbf{V}_L$; $\mathbf{I}_2$ may be found from $\mathbf{V}$; and $\mathbf{I}$ is the sum $\mathbf{I}_1 + \mathbf{I}_2$. Alternatively, the solution could have started with a trial $\mathbf{V}_R$ or a trial $\mathbf{V}_L$, but to start with $\mathbf{I}_2$ or $\mathbf{V}$ would not allow an easy construction for $\mathbf{I}_1$.

Occasionally, in more complicated networks, an impasse is reached, where no initial trial phasor leads simply back to the source. In this event, some numerical calculation is necessary, or more than a single rotation and stretching or compression of phasors on the diagram is expedient.

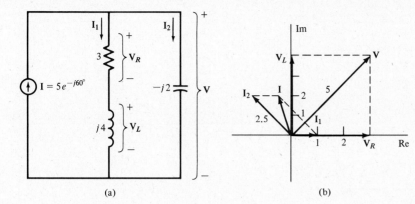

(a)

(b)

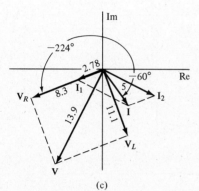

(c)

Figure 7-30   Another graphical network solution.
(a) Phasor network.
(b) Trial phasor diagram.
(c) Revised phasor diagram.

D7-24

Using entirely *graphical* methods, construct phasor diagrams for each of the following networks, showing every voltage and every current phasor:

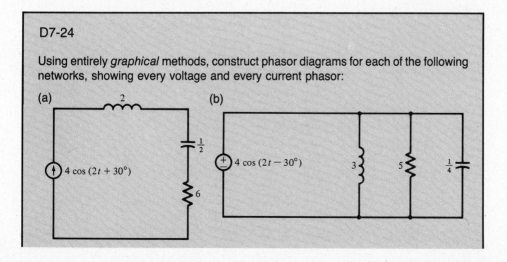

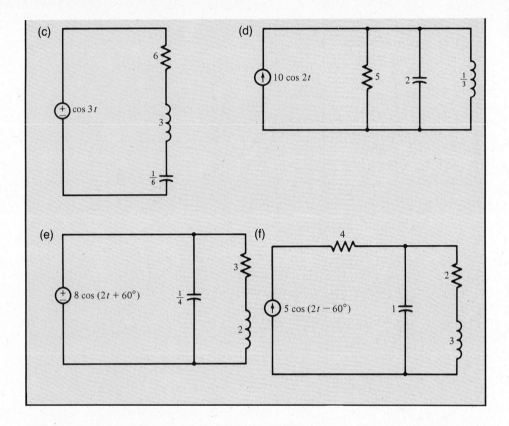

GRAPHICAL PHASOR METHODS

The sum of two sinusoidal functions of the same frequency may be found by adding their phasors to obtain the phasor for the sum.

The phasor relationships for the sink reference relation voltage and current for R, L, and C elements are as follows:

Resistor: Current is in phase with the voltage, with magnitude (1/R), times the voltage magnitude.

Inductor: Current lags the voltage by 90°, with magnitude (1/ωL), times the voltage magnitude.

Capacitor: Current leads the voltage by 90°, with magnitude (ωC), times the voltage magnitude.

The graphical solution of a single-source phasor network involves drawing a phasor diagram upon which are plotted the phasors for each network voltage and current. In trivial cases the other phasors are found by starting with the source phasor and proceeding through the network, using the R, L, and C phasor relationships.

In general, beginning with a trial voltage or current elsewhere in the network, the source phasor corresponding to the trial signal is computed. Then the resulting phasor diagram is rotated and scaled to give the source phasor and the true solution.

CHAPTER 7 PROBLEMS

Basic Problems

Sinusoidal Functions
1. Sketch plots of the following sinusoidal functions. Find their amplitude, radian frequency, phase angle (in terms of the cosine function), and period:
   (a) $-10 \cos (10^6 t + 1000)$
   (b) $9 \sin (60° - 7t)$
   (c) $60 \cos (-377t - 140°)$
   (d) $-100 \sin \left(8t - \dfrac{16\pi}{5} \right)$

2. Convert to the form $A \cos (\omega t + \theta)$ with $A \geq 0$, $\omega > 0$, and $-360° < \theta < 360°$:
   (a) $f(t) = 3 \sin (80t - 300°)$
   (b) $f(t) = 20 \cos 10t + 5 \sin 10t$
   (c) $f(t) = -8 \cos 377t - 9 \sin 377t$
   (d) $f(t) = 100 \cos (3t + 90°) - 50 \sin 3t$

Differential Equations with Sinusoidal Driving Functions
3. Find the general solutions of each of the following differential equations:

   (a) $\dfrac{dy}{dt} = 10 \cos 20t$

   (b) $2 \dfrac{dy}{dt} + y = -6 \sin 3t$

   (c) $\dfrac{dy}{dt} + 4y = 5 \cos (2t + 45°)$

   (d) $\dfrac{d^2y}{dt^2} + 5 \dfrac{dy}{dt} + 6y = 7 \sin 2t$

   (e) $3 \dfrac{dy}{dt} + 2y = 8 \cos (2t - 30°)$

   (f) $\dfrac{d^3y}{dt^3} = 10 \cos 3t - 5 \sin 3t$

   (g) $\dfrac{d^2y}{dt^2} + 2 \dfrac{dy}{dt} + 5y = 3 \cos 3t$

   (h) $\dfrac{d^3y}{dt^3} + 3 \dfrac{d^2y}{dt^2} + 2 \dfrac{dy}{dt} = 4 \cos (3t + 100°)$

Complex Algebra
4. Find the indicated complex numbers. Express in both rectangular form and in polar form, and sketch the location of each number on the complex plane:
   (a) $(3 - j4)(-5 + j6)$
   (b) $\dfrac{(4 + j5)}{(2 - j3)}$

(c) $6e^{j30°} + j7$

(d) $-3e^{j100°} + 5e^{-j40°}$

(e) $\dfrac{(6 - j7)(-3 - j8)}{(-50 + j40)}$

(f) $\dfrac{j4 - e^{j30°}}{-3 - j5}$

(g) $\dfrac{3 - j2}{2 - j3} + 1$

(h) $\dfrac{(5 - j2)(-2 + j2)}{1 + (3 - j)(-2 + j)}$

Method of Components

5. Use the method of components to find the indicated signals. There are two signals to find in each network.

(a)

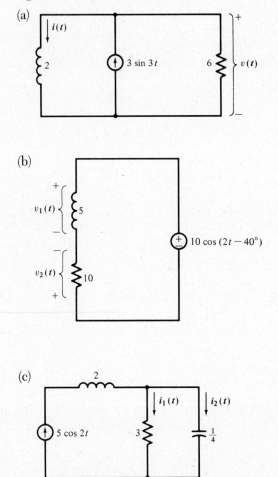

(b)

(c)

A high-power radio transmitter installation. The amplitude or frequency of a sinusoidal "carrier" signal is varied to represent the sound, picture, or data to be transmitted. The power generated, which is delivered to an antenna, is typically thousands of watts. (*Photo courtesy of the Radio Corporation of America.*)

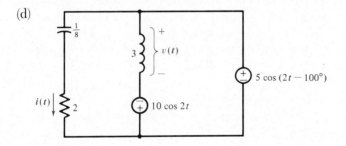

(d)

$v(t)$

$5 \cos (2t - 100°)$

$\frac{1}{8}$

3

$i(t)$

2

$10 \cos 2t$

Method of Sinors

6. Use the method of sinors to find the indicated signals. There are two signals to find in each network.

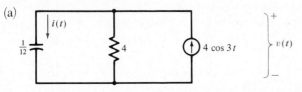

(a)

$i(t)$

$\frac{1}{12}$

4

$4 \cos 3t$

$+$

$v(t)$

$-$

(b)

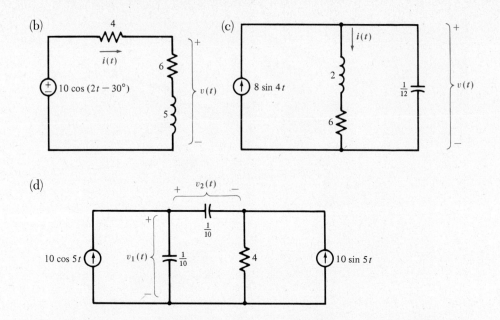

(c)

(d)

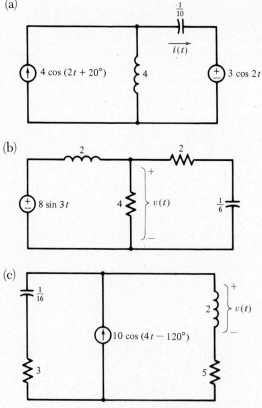

7. Use sinor solution methods and equivalent circuits to find the indicated signals:

(a)

(b)

(c)

Domestic communications satellite receiving station. Communications satellites use a wide band of frequencies in order to accommodate a large number of voice, data, and video channels. (*Photo courtesy of AT&T—Long Lines, San Francisco, CA.*)

(d)

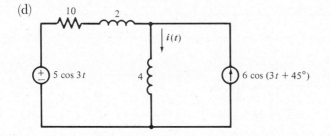

8. Use sinor methods and systematic simultaneous nodal equations to find the indicated signals:

(a)

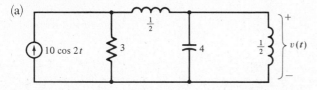

(b)

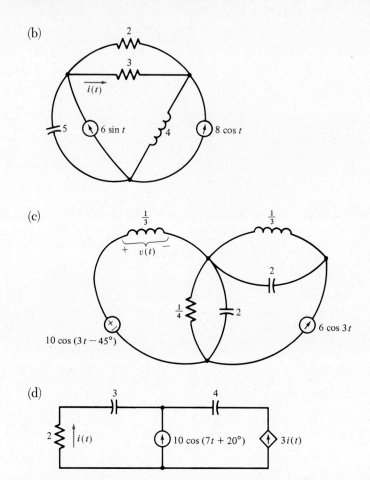

(c)

(d)

9. Use sinor methods and systematic simultaneous mesh equations to find the indicated signals:

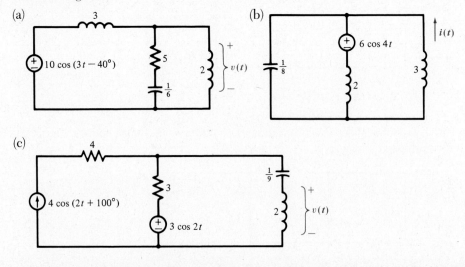

(d)

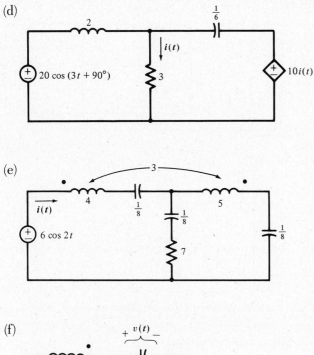

(e)

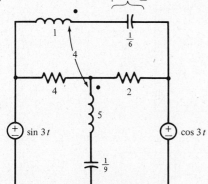

(f)

Phasor Solutions

10. Use the method of phasors to find the indicated signals. There are two signals to find in each network:

(a)

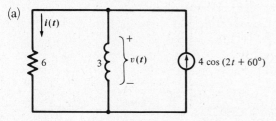

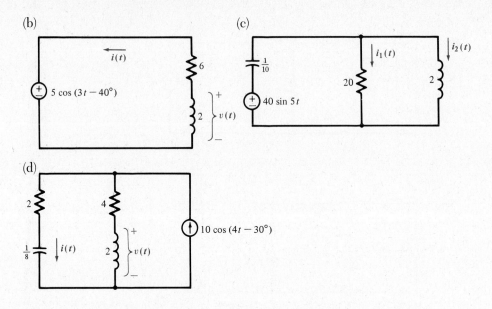

(b)

$5 \cos (3t - 40°)$

$i(t)$

6

2 $v(t)$

(c)

$\frac{1}{10}$

$40 \sin 5t$

20

2

$i_1(t)$

$i_2(t)$

(d)

2

4

$\frac{1}{8}$

$i(t)$

2 $v(t)$

$10 \cos (4t - 30°)$

11. Use phasor methods and equivalent circuits to find the indicated signals:

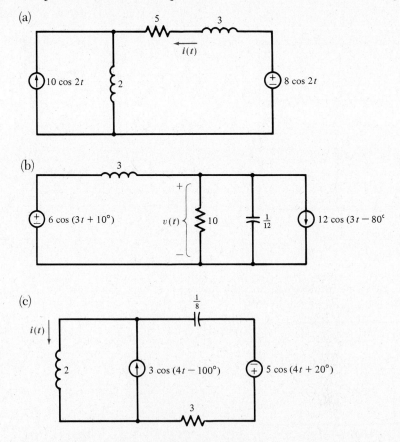

(a)

5

3

$i(t)$

$10 \cos 2t$

2

$8 \cos 2t$

(b)

3

$6 \cos (3t + 10°)$

$v(t)$

10

$\frac{1}{12}$

$12 \cos (3t - 80°)$

(c)

$\frac{1}{8}$

$i(t)$

2

$3 \cos (4t - 100°)$

$5 \cos (4t + 20°)$

3

(d)

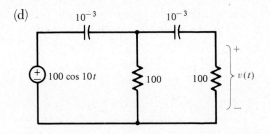

12. Use phasor methods and systematic simultaneous nodal equations to find the indicated signals:

(a)

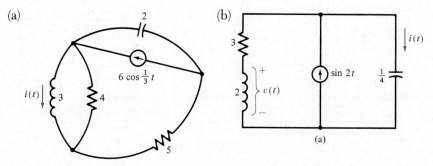

(b)

(a)

(c)

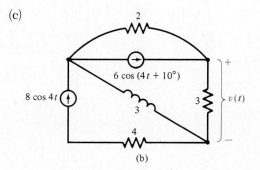

(b)

(d)

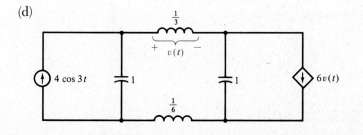

13. Use phasor methods and systematic simultaneous mesh equations to find the indicated signals:

(a)

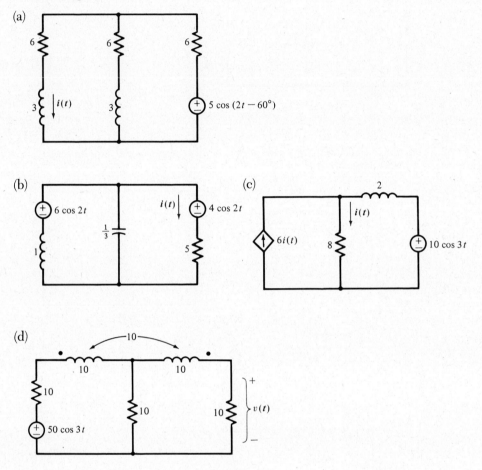

(b)

(c)

(d)

Sources with Different Frequencies

14. Use phasor methods to find the indicated signal:

(a)

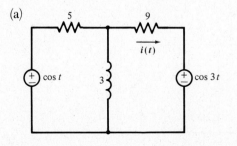

(b)

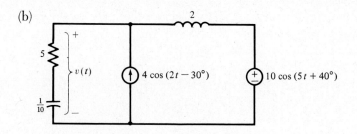

Graphical Phasor Methods

15. Using graphical methods with phasors, find the constants A, B, C, and θ:

(a) $9 \cos (10t + 130°) = A \cos 10t + B \sin 10t$

(b) $6 \cos (4t - 40°) - 7 \cos (4t + 80°) = A \cos 4t + B \sin 4t = C \cos (4t + \theta)$

(c) $5 \sin (10^4 t - 70°) - 3 \sin (10^4 t) = A \cos 10^4 t + B \sin 10^4 t$

16. Using entirely graphical methods, find the indicated signals:

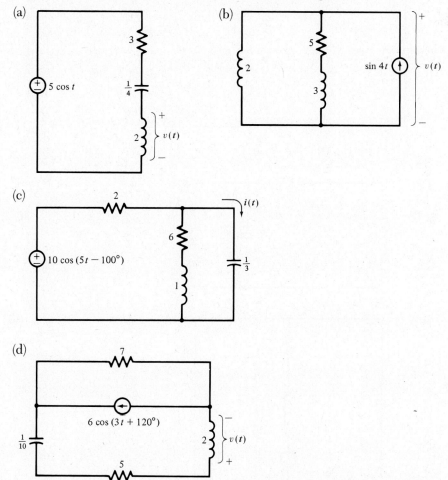

17. Using entirely graphical methods, construct phasor diagrams for the following networks, showing each of the indicated phasors:

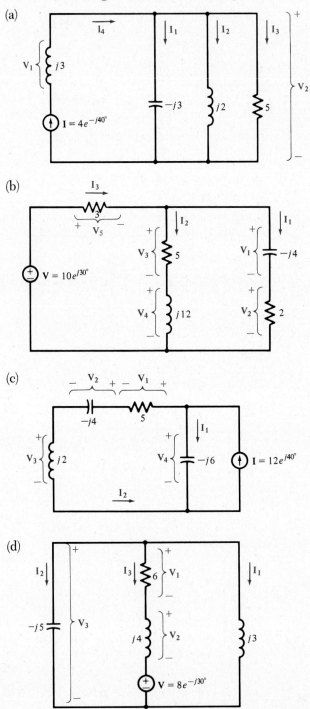

Switched Sinusoidal Networks

18. Find the indicated signals before and after time $t = 0$ and sketch them:

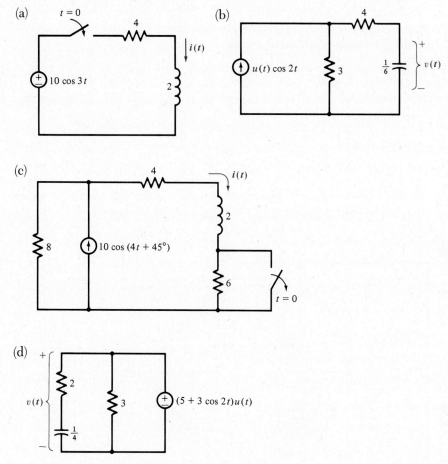

(a)

(b)

(c)

(d)

Practical Problems

AC Ammeters and Voltmeters

When a sinusoidal signal is applied to a dc meter such as a permanent magnet ammeter or voltmeter, the instrument will respond only if the frequency is very low. At higher frequencies, the inertia of the mechanical movement results in a deflection proportional to the average force, which, for a sinusoid, is zero.

Meters suitable for indicating the amplitude of sinusoidal currents and voltages average the square or the magnitude of the sinusoidal quantity and produce a reading of the corresponding sinusoidal amplitude. A *two-coil* meter uses the attraction between two current-carrying coils to produce a force proportional to the average of the square of the current. An *iron vane* instrument uses the attraction between a current-carrying coil and the induced magnetization of a piece of soft iron to produce a similar force. The scales of these meters tend to be compressed at the low end and spread out at high values of current or voltage because of the nonlinear force relation.

Most electronic ac meters respond to the average of the magnitude (or absolute value) of the measured voltage or current.

19. An ac ammeter responds with a deflection proportional to the average of the square of the current. If a constant applied current of 10 A produces a certain deflection, what must be the amplitude of sinusoidal current to produce the same deflection?

Impedance Bridge

20. The adjustable network shown is one type of impedance bridge. The resistor R and inductance or capacitance, L or C, are adjusted until $\mathbf{V} = 0$. The bridge is then said to be balanced. Show that, at balance, the impedance $(R + j\omega L)$ or $[R - (j/\omega C)]$ is proportional to the unknown impedance Z at the frequency of $\mathbf{V}_{in}$.

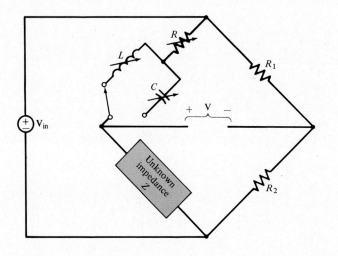

Phase-Difference Metering

21. The phase difference between two sinusoidal signals may be measured by adjusting the amplitudes of each to a fixed value A, adding them, and measuring the amplitude of their sum. The sum has amplitude $2A$ if the signals are in phase, and zero amplitude if the signals differ in phase by $180°$.

What is the relationship between the phase difference of the two signals and the amplitude of their sum?

22. Another method of phase-difference measurement is to pass one of the sinusoidal signals through an additional, adjustable, calibrated phase shift and adjust this phase shift until the resulting signals are exactly in phase. The phase difference is then the amount of added phase shift of one signal necessary to make the phases equal.

For the phase-shifting network below, show that

$$|\mathbf{V}_{out}| = \tfrac{1}{2}|\mathbf{V}_{in}|$$

at every frequency, and find the amount of phase shift,

$$\angle\mathbf{V}_{out} - \angle\mathbf{V}_{in}$$

as a function of R and the source radian frequency, ω.

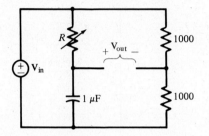

High Voltages and Currents in LC Networks

23. When an inductor and capacitor are connected in series across a voltage source, the individual inductor and capacitor voltages can have amplitudes much larger than the source voltage amplitude.

(a) Use graphical phasor methods to find the inductor and capacitor voltage amplitudes in the network of (a).

(b) In the network of (b), the inductor model includes its winding resistance. Use graphical phasor methods to find the inductor and capacitor voltage amplitudes.

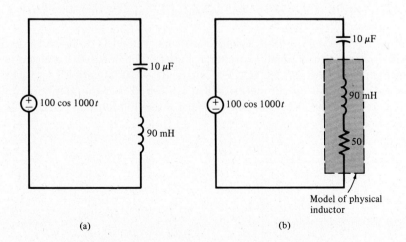

Model of physical
inductor

(a)   (b)

24. When an inductor and a capacitor in parallel are driven by a sinusoidal source, the individual inductor and capacitor current amplitudes can be large compared to their sum, the net current through the parallel combination.

(a) Use graphical phasor methods to find the inductor, capacitor, and net current amplitudes in the network of (a).

(b) In the network of (b), the inductor model includes its winding resistance. Find the inductor, capacitor, and net current amplitudes.

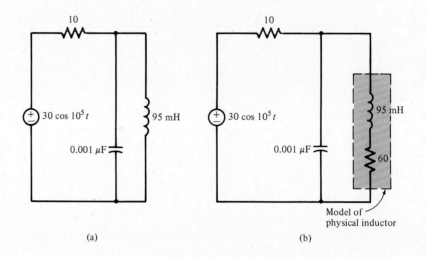

(a)   (b)

Model of physical inductor

Capacitive Voltage Divider

25. When a voltage divider network for sinusoidal signals is needed, capacitors are often used rather than resistors because, ideally, capacitors do not dissipate power as do resistors.

(a) For the network model of (a), find values of C_1 and C_2 such that $v_1(t) = (\frac{1}{6})v(t)$.

(b) Practically, capacitors exhibit leakage resistance, modeled by R_1 and R_2 in (b), so that arbitrarily small capacitor values cannot be used for accurate voltage division. If $R_1 = 10^6$ Ω, $C_1 = 1$ μF, $R_2 = 10^5$ Ω, and $C_2 = 5$ μF, what is the percent amplitude error if it is desired that the amplitude of $v_1(t)$ be one-fifth the amplitude of $v(t)$? What is the difference in phase angle between v_1 and v?

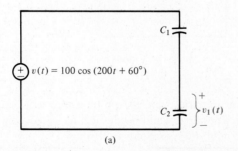

(a)

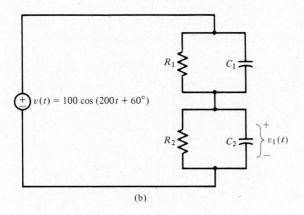

$v(t) = 100 \cos (200t + 60°)$

(b)

Advanced Problems

Phasors
26. The equation

$$f(t) = Re\,[\underline{f}(t)]$$

relates a sinor $\underline{f}(t)$ to the time function $f(t)$ that it represents. Find the equation that relates a phasor to the time function it represents.

27. Show that if the rectangular components of a phasor are

$$\mathbf{V} = a + jb$$

the quadrature components of the corresponding sinusoid are

$$v(t) = a \cos \omega t + (-b) \sin \omega t$$

Network Design
28. Design networks, to be driven by sinusoidal voltage sources of frequency 1000 Hz, for which another network voltage has
   (a) phase angle 45 degrees greater than that of the source
   (b) phase angle 45 degrees less than that of the source
   (c) phase angle 90 degrees greater than that of the source
   (d) phase angle 90 degrees less than that of the source

29. Design a network, to be driven by a sinusoidal voltage source of frequency 1000 Hz, for which another network voltage has amplitude one-half that of the source and phase angle 30 degrees greater than that of the source.

Approximate Differentiation and Integration
30. The network of (a) produces a voltage $v_1(t)$ that is, under certain circumstances, very nearly proportional to the derivative of the voltage $v(t)$. The voltage $v_2(t)$ in the network of (b) is, under the proper circumstances, approximately proportional to the integral of $v(t)$.

(a) For $R = 100 \text{ k}\Omega$ and $C = 0.01 \mu\text{F}$, find a range of frequency (in Hz) of *sinusoidal* source voltage $v(t)$ for which $v_1(t)$ is approximately proportional to the derivative of $v(t)$.

(b) For $R = 270 \text{ k}\Omega$ and $C = 2 \mu\text{F}$, find a range of Hertz frequency of sinusoidal source voltage for which $v_2(t)$ is nearly proportional to the integral of $v(t)$.

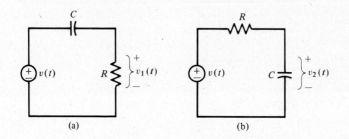

(a)   (b)

Components, Sinors, and Phasors for Differential Equations

31. Find the forced solution of the differential equation

$$\frac{d^2y}{dt^2} + 5\frac{dy}{dt} + 6y = 10 \cos (2t + 45°)$$

(a) by expanding the driving function into Euler components and superimposing them
(b) by replacing the driving function with the corresponding sinor and solving
(c) by using phasors

Chapter Eight
Resonance

A very little error in tuning, easily made by altering the
position of the slider, will make the circuits quite
unresponsive . . .

Sir Oliver Joseph Lodge
From the *Proceedings of the
Royal Society*, 1891

These are adjustable condensers [capacitors], preferably in
the form of two metallic tubes separated by a dielectric and
sliding telescopically on each other as in this way their
capacity can be readily varied with accuracy to tune the
circuits.

Guglielmo Marconi
From British Patent No. 7777
London, 1900

8.1 PREVIEW

We now apply the techniques of sinusoidal-response calculation to the analysis and
understanding of resonance in networks. The phenomenon of resonance is basic to
such applications as radio, radar, and telephone transmission, and it is also very
useful in a variety of other fields.

The chapter begins by considering resonance in series and parallel *RLC* networks. There then follows discussion of resonance in general, with more involved
examples. Both the magnitude extrema and the zero imaginary-part definitions of
resonance are included, and attention is given to bandwidth and quality factor Q,
particularly in high-Q networks. Frequency response of impedances and of other
transfer functions is discussed in quite some detail.

Decomposition of an impedance at a single frequency into series and parallel
resistive and reactive components, a concept of great practical and design importance, is then discussed and examples of obtaining network resonance in this way
are given.

When you complete this chapter, you should know—

1. the characteristics of resonance in series and in parallel *RLC* networks;
2. what frequency response plots are, and how to construct them;
3. the general definitions and meaning of resonance and bandwidth;
4. about the quality factor Q of a resonant network, and its relationship to bandwidth;
5. how to represent an impedance at a single frequency as a series combination and as a parallel combination of resistance and reactance;
6. how to connect a series or parallel inductor or capacitor to a network to make it resonant at a given frequency.

8.2 THE SERIES *RLC* NETWORK

8.2.1 Series Resonance

In two-terminal networks containing both inductors and capacitors, there are one or more sinusoidal frequencies for which the network impedance, $Z(s = j\omega)$, is entirely real, as it is for a resistor. This phenomenon is termed *resonance*. In the vicinity of such a resonant frequency, the impedance at the network terminals generally exhibits a peak or a dip in magnitude. Networks may thus be designed to emphasize frequencies near their resonance. In a radio receiver, for example, the resonant frequency of a network may be adjusted by varying an element value (often the value of a capacitor) to "tune in" the range of frequencies sent by a transmitting station, greatly attenuating other frequencies sent by other stations.

Perhaps the most simple resonant network is the series *RLC* one, diagramed in Figure 8-1. For sinusoidal signals with radian frequency ω, the impedance as viewed from the network terminals is

$$Z(s = j\omega) = R + j\omega L + \frac{1}{j\omega C} = R + j\left(\omega L - \frac{1}{\omega C}\right)$$

The frequency $\omega = \omega_r$, for which

$$\omega_r L = \frac{1}{\omega_r C} \qquad \omega_r^2 LC = 1$$

$$\omega_r = \frac{1}{\sqrt{LC}}$$

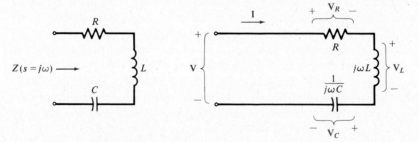

Figure 8-1 Series *RLC* network.

is called the network's *resonant frequency.* It is the frequency at which the capacitor and inductor impedances add to zero. That is, at the resonant frequency, the series combination of inductor and capacitor behave as a short circuit. The resonant frequency in hertz (cycles per second) for the series *RLC* network is

$$f_r = \frac{\omega_r}{2\pi} = \frac{1}{2\pi \sqrt{LC}}$$

8.2.2 Behavior as a Function of Frequency

At frequencies lower than the resonant frequency,

$$\frac{1}{\omega C} > \omega L$$

the series *RLC* network impedance is

$$Z(s = j\omega) = R + j\left(\omega L - \frac{1}{\omega C}\right) = R + j(\text{negative number})$$

At such frequencies, the sink-reference network current leads the network voltage, as in an *RC* network. At resonance,

$$Z(s = j\omega) = R + j(0) = R$$

and the network voltage and current are in phase with one another.

At frequencies higher than the resonant frequency,

$$\omega L > \frac{1}{\omega C}$$

the impedance is

$$Z(s = j\omega) = R + j\left(\omega L - \frac{1}{\omega C}\right) = R + j(\text{positive number})$$

The sink-reference network current lags the voltage, as in an *RL* network.

The magnitude of the impedance,

$$|Z(s = j\omega)| = \sqrt{R^2 + \left(\omega L - \frac{1}{\omega C}\right)^2}$$

is minimum at resonance, which is to say that for constant applied voltage amplitude, the resulting current is of greatest amplitude at the resonant frequency.

Phasor diagrams for the series *RLC* network, in terms of the current **I**, with angle taken to be 0°, are shown in Figure 8-2(a). Relations for a frequency below resonance, at resonant frequency, and at a frequency above resonance are shown. The real part of the network impedance is constant, while the imaginary part of the impedance is negative at low frequencies and positive at high frequencies, as shown in Figure 8-2(b).

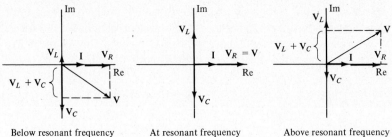

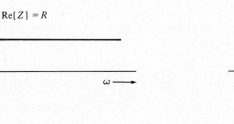

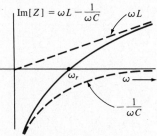

Below resonant frequency   At resonant frequency   Above resonant frequency

(a)

(b)

Figure 8-2 Phasor and impedance relations for the series *RLC* network.
(a) Phasor relations below, at, and above resonant frequency.
(b) Real and imaginary parts of the series network impedance as functions of frequency.

8.2.3 Numerical Example

The series *RLC* network of Figure 8-3, for example, has impedance, for sinusoidal signals with radian frequency ω,

$$Z(s = j\omega) = R + j\left(\omega L - \frac{1}{\omega C}\right) = 40 + j\left(15\omega - \frac{1000}{2\omega}\right)$$

It is resonant at the frequency

$$\omega_r = \frac{1}{\sqrt{LC}} = \frac{1}{\sqrt{30 \times 10^{-3}}} = 5.77 \text{ rad/s}$$

or

$$f_r = \frac{1}{2\pi \sqrt{LC}} = 0.92 \text{ Hz}$$

At the resonant frequency,

$$Z(s = j5.77) = R + j(0) = 40 \ \Omega$$

the inductor and capacitor impedances (which are purely imaginary with opposite algebraic signs) sum to zero. At a frequency below resonance, say one-half the resonant frequency,

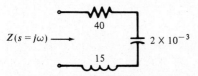

Figure 8-3 Series resonance example.

$$Z\left(s = j\,\tfrac{5.77}{2}\right) = 40 + j[43.3 - 173.3] = 40 - j130\ \Omega$$

The purely imaginary inductor impedance cancels part, but not all, of the capacitor's impedance.

At frequencies above resonance, the reverse occurs. At twice the resonant frequency, for example,

$$Z(s = j11.54) = 40 + j(173.3 - 43.3) = 40 + j130\ \Omega$$

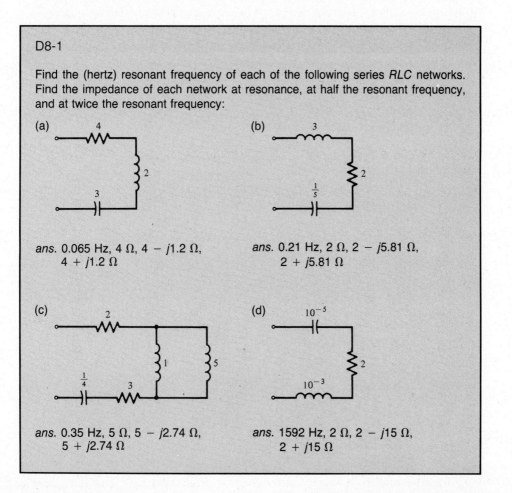

D8-1

Find the (hertz) resonant frequency of each of the following series *RLC* networks. Find the impedance of each network at resonance, at half the resonant frequency, and at twice the resonant frequency:

(a)

ans. 0.065 Hz, 4 Ω, 4 − *j*1.2 Ω,
4 + *j*1.2 Ω

(b)

ans. 0.21 Hz, 2 Ω, 2 − *j*5.81 Ω,
2 + *j*5.81 Ω

(c)

ans. 0.35 Hz, 5 Ω, 5 − *j*2.74 Ω,
5 + *j*2.74 Ω

(d)

ans. 1592 Hz, 2 Ω, 2 − *j*15 Ω,
2 + *j*15 Ω

SERIES *RLC* NETWORKS

The impedance for sinusoidal signals of a series *RLC* network is

$$Z = R + j\omega L + \frac{1}{j\omega C} = R + j\left(\omega L - \frac{1}{\omega C}\right)$$

where ω is the source radian frequency. The network is said to be resonant at the frequency

$$\omega_r = \frac{1}{\sqrt{LC}}, \qquad f_r = \frac{1}{2\pi \sqrt{LC}}$$

for which the series combination of inductor and capacitor is equivalent to a short circuit, and

$$Z = R$$

As a function of frequency, the sink-reference network voltage is in phase with the current at resonance. The current phase leads the voltage (as in an *RC* network) below the resonant frequency and lags it (as in an *RL* network) above resonance.

For constant network voltage amplitude, the sink reference network current is maximum and in phase with the voltage at resonance.

8.3 THE PARALLEL *RLC* NETWORK

8.3.1 Parallel Resonance

Another simple resonant network is the parallel *RLC* network, Figure 8-4. It has impedance, for sinusoidal signals,

$$Z(s = j\omega) = \frac{1}{\dfrac{1}{R} + j\left(\omega C - \dfrac{1}{\omega L}\right)}$$

The *admittance* of this network (admittance is the inverse of impedance),

$$Y(s = j\omega) = \frac{1}{Z(s = j\omega)} = \frac{1}{R} + j\left(\omega C - \frac{1}{\omega L}\right)$$

is of the same form as the impedance of the series *RLC* network, with the roles of L and C interchanged, and R replaced by $1/R$.

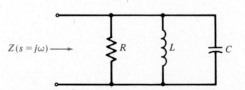

$Z(s = j\omega) \longrightarrow$ R L C

Figure 8-4   Parallel *RLC* network.

The frequency $\omega = \omega_r$, for which

$$\omega_r C = \frac{1}{\omega_r L} \qquad \omega_r = \frac{1}{\sqrt{LC}} \text{ rad/s}$$

is the network's resonant frequency, where

$$Z(s = j\omega_r) = \frac{1}{\dfrac{1}{R} + j(0)} = R$$

The purely imaginary parallel inductor and capacitor impedances combine to give an *open* circuit at the resonant frequency.

8.3.2 Behavior as a Function of Frequency

Multiplying the numerator and denominator of Z by the complex conjugate of the denominator places Z in rectangular form, where its real and imaginary parts are evident:

$$Z(s = j\omega) = \frac{1}{\dfrac{1}{R} + j\left(\omega C - \dfrac{1}{\omega L}\right)}$$

$$= \frac{\dfrac{1}{R} - j\left(\omega C - \dfrac{1}{\omega L}\right)}{\left[\dfrac{1}{R} + j\left(\omega C - \dfrac{1}{\omega L}\right)\right]\left[\dfrac{1}{R} - j\left(\omega C - \dfrac{1}{\omega L}\right)\right]}$$

$$= \frac{\dfrac{1}{R} + j\left(\dfrac{1}{\omega L} - \omega C\right)}{\left(\dfrac{1}{R}\right)^2 + \left(\omega C - \dfrac{1}{\omega L}\right)^2}$$

At frequencies below the resonant frequency,

$$\omega C < \frac{1}{\omega L}$$

the parallel *RLC* network impedance is

$$Z(s = j\omega) = \frac{\dfrac{1}{R} + j(\text{positive number})}{\text{positive number}}$$

At such frequencies, the sink-reference network current lags the network voltage, as in an *RL* network.

At frequencies above resonance,

$$\omega C > \frac{1}{\omega L}$$

the impedance is

$$Z(s = j\omega) = \frac{\dfrac{1}{R} + j(\text{negative number})}{\text{positive number}}$$

The sink-reference current leads the voltage, as in an RC network.

The magnitude of the impedance is maximum at resonance, since the magnitude of

$$Y(s = j\omega) = \frac{1}{R} + j\left(\omega C - \frac{1}{\omega L}\right) = \frac{1}{Z(s = j\omega)}$$

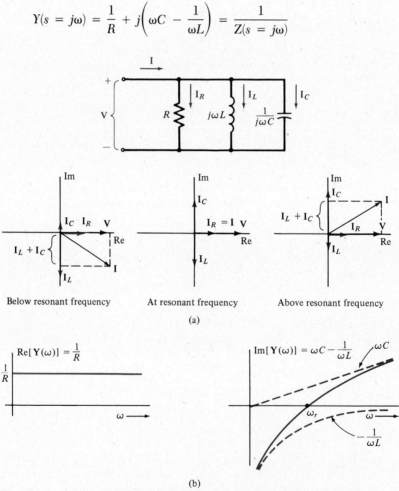

Below resonant frequency   At resonant frequency   Above resonant frequency

(a)

(b)

Figure 8-5   Phasor and admittance relations in the parallel RLC network.
(a) Phasor relations below, at, and above resonant frequency.
(b) Real and imaginary parts of the parallel network admittance as functions of frequency.

is minimum at resonance. For constant applied voltage amplitude, the resulting current has the smallest amplitude at the resonant frequency.

Phasor diagrams for the parallel *RLC* network are shown in Figure 8-5(a), in terms of the voltage **V** taken at angle 0°. The real part of the network admittance is constant, while the imaginary part of the admittance is negative at low frequencies and positive at high frequencies, as shown in Figure 8-5(b).

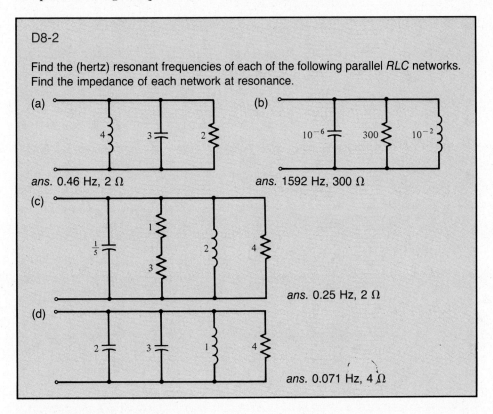

D8-2

Find the (hertz) resonant frequencies of each of the following parallel *RLC* networks. Find the impedance of each network at resonance.

(a)
ans. 0.46 Hz, 2 Ω

(b)
ans. 1592 Hz, 300 Ω

(c)
ans. 0.25 Hz, 2 Ω

(d)
ans. 0.071 Hz, 4 Ω

8.3.3 Numerical Example

The parallel *RLC* network in Figure 8-6 has impedance for sinusoidal signals

$$Z(s = j\omega) = \cfrac{1}{\cfrac{1}{R} + j\omega C + \cfrac{1}{j\omega L}} = \cfrac{1}{\cfrac{1}{50} + j\left(2 \times 10^{-3}\omega - \cfrac{1}{4\omega}\right)}$$

It is resonant at the frequency

$$\omega_r = \frac{1}{\sqrt{LC}} = \frac{1}{\sqrt{8 \times 10^{-3}}} = 11.2 \text{ rad/s},$$

or

$$f_r = \frac{1}{2\pi \sqrt{LC}} = 1.78 \text{ Hz}$$

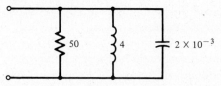

Figure 8-6   Parallel resonance example.

At the resonant frequency,

$$Z(s = j11.2) = \frac{1}{\frac{1}{R} + j(0)} = R = 50 \ \Omega$$

At a frequency below resonance, say one-half the resonant frequency,

$$Z(s = j5.6) = \frac{1}{0.02 + j(0.0112 - 0.0446)}$$

$$= \frac{1}{0.02 - j0.0335} = 13.1 + j22.0 \ \Omega$$

The sink-reference network current lags the voltage, as in an *RL* network.

At frequencies above resonance, current leads voltage, as in an *RC* network. At twice the resonant frequency,

$$Z(s = j22.4) = \frac{1}{0.02 + j(0.0448 - 0.01116)}$$

$$= \frac{1}{0.02 + j0.0335} = 13.1 - j22.0 \ \Omega$$

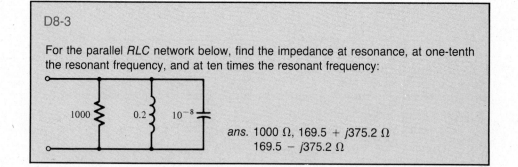

D8-3

For the parallel *RLC* network below, find the impedance at resonance, at one-tenth the resonant frequency, and at ten times the resonant frequency:

ans. 1000 Ω, 169.5 + j375.2 Ω
169.5 − j375.2 Ω

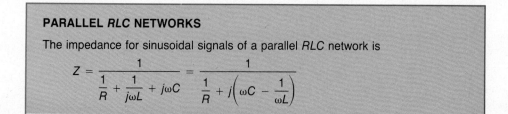

PARALLEL *RLC* NETWORKS

The impedance for sinusoidal signals of a parallel *RLC* network is

$$Z = \frac{1}{\frac{1}{R} + \frac{1}{j\omega L} + j\omega C} = \frac{1}{\frac{1}{R} + j\left(\omega C - \frac{1}{\omega L}\right)}$$

where ω is the source radian frequency. The network is resonant at the frequency

$$\omega_r = \frac{1}{\sqrt{LC}} \qquad f_r = \frac{1}{2\pi\sqrt{LC}}$$

where the parallel *LC* combination is equivalent to an open circuit, and

$$Z = R$$

At frequencies below resonance, the effect of the inductor dominates that of the capacitor, and the sink-reference voltage phase leads the current phase. Above resonance, the voltage lags the current.

For constant voltage amplitude, the sink-reference network current has minimum amplitude and is in phase with the voltage at resonance.

8.4 FREQUENCY RESPONSE

8.4.1 Amplitude Ratio and Phase Shift

For sinusoidal signals, the impedance of a two-terminal network, Figure 8-7(a), has magnitude that is the ratio of voltage amplitude to current amplitude:

$$|Z(s = j\omega)| = \frac{|\mathbf{V}|}{|\mathbf{I}|} = \frac{\text{amplitude of voltage}}{\text{amplitude of current}}$$

The angle of the impedance is the difference in phase, or *phase shift*, between the sink reference voltage and current:

$$\angle Z(s = j\omega) = \angle\mathbf{V} - \angle\mathbf{I}$$

$$= (\text{voltage phase angle}) - (\text{current phase angle})$$

Plots of $|Z(s = j\omega)|$ and $\angle Z(s = j\omega)$, as functions of ω, are called *frequency response plots* of the impedance.

For example, the specific network of Figure 8-7(b) has impedance

$$Z(s) = 1 + \frac{6s}{3 + 2s} = \frac{8s + 3}{2s + 3}$$

For sinusoidal signals, $s = j\omega$,

$$Z(s = j\omega) = \frac{8j\omega + 3}{2j\omega + 3}$$

The ratio of the terminal voltage to current amplitudes is, as a function of the sinusoidal radian frequency ω,

$$\frac{|\mathbf{V}|}{|\mathbf{I}|} = Z(s = j\omega) = \frac{\sqrt{64\omega^2 + 9}}{\sqrt{4\omega^2 + 9}}$$

The phase shift between the terminal voltage and current is, as a function of ω,

$$\angle Z(s = j\omega) = \arctan\frac{8\omega}{3} - \arctan\frac{2\omega}{3}$$

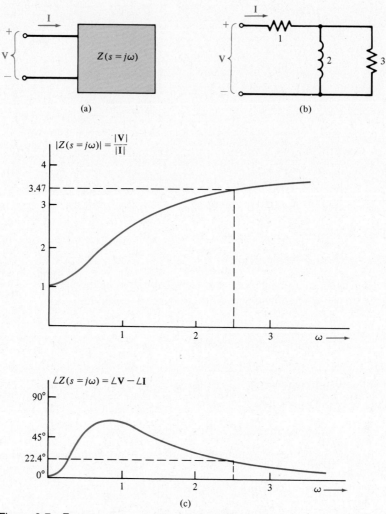

Figure 8-7   Frequency response of an impedance.

The frequency response plots for this impedance are given in Figure 8-7(c). At very low frequencies, the voltage-current amplitude ratio is near unity, while at higher frequencies, it approaches 4.0. The phase difference between the voltage and current is near zero at low frequencies, rises to a maximum of about 60° near $\omega = 0.8$, and approaches zero again at high frequencies. Using the plots, a sinusoidal current, for example, with radian frequency $\omega = 2.5$, will result in a voltage with amplitude 3.47 times the current amplitude and phase angle 22.4° larger than the current phase angle. The current

$$i(t) = 10 \cos (2.5t + 50°)$$

for instance, results in the voltage

$$v(t) = 10(3.47) \cos (2.5t + 50° + 22.4°) = 34.7 \cos (2.5t + 72.4°)$$

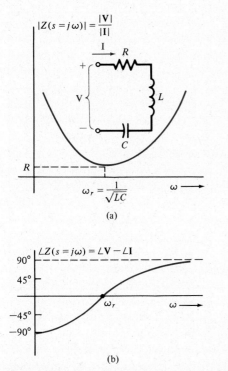

(a)

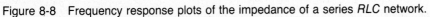

(b)

Figure 8-8   Frequency response plots of the impedance of a series *RLC* network.

8.4.2 Frequency Response of the Series *RLC* Network

For the series *RLC* network,

$$Z(s = j\omega) = R + j\left(\omega L - \frac{1}{\omega C}\right)$$

The ratio of amplitudes of the network voltage to the network current is

$$\frac{|\mathbf{V}|}{|\mathbf{I}|} = |Z(s = j\omega)| = \sqrt{R^2 + \left(\omega L - \frac{1}{\omega C}\right)^2}$$

a sketch of which is shown in Figure 8-8(a).

The difference in phase angle between the series *RLC* network's sink reference voltage and current is

$$\angle\mathbf{V} - \angle\mathbf{I} = \angle Z(s = j\omega) = \arctan\left(\frac{\omega L - \dfrac{1}{\omega C}}{R}\right)$$

which is sketched in Figure 8-8(b). For small values of ω, the effect of the series capacitor dominates that of the inductor, and the impedance angle approaches $-90°$. At resonance, it is $0°$; and for large ω, where the series inductor dominates, the impedance angle approaches $+90°$.

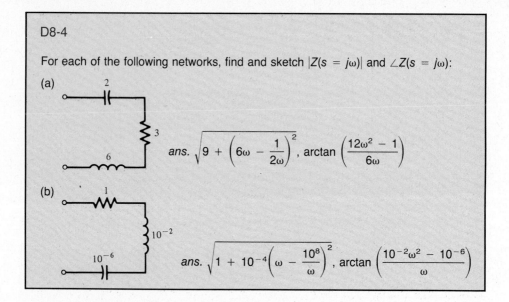

D8-4

For each of the following networks, find and sketch $|Z(s = j\omega)|$ and $\angle Z(s = j\omega)$:

(a)

ans. $\sqrt{9 + \left(6\omega - \dfrac{1}{2\omega}\right)^2}$, arctan $\left(\dfrac{12\omega^2 - 1}{6\omega}\right)$

(b)

ans. $\sqrt{1 + 10^{-4}\left(\omega - \dfrac{10^8}{\omega}\right)^2}$, arctan $\left(\dfrac{10^{-2}\omega^2 - 10^{-6}}{\omega}\right)$

8.4.3 Frequency Response of Other Signals

In general, frequency response may express the amplitude ratio and phase shift of any two sinusoidal signals in a network, as a function of frequency. One could plot the frequency response of an admittance, or the relation between two voltages or between two network currents.

For example, the ratio of the inductor voltage phasor to the total voltage phasor in the series RLC network in Figure 8-9(a) is

$$\frac{\mathbf{V}_L}{\mathbf{V}} = \frac{j\omega L}{R + j\omega L + \dfrac{1}{j\omega C}} = \frac{-\omega^2 LC}{-\omega^2 LC + j\omega RC + 1}$$

as given by the voltage divider rule. The amplitude ratio of these two voltages is

$$\frac{|\mathbf{V}_L|}{|\mathbf{V}|} = \frac{\omega^2 LC}{\sqrt{(1 - \omega^2 LC)^2 + \omega^2 R^2 C^2}}$$

and the phase shift is

$$\angle \mathbf{V}_L - \angle \mathbf{V} = 180° - \arctan\left(\frac{\omega RC}{1 - \omega^2 LC}\right)$$

Frequency response curves for this phasor ratio and for the ratio

$$\frac{\mathbf{V}_C}{\mathbf{V}} = \frac{\dfrac{1}{j\omega C}}{R + j\omega L + \dfrac{1}{j\omega C}} = \frac{1}{-\omega^2 LC + j\omega RC + 1}$$

in the series network are sketched in Figure 8-9(b).

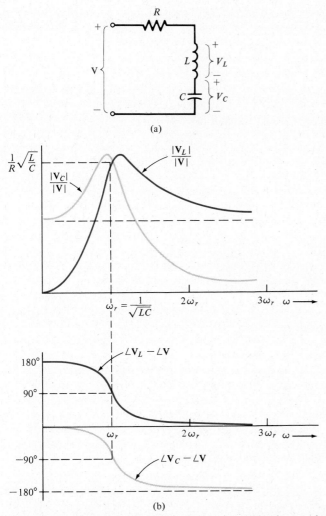

(a)

$$\omega_r = \frac{1}{\sqrt{LC}}$$

(b)

Figure 8-9 Frequency response of other quantities in a series *RLC* network.

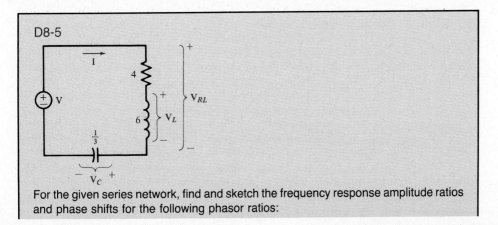

D8-5

For the given series network, find and sketch the frequency response amplitude ratios and phase shifts for the following phasor ratios:

(a) I/V

ans. $\dfrac{1}{\sqrt{16 + [6\omega - (3/\omega)]^2}}$, $\arctan\left(\dfrac{-6\omega^2 - 3}{4\omega}\right)$

(b) V_L/V

ans. $\dfrac{6\omega}{\sqrt{16 + [6\omega - (3/\omega)]^2}}$, $90° + \arctan\left(\dfrac{-6\omega^2 - 3}{4\omega}\right)$

(c) V_C/V

ans. $\dfrac{3}{\omega\sqrt{16 + [6\omega - (3/\omega)]^2}}$, $-90° + \arctan\left(\dfrac{-6\omega^2 - 3}{4\omega}\right)$

(d) V_{RL}/V

ans. $\dfrac{\sqrt{16 + 36\omega^2}}{\sqrt{16 + [6\omega - (3/\omega)]^2}}$, $\arctan\left(\dfrac{3\omega}{2}\right) + \arctan\left(\dfrac{-6\omega^2 - 3}{4\omega}\right)$

8.4.4 Parallel *RLC* Networks

The impedance of the parallel *RLC* network is

$$Z(s = j\omega) = \dfrac{1}{\dfrac{1}{R} + j\left(\omega C - \dfrac{1}{\omega L}\right)}$$

The amplitude ratio of voltage to current at the terminals is

$$|Z(s = j\omega)| = \dfrac{1}{\sqrt{\left(\dfrac{1}{R}\right)^2 + \left(\omega C - \dfrac{1}{\omega L}\right)^2}}$$

and the phase shift between the two is

$$\angle Z(s = j\omega) = -\arctan\left(\dfrac{\omega C - \left(\dfrac{1}{\omega L}\right)}{\left(\dfrac{1}{R}\right)}\right)$$

These frequency response curves are sketched in Figure 8-10. The amplitude curve is the inverse of a series *RLC* network amplitude curve, since the admittance of the parallel network has the same form as the impedance of a series network.

The ratio of the capacitor current phasor to the total current phasor, Figure 8-11(a), is given by the current divider rule:

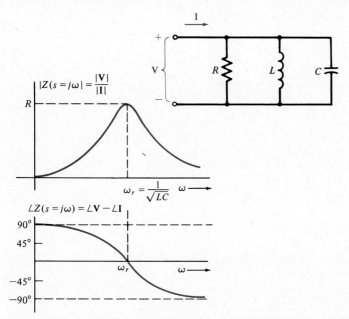

Figure 8-10   Frequency response of the impedance of a parallel *RLC* network.

$$\frac{\mathbf{I}_C}{\mathbf{I}} = \frac{\dfrac{j\omega RL}{R + j\omega L}}{\dfrac{1}{j\omega C} + \dfrac{j\omega RL}{R + j\omega L}} = \frac{-\omega^2 LC}{-\omega^2 LC + j\omega \dfrac{L}{R} + 1}$$

The amplitude ratio of these two currents is

$$\frac{|\mathbf{I}_C|}{|\mathbf{I}|} = \frac{\omega^2 LC}{\sqrt{(1 - \omega^2 LC)^2 + \omega^2 \left(\dfrac{L}{R}\right)^2}}$$

and their phase shift is

$$\angle \mathbf{I}_C - \angle \mathbf{I} = 180° - \arctan\left(\frac{\omega L/R}{1 - \omega^2 LC}\right)$$

This frequency response is sketched in Figure 8-11(b), as is the frequency response

$$\frac{\mathbf{I}_L}{\mathbf{I}} = \frac{\dfrac{\dfrac{R}{j\omega C}}{R + \dfrac{1}{j\omega C}}}{\dfrac{\dfrac{R}{j\omega C}}{R + \dfrac{1}{j\omega C}} + j\omega L} = \frac{1}{-\omega^2 LC + j\omega \dfrac{L}{R} + 1}$$

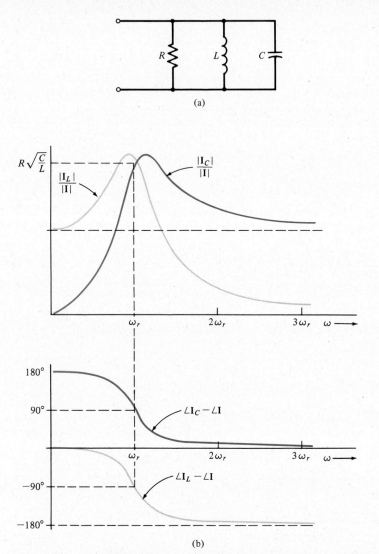

(a)

(b)

Figure 8-11   Frequency response of other quantities in a parallel *RLC* network.

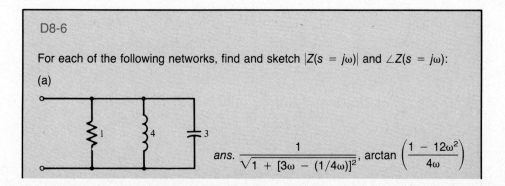

D8-6

For each of the following networks, find and sketch $|Z(s = j\omega)|$ and $\angle Z(s = j\omega)$:

(a)

ans. $\dfrac{1}{\sqrt{1 + [3\omega - (1/4\omega)]^2}}$, $\arctan\left(\dfrac{1 - 12\omega^2}{4\omega}\right)$

(b)

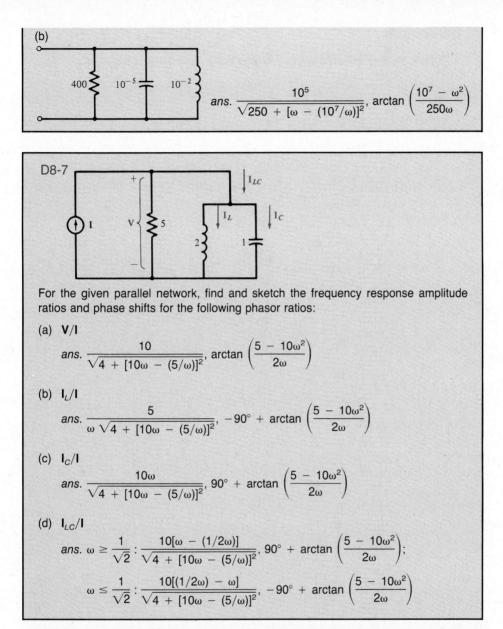

ans. $\dfrac{10^5}{\sqrt{250 + [\omega - (10^7/\omega)]^2}}$, arctan $\left(\dfrac{10^7 - \omega^2}{250\omega}\right)$

D8-7

For the given parallel network, find and sketch the frequency response amplitude ratios and phase shifts for the following phasor ratios:

(a) **V/I**

ans. $\dfrac{10}{\sqrt{4 + [10\omega - (5/\omega)]^2}}$, arctan $\left(\dfrac{5 - 10\omega^2}{2\omega}\right)$

(b) **I_L/I**

ans. $\dfrac{5}{\omega \sqrt{4 + [10\omega - (5/\omega)]^2}}$, $-90° + $ arctan $\left(\dfrac{5 - 10\omega^2}{2\omega}\right)$

(c) **I_C/I**

ans. $\dfrac{10\omega}{\sqrt{4 + [10\omega - (5/\omega)]^2}}$, $90° + $ arctan $\left(\dfrac{5 - 10\omega^2}{2\omega}\right)$

(d) **I_LC/I**

ans. $\omega \geq \dfrac{1}{\sqrt{2}} : \dfrac{10[\omega - (1/2\omega)]}{\sqrt{4 + [10\omega - (5/\omega)]^2}}$, $90° + $ arctan $\left(\dfrac{5 - 10\omega^2}{2\omega}\right)$;

$\omega \leq \dfrac{1}{\sqrt{2}} : \dfrac{10[(1/2\omega) - \omega]}{\sqrt{4 + [10\omega - (5/\omega)]^2}}$, $-90° + $ arctan $\left(\dfrac{5 - 10\omega^2}{2\omega}\right)$

FREQUENCY RESPONSE

For sinusoidal signals, the magnitude of an impedance is the ratio of voltage amplitude to current amplitude. The impedance angle is the phase angle of the voltage minus the phase angle of the current, the phase shift between voltage and current.

Frequency response plots consist of a plot of the ratio of the amplitudes of two sinusoidal signals in a network, as a function of frequency, and a plot of the difference in phase between the two signals, as a function of frequency.

8.5 BANDWIDTH

8.5.1 Series Network Half-Power Frequencies and Bandwidth

In the series RLC network at the frequency $\omega = \omega_1$ for which

$$\frac{1}{\omega_1 C} - \omega_1 L = R$$

the impedance is

$$Z(s = j\omega_1) = R + j\left(\omega_1 L - \frac{1}{\omega_1 C}\right) = R - jR$$

At this frequency,

$$|Z(s = j\omega_1)| = \sqrt{R^2 + (-R)^2} = R\sqrt{2} \quad \text{and} \quad \angle Z(s = j\omega_1) = -45°$$

At the frequency $\omega = \omega_2$, for which

$$\omega_2 L - \frac{1}{\omega_2 C} = R$$

$$Z(s = j\omega_2) = R + jR$$

and

$$|Z(s = j\omega_2)| = \sqrt{R^2 + R^2} = R\sqrt{2} \quad \angle Z(s = j\omega_2) = 45°$$

The frequencies ω_1 and ω_2 (or their hertz counterparts $f_1 = \omega_1/2\pi$ and $f_2 = \omega_2/2\pi$) are termed the *half-power frequencies* of the series RLC network. Since electrical power is related to the square of current, increasing the impedance by a factor of $\sqrt{2}$, at constant voltage, halves power. The half-power frequencies of the series RLC network are indicated in Figure 8-12.

The interval of frequency between the half-power frequencies is called the *bandwidth* of a resonant network. For the series RLC network, where

$$\frac{1}{\omega_1 C} - \omega_1 L = R \qquad \omega_1^2 LC + \omega_1 RC - 1 = 0$$

$$\omega_1 = \frac{-RC \pm \sqrt{R^2 C^2 + 4LC}}{2LC}$$

The solution for positive ω_1 is

$$\omega_1 = \frac{-RC + \sqrt{R^2 C^2 + 4LC}}{2LC}$$

Similarly, for ω_2,

$$\omega_2 L - \frac{1}{\omega_2 C} = R \qquad \omega_2^2 LC - \omega_2 RC - 1 = 0$$

$$\omega_2 = \frac{RC \pm \sqrt{R^2 C^2 + 4LC}}{2LC}$$

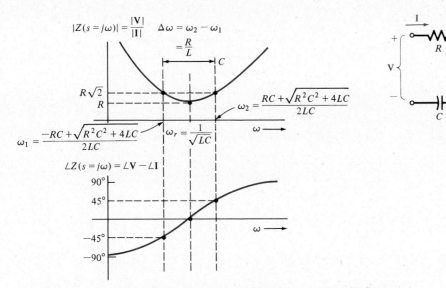

Figure 8-12   Half-power frequencies and bandwidth of a series RLC network.

and the solution for positive ω_2 is

$$\omega_2 = \frac{RC + \sqrt{R^2C^2 + 4LC}}{2LC}$$

The difference between ω_2 and ω_1 is

$$\Delta\omega = \omega_2 - \omega_1 = \frac{R}{L} \quad \text{or} \quad \Delta f = \frac{\Delta\omega}{2\pi} = \frac{R}{2\pi L}$$

as indicated in Figure 8-12.

8.5.2 Bandwidth of the Parallel Network

In the parallel RLC network, for which

$$Z(s = j\omega) = \frac{1}{\dfrac{1}{R} + j\left(\omega C - \dfrac{1}{\omega L}\right)}$$

the amplitude ratio $|Z(s = j\omega)|$ exhibits a maximum at resonance. The half-power points are then defined to be the frequencies for which the magnitude of Z is reduced by a factor of $\sqrt{2}$ from its value at resonance. This occurs for ω_1, for which

$$\frac{1}{\omega_1 L} - \omega_1 C = \frac{1}{R}$$

and for ω_2, for which

$$\omega_2 C - \frac{1}{\omega_2 L} = \frac{1}{R}$$

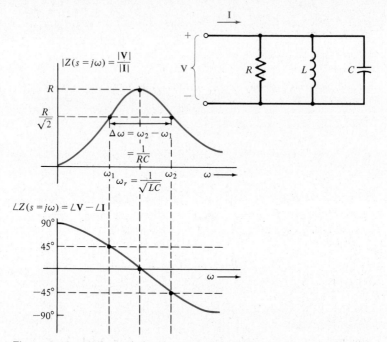

Figure 8-13   Half-power frequencies and bandwidth of a parallel *RLC* network.

The solutions for ω_1 and ω_2 are

$$\omega_1 = \frac{-\dfrac{L}{R} + \sqrt{\left(\dfrac{L}{R}\right)^2 + 4LC}}{2LC} \qquad \omega_2 = \frac{\dfrac{L}{R} + \sqrt{\left(\dfrac{L}{R}\right)^2 + 4LC}}{2LC}$$

and the bandwidth of this network is

$$\Delta\omega = \omega_2 - \omega_1 = \frac{1}{RC} \text{ rad/s} \qquad \text{or} \qquad \Delta f = \frac{\Delta\omega}{2\pi} = \frac{1}{2\pi RC} \text{ Hz}$$

The phase shift between voltage and current is $\pm 45°$ at the half-power frequencies, as illustrated in the sketch of Figure 8-13.

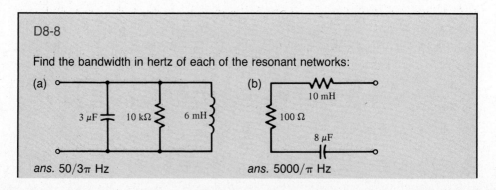

D8-8

Find the bandwidth in hertz of each of the resonant networks:

(a)

3 µF   10 kΩ   6 mH

(b)

10 mH   100 Ω   8 µF

ans. 50/3π Hz   *ans.* 5000/π Hz

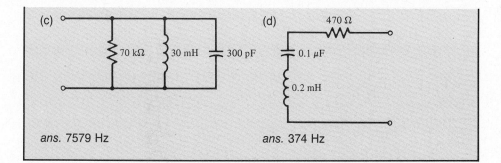

ans. 7579 Hz   ans. 374 Hz

BANDWIDTH

The bandwidth of a resonant minimum in frequency response amplitude is the frequency interval, extending from one side of the minimum to the other, for which the amplitude ratio is below $\sqrt{2}$ times its value at the minimum. The frequencies at which the amplitude ratio equals $\sqrt{2}$ times the minimum value are called *half-power frequencies*.

The series *RLC* network has bandwidth

$$\Delta\omega = \frac{R}{L} \qquad \Delta f = \frac{R}{2\pi L}$$

For a resonant maximum in frequency response amplitude, the half-power frequencies are those for which the amplitude ratio equals $(1/\sqrt{2})$ times the maximum value. The bandwidth is the difference between the half-power frequencies.

The parallel *RLC* network has bandwidth

$$\Delta\omega = \frac{1}{RC} \qquad \Delta f = \frac{1}{2\pi RC}$$

For the series and the parallel *RLC* networks, the phase shifts between network voltage and current are $\pm45°$ at the half-power frequencies.

8.6 RESONANCE IN GENERAL

More complicated *RLC* networks than the series and parallel ones also exhibit resonance. In fact, if there are several *L*s and *C*s, there may be two or more resonant frequencies. In the more involved networks, frequencies for which

$$\angle Z(s = j\omega) = 0$$

are not precisely the same frequencies for which $|Z(s = j\omega)|$ has maxima or minima. In practice, both the zero-angle and the magnitude extrema are used as definitions of resonance, the choice depending on the application at hand.

For example, the *RLC* network of Figure 8-14 commonly occurs when a physical inductor and capacitor are connected in parallel. The winding resistance of the inductor is not negligible and is represented by the resistance *R*. This network is resonant, but it is neither a series nor a parallel *RLC* network. For sinusoidal signals,

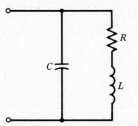

Figure 8-14   A series-parallel *RLC* network.

the impedance of this network at the terminals is

$$Z(s = j\omega) = \frac{(R + j\omega L)(1/j\omega C)}{R + j\omega L + (1/j\omega C)} = \frac{R + j\omega L}{1 - \omega^2 LC + j\omega RC}$$

The real and imaginary parts of Z are as follows:

$$Z(s = j\omega) = \frac{(R + j\omega L)(1 - \omega^2 LC - j\omega RC)}{(1 - \omega^2 LC + j\omega RC)(1 - \omega^2 LC - j\omega RC)}$$

$$= \frac{R - \omega^2 RLC + \omega^2 LRC}{(1 - \omega^2 LC)^2 + \omega^2 R^2 C^2} - \frac{j(\omega^3 L^2 C + \omega R^2 C - \omega L)}{(1 - \omega^2 LC)^2 + \omega^2 R^2 C^2}$$

The angle of Z is zero, and thus the network is zero-angle resonant for ω for which the imaginary part of Z is zero:

$$\omega^3 L^2 C + \omega(R^2 C - L) = 0$$

$$\omega = 0, \quad \pm \sqrt{\frac{1 - \left(\dfrac{R^2 C}{L}\right)}{LC}}$$

Ordinarily the obvious $\omega = 0$ solution for a network, corresponding to constant voltages and currents, is not considered to be a resonant frequency; only positive values of ω are sought, so

$$\omega_r = \sqrt{\frac{1 - \left(\dfrac{R^2 C}{L}\right)}{LC}} = \sqrt{1 - \left(\frac{R^2 C}{L}\right)}\left(\frac{1}{\sqrt{LC}}\right)$$

for this network.

For $R = 0$, the network is parallel LC, and

$$\omega_r = \frac{1}{\sqrt{LC}}$$

as expected. For significant R, the resonant frequency is lowered by the factor $\sqrt{1 - (R^2 C/L)}$.

Almost everyone encounters resonance at an early age. Children learn to pump a swing at a frequency for which the oscillations increase rapidly; and they learn to slosh the water back and forth in a bathtub at just the rate that puts most of it on the floor in minimum time.

In these and many other cases, the system's resonant frequency is very nearly the rate at which it tends to oscillate naturally. For example, one pumps a swing at just about the rate it swings when the rider is aboard but not pumping. The situation is similar for resonant electrical networks: Resonant frequencies tend to be very nearly the frequencies of a network's oscillatory natural response.

D8-9

Find the zero-angle resonant frequencies (in radians per second) for each of the following networks:

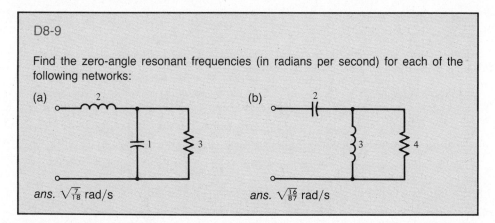

(a)

(b)

ans. $\sqrt{\frac{7}{18}}$ rad/s

ans. $\sqrt{\frac{16}{87}}$ rad/s

RESONANCE

There are two common definitions of resonance. In the first, a two-terminal network with impedance $Z(s = j\omega)$ is said to be resonant at any frequency for which

$$\angle Z(s = j\omega) = 0°$$

that is, any frequency at which Z is entirely real. At a resonant frequency, the sink reference network voltage and current are in phase, and the network is equivalent to a resistor. A network is resonant according to the second definition at any frequency for which $|Z(s = j\omega)|$ exhibits an extrema (a maximum or a minimum).

For the series and the parallel RLC networks, both definitions yield the same resonant frequency (which is the same $\omega_r = 1/\sqrt{LC}$ for each type of network). For more involved networks, the two definitions generally give differing results.

8.7 QUALITY FACTOR

The quality factor Q of a resonant network is a measure of the narrowness of its bandwidth, that is, the "sharpness" of the resonant maximum or minimum of $|Z(s = j\omega)|$. It is defined to be

$$Q = 2\pi \frac{\text{energy stored in the network at resonant frequency}}{\text{energy dissipated in the network per sinusoidal cycle at resonance}}$$

For the series RLC network, energy is stored in the inductor in amount

$$W_L(t) = \tfrac{1}{2}Li_L^2(t)$$

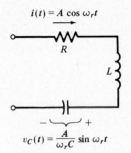

$i(t) = A \cos \omega_r t$

R

L

$v_C(t) = \dfrac{A}{\omega_r C} \sin \omega_r t$

Figure 8-15 Series *RLC* network at resonance.

and in the capacitor,

$$W_C(t) = \tfrac{1}{2} C v_C^2(t)$$

Let the resonant-frequency sinusoidal current in the network be

$$i(t) = A \cos (\omega_r t)$$

as in the drawing of Figure 8-15. The current $i(t)$ is the inductor current, so

$$W_L = \frac{1}{2} L i^2(t) = \frac{LA^2}{2} \cos^2 \omega_r t$$

The capacitor voltage is then

$$v_C(t) = \frac{A}{\omega_r C} \sin \omega_r t$$

giving

$$W_C = \frac{1}{2} C v_C^2 = \frac{A^2}{2\omega_r^2 C} \sin^2 \omega_r t$$

Substituting $\omega_r^2 = 1/LC$.

$$W_C = \frac{LA^2}{2} \sin^2 \omega_r t$$

The total stored energy is

$$W = W_L + W_C = \frac{LA^2}{2} (\sin^2 \omega_r t + \cos^2 \omega_r t) = \frac{LA^2}{2}$$

which is constant. Energy is alternately stored more in the inductor, then more in the capacitor, in such a way that the total amount of stored energy is constant.

At resonance, energy is dissipated in the resistor in the series network according to

$$p_{\text{into}}(t) = R i^2(t) = R A^2 \cos^2 \omega_r t$$

In one sinusoidal cycle, the net energy dissipated is

$$P = \int_0^{2\pi/\omega_r} R A^2 \cos^2 \omega_r t \, dt = R A^2 \int_0^{2\pi/\omega_r} \left(\frac{1}{2} + \frac{1}{2} \cos 2\omega_r t \right) dt$$

$$= \frac{RA^2}{2}\left[1 + \frac{\sin 2\omega_r t}{2\omega_r}\right]_0^{2\pi/\omega_r} = \frac{RA^2}{2}\left[\frac{2\pi}{\omega_r} + 0\right]$$

$$= \frac{\pi RA^2}{\omega_r}$$

The Q of a series RLC network is thus

$$Q = \frac{2\pi L A^2/2}{\pi R A^2/\omega_r} = \frac{\omega_r L}{R} = \sqrt{\frac{L}{R^2 C}}$$

A similar calculation of Q for RLC in parallel gives

$$Q = \frac{R}{\omega_r L} = \sqrt{\frac{R^2 C}{L}}$$

for that network. For both the series and the parallel RLC networks, bandwidth and Q are related by

$$\frac{\Delta\omega}{\omega_r} = \frac{\Delta f}{f_r} = \frac{1}{Q}$$

and this relationship applies to other resonant networks that have sufficiently high Q to be approximated well in their behavior near the resonant frequency by a series or parallel network.

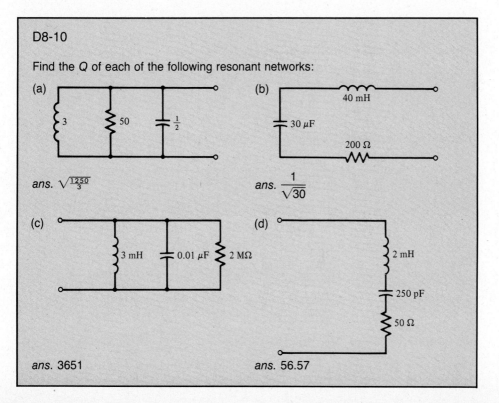

D8-10

Find the Q of each of the following resonant networks:

(a)

3   50 $\frac{1}{2}$

ans. $\sqrt{\frac{1250}{3}}$

(b)

40 mH

30 μF

200 Ω

ans. $\dfrac{1}{\sqrt{30}}$

(c)

3 mH   0.01 μF   2 MΩ

ans. 3651

(d)

2 mH

250 pF

50 Ω

ans. 56.57

QUALITY FACTOR

The quality factor Q of a resonant network is defined to be

$$Q = 2\pi \frac{\text{energy stored in the network}}{\text{energy dissipated in the network per sinusoidal cycle}}$$

at the resonant frequency.

For the series RLC network,

$$Q = \frac{\omega_r L}{R} = \sqrt{\frac{L}{R^2 C}}$$

For the parallel RLC network,

$$Q = \frac{R}{\omega_r L} = \sqrt{\frac{R^2 C}{L}}$$

In the series and the parallel RLC networks,

$$\frac{\Delta \omega}{\omega_r} = \frac{\Delta f}{f_r} = \frac{1}{Q}$$

This relationship is closely approximated in other resonant networks with sufficiently high Q.

8.8 RESISTIVE AND REACTIVE COMPONENTS OF AN IMPEDANCE

8.8.1 Series Resistance-Reactance Equivalent

At any fixed frequency, ω_0, the impedance of a two-terminal network is a complex number

$$Z(s = j\omega_0) = \mathcal{R} + j\mathcal{X}$$

At frequency ω_0, the same impedance could be produced by a resistor of resistance $\mathcal{R}$, in series with an inductor if $\mathcal{X}$ is positive, or a capicitor if $\mathcal{X}$ is negative.

As an example consider the network of Figure 8-16(a). At the frequency $\omega = 10$ rad/s, the impedance at the terminals is

$$Z(s = j10) = j5 + \frac{(10 + j20)(-j10)}{10 + j20 - j10} = j5 + \frac{(200 - j100)(10 - j10)}{(10 + j10)(10 - j10)}$$

$$= j5 + \frac{1000 - j3000}{200} = 5 - j10$$

At that frequency, $\omega = 10$, a 5-Ω resistor in series with a capacitor for which

$$-\frac{j}{10C} = -j10 \qquad C = \frac{1}{100}$$

will produce the same impedance. The original network and the RC network are thus equivalent for 10 rad/s sinusoidal signals, as indicated in Figure 8-16(b).

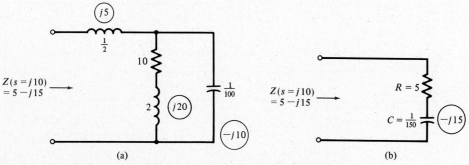

Figure 8-16   Network and series equivalent at a single frequency.

At another frequency, of course, the network would have an impedance with different real and imaginary parts, and the series equivalent would generally be different. For example, for $\omega = 2$,

$$Z(s = j2) = j1 + \frac{(10 + j4)(-j50)}{10 + j4 - j50} = j1 + \frac{(200 - j500)(10 + j46)}{(10 - j46)(10 + j46)}$$

$$= j1 + \frac{25{,}000 + j4200}{2216} = 11.28 + j2.9$$

At this frequency, the impedance is produced by an 11.28-Ω resistor in series with an inductor for which

$$j2L = j2.9 \qquad L = 1.45$$

as in Figure 8-17.

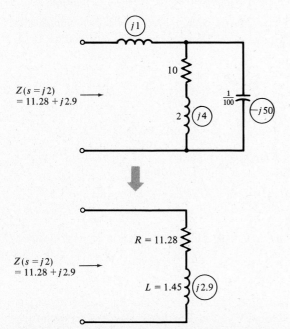

Figure 8-17   Series equivalent at another frequency.

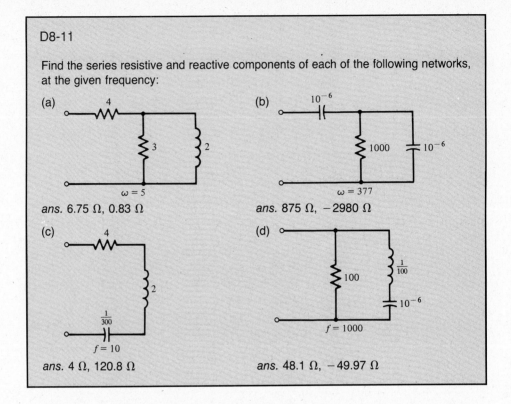

D8-11

Find the series resistive and reactive components of each of the following networks, at the given frequency:

(a)

4

3

2

$\omega = 5$

ans. 6.75 Ω, 0.83 Ω

(b)

10^{-6}

1000

10^{-6}

$\omega = 377$

ans. 875 Ω, -2980 Ω

(c)

4

2

$\frac{1}{300}$

$f = 10$

ans. 4 Ω, 120.8 Ω

(d)

100

$\frac{1}{100}$

10^{-6}

$f = 1000$

ans. 48.1 Ω, -49.97 Ω

8.8.2 Parallel Resistance-Reactance Equivalent

The term *reactance* is used to specify the entirely imaginary impedance for sinusoidal signals produced by inductors and capacitors. The reactance X of an inductor is

$$X = \omega L$$

and of a capacitor is

$$X = -\frac{1}{\omega C}$$

Dating from many years ago, the *inductive reactance* χ_L of an inductor and the *capacitive reactance* χ_C of a capacitor are defined according to

$$\chi_L = \omega L \qquad \chi_C = \frac{1}{\omega C}$$

In terms of these quantities, the net reactance of a series combination of inductor and capacitor is then

$$X = \chi_L - \chi_C$$

The symbol X will be used exclusively here, as is common, to denote the imaginary part of an impedance; the special symbol χ (chi) will indicate the positive quantities χ_L or χ_C.

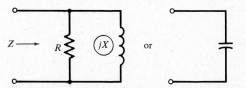

Figure 8-18   Parallel resistance and reactance.

At a given frequency, an arbitrary impedance may also be obtained from a resistance in *parallel* with a reactance, Figure 8-18. Denoting the reactance of an inductor or capacitor by X,

$$Z = \frac{R(jX)}{R + jX} = \frac{jRX(R - jX)}{(R + jX)(R - jX)} = \frac{RX^2}{R^2 + X^2} + j\frac{R^2X}{R^2 + X^2}$$

for the parallel connection.

For Z to have given real part $\mathcal{R}$ and given imaginary part $\mathcal{X}$,

$$\begin{cases} \dfrac{RX^2}{R^2 + X^2} = \mathcal{R} \\[3mm] \dfrac{R^2X}{R^2 + X^2} = \mathcal{X} \end{cases}$$

Solving for R and X, the parallel resistance and reactance in terms of the real and imaginary parts of Z is

$$\begin{cases} R = \dfrac{\mathcal{R}^2 + \mathcal{X}^2}{\mathcal{R}} \\[3mm] X = \dfrac{\mathcal{R}^2 + \mathcal{X}^2}{\mathcal{X}} \end{cases}$$

An impedance

$$Z = \mathcal{R} + j\mathcal{X}$$

may thus be obtained with the parallel connection of a resistance R and a reactance jX, where the values of R and X are given by these relations.

Consider the network in Figure 8-19, which, at the frequency $\omega = 4$, has impedance

$$Z(s = j4) = \frac{(7 + j8)(1 - j20)}{7 + j8 + 1 - j20} = \frac{(167 - j132)(8 + j12)}{(8 - j12)(8 + j12)}$$

$$= \frac{2920 + j948}{64 + 144} = 14.04 + j4.56 \ \Omega$$

The parallel connection of a resistor and an inductor of appropriate values will also have this impedance for $\omega = 4$.

Using

$$\mathcal{R} = 14.04 \ \Omega \qquad \mathcal{X} = 4.56 \ \Omega$$

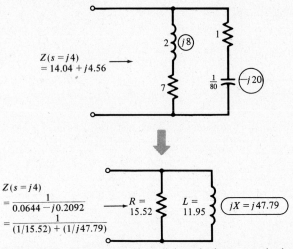

Figure 8-19   Network and parallel equivalent at a single frequency.

the equivalent parallel resistance and reactance are

$$R = \frac{\mathcal{R}^2 + \mathcal{X}^2}{\mathcal{R}} = \frac{197.1 + 20.79}{14.04} = 15.52 \ \Omega$$

$$X = \frac{\mathcal{R}^2 + \mathcal{X}^2}{\mathcal{X}} = \frac{217.9}{4.56} = 47.79 \ \Omega$$

The inductor with the required reactance at $\omega = 4$ has value

$$j4L = j47.79 \qquad L = 11.95 \text{ H}$$

as shown.

At a different frequency, $\omega = 10$ for example, the original network has impedance

$$Z(s = j10) = \frac{(7 + j20)(1 - j8)}{7 + j20 + 1 - j8} = \frac{(167 - j36)(8 - j12)}{(8 + j12)(8 - j12)}$$

$$= \frac{904 - j2292}{64 + 144} = 4.35 - j11 \ \Omega$$

At this frequency, the impedance has a negative imaginary part that will involve a capacitor instead of an inductor in the parallel equivalent. The equivalent parallel resistance and reactance are

$$R = \frac{\mathcal{R}^2 + \mathcal{X}^2}{\mathcal{R}} = \frac{(4.35)^2 + (-11)^2}{4.35} = 32.2 \ \Omega$$

$$X = \frac{\mathcal{R}^2 + \mathcal{X}^2}{\mathcal{X}} = \frac{(4.35)^2 + (-11)^2}{-11} = -12.7 \ \Omega$$

The capacitor with the required reactance at $\omega = 10$ has value

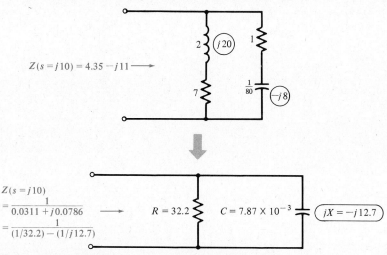

$Z(s = j10) = 4.35 - j11 \longrightarrow$

Figure 8-20   Parallel equivalent at another frequency.

$Z(s = j10)$
$$= \frac{1}{0.0311 + j0.0786} \longrightarrow$$
$$= \frac{1}{(1/32.2) - (1/j12.7)}$$

$R = 32.2$ $C = 7.87 \times 10^{-3}$ $jX = -j12.7$

$$\frac{-j}{10C} = -j12.7 \qquad C = 7.87 \times 10^{-3} \text{ F}$$

and is shown in Figure 8-20.

D8-12

Find the parallel resistive and reactive components of each of the following networks at the given frequency:

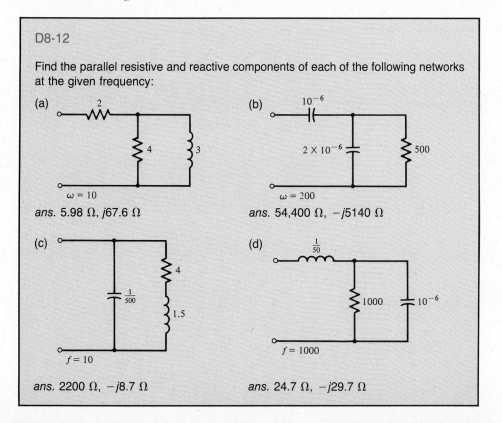

(a)

2

4   3

$\omega = 10$

ans. 5.98 Ω, j67.6 Ω

(b)

10^{-6}

2×10^{-6} 500

$\omega = 200$

ans. 54,400 Ω, −j5140 Ω

(c)

$\frac{1}{500}$

4

1.5

$f = 10$

ans. 2200 Ω, −j8.7 Ω

(d)

$\frac{1}{50}$

1000 10^{-6}

$f = 1000$

ans. 24.7 Ω, −j29.7 Ω

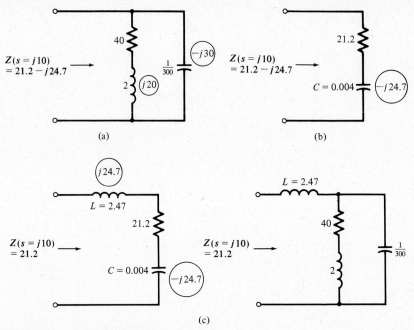

Figure 8-21   Series resonating a network.

8.8.3 Obtaining Series Resonance

At a single frequency, any network, with impedance $\mathcal{R} + j\mathcal{X}$ at that frequency, may be made zero-angle resonant by the addition of a series reactive impedance $-j\mathcal{X}$. For example, consider the behavior of the network in Figure 8-21(a) for sinusoidal signals of radian frequency $\omega = 10$. The impedance of this network is

$$Z(s = j10) = \frac{(40 + j20)(-j30)}{40 + j20 - j30} = \frac{600 - j1200}{40 - j10} = \frac{(600 - j1200)(40 + j10)}{(40 - j10)(40 + j10)}$$

$$= \frac{(24000 + 12000) - j(48000 - 6000)}{1600 + 100} = 21.2 - j24.7 \; \Omega$$

At this single frequency, $\omega = 10$, the same impedance would be produced by a resistor and capacitor in series, as shown in Figure 8-21(b). The addition of an inductor with impedance $j24.7 \; \Omega$, Figure 8-21(c), will make the series-equivalent network, and thus the original network, resonant at frequency $\omega = 10$. The required inductor value is given by

$$j\omega L = j24.7 \qquad L = \tfrac{24.7}{10} = 2.47 \text{ H}$$

D8-13

Find the values of additional series inductance or capacitance that will make each of the following networks resonant (with zero phase angle) at the given frequency:

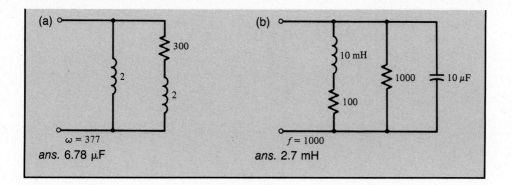

8.8.4 Obtaining Parallel Resonance

Using the parallel single-frequency equivalent, any network may also be made zero-angle resonant at that frequency by the addition of an appropriate parallel reactance. For the example of the previous section, at the given frequency $\omega = 10$, a parallel connection of a resistor and a reactance also has the same impedance as the given network. Using the series-parallel relations,

$$R = \frac{\mathscr{R}^2 + \mathscr{X}^2}{\mathscr{R}} = \frac{(21.2)^2 + (-24.7)^2}{21.2} = 50 \ \Omega$$

$$X = \frac{\mathscr{R}^2 + \mathscr{X}^2}{\mathscr{X}} = \frac{(21.2)^2 + (-24.7)^2}{-24.7} = -42.9 \ \Omega$$

The parallel equivalent, and thus the original network, will be resonant at the frequency $\omega = 10$ if a parallel inductor with reactance $+42.9 \ \Omega$ is placed across the terminals. The value of the added parallel inductor that causes resonance is given by

$$j\omega L = j42.9 \qquad L = \tfrac{42.9}{10} = 4.29 \text{ H}$$

as shown in Figure 8-22.

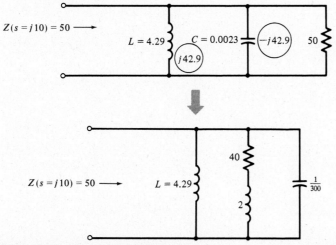

Figure 8-22   Parallel resonating a network.

D8-14

Find the values of additional parallel inductance or capacitance that will make each of the following networks resonant (with zero phase angle) at the given frequency:

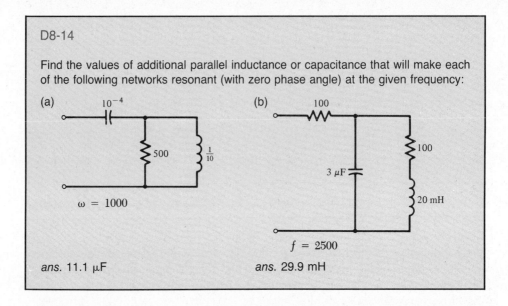

(a)

10^{-4}

500 $\frac{1}{10}$

$\omega = 1000$

ans. 11.1 μF

(b)

100

$3\ \mu F$

100

$20\ mH$

$f = 2500$

ans. 29.9 mH

RESISTIVE AND REACTIVE COMPONENTS OF AN IMPEDANCE

For sinusoidal signals at a single frequency, any impedance is a complex number

$$Z = \mathcal{R} + j\mathcal{X}$$

that could be produced by a resistor in series with an inductor or a capacitor, depending on the algebraic sign of $\mathcal{X}$.

The real part $\mathcal{R}$ of the impedance is called its *resistive* component, and the imaginary part $\mathcal{X}$ is called the *reactive* component of Z. $\mathcal{X}$ is said to be an inductive reactance if positive and a capacitive reactance if negative.

Alternatively, at a single frequency, any impedance may be produced by an appropriate resistor R in parallel with an inductor or a capacitor of reactance X, according to

$$\begin{cases} \mathcal{R} = \dfrac{RX^2}{R^2 + X^2} \\[3mm] \mathcal{X} = \dfrac{R^2 X}{R^2 + X^2} \end{cases}$$

or

$$\begin{cases} R = \dfrac{\mathcal{R}^2 + \mathcal{X}^2}{\mathcal{R}} \\[3mm] X = \dfrac{\mathcal{R}^2 + \mathcal{X}^2}{\mathcal{X}} \end{cases}$$

At a single frequency a reactance $-\mathcal{X}$ in series with an impedance

$$Z = \mathcal{R} + j\mathcal{X}$$

will make the combination resonant (with zero phase angle) at that frequency. Similarly the parallel combination of Z and the reactance

$$-X = -\frac{\mathcal{R}^2 + \mathcal{X}^2}{\mathcal{X}}$$

is resonant (with zero phase angle) at that frequency.

CHAPTER EIGHT PROBLEMS

Basic Problems

Series and Parallel RLC *Networks*
1. Find the resonant frequency of each of the following networks, their impedance, bandwidth, and Q at resonance:

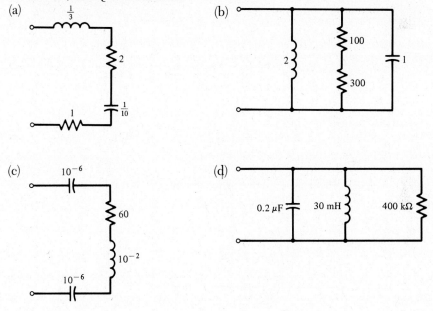

2. For a series RLC network with $R = 2\ \Omega$, $L = 100$ mH, and $C = 0.02\ \mu F$, find the impedance at resonance, one-fifth, one-half, twice, and five times the resonant frequency.
3. For a parallel RLC network with $R = 1000\ \Omega$, $L = 200$ mH, and $C = 0.3\ \mu F$, find the impedance at resonance, one-fifth, one-half, twice, and five times the resonant frequency.
4. For the network below, draw phasor diagrams showing the voltage and each current

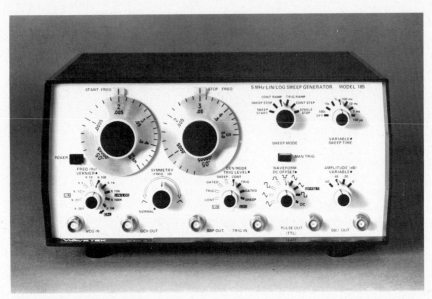

A sweep frequency function generator that can produce a sinusoidal signal with slowly increasing frequency for freqency-response measurements. (*Photo courtesy of Wavetek.*)

(a) at resonance.
(b) at one-fourth the resonant frequency.
(c) at four times the resonant frequency.

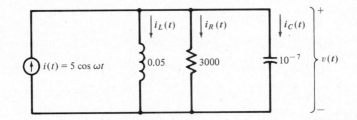

Frequency Response

5. For each of the following networks, find and sketch $|Z(s = j\omega)|$ and $\angle Z(s = j\omega)$:

(a)

(b)

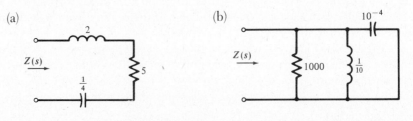

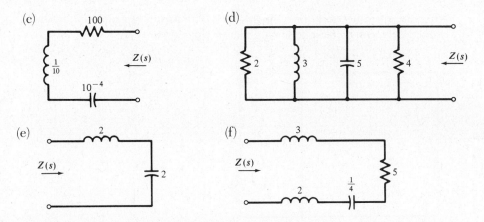

6. For the given series network, find and sketch the frequency response amplitude ratios and phase shifts for the following phasor ratios:

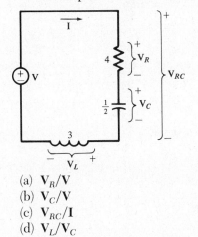

(a) $\mathbf{V}_R/\mathbf{V}$
(b) $\mathbf{V}_C/\mathbf{V}$
(c) $\mathbf{V}_{RC}/\mathbf{I}$
(d) $\mathbf{V}_L/\mathbf{V}_C$

7. For the given parallel network, find and sketch the frequency response amplitude ratios and phase shifts for the following phasor ratios:

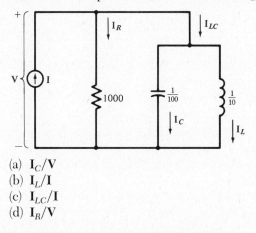

(a) $\mathbf{I}_C/\mathbf{V}$
(b) $\mathbf{I}_L/\mathbf{I}$
(c) $\mathbf{I}_{LC}/\mathbf{I}$
(d) $\mathbf{I}_R/\mathbf{V}$

8. For the given network, find and sketch the frequency response amplitude ratios and phase shifts for the following phasor ratios:

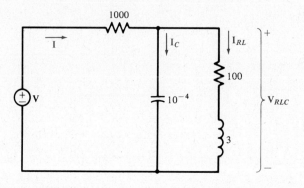

(a) $\mathbf{V}/\mathbf{I}$

(b) $\mathbf{I}_{RL}/\mathbf{I}$

(c) $\mathbf{I}_C/\mathbf{I}$

(d) $\mathbf{V}_{RLC}/\mathbf{V}$

9. Sketch frequency response plots for the following impedances:

(a) $Z(s) = 4 + 10s$

(b) $Z(s) = 4 + \dfrac{1}{10s}$

(c) $Z(s) = \dfrac{s}{s^2 + 3s + 2}$

(d) $Z(s) = \dfrac{s^2 + 2s + 3}{s}$

(e) $Z(s) = \dfrac{s + 1}{s + 10}$

(f) $Z(s) = \dfrac{s + 10}{s + 1}$

(g) $Z(s) = \dfrac{s^2 + 2s + 10}{s + 3}$

(h) $Z(s) = \dfrac{s + 2}{s^2 + s + 10}$

Resonant Network Design

10. Design a series *RLC* network with a resonant frequency of 1000 Hz and a Q of 30. Find the network bandwidth.

11. Design a parallel *RLC* network with a Q of 25 and 100-kHz resonant frequency. What is the bandwidth of the network?

12. Design a resonant series *RLC* network with a resonant frequency of 100 Hz and a bandwidth of 10 Hz. Find the Q of the network.

The frequency response of a network is measured, tabulated and displayed under the control of a specialized desktop computer system. (*Photo courtesy of Gould, Inc.*)

13. Design a resonant parallel *RLC* network with bandwidth 10 kHz and resonant frequency 1 MHz. What is the network's *Q*?

14. Design a series *RLC* network with resonant frequency 1 kHz, for which the magnitude of the network impedance at 1.2 kHz is twice that at resonance. Find the network bandwidth and *Q*.

15. Design a parallel *RLC* network to be resonant at 10 kHz and to have an impedance magnitude at 8 kHz that is one-third that at resonance. What are the network's bandwidth and *Q*?

General Resonance

16. Find the zero-angle resonant frequencies of each of the following networks:

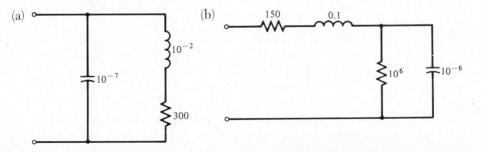

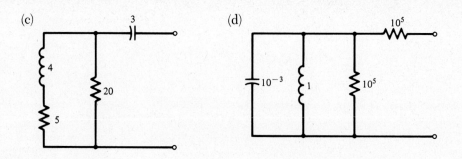

(c)   (d)

17. Find the frequencies for which each of the following networks exhibits a maximum or minimum of impedance magnitude:

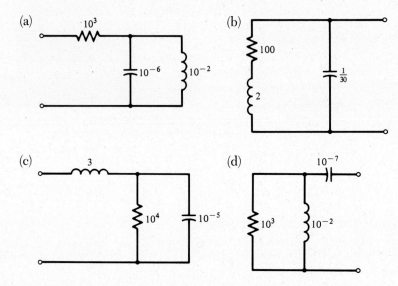

(a)   (b)

(c)   (d)

Resistive and Reactive Impedance Components
18. Find the series resistive and reactive components and the parallel resistive and reactive components of each of the following networks at the given frequency:

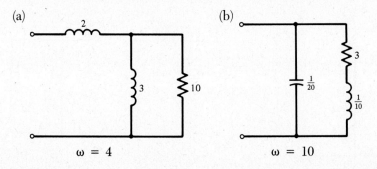

(a)   (b)

$\omega = 4$ $\omega = 10$

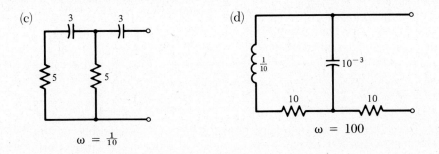

(c) $\omega = \frac{1}{10}$

(d) $\omega = 100$

19. Find the values of additional series inductance or capacitance and the values of additional parallel inductance or capacitance that will make each of the networks resonant (with zero phase angle) at the given frequency:

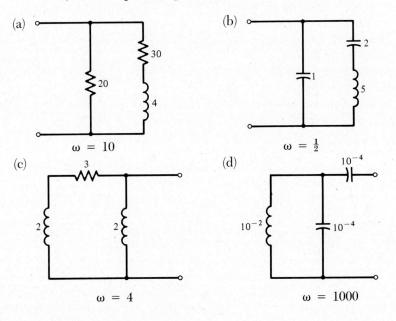

(a) $\omega = 10$

(b) $\omega = \frac{1}{2}$

(c) $\omega = 4$

(d) $\omega = 1000$

20. Design a network that has impedance

$$Z(s = j10) = 3 + j30$$

at radian frequency $\omega = 10$ and impedance

$$Z(s = j20) = 3 - j40$$

at radian frequency $\omega = 20$.

Practical Problems

Adjustable Tuning
21. It is desired to construct a parallel resonant network with a 15-mH inductor and an adjustable capacitor that may be adjusted to be resonant at any frequency

within the standard broadcast band of 550 to 1600 kHz. Over what range of capacitance must the capacitor be adjustable? The largest capacitance, C_{max}, will give resonance at 550 kHz, and the smallest capacitance, C_{min}, will give resonance at 1600 kHz. Find the resonant frequency for the midvalue of capacitance,

$$\frac{C_{max} - C_{min}}{2}$$

The tuning scales of many radio receivers are "spread out" at the lower frequencies because of the nonlinear relation between capacitance (or inductance) and resonant frequency.

Resonance Measurements

22. A physical inductor with an inductance of 10 mH and 30-Ω winding resistance is connected in series with a 0.01-μF capacitor and a voltage source of constant amplitude but adjustable frequency.
   (a) Will the inductor voltage be maximum at a frequency above or below the resonant frequency?
   (b) If the maximum network-current amplitude is 15 mA, what is the source-voltage amplitude?
   (c) At what frequency will the network current and the source voltage be in phase with one another?
   (d) Over how wide a range of frequency will the network current and source-voltage phases be within $\pm 45°$ of one another?

23. A physical inductor with an inductance of 20 mH and 500-Ω winding resistance is connected in parallel with a 300-pF capacitor with a 10-MΩ leakage resistance. Find the precise (zero-angle) resonant frequency of this series-parallel network.

Resonance of a Coil

A coil of wire, in addition to having inductance and resistance, has capacitances between each wire segment, as indicated schematically in the sketch of (a). At low frequencies, these stray capacitances, which depend on the construction of the coil, are negligible. At sufficiently high frequencies, the stray capacitances become important and may be modeled by a single capacitor in parallel with the coil inductance and resistance, (b). A coil of wire thus acts as a resonant circuit at very high frequencies, being inductive below resonance and *capacitive* above its resonance.

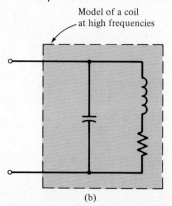

Model of a coil
at high frequencies

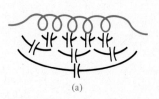

(a)

(b)

At still higher frequencies, a more complicated model, involving several inductors and capacitors, is needed.

24. A certain coil has an inductance of 10 mH and a resistance of 50 Ω, when measured at frequencies that are small compared to the coil's resonant frequency. What is the effective capacitance if the coil is self-resonant at 3 MHz?

25. A coil with negligible winding resistance has inductance 200 μH and is self-resonant at 25 MHz. What is the coil impedance at 20 MHz? At 30 MHz?

Tuned Coupled Circuits

26. A resonant circuit that is especially useful in radio receivers consists of two parallel *LC* networks that are magnetically coupled, as in (a). These are called I.F. (for "intermediate frequency") transformers. In practice, L_1C_1 and L_2C_2 are separately tuned (by moving the capacitor plates or by inserting ferromagnetic core material into the inductors) to slightly different frequencies, so that the network's frequency response **V/I** approximates the ideal rectangular shape shown dotted in (b). This is termed "stagger tuning."

For $L_1 = L_2 = 100$ mH, $C_1 = C_2 = 0.01$ μF, and $M = 50$ mH, find the frequency for which $|\mathbf{V}|/|\mathbf{I}|$ is maximum.

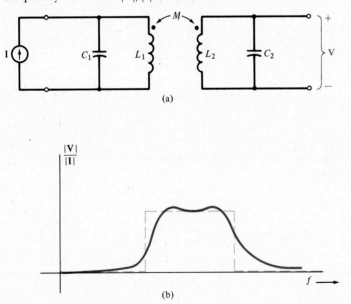

(a)

(b)

Advanced Problems

Resonance

27. Find and sketch plots of the zero-angle resonant frequency of the following network as a function of

   (a) *R*

   (b) *L*

   (c) *C*

Consider only positive values of R, L, and C.

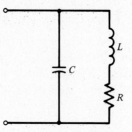

28. Find the impedance of the following network as a function of radian frequency ω. Find the zero-angle resonant frequency or frequencies.

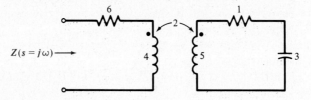

Quality Factor and Bandwidth

29. Show that the total stored energy in a parallel RLC network is constant at the resonant frequency.

30. Show that for a parallel RLC network

$$Q = \sqrt{\frac{R^2 C}{L}}$$

31. Using the definition of quality factor Q given in Section 8.7, find a formula in terms of R, L, and C for the Q of the following resonant network:

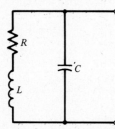

32. Using the definition of bandwidth given in Section 8.5, find a formula in terms of R, L, and C for the bandwidth of the following resonant network:

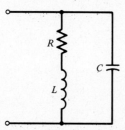

Resistive and Reactive Components

33. Starting with the relations

$$\begin{cases} \dfrac{RX^2}{R^2 + X^2} = \mathscr{R} \\[4mm] \dfrac{R^2 X}{R^2 + X^2} = \mathscr{X} \end{cases}$$

algebraically derive the relations

$$\begin{cases} R = \dfrac{\mathscr{R}^2 + \mathscr{X}^2}{\mathscr{R}} \\[4mm] X = \dfrac{\mathscr{R}^2 + \mathscr{X}^2}{\mathscr{X}} \end{cases}$$

Frequency Response

34. Design a network that has impedance

$$Z(s = j10) = 3 + j30$$

at radian frequency $\omega = 10$ and impedance

$$Z(s = j20) = 5 - j40$$

at radian frequency $\omega = 20$.

Chapter Nine
Power in Sinusoidally Driven Networks

The electric lighting company with which I am connected purchased some time ago the patents for a complete alternating [current] system and my protest against this action can be found upon its minute book. Up to the present, I have succeeded in inducing them not to offer this system to the public, nor will they do so with my consent.

Thomas A. Edison [1847–1931]
in "Dangers of Electrical Lighting"
North American Review, 1887

If Edison had a needle to find in a haystack, he would proceed at once with the diligence of the bee to examine straw after straw until he found the object of his search. I was a sorry witness to such doings, knowing that a little theory and calculation would have saved him ninety percent of his labor.

Nikola Tesla [1856–1943]
quoted in the *New York Times*, 1931

9.1 PREVIEW

We now focus upon electrical power flow in networks with sinusoidal voltages and currents. The relations developed are fundamental to the areas of power, electronic design, and communications and have applications in other areas, as well.

Once the key analytical results are obtained, concepts and applications closely related to power transmission are discussed. Motivation for using rms phasors develops as single-phase power system objectives and terminology are examined. Controlled source, ideal transformer, and tee network transformer models are developed. Special emphasis is given to the interpretation and solution of problems of the type commonly found on Engineer-in-Training examinations, where such details as polarities, rms values, and element models must often be inferred.

The basics of three-phase power transmission, including delta-wye source and load transformation, are given. Transmission efficiency and constancy of power flow of three-phase systems are emphasized, as the objective here is an introductory understanding upon which a later course in power systems can build.

When you complete this chapter, you should know—

1. how power flow in networks with sinusoidal signals may be expressed as an average part plus a fluctuating part;
2. relations for average power flow in the basic elements and in general;
3. what reactive power is, and reactive power relations for the basic elements and in general;
4. the concept of complex power and complex power relations;
5. how rms values are used in power calculations;
6. the configuration of a single-phase power transmission system and how to solve related problems;
7. basic transformer models;
8. ideal transformer properties, including voltage and current ratios and impedance reflection properties;
9. why power factor correction is important in electrical power transmission and how to use resonance methods to solve such problems;
10. what three-phase power transmission is, its advantages, and various source and load configurations;
11. how to use source and load delta-wye transformations to convert balanced three-phase power networks to equivalent single-phase problems.

9.2 SINUSOIDAL POWER

9.2.1 Sinusoidal Power in General

The electrical power flow into an element is the product of the element's sink-reference voltage and current. For sinusoidal voltage,

$$v(t) = A \cos (\omega t + \theta)$$

and a sinusoidal element current,

$$i(t) = B \cos (\omega t + \phi)$$

the power flow is

$$p_{\text{into}}(t) = v(t)i(t) = AB \cos (\omega t + \theta) \cos (\omega t + \phi)$$

Using the trigonometric identity

$$\cos x \cos y = \tfrac{1}{2} \cos (x - y) + \tfrac{1}{2} \cos (x + y)$$

$$p_{\text{into}}(t) = \underbrace{\frac{AB}{2} \cos (\theta - \phi)}_{\substack{\text{constant} \\ \text{component}}} + \underbrace{\frac{AB}{2} \cos (2\omega t + \theta + \phi)}_{\substack{\text{fluctuating} \\ \text{component}}}$$

The power consists of a constant component that depends on the difference in phase angles between voltage and current, $(AB/2) \cos (\theta - \phi)$, plus a sinusoidal, fluctuating component with frequency twice that of the voltage or current.

9.2.2 Sinusoidal Power in Resistors, Inductors, and Capacitors

In a resistor, the sink reference voltage and current are in phase. If

$$v_R(t) = A \cos (\omega t + \theta)$$

then

$$i_R(t) = \frac{v_R}{R} = \frac{A}{R} \cos (\omega t + \theta)$$

and

$$p_{\text{into } R}(t) = \frac{A^2}{R} \cos^2 (\omega t + \theta) = \frac{A^2}{2R} + \frac{A^2}{2R} \cos (2\omega t + 2\theta)$$

As shown in the sketch of typical voltage, current, and power in Figure 9-1, the power flow consists of equal-amplitude constant and fluctuating components so that $p_{\text{into } R}(t)$ is never negative.

In an inductor, sink reference voltage and current are 90° out of phase. If

$$v_L(t) = A \cos (\omega t + \theta)$$

the forced sinusoidal inductor current is

$$i_L(t) = \frac{A}{\omega L} \cos (\omega t + \theta - 90°)$$

and

$$p_{\text{into } L}(t) = \frac{A^2}{\omega L} \cos (\omega t + \theta) \cos (\omega t + \theta - 90°)$$

$$= \frac{A^2}{2\omega L} \cos 90° + \frac{A^2}{2\omega L} \cos (2\omega t + 2\theta - 90°)$$

$$= \frac{A^2}{2\omega L} \cos (2\omega t + 2\theta - 90°)$$

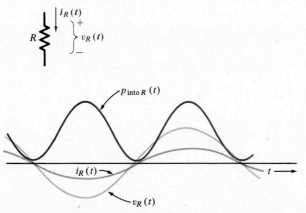

Figure 9-1   Voltage, current, and power for a resistor.

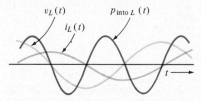

Figure 9-2   Voltage, current, and power for an inductor.

There is no constant component to the power flow, as indicated in the sketch of Figure 9-2. Power flows into and out of the inductor, back and forth, twice each cycle of the voltage or current.

The sink reference current leads the voltage by 90° in a capacitor. If

$$v_C(t) = A \cos(\omega t + \theta)$$

the forced sinusoidal capacitor current is

$$i_C(t) = A\omega C \cos(\omega t + \theta + 90°)$$

and

$$p_{\text{into}}(t) = A^2\omega C \cos(\omega t + \theta) \cos(\omega t + \theta + 90°)$$

$$= \frac{A^2\omega C}{2} \cos(-90°) + \frac{A^2\omega C}{2} \cos(2\omega t + 2\theta + 90°)$$

$$= \frac{A^2\omega C}{2} \cos(2\omega t + 2\theta + 90°)$$

As with the inductor, the capacitor power flow contains no constant component, as illustrated in Figure 9-3. Power is alternately stored in and released from the capacitor, twice during each cycle of the voltage or current.

9.2.3 Conservation of Energy

In a network, the net power flowing out of sources equals the net power flow into the other elements at every instant of time. For example, consider the network of Figure 9-4, in which all voltages and currents have been found and are indicated on the network diagram. The electrical power flow into the inductor is

$$p_{\text{into } L}(t) = v_1(t)i_1(t) = [9.92 \cos(2t + 38.3°)][2.48 \cos(2t - 51.7°)]$$

$$= 12.39 \cos(4t - 13.4°) \text{ W}$$

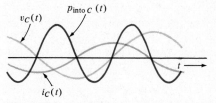

Figure 9-3   Voltage, current, and power for a capacitor.

The electrical power flow into the capacitor is

$$p_{\text{into } C}(t) = v_4(t)i_3(t) = [3.93 \cos (2t - 123.3°)][1.31 \cos (2t - 33.3°)]$$

$$= 2.57 \cos (4t - 156.6°) \text{ W}$$

The electrical power flow into the 5-Ω resistor is

$$p_{\text{into } 5}(t) = v_2(t)i_2(t) = [6.53 \cos (2t - 70.2°)][1.31 \cos (2t - 70.2°)]$$

$$= 4.28 + 4.28 \cos (4t - 140.4°) \text{ W}$$

and the electrical power flow into the 4-Ω resistor is

$$p_{\text{into } 4}(t) = v_3(t)i_3(t) = [5.22 \cos (2t - 33.3°)][1.31 \cos (2t - 33.3°)]$$

$$= 3.42 + 3.42 \cos (4t - 66.6°) \text{ W}$$

The electrical power flow out of the source,

$$p_{\text{out}}(t) = v(t)i_1(t) = [10 \cos 2t][2.48 \cos (2t - 51.7°)]$$

$$= 12.4 \cos (51.7°) + 12.4 \cos (4t - 51.7°)$$

$$= 7.7 + 12.4 \cos (4t - 51.7°) \text{ W}$$

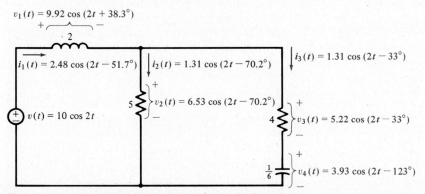

Figure 9-4   Network with sinusoidal signals indicated.

is equal to the sum of the power flow into the other four elements:

$$p_{\text{out}}(t) = p_{\text{into } L}(t) + p_{\text{into } C}(t) + p_{\text{into } 5}(t) + p_{\text{into } 4}(t)$$

$$= 12.39 \cos (4t - 13.4°) + 2.57 \cos (4t - 156.6°) + 4.28$$

$$+ 4.28 \cos (4t - 140.4°) + 3.42 + 3.42 \cos (4t - 66.6°) \text{ W}$$

That the instantaneous power flow out of the source equals the net power flow into the other elements may be verified as follows: The constant parts add to equal the constant part of $p_{\text{out}}(t)$, 7.7. The fluctuating part of this sum consists of the sum of four sinusoidal components (with frequencies twice that of the voltages and currents) with various amplitudes and phase angles. Decomposing each into quadrature components, there results

$$12.39 \cos (4t - 13.4°) = 12.05 \cos 4t + 2.86 \sin 4t$$

$$2.57 \cos (4t - 156.6°) = -2.35 \cos 4t + 1.02 \sin 4t$$

$$4.28 \cos (4t - 140.4°) = -3.3 \cos 4t + 2.74 \sin 4t$$

$$3.42 \cos (4t - 66.6°) = 1.35 \cos 4t + 3.15 \sin 4t$$

the sum of which is

$$7.73 \cos 4t + 9.77 \sin 4t = 12.4 \cos (4t - 51.7°)$$

which equals the corresponding component in the source power flow.

It should be noted that there are times when the power flow out of the source is negative; that is, there are intervals of time when energy is flowing from the rest of the network back into the source. This situation is typical of sinusoidally driven networks. Energy flows from the source to the other network elements. Energy that flows into resistors is completely dissipated. Energy that flows into reactive elements, Ls and Cs, is stored, then released, allowing a return of some energy to the source.

If a network has more than one source, energy may be exchanged by the sources, and it sometimes happens that one source mainly supplies energy while another source mainly absorbs it.

D9-1

Element voltages and currents for the network below have been found and appear on the network diagram. Find the electrical power flow into each element as a function of time and write the equation relating power flow out of sources to power flow into the other network elements.

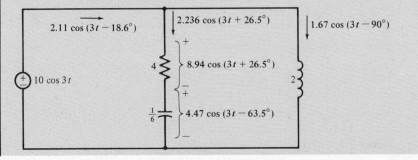

> *ans.* $p_{\text{into } R} = 19.99 \cos^2 (3t + 26.5°)$
> $\qquad\qquad = 10 + 10 \cos (6t + 53°) \text{ W}$
>
> $p_{\text{into } C} = 10 \cos (3t + 26.5°) \cos (3t - 63.5°)$
> $\qquad\qquad = 5 \cos (6t - 37°) \text{ W}$
>
> $p_{\text{into } L} = 16.7 \cos (3t - 90°) \cos (3t)$
> $\qquad\qquad = 8.3 \cos (6t - 90°) \text{ W}$
>
> $p_{\text{out of source}} = 21.1 \cos (3t - 18.6°) \cos (3t)$
> $\qquad\qquad = 10 + 10.6 \cos (6t - 18.6°) \text{ W}$

SINUSOIDAL POWER

The electrical power flow into an element is the product of its sink reference voltage and current.

The product of a sinusoidal voltage,

$$v(t) = A \cos (\omega t + \theta)$$

and a sinusoidal current,

$$i(t) = B \cos (\omega t + \phi)$$

is

$$p(t) = v(t)i(t) = \frac{AB}{2} \cos (\theta - \phi) + \frac{AB}{2} \cos (2\omega t + \theta + \phi)$$

The sum of electrical power flows out of sources equals the sum of electrical power flows into the other elements in a network at every instant of time.

9.3 AVERAGE POWER FLOW

9.3.1 Average Power in General

For sinusoidal, sink reference voltage and current,

$$v(t) = A \cos (\omega t + \theta) \qquad i(t) = B \cos (\omega t + \phi)$$

an element's power flow is

$$p_{\text{into}}(t) = v(t)i(t) = \frac{AB}{2} \cos (\theta - \phi) + \frac{AB}{2} \cos (2\omega t + \theta + \phi)$$

The average power flow is the constant term

$$P_{\text{into}} = \frac{AB}{2} \cos (\theta - \phi)$$

Average power is denoted by a capital letter P, to distinguish it from instantaneous power $p(t)$.

The phasors representing $v(t)$ and $i(t)$ are

$$\mathbf{V} = Ae^{j\theta} \qquad \mathbf{I} = Be^{j\phi}$$

and in terms of these, the average power is

$$P_{\text{into}} = \tfrac{1}{2}|\mathbf{V}|\,|\mathbf{I}|\cos(\angle\mathbf{V} - \angle\mathbf{I})$$

9.3.2 Average Power in Resistors, Inductors, and Capacitors

For a resistor, voltage and current are in phase, so

$$P_{\text{into }R} = \frac{AB}{2}\cos(0°) = \frac{AB}{2}$$

one-half the product of the voltage amplitude with the current amplitude. In terms of the phasors describing the resistor voltage and current,

$$P_{\text{into }R} = \tfrac{1}{2}|\mathbf{V}_R|\,|\mathbf{I}_R|$$

For an inductor, the current lags the voltage by 90°, giving

$$P_{\text{into }L} = \frac{AB}{2}\cos(90°) = 0$$

For a capacitor, where current leads voltage by 90°,

$$P_{\text{into }C} = \frac{AB}{2}\cos(-90°) = 0$$

Energy flows back and forth, into and out of inductors and capacitors, but the average flows are zero.

9.3.3 Conservation of Average Energy

In a network, the power flow out of sources equals the power flow into the other elements at each instant of time. The *average* power flow from sources thus equals the *average* power flow into the other elements. In many practical situations it is the average power rather than the detailed, instantaneous power that is of interest so far as element ratings, capabilities, and efficiency are concerned.

As a numerical example of average power flow relations, consider again the example network of Figure 9-4. The average electrical power flow into the 5-Ω resistor is

$$P_{\text{into }5} = \frac{(6.53)(1.31)}{2} = 4.28 \text{ W}$$

The 4-Ω resistor, similarly, has an average power flow of

$$P_{\text{into }4} = \frac{(5.22)(1.31)}{2} = 3.42 \text{ W}$$

The average power flows into the inductor and capacitor are each zero:

$$P_{\text{into } L} = 0 \qquad P_{\text{into } C} = 0$$

The average power flow out of the source is

$$P_{\text{out}} = \frac{(10)(2.48)}{2} \cos 51.7° = 7.7 \text{ W}$$

which equals the sum of the average power flows into the other network elements:

$$P_{\text{out}} = P_{\text{into } 5} + P_{\text{into } 4} + P_{\text{into } L} + P_{\text{into } C}$$

D9-2

Element voltages and currents have been found for the network below and appear on the network diagram. Find the average electrical power flow into each element and verify that the net average power flow out of sources equals the net average power flow into the other elements.

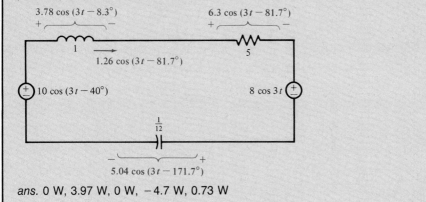

ans. 0 W, 3.97 W, 0 W, −4.7 W, 0.73 W

AVERAGE POWER

In general, the average power flow into an element with sink reference sinusoidal voltage and current

$$v(t) = A \cos (\omega t + \theta) \qquad \mathbf{V} = Ae^{j\theta}$$
$$i(t) = B \cos (\omega t + \phi) \qquad \mathbf{I} = Be^{j\phi}$$

is

$$P_{\text{into}} = \frac{AB}{2} \cos (\theta - \phi) = \frac{1}{2} |\mathbf{V}| \, |\mathbf{I}| \cos (\angle \mathbf{V} - \angle \mathbf{I})$$

Average power relations for the basic elements are as follows:

Resistor $\qquad P_{\text{into}} = \dfrac{AB}{2} = \dfrac{1}{2} |V_R| \, |I_R|$

Inductor	$P_{into} = 0$	
Capacitor	$P_{into} = 0$	

The sum of average electrical power flows out of sources equals the sum of average electrical power flows into the other elements of a network.

9.4 MAXIMUM POWER TRANSFER

Consider the network of Figure 9-5, for which it is desired to choose the impedance

$$Z_L = R_L + jX_L$$

for maximum average electrical power flow into Z_L. The impedance Z_L is commonly termed the "load impedance" in such a situation.

The average power flow into Z_L is

$$P = \frac{|\mathbf{I}|^2 R_L}{2}$$

and the current phasor is

$$\mathbf{I} = \frac{\mathbf{V}_T}{Z_T + Z_L} = \frac{\mathbf{V}_T}{(R_T + R_L) + j(X_T + X_L)}$$

giving

$$P = \frac{|\mathbf{V}_T|^2 R_L}{2[(R_T + R_L)^2 + (X_T + X_L)^2]}$$

Equating the partial derivatives of P with respect to R_L and X_L to zero, to find the minimum, gives the simultaneous equations

$$\begin{cases} \frac{\partial P}{\partial R_L} = \frac{2[(R_T + R_L)^2 + (X_T + X_L)^2]|\mathbf{V}_T|^2 - |\mathbf{V}_T|^2 R_L \cdot 4(R_T + R_L)}{2[(R_T + R_L)^2 + (X_T + X_L)^2]^2} \\ \\ = \frac{|\mathbf{V}_T|^2[R_T^2 - R_L^2 + (X_T + X_L)^2]}{2[(R_T + R_L)^2 + (X_T + X_L)^2]^2} = 0 \\ \\ \frac{\partial P}{\partial X_L} = \frac{-|\mathbf{V}_T|^2 R_L \cdot 4(X_T + X_L)}{2[(R_T + R_L)^2 + (X_T + X_L)^2]^2} = 0 \end{cases}$$

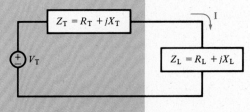

Thevenin equivalent of the rest
of the network

Figure 9-5   Maximum power transfer for sinusoidal signals.

These conditions are satisfied if and only if

$$X_L = -X_T \quad \text{and} \quad R_L = R_T$$

This is to say that to adjust a load impedance for maximum average power flow into it, choose the real part of the load impedance equal to the real part of the Thévenin impedance of the rest of the network. And choose the imaginary part of the load impedance to be the negative of the Thévenin impedance's imaginary part:

$$Z_L = R_T - jX_T = Z_T^*$$

This result may be visualized as follows: The series circuit is made resonant by canceling the Thévenin reactance with the load reactance. Then the load resistance is made equal to the Thévenin resistance, as in the result for resistive networks.

As a numerical example, consider the network in Figure 9-6(a), in which it is desired to choose the impedance Z for maximum average power flow into Z. The Thévenin impedance of the portion of the network connected to Z is computed by finding the terminal impedance, with the source set to zero, in Figure 9-6(b). The Thévenin voltage $\mathbf{V}_T$ could also be found, but it is not needed for the calculation at hand. In Figure 9-6(c) the network connected to Z is replaced by the Thévenin equivalent, which is equivalent so far as Z is concerned. Maximum average power will flow into Z if it is chosen to be $Z = 9 + j3 \ \Omega$.

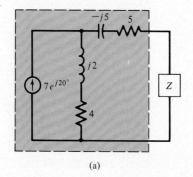

(a)

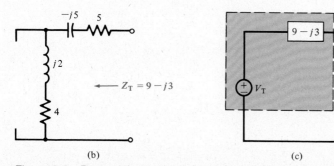

(b) (c)

Figure 9-6 Determining an impedance for maximum power transfer.

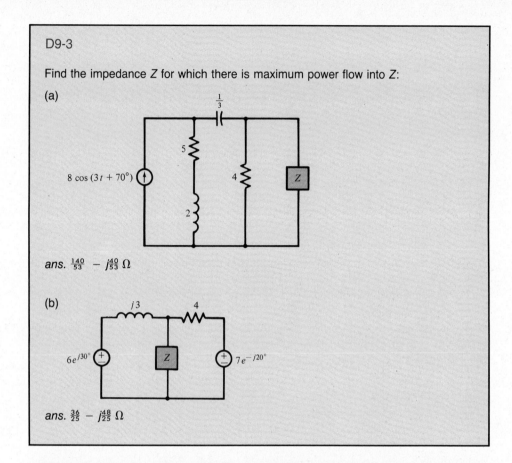

D9-3

Find the impedance Z for which there is maximum power flow into Z:

(a)

ans. $\frac{140}{53} - j\frac{40}{53}$ Ω

(b)

ans. $\frac{36}{25} - j\frac{48}{25}$ Ω

MAXIMUM POWER TRANSFER

For sinusoidal signals and an adjustable load impedance Z_L, maximum average power is transferred from a two-terminal network with Thévenin impedance Z_T when

$$Z_L = Z_T^*$$

9.5 REACTIVE POWER

9.5.1 Resistive and Reactive Components of Instantaneous Power

For sinusoidal voltage and current,

$$v(t) = A \cos (\omega t + \theta) \qquad i(t) = B \cos (\omega t + \phi)$$

a general power flow is

$$p(t) = v(t)i(t) = \frac{AB}{2} \cos (\theta - \phi) + \frac{AB}{2} \cos (2\omega t + \theta + \phi)$$

Using the trigonometric identity

$$\cos (x + y) = \cos x \cos y - \sin x \sin y$$

$$\cos (2\omega t + \theta + \phi) = \cos (\theta - \phi) \cos (2\omega t + 2\phi)$$

$$- \sin (\theta - \phi) \sin (2\omega t + 2\phi)$$

$$= \cos (\theta - \phi) \cos (2\omega t + 2\phi)$$

$$+ \sin (\theta - \phi) \cos (2\omega t + 2\phi + 90°)$$

gives

$$p(t) = \left[\frac{AB}{2} \cos (\theta - \phi)\right][1 + \cos (2\omega t + 2\phi)]$$

$$+ \left[\frac{AB}{2} \sin (\theta - \phi)\right] \cos (2\omega t + 2\phi + 90°)$$

Power flow is thus expressible as a sum,

$$p(t) = p_{\mathcal{R}}(t) + p_{\mathcal{X}}(t)$$

of a resistor power flow

$$p_{\mathcal{R}}(t) = \left[\frac{AB}{2} \cos (\theta - \phi)\right][1 + \cos (2\omega t + 2\phi)]$$

plus a reactive (inductor or capacitor) power flow

$$p_{\mathcal{X}}(t) = \left[\frac{AB}{2} \sin (\theta - \phi)\right] \cos (2\omega t + 2\phi + 90°)$$

The resistive component of power flow has average value

$$P = \frac{AB}{2} \cos (\theta - \phi)$$

as indicated in the sketch of Figure 9-7, whereas the reactive component of the instantaneous power has zero average value, and amplitude

$$Q = \frac{AB}{2} \sin (\theta - \phi)$$

as shown.

9.5.2 General Reactive Power

The reactive power flow Q into an element is defined as

$$Q_{\text{into}} = \tfrac{1}{2}AB \sin (\theta - \phi)$$

where the element's sink reference voltage and current are

$$v(t) = A \cos (\omega t + \theta) \qquad i(t) = B \cos (\omega t + \phi)$$

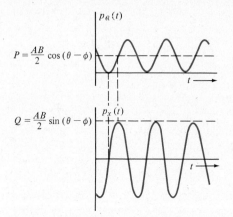

$P = \dfrac{AB}{2} \cos(\theta - \phi)$

$Q = \dfrac{AB}{2} \sin(\theta - \phi)$

Figure 9-7 Resistive and reactive components of instantaneous power flow.

Reactive power is the amplitude of the reactive component of the instantaneous power flow, which is the component that is alternately stored and returned to the rest of the network every half cycle of $v(t)$ and $i(t)$.

The SI unit of reactive power is watt/j = var, which stands for volt-amperes reactive. Reactive power Q should not be confused with the Q of a resonant circuit; the two are different quantities with the same symbol.

9.5.3 Reactive Power in Resistors, Inductors, and Capacitors

For a resistor the voltage and current are in phase and

$$Q_{\text{into } R} = \frac{AB}{2} \sin(0°) = 0$$

All of the energy that flows into a resistor is dissipated; none is alternately stored then released.

For an inductor, the current lags the voltage by 90°, so

$$Q_{\text{into } L} = \frac{AB}{2} \sin(90°) = \frac{AB}{2}$$

The reactive power is one-half the product of voltage amplitude with current amplitude. In terms of the phasors describing the inductor voltage and current,

$$Q_{\text{into } L} = \tfrac{1}{2}|\mathbf{V}_L|\,|\mathbf{I}_L|$$

For a capacitor, where current leads voltage by 90°,

$$Q_{\text{into } C} = \frac{AB}{2} \sin(-90°) = -\frac{AB}{2}$$

In terms of the voltage and current phasors,

$$Q_{\text{into} C} = -\tfrac{1}{2}|\mathbf{V}_C|\,|\mathbf{I}_C|$$

The negative nature of reactive power for a capacitor indicates a negative amplitude for the reactive component of instantaneous power. This is to say that a capacitor releases energy during the time interval an inductor stores it, and vice versa.

9.5.4 Conservation of Reactive Energy

The reactive power oscillations are $90°$ out of phase with the power oscillations associated with energy dissipation and represent energy that is interchanged between sources and energy-storing elements every half cycle of the voltage or current. The sum of the reactive powers out of sources in a network equals the sum of reactive powers into the other elements.

As a numerical example, consider the network of Figure 9-4 once again. The reactive power flow into the inductor is

$$Q_{\text{into } L} = \frac{(9.92)(2.48)}{2} = 12.3 \text{ var}$$

The capacitor has reactive power flow

$$Q_{\text{into } C} = -\frac{(3.93)(1.31)}{2} = -2.57 \text{ var}$$

The resistors each have zero reactive power flow:

$$Q_{\text{into } 5} = 0 \qquad Q_{\text{into } 4} = 0$$

The reactive power flow out of the source is

$$Q_{\text{out}} = \frac{(10)(2.48)}{2} \sin 51.7° = 9.73 \text{ var}$$

which equals the sum of reactive power flows into the other network elements:

$$Q_{\text{into } 5} + Q_{\text{into } 4} + Q_{\text{into } L} + Q_{\text{into } C} = Q_{\text{out}}$$

D9-4

Element voltages and currents have been found for the network below and appear on the network diagram. Find the reactive power flow into each element and verify that the net reactive power flow out of sources equals the net reactive power flow into the other elements.

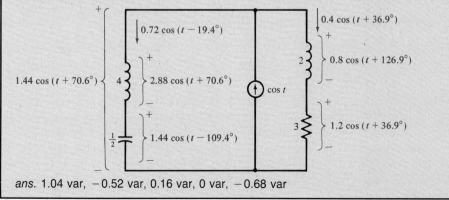

ans. 1.04 var, −0.52 var, 0.16 var, 0 var, −0.68 var

REACTIVE POWER

Reactive power Q (not to be confused with the Q of a resonant circuit) is a measure of the energy that flows into an element, is stored, and then flows out of the element each half cycle.

In general, the reactive power flow into an element with sink reference sinusoidal voltage and current

$$v(t) = A \cos(\omega t + \theta) \qquad \mathbf{V} = Ae^{j\theta}$$

$$i(t) = B \cos(\omega t + \phi) \qquad \mathbf{I} = Be^{j\phi}$$

is

$$Q_{into} = \frac{AB}{2} \sin(\theta - \phi) = \frac{1}{2} |\mathbf{V}| \, |\mathbf{I}| \sin(\angle\mathbf{V} - \angle\mathbf{I})$$

Reactive power relations for the basic elements are as follows:

Resistor $\qquad Q_{into} = 0$

Inductor $\qquad Q_{into} = \dfrac{AB}{2} = \dfrac{1}{2} |\mathbf{V}_L| \, |\mathbf{I}_L|$

Capacitor $\qquad Q_{into} = -\dfrac{AB}{2} = -\dfrac{1}{2} |\mathbf{V}_C| \, |\mathbf{I}_C|$

The sum of reactive power flows out of sources equals the sum of reactive power flows into the other elements of a network.

9.6 COMPLEX POWER

In performing power calculations, it is often helpful to combine the average and reactive power for an element into a complex quantity, which is termed *complex power:*

$$\mathscr{P} = P + jQ$$

If the element's sink reference voltage and current are

$$v(t) = A \cos(\omega t + \theta) \qquad i(t) = B \cos(\omega t + \phi)$$

then

$$\mathscr{P} = \frac{AB}{2} \cos(\theta - \phi) + j\frac{AB}{2} \sin(\theta - \phi) = \frac{AB}{2} e^{j(\theta - \phi)}$$

The real part of $\mathscr{P}$ is the average power, and the imaginary part of $\mathscr{P}$ is the reactive power. The angle of $\mathscr{P}$ is the phase difference between voltage and current.

Average power is sometimes called *real power* or *active power,* and reactive power is sometimes called *imaginary power,* to give a symmetry to these terms comparable to their mathematical symmetry.

In terms of the sink reference voltage and current phasors for an element, **V** and **I**, its complex power is

$$\mathcal{P} = P + jQ = \tfrac{1}{2}|\mathbf{V}|\,|\mathbf{I}|\cos(\angle\mathbf{V} - \angle\mathbf{I}) + j\tfrac{1}{2}|\mathbf{V}|\,|\mathbf{I}|\sin(\angle\mathbf{V} - \angle\mathbf{I})$$

$$= \tfrac{1}{2}|\mathbf{V}|\,|\mathbf{I}|e^{j(\angle\mathbf{V} - \angle\mathbf{I})} = \tfrac{1}{2}\mathbf{VI}^*$$

where **I*** denotes the complex conjugate of **I**. The SI unit for complex power, like real power, is the watt (W).

Conservation of average energy and conservation of reactive energy in a network are simultaneously expressed by stating that the net complex power flow out of sources equals the net complex power flow into the other network elements.

D9-5

Element voltages and currents have been found for the network below and appear on the network diagram. Find the complex power flow into each element and verify that the net complex power flow out of sources equals the net complex power into the other elements.

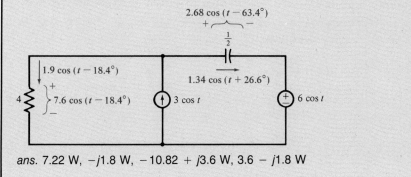

ans. 7.22 W, $-j1.8$ W, $-10.82 + j3.6$ W, $3.6 - j1.8$ W

COMPLEX POWER

With sink reference sinusoidal voltage and current

$$v(t) = A\cos(\omega t + \theta) \qquad \mathbf{V} = Ae^{j\theta}$$

$$i(t) = B\cos(\omega t + \phi) \qquad \mathbf{I} = Be^{j\phi}$$

the complex power flow into an element is

$$\mathcal{P}_{\text{into}} = P_{\text{into}} + jQ_{\text{into}}$$

$$= \tfrac{1}{2}AB\cos(\theta - \phi) + j\tfrac{1}{2}AB\sin(\theta - \phi)$$

$$= \tfrac{1}{2}ABe^{j(\theta - \phi)}$$

$$= \tfrac{1}{2}|\mathbf{V}|\,|\mathbf{I}|e^{j(\angle\mathbf{V} - \angle\mathbf{I})}$$

$$= \tfrac{1}{2}\mathbf{VI}^*$$

> Complex power is used to combine concisely average (or *active* or *real*) and reactive (or *imaginary*) power into a single mathematical quantity.
>
> The sum of complex power flows out of sources equals the sum of complex power flows into the other elements of a network.

9.7 POWER CALCULATIONS USING RMS VALUES

9.7.1 RMS Values

Over the years, considerable effort has been devoted to expressing the power relations as simply as possible. One concern has been the factors of one-half in the expressions for P and Q. For *constant* voltages and currents (termed dc or *direct current* conditions), power is the voltage-current product; but the power product involves a factor of one-half the amplitude product in the sinusoidal (ac or *alternating current*) case.

The root-mean-square (rms) of a sinusoid is related to its amplitude by

$$\text{(rms value)} = \frac{1}{\sqrt{2}} \text{(amplitude)}$$

so in terms of voltage and current

$$v(t) = A \cos(\omega t + \theta) \qquad i(t) = B \cos(\omega t + \phi)$$

the rms values are

$$V_{\text{rms}} = \frac{A}{\sqrt{2}} \qquad I_{\text{rms}} = \frac{B}{\sqrt{2}}$$

giving

$$P = \tfrac{1}{2}AB \cos(\theta - \phi) = V_{\text{rms}} I_{\text{rms}} \cos(\theta - \phi)$$

$$Q = \tfrac{1}{2}AB \sin(\theta - \phi) = V_{\text{rms}} I_{\text{rms}} \sin(\theta - \phi)$$

In a resistor, a constant (dc) voltage V and current $I = V/R$ produces electrical power conversion into heat in the amount

$$P_{\text{dc}} = VI$$

The same resistor will have an equal average electrical power flow and so will be heated the same amount by sinusoidal voltage and current with rms values equal to the dc values:

$$V_{\text{rms}} = V \qquad I_{\text{rms}} = V_{\text{rms}}/R = I$$

$$P_{\text{ac}} = V_{\text{rms}} I_{\text{rms}} = VI$$

The amplitudes of the corresponding sinusoids are $\sqrt{2}$ times as large as the rms values.

Root-mean-square (rms) means the square root of the average (or mean) of the square of a function. For a sinusoidal function such as

$$v(t) = A \cos (\omega t + \theta)$$

the average of the square is

$$\frac{1}{\text{period of } v(t)} \int_{\substack{\text{integral over} \\ \text{one period of } v(t)}} v^2(t) \, dt = \frac{\omega}{2\pi} \int_0^{2\pi/\omega} A^2 \cos^2 (\omega t + \theta) \, dt$$

$$= \frac{\omega}{2\pi} \int_0^{2\pi/\omega} \left[\frac{A^2}{2} + \frac{A^2}{2} \cos (2\omega t + 2\theta) \right] dt$$

$$= \frac{A^2}{2}$$

The root-mean-square (rms) of the sinusoid is the square root of the above, which is

$$V_{\text{rms}} = \frac{A}{\sqrt{2}}$$

9.7.2 RMS Phasors

In situations in which ac power calculations are used extensively, it is helpful to redefine phasors so that phasor magnitudes are all the rms values of the sinusoids they represent, instead of their amplitudes. The sinusoidal signal

$$v(t) = A \cos (\omega t + \theta)$$

is represented by

$$\mathbf{V}_{\text{rms}} = \frac{A}{\sqrt{2}} e^{j\theta}$$

instead of by

$$\mathbf{V} = A e^{j\theta}$$

Scaling all phasors by the factor $1/\sqrt{2}$ does not change the relations between them in a network; only the translation between phasor and time function is involved. An impedance relating two ordinary phasors also relates the rms phasors:

$$\mathbf{Z} = \frac{\mathbf{V}}{\mathbf{I}} = \frac{\mathbf{V}/\sqrt{2}}{\mathbf{I}/\sqrt{2}} = \frac{\mathbf{V}_{\text{rms}}}{\mathbf{I}_{\text{rms}}}$$

In practice, both ordinary and rms phasors are used in sinusoidal response calculation. The electric power industry uses rms phasors almost exclusively; whereas in communications, control systems, and electronics, ordinary (or peak) phasors seem to have the edge. It certainly is a bother to be dealing continually with factors of $\sqrt{2}$ if power calculations are not to be made.

9.7.3 Volt-Amperes and Power Factor

By defining volt-amperes (VA) and power factor (PF) as

$$VA = V_{rms}I_{rms}$$

and

$$PF = \cos(\angle \mathbf{V} - \angle \mathbf{I})$$

average power may be expressed as

$$P = V_{rms}I_{rms} \cos(\angle \mathbf{V} - \angle \mathbf{I}) = (VA)(PF)$$

The volt-amperes are simply the product of an element's rms voltage and current. Each may be read with ac meters, the scales of which read rms values for sinusoids. Volt-amperes is sometimes called *apparent power* because for constant (dc) voltages and currents, the power would simply be voltage times current. The power factor is a number between zero and unity that converts the "apparent power" to actual power.

Suppose a 220-V rms voltage is applied to an impedance

$$Z = 3 + j2 \ \Omega$$

The rms of the current is given by

$$I_{rms} = \frac{V_{rms}}{|Z|} = \frac{220}{\sqrt{3^2 + 2^2}} = \frac{220}{\sqrt{13}}$$

Hence the volt-amperes are

$$VA = V_{rms}I_{rms} = \frac{(220)^2}{\sqrt{13}}$$

The power factor is

$$PF = \cos(\angle \mathbf{V} - \angle \mathbf{I}) = \cos(\angle Z)$$

$$= \cos\left(\arctan\frac{2}{3}\right) = \frac{3}{\sqrt{13}}$$

The average power flow into the impedance is

$$P = (VA)(PF) = \frac{(220)^2 \cdot 3}{13} = 11,169 \ W$$

Although the phase difference between voltage and current in an element may have any value in a 360° range, negative power factors are not used in practice. Instead, one deals with power flows out of elements that supply average power and with power flows into elements that dissipate average power.

Even within a voltage-current phase difference range of ±90°, the algebraic sign of the phase difference cannot be determined from the power factor, since

$$\cos(\theta - \phi) = \cos(\phi - \theta)$$

To convey the sense of the phase difference, a power factor is said to be *inductive* or *lagging* (current lags voltage) if $\angle\mathbf{V} - \angle\mathbf{I}$ is positive, and *capacitive* or *leading* (current leads voltage) if $\angle\mathbf{V} - \angle\mathbf{I}$ is negative.

Let the power into an impedance with magnitude 10 Ω and power factor 0.5 leading be 600 W. The angle of the impedance is

$$\angle Z = -\cos^{-1} 0.5 = -60°$$

the minus sign applying because current leads voltage. Thus

$$Z = 10e^{-j60°} = 5 - j8.66 \ \Omega$$

In terms of rms values, the average power is

$$P = V_{rms}I_{rms}(PF)$$

$$600 = V_{rms}I_{rms}(0.5)$$

so that

$$V_{rms}I_{rms} = (VA) = 1200$$

The ratio of rms voltage to current is the magnitude of the impedance,

$$\frac{V_{rms}}{I_{rms}} = 10$$

and substituting into the power relation gives

$$V_{rms} = 109.5 \ \text{V} \qquad I_{rms} = 10.95 \ \text{A}$$

The reactive power is

$$Q = (VA) \sin (\angle Z) = 1200 \sin (-60°) = -1039.2 \ \text{var}$$

D9-6

The following are representative of a class of problems found on many engineer-in-training (EIT) examinations. Here, rms values, sink reference relations for impedances (loads), and source references for sources (generators) are implied unless otherwise stated. "Impedance" often means the magnitude of the impedance.

(a) A certain load has impedance 10 Ω and power factor 0.9, lagging. If the load voltage is 110, what are the current, the volt-amperes, the power, and the reactive power?

ans. 11 A, 1210 VA, 1089 W, 528.5 var

(b) A generator supplies 3000 VA to a load at 220 V with a capacitive power factor of 0.85. Find the current, the real power, and the reactive power.

ans. 13.64 A, 2550 W, −1576 var

(c) A load of $10 + j4$ is connected through wires of total resistance $2 \, \Omega$ to a 440-V generator. What are the power factor, the volt-amperes, and the power of the load and of the generator? What power is lost in the wires?

ans. 0.928, 13,032 VA, 12,100 W, 0.949, 15,305 VA, 14,250 W, 2420 W

(d) A 10.6-Ω impedance with capacitive power factor 0.7 is connected in parallel with 4.8 Ω with 0.9 power factor, lagging. What is the impedance and what is the power factor of the combination?

ans. 3.93 Ω, 0.9959 inductive

POWER CALCULATIONS WITH RMS VALUES

The rms value of a sinusoidal signal is $(1/\sqrt{2}) \cong 0.707$ times its amplitude. In terms of rms values,

$$P_{into} = V_{rms}I_{rms} \cos (\angle \mathbf{V} - \angle \mathbf{I})$$

$$Q_{into} = V_{rms}I_{rms} \sin (\angle \mathbf{V} - \angle \mathbf{I})$$

$$|Z| = \frac{|\mathbf{V}|}{|\mathbf{I}|} = \frac{V_{rms}}{I_{rms}}$$

When dealing extensively with rms values, as in the electric power industry, it is convenient to modify the definitions of phasors so that phasor magnitudes are rms values of the sinusoidal signals rather than their amplitudes (peak values).

Average power flow is commonly expressed in the form

$$P_{into} = \text{(volt-amperes)(power factor)} = \text{(VA)(PF)}$$

The volt-amperes of an element is the product of its rms voltage and current:

$$\text{VA} = V_{rms}I_{rms}$$

The power factor of an element is

$$\text{PF} = \cos (\angle \mathbf{V} - \angle \mathbf{I}) = \cos (\angle \mathbf{I} - \angle \mathbf{V})$$

A power factor is said to be *inductive* or *lagging* if $\angle \mathbf{V} - \angle \mathbf{I}$ is between $0°$ and $90°$. It is said to be *capacitive* or *leading* if $\angle \mathbf{V} - \angle \mathbf{I}$ is between $0°$ and $-90°$.

9.8 SINGLE-PHASE POWER TRANSMISSION

9.8.1 A Power System Model

A simple ac power transmission system consists of a generator, interconnecting wires or "lines," and the device or devices using the power, called the *load*. As illustrated in the drawing of Figure 9-8, the generator may be modeled by a Thévenin equivalent, the lines by an impedance, often resistive, and the load by another impedance. Most power systems in North America operate at a frequency of 60 Hz; 50 Hz is commonly used in other parts of the world.

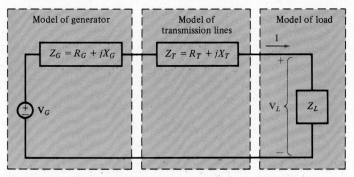

Figure 9-8   Single-phase power transmission model.

The *efficient* transmission of power is very different from the transfer of *maximum* power discussed in Section 9.4. Maximum power transfer is desirable in applications such as communications processing, where the signals involve a small amount of power and are competing with undesirable noise signals. Efficient transfer is important in power systems where large, expensive amounts of energy are involved. Energy losses in the generator and the transmission lines are wasted energy not delivered to the load.

In the remainder of this chapter, in keeping with electric power industry practice, phasors will be of the rms type, with magnitude equal to the rms value of the represented sinusoid. The phasor coordinate notation $A \angle \theta$ will be used rather than the polar form $Ae^{j\theta}$.

9.8.2 Power Factor Correction

For a given amount of average power flow into the load (using rms phasors),

$$P_{\text{into load}} = |\mathbf{V}_L| \, |\mathbf{I}| \cos{(\angle Z_L)}$$

and a fixed load voltage amplitude $|\mathbf{V}_L|$, the load-current amplitude is smallest when the power factor of the load is unity, so that the load voltage and current are in phase.

This load current flows through the lines and generator. The average power losses in the lines,

$$P_{\text{into lines}} = |\mathbf{I}|^2 R_T$$

and in the generator,

$$P_{\text{into generator}} = |\mathbf{I}|^2 R_G$$

are proportional to its square. Thus in the interest of reducing power loss in the generator and the lines, a purely resistive, unity power factor, load impedance is desirable.

In practice, many loads involving large amounts of power, as in an industrial installation, do not have a unity power factor. The usual situation is that of an inductive load, typical of induction motors. In these cases it is advantageous to modify the load impedance by adding a parallel reactance, so that the composite load has unity power factor. The problem of finding the desired parallel reactance is that of obtaining parallel resonance of the load at the power system frequency.

Consider the original inductive load impedance Z_L in Figure 9-9. A parallel capacitive reactance is to be added, so that the parallel combination is purely resistive. A direct solution for the desired reactance X is as follows:

$$Z_0 = \frac{(jX)(10 + j3)}{10 + j3 + jX} = \frac{-3X + j10X}{10 + j(3 + X)}$$

$$= \frac{(3X - j10X)[10 - j(3 + X)]}{100 + (3 + X)^2} = \frac{10X^2 - j(3X^2 + 109X)}{100 + (3 + X)^2}$$

Setting

$$3X^2 + 109X = 0$$

the nonzero reactance of

$$X = -\frac{109}{3}$$

will give unity power factor for the overall load. As the needed reactance is negative, it will be provided by a capacitor.

At a power line frequency of 60 Hz, the value of capacitance needed is given by

$$\frac{1}{\omega C} = \frac{109}{3}$$

$$C = \frac{3}{(6.28)(60)(109)} = 7.3 \times 10^{-5} = 73 \ \mu\text{F}$$

The parallel reactance for correction of the load power factor to unity may also be easily calculated by considering reactive powers. The reactive power flow into Z_L is

$$Q_{\text{into } Z_L} = |\mathbf{I}_L|^2 \cdot 3 = \left(\frac{|\mathbf{V}_L|}{|10 + j3|}\right)^2 \cdot 3 = \frac{3|\mathbf{V}_L|^2}{109}$$

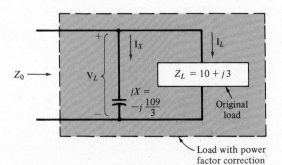

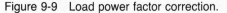

Figure 9-9 Load power factor correction.

The reactive power flow into the parallel capacitive reactance is

$$Q_{\text{into } X} = |\mathbf{I}_X|^2 \cdot X = \left(\frac{|\mathbf{V}_L|}{X}\right)^2 \cdot X = \frac{|\mathbf{V}_L|^2}{X}$$

When the overall load impedance $\mathbf{Z}_0$ is purely resistive, the net reactive power is zero, giving

$$\frac{3|\mathbf{V}_L|^2}{109} + \frac{|\mathbf{V}_L|^2}{X} = 0 \qquad X = -\frac{109}{3}$$

Power factor correction is thus seen to be the connection of additional reactance to the load to bring its net reactive power to zero. A nonzero load reactive power means that energy is being stored in the load then returned to the source, back and forth, twice each cycle, giving rise to additional power losses. The added reactance makes the load resonant at the power line frequency and confines this interchange of reactive power to the elements within the load.

If it is desired to correct a load power factor only partially, to say 0.95 instead of unity, the complex power $(P + jQ)$ is especially useful since the angle of $\mathcal{P}$ is the phase difference between voltage and current.

D9-7

The following are representative of a class of problems found on many EIT examinations:

(a) In a 440-V ac circuit, an inductive load draws 60 A at a power factor of 0.8. Find the volt-amperes, the power, the reactive volt-amperes, and the load impedance.

  ans. 26,400 VA, 21,120 W, 15,840 var, 5.87 + j4.4 Ω

(b) A circuit, consisting of a 3-Ω resistance in series with a 7-Ω reactance, is connected to a 110-V, 60-Hz source. Find the current, the real volt-amperes, the reactive volt-amperes, and the total volt-amperes of the load.

  ans. 14.44 A, 625.9 W, 1460 var, 1588 VA

(c) A 10-kW, 60-Hz, 120-V load has a power factor of 0.9, leading. Find the current, the volt-amperes, and the reactive volt-amperes.

  ans. 92.6 A, 11,111 VA, −4843 var

(d) Two loads are connected in parallel across a 120-V, 50-Hz ac line. The first load is 6000 VA at a power factor of 0.9, leading. The second load is 9000 VA with a 0.8 power factor, lagging. Find the total volt-amperes, the overall power factor, the line current, and the net real and reactive powers.

  ans. 12,904 VA, 0.976, 107.5 A, 12,600 W, 2785 var

(e) It is desired to correct the power factor of a 240-V, 5-kW, 6-kVA inductive load to unity. What reactance should be connected in parallel?

  ans. X = −17.37

(f) A 12-kW, 60-Hz, 220-V load has a power factor of 0.75, lagging. It is desired to correct the power factor to 0.9, lagging. What value of capacitor should be connected in parallel with the load?

ans. 2.62 × 10⁻⁴ F

SINGLE-PHASE POWER TRANSMISSION

The major elements of a single-phase power transmission system consist of the generator, modeled by a Thévenin equivalent, the lines, modeled by an impedance (often a pure resistance), and the load, modeled by another impedance.

   Correcting the power factor of a load impedance reduces power losses in the generator and lines.

   The most common power factor correction problem involves finding a capacitance to place in parallel with an inductive load such that their combination is entirely real.

9.9 TRANSFORMERS

9.9.1 Controlled Source Model

The term *transformer* generally means a set of coupled inductors with a relatively large degree of magnetic flux coupling between the coils, intended for efficient transmission of electrical power. One coil is termed the *primary* and the other is called the *secondary*, although which is called which is somewhat arbitrary. Usually the label "primary" is attached to the coil that is to be connected closest to the source of power, so that average electrical power flows from the primary to the secondary.

   For power applications, a more tractable transformer model than the coupled inductors, Figure 9-10(a), or their controlled source equivalent, Figure 9-10(b), may be derived. The equations for $\mathbf{V}_1$ and $\mathbf{V}_2$, in terms of $\mathbf{I}_1$ and $\mathbf{I}_2$, are

$$\begin{cases} \mathbf{V}_1 = j\omega L_1 \mathbf{I}_1 + j\omega M \mathbf{I}_2 \\ \mathbf{V}_2 = j\omega M \mathbf{I}_1 + j\omega L_2 \mathbf{I}_2 \end{cases} \tag{9-1}$$

Eliminating $\mathbf{I}_2$ between these two equations gives

$$\mathbf{V}_1 = \frac{M}{L_2} \mathbf{V}_2 + j\omega\left(\frac{L_1 L_2 - M^2}{L_2}\right)\mathbf{I}_1$$

Similarly, eliminating $\mathbf{I}_1$ between the same equations,

$$\mathbf{V}_2 = \frac{M}{L_1} \mathbf{V}_1 + j\omega\left(\frac{L_1 L_2 - M^2}{L_1}\right)\mathbf{I}_2$$

These two new equations are equivalent to the original two, and give the alternative representation of Figure 9-10(c), where it is seen that each winding involves some inductance in series with a voltage proportional to the voltage applied to the opposite winding.

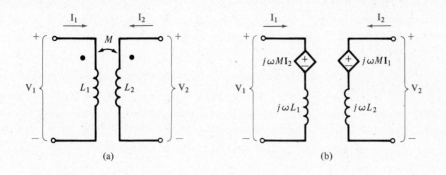

(a) (b)

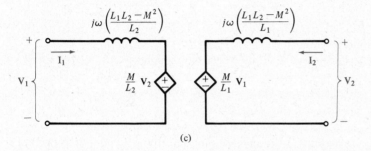

(c)

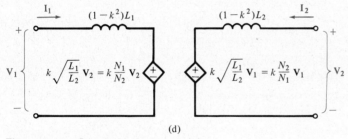

(d)

Figure 9-10   Coupled inductor models.

In terms of the coupling coefficient of the transformer,

$$k = \frac{M}{\sqrt{L_1 L_2}}$$

these equations are

$$\begin{cases} \mathbf{V}_1 = k\sqrt{\dfrac{L_1}{L_2}}\,\mathbf{V}_2 + j\omega(1 - k^2)L_1 \mathbf{I}_1 \\[2mm] \mathbf{V}_2 = k\sqrt{\dfrac{L_2}{L_1}}\,\mathbf{V}_1 + j\omega(1 - k^2)L_2 \mathbf{I}_2 \end{cases}$$

(9-2)

The quantity

$$\sqrt{\frac{L_2}{L_1}} = \frac{N_2}{N_1}$$

is called the *turns ratio* of the transformer; N_2/N_1 is approximately the ratio of the number of turns of the L_2 coil to the number of turns of the L_1 coil. The model is relabeled in terms of k and turns ratios in Figure 9-10(d).

9.9.2 Ideal Transformers

An idealized transformer model, for which the magnetic coupling coefficient k is maximum,

$$k = 1 \qquad M = \sqrt{L_1 L_2}$$

is a useful approximation for many practical transformers. It is called the *ideal transformer* and is symbolized by two inductor symbols and a turns ratio, as in the example of Figure 9-11. If the senses of the windings are of importance to the application, they should be indicated.

For the ideal transformer, $k = 1$, and the winding voltages are related by the turns ratio:

$$\frac{V_2}{V_1} = \sqrt{\frac{L_2}{L_1}} = \frac{N_2}{N_1}$$

In general, the ratio of currents in the two windings is, from the equations (9-2) of the previous section,

$$\frac{I_2}{I_1} = \frac{L_1}{L_2} \cdot \frac{V_2 - k\sqrt{\dfrac{L_2}{L_1}}\,V_1}{V_1 - k\sqrt{\dfrac{L_1}{L_2}}\,V_2} = -\sqrt{\frac{L_1}{L_2}} \left[\frac{V_2 - k\sqrt{\dfrac{L_2}{L_1}}\,V_1}{kV_2 - \sqrt{\dfrac{L_2}{L_1}}\,V_1} \right]$$

For k near unity, the factor in brackets approaches unity, giving

$$\frac{I_2}{I_1} = -\sqrt{\frac{L_1}{L_2}} = -\frac{N_1}{N_2}$$

for perfect coupling. The currents I_1 and I_2 are related by the negative inverse of the turns ratio. The minus sign indicates that I_1 and I_2 are in opposite senses.

In an ideal transformer when the voltage is stepped up or stepped down by the turns ratio,

$$V_2 = \frac{N_2}{N_1} V_1$$

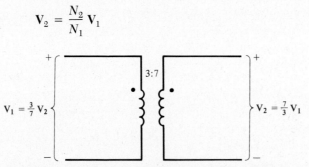

Figure 9-11   Voltage relations for an ideal transformer.

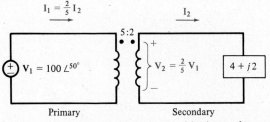

Figure 9-12 Network involving an ideal transformer.

the current is stepped in inverse proportion,

$$\mathbf{I}_2 = -\frac{N_1}{N_2}\mathbf{I}_1$$

so that the voltage-current product is unchanged. All of the electrical power flowing into one winding flows out of the other winding.

As an example of these relations, consider the network model of Figure 9-12, which contains an ideal transformer with turns ratio 5:2. The secondary voltage is two-fifths that of the primary, so the secondary current (with reference direction reversed from that used previously) is

$$\mathbf{I}_2 = \frac{\mathbf{V}_2}{4 + j2} = \frac{\frac{2}{5}\mathbf{V}_1}{4 + j2} = \frac{40\,\underline{/50°}}{4.47\,\underline{/26.6°}}$$

$$= 8.95\,\underline{/23.4°}$$

The primary current is two-fifths that of the secondary:

$$\mathbf{I}_1 = \tfrac{2}{5}\mathbf{I}_2 = 3.58\,\underline{/23.4°}$$

Transformers are commonly used to increase or decrease generated voltages in power systems. Power is generated by rotating machines at voltages (typically 5 to 50 kV) that are convenient for the generator construction. Transformers are used to step up the generator voltage, while stepping down the current, for long-distance transmission. Since the power dissipated in a transmission line is the line resistance times the square of the rms line current, the smaller the current, the smaller the losses. Long-distance power transmission is done at the highest practical voltage, typically 200 to 750 kV.

As the power transmission lines branch to localized distribution areas, transformers successively step down the line voltages to values suitable for the area, eventually reaching the low voltages common in residential distribution. Power lines servicing residential areas typically involve rms line voltages of 2 to 10 kV, which are stepped down to a center-tapped 220 V, allowing the home user to connect to either 110 V for low-power appliances or to 220 V for greater efficiency in supplying devices having larger power requirements. With the centertap grounded, supplied voltages do not exceed 110 V rms with respect to ground. The drawing of Figure 9-13 illustrates home power distribution and utilization.

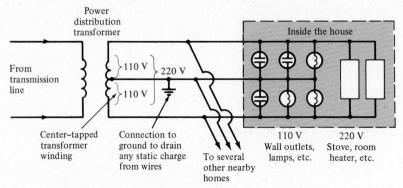

Figure 9-13   Residential power distribution.

9.9.3 Impedance Reflection

When the secondary of an ideal transformer is connected to an impedance, as in the diagram of Figure 9-14, the ratio of secondary phasor voltage to current is that impedance. The primary voltage and current are related to the secondary quantities by the turns ratio as shown, so the impedance at the primary winding terminals is

$$Z_1 = \frac{\mathbf{V}_1}{\mathbf{I}_1} = \frac{\dfrac{N_1}{N_2}\mathbf{V}_2}{\dfrac{N_2}{N_1}\mathbf{I}_2} = \left(\frac{N_1}{N_2}\right)^2 \frac{\mathbf{V}_2}{\mathbf{I}_2} = \left(\frac{N_1}{N_2}\right)^2 Z_2$$

Viewed through the transformer, the impedance connected to the secondary is magnified by the turns ratio squared. This result is independent of the transformer winding senses.

In communications systems, the impedance reflection property of transformers is used to match fixed load impedances to networks for maximum power transfer.

Impedance reflection is a useful technique in equivalent circuit solution of networks involving transformers. For example, the impedance as viewed from the primary terminals of the transformer in Figure 9-15(a) is

$$Z = (\tfrac{2}{3})^2(6 - j3) = \tfrac{8}{3} - j\tfrac{4}{3}$$

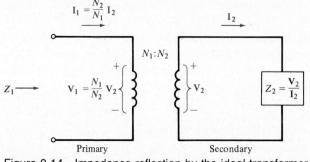

Figure 9-14   Impedance reflection by the ideal transformer.

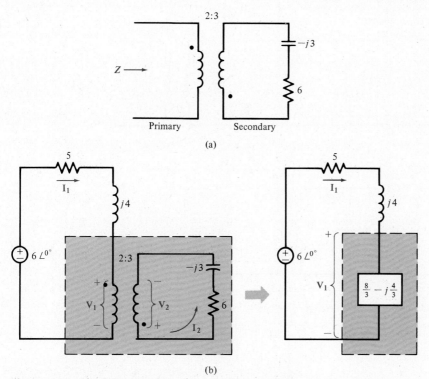

Figure 9-15 Network solution using impedance reflection.

If the primary of this transformer is connected as shown in Figure 9-15(b), so far as the rest of the network is concerned, the primary terminals present the impedance Z as in the equivalent network. So

$$\mathbf{I}_1 = \frac{6 \angle 0°}{\frac{23}{3} + j\frac{8}{3}} = \frac{6 \angle 0°}{8.1 \angle 19.2°} = 0.74 \angle 19.2°$$

$$\mathbf{V}_1 = 6 \angle 0° \frac{\frac{8}{3} - j\frac{4}{3}}{\frac{23}{3} + j\frac{8}{3}} = \frac{6 \angle 0° (2.98 \angle -26.6°)}{8.1 \angle 19.2°} = 2.2 \angle -45.8°$$

Using the turns ratio,

$$\mathbf{V}_2 = \tfrac{3}{2}\mathbf{V}_1 = \tfrac{3}{2}(2.2 \angle -45.8°) = 3.3 \angle -45.8°$$

and

$$\mathbf{I}_2 = \tfrac{2}{3}\mathbf{I}_1 = \tfrac{2}{3}(0.74 \angle -19.2°) = 0.49 \angle -19.2°$$

9.9.4 Transformer Models Using Ideal Transformers

Another type of transformer model that is commonly used in practice involves an ideal transformer as part of the model. In Figure 9-16(a), the ideal transformer voltage and current to the left are related to the transformer voltage and current to the right by

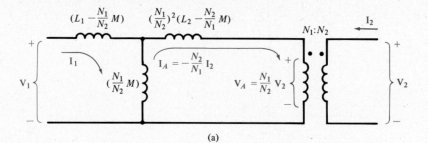

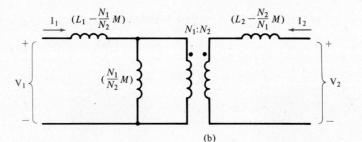

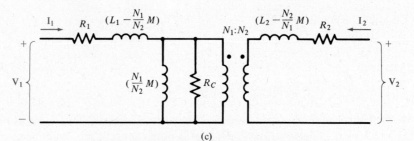

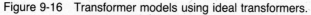

Figure 9-16   Transformer models using ideal transformers.
(a) Tee network followed by ideal transformer.
(b) Inductance reflected through the ideal transformer.
(c) Model with winding resistances and core loss resistance.

$$\mathbf{V}_A = \frac{N_1}{N_2}\,\mathbf{V}_2 \qquad \mathbf{I}_A = -\frac{N_2}{N_1}\,\mathbf{I}_2$$

as shown. Loop equations for the rest of this network are as follows

$$
\begin{cases}
j\omega L_1 \mathbf{I}_1 - j\omega\left(\dfrac{N_1}{N_2}\,M\right)\left(-\dfrac{N_2}{N_1}\,\mathbf{I}_2\right) = \mathbf{V}_1 \\[2ex]
-j\omega\left(\dfrac{N_1}{N_2}\,M\right)\mathbf{I}_1 + j\omega\left(\dfrac{N_1}{N_2}\right)^2 L_2\left(-\dfrac{N_2}{N_1}\,\mathbf{I}_2\right) = -\left(\dfrac{N_1}{N_2}\,\mathbf{V}_2\right)
\end{cases}
$$

or

$$
\begin{cases}
j\omega L_1\,\mathbf{I}_1 + j\omega M\,\mathbf{I}_2 = \mathbf{V}_1 \\[1ex]
j\omega M\,\mathbf{I}_1 + j\omega L_2\,\mathbf{I}_2 = \mathbf{V}_2
\end{cases}
$$

which are seen to be identical to equations (9-1) for any coupled inductors.

When the inductance involving L_2 is reflected through the ideal transformer windings, the model of Figure 9-16(b), also described by equations (9-1), results. The series inductances, using

$$M = k \sqrt{L_1 L_2} \qquad \sqrt{\frac{L_1}{L_2}} = \frac{N_1}{N_2}$$

are

$$L_1 - \frac{N_1}{N_2} M = (1 - k) L_1$$

$$L_2 - \frac{N_2}{N_1} M = (1 - k) L_2$$

which become zero for perfect magnetic coupling, $k = 1$, are termed *leakage inductances*, referring to imperfect coupling caused by leakage of magnetic flux. The inductance

$$L_M = \frac{N_1}{N_2} M$$

is termed the *magnetization inductance*.

In Figure 9-16(c), resistances R_1 and R_2 of the wires comprising the transformer windings have been added to the model. The added "core loss" resistance R_C is used to account for power losses in the magnetization of a transformer's ferromagnetic core.

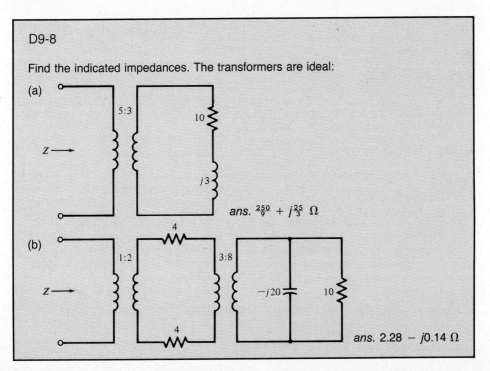

D9-8

Find the indicated impedances. The transformers are ideal:

(a)

5:3

10

$Z \longrightarrow$

$j3$

ans. $\frac{250}{9} + j\frac{25}{3}$ Ω

(b)

4

1:2   3:8

$Z \longrightarrow$

$-j20$ 10

4

ans. $2.28 - j0.14$ Ω

D9-9

For the following networks involving ideal transformers, find the indicated phasors:

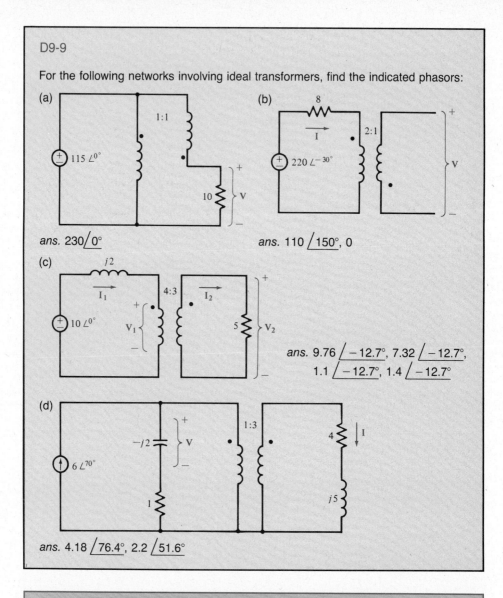

(a)

ans. $230\underline{/0°}$

(b)

ans. $110\underline{/150°}$, 0

(c)

ans. $9.76\underline{/-12.7°}$, $7.32\underline{/-12.7°}$,
$1.1\underline{/-12.7°}$, $1.4\underline{/-12.7°}$

(d)

ans. $4.18\underline{/76.4°}$, $2.2\underline{/51.6°}$

TRANSFORMERS

Transformers are coupled inductors with large coupling coefficients. The controlled source and leakage inductance model, in terms of the coupling coefficient and turns ratio, is useful for such devices, as are models involving ideal transformers.

An ideal transformer is a model of coupled inductors for which the magnetic coupling coefficient is maximum, $k = 1$.

For an ideal transformer, the primary voltage is the turns ratio times the secondary voltage. The primary current is the negative inverse of the turns ratio times the secondary current. The impedance as viewed from the primary of an ideal transformer is the secondary impedance times the square of the turns ratio.

9.10 THREE-PHASE POWER TRANSMISSION

9.10.1 Balanced Three-Phase Power Flow

Consider the three networks of Figure 9-17(a) which are identical except for the phases of the source voltages. The source voltage phase angles are located symmetrically at 120° intervals, as indicated on the typical phasor diagram of Figure 9-17(b). In consequence of the symmetry, $\mathbf{I}_1$, $\mathbf{I}_2$, and $\mathbf{I}_3$ differ only in that their phase angles are spaced 120° apart.

The instantaneous electrical power flow out of the first source is (using rms phasors)

$$p_1(t) = AB \cos (\theta - \phi) + AB \cos (2\omega t + \theta + \phi)$$

For the second and third sources, the power flows are

$$p_2(t) = AB \cos (\theta - \phi) + AB \cos (2\omega t + \theta + \phi + 240°)$$

$$p_3(t) = AB \cos (\theta - \phi) + AB \cos (2\omega t + \theta + \phi + 480°)$$

$$= AB \cos (\theta - \phi) + AB \cos (2\omega t + \theta + \phi + 120°)$$

The total instantaneous power flow from the three sources is

$$p(t) = p_1(t) + p_2(t) + p_3(t)$$

$$= 3AB \cos (\theta - \phi) + AB\{\cos (2\omega t + \theta + \phi)$$

$$+ \cos (2\omega t + \theta + \phi + 240°) + \cos (2\omega t + \theta + \phi + 120°)\}$$

$$= 3AB \cos (\theta - \phi)$$

which is *constant*, the terms in brackets summing to zero. Individually, the power flows vary with time, but the sum of the three powers is a constant. This *three-phase* arrangement is particularly advantageous for the generation and utilization of electrical power.

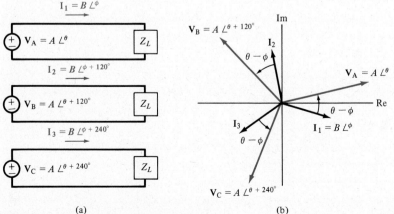

(a)	(b)

Figure 9-17   Three symmetrical circuits.

For single-phase power generation, the flow of mechanical power into a generator from, say, a turbine, varies with time as does $p_1(t)$. During part of the shaft rotation, a large amount of mechanical power must be converted to electrical power. During another part of the shaft rotation, there is less mechanical-electrical power conversion. In fact, electrical power is returned to the mechanical sources in any interval of time when $p_1(t)$ is negative. Not only is this time-varying power flow hard on the generator and turbine bearings, but it is inefficient since there are power losses in each conversion back and forth between electrical and mechanical energy.

If all three generators are connected to the same shaft (in practice, three symmetrical generator coils are placed in the same housing), the net power flow is constant. This result is true also of the load. The net power flow into a balanced three-phase load, which might consist of three motor coils on the same shaft, is constant.

9.10.2 Transmission Efficiency

A three-phase system also offers advantages in the transmission of electrical power. To reduce the number of wires involved in the distribution of power, three main conductors and a common, *neutral* conductor may be used, as indicated in Figure 9-18(a). The current in the neutral wire is the sum of the three load currents that, from a phasor diagram such as that of Figure 9-17(b), are seen to add to zero. Only three wires are really needed then, as indicated in Figure 9-18(b). It is common to draw diagrams for three-phase networks in the manner of Figure 9-18(b), which is suggestive of phasor positions.

A three-phase system can transmit three times the power over three wires that a single-phase system can deliver over two wires, using the same generator voltage

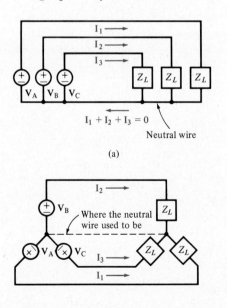

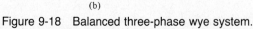

Figure 9-18   Balanced three-phase wye system.

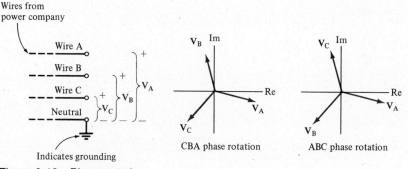

Figure 9-19 Phase rotation.

and line current. In practice, a neutral wire of relatively small diameter is often used to allow for small imbalances in the system. In any event, the neutral connection is wired to conductors buried in the earth (*grounded*), to prevent accumulation of static charge on the conductors.

Models with perfectly symmetrical three-phase sources and loads are referred to as *balanced* three-phase systems. If relatively small yet significant asymmetries are present, the system is called an *unbalanced* three-phase system. Unbalanced systems will not be considered further here, except to note that any network may be solved by conventional means; the symmetry of a balanced system allows simplified solutions.

Given the incoming wires of a three-phase power distribution system, it is often of the utmost importance to know the *phase sequence*. Labeling the wires A, B, and C, the line-to-neutral voltages, Figure 9-19, can crest in the order A–B–C or in the order C–B–A. The phase sequence makes quite a difference in many applications. For instance, interchanging connections will result in reversal of the direction of rotation of a three-phase induction motor.

9.10.3 Balanced Wye Systems

A simple model of a three-phase power system consists of wye connections of sources and loads, such as the example in Figure 9-20(a), in which each generator is modeled by a Thévenin equivalent, each wire by a resistance, and each load by an impedance. Provided the three-phase system is balanced, the neutral wire may be added, and the three equivalent single-phase networks solved, as in Figure 9-20(b). Of course it is really only necessary to calculate a solution of one of the single-phase networks; the other network solutions differ only in phase, by 120° and 240°.

The solution for $\mathbf{I}_1$ is

$$\mathbf{I}_1 = \frac{500\,\underline{/30°}}{5 + j5} = \frac{500\,\underline{/30°}}{5\sqrt{2}\,\underline{/45°}} = \frac{100}{\sqrt{2}}\,\underline{/-15°}$$

Then

$$\mathbf{I}_2 = \frac{100}{\sqrt{2}}\,\underline{/105°} \qquad \mathbf{I}_3 = \frac{100}{\sqrt{2}}\,\underline{/225°}$$

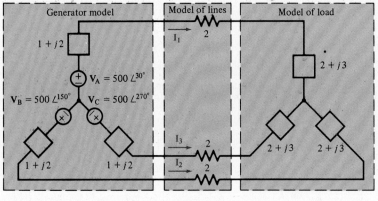

(a)

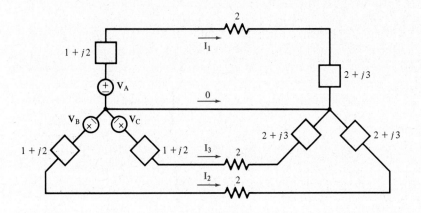

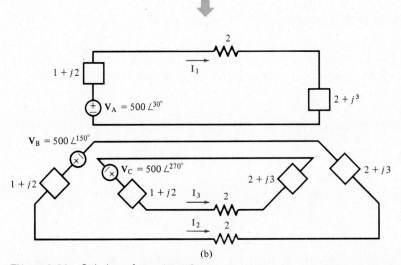

(b)

Figure 9-20   Solution of a wye system.

D9-10

Draw a phasor diagram showing every voltage and current in the following system:

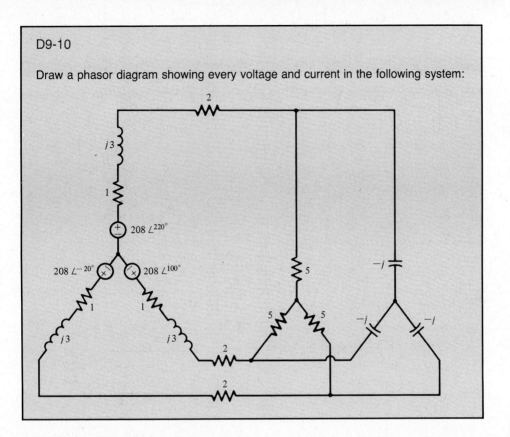

9.10.4 Balanced Delta Systems

The line-to-line voltages corresponding to a three-phase wye connection of sources, Figure 9-21(a), may be easily found, using a phasor diagram. The wye voltages are

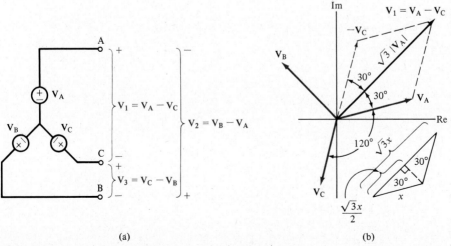

(a)   (b)

Figure 9-21   Line-to-line voltage in terms of wye voltages.

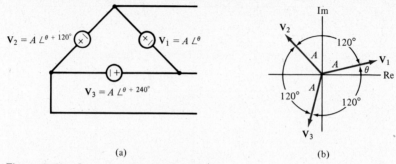

(a)   (b)

Figure 9-22   Sources connected in delta.

termed the *phase* voltages, to distinguish them. The *line-to-line* voltage

$$\mathbf{V}_1 = \mathbf{V}_A - \mathbf{V}_C$$

is seen from the phasor diagram construction of Figure 9-21(b) to lead $\mathbf{V}_A$ in phase by 30° and to have magnitude $\sqrt{3}$ times the magnitude of $\mathbf{V}_A$. The other two line voltages, $\mathbf{V}_2$ and $\mathbf{V}_3$, are of the same magnitude as $\mathbf{V}_1$ and are spaced at 120° intervals with $\mathbf{V}_1$, as shown on the phasor diagram.

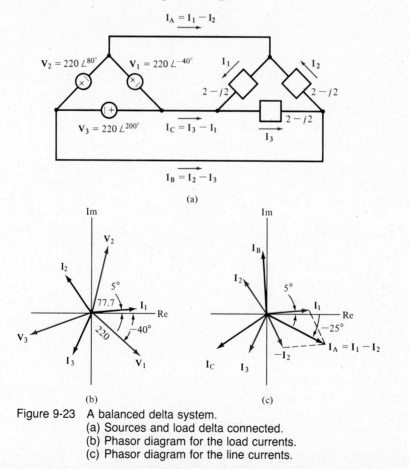

(a)

(b)   (c)

Figure 9-23   A balanced delta system.
        (a) Sources and load delta connected.
        (b) Phasor diagram for the load currents.
        (c) Phasor diagram for the line currents.

Another way to model a three-phase power system is with a delta connection of sources, such as that shown in Figure 9-22(a). The voltage phasors, Figure 9-22(b), being equal in magnitude and at 120° intervals in phase, add to zero around the loop of sources, as they must. Any one of the three sources is redundant and could be removed so far as network solution is concerned, although this is seldom done in practice. In this arrangement, the delta-connected sources provide the line-to-line voltages rather than the phase voltages.

If sources in delta are connected to a delta connection of impedances, the voltage across each impedance is a known source voltage, and the network solution is particularly easy, as in the example of Figure 9-23, for which

$$\mathbf{I}_1 = \frac{\mathbf{V}_1}{2 - j2} = \frac{220 \,/\!-40°}{2.83 \,/\!-45°} = 77.7 \,/\!5°$$

$$\mathbf{I}_2 = 77.7 \,/\!125° \qquad \mathbf{I}_3 = 77.7 \,/\!245°$$

The line currents, $\mathbf{I}_A$, $\mathbf{I}_B$, and $\mathbf{I}_C$ may be found from the load currents by phasor addition similar to that used for finding line voltages from phase voltages. From the phasor diagram of Figure 9-23(c),

$$\mathbf{I}_A = 77.7 \sqrt{3} \,/\!-25° \qquad \mathbf{I}_B = 77.7 \sqrt{3} \,/\!95° \qquad \mathbf{I}_C = 77.7 \sqrt{3} \,/\!215°$$

9.10.5 Delta-Wye Source Transformation

Transformation between wye and equivalent delta sources may be used to simplify the solution of balanced three-phase systems. In the network of Figure 9-24, for

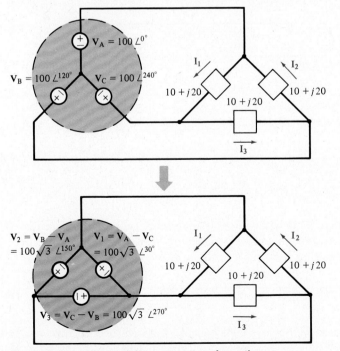

Figure 9-24   Wye-to-delta source transformation.

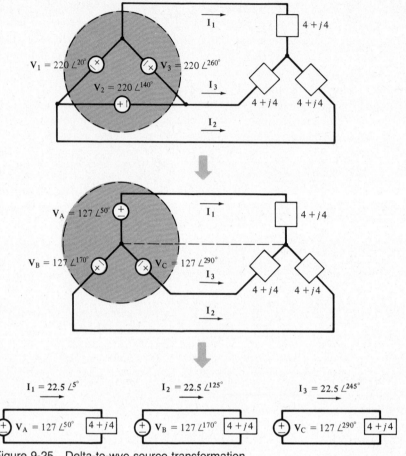

Figure 9-25 Delta-to-wye source transformation.

example, the sources are wye connected and the load is delta connected. Converting the wye sources to equivalent delta ones gives a balanced delta system for which the solution is simple:

$$\mathbf{I}_1 = \frac{\mathbf{V}_1}{10 + j20} = \frac{100\sqrt{3}\,\underline{/30^\circ}}{22.36\,\underline{/63.5^\circ}} = 7.75\,\underline{/-33.5^\circ}$$

Transformation from delta to wye sources is also useful. In the system of Figure 9-25, with delta-connected sources and wye-connected load, a delta-to-wye source transformation greatly simplifies the analysis.

D9-11

Perform the indicated delta-wye or wye-delta source transformations, and draw a phasor diagram that shows the six phasors involved in each case.

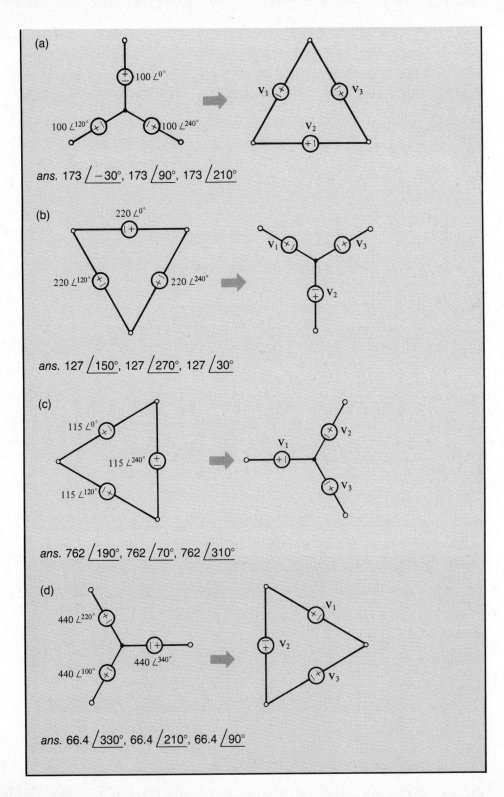

(a)

ans. 173 $\underline{/-30°}$, 173 $\underline{/90°}$, 173 $\underline{/210°}$

(b)

ans. 127 $\underline{/150°}$, 127 $\underline{/270°}$, 127 $\underline{/30°}$

(c)

ans. 762 $\underline{/190°}$, 762 $\underline{/70°}$, 762 $\underline{/310°}$

(d)

ans. 66.4 $\underline{/330°}$, 66.4 $\underline{/210°}$, 66.4 $\underline{/90°}$

D9-12

Use delta-wye or wye-delta source transformations to find the indicated phasors:

(a)

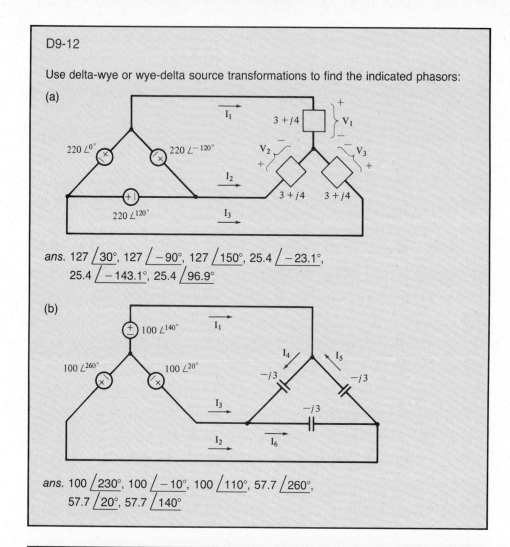

ans. 127 $\underline{/30°}$, 127 $\underline{/-90°}$, 127 $\underline{/150°}$, 25.4 $\underline{/-23.1°}$, 25.4 $\underline{/-143.1°}$, 25.4 $\underline{/96.9°}$

(b)

ans. 100 $\underline{/230°}$, 100 $\underline{/-10°}$, 100 $\underline{/110°}$, 57.7 $\underline{/260°}$, 57.7 $\underline{/20°}$, 57.7 $\underline{/140°}$

THREE-PHASE POWER TRANSMISSION

Three-phase power transmission has the advantage of conveying more power per wire, using three wires, than a single-phase system using two wires of the same capacity. The total power transferred in a balanced three-phase system is constant with time.

For analysis of a balanced wye-connected source and load, the neutral connection may be inserted if it is not already present, and the problem reduces to that of three symmetric single-phase systems. The neutral connection cannot be made if it is not already present in an *unbalanced* system.

For three-phase systems involving a mixture of delta- and wye-connected sources and loads, it is often useful to convert a set of wye sources to an equivalent set of delta sources. For balanced systems, the delta source voltages have amplitude $\sqrt{3}$ as large as the wye voltages and lead them by 30°.

9.11 DELTA-WYE LOAD TRANSFORMATION

It is often convenient to convert between equivalent delta and wye loads in analyzing a three-phase power system. In Chapter 3, it was shown that for any three resistances connected in delta, another set of wye-connected resistances can be found such that the two three-terminal elements are indistinguishable, and vice versa. The same equivalence carries over to impedances, of course.

For delta and wye impedances as in Figure 9-26, the equivalent wye impedances are related to the delta ones by

$$Z_A = \frac{Z_2 Z_3}{Z_1 + Z_2 + Z_3}$$

$$Z_B = \frac{Z_1 Z_3}{Z_1 + Z_2 + Z_3}$$

$$Z_C = \frac{Z_1 Z_2}{Z_1 + Z_2 + Z_3}$$

Each of these relations is of the form

$$Z_N = \frac{\text{product of the two impedances connected to terminal } i}{\text{sum of the three impedances}}$$

The delta impedances are related to the wye ones by

$$Z_1 = \frac{Z_A Z_B + Z_A Z_C + Z_B Z_C}{Z_A}$$

$$Z_2 = \frac{Z_A Z_B + Z_A Z_C + Z_B Z_C}{Z_B}$$

$$Z_3 = \frac{Z_A Z_B + Z_A Z_C + Z_B Z_C}{Z_C}$$

each of these equations being of the form

$$Z_i = \frac{\text{sum of products of impedances taken two at a time}}{\text{impedance connected to terminal opposite to } Z_i}$$

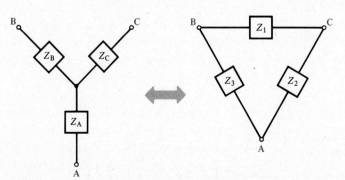

Figure 9-26   Delta-wye equivalence of impedances.

For *balanced* three-phase systems, where each of the three wye- or three delta-connected load impedances are equal, delta-wye conversion reduces to

$$Z_{wye} = \tfrac{1}{3} Z_{delta}$$

Figure 9-27 shows an example of the use of delta-wye transformation to simplify the solution of a balanced three-phase system. First, the delta load is converted to an equivalent wye load, and the series impedances are combined. Then the combined wye impedances are converted to an equivalent delta load, giving a balanced delta system for solution. Alternatively, the load could have been left as a wye and the delta sources converted to an equivalent wye.

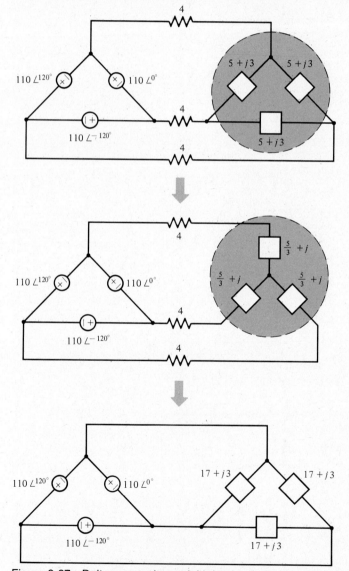

Figure 9-27 Delta-wye and wye-delta load transformations.

D9-13

Use delta-wye or wye-delta load transformations to find the indicated phasors:

(a)

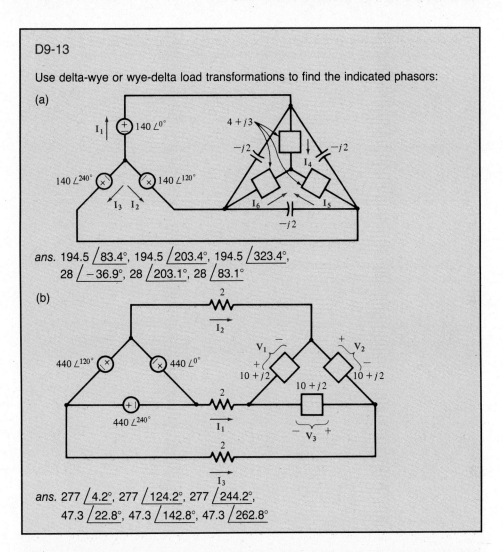

ans. 194.5 $\underline{/83.4°}$, 194.5 $\underline{/203.4°}$, 194.5 $\underline{/323.4°}$,
28 $\underline{/-36.9°}$, 28 $\underline{/203.1°}$, 28 $\underline{/83.1°}$

(b)

ans. 277 $\underline{/4.2°}$, 277 $\underline{/124.2°}$, 277 $\underline{/244.2°}$,
47.3 $\underline{/22.8°}$, 47.3 $\underline{/142.8°}$, 47.3 $\underline{/262.8°}$

D9-14

The following are typical of a class of EIT examination problems:

(a) A balanced, three-phase wye-connected 2400-W load has line-to-line voltages of 220 and a lagging power factor of 0.85. Find the line currents and the total volt-amperes of the load.

   ans. 7.41 A, 2824 VA

(b) A balanced, three-phase delta-connected load has phase voltages of 208 V and line currents of 15 A. Total power delivered to the load is 2200 W. Find the load impedance per phase and the power factor.

   ans. 24 $\underline{/66°}$, 0.407

(c) A balanced, three-phase system has phase voltages of 95 V at the generator and 90 V at the load. Each of the three elements of a delta-connected load require 600 VA at a power factor of 0.7. If the wires are resistive, what are the losses (per wire) in the lines?

ans. 31.6 W

(d) A three-phase generator supplies 3800 VA at a power factor of 0.95, lagging. Line losses are 100 W per wire, and 3000 W are delivered to the load. What are the losses in the generator?

ans. 310 W

DELTA-WYE LOAD TRANSFORMATION

Three-terminal delta- and wye-connected impedances are equivalent to one another when there are the following relations:

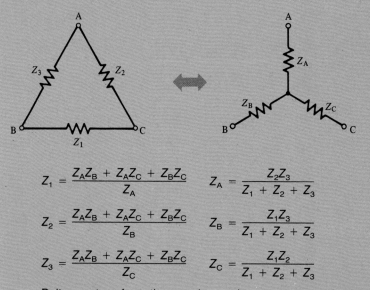

$$Z_1 = \frac{Z_A Z_B + Z_A Z_C + Z_B Z_C}{Z_A} \qquad Z_A = \frac{Z_2 Z_3}{Z_1 + Z_2 + Z_3}$$

$$Z_2 = \frac{Z_A Z_B + Z_A Z_C + Z_B Z_C}{Z_B} \qquad Z_B = \frac{Z_1 Z_3}{Z_1 + Z_2 + Z_3}$$

$$Z_3 = \frac{Z_A Z_B + Z_A Z_C + Z_B Z_C}{Z_C} \qquad Z_C = \frac{Z_1 Z_2}{Z_1 + Z_2 + Z_3}$$

Delta-wye transformation may be used to convert a wye load to an equivalent delta load, or vice versa. For balanced systems, the wye impedances are one-third the delta impedances.

CHAPTER NINE PROBLEMS

Basic Problems

Instantaneous, Average, and Reactive Power

1. For the following networks, find the instantaneous electrical power flow into each element as a function of time.

A large power transformer. This unit is three-phase and is designed to operate at 220kV. (*Photo courtesy of Southern California Edison Co.*)

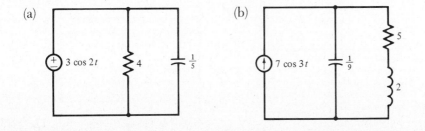

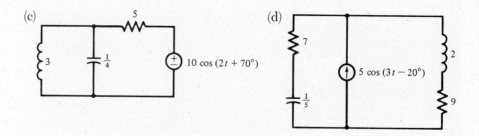

2. For the following networks, find the average electrical power flow into each element and verify that the net average power flow out of sources equals the net average power flow into the other elements:

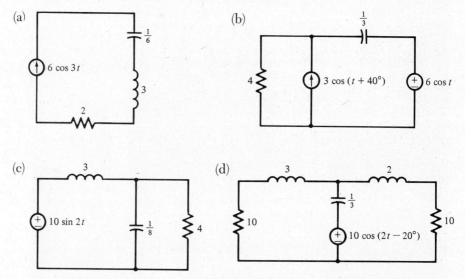

3. For the following networks, find the reactive power flow into each element and verify that the net reactive power flow out of sources equals the net reactive power flow into the other elements. The given phasors are rms phasors.

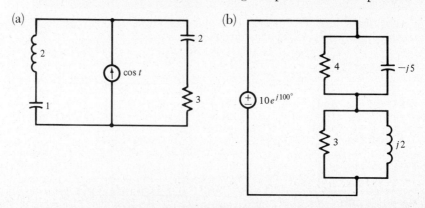

A huge turbine generator at a hydroelectric power plant. This machine will supply up to 35 megawatts of electric power. (*Photo courtesy of Southern California Edison Co.*)

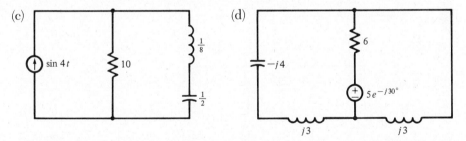

4. For the following networks, find the complex power flow into each element and verify that the net complex power flow out of sources equals the net complex power flow into the other elements. The given phasors are rms phasors.

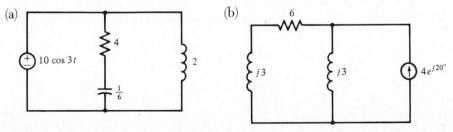

(c)

(d)

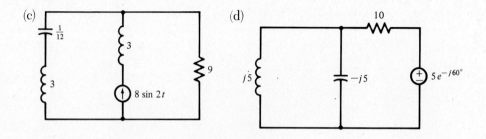

Maximum Power Transfer

5. For the networks below, find the value of Z for which the average power flow into Z is maximized:

(a)

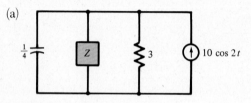

(b)

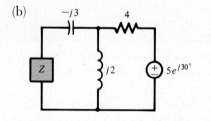

(c)

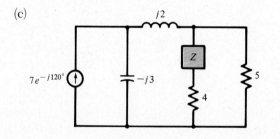

(d)

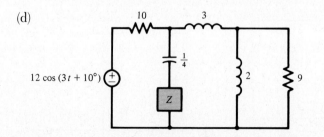

Single-Phase Power

6. Find the values of reactance X necessary to achieve unity power factor for each of the loads shown, at the indicated frequency. Also find the amount of inductance or capacitance needed to produce those reactances at the given frequency.

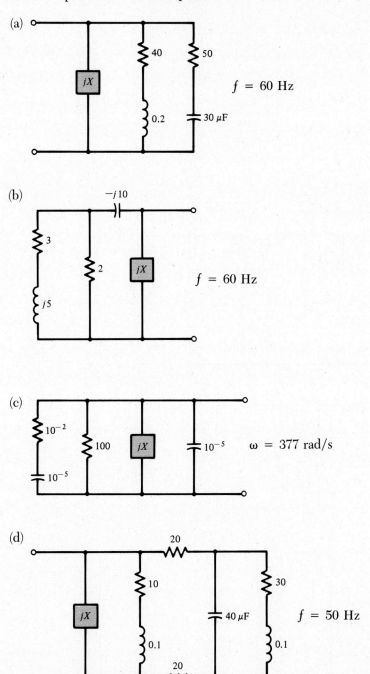

(a)

40   50

jX

f = 60 Hz

0.2   30 μF

(b)

$-j10$

3

2   jX

f = 60 Hz

$j5$

(c)

10^{-2}

100   jX 10^{-5} ω = 377 rad/s

10^{-5}

(d)

20

10   30

jX 40 μF f = 50 Hz

0.1   0.1

20

7. Draw a phasor diagram showing the indicated voltages and currents in the following system for $C = 0$. Then show the same quantities on a second phasor diagram for C chosen to give unity power factor at the load.

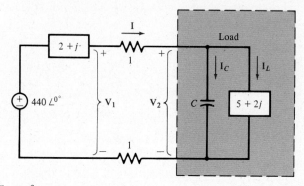

Transformers

8. Find the indicated phasors. The transformers are ideal, and there are two phasors to find in each problem:

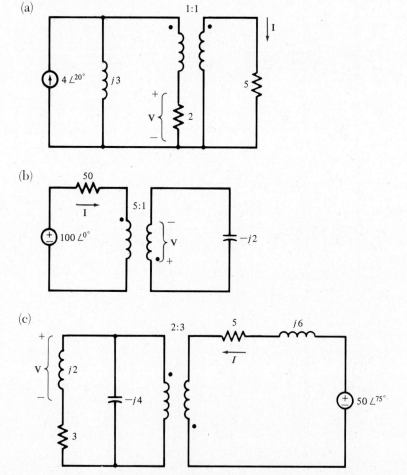

(d)

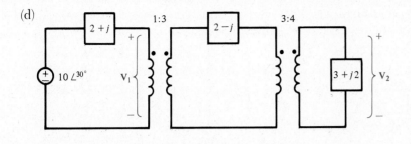

9. Find the ideal transformer turns ratio $N_1 : N_2$ and the reactance X for maximum power transfer into the resistor R.

(a)

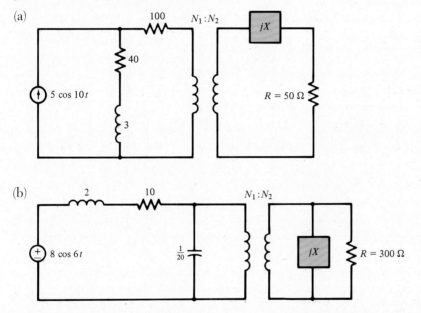

(b)

10. A capacitor is connected to the secondary terminals of a transformer. For the given transformer model parameters, find the impedance $Z(s = j\omega)$ at the primary terminals.

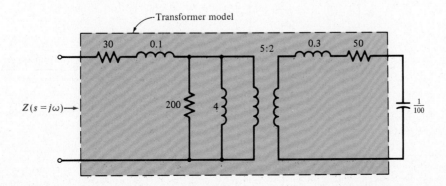

Delta-Wye Transformation

11. Convert the network to an equivalent wye:

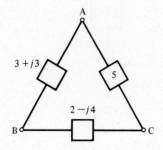

12. Convert the network to an equivalent delta:

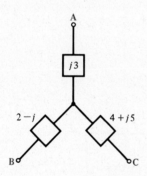

13. Use delta-wye transformations and series and parallel impedance combinations to find the impedance at the terminals of the following network:

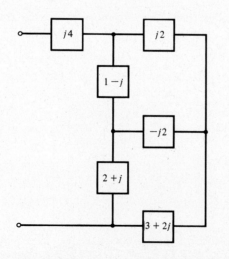

Three-Phase Power

14. Perform delta-wye transformations of the following three-phase sources. Label the A, B, and C terminals in the equivalent.

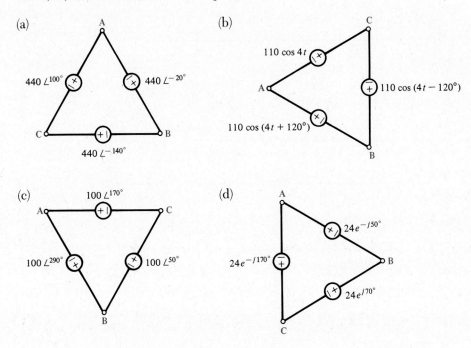

(a)

(b)

(c)

(d)

15. Perform wye-delta transformations of the following three-phase sources. Label the A, B, and C terminals in the equivalent.

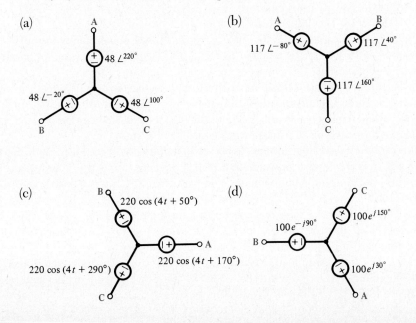

(a)

(b)

(c)

(d)

16. Draw phasor diagrams showing each indicated voltage and current in the following balanced three-phase system:

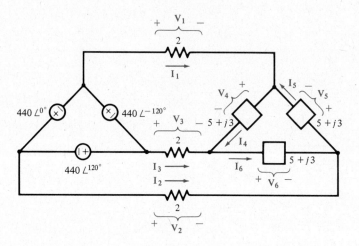

Practical Problems

The following problems are representative of a class found on many EIT examinations:

17. A certain load consists of $10 + j4 \ \Omega$ in parallel with $-j3 \ \Omega$. What is the overall power factor?

18. A generator supplies two parallel 115-V loads at a 0.86 power factor, lagging. One load is 4800 VA at a power factor of 0.9, lagging. Is the power factor of the other load leading or lagging? If the other load consumes 4200 W, what are its volt-amperes and its power factor?

19. An 1100-VA, 60-Hz, 220-V load has a power factor of 0.85, lagging. Find the value of parallel capacitance that will correct the load power factor to unity.

20. A 3000-W, 125-V load has a power factor of 0.8, lagging. What value of parallel reactance will correct the power factor to 0.95, lagging?

21. A balanced three-phase system has a line-to-line voltage of 440. One load, which is wye-connected, consumes 2000 W at a power factor of 0.9, leading; and a second, delta-connected load, consumes 3500 watts at 0.8 power factor, lagging. Find the line current and the power factor of the system.

22. A three-phase generator supplies 9600 VA to a delta-connected load with phase voltages of 255-V and a power factor of 0.8, lagging. Find the generator line currents, the generator line voltages, the load currents, the net power, and the net reactive power.

23. It is desired to correct the power factor of a three-phase 440 V, 45-kW, 55-kVA inductive load to unity. If the load is wye-connected, what reactances should be connected in parallel with each load impedance?

24. A balanced three-phase load consists of impedances $3 + 2j \ \Omega$ connected in wye and impedances $4 - j \ \Omega$ in delta. What is the power factor of the load, and what are the equivalent net wye load impedances?

25. A three-phase 60-Hz delta-connected load has 220-V line voltages and a power factor of 0.8, lagging. If the line currents are each 15 amperes, what value of capacitor should be placed between each pair of lines to correct the power factor to 0.95, lagging?

Advanced Problems

Sinusoidal Power

26. An element is said to be *passive* if for sinusoidal voltage and current of any frequency, the average electrical power flow is always *into* the element. Show that an element with impedance $Z(s)$ is passive if and only if

$$\text{Re}[Z(s = j\omega)] \geq 0$$

for all ω.

27. There are other numbers of balanced circuits besides three that have the property of constant net power flow. Find a two-phase system for which the total instantaneous power flow is constant. How many wires are necessary for power transmission in this system?

Wattmeters

28. A wattmeter is a meter that indicates average power. Mechanical wattmeters involve two coils, one for voltage, the other for current. The force developed between the coils is proportional to the voltage-current product. The average force, indicated by the meter scale, is the average of the voltage-current product, which is the average power.

   Wattmeter symbols are shown in (a). The voltage-sensing terminals are labeled *V* and are connected as one would connect a voltmeter, across the voltage to be sensed. The current sensing terminals *I* are connected as an ammeter is connected, in series with the current to be sensed.

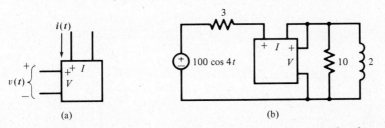

(a)   (b)

   The plus signs beside a voltage and a current terminal indicate the senses of positive voltage and current that produce an upward deflection of the meter pointer. For voltage and current

$$v(t) = A \cos (\omega t + \theta) \qquad i(t) = B \cos (\omega t + \phi)$$

with the reference senses shown in (a), the wattmeter will indicate the quantity

$$P = \frac{AB}{2} \cos (\theta - \phi)$$

   What will be the reading of the wattmeter in the network of (b)?

29. A wattmeter may be used to measure the power flow from a source in a three-phase system, as shown in (a) for sources in wye. For a perfectly balanced power system, the total power flow is three times this reading.

Show that in a balanced three-phase system, the wattmeter connection of (b) will produce a wattmeter reading of $\frac{3}{2} P \pm (\sqrt{3}/2)Q$, where P and Q are the average and the reactive power of one source, and where the algebraic sign depends on the phase rotation.

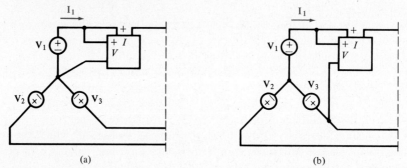

(a)   (b)

30. In an unbalanced three-phase power system, the total power flow could be measured by using three wattmeters, each responding to the power flow out of one of the sources, as indicated in (a). In terms of rms phasors, the total power is

$$P = |\mathbf{V}_1|\,|\mathbf{I}_1|\cos\theta_1 + |\mathbf{V}_2|\,|\mathbf{I}_2|\cos\theta_2 + |\mathbf{V}_3|\,|\mathbf{I}_3|\cos\theta_3$$

where

$$\theta_1 = \angle\mathbf{V}_1 - \angle\mathbf{I}_1$$

$$\theta_2 = \angle\mathbf{V}_2 - \angle\mathbf{I}_2$$

$$\theta_3 = \angle\mathbf{V}_3 - \angle\mathbf{I}_3$$

Show that the commonly used arrangement of *two* wattmeters in (b) produces readings whose sum is the total power flow out of the sources.

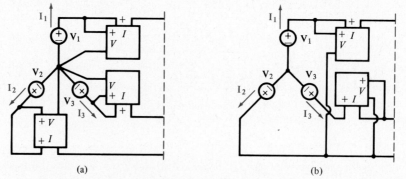

(a)   (b)

In this two-wattmeter method of power measurement, one of the wattmeter readings may be negative, meaning that the magnitude of its reading should be subtracted from that of the other wattmeter to give the total power flow.

Chapter Ten
Trigonometric Fourier Series

The series arranged according to sines or cosines of multiple arcs are always convergent; that is to say, on giving to the variable any value whatever that is not imaginary, the sum of the terms converges more and more to a single fixed limit, which is the value of the developed function.

Joseph Fourier
From *Analytical Theory of Heat*
Cambridge, 1878

10.1 PREVIEW

A Fourier series is a sum of harmonically related sinusoidal functions. Just as a function's Taylor series expansion is useful for many purposes, Fourier series expansions of periodic functions are highly useful for network analysis.

In this chapter are presented ideas of periodicity, series convergence, series approximation, and the use of series for solution of differential equations and the electrical networks they represent. For the trigonometric series, these are quickly understood and easily visualized. The trigonometric series is carefully developed so that when transitions are made to the complex exponential form of the series and to the Fourier transform in the next chapter, there will be considerably fewer new concepts to master at once.

Trigonometric Fourier series expansion is developed from the orthogonality relations. The mean square error (MSE) between two periodic functions is defined, and it is shown that a truncated trigonometric series gives an approximation with minimum MSE if its coefficients are the Fourier coefficients. MSE relations are also used to derive Parseval's relation and the expression for MSE of a truncated series approximation.

The Dirichlet conditions, convergence of the Fourier series, and the behavior of a truncated Fourier series approximation at a point of discontinuity of the original function are discussed. Series convergence is outlined but not proved, since a proof involves more advanced concepts than are appropriate at this stage.

Fourier series methods for the solution of differential equations with periodic driving functions and networks with periodic sources are examined, preparing a foundation for the more efficient solution methods in the next chapter.

When you complete this chapter, you should know—

1. the meaning of periodicity and properties of periodic functions;
2. how to calculate trigonometric Fourier series for periodic functions;
3. conditions for which a Fourier series converges;
4. how to apply Parseval's relation;
5. what the Gibbs phenomenon is;
6. how to use various properties, including those for even and odd functions, amplitude scaling, level shift, scale change, time shift, and derivative and integral relations, in finding Fourier series;
7. the Fourier series for such functions as square waves, sawtooth waves, half- and full-wave rectified sinusoids and triangular waves;
8. how to use trigonometric Fourier series to solve linear, time-invariant differential equations with periodic driving functions;
9. how to convert between quadrature, cosine and angle, and sine and angle forms of the series;
10. how to use trigonometric Fourier series to find network response to periodic sources.

10.2 PERIODIC FUNCTIONS AND TRIGONOMETRIC SERIES

10.2.1 Periodicity

Periodic functions repeat over and over, as do the examples of Figure 10-1. A function $f(t)$ is periodic with period $T > 0$ if

$$f(t) = f(t + T)$$

To specify a periodic function, it is only necessary to sketch or give a functional description for one period. Normally, the period of a periodic function means the smallest interval of repetition, T. Of course, if a function repeats in an interval T, it repeats in intervals $2T$, $3T$, and so on, also. One important periodic function that has no smallest interval of repetition is the constant function,

$$f = A, \text{ a constant}$$

shown in Figure 10-1(f). The constant is periodic with any interval of repetition. The sinusoidal functions $A \cos \omega t$ and $B \sin \omega t$ are periodic with period

$$T = \frac{2\pi}{\omega}$$

In terms of their period, these functions are

$$A \cos \frac{2\pi t}{T} \quad \text{and} \quad B \sin \frac{2\pi t}{T}$$

In every interval t of T units, the argument of these sinusoidal functions changes by 2π radians.

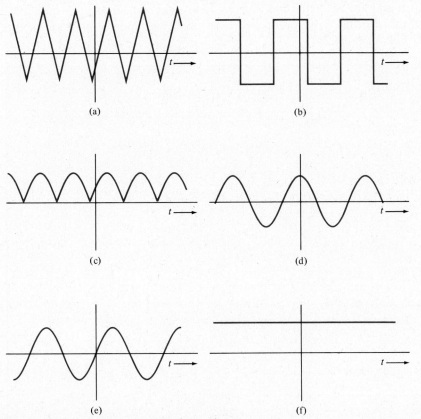

(a)

(b)

(c)

(d)

(e)

(f)

Figure 10-1   Examples of periodic functions.

Sums of sinusoidal functions of harmonically related frequencies are periodic. The most general such function is the trigonometric series

$$f(t) = d_0 + \left(a_1 \cos \frac{2\pi t}{T} + b_1 \sin \frac{2\pi t}{T} \right) + \left(a_2 \cos \frac{4\pi t}{T} + b_2 \sin \frac{4\pi t}{T} \right)$$

$$+ \cdots + \left(a_n \cos \frac{n2\pi t}{T} + b_n \sin \frac{n2\pi t}{T} \right) + \cdots$$

$$= d_0 + \sum_{n=1}^{\infty} \left(a_n \cos \frac{n2\pi t}{T} + b_n \sin \frac{n2\pi t}{T} \right)$$

where $d_0, a_1, b_1, a_2, b_2, \ldots$, are constants. This function consists of a constant term, d_0, a general sinusoid

$$a_1 \cos \omega_0 t + b_1 \sin \omega_0 t$$

of *fundamental* radian frequency

$$\omega_0 = \frac{2\pi}{T}$$

a general sinusoid of twice the fundamental frequency, and so on. The individual sinusoidal terms in the series are called *harmonics*; the fundamental term is also called the *first harmonic*.

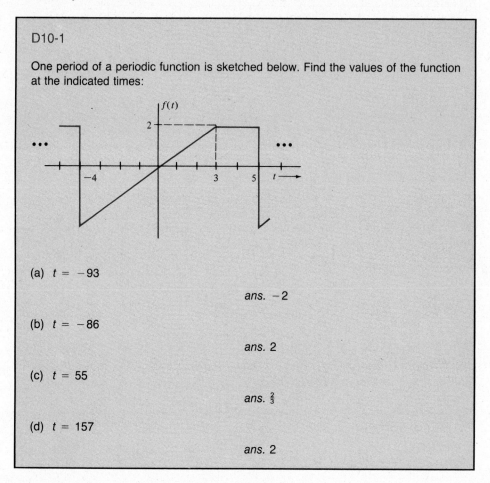

D10-1

One period of a periodic function is sketched below. Find the values of the function at the indicated times:

(a) $t = -93$

ans. -2

(b) $t = -86$

ans. 2

(c) $t = 55$

ans. $\frac{2}{3}$

(d) $t = 157$

ans. 2

10.2.2 Orthogonality Relations

The average of a periodic function is its average over one or any integral number of cycles. If $f(t)$ is periodic with period T,

$$\begin{bmatrix} \text{average value} \\ \text{of } f(t) \end{bmatrix} = \langle f(t) \rangle = \frac{1}{T} \int_T f(t) \, dt = \frac{1}{nT} \int_{nT} f(t) \, dt$$

where n is a positive integer. Instead of giving specific limits to the integrals involved in averages, the range of the limits may be indicated below the integral sign, as it is above. For a periodic function with period T, where the integration is started is of no consequence when the integration is over an interval of T or a multiple of T

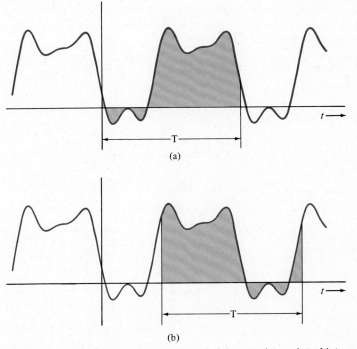

Figure 10-2 Illustrating independence of the starting point of integration over a period of a periodic function.

units. This property is illustrated in Figure 10-2. Some possible limits on integration over an interval T are as follows:

$$\int_T f(t) \, dt = \int_0^T f(t) \, dt = \int_{-(T/2)}^{(T/2)} f(t) \, dt = \int_{1.5}^{1.5+T} f(t) \, dt$$

The easiest set of limits for each situation should be chosen in practice.

The integral over a period or any integer number of periods of a sinusoidal function is zero. It is convenient to place this result in the following form:

$$\int_T \cos \frac{n2\pi t}{T} \, dt = 0, \quad n = 1, 2, 3, \ldots \tag{10-1}$$

$$\int_T \sin \frac{n2\pi t}{T} \, dt = 0, \quad n = 1, 2, 3, \ldots \tag{10-2}$$

These are two of a set of five equations known as the *orthogonality relations* for periodic sinusoidal functions.

A more interesting orthogonality relation is

$$\int_T \cos \frac{m2\pi t}{T} \sin \frac{n2\pi t}{T} \, dt = 0 \tag{10-3}$$

for any integers m and n. This is to say that the integral of the product of harmonically related sine and cosine functions, over a common interval of repetition T, is zero. To show (10-3), use the trigonometric identity

$$\cos x \sin y = \tfrac{1}{2} \sin (x + y) - \tfrac{1}{2} \sin (x - y)$$

which gives

$$\int_T \cos \frac{m2\pi t}{T} \sin \frac{n2\pi t}{T} \, dt$$

$$= \frac{1}{2} \int_T \sin \left[\frac{(m + n)2\pi t}{T} \right] dt - \frac{1}{2} \int_T \sin \left[\frac{(m - n)2\pi t}{T} \right] dt = 0$$

A similar relation is

$$\int_T \cos \frac{m2\pi t}{T} \cos \frac{n2\pi t}{T} \, dt = \begin{cases} 0 & m \neq n \\ \dfrac{T}{2} & m = n \end{cases} \tag{10-4}$$

for integers m and n. Using

$$\cos x \cos y = \tfrac{1}{2} \cos (x + y) + \tfrac{1}{2} \cos (x - y)$$

then

$$\int_T \cos \frac{m2\pi t}{T} \cos \frac{n2\pi t}{T} \, dt$$

$$= \frac{1}{2} \int_T \cos \left[\frac{(m + n)2\pi t}{T} \right] dt + \frac{1}{2} \int_T \cos \left[\frac{(m - n)2\pi t}{T} \right] dt$$

Each of the terms above, being the integral of a sinusoid over an integer number of cycles, is zero, provided $m \neq n$. For $m = n$, (10-4) becomes

$$\int_T \cos^2 \frac{n2\pi t}{T} \, dt = \frac{1}{2} \int_T \cos \frac{n4\pi t}{T} \, dt + \frac{1}{2} \int_T \cos (0) \, dt = \frac{T}{2}$$

Similarly,

$$\int_T \sin \frac{m2\pi t}{T} \sin \frac{n2\pi t}{T} \, dt = \begin{cases} 0 & m \neq n \\ \dfrac{T}{2} & m = n \end{cases} \tag{10-5}$$

for integers m and n.

PERIODIC FUNCTIONS AND TRIGONOMETRIC SERIES

A function $f(t)$ is *periodic* if

$$f(t + T) = f(t)$$

for some positive constant T. The smallest interval of repetition T is termed the *period* of a periodic function. The constant function is periodic but has no smallest interval of repetition. Sinusoidal functions

$$f(t) = A \cos(\omega t + \theta)$$

are periodic, with period

$$T = 2\pi/\omega$$

The *average* of a periodic function $f(t)$ is given by

$$\langle f(t) \rangle = \frac{1}{T} \int_T f(t)\, dt$$

The *orthogonality relations* are as follows:

$$\int_T \cos \frac{n2\pi t}{T}\, dt = 0 \tag{10-1}$$

$$\int_T \sin \frac{n2\pi t}{T}\, dt = 0 \tag{10-2}$$

$$\int_T \cos \frac{m2\pi t}{T} \sin \frac{n2\pi t}{T}\, dt = 0 \tag{10-3}$$

$$\int_T \cos \frac{m2\pi t}{T} \cos \frac{n2\pi t}{T}\, dt = \begin{cases} 0 & m \neq n \\ (T/2) & m = n \end{cases} \tag{10-4}$$

$$\int_T \sin \frac{m2\pi t}{T} \sin \frac{n2\pi t}{T}\, dt = \begin{cases} 0 & m \neq n \\ (T/2) & m = n \end{cases} \tag{10-5}$$

where m and n are integers.

10.3 FOURIER SERIES CALCULATION

10.3.1 Coefficients of the Series if it Converges

If it is possible to expand a periodic function $f(t)$ into a trigonometric series,

$$f(t) = d_0 + \sum_{n=1}^{\infty} \left(a_n \cos \frac{n2\pi t}{T} + b_n \sin \frac{n2\pi t}{T} \right) \tag{10-6}$$

the orthogonality relations show how the series coefficients must be related to the function. If the function and the series are everywhere equal, the series is said to *converge* to the function. Integrating both sides of the above,

$$\int_T f(t)\, dt = \int_T d_0\, dt + \sum_{n=1}^{\infty} \left[a_n \int_T \cos \frac{n2\pi t}{T}\, dt + b_n \int_T \sin \frac{n2\pi t}{T}\, dt \right]$$

The sinusoids integrate to zero, leaving

$$\int_T f(t)\, dt = T d_0$$

$$d_0 = \frac{1}{T} \int_T f(t)\, dt = \langle f(t) \rangle \tag{10-7}$$

The constant term in the series is the average value of the function. This result is certainly sensible; if $f(t)$ has a nonzero average value, it must be contained in d_0 because all the other terms in the series are sinusoids, with zero average value.

Multiplying both sides of the harmonic trigonometric expansion (10-6) by the first harmonic cosine function and integrating over a period T,

$$\int_T f(t) \cos \frac{2\pi t}{T} \, dt = d_0 \int_T \cos \frac{2\pi t}{T} \, dt + a_1 \int_T \cos^2 \left(\frac{2\pi t}{T}\right) dt$$

$$+ b_1 \int_T \sin \frac{2\pi t}{T} \cos \frac{2\pi t}{T} \, dt + a_2 \int_T \cos \frac{4\pi t}{T} \cos \frac{2\pi t}{T} \, dt$$

$$+ b_2 \int_T \sin \frac{4\pi t}{T} \cos \frac{2\pi t}{T} \, dt + \cdots$$

$$= d_0 \cdot 0 + a_1 \left(\frac{T}{2}\right) + b_1 \cdot 0 + a_2 \cdot 0 + b_2 \cdot 0 + \cdots$$

gives

$$a_1 = \frac{2}{T} \int_T f(t) \cos \frac{2\pi t}{T} \, dt$$

using the orthogonality relations. Multiplying by the second harmonic cosine and integrating over T similarly gives

$$a_2 = \frac{2}{T} \int_T f(t) \cos \frac{4\pi t}{T} \, dt$$

and in general

$$a_n = \frac{2}{T} \int_T f(t) \cos \frac{n2\pi t}{T} \, dt \qquad (10\text{-}8)$$

Similarly,

$$b_n = \frac{2}{T} \int_T f(t) \sin \frac{n2\pi t}{T} \, dt \qquad (10\text{-}9)$$

Relations (10-7), (10-8), and (10-9) give the series coefficients, provided the series converges, from the function. When the coefficients of a trigonometric series are chosen in this way, the series is termed a *Fourier series* for the periodic function $f(t)$. Joseph Fourier [1768–1830] was Professor of Mathematics at the Ecole Polytechnique in Paris. His solution of certain heat flow problems involved expanding arbitrary periodic functions into harmonic trigonometric series.

The constant term d_0 is sometimes written as $(a_0/2)$, so that the formula for a_n with $n = 0$ also applies to this term. The Fourier series then has the form

$$f(t) = \frac{a_0}{2} + \sum_{n=1}^{\infty} \left(a_n \cos \frac{n2\pi t}{T} + b_n \sin \frac{n2\pi t}{T}\right)$$

As a practical matter, however, it is seldom that the general integral for a_n can be used to obtain a_0. A separate integration usually must be done for the constant term in the Fourier series.

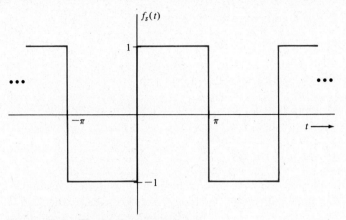

Figure 10-3   The basic square wave.

10.3.2 Fourier Series for a Square Wave

The periodic function $f_s(t)$ of Figure 10-3 is a square wave function. The specific function shown, which switches between values of plus and minus one and has period 2π seconds will be here called the *basic square wave*. Its period is $T = 2\pi$. If the basic square wave may be expanded into a Fourier series,

$$f_s(t) = d_0 + \sum_{n=1}^{\infty} (a_n \cos nt + b_n \sin nt)$$

its Fourier coefficients must be as follows:

$$d_0 = \frac{1}{2\pi} \int_{2\pi} f_s(t)\, dt = \frac{1}{2\pi} \int_{-\pi}^{0} (-1)\, dt + \frac{1}{2\pi} \int_{0}^{\pi} 1\, dt = -\frac{1}{2} + \frac{1}{2} = 0$$

$$a_n = \frac{2}{2\pi} \int_{2\pi} f_s(t) \cos nt\, dt = \frac{1}{\pi} \int_{-\pi}^{0} (-1) \cos nt\, dt + \frac{1}{\pi} \int_{0}^{\pi} (1) \cos nt\, dt$$

$$= \frac{1}{\pi} \left.\frac{-\sin nt}{n}\right|_{-\pi}^{0} + \frac{1}{\pi} \left.\frac{\sin nt}{n}\right|_{0}^{\pi} = \frac{1}{\pi}(0 - 0) + \frac{1}{\pi}(0 - 0) = 0$$

$$b_n = \frac{2}{2\pi} \int_{2\pi} f_s(t) \sin nt\, dt = \frac{1}{\pi} \int_{-\pi}^{0} (-1) \sin nt\, dt + \frac{1}{\pi} \int_{0}^{\pi} (1) \sin nt\, dt$$

$$= \frac{1}{\pi} \left.\frac{\cos nt}{n}\right|_{-\pi}^{0} + \frac{1}{\pi} \left.\frac{-\cos nt}{n}\right|_{0}^{\pi}$$

$$= \frac{1}{\pi n} [1 - \cos(-n\pi)] + \frac{1}{\pi n} [-\cos n\pi + 1]$$

$$= \frac{2}{\pi n} (1 - \cos n\pi)$$

For n even, $\cos n\pi = 1$. For n odd, $\cos n\pi = -1$, giving

$$
b_n = \begin{cases} 0, & n \text{ even} \\ \dfrac{4}{n\pi}, & n \text{ odd} \end{cases}
$$

The Fourier series for the basic square wave, assuming it converges, is thus

$$
f_s(t) = d_0 + \sum_{n=1}^{\infty} (a_n \cos nt + b_n \sin nt) = \sum_{\substack{n=1 \\ n \text{ odd}}}^{\infty} \frac{4}{n\pi} \sin nt
$$

In Figure 10-4 are plotted the first several terms of this series. As higher harmonics are added, the sum looks more and more like the square wave, so it appears that the basic square wave can indeed be expanded into a Fourier series.

10.3.3 Fourier Series for a Sawtooth Wave

As another example of Fourier series calculation, consider the sawtooth wave of Figure 10-5. The average value of this function is the area under one cycle divided by the period

$$
d_0 = \langle f_v(t) \rangle = \tfrac{1}{2}
$$

In the period between $t = 0$ and $t = 1$, this function is $f_v(t) = t$. An integration to find the average value,

$$
d_0 = \frac{1}{1} \int_0^1 t \, dt = \left. \frac{t^2}{t} \right|_0^1 = \frac{1}{2}
$$

obtains the same result.

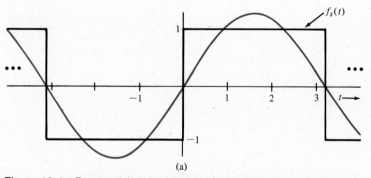

(a)

Figure 10-4   Representing the basic square wave with a Fourier series.
(a) First harmonic approximation.
(b) Approximation through the third harmonic.
(c) Approximation through the fifth harmonic.
(d) Approximation through the ninth harmonic.
(e) Approximation through the 19th harmonic.

Figure 10-4   (*Continued*)

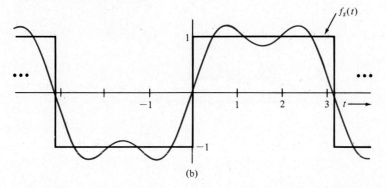

(b)

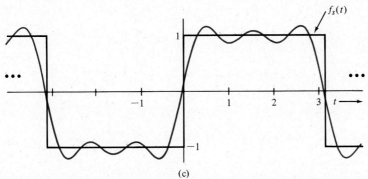

(c)

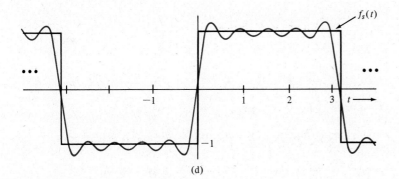

(d)

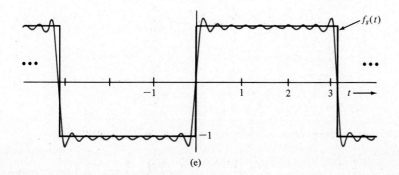

(e)

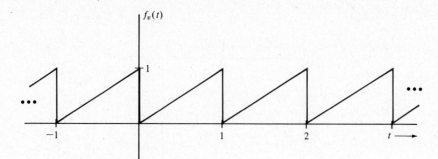

Figure 10-5 A sawtooth wave.

The other coefficients of the sawtooth wave's Fourier series are found as follows:

$$a_n = \frac{2}{1} \int_0^1 t \cos \frac{n2\pi t}{1} \, dt = 2\left[\frac{\cos n2\pi t}{4n^2\pi^2} + \frac{t \sin n2\pi t}{n2\pi} \right]_0^1$$

$$= 2\left[\frac{1-1}{4^2\pi^2} + \frac{0-0}{n2\pi} \right] = 0$$

Similarly,

$$b_n = \frac{2}{1} \int_0^1 t \sin n2\pi t \, dt = 2\left[\frac{\sin n2\pi t}{4n^2\pi^2} - \frac{t \cos n2\pi t}{n2\pi} \right]_0^1$$

$$= 2\left[\frac{0-0}{4n^2\pi^2} - \frac{1-0}{n2\pi} \right] = \frac{-1}{n\pi}$$

so that

$$f_v(t) = \frac{1}{2} + \sum_{n=1}^{\infty} \frac{-1}{n\pi} \sin n2\pi t$$

Partial sums of the series are plotted in Figure 10-6, and the series appears to approach convergence as more and more terms are added.

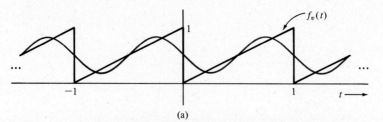

(a)

Figure 10-6 Representing a sawtooth wave with a Fourier series.
(a) Constant and first harmonic approximation.
(b) Approximation through the second harmonic.
(c) Approximation through the third harmonic.
(d) Approximation through the fifth harmonic.
(e) Approximation through the 16th harmonic.

Figure 10-6   (*Continued*)

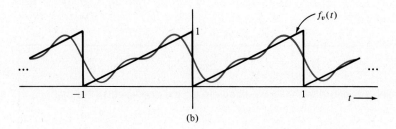

(b)

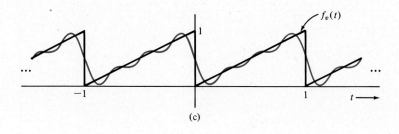

(c)

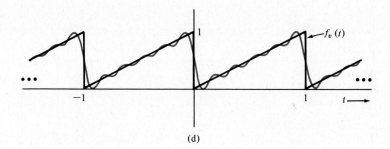

(d)

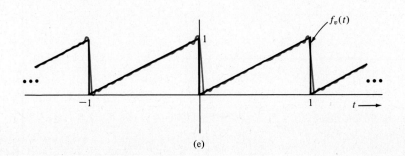

(e)

The two series calculated so far have had zero coefficients for the cosine terms. This simplification is due to the symmetries of the simple functions involved and will be explored in Section 10.5.

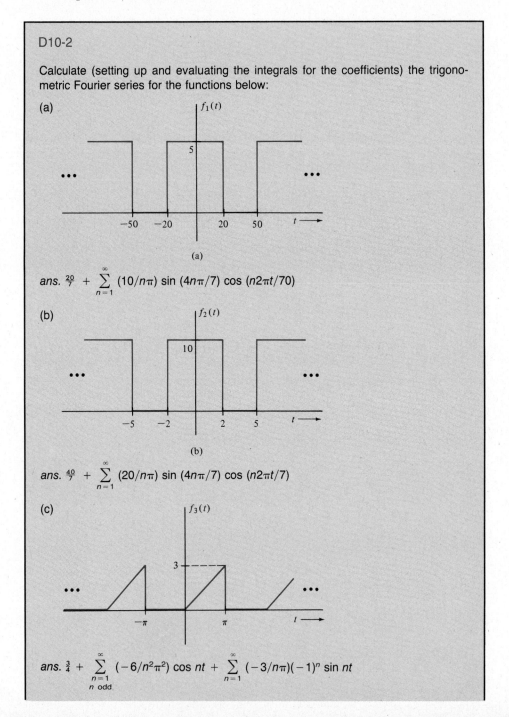

D10-2

Calculate (setting up and evaluating the integrals for the coefficients) the trigonometric Fourier series for the functions below:

(a)

(a)

ans. $\frac{20}{7} + \sum\limits_{n=1}^{\infty} (10/n\pi) \sin(4n\pi/7) \cos(n2\pi t/70)$

(b)

(b)

ans. $\frac{40}{7} + \sum\limits_{n=1}^{\infty} (20/n\pi) \sin(4n\pi/7) \cos(n2\pi t/7)$

(c)

ans. $\frac{3}{4} + \sum\limits_{\substack{n=1 \\ n\ \text{odd}}}^{\infty} (-6/n^2\pi^2) \cos nt + \sum\limits_{n=1}^{\infty} (-3/n\pi)(-1)^n \sin nt$

(d)

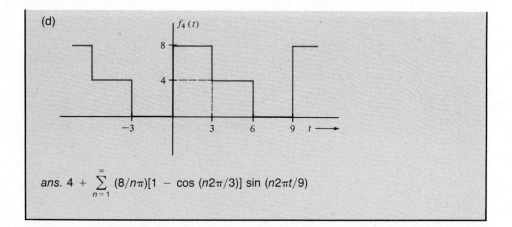

ans. $4 + \sum\limits_{n=1}^{\infty} (8/n\pi)[1 - \cos (n2\pi/3)] \sin (n2\pi t/9)$

FOURIER SERIES CALCULATION

The orthogonality relations may be used to show that if a function $f(t)$ equals a harmonic trigonometric series,

$$f(t) = d_0 + \sum_{n=1}^{\infty} \left(a_n \cos \frac{n2\pi t}{T} + b_n \sin \frac{n2\pi t}{T} \right) \qquad (10\text{-}6)$$

then the series coefficients are related to the function by

$$d_0 = \frac{1}{T} \int_T f(t) \, dt = \langle f(t) \rangle \qquad (10\text{-}7)$$

$$a_n = \frac{2}{T} \int_T f(t) \cos \frac{n2\pi t}{T} \, dt = 2\left\langle f(t) \ \cos \frac{n2\pi t}{T} \right\rangle \qquad (10\text{-}8)$$

$$b_n = \frac{2}{T} \int_T f(t) \sin \frac{n2\pi t}{T} \, dt = 2\left\langle f(t) \sin \frac{n2\pi t}{T} \right\rangle \qquad (10\text{-}9)$$

A harmonic series with coefficients chosen in this manner is termed the *Fourier series* for the function $f(t)$.

Some basic Fourier series are as follows:

Basic Square Wave

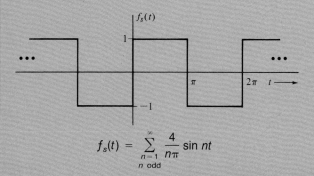

$$f_s(t) = \sum_{\substack{n=1 \\ n \text{ odd}}}^{\infty} \frac{4}{n\pi} \sin nt$$

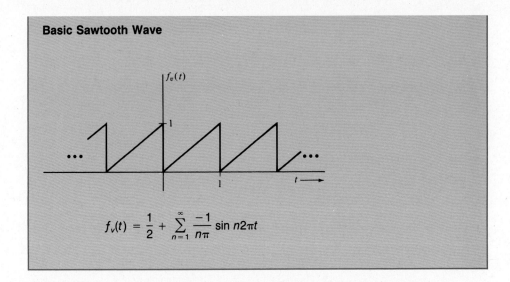

Basic Sawtooth Wave

$$f_v(t) = \frac{1}{2} + \sum_{n=1}^{\infty} \frac{-1}{n\pi} \sin n2\pi t$$

10.4 SERIES CONVERGENCE AND PARSEVAL'S RELATION

10.4.1 Minimization of MSE by Truncated Series

The mean square error (MSE) between two periodic functions, $f(t)$ and $g(t)$, each with period T, is

$$\text{MSE} = \frac{1}{T} \int_T [f(t) - g(t)]^2 \, dt$$

The MSE is never negative, and if $f(t) = g(t)$, MSE = 0. The amount of MSE is a good measure of error between two periodic functions. Because the difference between the functions is squared, positive errors cannot be cancelled by negative ones. The square relationship also tends to give greater weight to large errors than to small ones. Other measures of the error between two functions can be used, but MSE is particularly attractive because it gives very useful results relatively easily.

Suppose a periodic function $f(t)$ is approximated by a trigonometic series truncated beyond the Nth harmonic:

$$g_N(t) = \delta_0 + \sum_{n=1}^{N} \left(\alpha_n \cos \frac{n2\pi t}{T} + \beta_N \sin \frac{n2\pi t}{T} \right)$$

The coefficients of the approximating truncated series are not necessarily the Fourier coefficients, so they are here given different symbols.

The MSE between the function and the approximating series is

$$\text{MSE} = \frac{1}{T} \int_T [f(t) - G_N(t)]^2 \, dt$$

(10-10)

$$= \frac{1}{T} \int_T f^2(t) \, dt - \frac{2}{T} \int_T f(t)g_N(t) \, dt + \frac{1}{T} \int_T g_N^2(t) \, dt$$

Now

$$\frac{2}{T} \int_T f(t) g_N(t) \, dt$$

$$= \frac{2}{T} \int_T f(t) \left[\delta_0 + \sum_{n=1}^{N} \left(\alpha_n \cos \frac{n 2\pi t}{T} + \beta_n \sin \frac{n 2\pi t}{T} \right) \right] dt = 2\delta_0 \left[\frac{1}{T} \int_T f(t) \, dt \right]$$

$$+ \alpha_1 \left[\frac{2}{T} \int_T f(t) \cos \frac{2\pi t}{T} \, dt \right] + \beta_1 \left[\frac{2}{T} \int_T f(t) \sin \frac{2\pi t}{T} \, dt \right] + \cdots$$

$$+ \alpha_N \left[\frac{2}{T} \int_T f(t) \cos \frac{N 2\pi t}{T} \, dt \right] + \beta_N \left[\frac{2}{T} \int_T f(t) \cos \frac{N 2\pi t}{T} \, dt \right]$$

$$= 2\delta_0 d_0 + \alpha_1 a_1 + \beta_1 b_1 + \cdots + \alpha_N a_N + \beta_N b_N$$

where $a_0, a_1, a_2, \ldots, a_N, b_1, b_2, \ldots, b_N$ are the Fourier coefficients for $f(t)$. The third term in (10-10) is

$$\frac{1}{T} \int_T g_N^2(t) \, dt = \frac{1}{T} \int_T \left[\delta_0 + \sum_{n=1}^{N} \left(\alpha_n \cos \frac{n 2\pi t}{T} + \beta_n \sin \frac{n 2\pi t}{T} \right) \right]$$

$$\left[\delta_0 + \sum_{n=1}^{N} \left(\alpha_n \cos \frac{n 2\pi t}{T} + \beta_n \sin \frac{n 2\pi t}{T} \right) \right] dt$$

$$= \delta_0^2 + \frac{1}{2} \alpha_1^2 + \frac{1}{2} \beta_1^2 + \cdots + \frac{1}{2} \alpha_N^2 + \frac{1}{2} \beta_N^2$$

giving

$$\text{MSE} = \frac{1}{T} \int_T f^2(t) \, dt - 2\delta_0 d_0 - \alpha_1 a_1 - \beta_1 b_1 + \cdots - \alpha_N a_N - \beta_1 b_N$$

$$+ \delta_0^2 + \frac{1}{2} \alpha_1^2 + \frac{1}{2} \beta_1^2 + \cdots + \frac{1}{2} \alpha_N^2 + \frac{1}{2} \beta_N^2 \qquad (10\text{-}11)$$

The values of $\delta_0, \alpha_1, \beta_1, \ldots, \alpha_N, \beta_N$ that give minimal MSE may be found by inspection once the MSE expression is placed in a certain form. Completing the square by adding and subtracting

$$d_0^2 + \tfrac{1}{2} a_1^2 + \tfrac{1}{2} b_1^2 + \cdots + \tfrac{1}{2} \alpha_N^2 + \tfrac{1}{2} \beta_N^2$$

gives

$$\text{MSE} = \frac{1}{T} \int_T f^2(t) \, dt + \frac{1}{2} \left[2(\delta_0 - d_0)^2 + (\alpha_1 - a_1)^2 + (\beta_1 - b_1)^2 \right.$$

$$+ \cdots + (\alpha_N - a_N)^2 + (\beta_N - b_N)^2]$$

$$- d_0^2 - \frac{1}{2} a_1^2 - \frac{1}{2} b_1^2 - \cdots - \frac{1}{2} a_N^2 - \frac{1}{2} b_N^2$$

The values of the integral and the Fourier coefficients are fixed numbers. Taking $\delta_0, \alpha_1, \beta_1, \ldots, \alpha_N, \beta_N$ to be adjustable, the MSE is minimized by choosing

$$\delta_0 = d_0,$$

$$\alpha_n = a_n \qquad n = 1, 2, \ldots, N$$

$$\beta_n = b_n \qquad n = 1, 2, \ldots, N$$

That is, the MSE is minimized when the truncated series coefficients are chosen to be the Fourier coefficients. The minimum error is then given by

$$\text{MSE} = \frac{1}{T} \int_T f^2(t)\, dt - d_0^2 - \frac{1}{2} a_1^2 - \frac{1}{2} b_1^2 - \cdots - \frac{1}{2} a_N^2 - \frac{1}{2} b_N^2$$

$$= \frac{1}{T} \int_T f^2(t)\, dt - d_0^2 - \frac{1}{2} \sum_{n=1}^{N} (a_n^2 + b_n^2) \tag{10-12}$$

That the choice of Fourier coefficients gives minimum MSE can be shown also by equating the partial derivatives of MSE (10-11) with respect to $\delta_0, \alpha_1, \beta_1, \ldots, \alpha_N, \beta_N$ to zero:

$$\frac{\partial \text{MSE}}{\partial \delta_0} = -2d_0 + 2\delta_0 = 0$$

$$\frac{\partial \text{MSE}}{\partial \alpha_1} = -a_1 + \alpha_1 = 0$$

$$\frac{\partial \text{MSE}}{\partial \beta_1} = -b_1 + \beta_1 = 0$$

$$\vdots$$

$$\frac{\partial \text{MSE}}{\partial \beta_N} = -b_N + \beta_N = 0$$

With this approach, it is also necessary to verify that the solution obtained is a minimum.

10.4.2 Convergence of the Series

A periodic function $f(t)$ is said to satisfy the *Dirichlet conditions* if—

1. $f(t)$ has a finite number of finite maxima and minima in each period, and
2. $f(t)$ has a finite number of finite discontinuities in each period.

Suppose $f(t)$ has period T, satisfies the Dirichlet conditions, and is approximated by a Fourier series that is truncated above the Nth harmonic:

$$f_N(t) = d_0 + \sum_{n=1}^{N} \left(a_n \cos \frac{n2\pi t}{T} + b_n \sin \frac{n2\pi t}{T} \right)$$

The MSE of the truncated series approximation is

$$\text{MSE} = \frac{1}{T} \int_T [f(t) - f_N(t)]^2 \, dt$$

It can be shown that, as higher and higher harmonics are included in the truncated series, the MSE goes to zero:

$$\underset{N \to \infty}{\text{limit}} \; \text{MSE} = 0$$

This is to say that the MSE between a periodic function satisfying the Dirichlet conditions and its infinite Fourier series is zero.

If the MSE between two functions,

$$\text{MSE} = \frac{1}{T} \int_T [f(t) - g(t)]^2 \, dt$$

is zero, the functions are not necessarily equal everywhere, however. They may differ by finite amounts at a finite number of isolated points in each period T.

Letting $f(t)$ represent a periodic function satisfying the Dirichlet conditions and $\hat{f}(t)$ its (infinite) Fourier series, it can be shown that

$$\hat{f}(t) = f(t)$$

at every point where $f(t)$ is continuous. At points of discontinuity of $f(t)$, the function and the series may differ, however. At a point where $f(t)$ is discontinuous, the series will converge to the midpoint of the discontinuity:

$$\hat{f}(t_0) = \frac{1}{2} \left[\underset{t \to t_0^-}{\text{limit}} \, f(t) + \underset{t \to t_0^+}{\text{limit}} \, f(t) \right]$$

Convergence, except possibly at points of discontinuity, is commonly indicated by a dot over the equal sign:

$$f(t) \doteq d_0 + \sum_{n=1}^{\infty} \left(a_n \cos \frac{n2\pi t}{T} + b_n \sin \frac{n2\pi t}{T} \right)$$

The dot, emphasizing that a function and its series may differ at isolated points of discontinuity, will not be included further here.

The Dirichlet conditions are not absolutely necessary for convergence of the Fourier series, but they are sufficient to ensure convergence.

D10-3

For each of the following periodic functions, determine whether or not the Dirichlet conditions are satisfied. If they are, determine the function to which the corresponding Fourier series converges.

(a) $f(t) = \tan t$

ans. Function is not everywhere finite

(b) One period of a periodic function $f(t)$ is the interval $(-2, 2]$, where

$$f(t) = \begin{cases} t & -1 \le t \le 1 \\ 0 & \text{otherwise} \end{cases}$$

$$\text{ans.} \begin{cases} t & -1 < t < 1 \\ -\frac{1}{2} & t = -1 \\ \frac{1}{2} & t = 1 \\ 0 & \text{otherwise} \end{cases}$$

(c) One period of the periodic function $f(t)$ is the interval $(0, 10]$, where

$$f(t) = \sin(1/t)$$

ans. Function has an infinite number of maxima and minima in a period

(d) $f(t) = \begin{cases} 1 & t \text{ a rational number} \\ -1 & t \text{ an irrational number} \end{cases}$

over the period $(0, 1]$

ans. Function has an infinite number of maxima and minima in a period

(e) $f(t) = \begin{cases} 10 & t \ne k \quad k \text{ an integer} \\ 20 & t = k \quad k \text{ an integer} \end{cases}$

over the period $(-5, 5]$

ans. 10 everywhere

10.4.3 Parseval's Relation

When a function $f(t)$ satisfying the Dirichlet conditions is represented by a Fourier series, the MSE between the function and the (infinite) series is zero. From (10-10) with $N = \infty$,

$$\text{MSE} = \frac{1}{T} \int_T f^2(t)\, dt - d_0^2 - \frac{1}{2} \sum_{n=1}^{\infty} (a_n^2 + b_n^2) = 0$$

$$\frac{1}{T} \int_T f^2(t)\, dt = d_0^2 + \frac{1}{2} \sum_{n=1}^{\infty} (a_n^2 + b_n^2) \tag{10-13}$$

This result, known as *Parseval's relation*, expresses the average of the square of the periodic function in terms of its Fourier coefficients.

For example, the sawtooth wave,

$$f_v(t) = \frac{1}{2} + \sum_{n=1}^{\infty} \frac{1}{n\pi} \sin n2\pi t$$

has average square given by Parseval's relation as follows:

$$\frac{1}{1} \int_0^1 f_v^2(t)\, dt = \left(\frac{1}{2}\right)^2 + \frac{1}{2} \sum_{n=1}^{\infty} \left(\frac{1}{n\pi}\right)^2 = \frac{1}{4} + \frac{1}{2\pi^2} \sum_{n=1}^{\infty} \frac{1}{n^2}$$

Using the relation

$$\sum_{n=1}^{\infty} \frac{1}{n^2} = \frac{\pi^2}{6}$$

$$\int_0^1 f_v^2(t) \, dt = \frac{1}{4} + \frac{1}{2\pi^2} \left(\frac{\pi^2}{6} \right) = \frac{1}{3}$$

a result that is identical to that obtainable by directly integrating $f_v^2(t)$.

Parseval's relation shows a number of useful properties of the Fourier series, for example, that

$$\lim_{n \to \infty} a_n = 0 \qquad \text{and} \qquad \lim_{n \to \infty} b_n = 0$$

since the infinite sum on the right of (10-13) must converge. In fact, for convergence, a_n^2 and b_n^2 must eventually decrease with n faster than $(1/n)$.

The root-mean-square value (rms) of a periodic function is the square root of the average (or mean) of its square over one period. It is simply related to the function's Fourier coefficients:

$$\text{RMS}[f(t)] = \sqrt{\frac{1}{T} \int_T f^2(t) \, dt} = \sqrt{d_0^2 + \frac{1}{2} \sum_{n=1}^{\infty} (a_n^2 + b_n^2)}$$

10.4.4 MSE from Series Truncation

For a periodic function $f(t)$ satisfying the Dirichlet conditions,

$$\frac{1}{T} \int_T f^2(t) \, dt = d_0^2 + \frac{1}{2} \sum_{n=1}^{\infty} (a_n^2 + b_n^2)$$

A truncated Fourier series approximation to $f(t)$,

$$f_N(t) = d_0 + \sum_{n=1}^{N} \left(a_n \cos \frac{n2\pi t}{T} + b_n \sin \frac{n2\pi t}{T} \right)$$

has

$$\text{MSE} = \frac{1}{T} \int_T [f(t) - f_N(t)]^2 \, dt = \frac{1}{T} \int_T f^2(t) \, dt - d_0^2 - \frac{1}{2} \sum_{n=1}^{N} (a_n^2 + b_n^2)$$

according to (10-12). Substituting for the integral from (10-13),

$$\text{MSE} = d_0^2 + \frac{1}{2} \sum_{n=1}^{\infty} (a_n^2 + b_n^2) - d_0^2 - \frac{1}{2} \sum_{n=1}^{N} (a_n^2 + b_n^2)$$

$$= \frac{1}{2} \sum_{n=N+1}^{\infty} (a_n^2 + b_n^2) \qquad\qquad (10\text{-}14)$$

The MSE between a periodic function and a truncated Fourier series approximation is thus expressed as one-half the sum of the squares of the Fourier coefficients of the truncated terms.

For example, if a function with Fourier series

$$f(t) = 10 + \sum_{n=1}^{\infty} \left[\left(\frac{3}{n} \right) \cos 8nt + \left(\frac{-4}{n^2} \right) \sin 8nt \right]$$

is approximated by a series truncated beyond the fifth harmonic,

$$f_5(t) = 10 + \sum_{n=1}^{5} \left[\left(\frac{3}{n} \right) \cos 8nt + \left(\frac{-4}{n^2} \right) \sin 8nt \right]$$

the MSE between the function and the approximation will be

$$\text{MSE} = \frac{1}{T} \int_T [f(t) - f_5(t)]^2 \, dt = \frac{1}{2} \sum_{n=6}^{\infty} \left[\left(\frac{3}{n} \right)^2 + \left(\frac{-4}{n^2} \right)^2 \right]$$

$$= 0.13 + 0.095 + 0.0724 + \cdots \cong 0.92$$

10.4.5 Gibbs' Phenomenon

When the Fourier series for a discontinuous function is truncated, the approximate function exhibits an oscillatory behavior in the vicinity of the original function's discontinuity. The plots of truncated series approximations for square waves in Figure 10-7 illustrate this property, which is known as *Gibbs' phenomenon*. As the number of harmonics in the truncated series approximation to a discontinuous function is increased, the Gibbs' oscillations become more concentrated and more rapid. The overshoot of the truncated approximation, however, remains at about 9 percent of amount of the discontinuity, regardless of the (finite) number of harmonics present.

D10-4

Consider the periodic function

$$f(t) = 3 + \sum_{n=1}^{\infty} \left[\left(\frac{4}{n^2 + 1} \right) \cos 10nt + \left(\frac{1}{n^3} \right) \sin 10nt \right]$$

and an approximation to that function by the truncated series

$$f_{10}(t) = 3 + \sum_{n=1}^{10} \left[\left(\frac{4}{n^2 + 1} \right) \cos 10nt + \left(\frac{1}{n^3} \right) \sin 10nt \right]$$

(a) Find the RMS value of $f(t)$.

$$\text{ans.} \ \sqrt{9 + \frac{1}{2} \sum_{n=1}^{\infty} \left[\frac{16}{(n^2 + 1)^2} + \frac{1}{n^6} \right]} \cong \sqrt{11.96} = 3.46$$

(b) Find the average square error between the function and the approximation.

$$\text{ans.} \ \frac{1}{2} \sum_{n=11}^{\infty} \left[\left(\frac{4}{n^2 + 1} \right)^2 + \left(\frac{1}{n^3} \right)^2 \right] \cong 0.00267$$

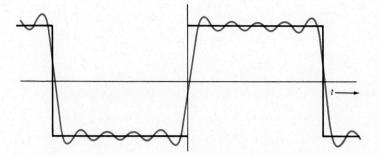

(a) Series truncated after the ninth harmonic

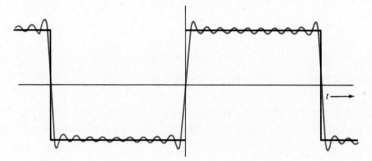

(b) Series truncated after the 19th harmonic

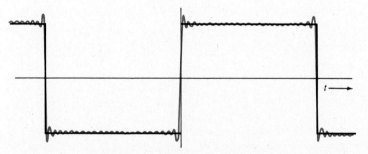

(c) Series truncated after the 49th harmonic

Figure 10-7 Gibbs' phenomenon for a square wave. Overshoot is approximately 9 percent of the amount of the discontinuity regardless of the degree of truncation.

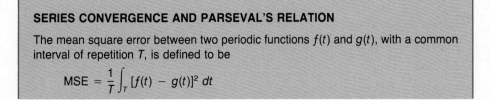

SERIES CONVERGENCE AND PARSEVAL'S RELATION

The mean square error between two periodic functions $f(t)$ and $g(t)$, with a common interval of repetition T, is defined to be

$$MSE = \frac{1}{T}\int_T [f(t) - g(t)]^2\, dt$$

When a periodic function $f(t)$ is approximated by a *truncated* harmonic trigonometric series,

$$f(t) \cong g_N(t) = \delta_0 + \sum_{n=1}^{N} \left(\alpha_n \cos \frac{n2\pi t}{T} + \beta_n \sin \frac{n2\pi t}{T} \right)$$

the MSE of the approximation is minimized if the series coefficients are chosen to be the Fourier coefficients:

$$\delta_0 = d_0 = \frac{1}{T} \int_T f(t) \, dt$$

$$\alpha_n = a_n = \frac{2}{T} \int_T f(t) \cos \frac{n2\pi t}{T} \, dt$$

$$\beta_n = b_n = \frac{2}{T} \int_T f(t) \sin \frac{n2\pi t}{T} \, dt$$

For a periodic function satisfying the *Dirichlet* conditions of having

1. a finite number of finite maxima and minima in a period
2. a finite number of finite discontinuities in each period,

the function and its (infinite) Fourier series are equal, except possibly at points of discontinuity of $f(t)$, where the series converges to the midpoint of the discontinuity.
For a function satisfying the Dirichlet conditions, *Parseval's relation* applies:

$$\frac{1}{T} \int_T f^2(t) \, dt = d_0^2 + \frac{1}{2} \sum_{n=1}^{\infty} (a_n^2 + b_n^2) \qquad (10\text{-}13)$$

The MSE between a periodic function and a truncated Fourier series approximation,

$$f(t) \cong f_N(t) = d_0 + \sum_{n=1}^{N} \left(a_n \cos \frac{n2\pi t}{T} + b_n \sin \frac{n2\pi t}{T} \right)$$

is given by

$$\text{MSE} = \frac{1}{T} \int_T [f(t) - f_N(t)]^2 \, dt = \frac{1}{2} \sum_{n=N+1}^{\infty} (a_n^2 + b_n^2) \qquad (10\text{-}14)$$

When a discontinuous function is approximated by a truncated Fourier series, the series approximation exhibits an overshoot of about 9 percent in the vicinity of the discontinuity, a behavior known as *Gibbs' phenomenon.*

10.5 SERIES PROPERTIES AND MANIPULATION

10.5.1 Even and Odd Functions

A function $f(t)$ is termed *even* if it has the property

$$f(-t) = f(t)$$

Figure 10-8(a) shows several examples of even functions. The cosine function and the constant are even. A function $f(t)$ is termed *odd* if

$$f(-t) = -f(t)$$

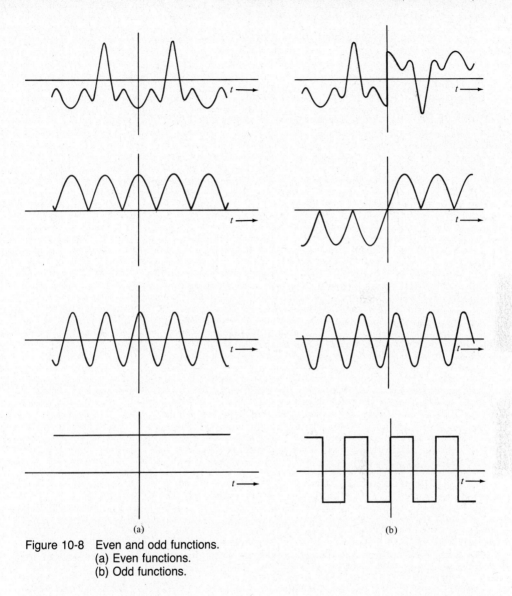

Figure 10-8 Even and odd functions.
(a) Even functions.
(b) Odd functions.

Representative odd functions are sketched in Figure 10-8(b). The sine function is odd.

Any function may be decomposed into the sum of any even part plus an odd part as follows:

$$f(t) = \underbrace{\tfrac{1}{2}[f(t) + f(-t)]}_{\text{even part}} + \underbrace{\tfrac{1}{2}[f(t) - f(-t)]}_{\text{odd part}}$$

Sums and products of even and odd functions have the following properties:

(even) + (even) = (even)   (odd) + (odd) = (odd)

(even)(even) = (even)   (even)(odd) = (odd)   (odd)(odd) = (even)

When an odd function is integrated over symmetrical limits, the result is zero:

$$\int_{-T/2}^{T/2} (\text{odd}) \, dt = 0$$

When an even function is integrated over symmetrical limits, the result is twice the value of the integral over the positive (or negative) domain:

$$\int_{-T/2}^{T/2} (\text{even}) \, dt = 2 \int_{0}^{T/2} (\text{even}) \, dt$$

The Fourier series consists of easily identified even and odd parts:

$$f(t) = \underbrace{d_0 + \sum_{n=1}^{\infty} a_n \cos \frac{n2\pi t}{T}}_{\text{even part}} + \underbrace{\sum_{n=1}^{\infty} b_n \sin \frac{n2\pi t}{T}}_{\text{odd part}}$$

When an even periodic function is expanded in a Fourier series, only the even terms appear in the series; all the b_ns are zero. Similarly, d_0 and all the a_ns are zero if an odd function is expanded into a Fourier series. If $f(t)$ is odd,

$$d_0 = \frac{1}{T} \int_{T} (\text{odd}) \, dt = \frac{1}{T} \int_{-T/2}^{T/2} (\text{odd}) \, dt = 0$$

$$a_n = \frac{2}{T} \int_{T} \underbrace{f(t)}_{\text{odd}} \underbrace{\cos \frac{n2\pi t}{T}}_{\text{even}} \, dt = \frac{2}{T} \int_{-T/2}^{T/2} (\text{odd}) \, dt = 0$$

If $f(t)$ is even,

$$b_n = \frac{2}{T} \int_{T} \underbrace{f(t)}_{\text{even}} \underbrace{\sin \frac{n2\pi t}{T}}_{\text{odd}} \, dt = \frac{2}{T} \int_{-T/2}^{T/2} (\text{odd}) \, dt = 0$$

The basic square wave, for example, is an odd function and has a Fourier series involving only sine terms. The coefficients of the cosine terms were thus found to be zero. Other, more involved symmetries than evenness and oddness result in Fourier series with other sets of coefficients zero. The basic square wave, for example, is not only odd, but is symmetric about $t = \pi/2$. (This is known as *odd half-wave symmetry*.) As a consequence, the Fourier series involves only odd harmonic terms; every other b_n is zero.

10.5.2 Amplitude Change, Level Shift, and Sums of Series

Given the series for a function, the series for the function that is a constant K times the original function has Fourier coefficients that are K times the original coefficients.

Changing the average value of a periodic function changes only the constant term, d_0, in its Fourier series. If

$$g(t) = f(t) + K$$

where K is a constant, the cosine coefficients for $g(t)$ are

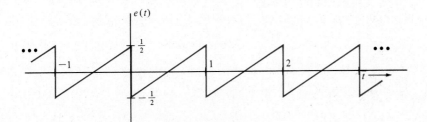

Figure 10-9   A level shifted sawtooth wave.

$$a_n = \frac{2}{T} \int_T [f(t) + K] \cos \frac{n2\pi t}{T} \, dt$$

$$= \frac{2}{T} \int_T f(t) \cos \frac{n2\pi t}{T} \, dt + \frac{2}{T} \int_T K \cos \frac{n2\pi t}{T} \, dt$$

$$= \frac{2}{T} \int_T f(t) \cos \frac{n2\pi t}{T} \, dt$$

which are those for $f(t)$, and similarly for the sine coefficients.

For the level-shifted sawtooth wave of Figure 10-9.

$$e(t) = f_v(t) - \frac{1}{2} = \sum_{n=1}^{\infty} \frac{-1}{n\pi} \sin n2\pi t$$

It is thus evident why the sawtooth has only sine harmonic terms. Except for a level change, the original basic sawtooth wave would be an odd function.

A periodic function that is the sum of two simpler periodic functions has a Fourier series that is the sum of the series for the individual functions.

10.5.3 Time Scale Change

When the time scale of a periodic function is changed, as in the examples of Figure 10-10, the Fourier coefficients do not change. Only the fundamental frequency and the harmonic frequencies change to fit the function's period. The average value, d_0, of the function is unchanged by a change of time scale. The other Fourier coefficients, the as and the bs, are averages, too, which are unaffected by the time scale:

$$a_n = \frac{2}{T} \int_T f(t) \cos \frac{n2\pi t}{T} \, dt = \left\langle 2f(t) \cos \frac{n2\pi t}{T} \right\rangle$$

$$b_n = \frac{2}{T} \int_T f(t) \sin \frac{n2\pi t}{T} \, dt = \left\langle 2f(t) \sin \frac{n2\pi t}{T} \right\rangle$$

The function $e(t)$ in Figure 10-11 is 9 times a basic sawtooth function, level shifted by 7 units. The time scale has been changed so that the function has a period of 10^{-3}. Its Fourier series is found as follows:

$$e(t) = 7 + 9\left[\frac{1}{2} + \sum_{n=1}^{\infty} \frac{-1}{n\pi} \sin \frac{n2\pi t}{10^{-3}} \right] = \frac{23}{2} + \sum_{n=1}^{\infty} \frac{-9}{n\pi} \sin (10^3 n2\pi t)$$

This time scaling property means, too, that any convenient time scale can be chosen for the calculation of the Fourier coefficients.

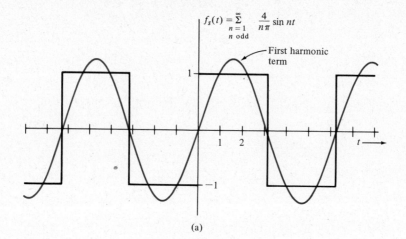

$$f_s(t) = \sum_{\substack{n=1 \\ n \text{ odd}}}^{\infty} \frac{4}{n\pi} \sin nt$$

First harmonic term

(a)

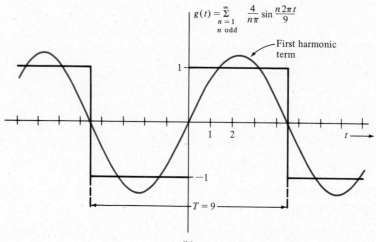

$$g(t) = \sum_{\substack{n=1 \\ n \text{ odd}}}^{\infty} \frac{4}{n\pi} \sin \frac{n2\pi t}{9}$$

First harmonic term

$T = 9$

(b)

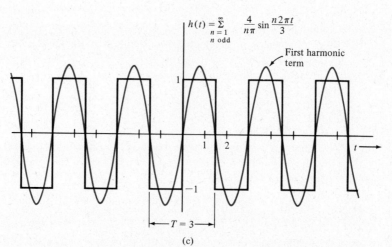

$$h(t) = \sum_{\substack{n=1 \\ n \text{ odd}}}^{\infty} \frac{4}{n\pi} \sin \frac{n2\pi t}{3}$$

First harmonic term

$T = 3$

(c)

Figure 10-10 Changing the time scale of a periodic function.

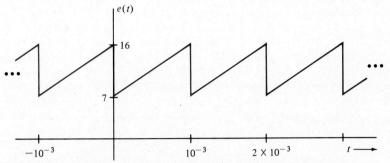

Figure 10-11   Sawtooth function with time scale and other changes.

10.5.4 Time Origin Shift

If the origin of the time scale of a periodic function is shifted, as in Figure 10-12, each harmonic sinusoid in the function's Fourier series is time shifted that amount:

$$f(t) = d_0 + \sum_{n=1}^{\infty} \left(a_n \cos \frac{n2\pi t}{T} + b_n \sin \frac{n2\pi t}{T} \right)$$

$$g(t) = f(t - \tau) = d_0 + \sum_{n=1}^{\infty} \left[a_n \cos \frac{n2\pi}{T}(t - \tau) + b_n \sin \frac{n2\pi}{T}(t - \tau) \right]$$

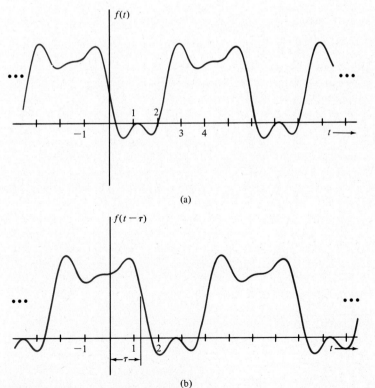

Figure 10-12   Function time shifting.

The trigonometric identities

$$\cos{(x \pm y)} = \cos{x}\cos{y} \mp \sin{x}\sin{y}$$

$$\sin{(x \pm y)} = \sin{x}\cos{y} \pm \cos{x}\sin{y}$$

may be used to place a time-shifted series into the quadrature form:

$$g(t) = d_0 + \sum_{n=1}^{\infty}\left(a_n\cos\frac{n2\pi t}{T}\cos\frac{n2\pi\tau}{T} + a_n\sin\frac{n2\pi t}{T}\sin\frac{n2\pi\tau}{T}\right.$$

$$\left. + b_n\sin\frac{n2\pi t}{T}\cos\frac{n2\pi\tau}{T} - b_n\cos\frac{n2\pi t}{T}\sin\frac{n2\pi\tau}{T}\right)$$

$$= d_0 + \sum_{n=1}^{\infty}\left[\left(a_n\cos\frac{n2\pi\tau}{T} - b_n\sin\frac{n2\pi\tau}{T}\right)\cos\frac{n2\pi t}{T}\right.$$

$$\left. + \left(a_n\sin\frac{n2\pi\tau}{T} + b_n\cos\frac{n2\pi\tau}{T}\right)\sin\frac{n2\pi t}{T}\right]$$

The time-shifted square wave of Figure 10-13 provides an important example of Fourier series time shifting. The function shown is related to the basic square wave, $f_s(t)$, by

$$g(t) = f_s\left(t + \frac{\pi}{2}\right) = \sum_{\substack{n=1 \\ n\ \text{odd}}}^{\infty}\frac{4}{n\pi}\sin n\left(t + \frac{\pi}{2}\right)$$

To place the Fourier series for $g(t)$ in the quadrature form, the trigonometric identity

$$\sin{(x + y)} = \sin{x}\cos{y} + \cos{x}\sin{y}$$

gives

$$g(t) = \sum_{\substack{n=1 \\ n\ \text{odd}}}^{\infty}\left[\left(\frac{4}{n\pi}\sin\frac{n\pi}{2}\right)\cos nt + \left(\frac{4}{n\pi}\cos\frac{n\pi}{2}\right)\sin nt\right]$$

For n odd,

$$\sin\frac{n\pi}{2} = (-1)^{(n-1)/2} \qquad \text{and} \qquad \cos\frac{n\pi}{2} = 0$$

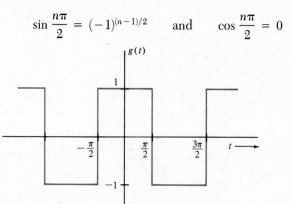

Figure 10-13   A time-shifted square wave.

giving

$$g(t) = \sum_{\substack{n=1 \\ n \text{ odd}}}^{\infty} \frac{4(-1)^{(n-1)/2}}{n\pi} \cos nt$$

the series for an even function.

10.5.5 Series for Derivatives and Integrals of Periodic Functions

Provided the functions involved are periodic and have Fourier series that converge, series for derivatives and integrals of periodic functions may be found by differentiating or integrating the original series term by term. For a function with Fourier series

$$f(t) = d_0 + \sum_{n=1}^{\infty} \left(a_n \cos \frac{n2\pi t}{T} + b_n \sin \frac{n2\pi t}{T} \right)$$

the derivative function has Fourier series

$$g(t) = \frac{df}{dt} = \sum_{n=1}^{\infty} \left(\frac{-n2\pi}{T} a_n \sin \frac{n2\pi t}{T} + \frac{n2\pi}{T} b_n \cos \frac{n2\pi t}{T} \right)$$

$$= \sum_{n=1}^{\infty} \left[\left(\frac{n2\pi b_n}{T} \right) \cos \frac{n2\pi t}{T} + \left(\frac{-n2\pi a_n}{T} \right) \sin \frac{n2\pi t}{T} \right]$$

For the integral of a periodic function to be periodic, the original function's average value, d_0, must be zero. Otherwise, the constant term in the integrated function gives rise to a ramp term in the integral, which is not periodic. If

$$f(t) = \sum_{n=1}^{\infty} \left(a_n \cos \frac{n2\pi t}{T} + b_n \sin \frac{n2\pi t}{T} \right)$$

$$h(t) = \int_T f(t) \, dt = K + \sum_{n=1}^{\infty} \left(\frac{T}{n2\pi} a_n \sin \frac{n2\pi t}{T} - \frac{T}{n2\pi} b_n \cos \frac{n2\pi t}{T} \right)$$

where K is the constant of integration.

Consider the basic triangular wave of Figure 10-14. This function is the integral

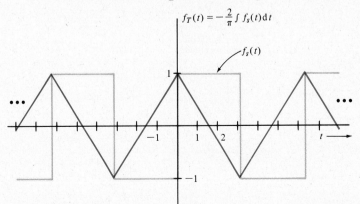

Figure 10-14   Basic triangular wave.

of a scaled and inverted square wave, with the constant of integration $K = 0$:

$$f_T(t) = -\frac{2}{\pi} \int f_s(t)\, dt = -\frac{2}{\pi} \int \left[\sum_{\substack{n=1 \\ n \text{ odd}}}^{\infty} \frac{4}{n\pi} \sin nt \right] dt$$

$$= \sum_{\substack{n=1 \\ n \text{ odd}}}^{1} \frac{8}{n^2\pi^2} \cos nt$$

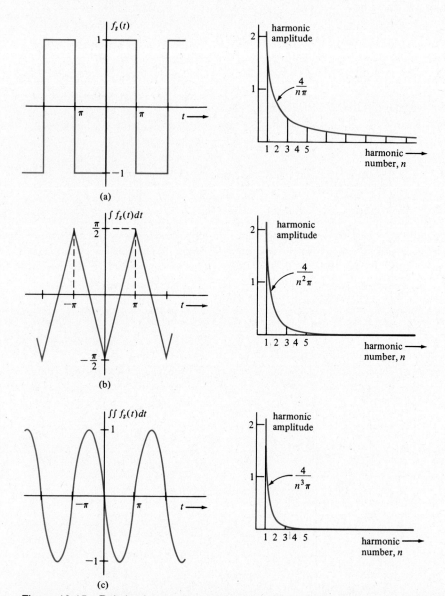

Figure 10-15   Relative harmonic content of some representative periodic functions.

The derivative and integral relations show the following qualitative relations between periodic functions: Smooth curves have high harmonics with much smaller relative amplitudes than do curves with discontinuities. The relative harmonic content of representative discontinuous, discontinuous-slope, and continuous-slope functions is shown in Figure 10-15.

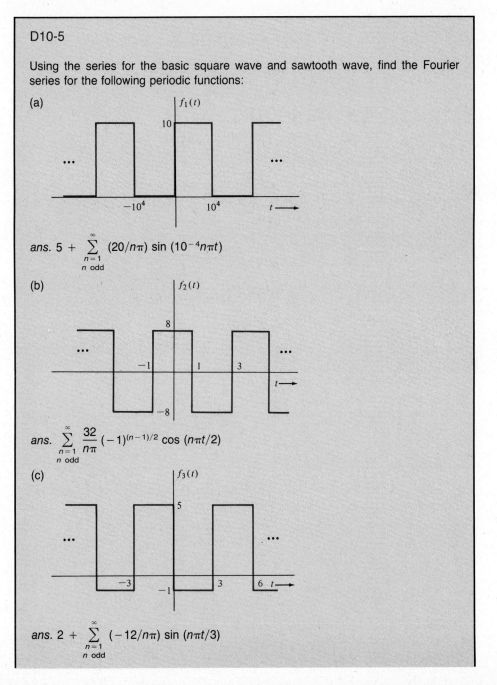

D10-5

Using the series for the basic square wave and sawtooth wave, find the Fourier series for the following periodic functions:

(a)

$f_1(t)$

10

-10^4 10^4 $t \longrightarrow$

ans. $5 + \sum_{\substack{n=1 \\ n \text{ odd}}}^{\infty} (20/n\pi) \sin (10^{-4}n\pi t)$

(b)

$f_2(t)$

8

-1 1   3 $t \longrightarrow$

-8

ans. $\sum_{\substack{n=1 \\ n \text{ odd}}}^{\infty} \frac{32}{n\pi} (-1)^{(n-1)/2} \cos (n\pi t/2)$

(c)

$f_3(t)$

5

-3 -1 3 6 $t \longrightarrow$

ans. $2 + \sum_{\substack{n=1 \\ n \text{ odd}}}^{\infty} (-12/n\pi) \sin (n\pi t/3)$

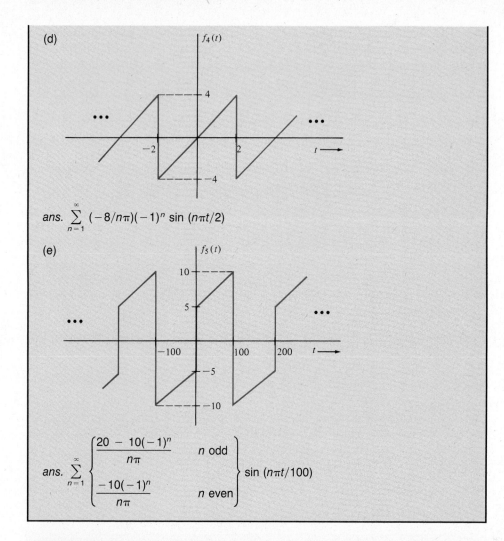

(d)

ans. $\displaystyle\sum_{n=1}^{\infty} (-8/n\pi)(-1)^n \sin (n\pi t/2)$

(e)

ans. $\displaystyle\sum_{n=1}^{\infty} \left\{ \begin{array}{ll} \dfrac{20 - 10(-1)^n}{n\pi} & n \text{ odd} \\[3mm] \dfrac{-10(-1)^n}{n\pi} & n \text{ even} \end{array} \right\} \sin (n\pi t/100)$

SERIES PROPERTIES AND MANIPULATION

An even function $f_E(t)$ has the property

$$f_E(-t) = f_E(t)$$

and an odd function $f_0(t)$ satisfies

$$f_0(-t) = -f_0(t)$$

Any function may be decomposed into even and odd parts:

$$f(t) = f_E(t) + f_0(t) = \underbrace{\tfrac{1}{2}[f(t) + f(-t)]}_{\text{even}} + \underbrace{\tfrac{1}{2}[f(t) - f(-t)]}_{\text{odd}}$$

The even and odd parts of a periodic function are, in terms of its Fourier series, as follows:

$$f(t) = \underbrace{d_0 + \sum_{n=1}^{\infty} a_n \cos \frac{n2\pi t}{T}}_{\text{even}} + \underbrace{\sum_{n=1}^{\infty} b_n \sin \frac{n2\pi t}{T}}_{\text{odd}}$$

If $f(t)$ is even, each b_n is zero; if $f(t)$ is odd, d_0 and each a_n is zero.

The periodic function $Kf(t)$, where K is a constant, has Fourier series with coefficients that are the coefficients of the series for $f(t)$, multiplied by K. The periodic function $[f(t) + K]$, where K is a constant, has the same Fourier series as that for $f(t)$, except that d_0 in the $f(t)$ series is replaced by $(d_0 + K)$ in the new series.

Time scaled periodic functions have unchanged Fourier coefficients; their harmonic frequencies are changed to correspond to the new period. Time-shifted periodic functions have time shifted Fourier series, which may be placed in the quadrature form using trigonometric identities.

Provided $f(t)$ and its derivative df/dt satisfy the Dirichlet conditions, the Fourier series for df/dt may be obtained from that for $f(t)$ by term-by-term differentiation:

$$f(t) = d_0 + \sum_{n=1}^{\infty} \left(a_n \cos \frac{n2\pi t}{T} + b_n \sin \frac{n2\pi t}{T} \right)$$

$$\frac{df}{dt} = \sum_{n=1}^{\infty} \left[\left(\frac{n2\pi b_n}{T} \right) \cos \frac{n2\pi t}{T} + \left(\frac{-n2\pi a_n}{T} \right) \sin \frac{n2\pi t}{T} \right]$$

For the (indefinite) integral of a periodic function to be periodic, the function must have zero average value: $d_0 = 0$. Provided the integral is periodic, its Fourier series may be obtained from that for $f(t)$ by term-by-term integration:

$$f(t) = \sum_{n=1}^{\infty} \left(a_n \cos \frac{n2\pi t}{T} + b_n \sin \frac{n2\pi t}{T} \right)$$

$$\int f(t)\, dt = (\text{arbitrary constant})$$
$$+ \sum_{n=1}^{\infty} \left[\left(\frac{-T b_n}{n2\pi} \right) \cos \frac{n2\pi t}{T} + \left(\frac{T a_n}{n2\pi} \right) \sin \frac{n2\pi t}{T} \right]$$

The harmonic amplitudes of continuous functions decay with harmonic number, n, for large n, more rapidly than do functions with discontinuities.

Basic Triangular Wave

$$f_T(t) = -\frac{2}{\pi} \int f_s(t)\, dt$$

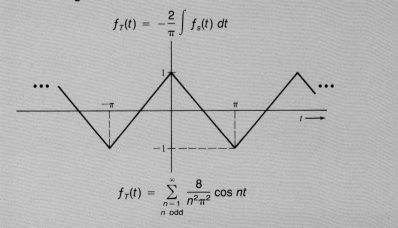

$$f_T(t) = \sum_{\substack{n=1 \\ n \text{ odd}}}^{\infty} \frac{8}{n^2 \pi^2} \cos nt$$

10.6 TRIGONOMETRIC SOLUTION OF DIFFERENTIAL EQUATIONS WITH PERIODIC DRIVING FUNCTIONS

The Fourier series allows solution of linear, time-invariant differential equations with periodic driving functions. When the driving function is expanded into a Fourier series, the solution component due to each term separately may be found, then summed to find the response to the periodic function.

Consider the following equation, where the driving function is the level-shifted square wave of Figure 10-16.

$$\frac{d^2y}{dt^2} + 3\frac{dy}{dt} + 2y = f(t) = 5 + \sum_{\substack{n=1 \\ n \text{ odd}}}^{\infty} \frac{20}{n\pi} \sin n2\pi t$$

The term in the forced solution corresponding to the constant term in the driving function is the forced solution to

$$\frac{d^2y_0}{dt^2} + 3\frac{dy_0}{dt} + 2y_0 = 5$$

which is

$$y_0 = \frac{5}{2}$$

The forced solution component due to the driving function's first harmonic is the forced solution to

$$\frac{d^2y_1}{dt^2} + 3\frac{dy_1}{dt} + 2y_1 = \frac{20}{\pi} \sin 2\pi t$$

Substituting a general sinusoid of radian frequency 2π

$$y_1(t) = a_1 \cos 2\pi t + b_1 \sin 2\pi t$$

then

$$- 4\pi^2 a_1 \cos 2\pi t - 4\pi^2 b_1 \sin 2\pi t - 6\pi a_1 \sin 2\pi t$$

$$+ 6\pi b_1 \cos 2\pi t + 2a_1 \cos 2\pi t + 2b_1 \sin 2\pi t$$

$$= (20/\pi) \sin 2\pi t$$

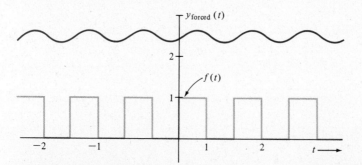

Figure 10-16 Trigonometric Fourier series solution of a differential equation with a periodic driving function.

$$\begin{cases} (2 - 4\pi^2)a_1 + & 6\pi b_1 = 0 \\ -6\pi a_1 + (2 - 4\pi^2)b_1 = (20/\pi) \end{cases}$$

$$a_1 = \frac{\begin{vmatrix} 0 & 6\pi \\ (20/\pi) & (2 - 4\pi^2) \end{vmatrix}}{\begin{vmatrix} (2 - 4\pi^2) & 6\pi \\ -6\pi & (2 - 4\pi^2) \end{vmatrix}} = \frac{-120}{4 - 8\pi^2 + 16\pi^4 + 36\pi^2}$$

$$= \frac{-120}{4 + 28\pi^2 + 16\pi^4} = -0.065$$

$$b_1 = \frac{\begin{vmatrix} (2 - 4\pi^2) & 0 \\ -6\pi & (20/\pi) \end{vmatrix}}{4 + 28\pi^2 + 16\pi^4} = \frac{80 - (40/\pi)}{4 + 28\pi^2 + 16\pi^4} = 0.13$$

Thus

$$y_1(t) = -0.065 \cos 2\pi t + 0.13 \sin 2\pi t$$

As the driving function's second harmonic is zero, the corresponding second harmonic forced solution component is zero:

$$y_2(t) = 0.$$

The forced solution due to the nth nonzero driving function harmonic is given by

$$\frac{d^2 y_n}{dt^2} + 3 \frac{dy_n}{dt} + 2y_n = \frac{4}{n\pi} \sin n2\pi t$$

Substituting a general sinusoid of radian frequency $n2\pi$

$$y_n(t) = a_n \cos n2\pi t + b_n \sin n2\pi t$$

there results

$$-4n^2\pi^2 a_n \cos n2\pi t - 4n^2\pi^2 b_n \sin n2\pi t - 6n\pi a_n \sin n2\pi t$$

$$+ 6n\pi b_n \cos n2\pi t + 2a_n \cos n2\pi t + 2b_n \sin n2\pi t$$

$$= (20/n\pi)$$

$$\begin{cases} (2 - 4n^2\pi^2)a_n + & 6n\pi b_n = 0 \\ -6n\pi a_n + (2 - 4n^2\pi^2)b_n = (20/n\pi) \end{cases}$$

$$a_n = \frac{\begin{vmatrix} 0 & 6n \\ (20/n\pi) & (2 - 4n^2\pi^2) \end{vmatrix}}{\begin{vmatrix} (2 - 4n^2\pi^2) & 6n\pi \\ -6n\pi & (2 - 4n^2\pi^2) \end{vmatrix}} = \frac{-120}{4 + 28n^2\pi^2 + 16n^4\pi^4}$$

$$b_n = \frac{\begin{vmatrix} (2 - 4n^2\pi^2) & 0 \\ -6n\pi & (20/n\pi) \end{vmatrix}}{4 + 28n^2\pi^2 + 16n^4\pi^4} = \frac{80n\pi - (40/n\pi)}{4 + 28n^2\pi^2 + 16n^4\pi^4}$$

The entire forced solution of the differential equation is then

$$y_{forced}(t) = y_0 + y_1(t) + y_2(t) + y_3(t) + \cdots$$

$$= \frac{5}{2} + \sum_{\substack{n=1 \\ n \text{ odd}}}^{\infty} \left[\left(\frac{-120}{4 + 28n^2\pi^2 + 16n^4\pi^4} \right) \cos n2\pi t \right.$$

$$\left. + \left(\frac{80n\pi - (40/n\pi)}{4 + 28n^2\pi^2 + 16n^4\pi^4} \right) \sin n2\pi t \right]$$

A plot of this solution is also given in Figure 10-16. Because the harmonic amplitudes in the Fourier series for $y_{forced}(t)$ decrease so rapidly with harmonic number n, low order truncations of the series will be relatively accurate. From the plot, dominance of the constant and first harmonic term is evident.

Occasionally, a Fourier series solution may be recognized as being simply related to known series. More often, the series is not a simple one, and the shape of the periodic function it represents is not easily recognized from its formula. Then, a plot of the series, truncated appropriately, helps in visualizing the function.

D10-6

Find the Fourier series for the forced solutions of the following differential equations with periodic driving functions:

(a) $\dfrac{dy}{dt} + 5y = \displaystyle\sum_{n=1}^{\infty} \dfrac{7}{n^2 + 3} \sin nt$

ans. $\displaystyle\sum_{n=1}^{\infty} \dfrac{1}{(n^2 + 3)(n^2 + 25)} [-7n \cos nt + 35 \sin nt]$

(b) $2\dfrac{dy}{dt} + 3y = f(t)$

where $f(t)$ is the basic square wave with period 2 seconds.

ans. $\displaystyle\sum_{\substack{n=1 \\ n \text{ odd}}}^{\infty} \left[\dfrac{-8}{9 + 4n^2\pi^2} \cos n\pi t + \dfrac{12}{n\pi(9 + 4n^2\pi^2)} \sin n\pi t \right]$

TRIGONOMETRIC SOLUTION OF DIFFERENTIAL EQUATIONS WITH PERIODIC DRIVING FUNCTIONS

Forced solutions to linear differential equations with periodic driving functions may be found by expanding the driving function into a Fourier series and superimposing the individual terms. The sum of the forced solution due to the constant driving function term and the solutions for each harmonic term results in another Fourier series.

10.7 OTHER TRIGONOMETRIC SERIES FORMS

10.7.1 Cosine and Angle Form

To this point, the quadrature (cosine and sine) form of the Fourier series has been emphasized. The alternative cosine and angle form for the series, and the sine and angle form, are now developed. These other forms are sometimes easier to use in practice and are helpful in visualizing the harmonic components of a periodic function. Using the relation

$$D \cos \omega t + E \sin \omega t = A \cos (\omega t + \theta)$$

where

$$A = \sqrt{D^2 + E^2} \qquad \theta = -\arctan \frac{E}{D}$$

for converting a sinusoidal function in quadrature form to the cosine and angle form (Section 7.2), a Fourier series may be expressed as

$$f(t) = d_0 + \sum_{n=1}^{\infty} \left(a_n \cos \frac{n2\pi t}{T} + b_n \sin \frac{n2\pi t}{T} \right)$$

$$= d_0 + \sum_{n=1}^{\infty} c_n \cos \left(\frac{n2\pi t}{T} + \theta_n \right)$$

where

$$\begin{cases} c_n = \sqrt{a_n^2 + b_n^2} \\ \theta_n = -\arctan \dfrac{b_n}{a_n} \end{cases}$$

For example, the series

$$f(t) = 10 + \sum_{n=1}^{\infty} \left(\frac{4}{n^2} \cos \frac{nt}{5} + \frac{3}{n} \sin \frac{nt}{5} \right)$$

$$= 10 + \sum_{n=1}^{\infty} \sqrt{\left(\frac{4}{n^2} \right)^2 + \left(\frac{3}{n} \right)^2} \cos \left(\frac{nt}{5} - \arctan \frac{3n}{4} \right)$$

10.7.2 Sine and Angle Form

The sine and angle form of a Fourier series is simply related to the cosine and angle form, as follows:

$$f(t) = d_0 + \sum_{n=1}^{\infty} \left(a_n \cos \frac{n2\pi t}{T} + b_n \sin \frac{n2\pi t}{T} \right)$$

$$= d_0 + \sum_{n=1}^{\infty} c_n \cos \left(\frac{n2\pi t}{T} + \theta_n \right) = d_0 + \sum_{n=1}^{\infty} c_n \sin \left(\frac{n2\pi t}{T} + \phi_n \right)$$

where

$$c_n = \sqrt{a_n^2 + b_n^2}$$

$$\phi_n = \theta_n + 90° = 90° - \arctan \frac{b_n}{a_n} = \arctan \frac{a_n}{b_n}$$

Time shifts of simple functions often give cosine and angle or sine and angle forms of the series. For example, if the basic square wave is delayed one second, the easiest form of the Fourier series for the result is

$$g(t) = \sum_{\substack{n=1 \\ n \text{ odd}}}^{\infty} \frac{4}{n\pi} \sin n(t - 1) = \sum_{\substack{n=1 \\ n \text{ odd}}}^{\infty} \frac{4}{n\pi} \sin (nt - n)$$

which is in sine and angle form.

10.7.3 Parseval's Relation

Parseval's relation (10-13) in terms of the harmonic amplitudes, is

$$\frac{1}{T} \int_T f^2(t) \, dt = d_0^2 + \frac{1}{2} \sum_{n=1}^{\infty} (a_n^2 + b_n^2) = d_0^2 + \frac{1}{2} \sum_{n=1}^{\infty} c_n^2$$

It is here seen that it is the *amplitudes* of the component harmonics that affect the average of the square of a periodic function; the harmonic phase angles have no effect. The rms value of $f(t)$ is thus

$$\text{RMS}[f(t)] = \sqrt{\frac{1}{T} \int_T f^2(t) \, dt} = \sqrt{d_0^2 + \frac{1}{2} \sum_{n=1}^{\infty} c_n^2}$$

The expression (10-14) for the mean square error of a truncated Fourier series in terms of the coefficients is, in terms of the harmonic amplitudes,

$$\text{MSE} = \frac{1}{2} \sum_{n=N+1}^{\infty} (a_n^2 + b_n^2) = \frac{1}{2} \sum_{n=N+1}^{\infty} c_n^2$$

For example, approximating a function with Fourier series

$$f(t) = 8 + \sum_{n=1}^{\infty} \frac{3}{n^3} \cos \left(10^6 t - \frac{n\pi}{16} \right)$$

by a series truncated beyond the tenth harmonic gives

$$\text{MSE} = \frac{1}{2} \sum_{n=11}^{\infty} \left(\frac{3}{n^3} \right)^2 = \frac{9}{2} \sum_{n=11}^{\infty} \frac{1}{n^6}$$

$$= 2.5 \times 10^{-6} + 1.5 \times 10^{-6} + 0.9 \times 10^{-6} + \cdots \cong 6.4 \times 10^{-6}$$

D10-7

Find the indicated series form from the given Fourier series:

(a) Find the cosine and angle form of the series

$$f(t) = -2 + \sum_{n=1}^{\infty} \left[\frac{1}{n^2 + 1} \cos n\pi t + \frac{n}{n^2 + 1} \sin n\pi t \right]$$

ans. $-2 + \sum_{n=1}^{\infty} \frac{1}{\sqrt{n^2 + 1}} \cos (n\pi t - \arctan n)$

(b) Find the sine and angle form of the series

$$f(t) = 6 + \sum_{\substack{n=1 \\ n \text{ odd}}}^{\infty} \left[\frac{\pi n}{n^2 + 4} \cos 5nt + \frac{1}{n^2 + 4} \sin 5nt \right]$$

ans. $6 + \sum_{\substack{n=1 \\ n \text{ odd}}}^{\infty} \frac{\sqrt{\pi^2 n^2 + 1}}{n^2 + 4} \sin [5nt + 90° - \arctan (1/\pi n)]$

(c) Find the quadrature form of the series

$$f(t) = -2 + \sum_{\substack{n=1 \\ n \text{ odd}}}^{\infty} \frac{1}{n^2 + 1} \sin \left(\frac{nt}{8} + \frac{n\pi}{4} \right)$$

ans. $-2 + \sum_{\substack{n=1 \\ n \text{ odd}}}^{\infty} \left[\frac{\sin (n\pi/4)}{n^2 + 1} \cos (nt/8) + \frac{\cos (n\pi/4)}{n^2 + 1} \sin (nt/8) \right]$

(d) Find the quadrature form of the series

$$f(t) = \sum_{n=1}^{\infty} \frac{1}{n^2 + 3} \cos \left(100nt - \frac{n\pi}{8} \right)$$

ans. $\sum_{n=1}^{\infty} \left(\frac{\cos (n\pi/8)}{n^2 + 3} \cos 100nt + \frac{\sin (n\pi/8)}{n^2 + 3} \sin 100nt \right)$

OTHER TRIGONOMETRIC SERIES FORMS

Alternative trigonometric Fourier series forms are:

$$f(t) = d_0 + \sum_{n=1}^{\infty} \left(a_n \cos \frac{n2\pi t}{T} + b_n \sin \frac{n2\pi t}{T} \right)$$

$$= d_0 + \sum_{n=1}^{\infty} c_n \cos \left(\frac{n2\pi t}{T} + \theta_n \right) = d_0 + \sum_{n=1}^{\infty} c_n \sin \left(\frac{n2\pi t}{T} + \phi_n \right)$$

Relations between the various constants are as follows:

$$\begin{cases} c_n = \sqrt{a_n^2 + b_n^2} \\ \theta_n = -\arctan (b_n/a_n) \end{cases} \quad \begin{cases} a_n = c_n \cos \theta_n \\ b_n = -c_n \sin \theta_n \end{cases}$$

$$\phi_n = \theta_n + 90°$$

In terms of the coefficients, Parseval's relation is

$$\frac{1}{T} \int_T f^2(t) \, dt = d_0^2 + \frac{1}{2} \sum_{n=1}^{\infty} (a_n^2 + b_n^2) = d_0^2 + \frac{1}{2} \sum_{n=1}^{\infty} c_n^2$$

10.8 TRIGONOMETRIC SOLUTION OF NETWORKS WITH PERIODIC SOURCES

10.8.1 Forced Periodic Solution

Consider the network of Figure 10-17(a), which has the sawtooth wave periodic source sketched in Figure 10-17(b). Superimposing the harmonic source components

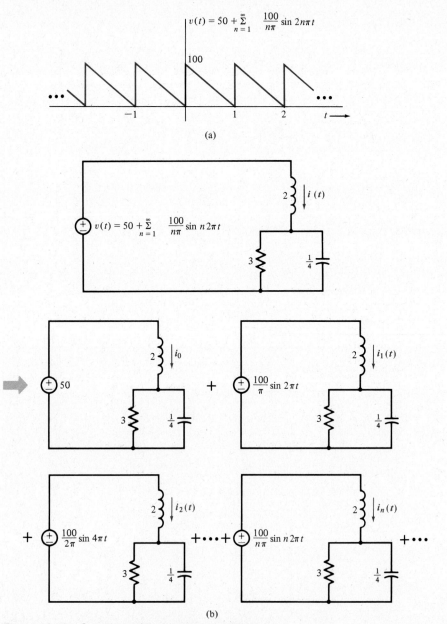

(a)

(b)

Figure 10-17 Superimposing source components.

gives the series of component sinusoidal source problems shown. The component sinusoidal problem for the constant source term is solved in Figure 10-18(a). For the constant source, $s = 0$ and the inductor impedance is zero while the capacitor impedance is infinite, so

$$i_0 = \tfrac{50}{3}$$

In the first harmonic component problem, Figure 10-18(b), the sinusoidal source $(1/\pi) \sin 2\pi t$ is replaced by the sinor $(1/\pi)e^{j2\pi t}$. Impedances are evaluated at $s = j2\pi$, giving a sinor current

$$\underline{i}_1(t) = \cfrac{\dfrac{100}{\pi}e^{j2\pi t}}{j4\pi - \cfrac{6j}{3\pi - 2j}} = \cfrac{\left(300 - \dfrac{200}{\pi}j\right)e^{j2\pi t}}{j12\pi^2 + 8\pi - 6j} = \cfrac{(307e^{-j12^\circ})e^{j2\pi t}}{115.2e^{j77^\circ}} = 2.66e^{j(2\pi t - 89^\circ)}$$

corresponding to a sinusoidal current in the original network of

$$i_1(t) = 2.66 \sin (2\pi t - 89^\circ)$$

Phasor methods would, of course, give the same result; the exponential time factors are dropped from the sinors to form the corresponding phasors.

The second harmonic component problem, Figure 10-18(c), is similar to that for the first harmonic. For it,

$$\underline{i}_2(t) = \cfrac{(100/2\pi)e^{j4\pi t}}{j8\pi - \cfrac{3j}{3\pi - j}} = \cfrac{[150 - (j50/\pi)]e^{j4\pi t}}{j24\pi^2 + 8\pi - 3j}$$

$$= \cfrac{(151e^{-j6^\circ})e^{j4\pi t}}{235e^{j83^\circ}} = 0.64e^{j(4\pi t - 89^\circ)}$$

$$i_2(t) = 0.64 \sin (4\pi t - 89^\circ)$$

For the nth harmonic current component, Figure 10-18(d),

$$\underline{i}_n(t) = \cfrac{\dfrac{100}{n\pi}e^{jn2\pi t}}{j4n\pi - \cfrac{6j}{3n\pi - 2j}} = \cfrac{\left(300 - \dfrac{200j}{n\pi}\right)e^{jn2\pi t}}{j12n^2\pi^2 + 8n\pi - 6j}$$

$$= \cfrac{100\left(\sqrt{9 + \dfrac{4}{n^2\pi^2}}\,e^{-j\,\arctan\frac{2}{3n\pi}}\right)e^{jn2\pi t}}{\sqrt{64n^2\pi^2 + (12n^2\pi^2 - 6)^2}\,e^{j\,\arctan\frac{12n^2\pi^2 - 6}{8n\pi}}}$$

$$i_n(t) = \cfrac{100}{n\pi}\sqrt{\cfrac{9n^2\pi^2 + 4}{144n^4\pi^4 - 12n^2\pi^2 + 36}}$$

$$\cdot \sin\left(n2\pi t - \arctan\frac{2}{3n\pi} - \arctan\frac{12n^2\pi^2 - 6}{8n\pi}\right)$$

The forced network solution is thus

$$i(t) = i_0 + \sum_{n=1}^{\infty} i_n(t)$$

which is expressed here as a Fourier series in the sine and angle form. A plot of this solution is given in Figure 10-19.

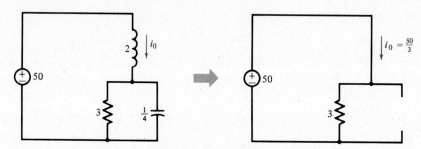

(a) Solution for the constant component of the network current

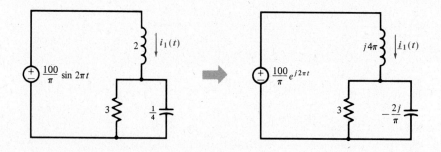

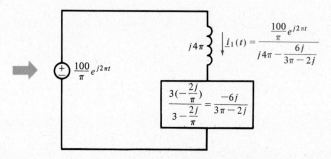

(b) First harmonic solution using sinors

Figure 10-18   Solutions for harmonic network current components.

Figure 10-18 *(Continued)*

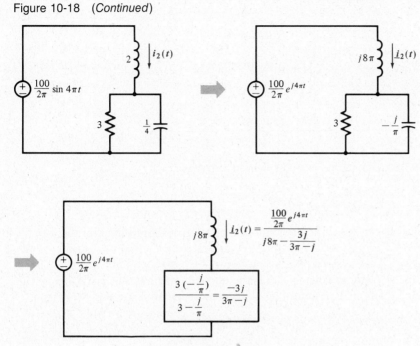

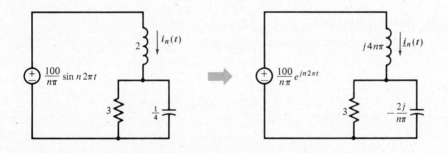

(c) Second harmonic solution

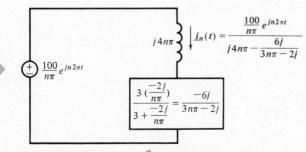

(d) n^{th} harmonic solution

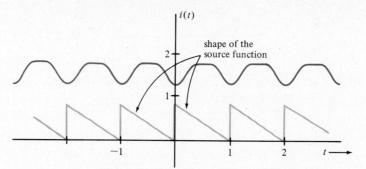

Figure 10-19   Solution for the example network current.

10.8.2 Periodic Power Flow

The electrical power flow into an element in an electrical network is of the form

$$p(t) = v(t)i(t)$$

where $v(t)$ is the element voltage, $i(t)$ is the element current, and v and i have the sink reference relation. For sinusoidal voltage and current of the same frequency,

$$v(t) = E \cos (\omega t + \gamma) \qquad i(t) = C \cos (\omega t + \theta)$$

$$p(t) = v(t)i(t) = CE \cos (\omega t + \gamma) \cos (\omega t + \theta)$$

$$= \frac{CE}{2} \cos (\gamma - \theta) + \frac{CE}{2} \cos (2\omega t + \gamma + \theta)$$

The average power flow is

$$P = \langle p(t) \rangle = \frac{CE}{2} \cos (\gamma - \theta)$$

the other sinusoidal term in $p(t)$ at radian frequency 2ω averaging to zero.

For periodic voltage and current, each with the same period, as occurs in networks with periodic sources,

$$v(t) = k_0 + \sum_{n=1}^{\infty} e_n \cos \left(\frac{n2\pi t}{T} + \gamma_n \right)$$

$$i(t) = d_0 + \sum_{n=1}^{\infty} c_n \cos \left(\frac{n2\pi t}{T} + \theta_n \right)$$

and

$$p(t) = \left[k_0 + \sum_{n=1}^{\infty} e_n \cos \left(\frac{n2\pi t}{T} + \gamma_n \right) \right] \left[d_0 + \sum_{n=1}^{\infty} c_n \cos \left(\frac{n2\pi t}{T} + \theta_n \right) \right]$$

The average power flow,

$$P = \langle p(t) \rangle = \frac{1}{T} \int_T v(t)i(t) \, dt$$

involves the averages of products of sinusoids. Using the orthogonality relations, only the integrals of the products of the constant terms and same-frequency terms are nonzero. The average power flow is thus

$$P = d_0 k_0 = \frac{1}{2} \sum_{n=1}^{\infty} e_n c_n \cos (\gamma_n - \theta_n)$$

For example, consider the network of Figure 10-20 with the given square wave source. The entire forced solution for $i(t)$ is

$$i(t) = i_1(t) + i_3(t) + i_5(t) + \cdots$$

$$= 0.19 \cos (t - 27°) + 0.039 \cos (3t + 124°)$$

$$+ 0.016 \cos (5t - 68°) + \cdots$$

and the instantaneous power flow out of the source is

$$p(t) = v(t)i(t)$$

$$= \frac{1}{2} \left[\frac{4}{\pi} \cos t + \frac{4}{3\pi} \cos (3t + 180°) + \frac{4}{5\pi} \cos 5t + \cdots \right]$$

$$\cdot [0.19 \cos (t - 27°) + 0.039 \cos (3t + 124°)$$

$$+ 0.016 \cos (5t - 68°) + \cdots]$$

The average power flow out of the source is thus

$$P = \frac{1}{2}\left(\frac{4}{\pi}\right)(0.19) \cos 27° + \frac{1}{2}\left(\frac{4}{3\pi}\right)(0.039) \cos 56°$$

$$+ \frac{1}{2}\left(\frac{4}{5\pi}\right)(0.016) \cos 68° + \cdots$$

$$= 0.108 + 0.00463 + 0.000764 + \cdots$$

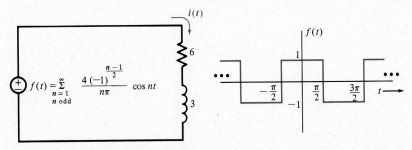

Figure 10-20   Finding network power flow.

D10-8

Find the trigonometric Fourier series for the indicated signals:

(a) $f_1(t) = -4 + \sum\limits_{n=1}^{\infty} \dfrac{3}{n^3 + 1} \cos\left(2nt + \dfrac{n\pi}{8}\right)$

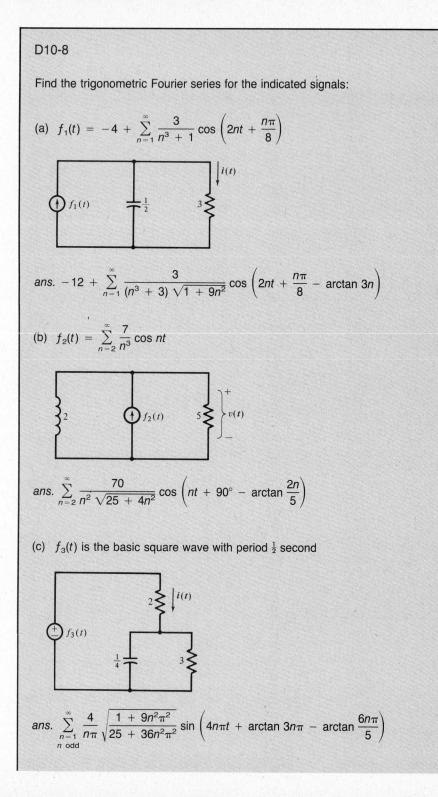

ans. $-12 + \sum\limits_{n=1}^{\infty} \dfrac{3}{(n^3 + 3)\sqrt{1 + 9n^2}} \cos\left(2nt + \dfrac{n\pi}{8} - \arctan 3n\right)$

(b) $f_2(t) = \sum\limits_{n=2}^{\infty} \dfrac{7}{n^3} \cos nt$

ans. $\sum\limits_{n=2}^{\infty} \dfrac{70}{n^2 \sqrt{25 + 4n^2}} \cos\left(nt + 90° - \arctan \dfrac{2n}{5}\right)$

(c) $f_3(t)$ is the basic square wave with period $\frac{1}{2}$ second

ans. $\sum\limits_{\substack{n=1 \\ n \text{ odd}}}^{\infty} \dfrac{4}{n\pi} \sqrt{\dfrac{1 + 9n^2\pi^2}{25 + 36n^2\pi^2}} \sin\left(4n\pi t + \arctan 3n\pi - \arctan \dfrac{6n\pi}{5}\right)$

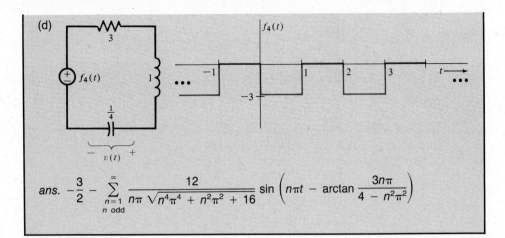

(d)

$$ans. \quad -\frac{3}{2} - \sum_{\substack{n=1 \\ n \text{ odd}}}^{\infty} \frac{12}{n\pi \sqrt{n^4\pi^4 + n^2\pi^2 + 16}} \sin\left(n\pi t - \arctan \frac{3n\pi}{4 - n^2\pi^2}\right)$$

D10-9

The source voltage $v(t)$ is a basic square wave with amplitude 10 and period $\frac{1}{8}$ second. Find the average power flow out of the source, expressed as an infinite series.

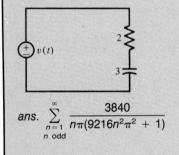

$$ans. \quad \sum_{\substack{n=1 \\ n \text{ odd}}}^{\infty} \frac{3840}{n\pi(9216n^2\pi^2 + 1)}$$

TRIGONOMETRIC SOLUTION OF NETWORKS WITH PERIODIC SOURCES

Forced solutions for voltages and currents in electrical networks with periodic driving functions may be found by expanding the source function(s) into a Fourier series and superimposing the individual terms. In practice, one finds the forced signals due to the constant term and for each harmonic term, resulting in another Fourier series.

The electrical power flow into a network element is

$$p_{\text{into}}(t) = v(t)i(t),$$

where $v(t)$ and $i(t)$ are the element's sink reference voltage and current. If the voltage and current are each periodic, with period T,

$$v(t) = k_0 + \sum_{n=1}^{\infty} e_n \cos\left(\frac{n2\pi t}{T} + \gamma_n\right)$$

$$i(t) = d_0 + \sum_{n=1}^{\infty} c_n \cos\left(\frac{n2\pi t}{T} + \theta_n\right)$$

the *average* element power flow is

$$P_{into} = \langle p_{into}(t) \rangle = \frac{1}{T} \int_T p_{into}(t) \, dt$$

$$= d_0 k_0 + \frac{1}{2} \sum_{n=1}^{\infty} e_n c_n \cos(\gamma_n - \theta_n)$$

The average power is the sum of the individual average powers, computed as if no other source components were present.

10.9 SERIES FOR RECTIFIED SINUSOIDS

10.9.1 Fourier Series for a Full-Wave Rectified Sinusoid

Rectified sinusoidal functions are important in many practical applications. Basic half- and full-wave functions are defined now, and the Fourier series for each is found. The function sketched in Figure 10-21 will be here referred to as the *basic full-wave rectified sinusoid*. It is an even function and has period $T = \pi$.

The Fourier series coefficients for $f_F(t)$ are given by

$$d_0 = \frac{1}{\pi} \int_{-\pi/2}^{\pi/2} \cos t \, dt = \frac{1}{\pi} \sin t \, \Big|_{-\pi/2}^{\pi/2} = \frac{2}{\pi}$$

$$a_n = \frac{2}{\pi} \int_{-\pi/2}^{\pi/2} \cos t \cos n2t \, dt$$

Using the trigonometric identity

$$\cos x \cos y = \tfrac{1}{2} \cos(x + y) + \tfrac{1}{2} \cos(x - y)$$

$$a_n = \frac{2}{\pi} \int_{-\pi/2}^{\pi/2} \left[\frac{1}{2} \cos(2n + 1)t + \frac{1}{2} \cos(2n - 1)t \right] dt$$

$$= \frac{\sin(2n + 1)t}{\pi(2n + 1)} \Big|_{-\pi/2}^{\pi/2} + \frac{\sin(2n - 1)t}{\pi(2n - 1)} \Big|_{-\pi/2}^{\pi/2}$$

$$= \frac{2 \sin \dfrac{(2n + 1)\pi}{2}}{\pi(2n + 1)} + \frac{2 \sin \dfrac{(2n - 1)\pi}{2}}{\pi(2n - 1)}$$

Some representative values are tabulated in Table 10-1, from which it is seen that

$$a_n = \begin{cases} \dfrac{-2}{\pi(2n + 1)} + \dfrac{2}{\pi(2n - 1)} = \dfrac{4}{\pi(4n^2 - 1)} & n = 1, 3, 5, \ldots \\[4mm] \dfrac{2}{\pi(2n + 1)} + \dfrac{-2}{\pi(2n - 1)} = \dfrac{-4}{\pi(4n^2 - 1)} & n = 2, 4, 6, \ldots \end{cases}$$

Because the function is even,

$$b_n = 0$$

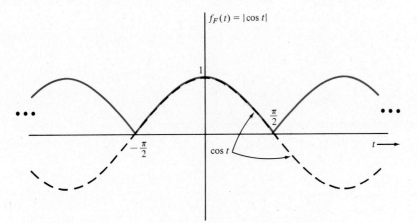

Figure 10-21 The basic full-wave rectified sinusoidal function.

Its Fourier series is thus

$$f_F(t) = \frac{2}{\pi} + \frac{4}{3\pi}\cos 2t - \frac{4}{15\pi}\cos 4t + \frac{4}{35\pi}\cos 6t - \frac{4}{143\pi}\cos 8t + \cdots$$

$$= \frac{2}{\pi} + \sum_{n=1}^{\infty} \frac{4(-1)^{n+1}}{\pi(4n^2 - 1)}\cos 2nt$$

The period of the full-wave rectified sinusoid is one-half that of the original sinusoidal function.

10.9.2 Fourier Series for a Half-Wave Rectified Sinusoid

The function of Figure 10-22(a) will be here termed the *basic half-wave rectified sinusoid*. It is the positive part of a cosine function of period $T = 2\pi$. The basic

Table 10-1 TERMS RELATED TO THE FOURIER SERIES OF A FULL-WAVE RECTIFIED SINUSOID

n	$\dfrac{(2n + 1)\pi}{2}$	$\sin\dfrac{(2n + 1)\pi}{2}$	$\dfrac{(2n - 1)\pi}{2}$	$\sin\dfrac{(2n - 1)\pi}{2}$
1	$\dfrac{3\pi}{2}$	-1	$\dfrac{\pi}{2}$	1
2	$\dfrac{5\pi}{2}$	1	$\dfrac{3\pi}{2}$	-1
3	$\dfrac{7\pi}{2}$	-1	$\dfrac{5\pi}{2}$	1
4	$\dfrac{9\pi}{2}$	1	$\dfrac{7\pi}{2}$	-1
5	$\dfrac{11\pi}{2}$	-1	$\dfrac{9\pi}{2}$	1

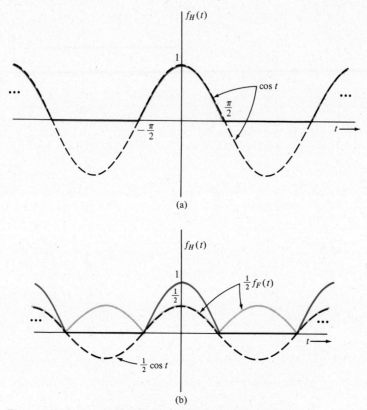

(a)

(b)

Figure 10-22   Basic half-wave rectified sinusoidal function.
(a) The half-wave rectified sinusoid.
(b) Forming the half-wave rectified sinusoid from the sum of the sinusoidal and full-wave rectified sinusoidal functions.

half-wave rectified sinusoid $f_H(t)$ is simply related to the basic full-wave rectified sinusoid $f_F(t)$ by

$$f_H(t) = \tfrac{1}{2} f_F(t) + \tfrac{1}{2} \cos t$$

as sketched in Figure 10-22(b).

So,

$$
\begin{aligned}
f_H(t) &= \frac{1}{2} \left[\frac{2}{\pi} + \sum_{n=1}^{\infty} \frac{4(-1)^{n+1}}{\pi(4n^2 - 1)} \cos 2nt \right] + \frac{1}{2} \cos t \\
&= \frac{1}{\pi} + \frac{1}{2} \cos t + \sum_{n=1}^{\infty} \frac{2(-1)^{n+1}}{\pi(4n^2 - 1)} \cos 2nt \\
&= \frac{1}{\pi} + \frac{1}{2} \cos t + \frac{2}{5\pi} \cos 2t - \frac{2}{17\pi} \cos 4t + \frac{2}{37\pi} \cos 6t \\
&\quad - \frac{2}{65\pi} \cos 8t + \cdots
\end{aligned}
$$

Replacing n by $(n/2)$ in the summation places this series in the standard form where n is the harmonic number:

$$f_H(t) = \frac{1}{\pi} + \frac{1}{2} \cos t + \sum_{\substack{n=2 \\ n \text{ even}}}^{\infty} \frac{2(-1)^{[(n/2)+1]}}{\pi(n^2 - 1)} \cos nt$$

The period of $f_H(t)$ is 2π units because it is composed of $f_F(t)$ with period π plus $\frac{1}{2}$ $\cos t$, which has period 2π.

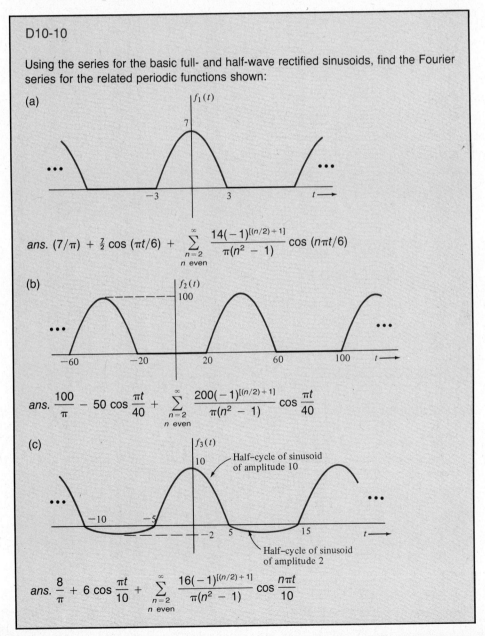

D10-10

Using the series for the basic full- and half-wave rectified sinusoids, find the Fourier series for the related periodic functions shown:

(a)

ans. $(7/\pi) + \frac{7}{2} \cos (\pi t/6) + \sum_{\substack{n=2 \\ n \text{ even}}}^{\infty} \frac{14(-1)^{[(n/2)+1]}}{\pi(n^2 - 1)} \cos (n\pi t/6)$

(b)

ans. $\dfrac{100}{\pi} - 50 \cos \dfrac{\pi t}{40} + \sum_{\substack{n=2 \\ n \text{ even}}}^{\infty} \dfrac{200(-1)^{[(n/2)+1]}}{\pi(n^2 - 1)} \cos \dfrac{\pi t}{40}$

(c)

Half–cycle of sinusoid of amplitude 10

Half–cycle of sinusoid of amplitude 2

ans. $\dfrac{8}{\pi} + 6 \cos \dfrac{\pi t}{10} + \sum_{\substack{n=2 \\ n \text{ even}}}^{\infty} \dfrac{16(-1)^{[(n/2)+1]}}{\pi(n^2 - 1)} \cos \dfrac{n\pi t}{10}$

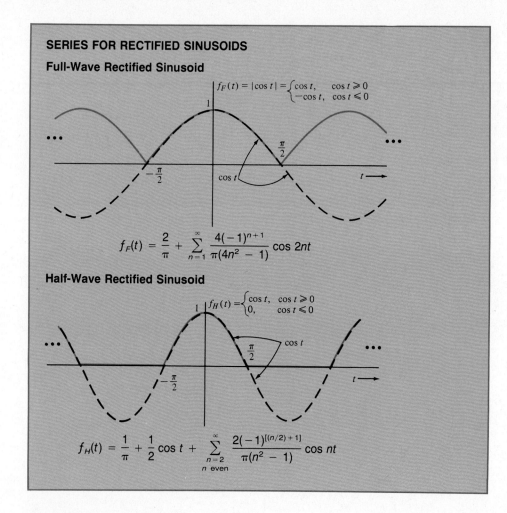

SERIES FOR RECTIFIED SINUSOIDS

Full-Wave Rectified Sinusoid

$$f_F(t) = |\cos t| = \begin{cases} \cos t, & \cos t \geqslant 0 \\ -\cos t, & \cos t \leqslant 0 \end{cases}$$

$$f_F(t) = \frac{2}{\pi} + \sum_{n=1}^{\infty} \frac{4(-1)^{n+1}}{\pi(4n^2 - 1)} \cos 2nt$$

Half-Wave Rectified Sinusoid

$$f_H(t) = \begin{cases} \cos t, & \cos t \geqslant 0 \\ 0, & \cos t \leqslant 0 \end{cases}$$

$$f_H(t) = \frac{1}{\pi} + \frac{1}{2}\cos t + \sum_{\substack{n=2 \\ n \text{ even}}}^{\infty} \frac{2(-1)^{[(n/2)+1]}}{\pi(n^2 - 1)} \cos nt$$

CHAPTER TEN PROBLEMS

Basic Problems

Periodic Functions

1. One period of the periodic function $f(t)$ is sketched. Find $f(-95)$.

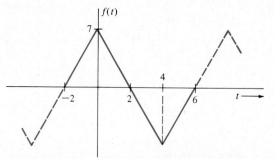

2. Find the periods of the following periodic functions:
  (a) $f(t) = -2 \sin 6t + 3 \cos 6t + 8 \sin 10t - 4 \cos 2t$
  (b) $f(t) = 3 \cos 7t + 4 \cos 8t$
  (c) $f(t) = 10 + 3 \cos \frac{8}{3} t - 4 \sin \frac{7}{2} t$

3. What is the form of a harmonic trigonometric series that completes three cycles every seven seconds?

Orthogonality Relations

4. Show, using the trigonometric identities

$$\cos (x \pm y) = \cos x \cos y \mp \sin x \sin y$$

that

$$\int_T \cos \frac{6\pi t}{T} \cos \frac{8\pi t}{T} \, dt = 0$$

Series Calculation

5. Calculate (setting up and evaluating the integrals for the coefficients) the trigonometric Fourier series for the following periodic functions:

(a)

(b)

(c)

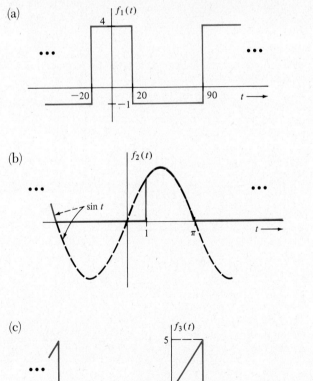

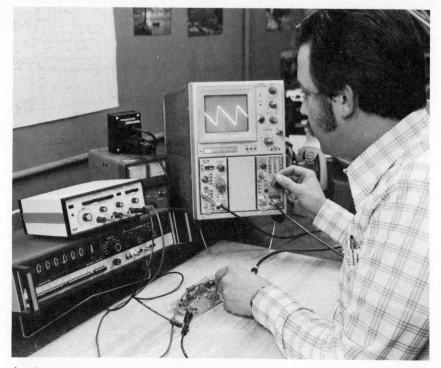

A voltage with a reversed sawtooth waveshape is used to test the performance of a filter network. (*Photo courtesy of Exact Electronics, Inc.*)

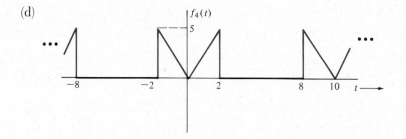

(d)

6. Find the Fourier series for the function $f(t)$, one period of which is the interval $(-2, 2]$, in which

$$f(t) = \begin{cases} 2t & 0 < t \le 2 \\ 0 & \text{otherwise} \end{cases}$$

7. Find the Fourier series for the periodic function

$$f(t) = \tfrac{1}{2} \cos^2 10t + \tfrac{1}{3} \cos^3 10t$$

8. Find the average value of the periodic function by
   (a) using geometric relationships
   (b) integration

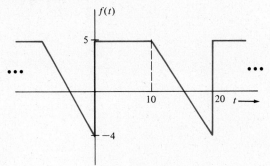

Parseval's Relation and Mean Square Error

9. Find the mean square error between the functions

$$f_1(t) = 3 \sin t \quad \text{and} \quad f_2(t) = 4 \cos t$$

10. Use Parseval's relation to find the average of the square of the basic square wave.

11. The Fourier series for a function is

$$f(t) = 10 + \sum_{n=1}^{\infty} \frac{1}{n^2} \cos\left(nt + \frac{n\pi}{4}\right)$$

Find the mean square error between the function and the approximation if $f(t)$ is approximated by

$$\hat{f}(t) = 10 + \sum_{n=1}^{3} \frac{1}{n^2} \cos\left(nt + \frac{n\pi}{4}\right)$$

12. Find the mean square error between the basic sawtooth wave and a Fourier series approximation that is truncated beyond the fourth harmonic.

13. One period of a periodic function with period 3 seconds is

$$f(t) = \begin{cases} 2t & 0 < t \le 2 \\ -4 & 2 < t \le 3 \end{cases}$$

To what function does the Fourier series for $f(t)$ converge?

Series Properties

14. Using the series for the basic square wave, basic sawtooth wave, and basic triangular wave, find the Fourier series for the following periodic functions:

(a)

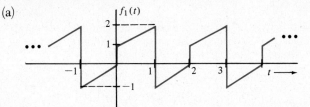

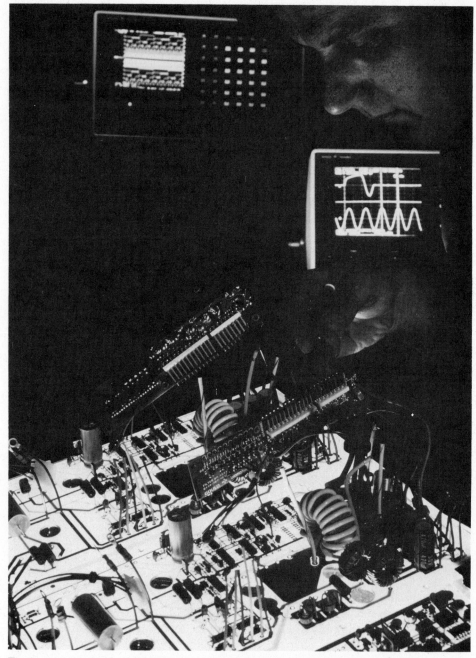

Testing a partially assembled network. The coils of wire are inductors and the larger tubular components are capacitors. The display instruments in the background show the shapes of several network voltages. (*Photo courtesy of Gould, Inc.*)

(b)

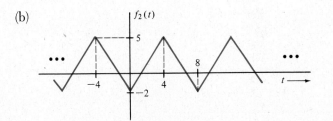

(c)

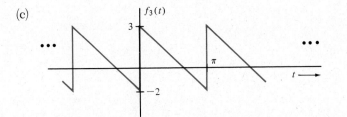

(d)

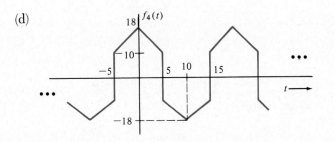

15. Find the even part and the odd part of the function

$$f(t) = \begin{cases} 2t, & t \geq 0 \\ 0, & t \leq 0 \end{cases}$$

and sketch them.

16. Show that if

$$f(t) = d_0 + \sum_{n=1}^{\infty} \left(a_n \cos \frac{n2\pi t}{T} + b_n \sin \frac{n2\pi t}{T} \right)$$

then

$$f(-t) = d_0 + \sum_{n=1}^{\infty} \left[a_n \cos \frac{n2\pi t}{T} + (-b_n) \sin \frac{n2\pi t}{T} \right]$$

Differential Equations with Periodic Driving Functions

17. Find the Fourier series for the forced solutions of the following differential equations:

(a) $2\dfrac{dy}{dt} + y = f(t)$

where $f(t)$ is the basic square wave with period 10 seconds.

(b) $\dfrac{dy}{dt} + 3y = g(t)$

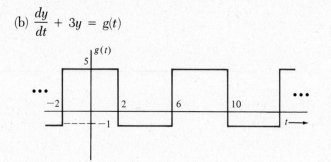

(c) $\dfrac{d^2y}{dt^2} + 4y = h(t)$

where $h(t)$ is the basic triangular wave.

(d) $\dfrac{d^2y}{dt^2} + 2\dfrac{dy}{dt} + 2y = 1 + \displaystyle\sum_{n=1}^{\infty} \dfrac{1}{n^2}\cos 3nt$

18. In the differential equation

$$\dfrac{dy}{dt} + 6y = f(t)$$

the driving function $f(t)$ is unknown before time $t = 0$, but is

$$f(t) = 4 + \sum_{\substack{n=2 \\ n \text{ even}}}^{\infty} \dfrac{3}{n^2 + 5}\sin \pi nt$$

after $t = 0$. Find the *general* solution for $y(t)$ after $t = 0$.

Periodically Driven Networks
19. Find the trigonometric Fourier series for the indicated signals:
   (a)

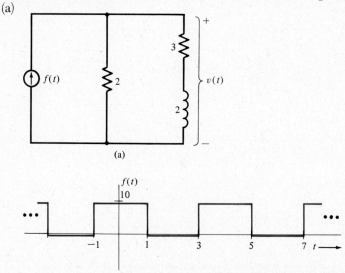

(a)

(b)

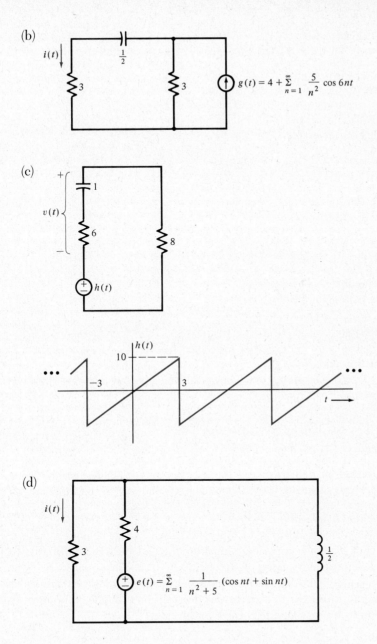

(c)

(d)

20. Find the average power dissipated in the resistor, expressed as an infinite series:

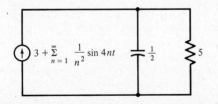

Rectified Sinusoids

21. Using the series for the basic full-wave rectified sinusoid, find the Fourier series for

$$f(t) = |\sin t|$$

Other Series Forms

22. Convert the Fourier series

$$f(t) = \sum_{n=0}^{\infty} \frac{3}{n^3 - 2} \cos\left(100nt + \frac{\pi n}{8}\right)$$

to the quadrature (cosine and sine) form.

23. Convert the Fourier series

$$f(t) = 4 + \sum_{n=1}^{\infty} \left(\frac{1}{1 + n^2} \cos nt + \frac{1}{n^2} \sin nt\right)$$

(a) to a cosine and angle form
(b) to a sine and angle form

Practical Problems

Finding the Function from the Series

24. Occasionally, a Fourier series solution may be recognized as being simply related to known series. The series

$$f(t) = \sum_{\substack{n=1 \\ n \text{ odd}}}^{\infty} \left(\frac{3}{n^2} \cos 10nt - \frac{1}{n} \sin 10nt\right)$$

is a sum of a series for a square wave, plus a series for a triangular wave. Express the series in terms of these two component parts and sketch the function $f(t)$.

Finding Coefficients Graphically

Calculation of the Fourier coefficients for a periodic function may be done graphically or numerically. For example, a periodic function $f(t)$ and the product of $f(t)$ with the first harmonic cosine function are plotted in (a). In (b), the area under one period of the product curve is found by subdivision of the integration interval and rectangular approximation, a process that is easy to program on a digital computer. The first harmonic cosine Fourier coefficient is then given by

$$a_1 = \frac{2}{T} \left(\begin{array}{l}\text{area under the } f(t) \cos \dfrac{2\pi t}{T} \\ \text{curve from 0 to } T\end{array}\right)$$

Similarly, calculation of the Fourier coefficient b_3 is indicated in (b), where the product of the periodic function with the third harmonic sine function is plotted. In (c), the area under the product curve, from 0 to T is approximated. Then

$$b_3 = \frac{2}{T} \left(\begin{array}{l}\text{area under the } f(t) \sin \dfrac{6\pi t}{T} \\ \text{curve from 0 to } T\end{array}\right)$$

For higher order coefficients, the product curve will fluctuate more rapidly, and a finer subdivision to approximate the area under the product curve will be desirable. Higher harmonic coefficients require more and more of the fine detail of $f(t)$, its minute fluctuations, for accurate calculation.

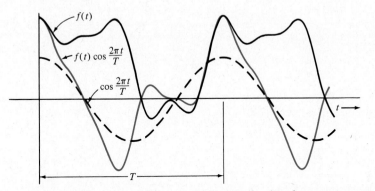

(a) Periodic function $f(t)$ and its product with the first harmonic cosine function

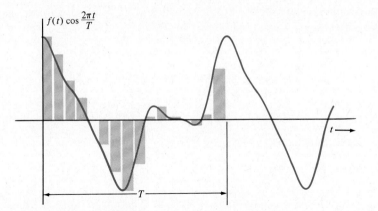

(b) Finding the approximate area of the product

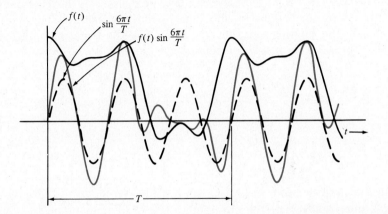

(c) Periodic function $f(t)$ and its product with the third harmonic sine function

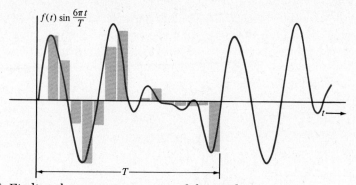

(d) Finding the approximate area of the product

25. By estimating the areas

$$\int_{-\pi}^{\pi} f(t) \cos nt$$

find approximate values for the Fourier coefficients d_0, a_1, and a_2 for the periodic function $f(t)$:

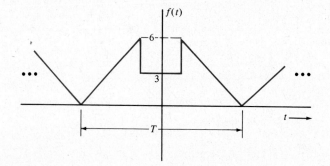

Nonlinearity

When a sinusoidal signal is passed through a nonlinear algebraic device, a periodic but generally nonsinusoidal signal results. If the nonlinear characteristic is expressed as a polynomial, the Fourier series for the output may be found by applying the trigonometric identities for powers of sinusoids:

$$\cos^2 x = \tfrac{1}{2}(1 + \cos 2x) \qquad \cos^3 x = \tfrac{1}{4}(3 \cos x + \cos 3x)$$

$$\sin^2 x = \tfrac{1}{2}(1 - \cos 2x) \qquad \sin^3 x = \tfrac{1}{4}(3 \sin x - \sin 3x)$$

$$\cos x \cos y = \tfrac{1}{2} \cos (x - y) + \tfrac{1}{2} \cos (x + y)$$

$$\sin x \cos y = \tfrac{1}{2} \sin (x + y) + \tfrac{1}{2} \sin (x - y)$$

$$\sin x \sin y = \tfrac{1}{2} \cos (x - y) - \tfrac{1}{2} \cos (x + y)$$

Nonlinear devices are used to generate precisely harmonic frequencies from an incoming sinusoidal signal. The distortion produced by unintentional nonlinearities in a system is commonly measured by comparing the rms harmonic content to the rms fundamental in the system output when the input signal is sinusoidal.

26. For an input

$$x(t) = 3 \cos 6t$$

to a nonlinear device with

$$y = x^4 + 2x^2 - 3x$$

use the trigonometric identities for powers of sinusoids to find the output $y(t)$.

27. A nonlinear device

$$y(x) = 4x^2$$

has input

$$x(t) = \sum_{\substack{n=1 \\ n \text{ odd}}}^{5} \frac{4}{n^2\pi^2} \cos nt$$

Find the amplitudes of the fifth and sixth harmonic components in $y(t)$.

Sampling of Bandlimited Functions

For a periodic function bandlimited above the Nth harmonic,

$$f(t) = d_0 + \sum_{n=1}^{N} \left(a_n \cos \frac{n2\pi t}{T} + b_n \sin \frac{n2\pi t}{T} \right),$$

$2N + 1$ samples of $f(t)$, evenly spaced across its period,

$$f(0), f\left(\frac{T}{2N + 1}\right), f\left(\frac{2T}{2N + 1}\right), \dots, f\left(\frac{2NT}{2N + 1}\right)$$

give $2N + 1$ independent linear algebraic equations in the $2N + 1$ Fourier coefficients. The function $f(t)$ is thus completely determined from the samples.

28. A function of the form

$$f(t) = d_0 + a_1 \cos t + b_1 \sin t$$

has samples

$$f(0) = -2 \qquad f\left(\frac{2\pi}{3}\right) = 1 \qquad f\left(\frac{4\pi}{3}\right) = 3$$

Find $f(t)$.

29. A periodic function $f(t)$ has period 2π and is bandlimited beyond the second harmonic. Find $f(t)$ if it has the samples

$$f(0) = 2 \qquad f\left(\frac{2\pi}{5}\right) = -1 \qquad f\left(\frac{4\pi}{5}\right) = 0 \qquad f\left(\frac{6\pi}{5}\right) = 0 \qquad f\left(\frac{8\pi}{5}\right) = 1$$

Advanced Problems

Periodicity
The sum of two periodic functions is periodic only if the functions share a common interval of repetition. If $f_1(t)$ is periodic with period T_1, and $f_2(t)$ is periodic with period T_2, then

$$f(t) = f_1(t) + f_2(t)$$

is periodic only if there are positive integers m and n such that

$$mT_1 = nT_2.$$

30. Is the function

$$f(t) = \sin t + \sin \sqrt{2}\, t$$

periodic?

31. Certain functions have Fourier series that contain only *odd* harmonics. Why doesn't it make sense to say a series contains only *even* harmonics?

Parseval's Relation
32. Using Parseval's relation, prove *Bessel's inequality:*

$$\frac{1}{T} \int_T f^2(t)\, dt \geq d_0^2 + \frac{1}{2} \sum_{n=1}^{N} (a_n^2 + b_n^2)$$

Even and Odd Periodic Expansions in an Interval
A function $f(t)$ satisfying the Dirichlet conditions in the interval $[0, L]$ may be expanded in the harmonic cosine form

$$f(t) = d_0 + \sum_{n=1}^{\infty} a_n \cos \frac{n\pi t}{L} \qquad 0 \leq t \leq L$$

in that interval, where

$$d_0 = \frac{1}{L} \int_0^L f(t)\, dt, \qquad a_n = \frac{2}{L} \int_0^L f(t) \cos \frac{n\pi t}{L}\, dt$$

Outside the interval, the series is the even periodic expansion of $f(t)$.
Alternatively, such a function may be expanded in harmonic sines,

$$f(t) = \sum_{n=1}^{\infty} b_n \sin \frac{n\pi t}{L} \qquad 0 < t < L$$

where

$$b_n = \frac{2}{L} \int_0^L f(t) \sin \frac{n\pi t}{L}\, dt$$

Outside the interval, the series is the odd periodic expansion of $f(t)$.

33. Expand the function $f(t)$
    (a) in harmonic cosines in the interval $(0,1)$
    (b) in harmonic sines in the interval $(0,1)$

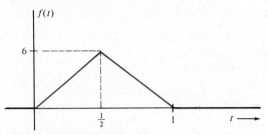

Windowing of Fourier Series

Improved truncated series approximations for periodic functions, particularly those with discontinuities, may be obtained by modifying the Fourier coefficients. By reducing the amplitude of the higher frequency components, the Gibbs' oscillations may be greatly reduced.

The truncation of a function's Fourier series may be imagined as passing the full series through a filter that changes the coefficients as follows: All coefficients for the Nth and lower harmonics are multiplied by unity. All coefficients for the $(N + 1)$th and higher harmonics are multiplied by zero. The process of modifying the amplitudes of a Fourier series is termed *windowing,* and the series truncation is an operation involving a rectangular window, as indicated in (a).

More gradual reductions of the Fourier harmonics results in less Gibbs' oscillation and more pleasing truncated series approximations. Three commonly used window functions are given in (b) and (c). The modified rectangular window simply reduces the amplitude of the highest nonzero harmonic by a factor of $\frac{1}{2}$. The effect of this modification may be seen in the figure, where comparisons of the series approximations are made for a square wave.

The von Hann window more severely modifies all but the constant series term, reducing the Gibbs' oscillations greatly.

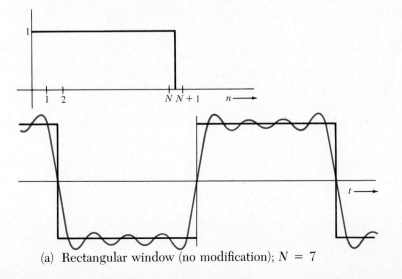

(a) Rectangular window (no modification); $N = 7$

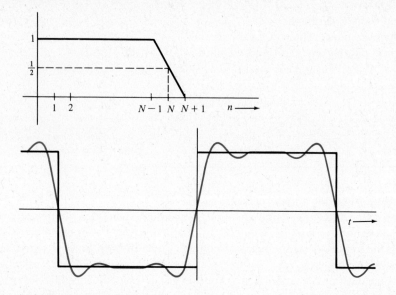

(b) Modified rectangular window (highest harmonic reduced in amplitude by $\frac{1}{2}$); $N = 7$

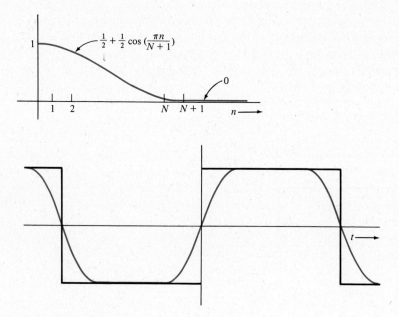

(c) Von Hann window; $N = 7$

34. From your acquaintance with truncated Fourier series so far, sketch your best guess of the shape of the series approximation of the periodic function $g(t)$, truncated beyond the fifth harmonic. Be sure to include the Gibbs' phenomena.

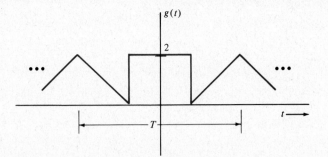

35. Find the harmonic trigonometric series for the basic sawtooth wave, windowed as follows:

(a) modified rectangular, with $N = 5$

(b) Von Hann, with $N = 5$

(c) Von Hann, with $N = 9$

Chapter Eleven
Exponential Fourier Series and the Fourier Transform

When mathematics is cleared away from physics, it becomes
set in logical form. But it is to be remembered that the men
who have in the past initiated great advances in mathematics
have usually been men who were employed in working out
physical questions. They supplied the purists with raw
material to be made coherent and elaborated.

Oliver Heavyside [1850–1925]
From *Electromagnetic Theory*
London, 1899

11.1 PREVIEW

The exponential form of the Fourier series, developed in this chapter, offers several
major advantages over the trigonometric series. Its expression is compact, and there
is a single formula for all of its coefficients. Most important, system response calculations are simplified considerably, since the expansion of a periodic function is
then into exponential components, to which impedance methods apply directly.

The exponential form of the Fourier series leads easily to the Fourier transform, an elegant, general tool for network analysis. Basic Fourier transform properties are examined and Fourier transform spectra are studied. The impulse is defined as a limiting behavior, and transforms of periodic functions are expressed in
terms of impulses. Response of differential equations and electrical networks to
nonperiodic sources are then found, using the Fourier transform.

The beginning student has a great deal to learn here, so the emphasis is on a
careful progression of topics, building on the more intuitive approach of the previous
chapter. Transitions are made from sinusoids to complex exponentials, then from
the discrete series to the continuous transform.

Because partial fraction expansion is developed in detail in the next chapter
in connection with the Laplace transform, there is little emphasis on Fourier transform inversion here. Although simple transforms are inverted, it is the process of
obtaining the solution transform that is stressed now.

When you complete this chapter, you should know—

1. the relations between the trigonometric and the exponential forms of the
   Fourier series;

2. how to calculate the coefficients of the exponential form of the Fourier series directly;
3. what the spectrum of a periodic function is and how to use it;
4. important properties of the exponential form of the series;
5. how to solve linear, time-invariant differential equations with periodic driving functions, using the exponential series;
6. how to use the exponential series to find network response to periodic sources;
7. the definition, meaning, and basic properties of the Fourier transform;
8. the meaning and uses of the spectrum of a nonperiodic function;
9. how to use the Fourier transform to solve linear, time-invariant integro-differential equations;
10. how to use the Fourier transform to solve electrical networks;
11. about impulses and how they are used to simplify calculations.

11.2 CONVERSIONS BETWEEN TRIGONOMETRIC AND EXPONENTIAL SERIES FORMS

11.2.1 Exponential Form of the Series

Using the Euler expansion of sinusoidal functions,

$$\cos x = \frac{1}{2}e^{jx} + \frac{1}{2}e^{-jx}$$

$$\sin x = \frac{e^{jx} - e^{-jx}}{2j} = -\frac{j}{2}e^{jx} + \frac{j}{2}e^{-jx}$$

an alternative Fourier series form is the following:

$$f(t) = d_0 + \sum_{n=1}^{\infty}\left(a_n \cos\frac{n2\pi t}{T} + b_n \sin\frac{n2\pi t}{T}\right)$$

$$= d_0 + \sum_{n=1}^{\infty}\left[\frac{a_n}{2}e^{j\frac{n2\pi t}{T}} + \frac{a_n}{2}e^{-j\frac{n2\pi t}{T}} - \frac{jb_n}{2}e^{j\frac{n2\pi t}{T}} + \frac{jb_n}{2}e^{-j\frac{n2\pi t}{T}}\right]$$

$$= d_0 + \sum_{n=1}^{\infty}\left[\left(\frac{a_n - jb_n}{2}\right)e^{j\frac{n2\pi t}{T}} + \left(\frac{a_n + jb_n}{2}\right)e^{-j\frac{n2\pi t}{T}}\right]$$

This expansion may be put into a more compact form by separating exponential terms and changing the algebraic sign of one of the summation indices. Then, recognizing that d_0 is a term of the same form, a single summation results.

$$f(t) = d_0 + \sum_{n=1}^{\infty}\left(\frac{a_n - jb_n}{2}\right)e^{j\frac{n2\pi t}{T}} + \sum_{n=1}^{\infty}\left(\frac{a_n + jb_n}{2}\right)e^{-j\frac{n2\pi t}{T}}$$

$$= d_0 + \sum_{n=1}^{\infty}\left(\frac{a_n - jb_n}{2}\right)e^{j\frac{n2\pi t}{T}} + \sum_{n=-1}^{-\infty}\left(\frac{a_{-n} + jb_{-n}}{2}\right)e^{j\frac{n2\pi t}{T}}$$

$$= d_0 \, e^{j\frac{0\cdot2\pi t}{T}} + \sum_{n=1}^{\infty} \left(\frac{a_n - jb_n}{2} \right) e^{j\frac{n2\pi t}{T}} + \sum_{n=-1}^{-\infty} \left(\frac{a_{-n} + jb_{-n}}{2} \right) e^{j\frac{n2\pi t}{T}}$$

$$= \sum_{n=-\infty}^{\infty} d_n \, e^{j\frac{n2\pi t}{T}}$$

where

$$d_n = \begin{cases} \dfrac{a_n - jb_n}{2} & n = 1, 2, 3, \ldots \\ d_0 & n = 0 \\ \dfrac{a_{-n} + jb_{-n}}{2} & n = -1, -2, -3, \ldots \end{cases}$$

This latter expansion of a periodic function into complex exponentials is termed the *exponential form* of the Fourier series. The coefficients of the exponential form, the d_ns, are seen to be a joining of the a_ns and b_ns into single complex numbers.

The coefficients, d_n, of the exponential Fourier series, which in general are complex numbers, have the property

$$d_{-n} = d_n^*$$

for each n. The average value, d_0, is always a real number.

As an example of conversion of a specific trigonometric Fourier series to the exponential form, consider the basic square wave:

$$f_s(t) = \sum_{\substack{n=1 \\ n \text{ odd}}}^{\infty} \frac{4}{n\pi} \sin nt = \sum_{\substack{n=1 \\ n \text{ odd}}}^{\infty} \frac{4}{n\pi} \left(\frac{-j}{2} e^{jnt} + \frac{j}{2} e^{-jnt} \right)$$

$$= \sum_{\substack{n=1 \\ n \text{ odd}}}^{\infty} \frac{-2j}{n\pi} e^{jnt} + \sum_{\substack{n=1 \\ n \text{ odd}}}^{\infty} \frac{2j}{n\pi} e^{-jnt} = \sum_{\substack{n=1 \\ n \text{ odd}}}^{\infty} \frac{-2j}{n\pi} e^{jnt} + \sum_{\substack{n=-1 \\ n \text{ odd}}}^{-\infty} \frac{-2j}{n\pi} e^{jnt}$$

$$= \sum_{\substack{n=-\infty \\ n \text{ odd}}}^{\infty} \frac{-2j}{n\pi} e^{jnt}$$

11.2.2 Relations Between Coefficients

The coefficients of the trigonometric form of the series are related to the d_ns by

$$\begin{cases} a_n = 2\text{Re}\,[d_n] = 2\text{Re}\left[\dfrac{a_n - jb_n}{2} \right] & n = 1, 2, 3, \ldots \\ b_n = -2\text{Im}\,[d_n] = -2\text{Im}\left[\dfrac{a_n - jb_n}{2} \right] & n = 1, 2, 3, \ldots \end{cases}$$

A series in exponential form

$$f(t) = \sum_{n=-\infty}^{\infty} \frac{2 + 4jn}{3n^2 + 1} e^{j8nt}$$

has equivalent quadrature trigonometric form with

$$d_0 = 2$$

$$a_n = 2\text{Re}\left[\frac{2 + 4jn}{3n^2 + 1}\right] = \frac{4}{3n^2 + 1}$$

$$b_n = -2\text{Im}\left[\frac{2 + 4jn}{3n^2 + 1}\right] = \frac{-8n}{3n^2 + 1}$$

giving

$$f(t) = 2 + \sum_{n=1}^{\infty}\left(\frac{4}{3n^2 + 1}\cos 8nt - \frac{8n}{3n^2 + 1}\sin 8nt\right)$$

The coefficients of the trigonometric cosine and angle series form are also simply related to the d_ns:

$$\begin{cases} c_n = \sqrt{a_n^2 + b_n^2} = 2|d_n| & n = 1, 2, 3, \ldots \\ \\ \theta_n = -\arctan\dfrac{b_n}{a_n} = \angle d_n & n = 1, 2, 3, \ldots \end{cases}$$

and

$$d_n = \begin{cases} \dfrac{1}{2}c_n e^{j\theta_n} & n = 1, 2, 3, \ldots \\ \\ d_0 & n = 0 \\ \\ \dfrac{1}{2}c_n e^{-j\theta_n} & n = 1, 2, 3, \ldots \end{cases}$$

The series

$$f(t) = 10 + \sum_{n=1}^{\infty}\frac{6}{n^2}\cos\left(100n\pi t + \frac{n\pi}{6}\right)$$

for example, has exponential form

$$f(t) = 10 + \sum_{\substack{n=-\infty \\ n \neq 0}}^{\infty}\left(\frac{3}{n^2}e^{j\frac{n\pi}{6}}\right)e^{j100n\pi t}$$

D11-1

Convert the following Fourier series to the indicated forms:

(a) $f(t) = 2 + \displaystyle\sum_{n=1}^{\infty}\left(\frac{1}{n^2 + 1}\right)\cos 8\pi nt$

to exponential form

ans. $2 + \displaystyle\sum_{\substack{n=-\infty \\ n \neq 0}}^{\infty}\frac{1}{2(n^2 + 1)}e^{j8\pi nt}$

(b) $f(t) = 4 + \sum\limits_{\substack{n=1 \\ n \text{ odd}}}^{\infty} \dfrac{2}{n^2 + 8} \cos\left(nt + \dfrac{n\pi}{8}\right)$

to exponential form

ans. $4 + \sum\limits_{\substack{n=-\infty \\ n \text{ odd}}}^{\infty} \left(\dfrac{1}{n^2 + 8} e^{j\frac{n\pi}{8}}\right) e^{jnt}$

(c) $f(t) = \sum\limits_{n=-\infty}^{\infty} \dfrac{3 + jn}{n^2 + 4} e^{j5nt}$

to the cosine and sine form

ans. $\dfrac{3}{4} + \sum\limits_{n=1}^{\infty} \left(\dfrac{6}{n^2 + 4} \cos 5nt - \dfrac{2n}{n^2 + 4} \sin 5nt\right)$

(d) $f(t) = \sum\limits_{\substack{n=-\infty \\ n \neq 0}}^{\infty} \dfrac{-6 + jn}{n^2} e^{j4\pi nt}$

to the quadrature (sine and cosine) form

ans. $\sum\limits_{\substack{n=-\infty \\ n \neq 0}}^{\infty} \left(\dfrac{-12}{n^2} \cos 4\pi nt - \dfrac{2}{n} \sin 4\pi nt\right)$

CONVERSIONS BETWEEN TRIGONOMETRIC AND EXPONENTIAL SERIES FORMS

Another Fourier series form is the exponential one,

$$f(t) = d_0 + \sum_{n=1}^{\infty} \left(a_n \cos \frac{n2\pi t}{T} + b_n \sin \frac{n2\pi t}{T}\right)$$

$$= \sum_{n-\infty}^{\infty} d_n e^{\frac{jn2\pi t}{T}}$$

where

$$d_n = \begin{cases} \dfrac{a_n - jb_n}{2} & n = 1, 2, 3, \ldots \\ d_0 & n = 0 \\ \dfrac{a_{-n} + jb_{-n}}{2} & n = -1, -2, -3, \ldots; \end{cases}$$

and

$a_n = 2\text{Re}[d_n] \qquad n = 1, 2, 3, \ldots$

$b_n = -2\text{Im}[d_n] \qquad n = 1, 2, 3, \ldots$

$c_n = \sqrt{a_n^2 + b_n^2} = 2|d_n| = 2|d_{-n}| \qquad n = 1, 2, 3, \ldots$

$\theta_n = -\arctan \dfrac{b_n}{a_n} = \underline{/d_n} = -\underline{/d_{-n}} \qquad n = 1, 2, 3, \ldots$

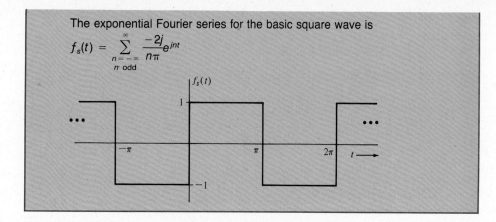

The exponential Fourier series for the basic square wave is

$$f_s(t) = \sum_{\substack{n=-\infty \\ n \text{ odd}}}^{\infty} \frac{-2j}{n\pi} e^{jnt}$$

11.3 EXPONENTIAL FOURIER SERIES CALCULATION

11.3.1 Integrals for the Coefficients

The coefficients for the exponential form of the Fourier series may be obtained directly by integration. As will be shown, each of the d_ns is given by the same integral formula, an improvement over the three formulas (for d_0, a_n, and b_n) involved with the trigonometric form of the series.

For $n = 1, 2, 3, \ldots$,

$$d_n = \frac{a_n - jb_n}{2} = \frac{1}{T}\int_T f(t) \cos \frac{n2\pi t}{T}\, dt - \frac{j}{T}\int_T f(t) \sin \frac{n2\pi t}{T}\, dt$$

$$= \frac{1}{T}\int_T f(t)\left(\cos \frac{n2\pi t}{T} - j\sin \frac{n2\pi t}{T}\right) dt = \frac{1}{T}\int_T f(t)e^{-j\frac{n2\pi t}{T}}\, dt$$

For $n = -1, -2, -3, \ldots$

$$d_n = \frac{a_{-n} + jb_{-n}}{2} = \frac{1}{T}\int_T f(t) \cos\left(\frac{-n2\pi t}{T}\right) dt + \frac{j}{T}\int_T f(t) \sin\left(\frac{-n2\pi t}{T}\right) dt$$

$$= \frac{1}{T}\int_T f(t)\left(\cos \frac{n2\pi t}{T} - j\sin \frac{n2\pi t}{T}\right) dt = \frac{1}{T}\int_T f(t)e^{-j\frac{n2\pi t}{T}}\, dt$$

which is the same relation as for positive n. The average value is also given by this formula, for $n = 0$,

$$d_0 = \frac{1}{T}\int_T f(t)\, dt = \frac{1}{T}\int_T f(t)e^{-j\frac{0\cdot 2\pi t}{T}}\, dt$$

so all of the exponential series coefficients may be calculated using

$$d_n = \frac{1}{T}\int_T f(t)e^{-j\frac{n2\pi t}{T}}\, dt$$

Consider the basic periodic pulse of Figure 11-1. The *duty cycle* of a periodic pulse is defined to be the ratio of "up time" to the period. For this pulse, the duty

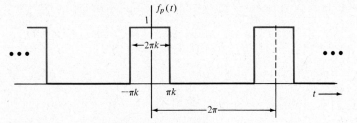

Figure 11-1 The basic periodic pulse.

cycle is k, and it is assumed, of course, that k is less than unity. The value of k is not yet specified, so that later results will apply to periodic pulses with any duty cycle. The exponential Fourier series coefficients for the basic pulse are given by

$$
d_n = \frac{1}{T} \int_T f_p(t) e^{-j\frac{n2\pi t}{T}} \, dt = \frac{1}{2\pi} \int_{-\pi k}^{\pi k} e^{-jnt} \, dt = \frac{1}{2\pi} \left[\frac{e^{-jnt}}{-jn} \right]_{-\pi k}^{\pi k}
$$

$$
= \frac{1}{n\pi} \left(\frac{e^{-jnk\pi} - e^{jnk\pi}}{-2j} \right) = \frac{\sin nk\pi}{n\pi}
$$

This relation is indeterminate for $n = 0$, so d_0 must be found separately, as follows:

$$
d_0 = \frac{1}{T} \int_T f_p(t) \, dt = \frac{1}{2\pi} \int_{-\pi k}^{\pi k} dt = k
$$

The basic periodic pulse thus has exponential Fourier series

$$
f_p(t) = \cdots + \left(\frac{\sin 2k\pi}{2\pi} \right) e^{-j2t} + \left(\frac{\sin k\pi}{\pi} \right) e^{-jt} + k
$$

$$
+ \left(\frac{\sin k\pi}{\pi} \right) e^{jt} + \left(\frac{\sin 2k\pi}{2\pi} \right) e^{j2t} + \cdots
$$

$$
= k + \sum_{\substack{n=-\infty \\ n \neq 0}}^{\infty} \frac{\sin nk\pi}{n\pi} e^{jnt} = k \operatorname{sinc}(nk\pi)
$$

where the sinc function is defined as

$$
\operatorname{sinc} x =
\begin{cases}
\dfrac{\sin x}{x} & x \neq 0 \\[2mm]
1 & x = 0
\end{cases}
$$

11.3.2 Parseval's Relation and Mean Square Error

In terms of the exponential series coefficients, Parseval's relation is

$$
\frac{1}{T} \int_T f^2(t) \, dt = d_0^2 + \frac{1}{2} \sum_{n=1}^{\infty} (a_n^2 + b_n^2) = d_0^2 + \frac{1}{2} \sum_{n=1}^{\infty} c_n^2
$$

$$
= d_0^2 + \frac{1}{2} \sum_{n=1}^{\infty} (2|d_n|)^2 = d_0^2 + 2 \sum_{n=1}^{\infty} |d_n|^2 = \sum_{n=-\infty}^{\infty} |d_n|^2
$$

A truncated Fourier series approximation, $g_N(t)$, to a periodic function

$$f(t) = d_0 + \sum_{n=1}^{\infty} \left(a_n \cos \frac{n2\pi t}{T} + b_n \sin \frac{n2\pi t}{T} \right) = \sum_{n=-\infty}^{\infty} d_n e^{j\frac{n2\pi t}{T}}$$

is given by

$$g_N(t) = d_0 + \sum_{n=1}^{N} \left(a_n \cos \frac{n2\pi t}{T} + b_n \sin \frac{n2\pi t}{T} \right) = \sum_{n=-N}^{N} d_n e^{j\frac{n2\pi t}{T}}$$

The mean square error of the approximation is

$$\text{MSE} = \frac{1}{T} \int_T [f(t) - g_N(t)]^2 \, dt = \frac{1}{2} \sum_{n=N+1}^{\infty} (a_n^2 + b_n^2) = \frac{1}{2} \sum_{n=N+1}^{\infty} c_n^2$$

$$= \frac{1}{2} \sum_{n=N+1}^{\infty} (2|d_n|)^2 = 2 \sum_{n=N+1}^{\infty} |d_n|^2$$

The MSE may be expressed as the sum of the squares of the coefficients of the truncated terms, if desired:

$$\text{MSE} = 2 \sum_{n=N+1}^{\infty} |d_n|^2 = \sum_{n=-(N+1)}^{-\infty} |d_n|^2 + \sum_{n=N+1}^{\infty} |d_n|^2$$

D11-2

Calculate exponential Fourier series for the following functions:

(a)

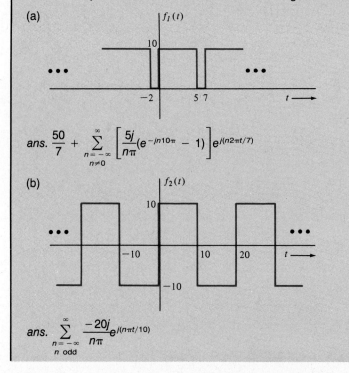

ans. $\dfrac{50}{7} + \displaystyle\sum_{\substack{n=-\infty \\ n \neq 0}}^{\infty} \left[\dfrac{5j}{n\pi}(e^{-jn10\pi} - 1) \right] e^{j(n2\pi t/7)}$

(b)

ans. $\displaystyle\sum_{\substack{n=-\infty \\ n \text{ odd}}}^{\infty} \dfrac{-20j}{n\pi} e^{j(n\pi t/10)}$

(c)

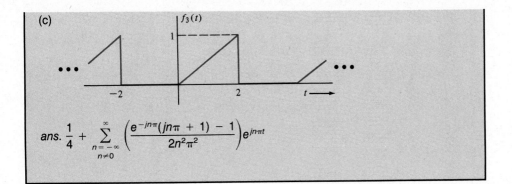

ans. $\dfrac{1}{4} + \displaystyle\sum_{\substack{n=-\infty \\ n\neq 0}}^{\infty} \left(\dfrac{e^{-jn\pi}(jn\pi + 1) - 1}{2n^2\pi^2} \right) e^{jn\pi t}$

EXPONENTIAL FOURIER SERIES CALCULATION

The coefficients of the exponential form of the Fourier series are each given by

$$d_n = \frac{1}{T} \int_T f(t) e^{\frac{-jn2\pi t}{T}} \, dt$$

Parseval's relation is, in terms of the d_ns,

$$\frac{1}{T} \int_T f^2(t) \, dt = \sum_{n=-\infty}^{\infty} |d_n|^2 = d_0^2 + 2 \sum_{n=1}^{\infty} |d_n|^2$$

and the mean square error of a truncated series approximation,

$$f(t) \cong f_N(t) = \sum_{n=-N}^{N} d_n e^{\frac{jn2\pi t}{T}}$$

is

$$\text{MSE} = \frac{1}{T} \int_T [f(t) - f_N(t)]^2 \, dt = 2 \sum_{n=N+1}^{\infty} |d_n|^2$$

The *duty cycle* of a periodic pulse is the ratio of "up time" to the period. For the basic periodic pulse with duty cycle k,

$$f_p(t) = k + \sum_{\substack{n=-\infty \\ n\neq 0}}^{\infty} \frac{\sin nk\pi}{\pi n} e^{jnt}$$

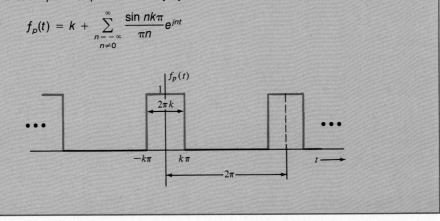

11.4 EXPONENTIAL SERIES LINE SPECTRA

11.4.1 The Real and Imaginary Part Spectrum

In this section, graphical representations of exponential Fourier series are developed. These portrayals, termed *line spectrum plots*, provide an important capability for the visualization of periodic signal properties.

The *spectrum* of a periodic function

$$f(t) = \sum_{n=-\infty}^{\infty} d_n e^{j\frac{n2\pi t}{T}}$$

is the set of coefficients d_n. A *spectral plot* for a periodic function consists of graphs of the d_ns versus the frequency of the corresponding complex exponential term. As the d_ns are generally complex numbers, two graphs are most convenient, one of the real parts of the d_ns versus frequency, and the other of the imaginary parts of the d_ns versus frequency.

For example, a periodic function with exponential series

$$f(t) = \sum_{n=-\infty}^{\infty} \frac{3 + jn}{n^2 + 4} e^{j10nt}$$

has

$$d_n = \frac{3 + jn}{n^2 + 4} \qquad \mathrm{Re}[d_n] = \frac{3}{n^2 + 4} \qquad \mathrm{Im}[d_n] = \frac{n}{n^2 + 4}$$

Real and imaginary part plots for the d_ns versus complex harmonic number n are shown in Figure 11-2(a).

True spectral plots show the d_ns versus the frequencies of the corresponding exponential terms,

$$\omega = \frac{n2\pi}{T}$$

as in Figure 11-2(b). That way, the series may be recovered from the plots. If desired, the hertz (cycle per second) frequency

$$f = \frac{\omega}{2\pi} = \frac{n}{T}$$

may be used, as in Figure 11-2(c).

For a real periodic function $f(t)$,

$$\mathrm{Re}[d_{-n}] = \mathrm{Re}[d_n] \qquad \text{and} \qquad \mathrm{Im}[d_{-n}] = -\mathrm{Im}[d_n]$$

The real parts of the d_ns are an even function of n, while the imaginary parts are odd in n. The real-part spectral plot is an even function of frequency and the imaginary-part plot is an odd function of frequency.

Consider the spectral plots of Figure 11-3. The real and imaginary parts of the first several d_ns may be read from the plots, along with the corresponding complex

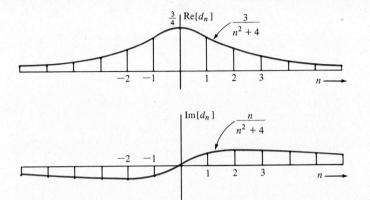

(a) Exponential Fourier series coefficients versus complex harmonic number

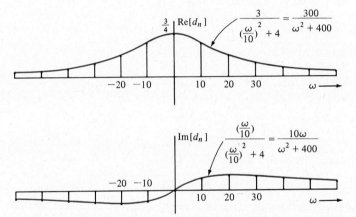

(b) Spectral plot: Real and imaginary parts of the exponential Fourier series coefficients are plotted versus the radian frequency of the corresponding complex exponential term

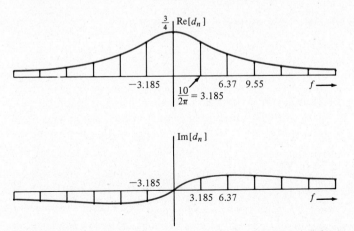

(c) Spectral plots in terms of the Hertz (cycles per second) frequency

Figure 11-2   Real and imaginary part spectral plots for an example periodic function.

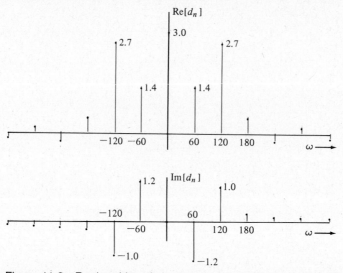

Figure 11-3   Real and imaginary part spectral plots for a periodic function.

exponential frequencies. The periodic function represented by these plots has Fourier series terms

$$f(t) = \cdots + (2.7 - j1.0)e^{-j120t} + (1.4 + j1.2)e^{-j60t} + (3.0)$$
$$+ (1.4 - j1.2)e^{j60t} + (2.7 + j1.0)e^{j120t} + \cdots$$

The corresponding trigonometric series terms are as follows:

$$f(t) = 3.0 + [(1.4 - j1.2)e^{j60t} + (1.4 + j1.2)e^{-j60t}]$$

$$+ [(2.7 + j1.0)e^{j120t} + (2.7 - j1.0)e^{-120t}] + \cdots$$

$$= 3.0 + 1.4(e^{j60t} + e^{-j60t}) + 1.2(-je^{j60t} + je^{-60t})$$

$$+ 2.7(e^{j120t} + e^{-j120t}) + 1.0(je^{j120t} - j3^{-j120t}) + \cdots$$

$$= 3.0 + 2.8 \cos 60t + 2.4 \sin 60t + 5.4 \cos 120t$$

$$- 2.0 \sin 120t + \cdots$$

11.4.2 The Magnitude and Angle Spectrum

An alternative form for spectral plots of a periodic function consists of graphs of the magnitudes and angles of the d_ns versus exponential radian or hertz frequency. For the periodic function

$$f(t) = \sum_{n=-\infty}^{\infty} \left(\frac{n^2 - 6jn}{n^4 + 8} \right) e^{j30\pi nt}$$

$$d_n = \frac{n^2 - 6jn}{n^4 + 8}$$

$$|d_n| = \frac{\sqrt{(n^2)^2 + (-6n)^2}}{n^4 + 8} = \frac{\sqrt{n^4 + 36n^2}}{n^4 + 8} \qquad \angle d_n = \arctan \frac{-6n}{n^2}$$

Magnitude and angle plots for the d_ns versus complex harmonic number n are shown in Figure 11-4(a). Spectral plots, using radian frequency and hertz frequency, are given in Figure 11-4(b) and (c).

For a real periodic function,

$$|d_{-n}| = |d_n| \quad \text{and} \quad \underline{/d_{-n}} = -\underline{/d_n}$$

The magnitude spectral plot is an even function of frequency, and the angle plot (except possibly for the d_0 term) may be expressed as an odd function of frequency.

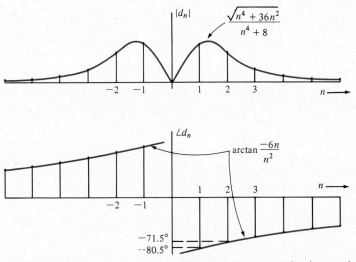

(a) Exponential Fourier series coefficients versus complex harmonic number

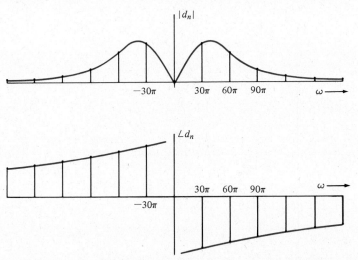

(b) Spectral plots: Magnitude and angle of the exponential Fourier series coefficients are plotted versus the radian frequency of the corresponding complex exponential term

Figure 11-4   Magnitude and angle spectral plots for an example periodic function. (Part (c) follows on p. 626.)

Figure 11-4 *(Continued)*

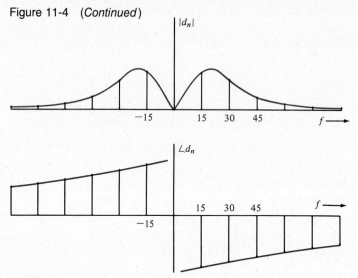

(c) Spectral plots in terms of Hertz frequency

There are many possibilities for the angle plot, since adding or subtracting a multiple of 360° to the angle is of no consequence.

An example of recovering series terms from spectral plots of magnitude and angle is given in the example of Figure 11-5, for which

$$f(t) = \cdots + (1.3e^{j65°})e^{-j50t} + (2.4e^{-j90°})e^{-j25t} + (1.8e^{j180°})$$
$$+ (2.4e^{j90°})e^{j25t} + (1.3e^{-j65°})e^{j50t} + \cdots$$

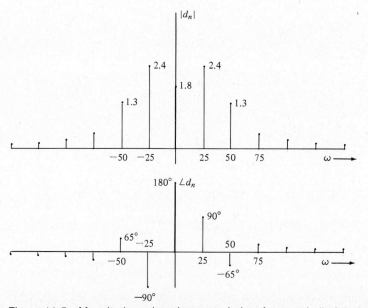

Figure 11-5 Magnitude and angle spectral plots for a periodic function.

The corresponding terms in the cosine and angle form of the trigonometric series for this function are as follows:

$$f(t) = -1.8 + 2.4[e^{j(25t+90°)} + e^{-j(25t+90°)}]$$
$$+ 1.3[e^{j(50t-65°)} + e^{-j(50t-65°)}] + \cdots$$
$$= -1.8 + 4.8 \cos(2t + 90°) + 2.6 \cos(50t - 65°) + \cdots$$

As, strictly speaking, the magnitude of a complex number is nonnegative, the negative average value of $f(t)$,

$$d_0 = -1.8$$

is represented by magnitude 1.8 and angle 180°.

The choice of magnitude and angle or real and imaginary part plots is made in practice according to which is easiest to construct or most useful to display. Radian frequency is generally used in analytical work because its use eliminates factors of 2π here and there. Hertz frequency is used most often in experimental work.

D11-3

Find and sketch the indicated type of spectra for each of the following periodic functions:

(a) Real and imaginary part:

$$f(t) = \sum_{n=-\infty}^{\infty} \frac{1 + 3jn}{n^2 + 1} e^{j4nt}$$

(b) Real and imaginary part:

$$f(t) = 10 + \sum_{\substack{n=-\infty \\ n \neq 0}}^{\infty} \frac{1}{1 - jn} e^{j\pi nt}$$

(c) Magnitude and angle:

$$f(t) = \sum_{n=-\infty}^{\infty} \frac{-2 + j3n}{n^2 + 4} e^{j100nt}$$

(d) Magnitude and angle:

$$f(t) = -4 + \sum_{\substack{n=-\infty \\ n \neq 0}}^{\infty} \frac{1}{n^4} e^{j[3nt + (n\pi/8)]}$$

D11-4

Find the exponential Fourier series of the functions with the following spectra:

(a)

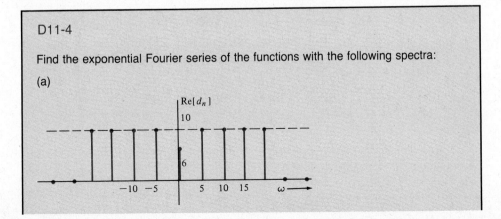

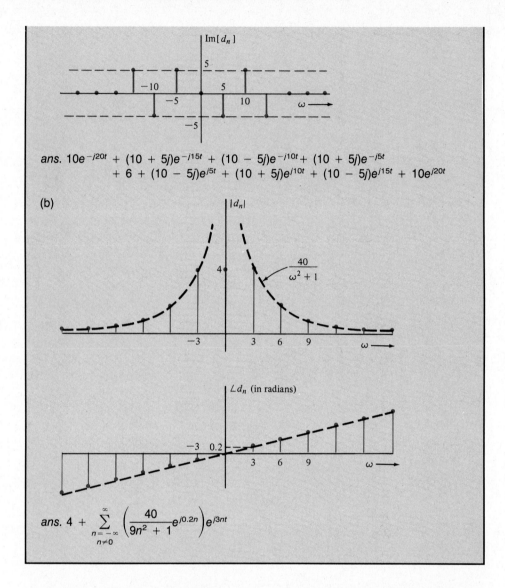

ans. $10e^{-j20t} + (10 + 5j)e^{-j15t} + (10 - 5j)e^{-j10t} + (10 + 5j)e^{-j5t}$
$+ 6 + (10 - 5j)e^{j5t} + (10 + 5j)e^{j10t} + (10 - 5j)e^{j15t} + 10e^{j20t}$

(b)

ans. $4 + \displaystyle\sum_{\substack{n=-\infty \\ n \neq 0}}^{\infty} \left(\frac{40}{9n^2 + 1} e^{j0.2n} \right) e^{j3nt}$

EXPONENTIAL SERIES SPECTRA

The *spectrum* of a periodic function

$$f(t) = \sum_{n=-\infty}^{\infty} d_n e^{\frac{jn2\pi t}{T}}$$

is the set of coefficients d_n as a function of radian frequency $\omega = (n2\pi/T)$ or hertz $f = (n/T)$. *Spectral plots* consist of either

(1) a graph of the real parts of the d_ns versus frequency and a graph of the imaginary parts of the d_ns versus frequency

or

(2) a graph of the magnitude of the d_ns versus frequency and a graph of the angle of the d_ns versus frequency.

The coefficients of the exponential Fourier series have the property

$$d_{-n} = d_n^*$$

Thus, the real part plot is even in ω, and the imaginary part plot is odd in ω. The magnitude plot is even in ω, and the angle plot may be arranged as an odd function of ω except possibly for the d_0 term at $\omega = 0$ which could have an angle of 180°.

11.5 MANIPULATING EXPONENTIAL SERIES

11.5.1 Even and Odd Functions

For an even periodic function, the d_ns are each real numbers:

$$d_n = \begin{cases} \dfrac{a_n - jb_n}{2} = \dfrac{a_n}{2} & n = 1, 2, 3, \ldots \\[3mm] \dfrac{a_{-n} + jb_{-n}}{2} = \dfrac{a_{-n}}{2} & n = -1, -2, -3, \ldots \end{cases}$$

The basic triangular wave, Figure 11-6(a), is an example of an even periodic function. Its trigonometric Fourier series was found earlier to be

$$f(t) = \sum_{\substack{n=1 \\ n \text{ odd}}}^{\infty} \frac{4}{n^2\pi^2} \cos nt$$

Its exponential Fourier series has entirely real coefficients d_n, as follows:

$$f(t) = \sum_{\substack{n=1 \\ n \text{ odd}}}^{\infty} \frac{4}{n^2\pi^2}\left(\frac{e^{jnt} + e^{-jnt}}{2}\right) = \sum_{\substack{n=1 \\ n \text{ odd}}}^{\infty} \frac{4}{n^2\pi^2}e^{jnt} + \sum_{\substack{n=1 \\ n \text{ odd}}}^{\infty} \frac{4}{n^2\pi^2}e^{-jnt}$$

$$= \sum_{\substack{n=1 \\ n \text{ odd}}}^{\infty} \frac{2}{n^2\pi^2}e^{jnt} + \sum_{n=-1}^{-\infty} \frac{2}{n^2\pi^2}e^{jnt} = \sum_{\substack{n=1 \\ n \text{ odd}}}^{\infty} \frac{2}{n^2\pi^2}e^{jnt}$$

Real and imaginary part spectral plots are easiest to sketch for an even periodic function, the imaginary part of the spectrum being zero. These are shown for the basic triangular wave in Figure 11-6(b).

For an odd periodic function, $d_0 = 0$ and

$$d_n = \begin{cases} \dfrac{a_n - jb_n}{2} = -\dfrac{jb_n}{2} & n = 1, 2, 3, \ldots \\[3mm] \dfrac{a_{-n} + jb_{-n}}{2} = \dfrac{jb_{-n}}{2} & n = -1, -2, -3, \ldots \end{cases}$$

The d_ns are each purely imaginary numbers.

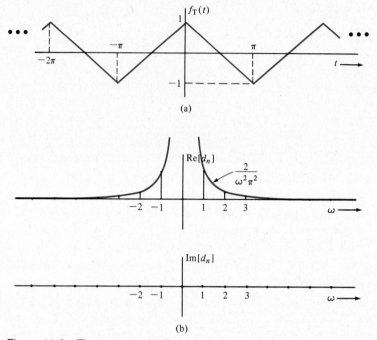

Figure 11-6 The exponential Fourier series coefficients for an even periodic function are entirely real.
(a) Basic triangular wave.
(b) Spectral plots for this even periodic function.

The level shifted sawtooth wave of Figure 11-7(a) is an odd periodic function and has trigonometric Fourier series

$$g(t) = \sum_{n=1}^{\infty} \frac{1}{n\pi} \sin n2\pi t$$

The exponential Fourier series for the function is as follows:

$$g(t) = \sum_{n=1}^{\infty} \frac{1}{n\pi} \left(\frac{e^{jn2\pi t} - e^{-jn2\pi t}}{2j} \right) = \sum_{n=1}^{\infty} \frac{-j}{2n\pi} e^{jn2\pi t} + \sum_{n=1}^{\infty} \frac{j}{2n\pi} e^{-jn2\pi t}$$

$$= \sum_{n=1}^{\infty} \frac{-j}{2n\pi} e^{jn2\pi t} + \sum_{n=-1}^{-\infty} \frac{j}{-2n\pi} e^{jn2\pi t} = \sum_{\substack{n=-\infty \\ n \neq 0}}^{\infty} \frac{-j}{2n\pi} e^{jn2\pi t}$$

Real and imaginary part spectral plots for this function are given in Figure 11-7(b).

For a function that is neither even nor odd, the real parts of the d_ns are the Fourier coefficients of the even part of the function, and the imaginary parts of the d_ns correspond to the odd part of the function.

11.5.2 Amplitude, Level, and Time Shifts

The series for a function that is a constant A times a periodic function has exponential series coefficients that are A times those of the original function. As with the trig-

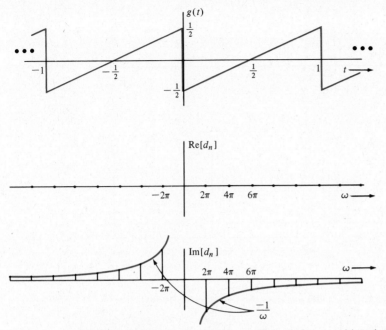

Figure 11-7   The exponential Fourier series coefficients for an odd periodic function are entirely imaginary.
(a) Sawtooth wave.
(b) Spectral plots for this odd periodic function.

onometric form of the series, shifting the level of a periodic function changes only the constant term in the series.

A time scale change does not affect the Fourier coefficients d_n. The exponential frequencies are changed to those associated with the new period, of course. For example, the periodic pulse function of Figure 11-8(a) has exponential series

$$f_p(t) = \frac{1}{10} + \sum_{\substack{n=-\infty \\ n \neq 0}}^{\infty} \frac{\sin \dfrac{n\pi}{10}}{\pi n} e^{jnt}$$

The related function of Figure 11-8(b) is

$$g(t) = -1 + 6f_p(t) = -\frac{4}{10} + \sum_{\substack{n=-\infty \\ n \neq 0}}^{\infty} \frac{6 \sin \dfrac{n\pi}{10}}{\pi n} e^{jnt}$$

The time scaled function in Figure 11-8(c) has series

$$h(t) = g(2\pi \times 10^6 t) = -\frac{4}{10} + \sum_{\substack{n=-\infty \\ n \neq 0}}^{\infty} \frac{6 \sin \dfrac{n\pi}{10}}{\pi n} e^{j2 \times 10^6 n\pi t}$$

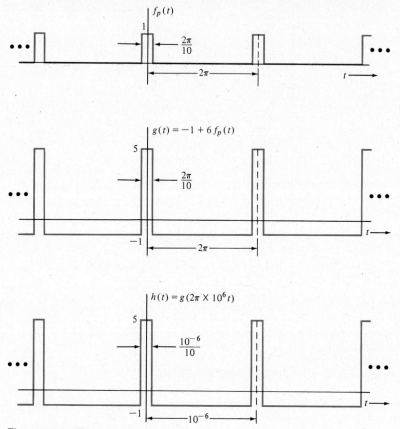

Figure 11-8   Periodic pulse functions.

When a periodic function

$$f(t) = \sum_{n=-\infty}^{\infty} d_n e^{\frac{jn2\pi t}{T}}$$

is time shifted, the exponential series coefficients are modified as follows:

$$g(t) = f(t-\tau) = \sum_{n=-\infty}^{\infty} d_n e^{\frac{jn2\pi(t-\tau)}{T}} = \sum_{n=-\infty}^{\infty} (d_n e^{-j\frac{n2\pi\tau}{T}}) e^{j\frac{n2\pi t}{T}}$$

$$= \sum_{n=-\infty}^{\infty} d'_n e^{j\frac{n2\pi t}{T}}$$

The coefficients for the time-shifted function are unchanged in magnitude. Their complex angles are changed by an amount proportional to the frequency of the term:

$$|d'_n| = |d_n| \, |e^{-j\frac{n2\pi\tau}{T}}| = |d_n|$$

$$\angle d'_n = \angle d_n + \angle e^{-j\frac{n2\pi\tau}{T}} = \angle d_n - \frac{n2\pi\tau}{T}$$

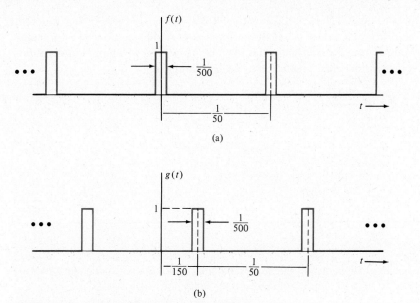

(a)

(b)

Figure 11-9   Time-shifting a periodic pulse.

The periodic pulse of Figure 11-9(a),

$$
f(t) = \frac{1}{10} + \sum_{\substack{n=-\infty \\ n \neq 0}}^{\infty} \frac{\sin \dfrac{n\pi}{10}}{\pi n} e^{j100\pi nt}
$$

is time-shifted to form the function in Figure 11-9(b), which has the following exponential series:

$$
g(t) = f(t - \tfrac{1}{150}) = \frac{1}{10} + \sum_{\substack{n=-\infty \\ n \neq 0}}^{\infty} \frac{\sin \dfrac{n\pi}{10}}{\pi n} e^{j100\pi n[t-(1/150)]}
$$

$$
= \frac{1}{10} + \sum_{\substack{n=-\infty \\ n \neq 0}}^{\infty} \left(\frac{\sin \dfrac{n\pi}{10}}{\pi n} e^{-j\frac{2n\pi}{3}} \right) e^{j100\pi nt}
$$

11.5.3 Derivatives and Integrals of Periodic Functions

The derivative of a periodic function, if its Fourier series exists, consists of terms that are the derivatives of the original series terms: For

$$
f(t) = \sum_{n=-\infty}^{\infty} d_n e^{j\frac{n2\pi t}{T}}
$$

$$
g(t) = \frac{df}{dt} = \sum_{n=-\infty}^{\infty} \left(j\frac{n2\pi}{T} d_n \right) e^{j\frac{n2\pi t}{T}}
$$

For the integral of a periodic function to be periodic, the original function must have zero average value, d_0. If the integral is periodic, its Fourier series may be found by integrating the original series term by term. The constant of integration, K, is the average value term in the series for the integral:

$$f(t) = \sum_{\substack{n=-\infty \\ n\neq 0}}^{\infty} d_n e^{j\frac{n2\pi t}{T}}$$

$$h(t) = \int f(t)\, dt = K + \sum_{\substack{n=-\infty \\ n\neq 0}}^{\infty} \left(\frac{-jT}{n2\pi} d_n\right) e^{j\frac{n2\pi t}{T}}$$

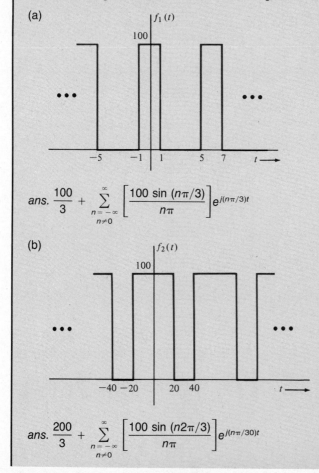

D11-5

Using the exponential series for the basic periodic pulse, find the exponential series for the following functions and sketch their magnitude and angle spectra:

(a)

ans. $\dfrac{100}{3} + \sum_{\substack{n=-\infty \\ n\neq 0}}^{\infty} \left[\dfrac{100 \sin{(n\pi/3)}}{n\pi}\right] e^{j(n\pi/3)t}$

(b)

ans. $\dfrac{200}{3} + \sum_{\substack{n=-\infty \\ n\neq 0}}^{\infty} \left[\dfrac{100 \sin{(n2\pi/3)}}{n\pi}\right] e^{j(n\pi/30)t}$

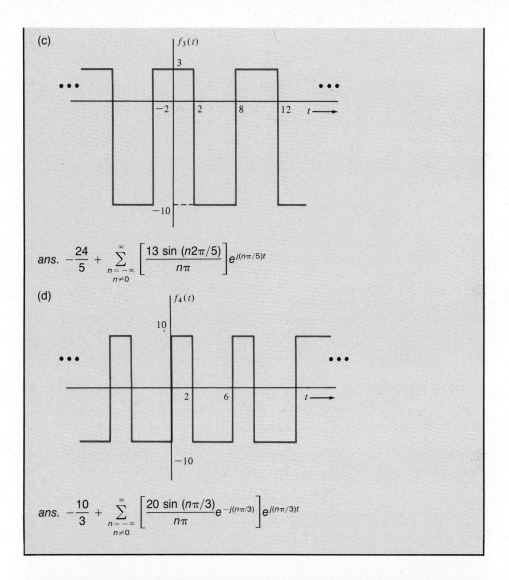

(c)

$f_3(t)$

ans. $-\dfrac{24}{5} + \displaystyle\sum_{\substack{n=-\infty \\ n\neq 0}}^{\infty} \left[\dfrac{13\sin(n2\pi/5)}{n\pi}\right] e^{j(n\pi/5)t}$

(d)

$f_4(t)$

ans. $-\dfrac{10}{3} + \displaystyle\sum_{\substack{n=-\infty \\ n\neq 0}}^{\infty} \left[\dfrac{20\sin(n\pi/3)}{n\pi}e^{-j(n\pi/3)}\right] e^{j(n\pi/3)t}$

MANIPULATING EXPONENTIAL SERIES

An even periodic function has an exponential Fourier series with coefficients d_n that are real. The coefficients for an odd function are purely imaginary.

The periodic function $Af(t)$, where A is a constant, has Fourier series coefficients that are the coefficients of the series for $f(t)$, multiplied by A. The periodic function $[f(t) + K]$, where K is a constant, has the same Fourier series as that for $f(t)$ except d_0 in the $f(t)$ series is replaced by $(d_0 + K)$ in the new series.

If the periodic function $f(t)$ has exponential Fourier series

$$f(t) = \sum_{n=-\infty}^{\infty} d_n e^{j\frac{n2\pi t}{T}}$$

the time shifted function has series

$$f(t - \tau) = \sum_{n=-\infty}^{\infty} (e^{-j\frac{n2\pi\tau}{T}} d_n) e^{j\frac{n2\pi t}{T}}$$

Provided it satisfies the Dirichlet conditions, the derivative of a periodic function has exponential series

$$\frac{df}{dt} = \sum_{n=-\infty}^{\infty} \left(\frac{jn2\pi d_n}{T}\right) e^{j\frac{n2\pi t}{T}}$$

For the (indefinite) integral of a periodic function to be periodic, the function must have zero average value: $d_0 = 0$.

$$f(t) = \sum_{\substack{n=-\infty \\ n\neq 0}}^{\infty} d_n e^{j\frac{n2\pi t}{T}}$$

The integral is then

$$\int f(t)\, dt = \text{(arbitrary constant)} + \sum_{\substack{n=-\infty \\ n\neq 0}}^{\infty} \frac{d_n T}{jn2\pi} e^{j\frac{n2\pi t}{T}}$$

11.6 EXPONENTIAL SERIES SOLUTION OF DIFFERENTIAL EQUATIONS WITH PERIODIC DRIVING FUNCTIONS

Consider again the differential equation with square wave driving function of Section 10.6, which was solved there using the trigonometric form of the Fourier series:

$$\frac{d^2y}{dt^2} + 3\frac{dy}{dt} + 2y = f(t) = \sum_{\substack{n=1 \\ n \text{ odd}}}^{\infty} \frac{20}{n\pi} \sin n2\pi t = 5 + \sum_{\substack{n=-\infty \\ n=\text{odd}}}^{\infty} \frac{-10j}{n\pi} e^{jn2\pi t}$$

As will be seen, the solution is obtained with considerably less effort using the exponential form of the series.

The forced solution component due to the driving function's constant term is the particular solution to

$$\frac{d^2y_0}{dt^2} + 3\frac{dy_0}{dt} + 2y_0 = 5 \qquad y_0 = \frac{5}{2}$$

The forced solution component to the nth nonzero exponential term in the series is the solution to

$$\frac{d^2y_n}{dt^2} + 3\frac{dy_n}{dt} + 2y_n = \frac{-10j}{n\pi} e^{jn2\pi t}$$

Substituting an exponential particular solution,

$$y_n(t) = d_n e^{jn2\pi t}$$

there results

$$(jn2\pi)^2 d_n e^{jn2\pi t} + 3(jn2\pi) d_n e^{jn2\pi t} + 2d_n e^{jn2\pi t} = \frac{-10j}{n\pi} e^{jn2\pi t}$$

$$(2 - 4n^2\pi^2 + 6jn\pi) d_n = \frac{-10j}{n\pi}$$

$$d_n = \frac{-10j}{n\pi(2 - 4n^2\pi^2 + 6jn\pi)}$$

$$y_n(t) = \frac{-10je^{jn2\pi t}}{n\pi(2 - 4n^2\pi^2 + 6jn\pi)}$$

The forced solution to the entire equation is then

$$y_f(t) = \frac{5}{2} + \sum_{\substack{n=-\infty \\ n=\text{odd}}}^{\infty} \frac{-10j}{n\pi(2 - 4n^2\pi^2 + 6jn\pi)} e^{jn2\pi t}$$

which is here expressed in the form of another exponential Fourier series. The general solution of the differential equation then consists of the sum of the natural and the forced components:

$$y(t) = K_1 e^{-t} + K_2 e^{-2t} + \frac{5}{2} + \sum_{\substack{n=-\infty \\ n=\text{odd}}}^{\infty} \frac{-10j}{n\pi(2 - 4n^2\pi^2 + 6jn\pi)} e^{jn2\pi t}$$

D11-6

For the following differential equations and periodic driving functions, find the exponential Fourier series for the forced solution $y(t)$:

(a) $\dfrac{d^2y}{dt^2} + 3\dfrac{dy}{dt} + 2y = 4f(t)$

where $f(t)$ is the basic square wave with period $\frac{1}{2}$ second.

ans. $\displaystyle\sum_{\substack{n=-\infty \\ n \text{ odd}}}^{\infty} \frac{-4j}{n\pi(1 - 8n^2\pi^2 + j6n\pi)} e^{j4n\pi t}$

(b) $\dfrac{dy}{dt} + 6y = \displaystyle\sum_{n=-\infty}^{\infty} \frac{3 - jn}{n^2 + 4} e^{j2nt}$

ans. $\displaystyle\sum_{n=-\infty}^{\infty} \frac{3 - jn}{(n^2 + 4)(6 + j2n)} e^{j2nt}$

(c) $3\dfrac{d^2y}{dt^2} + 2\dfrac{dy}{dt} + 4y = g(t)$

where $g(t)$ is the basic periodic pulse with period $\frac{1}{2}$ and duty cycle $\frac{1}{5}$.

ans. $\dfrac{1}{20} + \displaystyle\sum_{\substack{n=-\infty \\ n\neq 0}}^{\infty} \frac{\sin\dfrac{n\pi}{5}}{\pi n(4 - 48n^2\pi^2 + j8n\pi)} e^{j4n\pi t}$

EXPONENTIAL SERIES SOLUTION OF DIFFERENTIAL EQUATIONS WITH PERIODIC DRIVING FUNCTIONS

Particular solutions to linear differential equations with periodic driving functions may be found by expanding the driving function into an exponential Fourier series and superimposing the individual terms.

The component of the solution due to the nth exponential driving function component is exponential and may be found by substitution.

11.7 EXPONENTIAL SERIES SOLUTION OF NETWORKS WITH PERIODIC SOURCES

Network solution, too, is greatly simplified by use of the exponential form of the Fourier series because impedance methods apply directly to each series component. The network of Figure 11-10 was solved in Section 10.8, using trigonometric Fourier series. Using the exponential form of the series for the source,

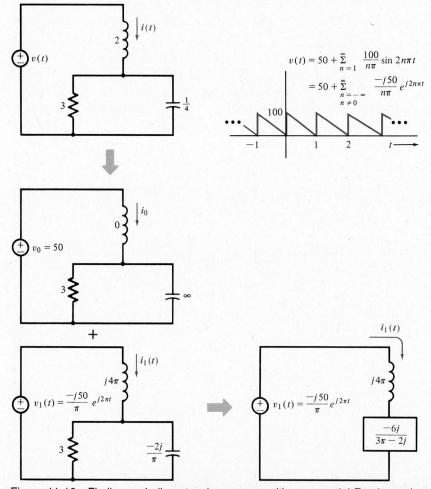

$$v(t) = 50 + \sum_{n=1}^{\infty} \frac{100}{n\pi} \sin 2n\pi t$$

$$= 50 + \sum_{\substack{n=-\infty \\ n \neq 0}}^{\infty} \frac{-j50}{n\pi} e^{j2n\pi t}$$

Figure 11-10   Finding periodic network response with exponential Fourier series.

Figure 11-10 *(Continued)*

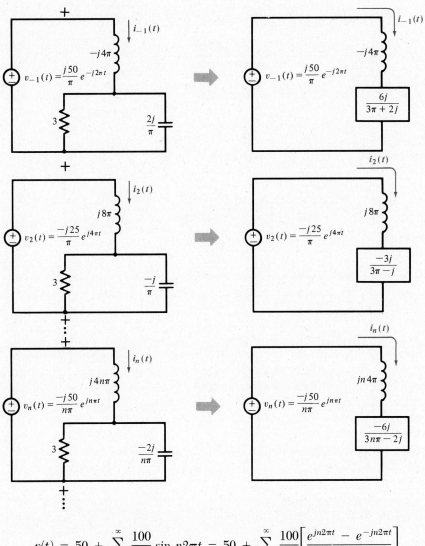

$$v(t) = 50 + \sum_{n=1}^{\infty} \frac{100}{n\pi} \sin n2\pi t = 50 + \sum_{n=1}^{\infty} \frac{100}{n\pi}\left[\frac{e^{jn2\pi t} - e^{-jn2\pi t}}{2j}\right]$$

$$= 50 + \sum_{\substack{n=-\infty \\ n \neq 0}}^{\infty} \frac{-50j}{n\pi} e^{jn2\pi t}$$

the first several corresponding components of $i(t)$ are as follows:

$$i_0 = \frac{50}{3}$$

$$i_1(t) = \frac{-j50e^{j2\pi t}}{\pi\left(j4\pi - \dfrac{6j}{3\pi - 2j}\right)} = \frac{-j50(3\pi + 2j)}{\pi(j12\pi^2 - j6 + 8\pi)}e^{j2\pi t}$$

$$i_{-1}(t) = \frac{j50e^{-j2\pi t}}{\pi\left(-j4\pi + \dfrac{6j}{3\pi + 2j}\right)} = \frac{j50(3\pi - 2j)}{\pi(-j12\pi^2 + j6 + 8\pi)}e^{-j2\pi t}$$

$$i_2(t) = \frac{-j25e^{j4\pi t}}{\pi\left(j8\pi - \dfrac{3j}{3\pi - j}\right)} = \frac{-j25(3\pi - j)}{\pi(j24\pi^2 - j3 + 8\pi)}e^{j4\pi t}$$

Impedance methods apply directly, since for each component problem, the source is exponential.

The nth component of $i(t)$, due to the nth exponential component of the source, is given in terms of n (except for $n = 0$) by

$$i_n(t) = \frac{-j50e^{jn2\pi t}}{n\pi\left(j4n\pi - \dfrac{6j}{3n\pi - 2j}\right)} = \frac{-j50(3n\pi - 2j)}{n\pi(j12n^2\pi^2 - j6 + 8n\pi)}e^{jn2\pi t}$$

The exponential Fourier series for $i(t)$ is thus as follows:

$$i(t) = i_0 + i_1 + i_{-1} + i_2 + i_{-2} + \cdots$$

$$= \frac{50}{3} + \sum_{\substack{n=-\infty \\ n\neq 0}}^{\infty} \frac{-j50(3n\pi - 2j)}{n\pi(j12n^2\pi^2 - j6 + 8n\pi)}e^{jn2\pi t}$$

In general, to find the forced response of a network with a periodic source, expand the source function into an exponential Fourier series and superimpose the individual terms. Because each of the individual terms is exponential, impedance methods apply directly, with no need for conversion to the equivalent sinor or phasor problems, as is the case with the trigonometric form of the series. It often happens that the components of the solution, like those of the source, may be easily expressed in terms of the component number n. If one or more of the source components are not given by the same formula as the rest of the terms, they must be found by separate calculations, as was the case with the constant term in the example network above.

D11-7

For the following networks with periodic sources, find the exponential Fourier series for the indicated signal:

(a) The source function $f(t)$ has the shape of the basic square wave with unit amplitude and 3-second period.

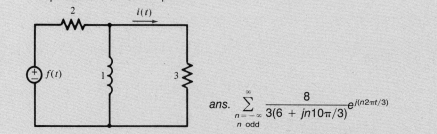

ans. $\displaystyle\sum_{\substack{n=-\infty \\ n\text{ odd}}}^{\infty} \frac{8}{3(6 + jn10\pi/3)}e^{j(n2\pi t/3)}$

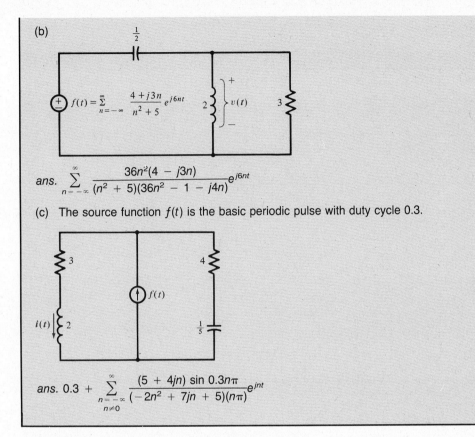

(b)

$$f(t) = \sum_{n=-\infty}^{\infty} \frac{4 + j3n}{n^2 + 5} e^{j6nt}$$

ans. $\displaystyle\sum_{n=-\infty}^{\infty} \frac{36n^2(4 - j3n)}{(n^2 + 5)(36n^2 - 1 - j4n)} e^{j6nt}$

(c) The source function $f(t)$ is the basic periodic pulse with duty cycle 0.3.

ans. $\displaystyle 0.3 + \sum_{\substack{n=-\infty \\ n \neq 0}}^{\infty} \frac{(5 + 4jn) \sin 0.3n\pi}{(-2n^2 + 7jn + 5)(n\pi)} e^{jnt}$

EXPONENTIAL SERIES SOLUTION OF NETWORKS WITH PERIODIC SOURCES

Forced solutions for voltages and currents in electrical networks with periodic driving functions may be found by expanding the source function(s) into exponential Fourier series and superimposing the individual source terms. As the source terms are exponential, impedance methods apply.

11.8 FOURIER TRANSFORM CALCULATION

11.8.1 Development of the Transformation

To accommodate functions $f(t)$ that are not periodic, $f(t)$ may be considered to be periodic with an extremely long period. Figure 11-11 shows a nonperiodic function $f(t)$ and several periodic approximations of it with increasingly longer periods, T. An approximation of this sort gets better and better in the period T centered on the origin, provided that the original function $f(t)$ approaches zero for large positive and negative t.

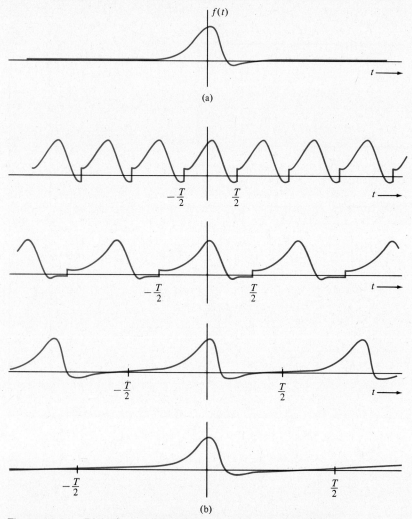

Figure 11-11 Periodic approximations of a nonperiodic function.
(a) The original function $f(t)$.
(b) Periodic approximations to $f(t)$ with various periods, T.

The results to follow will depend upon periodic approximations to $f(t)$ having Fourier series, which is guaranteed if $f(t)$ satisfies the Dirichlet conditions for all t:

(1) $f(t)$ has a finite number of finite maxima and minima
(2) $f(t)$ has a finite number of finite discontinuities

The results will also depend upon $f(t)$ approaching zero as $t \to \pm\infty$, one mathematical expression of which is the requirement that

$$\int_{-\infty}^{\infty} |f(t)|\, dt$$

be finite. These conditions are sufficient to guarantee the results but are not always necessary.

As a function of the period T, the exponential Fourier series coefficients for the periodic approximation of $f(t)$ are

$$d_n = \frac{1}{T} \int_{-T/2}^{T/2} f(t) e^{\frac{-jn2\pi t}{T}} \, dt$$

and, in the interval T centered on the origin,

$$f(t) = \sum_{n=-\infty}^{\infty} d_n \, e^{\frac{jn2\pi t}{T}} \qquad -\frac{T}{2} < t < \frac{T}{2}$$

As a function of radian frequency

$$\omega = \frac{n2\pi}{T}$$

the coefficients and the series approximation are given by

$$d(\omega) = \frac{1}{T} \int_{-T/2}^{T/2} f(t) e^{-j\omega t} \, dt$$

and

$$f(t) = \cdots + d\!\left(\omega = \frac{-4\pi}{T}\right) e^{\frac{-j4\pi t}{T}} + d\!\left(\omega = \frac{-2\pi}{T}\right) e^{\frac{-j2\pi t}{T}}$$

$$+ \, d(\omega = 0) + d\!\left(\omega = \frac{2\pi}{T}\right) e^{\frac{j2\pi t}{T}} + d\!\left(\omega = \frac{4\pi}{T}\right) e^{\frac{j4\pi t}{T}} + \cdots,$$

$$= \sum_{n=-\infty}^{\infty} d\!\left(\omega = \frac{n2\pi}{T}\right) e^{j\left(\frac{n2\pi}{T}\right)t} \qquad -\frac{T}{2} < t < \frac{T}{2}$$

The function $d(\omega)$ is the *spectral envelope* for the corresponding periodic function that has coefficients d_n equal to $d(\omega)$ at the evenly spaced discrete values of ω,

$$\omega = \ldots, \; -\frac{4\pi}{T}, \; -\frac{2\pi}{T}, \; 0, \; \frac{2\pi}{T}, \; \frac{4\pi}{T}, \ldots, \; \frac{n2\pi}{T}, \ldots$$

The periodic function represented by $d(\omega)$ is expressed as a sum of terms of the form

$$d(\omega)e^{j\omega t}$$

for these discrete values of ω.

Figure 11-12 illustrates how the spectrum of the periodic approximation of a nonperiodic function $f(t)$ changes as the approximating period is increased. For simplicity, an even function $f(t)$ has been chosen so that the spectrum has entirely real values and can be described by a single plot. In general, both the real and the imaginary parts of the spectrum have the character shown.

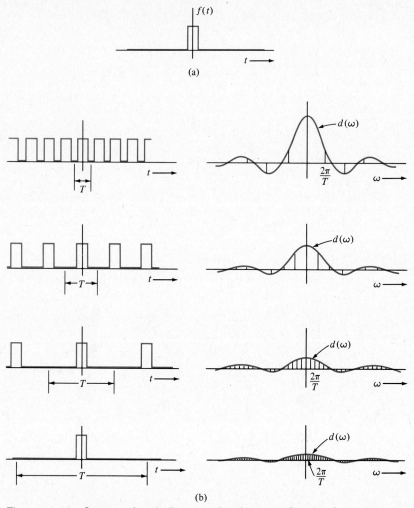

Figure 11-12   Spectra of periodic approximations to a function for various approximation
periods.
(a) Original function.
(b) Periodic approximations and spectra.

When a substantial nonzero portion of $f(t)$ lies outside the period T, the shape
of the spectral envelope is highly dependent on the size of T. When the period T
of the approximation is sufficiently large to include virtually all of the range of t for
which $f(t)$ is nonzero, the integral

$$\int_{-T/2}^{T/2} f(t)e^{-j\omega t} \, dt$$

changes very little with increase in T, and the spectral envelope function

$$d(\omega) = \frac{1}{T} \int_{-T/2}^{T/2} f(t)e^{-j\omega t} \, dt$$

maintains the same shape but is reduced in amplitude as T is increased. The discrete frequencies

$$\omega = \frac{n2\pi}{T}$$

in the spectrum of the periodic approximation are widely spaced for small T and become increasingly close to one another as T is increased.

Although the spectral envelope function gets smaller and smaller as the period T of the approximation is increased, the quantity

$$Td(\omega) = \int_{-T/2}^{T/2} f(t)e^{-j\omega t} \, dt,$$

approaches a nonzero limit

$$F(\omega) = \lim_{T \to \infty} Td(\omega) = \int_{-\infty}^{\infty} f(t)e^{-j\omega t} \, dt$$

This function, which is the limiting behavior of T times the spectral envelope function, is the Fourier transform of the nonperiodic function $f(t)$. Fourier transforms are commonly denoted by $\mathcal{F}$ or by the upper case of the symbol for the time function:

$$\mathcal{F}[f(t)] = F(\omega) = \int_{-\infty}^{\infty} f(t)e^{-j\omega t} \, dt$$

The Fourier transform of a function represents the density of the frequencies present in the function.

If the limit defining the Fourier transform exists for a function $f(t)$, the transform $F(\omega)$ is unique and is said to exist or to converge, and $f(t)$ is termed *Fourier transformable*. When the transform exists, $f(t)$ may be recovered from $F(\omega)$ through the process of inversion or inverse transformation, to be developed later in this section. The function $f(t)$ can be recovered approximately by using $F(\omega)$ to form a periodic approximation with a large period.

11.8.2 Fourier Transform of a Pulse

As an example of Fourier transform calculation, consider the function of Figure 11-13(a), which will here be called the *basic pulse*. It is a pulse of unit amplitude, centered on the origin, with width b. Its Fourier transform is calculated below in terms of b so that the result applies to pulses of any width:

$$F_B(\omega) = \int_{-\infty}^{\infty} f_B(t)e^{-j\omega t} \, dt = \int_{-(b/2)}^{(b/2)} e^{-j\omega t} \, dt = \left. \frac{e^{-j\omega t}}{-j\omega} \right|_{-\frac{b}{2}}^{\frac{b}{2}} = \frac{e^{-j\frac{\omega b}{2}} - e^{j\frac{\omega b}{2}}}{-j\omega}$$

$$= \frac{2}{\omega} \left[\frac{e^{j\frac{\omega b}{2}} - e^{-j\frac{\omega b}{2}}}{2j} \right] = \frac{2 \sin\left(\frac{\omega b}{2}\right)}{\omega} = b \operatorname{sinc}\left(\frac{\omega b}{2}\right)$$

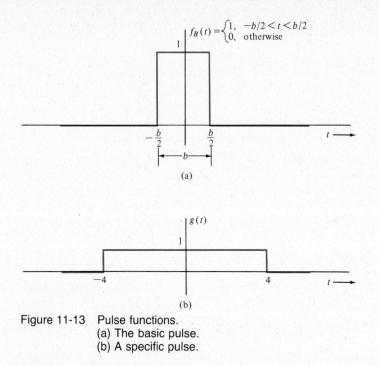

Figure 11-13   Pulse functions.
(a) The basic pulse.
(b) A specific pulse.

The specific pulse of Figure 11-13(b) has width $b = 8$, so that the Fourier transform of this function is

$$G(\omega) = \frac{2 \sin 4\omega}{\omega} = 8 \text{ sinc } 4\omega$$

11.8.3 Fourier Transforms of Switched Exponentials

Another very useful function that has an easily calculated Fourier transform is the basic switched-on exponential of Figure 11-14(a). For a positive constant a, the function satisfies the Dirichlet conditions and approaches zero for large positive and large negative t.

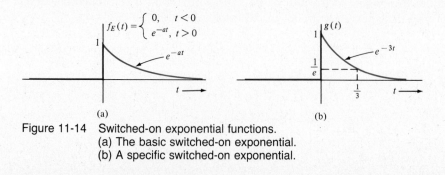

Figure 11-14   Switched-on exponential functions.
(a) The basic switched-on exponential.
(b) A specific switched-on exponential.

Its Fourier transform is given by

$$F_E(\omega) = \int_{-\infty}^{\infty} f_E(t)e^{-j\omega t}\,dt = \int_0^{\infty} e^{-at}e^{-j\omega t}\,dt = \int_0^{\infty} e^{-(a+j\omega)t}\,dt$$

$$= \frac{-1}{a+j\omega}\Big[e^{-(a+j\omega)t}\Big]_0^{\infty}$$

The upper limit in the integral evaluation involves

$$\lim_{t\to\infty} e^{-(a+j\omega)t} = \lim_{t\to\infty} e^{-at}e^{-j\omega t}$$

Now $e^{-j\omega t}$ is a complex function of t that has unity magnitude for any t. For a positive exponential constant a,

$$\lim_{t\to\infty} e^{-at} = 0$$

so

$$\lim_{t\to\infty} e^{-at}e^{-j\omega t} = 0$$

because the above quantity involves the product of a bounded (unity magnitude) term and a term that approaches zero. The Fourier transform is thus

$$F_E(\omega) = \frac{-1}{a+j\omega}[0 - 1] = \frac{1}{a+j\omega}$$

provided $a > 0$. The specific switched-on exponential of Figure 11-14(b) is of the form of $f_E(t)$ with $a = 3$ and so has transform

$$G(\omega) = \frac{1}{3+j\omega}$$

The switched-off exponential, Figure 11-15(a), consists of an expanding exponential before $t = 0$, then zero after $t = 0$. For a positive constant a, the function satisfies the Dirichlet conditions and dies out both ways in time. Its Fourier transform is given by

$$F_0(\omega) = \int_{-\infty}^{\infty} f_0(t)e^{-j\omega t}\,dt = \int_{-\infty}^{0} e^{at}e^{-j\omega t}\,dt$$

$$= \int_{-\infty}^{0} e^{(a-j\omega)t}\,dt = \frac{1}{a-j\omega}\Big[e^{(a-j\omega)t}\Big]_0^{\infty}$$

The lower limit in the integral evaluation is

$$\lim_{t\to-\infty} e^{at}e^{-j\omega t}$$

which, for positive a, is zero. The e^{at} term goes to zero, while the complex exponential has bounded magnitude. Thus

$$F_0(\omega) = \frac{1}{a-j\omega}[1 - 0] = \frac{1}{a-j\omega}$$

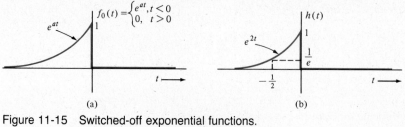

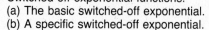

(a)   (b)

Figure 11-15 Switched-off exponential functions.
(a) The basic switched-off exponential.
(b) A specific switched-off exponential.

The specific switched-off exponential $h(t)$ in Figure 11-15(b) is of the form of $f_0(t)$ with $a = 2$ and so has transform

$$H(\omega) = \frac{1}{2 - j\omega}$$

The switched-on and the switched-off exponential functions have Fourier transforms that are distinguishable from one another by the relative signs of the real and imaginary parts of their denominators.

11.8.4 The Inverse Transform and Transform Pairs

To recover a function $f(t)$ from its Fourier transform, a periodic approximation to $f(t)$, with large period T centered at $t = 0$, may be formed from the transform $F(\omega)$:

$$f(t) = \sum_{n=-\infty}^{\infty} d_n \, e^{\frac{jn2\pi t}{T}} = \sum_{n=-\infty}^{\infty} d\left(\omega = \frac{n2\pi}{T}\right) e^{\frac{jn2\pi t}{T}}$$

$$\cong \sum_{n=-\infty}^{\infty} \frac{F\left(\omega = \dfrac{n2\pi}{T}\right)}{T} e^{\frac{jn2\pi t}{T}} \qquad -\frac{T}{2} < t < \frac{T}{2}$$

For larger and larger T, the approximation is increasingly good in the interval T centered at $t = 0$. In the limit as $T \to \infty$, except possibly at points of discontinuity, the approximation equals the function:

$$f(t) = \lim_{T \to \infty} \sum_{n=-\infty}^{\infty} \frac{F\left(\omega = \dfrac{n2\pi}{T}\right)}{T} e^{\frac{jn2\pi t}{T}}$$

Letting

$$\Delta\omega = \frac{2\pi}{T} \qquad T = \frac{2\pi}{\Delta\omega}$$

then

$$f(t) = \lim_{\Delta\omega \to 0} \sum_{n=-\infty}^{\infty} \frac{\Delta\omega}{2\pi} F(n\,\Delta\omega)e^{jn\,\Delta\omega t} = \lim_{\Delta\omega \to 0} \frac{1}{2\pi} \sum_{n=-\infty}^{\infty} F(n\,\Delta\omega)e^{j(n\,\Delta\omega)t}\,\Delta\omega$$

Since the limit of a sum of the form

$$\lim_{\Delta x \to 0} \sum_{n} y(n\,\Delta x)\,\Delta x = \int y(x)\,dx$$

is the defining relation for an integral, then

$$f(t) = \frac{1}{2\pi} \int_{-\infty}^{\infty} F(\omega)e^{j\omega t}\,d\omega$$

This integral relation is known as the inverse Fourier transform and is denoted by

$$f(t) = \mathscr{F}^{-1}[F(\omega)]$$

For functions for which the Fourier transform converges, the inverse transform recovers the original function from its transform. Like the Fourier series from which it is derived, if the original function is discontinuous, the recovered function converges to the midpoint of each discontinuity.

In practice, one makes use of tables of functions and their Fourier transforms whenever possible, looking up the transform from the time function or the time function from its transform. The *Fourier transform pair*,

$$F(\omega) = \int_{-\infty}^{\infty} f(t)e^{-j\omega t}\,dt \qquad f(t) = \frac{1}{2\pi} \int_{-\infty}^{\infty} F(\omega)e^{j\omega t}\,d\omega$$

define a transformation and inverse transformation that are highly useful for the solution of differential equations (and the electrical networks they represent), in much the same way as the transformation to and from logarithms is useful in solving arithmetic problems. Instead of dealing with time functions directly, as has been done here previously, differential equations can be expressed in terms of Fourier transforms. The transform of the solution can be found relatively easily and, from it, the solution as a function of time is found by inverse transforming. It is common to speak of Fourier transformed quantities as being in the *transform domain* or *frequency domain*, while the time functions they represent are in the *time domain*.

11.8.5 Parseval's Relation

For exponential Fourier series, Parseval's relation is

$$\frac{1}{T} \int_{T} f^2(t)\,dt = \sum_{n=-\infty}^{\infty} |d_n|^2$$

It relates the average of the square of a periodic function to the sum of squares of the magnitudes of its Fourier coefficients. As the period of a periodic function is made increasingly large to form the Fourier transform, a similar relation applies for transformable functions that are not periodic.

Multiplying by the period T, and integrating symmetrically about the origin, there results

$$\int_{-T/2}^{T/2} f^2(t)\, dt = T \sum_{n=-\infty}^{\infty} |d_n|^2 = \frac{1}{T} \sum_{n=-\infty}^{\infty} |T\, d_n|^2$$

$$\cong \frac{1}{T} \sum_{n=-\infty}^{\infty} \left| F\left(\omega = \frac{n2\pi}{T}\right) \right|^2$$

In the limit as T becomes very large, the approximation of Td_n approaches $F(\omega)$. With $T = 2\pi/\Delta\omega$,

$$\lim_{T\to\infty} \int_{-T/2}^{T/2} f^2(t)\, dt = \lim_{T\to\infty} \frac{1}{T} \sum_{n=-\infty}^{\infty} \left| F\left(\omega = \frac{n2\pi}{T}\right) \right|^2$$

$$\int_{-\infty}^{\infty} f^2(t)\, dt = \lim_{\Delta\omega\to 0} \frac{\Delta\omega}{2\pi} \sum_{n=-\infty}^{\infty} |F(n\,\Delta\omega)|^2 = \frac{1}{2\pi} \int_{-\infty}^{\infty} |F(\omega)|^2\, d\omega$$

This is the Fourier transform form of Parseval's relation. It states that, within the constant of proportionality 2π, the areas under the $f^2(t)$ and the corresponding $|F(\omega)|^2$ curves are equal. The squares of network voltages and currents are related to electrical power and energy, and here it is seen that power and energy also involve the squared magnitudes of the voltage and current transforms.

D11-8

Calculate (setting up and evaluating the integrals involved) the Fourier transforms of the indicated functions:

(a)

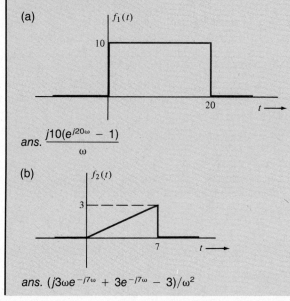

ans. $\dfrac{j10(e^{j20\omega} - 1)}{\omega}$

(b)

ans. $(j3\omega e^{-j7\omega} + 3e^{-j7\omega} - 3)/\omega^2$

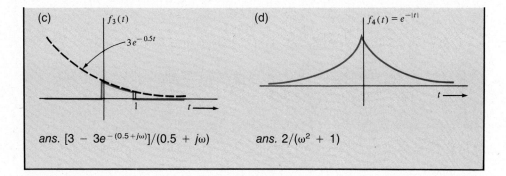

(c) $f_3(t)$

$3e^{-0.5t}$

1

$t \longrightarrow$

ans. $[3 - 3e^{-(0.5+j\omega)}]/(0.5 + j\omega)$

(d) $f_4(t) = e^{-|t|}$

$t \longrightarrow$

ans. $2/(\omega^2 + 1)$

FOURIER TRANSFORM CALCULATION

The periodic approximation, with period T, to a function $f(t)$ is the function $f(t)$ within the interval $(-T/2, T/2)$, repeated periodically in adjacent intervals. The Fourier transform of a function is the limit as the period T is made very large, of

$$F(\omega) = \lim_{T \to \infty} T \, d(\omega)$$

where $d(\omega)$ is the spectral envelope function of the periodic approximation to $f(t)$ with period T.

In terms of $f(t)$,

$$\mathscr{F}[f(t)] = F(\omega) = \int_{-\infty}^{\infty} f(t)e^{-j\omega t} \, dt$$

Sufficient conditions for convergence are that $f(t)$ satisfy the Dirichlet conditions and that

$$\int_{-\infty}^{\infty} |f(t)| \, dt$$

be finite.

Some basic Fourier transforms are as follows:

Basic Pulse

$$f_B(t) = \begin{cases} 1 & -(b/2) < t < (b/2) \\ 0 & \text{otherwise} \end{cases}$$

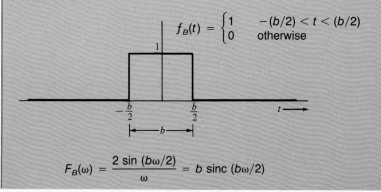

1

$-\dfrac{b}{2}$ $\dfrac{b}{2}$ $t \longrightarrow$

$\longleftarrow b \longrightarrow$

$$F_B(\omega) = \frac{2 \sin (b\omega/2)}{\omega} = b \text{ sinc } (b\omega/2)$$

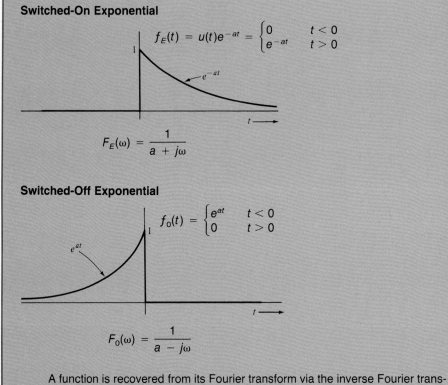

Switched-On Exponential

$$f_E(t) = u(t)e^{-at} = \begin{cases} 0 & t < 0 \\ e^{-at} & t > 0 \end{cases}$$

$-e^{-at}$

$$F_E(\omega) = \frac{1}{a + j\omega}$$

Switched-Off Exponential

$$f_0(t) = \begin{cases} e^{at} & t < 0 \\ 0 & t > 0 \end{cases}$$

e^{at}

$$F_0(\omega) = \frac{1}{a - j\omega}$$

A function is recovered from its Fourier transform via the inverse Fourier transformation:

$$f(t) = \mathscr{F}^{-1}[F(\omega)] = \frac{1}{2\pi} \int_{-\infty}^{\infty} F(\omega)e^{j\omega t}\, d\omega$$

At points of discontinuity of the original function $f(t)$, the inverse transformation converges to the midpoint of the discontinuity.

Parseval's relation is

$$\int_{-\infty}^{\infty} f^2(t)\, dt = \frac{1}{2\pi} \int_{-\infty}^{\infty} |F(\omega)|^2\, d\omega$$

11.9 FOURIER TRANSFORM PROPERTIES

11.9.1 Transform Manipulation

The Fourier transform of the sums of functions is the sum of the individual transforms,

$$\mathscr{F}[f_1(t) + f_2(t)] = \int_{-\infty}^{\infty} [f_1(t) + f_2(t)]e^{-j\omega t}\, dt$$

$$= \int_{-\infty}^{\infty} f_1(t)e^{-j\omega t}\, dt + \int_{-\infty}^{\infty} f_2(t)e^{-j\omega t}\, dt = F_1(\omega) + F_2(\omega)$$

and the transform of a constant A times a function is the constant times the transform:

$$\mathcal{F}[Af(t)] = \int_{-\infty}^{\infty} Af(t)e^{-j\omega t}\, dt = A \int_{-\infty}^{\infty} f(t)e^{-j\omega t}\, dt = AF(\omega)$$

These two properties establish the linearity of the Fourier transform operation. Generally, the Fourier transform of the product of two functions is *not* the product of the individual transforms.

If a function's time scale is changed, with the variable t replaced by (at) where a is a constant, the Fourier transform of the time-scaled function is changed as follows:

$$\mathcal{F}[f(at)] = \int_{-\infty}^{\infty} f(at)e^{-j\omega t}\, dt$$

Letting

$$t' = at \qquad dt' = a\, dt$$

$$\mathcal{F}[f(at)] = \int_{-\infty}^{\infty} f(t')e^{-\frac{j\omega t'}{a}} \frac{dt'}{a} = \frac{1}{a}\int_{-\infty}^{\infty} f(t')e^{-j\left(\frac{\omega}{a}\right)t'}\, dt' = \frac{1}{a}F\left(\frac{\omega}{a}\right)$$

The function of Figure 11-16(a), for example, has Fourier transform

$$F(\omega) = \int_{0}^{1} (1)e^{-j\omega t}\, dt = \frac{e^{-j\omega t}}{-j\omega}\bigg|_{0}^{1} = \frac{j(e^{-j\omega} - 1)}{\omega}$$

The related function of Figure 11-16(b),

$$g(t) = f(\tfrac{3}{2}t)$$

has Fourier transform

$$G(\omega) = \frac{2}{3}F(\tfrac{2}{3}\omega) = \frac{2}{3}\frac{j(e^{-j\frac{2\omega}{3}} - 1)}{\tfrac{2}{3}\omega} = \frac{j(e^{-j\frac{2\omega}{3}} - 1)}{\omega}$$

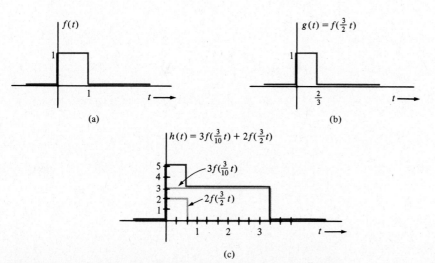

(a)   (b)

(c)

Figure 11-16   Forming Fourier transforms of time-scaled functions.

Compression of the time scale of $f(t)$, as in this example, results in an expansion in the ω scale and reduction of amplitude of the transform. The function of Figure 11-16(c) is related to $f(t)$ by

$$h(t) = 3f(\tfrac{3}{10}t) + 2f(\tfrac{3}{2}t)$$

Its Fourier transform is then given by

$$H(\omega) = 3\left[\frac{10}{3}\frac{j(e^{-j\frac{10\omega}{3}} - 1)}{\frac{10}{3}\omega}\right] + 2\left[\frac{2}{3}\frac{j(e^{-j\frac{2\omega}{3}} - 1)}{\frac{2}{3}\omega}\right]$$

$$= \frac{j}{\omega}(3e^{-j\frac{10\omega}{3}} + 2e^{-j\frac{2\omega}{3}} - 5)$$

When a function is shifted in time, as in Figure 11-17(a), the two Fourier transforms are related as follows:

$$\mathcal{F}[f(t - a)] = \int_{-\infty}^{\infty} f(t - a)e^{-j\omega t}\, dt$$

Letting

$$t' = t - a \qquad dt' = dt$$

$$\mathcal{F}[f(t - a)] = \int_{-\infty}^{\infty} f(t')e^{-j\omega(t' + a)}\, dt' = \int_{-\infty}^{\infty} f(t')e^{-j\omega t'}e^{-j\omega a}\, dt'$$

$$= e^{-j\omega a}\int_{-\infty}^{\infty} f(t')e^{-j\omega t'}\, dt' = e^{-j\omega a}F(\omega)$$

For example, the function $g(t)$ in Figure 11-17(b) is a switched-on exponential, delayed 5 seconds in time. The basic switched-on exponential

$$f_E(t) = \begin{cases} 0 & t < 0 \\ e^{-3t} & t > 0 \end{cases}$$

has Fourier transform

$$F_E(\omega) = \frac{1}{3 + j\omega}$$

Since

$$g(t) = f_E(t - 5)$$

then

$$G(\omega) = e^{-5j\omega}F_E(\omega) = \frac{e^{-5j\omega}}{3 + j\omega}$$

In the example of Figure 11-17(c),

$$h(t) = 3f_B(t - 1) - 3f_B(t + 1)$$

where $f_B(t)$ is the basic pulse with width 2 units. Since for this width of basic pulse

$$F_B(\omega) = \frac{2 \sin \omega}{\omega}$$

then

$$H(\omega) = 3e^{-j\omega}\left(\frac{2 \sin \omega}{\omega}\right) - 3e^{j\omega}\left(\frac{2 \sin \omega}{\omega}\right)$$

11.9.2 Transforms of Derivatives and Integrals

The Fourier transform of the time derivative of a function is simply related to the transform itself, provided of course that the derivative is Fourier transformable:

$$\mathscr{F}\left[\frac{df}{dt}\right] = \int_{-\infty}^{\infty} \underbrace{\frac{df}{dt}}_{u} \underbrace{e^{-j\omega t} \, dt}_{dv}$$

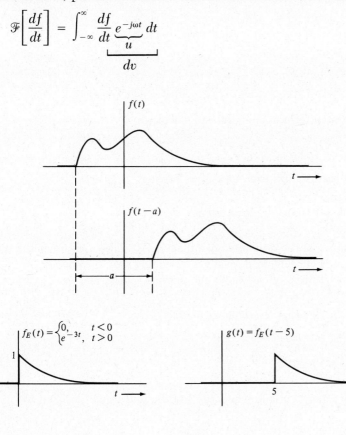

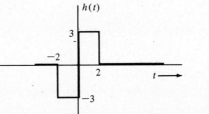

Figure 11-17   Forming Fourier transforms of time-shifted functions.

Integrating by parts,

$$\int_{-\infty}^{\infty} u \, dv = uv \Big|_{-\infty}^{\infty} - \int_{-\infty}^{\infty} v \, du$$

with

$$u = e^{-j\omega t}, \qquad du = -j\omega e^{-j\omega t} \, dt$$

$$dv = \frac{df}{dt} \, dt, \qquad v = f$$

there results

$$\mathcal{F}\left[\frac{df}{dt}\right] = f(t)e^{-j\omega t} \Big|_{-\infty}^{\infty} - \int_{-\infty}^{\infty} f(t)[-j\omega e^{-j\omega t}] \, dt$$

Since

$$\left|e^{-j\omega t}\right| = 1$$

so that the exponential term is bounded and since

$$\lim_{t \to \pm\infty} f(t) = 0$$

the evaluations of the uv term each give zero, leaving

$$\mathcal{F}\left[\frac{df}{dt}\right] = j\omega \int_{-\infty}^{\infty} f(t)e^{-j\omega t} \, dt = j\omega F(\omega)$$

The Fourier transform of the derivative of a function is simply $(j\omega)$ times the transform of the function itself. Higher derivatives, assuming they are Fourier transformable, involve higher powers of $(j\omega)$ times the original transform:

$$\mathcal{F}\left[\frac{d^2 f}{dt^2}\right] = (j\omega)^2 F(\omega) = -\omega^2 F(\omega)$$

$$\mathcal{F}\left[\frac{d^n f}{dt^n}\right] = (j\omega)^n F(\omega)$$

The running integral of a function $f(t)$ and the function are related by

$$\frac{d}{dt} \int_{-\infty}^{t} f(t) \, dt = f(t)$$

Assuming the integral is Fourier transformable,

$$j\omega \, \mathcal{F}\left[\int_{-\infty}^{t} f(t) \, dt\right] = F(\omega)$$

$$\mathcal{F}\left[\int_{-\infty}^{t} f(t) \, dt\right] = \frac{1}{j\omega} F(\omega)$$

Higher integrals, if they are Fourier transformable, are related by

$$\mathscr{F}\left[\int_{-\infty}^{t}\int_{-\infty}^{t} f(t)\, d^2t\right] = \frac{1}{(j\omega)^2}\, F(\omega)$$

$$\mathscr{F}\left[\underbrace{\int_{-\infty}^{t}\int_{-\infty}^{t}\cdots\int_{-\infty}^{t}}_{k\ \text{integrals}} f(t)\, d^k t\right] = \frac{1}{(j\omega)^k}\, F(\omega)$$

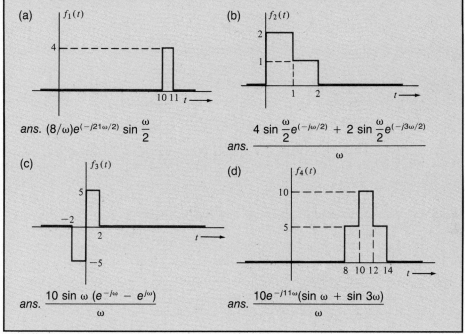

D11-9

Using the Fourier transform for the basic pulse, find the Fourier transforms of the following functions:

(a) $f_1(t)$

ans. $(8/\omega)e^{(-j21\omega/2)}\sin\dfrac{\omega}{2}$

(b) $f_2(t)$

ans. $\dfrac{4\sin\dfrac{\omega}{2}e^{(-j\omega/2)} + 2\sin\dfrac{\omega}{2}e^{(-j3\omega/2)}}{\omega}$

(c) $f_3(t)$

ans. $\dfrac{10\sin\omega\,(e^{-j\omega} - e^{j\omega})}{\omega}$

(d) $f_4(t)$

ans. $\dfrac{10e^{-j11\omega}(\sin\omega + \sin 3\omega)}{\omega}$

11.9.3 Fourier Transform Spectra

The spectrum of a Fourier transformable function $f(t)$ consists of the real and imaginary part functions

$$R(\omega) = \text{Re}[F(\omega)] \qquad X(\omega) = \text{Im}[F(\omega)]$$

or the magnitude and angle functions

$$M(\omega) = |F(\omega)| \qquad \Phi(\omega) = \underline{/F(\omega)}$$

When plotted versus radian frequency ω or Hertz frequency $f = \omega/2\pi$, these are termed *spectral plots* of the time function. The basic pulse of Figure 11-18(a), for

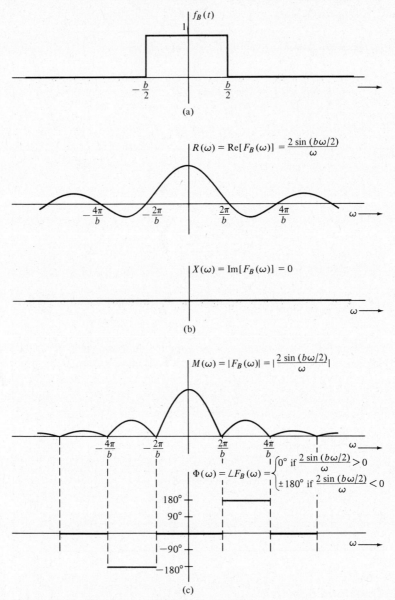

Figure 11-18   The basic pulse and its spectrum.

example, has Fourier transform

$$F_B(\omega) = \frac{2 \sin (b\omega/2)}{\omega}$$

in terms of the pulse width b. The real and imaginary part spectral plots for this function are given in Figure 11-18(b), the imaginary part of the transform being identically zero. Sketches of the spectrum magnitude and angle functions are given in Figure 11-18(c). As the magnitude of a complex quantity is, strictly speaking, never negative, when $F_B(\omega)$ is negative, it has been plotted with a positive magnitude and an angle of 180°.

Spectral plots for the specific switched-on exponential

$$g(t) = \begin{cases} 0 & t < 0 \\ e^{-2t} & t > 0 \end{cases}$$

are sketched in Figure 11-19. With

$$G(\omega) = \frac{10}{2 + j\omega} = \frac{10(2 - j\omega)}{(2 + j\omega)(2 - j\omega)} = \frac{20 - j10\omega}{4 + \omega^2}$$

$$R(\omega) = \text{Re}[G(\omega)] = \frac{20}{4 + \omega^2} \qquad X(\omega) = \text{Im}[G(\omega)] = \frac{-10\omega}{4 + \omega^2}$$

and these real and imaginary part functions are sketched in Figure 11-19(b). The magnitude and angle functions, sketched in Figure 11-19(c), are

$$M(\omega) = |G(\omega)| = \frac{10}{|2 + j\omega|} = \frac{10}{\sqrt{4 + \omega^2}}$$

$$\Phi(\omega) = \angle G(\omega) = -\arctan\left(\frac{\omega}{2}\right)$$

Like the d_ns of an exponential Fourier series, the Fourier transform of a real function of time has a real part that is even in ω and an imaginary part that is odd in ω:

$$F(\omega) = R(\omega) + jX(\omega)$$

$$R(-\omega) = R(\omega) \qquad X(-\omega) = -X(\omega)$$

These properties are easily demonstrated by comparing

$$F(\omega) = \int_{-\infty}^{\infty} f(t)e^{-j\omega t}\, dt$$

and

$$F(-\omega) = \int_{-\infty}^{\infty} f(t)e^{j\omega t}\, dt = F^*(\omega)$$

The Fourier transform magnitude function is even in ω:

$$F(\omega) = M(\omega)e^{j\Phi(\omega)}$$

$$M(-\omega) = M(\omega)$$

Its angle function is odd, except possibly for the point at $\omega = 0$, which could be 180°:

$$\Phi(-\omega) = -\Phi(\omega) \qquad \text{except possibly at } \omega = 0$$

The transform

$$F(\omega) = \frac{-10}{2 + \omega}$$

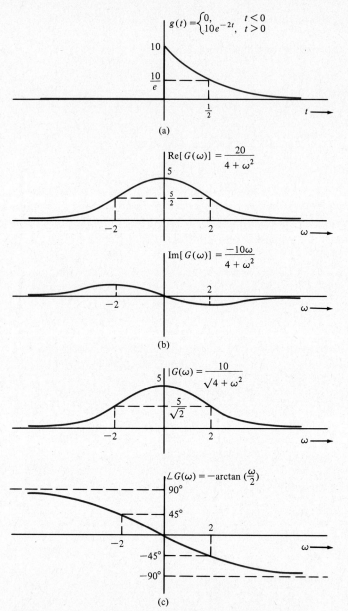

(a)

(b)

(c)

Figure 11-19 A switched-on exponential and its spectrum.

for example has

$$M(\omega = 0) = 5 \qquad \Phi(\omega = 0) = 180°$$

An even function $E(t)$ has a Fourier transform that is entirely real, since

$$\mathscr{F}[E(t)] = \int_{-\infty}^{\infty} E(t)e^{-j\omega t}\, dt = \int_{-\infty}^{\infty} E(t)[\cos \omega t - j \sin \omega t]\, dt$$

$$= \int_{-\infty}^{\infty} \underbrace{\underbrace{E(t)}_{\text{even}} \underbrace{\cos \omega t}_{\text{even}}}_{\text{even}} dt + \int_{-\infty}^{\infty} \underbrace{\underbrace{-jE(t)}_{\text{even}} \underbrace{\sin \omega t}_{\text{odd}}}_{\text{odd}} dt$$

$$= \int_{-\infty}^{\infty} E(t) \cos \omega t \, dt$$

Its imaginary part function is zero. Similarly, an odd function of time, $0(t)$ has a Fourier transform that is purely imaginary:

$$\mathscr{F}[0(t)] = \int_{-\infty}^{\infty} 0(t)e^{-j\omega t} \, dt = \int_{-\infty}^{\infty} 0(t) \left[\cos \omega t - j \sin \omega t \right] dt$$

$$= \int_{-\infty}^{\infty} \underbrace{\underbrace{0(t)}_{\text{odd}} \underbrace{\cos \omega t}_{\text{even}}}_{\text{odd}} dt - j \int_{-\infty}^{\infty} \underbrace{\underbrace{0(t)}_{\text{odd}} \underbrace{\sin \omega t}_{\text{odd}}}_{\text{even}} dt$$

$$= -j \int_{-\infty}^{\infty} 0(t) \sin \omega t \, dt$$

The real part and j times the imaginary parts of a function's Fourier transform are, respectively, the transforms of the even part and the odd part of the time function.

Real and imaginary part spectral plots are thus easiest for an even or an odd function because for these, one of the two plots is identically zero. Time-shift relationships are generally easiest to visualize with spectral plots of magnitude and angle since the time shift of a function $f(t)$ by amount a changes its Fourier transform by the factor $e^{-j\omega a}$:

$$G(\omega) = [f(t - a)] = e^{-j\omega a}F(\omega)$$

The two functions have the same transform magnitude

$$|G(\omega)| = |e^{-j\omega a}| \, |F(\omega)| = |F(\omega)|$$

and the time-shifted function's transform simply has an added term in its angle proportional to frequency:

$$\underline{/G(\omega)} = \underline{/e^{-j\omega a}} + \underline{/F(\omega)} = -\omega a + \underline{/F(\omega)}$$

D11-10

Find and sketch the real part and imaginary part and the magnitude and angle spectra of the functions with the following Fourier transforms:

(a) $F(\omega) = \dfrac{1}{j\omega}$

   ans. $0; \dfrac{-1}{\omega}; \dfrac{1}{|\omega|}; \omega > 0: -90°, \omega < 0: 90°$

(b) $F(\omega) = \dfrac{1}{5 - 2j\omega}$

ans. $\dfrac{5}{4\omega^2 + 25}, \dfrac{2\omega}{4\omega^2 + 25}, \dfrac{1}{\sqrt{4\omega^2 + 25}}$, arctan $(2\omega/5)$

(c) $F(\omega) = \dfrac{-3 + j\omega}{\omega^2 + 4}$

ans. $\dfrac{-3}{\omega^2 + 4}, \dfrac{\omega}{\omega^2 + 4}, \dfrac{\sqrt{\omega^2 + 9}}{\omega^2 + 4}$, $180° -$ arctan$(\omega/3)$

(d) $F(\omega) = \dfrac{1 - e^{-j\omega}}{\omega^2}$

ans. $\dfrac{1 - \cos \omega}{\omega^2}, \dfrac{\sin \omega}{\omega^2}, \dfrac{\sqrt{2 - 2\cos \omega}}{\omega^2}$, arctan $\dfrac{\sin \omega}{(1 - \cos \omega)}$

FOURIER TRANSFORM PROPERTIES

The following are important basic properties of the Fourier transform:

(1) $\mathcal{F}[Af(t)] = AF(\omega)$, where A is a constant.

(2) $\mathcal{F}[f_1(t) + f_2(t)] = F_1(\omega) + F_2(\omega)$

(3) $\mathcal{F}[f(at)] = \dfrac{1}{a} F\left(\dfrac{\omega}{a}\right)$, where a is a constant.

(4) $\mathcal{F}[f(t - a)] = e^{-j\omega a}F(\omega)$, where a is a constant.

(5) $\mathcal{F}\left[\dfrac{df}{dt}\right] = j\omega F(\omega)$, provided $\dfrac{df}{dt}$ is Fourier transformable.

Using this result, $\mathcal{F}\left[\dfrac{d^n f}{dt^n}\right] = (j\omega)^n F(\omega)$

(6) $\mathcal{F}\left[\displaystyle\int_{-\infty}^{t} f(t)\, dt\right] = \dfrac{1}{j\omega} F(\omega)$, provided $\displaystyle\int_{-\infty}^{t} f(t)\, dt$ is Fourier transformable.
Using this result,

$$\mathcal{F}\left[\underbrace{\int_{-\infty}^{t}\int_{-\infty}^{t}\cdots\int_{-\infty}^{t}}_{k \text{ integrals}} f(t)\, d^k t\right] = \left(\dfrac{1}{j\omega}\right)^k F(\omega)$$

The Fourier transform of a function is commonly termed its *spectrum*. Spectral plots for a function $f(t)$ consists of either

(1) Graphs of Re$[F(\omega)]$ and of Im$[F(\omega)]$ versus radian frequency ω or frequency in hertz, $f = \omega/2\pi$

or

(2) Graphs of $|F(\omega)|$ and of $\underline{/F(\omega)}$ versus frequency.

The Fourier transform of a real function $f(t)$ has the property

$$F(-\omega) = F^*(\omega).$$

Thus $\text{Re}[F(\omega)]$ is even in ω and $\text{Im}[F(\omega)]$ is odd in ω. $|F(\omega)|$ is an even function of ω, and $\underline{/F(\omega)}$ may be arranged as an odd function of ω, except possibly for the point at $\omega = 0$, which could have an angle of $180°$.

11.10 IMPULSES

11.10.1 Impulsive Functions

In the description and analysis of physical problems, it often happens that the detailed behavior of a function of time is unimportant to the result. For example, when a mass is struck with a hammer, it is the momentum transfer, not the highly variable details of the force applied to the mass on a microsecond time scale that is usually of interest.

A function that is concentrated in an interval of t that is small compared to any other interval of interest is called an *impulsive* function. Impulsive functions may be approximated by a limiting behavior whereby the function is nonzero only in the vicinity of a single point. The unit area rectangular pulse of Figure 11-20 is impulsive for sufficiently small widths T. In the limit as T approaches zero, an impulsive function that is concentrated at the origin approaches zero everywhere but at the origin, where it is infinite. As this impulsive function has unit area for every positive value of T, it also has unit area in the limit.

The symbol $\delta(t)$ is used to denote the limiting behavior of unit area impulsive functions as their areas become concentrated in the vicinity of the origin. $\delta(t)$ is called the *unit impulse*. Whenever it is used in an equation, it is implied that the equation is true in the limit, when $\delta(t)$ is replaced by any unit impulsive function, as the impulsive function becomes arbitrarily narrow and high near $t = 0$. The equations

$$\delta(t) = \begin{cases} 0 & t \neq 0 \\ \infty & t = 0 \end{cases}$$

and

$$\int_{-\epsilon}^{\epsilon} \delta(t)\, dt = 1$$

that state fundamental properties of the unit impulse mean that

$$\lim_{T \to 0} \xi(t, T) = \begin{cases} 0 & t \neq 0 \\ \infty & t = 0 \end{cases}$$

and

$$\lim_{T \to 0} \int_{-\epsilon}^{\epsilon} \xi(t, T)\, dt = 1$$

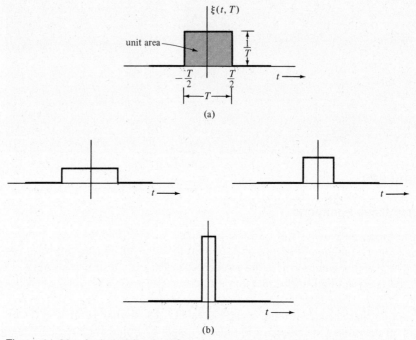

Figure 11-20 An impulsive function.
(a) Rectangular pulse with adjustable period T and unit area.
(b) The rectangular pulse for increasingly smaller values of T.

where $\xi(t,T)$ is any unit area impulsive function, such as the rectangular pulse of width T and height $1/T$.

The area of an impulse is called its *strength*. The unit impulse has unit strength, while $A\delta(t)$ has area A and thus strength A.

Many other functions besides the rectangular pulse are impulsive. Several others are sketched in Figure 11-21. Each function shown has unit area and becomes more and more concentrated near $t = 0$, as a parameter (in this case k) approaches zero. Each such function becomes an impulse, indistinguishable from any other impulse in the limit.

An arrow is used to represent an impulse on a sketch, as is indicated in Figure 11-22 for the function

$$f(t) = 5\delta(t + 3) - 3\delta(t - 2) + 2u(t + 4)$$

The senses and lengths of the arrows representing impulses are used to indicate relative impulse strengths, and the strengths may also be written next to the arrow. The scale used for the arrow lengths need not be the same as that for the rest of the function.

11.10.2 Impulse Properties

The product of a finite function $f(t)$ and an impulse,

$$h(t) = f(t)\,\delta(t)$$

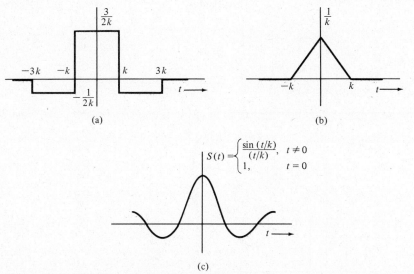

Figure 11-21   Some other impulsive functions.

is everywhere zero except where the impulse is concentrated, since in the limit, the impulse is zero everywhere else. In the limit as an impulsive function becomes narrower and higher, as illustrated in Figure 11-23, the product $h(t)$ becomes virtually $f(0)$ times the impulse. Provided that $f(t)$ is continuous at $t = 0$,

$$f(t) \, \delta(t) = f(0) \, \delta(t)$$

If the impulse is shifted so that it is concentrated at $t = a$, and if $f(t)$ is continuous at $t = a$, then

$$f(t) \, \delta(t - a) = f(a) \, \delta(t - a).$$

When the product of a finite function $f(t)$ and an impulse is integrated,

$$\int f(t) \, \delta(t) \, dt$$

there is no contribution to the integral from the product, except where the impulse is concentrated. If the integral includes $t = 0$ where the impulse is concentrated, and if $f(t)$ is continuous at $t = 0$, then

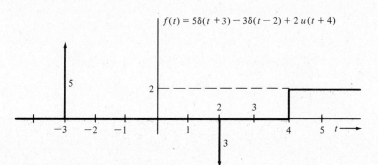

Figure 11-22   Representing impulses.

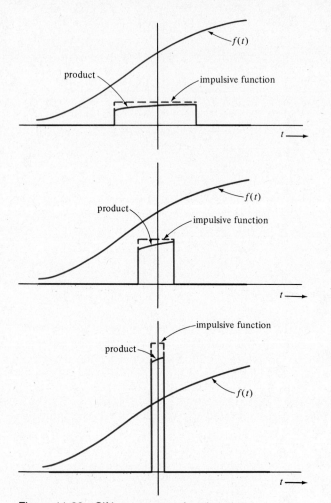

Figure 11-23 Sifting property of an impulse.

$$\int_{-\epsilon}^{\epsilon} f(t)\, \delta(t)\, dt \;=\; \int_{-\epsilon}^{\epsilon} f(0)\, \delta(t)\, dt \;=\; f(0) \int_{-\epsilon}^{\epsilon} \delta(t)\, dt \;=\; f(0)$$

This is known as the *sifting* property of the impulse.

The integral of the product of a bounded function and a time-shifted impulse sifts the value of the function at the point where the impulse is concentrated, assuming of course that the integral extends over that point, and provided that the function is continuous at that point:

$$\int_{a-\epsilon}^{a+\epsilon} f(t)\, \delta(t - a)\, dt \;=\; f(a)$$

Using the impulse sifting property, the Fourier transform of an impulse is

$$\mathscr{F}[\delta(t)] \;=\; \int_{-\infty}^{\infty} \delta(t) e^{-j\omega t}\, dt \;=\; e^{-j\omega(0)} \;=\; 1$$

As the Fourier transform is the density of frequencies in the corresponding time function, the impulse, which is an idealization of a sharp, narrow pulse, is seen to be composed of all frequencies, all with zero phase angle, in equal amounts.

11.10.3 Steps As Integrals of Impulses

When a running integral passes over an impulse, the value of the integral changes abruptly, by the amount of the impulse strength. Figure 11-24(a) shows drawings of the example function

$$g(t) = 2\delta(t + 2) + tu(t) + \delta(t - 3) - 3\delta(t - 4)$$

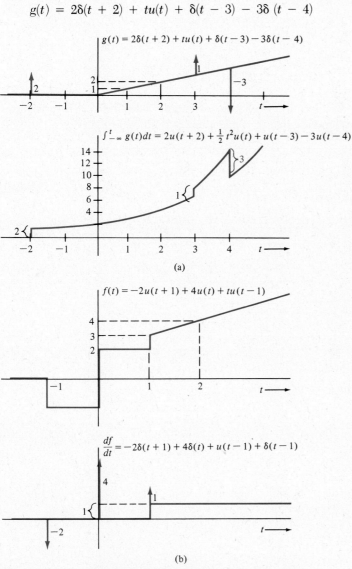

(a)

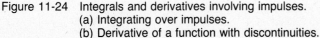

(b)

Figure 11-24   Integrals and derivatives involving impulses.
(a) Integrating over impulses.
(b) Derivative of a function with discontinuities.

and its integral

$$f(t) = \int_{-\infty}^{t} g(t)\, dt = 2u(t + 2) + \tfrac{1}{2}t^2 u(t) + u(t - 3) - 3u(t - 4)$$

Since a function $f(t)$ and its derivative are related by

$$f(t) = \int_{-\infty}^{t} \frac{df}{dt}\, dt$$

discontinuities in $f(t)$ give rise to impulses in df/dt at each of the discontinuity locations. An example of a discontinuous function

$$f(t) = -2u(t + 1) + 4u(t) + tu(t - 1)$$

and its derivative

$$\frac{df}{dt} = -2\delta(t + 1) + 4\delta(t) + u(t - 1) + \delta(t - 1)$$

are sketched in Figure 11-24(b). Each discontinuous "jump" in $f(t)$ results in an impulse in df/dt of strength equal to the amount of the discontinuity.

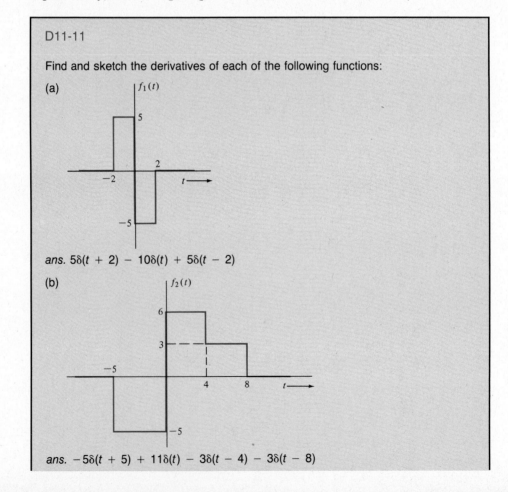

D11-11

Find and sketch the derivatives of each of the following functions:

(a)

ans. $5\delta(t + 2) - 10\delta(t) + 5\delta(t - 2)$

(b)

ans. $-5\delta(t + 5) + 11\delta(t) - 3\delta(t - 4) - 3\delta(t - 8)$

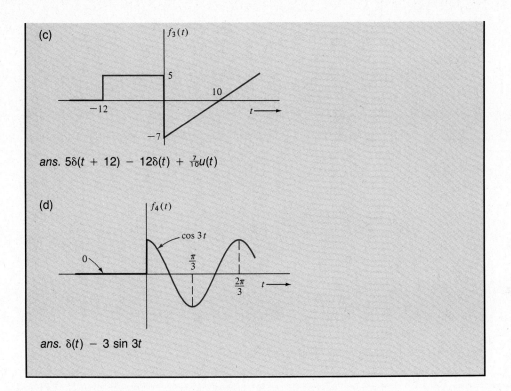

ans. $5\delta(t + 12) - 12\delta(t) + \frac{7}{10}u(t)$

ans. $\delta(t) - 3 \sin 3t$

11.10.4 Fourier Transforms of Periodic Functions

A Fourier *transform* that is an impulse,

$$F(\omega) = \delta(\omega)$$

represents the time function

$$\mathscr{F}^{-1}[\delta(\omega)] = \frac{1}{2\pi} \int_{-\infty}^{\infty} \delta(\omega)e^{j\omega t} \, d\omega = \frac{1}{2\pi}e^{j(0)t} = \frac{1}{2\pi}$$

a constant. This is sensible because the Fourier transform is the density of sinusoidal components to a signal, and a constant is a zero-frequency component with infinite density. The Fourier transform of a constant k is then

$$\mathscr{F}[k] = 2\pi k \, \delta(\omega)$$

Periodic signals such as the constant do not approach zero for large positive and large negative t. For these exceptions to the sufficient (but not necessary) conditions for convergence, the Fourier transform does converge.

Level-shifting a function $f(t)$ to form

$$g(t) = f(t) + k$$

where k is a constant, adds an impulse of strength $2\pi k$ to the transform, as is illustrated in Figure 11-25.

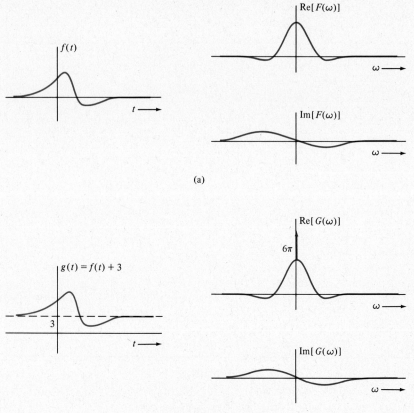

Figure 11-25 Fourier transform of a level-shifted function.
(a) Original function and spectrum.
(b) Level-shifted function and spectrum.

A frequency-shifted impulse in a Fourier transform

$$F(\omega) = \delta(\omega - b)$$

corresponds to the complex-exponential time function

$$f(t) = \frac{1}{2\pi} \int_{-\infty}^{\infty} \delta(\omega - b)e^{j\omega t} \, d\omega = \frac{1}{2\pi}e^{jbt}$$

so that sinusoidal time functions have Fourier transforms as follows:

$$\mathscr{F}\left[\cos bt = \frac{1}{2}e^{jbt} + \frac{1}{2}e^{-jbt} \right] = \pi \, \delta(\omega - b) + \pi \, \delta(\omega + b)$$

$$\mathscr{F}\left[\sin bt = \frac{-j}{2}e^{jbt} + \frac{j}{2}e^{-jbt} \right] = -j\pi \, \delta(\omega - b) + j\pi \, \delta(\delta + b)$$

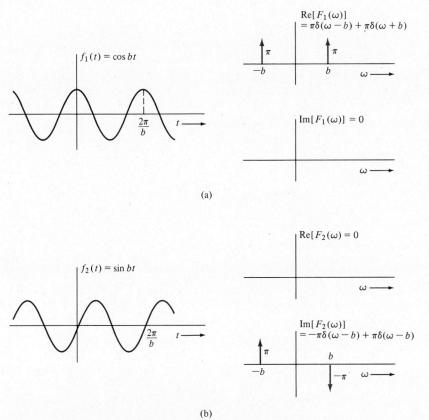

(a)

(b)

Figure 11-26 Fourier transforms of periodic functions.
(a) Transform of cos bt.
(b) Transform of sin bt.

Each sinusoid involves a pair of frequency-domain impulses, one at frequency b and the other at frequency $-b$, as shown in Figure 11-26(a) and (b).

A periodic function with exponential Fourier series

$$g(t) = \sum_{n=-\infty}^{\infty} d_n e^{\frac{jn2\pi t}{T}}$$

has Fourier transform

$$\mathcal{F}[g(t)] = G(\omega) = \sum_{n=-\infty}^{\infty} 2\pi d_n \delta\left(\omega - \frac{n2\pi}{T}\right)$$

The discrete spectrum of the series becomes a corresponding set of impulses in frequency in terms of frequency density, which is the Fourier transform. In this way, through the use of impulses, Fourier series can be included as part of the Fourier transform if desired.

D11-12

Find the Fourier transforms of the following functions:

(a) $f_1(t) = \begin{cases} 10 + e^{2t} & t < 0 \\ 10 & t > 0 \end{cases}$

$$ans. \quad \frac{1}{j\omega + 2} + 20\pi\, \delta(\omega)$$

(b) $f_2(t) = \displaystyle\sum_{n=-\infty}^{\infty} \frac{4}{2n^2 - 3} e^{j10nt}$

$$ans. \quad \sum_{n=-\infty}^{\infty} \frac{8\pi\, \delta(\omega - 10n)}{2n^2 - 3}$$

(c) $f_3(t)$ is the basic periodic pulse with duty cycle $1/3$.

$$ans. \quad (2\pi/3) + \sum_{\substack{n=-\infty \\ n\neq 0}}^{\infty} \frac{2\sin\dfrac{n\pi}{3}\, \delta(\omega - n)}{n}$$

(d) $f_4(t) = -4 + \displaystyle\sum_{\substack{n=0 \\ n\ \text{odd}}}^{\infty} \frac{1}{n^2} \cos n\pi t$

$$ans. \quad -8\pi\, \delta(\omega) + \sum_{\substack{n=-\infty \\ n\ \text{odd}}}^{\infty} \frac{\delta(\omega - n\pi)}{2n^2}$$

IMPULSES

An impulsive function is a function that is concentrated in a region that is small compared to any other interval of interest. It is particularly convenient in many applications to visualize the rectangular pulse, which is impulsive for sufficiently small values of the parameter T:

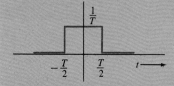

The symbol $\delta(t)$ is used to denote the limiting behavior of impulsive functions:

$$\delta(t) = \begin{cases} 0 & t \neq 0 \\ \infty & t = 0 \end{cases}$$

$$\int_{-\epsilon}^{\epsilon} \delta(t)\, dt = 1$$

When $\delta(t)$ appears in an equation, it is implied that the expression is true in the limit, as the area of an impulsive function in place of $\delta(t)$ becomes more and more concentrated near the origin. The sifting property of the impulse is as follows:

$$\int_{-\epsilon}^{\epsilon} f(t)\,\delta(t)\,dt = f(0) \qquad \int_{a-\epsilon}^{a+\epsilon} f(t)\,\delta(t-a)\,dt = f(a)$$

The Fourier transform of an impulse is the constant unity:

$$\mathcal{F}[\delta(t)] = 1$$

The constant function has Fourier transform

$$\mathcal{F}[k] = 2\pi k\,\delta(\omega),$$

hence the transform of a level-shifted function satisfies

$$\mathcal{F}[f(t) + k] = F(\omega) + 2\pi k\,\delta(\omega)$$

The Fourier transforms of sinusoidal functions are

$$\mathcal{F}[\cos bt] = \pi[\delta(\omega - b) + \delta(\omega + b)]$$

$$\mathcal{F}[\sin bt] = j\pi[-\delta(\omega - b) + \delta(\omega + b)]$$

and a periodic function, $f(t) = \sum\limits_{n=-\infty}^{\infty} d_n\, e^{\frac{jn2\pi t}{T}}$

has Fourier transform $F(\omega) = \sum\limits_{n=-\infty}^{\infty} 2\pi d_n\, \delta\left(\omega - \frac{n2\pi}{T}\right)$

11.11 FOURIER TRANSFORM SOLUTION OF DIFFERENTIAL EQUATIONS

When a differential equation is Fourier transformed, the result is an algebraic equation involving transforms. For example, in Fourier transforming

$$\frac{dy}{dt} + 4y = 3\delta(t)$$

there results

$$j\omega Y(\omega) + 4Y(\omega) = 3 \qquad Y(\omega) = \frac{3}{j\omega + 4}$$

The inverse transform of $Y(\omega)$ is the switched-on exponential

$$y(t) = 3u(t)e^{-4t}$$

This solution method requires that the differential equation driving function be known for all time and be Fourier transformable. The solution, which must also be Fourier transformable, also holds for all time.

Fourier transforming the differential equation

$$\frac{d^2y}{dt^2} + 5\frac{dy}{dt} + 6y = u(t)e^{-2t}$$

gives

$$(j\omega)^2Y(\omega) + 5j\omega Y(\omega) + 6Y(\omega) = \frac{1}{j\omega + 2}$$

$$(-\omega^2 + 5j\omega + 6)Y(\omega) = \frac{1}{j\omega + 2}$$

$$Y(\omega) = \frac{1}{(-\omega^2 + 5j\omega + 6)(j\omega + 2)}$$

To find

$$y(t) = \mathscr{F}^{-1}[Y(\omega)]$$

one would look up $Y(\omega)$ in a table of Fourier transform pairs in much the same way as a table of integrals is used. For a transform as complicated as the one above, some manipulation to express this function as a sum of simpler *partial fraction* terms will probably first be necessary, because transform tables (like integral tables) are seldom extensive enough to apply directly to involved functions. Identical methods are used for the inversion of Laplace transforms in the next chapter, and since our main concern at this time is with the solution process, the details of finding complicated functions from their transforms will be delayed until later.

Equations involving integrals may also be solved in this manner:

$$3\frac{dy}{dt} + 7y + 2\int_{-\infty}^{t} y \, dt = f(t) = \begin{cases} 4 & -1 \le t \le 1 \\ 0 & \text{otherwise} \end{cases}$$

$$3j\omega Y(\omega) + 7Y(\omega) + \frac{2}{j\omega}Y(\omega) = \frac{8 \sin \omega}{\omega}$$

$$\left(3j\omega + 7 + \frac{2}{j\omega}\right)Y(\omega) = \frac{8 \sin \omega}{\omega}$$

$$Y(\omega) = \frac{8j \sin \omega}{-3\omega^2 + 7j\omega + 2}$$

And, simultaneous integrodifferential equations Fourier transform to simultaneous algebraic equations that may be solved for the transforms of the variables.

D11-13

Use Fourier transform methods to find $y(t)$:

(a) $\dfrac{dy}{dt} + 3y = \delta(t)$

ans. $u(t)e^{-3t}$

(b) $2\dfrac{dy}{dt} + 7y = 5\delta(t)$

ans. $\frac{5}{2}u(t)e^{-(7/2)t}$

D11-14

Find the Fourier transform $Y(\omega)$:

(a) $3\dfrac{dy}{dt} + 7y = \begin{cases} 5e^{-6t} & t > 0 \\ 0 & t < 0 \end{cases}$ ans. $\dfrac{5}{(j\omega + 6)(3j\omega + 7)}$

(b) $\begin{cases} \dfrac{dx}{dt} = -2x + y + \delta(t) \\[2mm] \dfrac{dy}{dt} = -3x + 4\delta(t) \end{cases}$ ans. $\dfrac{4j\omega + 5}{-\omega^2 + 2j\omega - 3}$

(c) $3\dfrac{dy}{dt} + 4y + 5\displaystyle\int_{-\infty}^{t} y\,dt = 6f(t) - 7\dfrac{df}{dt}$

where $f(t)$ is a rectangular pulse, centered on the origin, with amplitude 2 and unit width.

ans. $\dfrac{(28\omega + j24)\sin\dfrac{\omega}{2}}{-3\omega^2 + 4j\omega + 5}$

(d) $9\dfrac{d^2y}{dt^2} + 3y = 5e^{-|3t|}$ ans. $\dfrac{30}{(9 + \omega^2)(-9\omega^2 + 3)}$

FOURIER TRANSFORM SOLUTION OF DIFFERENTIAL EQUATIONS

Provided that the solution and the driving function are Fourier transformable, a linear, time-invariant differential equation may be solved by Fourier transforming, solving for the transform of the solution, then inverse transforming. The solution thus obtained is the one for which

$$\lim_{t \to -\infty} y(t) = 0$$

Inversion of transforms is done by manipulation and use of tables of transform pairs.

11.12 FOURIER TRANSFORM NETWORK SOLUTION

11.12.1 Transformed Equations

Fourier transformation of integrodifferential network equations results in linear algebraic equations that may be solved for the transforms of the signals of interest. Once a signal transform is found, inversion to find the time function that is represented is straightforward.

The network of Figure 11-27(a) is described by the differential equation

$$2\frac{di}{dt} + 5i = 4\delta(t)$$

Taking the Fourier transform of both sides of the equation, there results

$$2j\omega I(\omega) + 5I(\omega) = 4$$

where

$$I(\omega) = \frac{4}{5 + 2j\omega} = \frac{2}{j\omega + (5/2)}$$

is the transform of $i(t)$. This is the transform of a switched-on exponential, so

$$i(t) = 2u(t)e^{-(5/2)t}$$

For the network of Figure 11-27(b), systematic simultaneous mesh equations are as follows:

$$\begin{cases} \left(3\frac{d}{dt} + 2\int_{-\infty}^{t} dt\right)i_1 - \left(2\int_{-\infty}^{t} dt\right)i_2 = 7u(t)e^{-8t} \\ -\left(2\int_{-\infty}^{t} dt\right)i_1 + \left(4\frac{d}{dt} + 5 + 2\int_{-\infty}^{t} dt\right)i_2 = 6\delta(t-9) \end{cases}$$

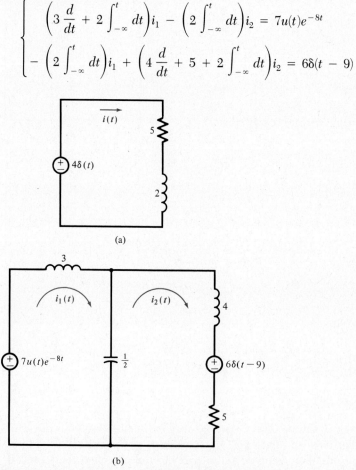

(a)

(b)

Figure 11-27 Fourier transforming network equations.

Fourier transforming these equations gives

$$\begin{cases} \left(3j\omega + \dfrac{2}{j\omega} \right) I_1(\omega) - \left(\dfrac{2}{j\omega} \right) I_2(\omega) = \dfrac{7}{j\omega + 8} \\[3mm] -\left(\dfrac{2}{j\omega} \right) I_1(\omega) + \left(4j\omega + 5 + \dfrac{2}{j\omega} \right) I_2(\omega) = 6e^{-j9\omega} \end{cases}$$

and

$$I_1(\omega) = \frac{\begin{vmatrix} \dfrac{7}{j\omega + 8} & -\dfrac{2}{j\omega} \\[4mm] 6e^{-j9\omega} & 4j\omega + 5 + \dfrac{2}{j\omega} \end{vmatrix}}{\begin{vmatrix} 3j\omega + \dfrac{2}{j\omega} & -\dfrac{2}{j\omega} \\[4mm] -\dfrac{2}{j\omega} & 4j\omega + 5 + \dfrac{2}{j\omega} \end{vmatrix}}$$

$$= \frac{(-3\omega^2 + 2)(j\omega + 8)6e^{-j9\omega} + 14}{(j\omega + 8)(-12j\omega^3 - 15\omega^2 + 14j\omega + 10)}$$

$$I_2(\omega) = \frac{\begin{vmatrix} 3j\omega + \dfrac{2}{j\omega} & \dfrac{7}{j\omega + 8} \\[4mm] -\dfrac{2}{j\omega} & 6e^{-j9\omega} \end{vmatrix}}{\begin{vmatrix} 3j\omega + \dfrac{2}{j\omega} & -\dfrac{2}{j\omega} \\[4mm] -\dfrac{2}{j\omega} & 4j\omega + 5 + \dfrac{2}{j\omega} \end{vmatrix}}$$

$$= \frac{12(j\omega + 8)e^{-j9\omega} - 28\omega^2 + 35j\omega + 2}{(j\omega + 8)(-12j\omega^3 - 15\omega^2 + 14j\omega + 10)}$$

These transforms are too complicated to invert at this time; however, similar inversions involving functions of comparable complexity will be done in the next chapter, in connection with the Laplace transform.

The Fourier transform solution method requires that all signals involved be Fourier transformable and that the driving functions be known for all time. The solution obtained in this manner holds for all time.

11.12.2 Transformed Networks

Fourier transformed equations may be written directly from a network diagram by replacing all signals by their Fourier transforms and representing all elements by their impedances for $s = j\omega$. The example network of Figure 11-28(a) is expressed in terms of the Fourier transform quantities in Figure 11-28(b). Systematic simultaneous nodal equations, in terms of the indicated node voltage transforms, are

$$\left(\frac{1}{2} + 3j\omega\right)V_1(\omega) - 3j\omega V_2(\omega) = 10 + \frac{1}{9 + j\omega}$$

$$-3j\omega V_1(\omega) + \left(3j\omega + \frac{1}{4j\omega} + \frac{1}{5} + \frac{1}{6}\right)V_2(\omega) - \left(\frac{1}{4j\omega} + \frac{1}{5}\right)V_3(\omega) = -\frac{1}{9 + j\omega}$$

$$-\left(\frac{1}{4j\omega} + \frac{1}{5}\right)V_2(\omega) + \left(\frac{1}{4j\omega} + \frac{1}{6j\omega} + \frac{1}{5} + 8j\omega\right)V_3(\omega) = 0$$

These are exactly the equations that would be obtained by writing the corresponding integrodifferential equations for the original network of Figure 11-28(a), then Fourier transforming the equations.

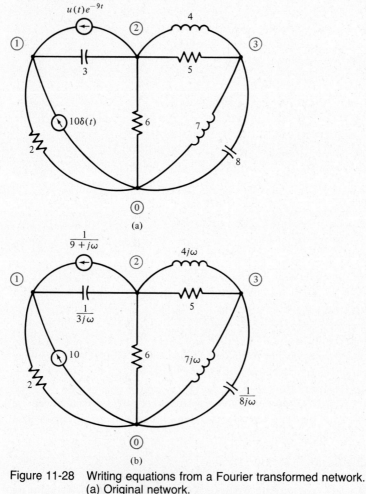

(a)

(b)

Figure 11-28 Writing equations from a Fourier transformed network.
(a) Original network.
(b) Fourier transformed network.

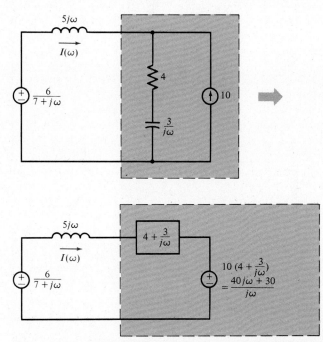

Figure 11-29   Using equivalent circuits with a Fourier transformed network.

Equivalent circuits apply to Fourier transformed networks because they represent manipulations of the equations describing the network that apply equally to the transformed equations. For example, equivalent circuits are used on the Fourier transformed network of Figure 11-29 to obtain

$$\left(5j\omega + 4 + \frac{3}{j\omega}\right)I(\omega) = \frac{6}{7 + j\omega} - \frac{40j\omega + 30}{j\omega}$$

$$(-5\omega^2 + 4j\omega + 3)I(\omega) = \frac{6j\omega - (40j\omega + 30)(7 + j\omega)}{7 + j\omega}$$

$$I(\omega) = \frac{40\omega^2 - 152j\omega - 210}{(-5\omega^2 + 4j\omega + 3)(7 + j\omega)}$$

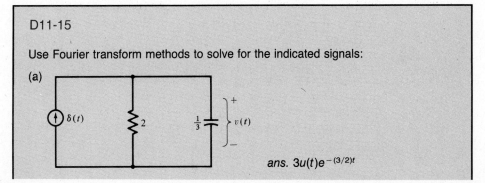

D11-15

Use Fourier transform methods to solve for the indicated signals:

(a)

ans. $3u(t)e^{-(3/2)t}$

(b)

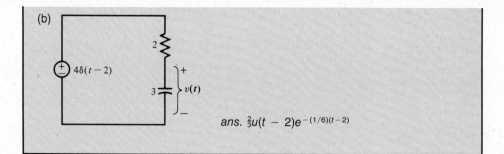

ans. $\frac{2}{3}u(t - 2)e^{-(1/6)(t-2)}$

D11-16

Find the Fourier transforms of the indicated signals:

(a)

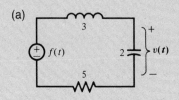

$f(t)$ is a rectangular pulse of unit length, centered on the origin, with width $\frac{1}{4}$.

ans. $\dfrac{4 \sin (\omega/8)}{\omega(-3\omega^2 + 5j\omega + 2)}$

(b)

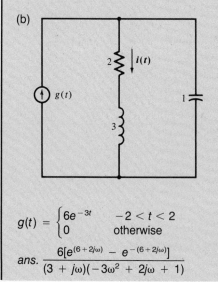

$$g(t) = \begin{cases} 6e^{-3t} & -2 < t < 2 \\ 0 & \text{otherwise} \end{cases}$$

ans. $\dfrac{6[e^{(6+2j\omega)} - e^{-(6+2j\omega)}]}{(3 + j\omega)(-3\omega^2 + 2j\omega + 1)}$

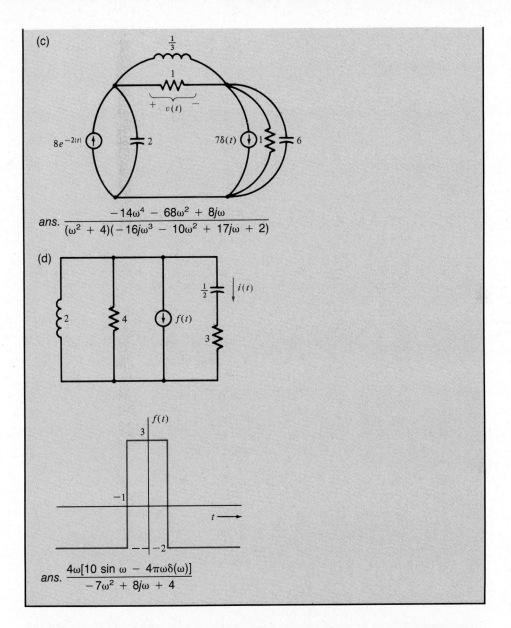

(c)

ans. $\dfrac{-14\omega^4 - 68\omega^2 + 8j\omega}{(\omega^2 + 4)(-16j\omega^3 - 10\omega^2 + 17j\omega + 2)}$

(d)

ans. $\dfrac{4\omega[10\sin\omega - 4\pi\omega\delta(\omega)]}{-7\omega^2 + 8j\omega + 4}$

FOURIER TRANSFORM NETWORK SOLUTION

If all signals involved are Fourier transformable, a network may be solved by solving the Fourier transformed network for the transform of the signal of interest, then inverting to find the signal as a function of time. In the transformed network, all signals are replaced by their Fourier transforms, and all elements by their impedances, evaluated at $s = j\omega$.

CHAPTER ELEVEN PROBLEMS

Basic Problems

Series Conversions

1. Convert to trigonometric form:

(a) $f_1(t) = \displaystyle\sum_{n=-\infty}^{\infty} \frac{3 + j2n}{n^2 + 4} e^{j\frac{3n}{2}t}$

(b) $f_2(t) = \displaystyle\sum_{\substack{n=-\infty \\ n \neq 0}}^{\infty} \frac{3}{2 - jn} e^{j4\pi nt}$

(c) $f_3(t) = \displaystyle\sum_{\substack{n=-\infty \\ n \text{ odd}}}^{\infty} \frac{e^{j\left(nt - \frac{n\pi}{8}\right)}}{6n}$

2. Find the exponential forms of the Fourier series for each of the following by converting previously derived trigonometric series:
   (a) Basic triangular wave
   (b) Basic full-wave rectified sinusoid
   (c) Basic half-wave rectified sinusoid

Exponential Series Calculation

3. Calculate (setting up and evaluating the integrals involved) the exponential Fourier series for the following functions:

(a)

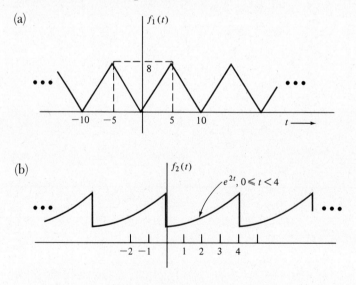

(b)

(c) The periodic function given by

$$f_3(t) = 6e^{-t} \qquad 0 \leq t < 2$$

and having period 2 seconds.

Series Spectra

4. Sketch both real and imaginary part and sketch magnitude and angle line spectra for the following periodic functions:
   (a) Basic periodic pulse with duty cycle one-half (which is a square wave centered at $t = 0$).
   (b) Basic periodic pulse with duty cycle one-third.
   (c) Basic periodic pulse with duty cycle two-thirds.
   (d) Basic periodic pulse with duty cycle one-sixth.

5. Find the exponential Fourier series of the periodic functions with the following spectra:

(a)

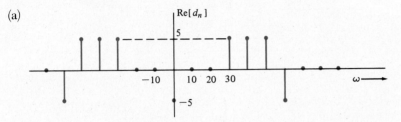

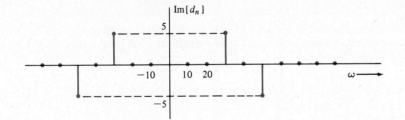

(b)

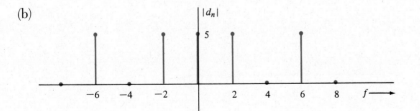

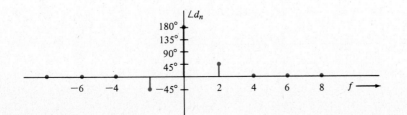

A two-arm assembly robot in a manufacturing plant is being programmed to execute the complicated series of operations needed to assemble a small gear train. (*Photo courtesy of Westinghouse Electric Corp.*)

Series Manipulation

6. Using the series for the basic periodic pulse, find the exponential Fourier series for the following functions and sketch their magnitude and angle spectra:

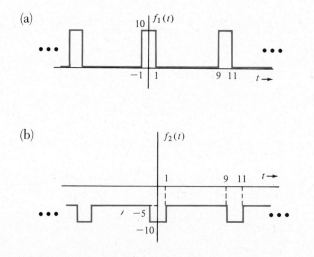

(c)

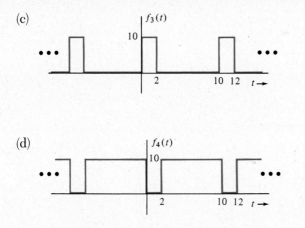

(d)

Differential Equations with Periodic Driving Functions

7. For the following differential equations and periodic driving functions, find the exponential Fourier series for the forced solution $y(t)$:

(a) $\dfrac{dy}{dt} + 6y = f_p(t)$

where $f_p(t)$ is the basic periodic pulse with duty cycle $\frac{1}{4}$.

(b) $7\dfrac{dy}{dt} + y = \displaystyle\sum_{n=-\infty}^{\infty} \dfrac{3}{n^4 + 1} e^{\frac{jnt}{8}}$

(c) $\dfrac{d^2y}{dt^2} - 4y = 8 + \displaystyle\sum_{\substack{n=-\infty \\ n \text{ odd}}}^{\infty} \dfrac{j}{n^3} e^{jn\pi t}$

(d) $\dfrac{d^2y}{dt^2} + 4\dfrac{dy}{dt} + 5y = f_S(t)$

where $f_S(t)$ is the basic square wave.

Networks with Periodic Sources

8. For the following networks with periodic sources, find the exponential Fourier series for the indicated signals:

(a)

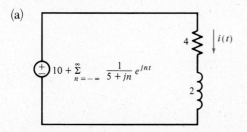

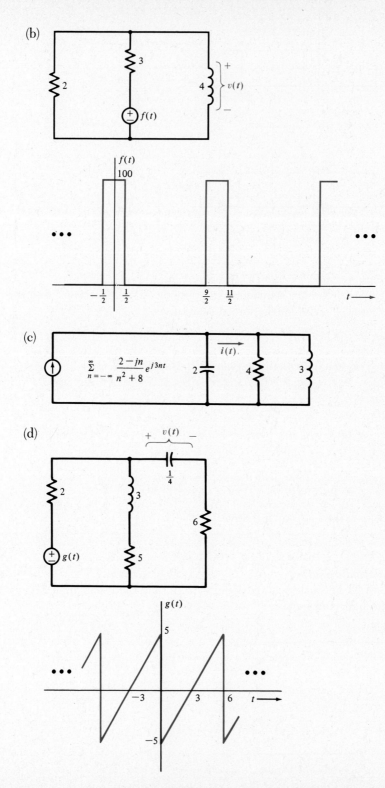

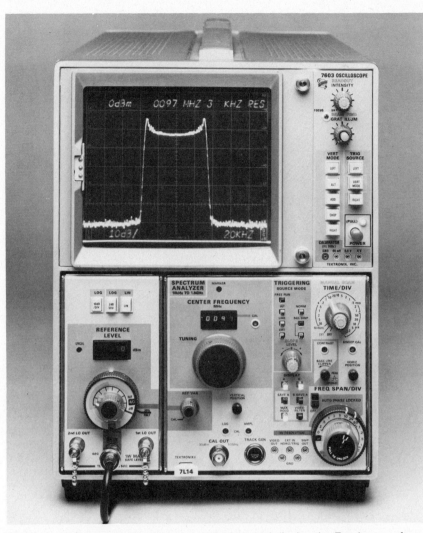

Spectrum analyzers such as this one calculate and display the Fourier transform magnitude of a signal voltage over a selected range of frequency. (*Photo courtesy of Tektronix, Inc.*)

Fourier Transform Calculation

9. Calculate (setting up and evaluating the integral involved) the Fourier transforms of the following functions:

(a)

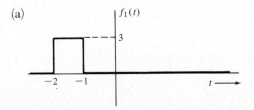

(b)

$$f_2(t) = \begin{cases} -3 & -2 \le t < -1 \\ 5 & -1 \le t < 1 \\ -3 & 1 \le t < 2 \\ 0 & \text{otherwise} \end{cases}$$

(c)

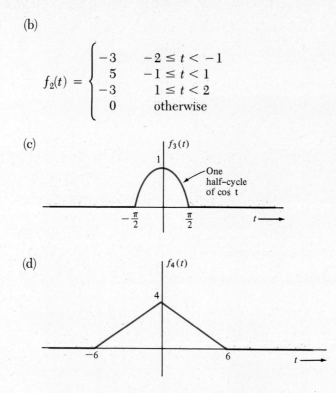

$f_3(t)$

1

One half-cycle of cos t

$-\dfrac{\pi}{2}$ $\dfrac{\pi}{2}$ $t \longrightarrow$

(d)

$f_4(t)$

4

-6 6 $t \longrightarrow$

Fourier Transform Properties
10. Using the Fourier transforms for the basic pulse and the switched exponentials, find the transforms of the following functions:

(a)

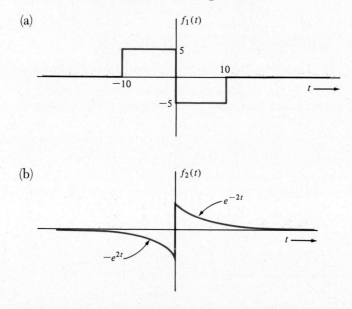

$f_1(t)$

5

-10 10 $t \longrightarrow$

-5

(b)

$f_2(t)$

e^{-2t}

$t \longrightarrow$

$-e^{2t}$

(c)

$$f_3(t) = \begin{cases} 0 & t < -5 \\ e^{-2t} & t \geq -5 \end{cases}$$

(d)

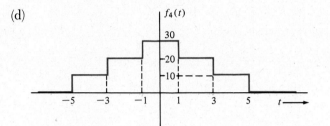

11. Consider the function

$$f(t) = \begin{cases} -2t & -3 \leq t \leq 3 \\ 0 & \text{otherwise} \end{cases}$$

Find the function

$$g(t) = \mathscr{F}^{-1}[F(\omega)]$$

Pay particular attention to the points of discontinuity.

12. Use the Fourier transform derivative relation to find $\mathscr{F}[df/dt]$ where

$$f(t) = \begin{cases} e^{-3t} & t \geq 0 \\ e^{3t} & t \leq 0 \end{cases}$$

Parseval's Transform Relation

13. For

$$f(t) = \begin{cases} t & -2 \leq t \leq 2 \\ 0 & \text{otherwise} \end{cases}$$

find

$$I = \int_{-\infty}^{\infty} |F(\omega)|^2 \, d\omega$$

14. For

$$F(\omega) = \begin{cases} j\omega & -10 \leq \omega \leq 10 \\ 0 & \text{otherwise} \end{cases}$$

find

$$I = \int_{-\infty}^{\infty} f^2(t) \, dt$$

Fourier Transform Spectra

15. Find and sketch the real part and imaginary part and the magnitude and angle spectra of the functions with the following Fourier transforms:

(a) $F_1(\omega) = \dfrac{j\omega}{\omega^2 + 4}$

(b) $F_2(\omega) = \dfrac{3}{j\omega + 1}$

(c) $F_3(\omega) = \dfrac{j\omega + 1}{(j\omega + 2)(j\omega + 3)}$

16. Find the function $f(t)$ that has the spectrum sketched:

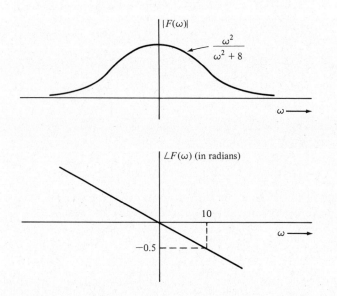

17. For the function with magnitude and angle spectra

$$F(\omega) = \dfrac{10\omega^2}{\omega^4 + 2\omega^2 + 10} \qquad \underline{/F(\omega)} = -3\omega$$

find and sketch the real and imaginary part spectra.

Impulses

18. Find the Fourier transform of the function

$$f(t) = \delta(t + 2) - \delta(t - 2)$$

19. Find

$$f(t) = \int_{-\infty}^{t} 3\delta(t - 4)\, dt$$

and sketch it.

20. Find the Fourier transform of the derivative of the function

$$f(t) = \begin{cases} 10e^{-3t} & t > 0 \\ 0 & t < 0 \end{cases}$$

Transforms of Periodic Functions

21. A function $f(t)$ has Fourier transform

$$F(\omega) = \frac{3}{\omega^2 + 4}$$

Find the Fourier transform of

$$g(t) = f(t) + 5$$

22. Find the Fourier transform of $f(t)$:

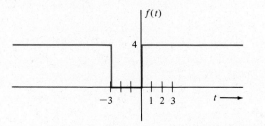

23. Find the time function that has Fourier transform

$$F(\omega) = 3\,\delta(\omega - 2) + \delta(\omega) + 3\,\delta(\omega + 2)$$

24. Find the Fourier transform of the basic periodic square wave with amplitude unity and period 1 millisecond.

Transform Solution of Differential Equations

25. Find the Fourier *transforms* of $y(t)$ in each case:

(a) $\dfrac{dy}{dt} + 3y = 2\,\delta(t)$

(b) $\dfrac{d^2y}{dt^2} + 3\dfrac{dy}{dt} + y = u(t)e^{-t}$

(c) $\dfrac{d^2y}{dt^2} + 2\dfrac{dy}{dt} + 3y = 4e^{-5|t|}$

(d) $\begin{cases} \dfrac{dy}{dt} = -3y + x - \delta(t - 4) \\[2mm] \dfrac{dx}{dt} = y - 5x + \delta(t + 6) \end{cases}$

Transform Solution of Networks
26. Use Fourier transform methods to solve for the indicated signals:

(a)

(b)

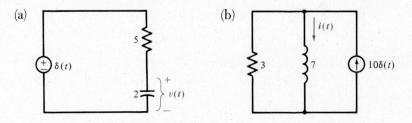

27. Find the Fourier *transforms* of the indicated signals:

(a)

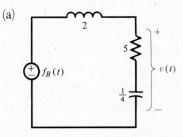

$f_B(t)$ is the basic pulse with duration $b = \frac{1}{2}$ s.

(b)

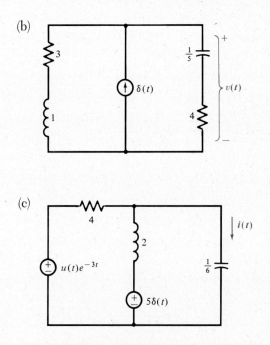

(c)

(d)

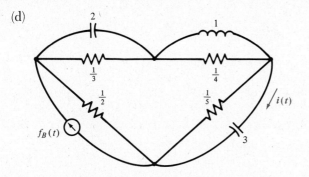

$f_B(t)$ is the basic pulse with duration $b = 6$ s.

(e)

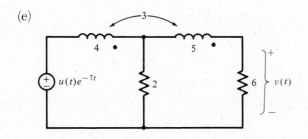

Practical Problems

Periodic Approximation

28. Approximate the function with Fourier transform

$$F(\omega) = \frac{2 - j\omega}{\omega^2 + 3}$$

by a periodic function with period 100 seconds.

Amplitude Modulation

29. Amplitude modulation (AM) is the multiplication of a sinusoid by a modulating signal, $f(t)$:

$$a(t) = f(t) \cos \omega_c t$$

Show that

$$\mathscr{F}[f(t) \cos \omega_c t] = \tfrac{1}{2}F(\omega - \omega_c) + \tfrac{1}{2}F(\omega + \omega_c)$$

The above relation demonstrates that AM involves positive- and negative-frequency translations of the spectrum of the modulating signal. Each frequency in $f(t)$ results in sum and difference frequencies with ω_c in the AM signal.

Advanced Problems

Average Power

30. Show that for periodic voltage and current

$$v(t) = \sum_{n=-\infty}^{\infty} \delta_n e^{j\frac{n2\pi t}{T}}$$

$$i(t) = \sum_{n=-\infty}^{\infty} d_n e^{j\frac{n2\pi t}{T}}$$

average power may be expressed as

$$P = \langle v(t) \, i(t) \rangle = \frac{1}{T}\int_T v(t) \, i(t) \, dt$$

$$= \delta_0 d_0 + 2 \sum_{n=1}^{\infty} \text{Re}\, [\delta_n d_n{}^*]$$

$$= \text{Re}\left[\sum_{n=-\infty}^{\infty} \delta_n d_n{}^* \right]$$

Inverse Transforms

31. Calculate the inverse Fourier transforms of the following:

(a) $F(\omega) = e^{-|\omega|}$

(b) $F(\omega) = \begin{cases} 1 - j & -1 < \omega < 0 \\ 1 + j & 0 < \omega < 1 \\ 0 & \text{otherwise} \end{cases}$

Fourier Transform Properties

32. Show that if

$$\mathcal{F}[f(t)] = F(\omega)$$

then

$$\mathcal{F}[f(-t)] = F^*(\omega)$$

where F^* denotes the complex conjugate of F.

Transforms of Periodic Functions

33. Show that the Fourier transform of a train of impulses in the time domain,

$$f(t) = \sum_{n=-\infty}^{\infty} \delta(t - n\,\Delta t)$$

where Δt is a constant, is a train of impulses in the frequency domain.

34. Find the time function that has periodic Fourier transform

$$F(\omega) = 4\cos 3\omega$$

Transforms of Complex Functions
A complex function of time

$$f(t) = x(t) + jy(t)$$

has Fourier transform

$$F(\omega) = X(\omega) + jY(\omega)$$

The even part of $\text{Re}[F(\omega)]$ is contributed by $X(\omega)$, and the odd part of $\text{Re}[F(\omega)]$ is contributed by $jY(\omega)$. The odd part of $\text{Im}[F(\omega)]$ is contributed by $X(\omega)$, and $jY(\omega)$ contributes the even part of $\text{Im}[F(\omega)]$.

35. Calculate $F(\omega)$ where

$$f(t) = \begin{cases} e^{j7t} & 0 < t < 1 \\ 0 & \text{otherwise} \end{cases}$$

Then sketch its real and imaginary part spectral plots.

36. Find $f(t) = \mathcal{F}^{-1}\left[\dfrac{3 - 2j}{j\omega + 4}\right]$ and sketch the real and imaginary parts of $f(t)$.

Multidimensional Fourier Transforms
Functions of several variables,

$$f(x, y, z, \ldots)$$

are Fourier transformed with respect to each of the variables according to

$$\mathcal{F}[f(x, y, z, \ldots)] = F(\omega_x, \omega_y, \omega_z, \ldots)$$

$$= \int_{-\infty}^{\infty} \int_{-\infty}^{\infty} \cdots \int_{-\infty}^{\infty} f(x, y, z, \ldots) e^{-j\omega_x x} e^{-j\omega_y y} e^{-j\omega_z z} \cdots dx\, dy\, dz \cdots$$

$$= \int_{-\infty}^{\infty} \int_{-\infty}^{\infty} \cdots \int_{-\infty}^{\infty} f(x, y, z, \ldots) e^{-j(\omega_x x + \omega_y y + \omega_z z + \cdots)} \, dx\, dy\, dz \ldots$$

The inverse transform is

$$f(x, y, z, \ldots) = \mathcal{F}^{-1}[F(\omega_x, \omega_y, \omega_z, \ldots)]$$

$$= \left(\frac{1}{2\pi}\right)^k \int_{-\infty}^{\infty} \int_{-\infty}^{\infty} \cdots \int_{-\infty}^{\infty} F(\omega_x, \omega_y, \omega_z, \ldots) \cdot$$

$$\cdot\, e^{j\omega_x x} e^{j\omega_y y} e^{j\omega_z z} \cdots d\omega_x\, d\omega_y\, d\omega_z \cdots$$

$$= \left(\frac{1}{2\pi}\right)^k \int_{-\infty}^{\infty} \int_{-\infty}^{\infty} \cdots \int_{-\infty}^{\infty} F(\omega_x, \omega_y, \omega_z, \ldots) \cdot$$

$$\cdot\, e^{j(\omega_x x + \omega_y y + \omega_z z + \cdots)} \, d\omega_x\, d\omega_y\, d\omega_z \ldots$$

where k is the number of transform variables.

Multidimensional Fourier transforms are commonly used in such fields as electromagnetics, optics, and image processing.

37. Show that the two dimensional Fourier transform of

$$f(x, y) = \begin{cases} e^{-(x+y)} & x > 0 \text{ and } y > 0 \\ 0 & \text{otherwise} \end{cases}$$

is

$$\frac{1}{(j\omega_x + 1)(j\omega_y + 1)}$$

Chapter Twelve
The Laplace Transform and Network Solutions

> The efforts of great men have brought about the discovery of the theorems and methods which together constitute the Science of Mathematics. . . . [Its] domain depends upon how much it is applied to natural phenomena which are the mathematical results of a number of invariate laws. While improving the natural sciences, mathematical models have opened new avenues for analysis. It is therefore evident that interaction among the sciences generates mutual support.
>
> Pierre-Simon Laplace
> From *Journal de l'École Polytechnic*, 1812

12.1 PREVIEW

The Laplace transformation gives a systematic, general method of switched network solution with routine incorporation of initial conditions. Like the Fourier transformation, to which it is closely related, it allows algebraic-equation manipulation in place of the complications involved when derivatives and integrals are dealt with directly. And Laplace transform is a potent mathematical tool with many other applications, as well.

The Laplace transform is now derived from the Fourier transform and its basic properties are developed. Initial conditions for the Laplace transform are defined at $t = 0^-$, as is most sensible for network problems. That way, the effects of impulses at $t = 0$ are included in the transform; for switched networks, the conditions immediately prior to the switching are those that are incorporated.

Because of their importance in application, sections on derivatives of discontinuous functions, derivatives of impulses, and transforms of piecewise polynomial functions are given. Efficient partial fraction coefficient evaluation methods are shown for distinct, repeated, and complex conjugate pair roots.

The Laplace transformation of integrodifferential equations easily leads to Laplace transformed networks in the case of zero initial conditions. Although nonzero initial conditions can be handled from the standpoint of the network equations, it is unnecessarily involved to do so; so source models equivalent to initial capacitor voltages and inductor currents are used.

Convolution is introduced in the concluding section of the chapter.

When you complete this chapter you should know—

1. how the Laplace transformation is related to that of Fourier, and how to calculate Laplace transforms of functions;
2. about the inverse Laplace transform and how it is used;
3. basic properties of the Laplace transform and how they may be used in transform calculation;
4. how to expand rational functions in partial fractions in order to find inverse Laplace transforms;
5. the special properties of partial fraction expansion involving complex roots;
6. the initial and final value theorems;
7. how to use the Laplace transform to solve single and simultaneous differential and integrodifferential equations;
8. how to use the Laplace transform to solve switched electrical networks with zero initial conditions;
9. how nonzero initial capacitor voltages and inductor currents can be routinely represented as equivalent sources, allowing easy Laplace transform solution of switched networks in general;
10. about convolution and how it is related to the Laplace transform and system response calculation, and how to perform it graphically.

12.2 LAPLACE TRANSFORM CALCULATION

12.2.1 Derivation of the Transform

The Fourier transformation is very useful for finding the response of systems to a large class of driving functions. Yet there are still many functions $f(t)$ that occur in practice for which the Fourier transform does not converge; that is,

$$\int_{-\infty}^{\infty} |f(t)|\ dt$$

is not finite, and the transform cannot be calculated (in the usual way) or $f(t)$ recovered. Typical of these functions are the switched-on ones, such as the step function and the others sketched in Figure 12-1. To derive the Laplace transformation, we consider a related function

$$g(t) = f(t)u(t)e^{-\sigma t}$$

which is the function of interest, $f(t)$, multiplied by a unit step, $u(t)$, and a decaying exponential, $e^{-\sigma t}$. For all but the most pathological of functions, "switching on" the function $f(t)$ by multiplying it by the unit step function and multiplying it by a sufficiently rapidly decaying exponential will result in a new function, $g(t)$, which converges. The function $g(t)$ is zero prior to $t = 0$ because $u(t)$ is zero there, and approaches zero for large t because of the decaying exponential factor, $e^{-\sigma t}$.

Virtually every switched-on function of interest to the description of physical processes may be made to converge when multiplied by a decaying exponential. For

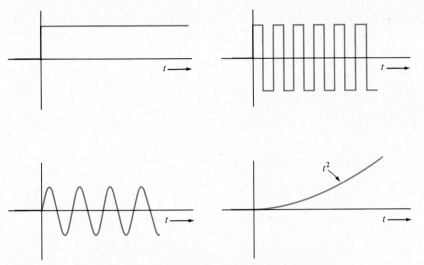

Figure 12-1   Some functions that are not Fourier transformable.

example, any power of t converges when multiplied by any decaying exponential. An expanding exponential such as

$$f(t) = e^{100t}$$

converges for $\sigma > 100$:

$$\int_0^\infty \left| e^{100t} e^{-\sigma t} \right| dt = \int_0^\infty e^{(100-\sigma)t} dt$$

Generally, convergence occurs for any convergence constant σ greater than some minimum value. Fortunately, it is not necessary in what follows to specify σ, except that it is to be any positive constant large enough to give convergence. Multiplication by the step function is important, too, since the exponential convergence factor "blows up" as one goes backward in time.

The Laplace transformation of a function $f(t)$ is the Fourier transform of the function

$$g(t) = f(t)u(t)e^{-\sigma t}$$

where σ is a constant that is sufficiently large for convergence of $g(t)$. The unit step function $u(t)$ is defined as

$$u(t) = \begin{cases} 0 & t < 0 \\ 1 & t \geq 0 \end{cases}$$

with $u(0) = 1$, so that the behavior of $f(t)$ at $t = 0$ is included in the transform. The Fourier transform of $g(t)$ is by definition

$$\mathcal{F}[g(t)] = \int_{-\infty}^{\infty} f(t)u(t)e^{-\sigma t}e^{-j\omega t}\, dt$$

Combining the exponentials and using the fact that $f(t)u(t)$ is zero prior to $t = 0$,

$$\mathscr{F}[g(t)] = \int_{0-}^{\infty} f(t)e^{-(\sigma+j\omega)t}\, dt$$

The 0^- limit indicates that, should it make a difference, the integration is to start just before time $t = 0$.

Letting the variable

$$s = \sigma + j\omega$$

the Laplace transform of $f(t)$ becomes

$$\mathscr{L}[f(t)] = F(s) = \mathscr{F}[f(t)u(t)e^{-\sigma t}]$$

$$= \int_{0-}^{\infty} f(t)e^{-st}\, dt$$

Laplace transformation is indicated by $\mathscr{L}[\ ]$ or by the upper case of the symbol for the time function, for example, $F(s)$. Pierre Simon Laplace (1749–1827), for whom the transform is named, was a French mathematical astronomer.

12.2.2 Fundamental Transforms

The unit impulse, represented in Figure 12-2(a), has Laplace transform

$$\mathscr{L}[\delta(t)] = \int_{0-}^{\infty} \delta(t)e^{-st}\, dt = e^0 = 1$$

That the lower limit of the transform integral is 0^- is crucial to this result. Were the limit $t = 0$, the impulse would be "cut in half" in some sense. A lower limit of 0^+ would exclude the effect of the impulse altogether, giving zero as the transform.

The unit step function, Figure 12-2(b), has Laplace transform given by

$$\mathscr{L}[u(t)] = \int_{0}^{\infty} 1e^{-st}\, dt = \left.\frac{e^{-st}}{-s}\right|_{0}^{\infty}$$

Now

$$\lim_{t\to\infty} e^{-st} = \lim_{t\to\infty} e^{-\sigma t}e^{-j\omega t} = 0$$

since

$$\left|e^{-j\omega t}\right| = 1$$

for all t and because $e^{-\sigma t}$, with σ positive, approaches zero for large t. Then

$$\mathscr{L}[u(t)] = \frac{0 - 1}{-s} = \frac{1}{s}$$

Switched-on exponential functions are shown in Figure 12-2(c) and (d). A switched-on exponential has Laplace transform

$$\mathscr{L}[e^{-at}] = \int_{0}^{\infty} e^{-at}e^{-st}\, dt = \int_{0}^{\infty} e^{-(s+a)t}\, dt = \left.\frac{e^{-(s+a)t}}{-(s+a)}\right|_{0}^{\infty}$$

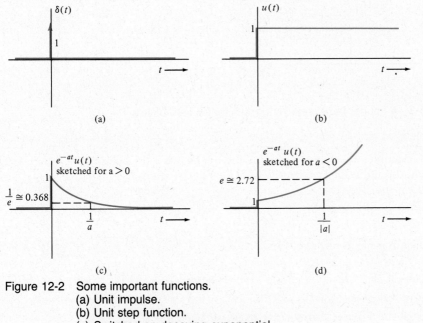

Figure 12-2   Some important functions.
(a) Unit impulse.
(b) Unit step function.
(c) Switched-on decaying exponential.
(d) Switched-on expanding exponential.

As with the case of the step function,

$$\lim_{t\to\infty} e^{-(s+a)t} = \lim_{t\to\infty} e^{-(\sigma+a)t}e^{-j\omega t} = 0$$

For convergence, σ must be sufficiently large so that the quantity $(\sigma + a)$ is positive. Thus

$$\mathscr{L}[e^{-at}] = \frac{1}{s + a}$$

12.2.3 Laplace Transforms from Fourier Transforms

Any Fourier transformable function $f(t)$ that is zero prior to time $t = 0$ has a Fourier transform given by

$$\mathscr{F}[f(t)] = \int_{0-}^{\infty} f(t)e^{-j\omega t}\, dt$$

Comparing with that function's Laplace transform,

$$\mathscr{L}[f(t)] = \int_{0-}^{\infty} f(t)e^{-st}\, dt$$

it is seen that one transform may be obtained from the other by interchanging $j\omega$ and s. If $f(t)$ is nonzero before $t = 0$ or if the Fourier transform of $f(t)$ does not converge, then of course this relation does not hold.

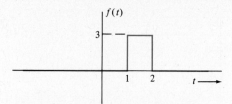

Figure 12-3 A Fourier transformable function that is zero prior to $t = 0$.

For example, the function $f(t)$ in Figure 12-3 is zero before $t = 0$ and has both a Fourier and a Laplace transform. The Fourier transform of $f(t)$ is

$$F(\omega) = \int_1^2 3e^{-j\omega t} \, dt = \left. \frac{3e^{-j\omega t}}{-j\omega} \right|_1^2 = \frac{3e^{-j\omega} - 3e^{-2j\omega}}{j\omega}$$

The Laplace transform of $f(t)$ is

$$F(s) = \int_1^2 3e^{-st} \, dt = \left. \frac{3e^{-st}}{-s} \right|_1^2 = \frac{3e^{-s} - 3e^{-2s}}{s}$$

which is just the Fourier transform with $j\omega$ replaced by s.

12.2.4 Inverse Laplace Transformation

To recover the function from its transformation, the inverse Fourier transform relation gives

$$\mathscr{F}[f(t)u(t)e^{-\sigma t}] = F(s)$$

$$f(t)u(t)e^{-\sigma t} = \mathscr{F}^{-1}[F(s)] = \frac{1}{2\pi} \int_{-\infty}^{\infty} F(s)e^{j\omega t} \, d\omega$$

$$f(t)u(t) = \frac{e^{\sigma t}}{2\pi} \int_{-\infty}^{\infty} F(s)e^{j\omega t} \, d\omega = \frac{1}{2\pi} \int_{-\infty}^{\infty} F(s)e^{(\sigma + j\omega)t} \, d\omega$$

$$= \frac{1}{2\pi} \int_{-\infty}^{\infty} F(s)e^{st} \, d\omega$$

Inverse Laplace transform is denoted by

$$\mathscr{L}^{-1}[F(s)] = f(t)u(t) = \begin{cases} f(t) & t \geq 0 \\ 0 & t < 0 \end{cases}$$

$$= \frac{1}{2\pi} \int_{-\infty}^{\infty} F(s)e^{st} \, d\omega$$

The inverse transformation recovers the transformed function $f(t)$ for $t \geq 0$ and gives zero for $t < 0$.

The calculation indicated by the inverse transformation is that of replacing s by $\sigma + j\omega$, where σ is a sufficiently large constant for convergence, and integrating with respect to ω. In practice, the integrals involved are often difficult, so tables of

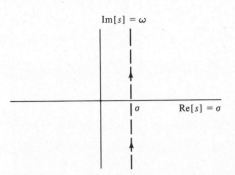

Figure 12-4 Path of integration on the complex plane for the inverse Laplace transform.

Laplace transform pairs are commonly developed by transform rather than inverse transform calculation, just as many entries in integral tables are most easily developed by differentiation.

The inverse Laplace transformation may be viewed as an integration on the complex plane, as indicated in Figure 12-4. The path of integration involves holding σ, the real part of s, constant and varying ω, the imaginary part of s. If desired, the inverse transform may be expressed in this manner by substituting

$$ds = d\sigma + jd\omega = jd\omega$$

The convergence factor σ is constant, so $d\sigma = 0$.

$$\mathcal{L}[F(s)] = \frac{1}{2\pi} \int_{-\infty}^{\infty} F(s)e^{st}\, d\omega = \frac{1}{2\pi j} \int_{\sigma-j\infty}^{\sigma+j\infty} F(s)e^{st}\, ds$$

The limits on the ds integral are the limits on s, which are $\sigma \pm j\infty$.

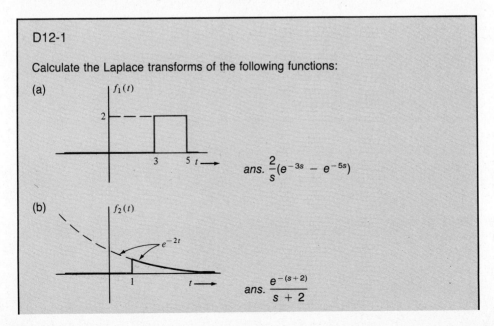

D12-1

Calculate the Laplace transforms of the following functions:

(a)

ans. $\dfrac{2}{s}(e^{-3s} - e^{-5s})$

(b)

ans. $\dfrac{e^{-(s+2)}}{s+2}$

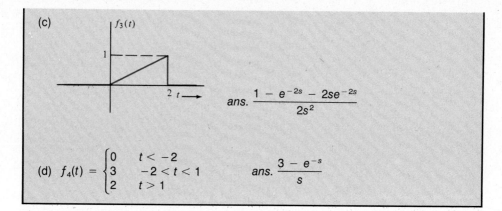

(c)

$$ans. \quad \frac{1 - e^{-2s} - 2se^{-2s}}{2s^2}$$

(d) $f_4(t) = \begin{cases} 0 & t < -2 \\ 3 & -2 < t < 1 \\ 2 & t > 1 \end{cases}$ $\qquad ans. \quad \dfrac{3 - e^{-s}}{s}$

LAPLACE TRANSFORM CALCULATION

The Laplace transform of a function $f(t)$ is

$$F(s) = \int_{0^-}^{\infty} f(t)e^{-st}\, dt$$

where $s = \sigma + j\omega$, and σ is a constant, termed the *convergence factor*, that is sufficiently large for convergence. Some basic Laplace transforms are as follows:

	$f(t) \quad t \geq 0$	$F(s)$
(1)	$\delta(t)$	2
(2)	1	$\dfrac{1}{s}$
(3)	e^{-at}	$\dfrac{1}{s + a}$

For a Fourier transformable function that is zero prior to $t = 0$, the Laplace transform is the Fourier transform with $j\omega$ replaced by s.

A function that is zero prior to $t = 0$ is recovered from its Laplace transform via the inverse transformation:

$$f(t)u(t) = \begin{cases} f(t) & t \geq 0 \\ 0 & t < 0 \end{cases} = \frac{1}{2\pi} \int_{-\infty}^{\infty} F(s)e^{st}\, d\omega$$

where

$$s = \sigma + j\omega$$

The inverse transformation may be written entirely in terms of the complex variable s as

$$f(t)u(t) = \frac{1}{2\pi j} \int_{\sigma - j\infty}^{\sigma + j\infty} F(s)e^{st}\, ds$$

12.3 SOME LAPLACE TRANSFORM PROPERTIES

12.3.1 Linearity

Basic Laplace transform properties are now derived. These are similar to the corresponding Fourier transform properties found previously. The Laplace transform is linear, in that the transform of a sum is the sum of the individual transforms,

$$\mathcal{L}[f_1(t) + f_2(t)] = \int_{0^-}^{\infty} [f_1(t) + f_2(t)]e^{-st}\,dt$$

$$= \int_{0^-}^{\infty} f_1(t)e^{-st}\,dt + \int_{0^-}^{\infty} f_2(t)e^{-st}\,dt = F_1(s) + F_2(s)$$

and the transform of a constant times a function is the constant times the function's transform:

$$\mathcal{L}[Af(t)] = \int_{0^-}^{\infty} Af(t)e^{-st}\,dt = A\int_{0^-}^{\infty} f(t)e^{-st}\,dt$$

$$= A\mathcal{L}[f(t)] = AF(s)$$

The Laplace transform of a product of two functions is generally *not* the product of the individual transforms.

The switched-on cosine function, Figure 12-5(a), has a Laplace transform that may be derived easily by expanding the cosine into a sum of Euler components:

$$\mathcal{L}[\cos bt] = \mathcal{L}[\tfrac{1}{2}e^{jbt} + \tfrac{1}{2}e^{-jbt}] = \mathcal{L}[\tfrac{1}{2}e^{jbt}] + \mathcal{L}[\tfrac{1}{2}e^{-jbt}]$$

$$= \frac{\tfrac{1}{2}}{s - jb} + \frac{\tfrac{1}{2}}{s + jb} = \frac{\tfrac{1}{2}s - \tfrac{1}{2}jb + \tfrac{1}{2}s + \tfrac{1}{2}jb}{s^2 + b^2}$$

$$= \frac{s}{s^2 + b^2}$$

Similarly, the switched-on sine, Figure 12-5(b), has transform

$$\mathcal{L}[\sin bt] = \mathcal{L}\left[\frac{-j}{2}e^{jbt}\right] + \mathcal{L}\left[\frac{j}{2}e^{-jbt}\right]$$

$$= \frac{-j/2}{s - jb} + \frac{j/2}{s + jb} = \frac{b}{s^2 + b^2}$$

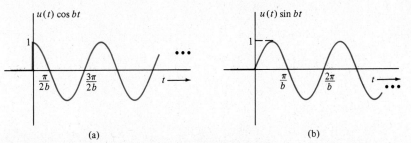

Figure 12-5   Switched-on cosine and sine functions.

When expressed in quadrature form, a general sinusoid

$$f(t) = C \cos (bt + \theta) = A \cos bt + B \sin bt$$

is seen to have transform

$$F(s) = \frac{As + Bb}{s^2 + b^2}$$

12.3.2 Exponential Weighting

A function $f(t)$ that is weighted by an exponential, e^{-at}, has a Laplace transform that is simply related to the transform of the function:

$$\mathscr{L}[f(t)e^{-at}] = \int_{0-}^{\infty} f(t)e^{-at}e^{-st} \, dt$$

$$= \int_{0-}^{\infty} f(t)e^{-(s+a)t} \, dt = F(s + a)$$

It is the transform of the function $f(t)$, with each variable s replaced by $(s + a)$. The exponentially weighted sinusoids, Figure 12-6, have the following transforms:

$$\mathscr{L}[e^{-at} \cos bt] = \frac{(s + a)}{(s + a)^2 + b^2} \qquad \mathscr{L}[e^{-at} \sin bt] = \frac{b}{(s + a)^2 + b^2}$$

$$\mathscr{L}[Ce^{-at} \cos (bt + \theta)] = \mathscr{L}[e^{-at}(A \cos bt + B \sin bt)]$$

$$= \frac{A(s + a) + Bb}{(s + a)^2 + b^2}$$

For example,

$$Y(s) = \frac{-6s + 5}{s^2 + 2s + 10} = \frac{-6(s + 1) + (11/3)(3)}{(s + 1)^2 + (3)^2}$$

is the Laplace transform of

$$y(t) = e^{-t}\left(-6 \cos 3t + \tfrac{11}{3} \sin 3t\right)$$

The function

$$x(t) = 9e^{3t} \cos (7t + 30°) = e^{3t}(9 \cos 30° \cos 7t - 9 \sin 30° \sin 7t)$$

$$= e^{3t}(7.8 \cos 7t - 4.5 \sin 7t)$$

has Laplace transform

$$X(s) = \frac{7.8(s - 3) - 4.5(7)}{(s - 3)^2 + 7^2} = \frac{7.8s - 54.9}{s^2 - 6s + 58}$$

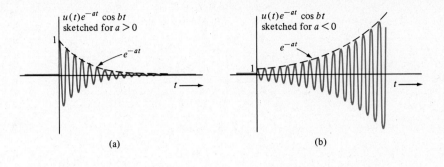

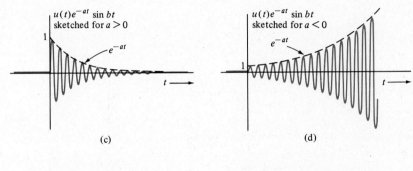

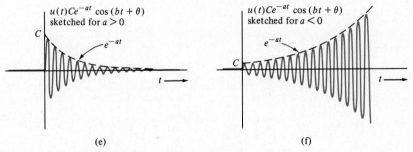

Figure 12-6   Exponentially weighted sinusoids.

12.3.3 Time Weighting

A function $f(t)$ multiplied by the variable t is termed time-weighted. The Laplace transform of a time-weighted function is related to the function itself, as follows:

$$\mathcal{L}[tf(t)] = \int_{0^-}^{\infty} tf(t)e^{-st}\, dt$$

$$= \int_{0^-}^{\infty} f(t)[te^{-st}]\, dt = \int_{0^-}^{\infty} f(t)\left[-\frac{d}{ds}e^{-st}\right] dt$$

$$= -\frac{d}{ds}\int_{0^-}^{\infty} f(t)e^{-st}\, dt = -\frac{dF(s)}{ds}$$

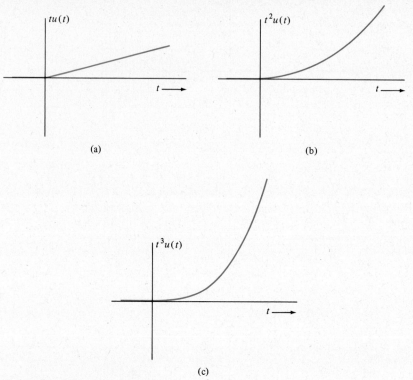

(a)

(b)

(c)

Figure 12-7   Powers of time.

Applying this property to the Laplace transform of a constant gives the transform for the ramp function, t, of Figure 12-7:

$$\mathcal{L}[t \cdot 1] = -\frac{d}{ds}\,\mathcal{L}[1] = -\frac{d}{ds}\!\left(\frac{1}{s}\right) = \frac{1}{s^2}$$

Similarly, higher powers of t, as in Figure 12-7, have the following transforms:

$$\mathcal{L}[t^2] = \mathcal{L}[t \cdot t] = -\frac{d}{ds}\!\left(\frac{1}{s^2}\right) = \frac{2}{s^3}$$

$$\mathcal{L}[t^3] = \mathcal{L}[t \cdot t^2] = -\frac{d}{ds}\!\left(\frac{2}{s^3}\right) = \frac{6}{s^4}$$

In general,

$$\mathcal{L}[t^n] = \frac{1 \cdot 2 \cdot 3 \ldots n}{s^{n+1}} = \frac{n!}{s^{n+1}} \qquad n = 0, 1, 2, \ldots$$

A time-weighted exponential, Figure 12-8, has transform

$$\mathcal{L}[te^{-at}] = -\frac{d}{ds}\,\mathcal{L}[e^{-at}] = -\frac{d}{ds}\!\left(\frac{1}{s + a}\right) = \frac{1}{(s + a)^2}$$

Higher powers of t times exponentials, also sketched in Figure 12-8, have transforms

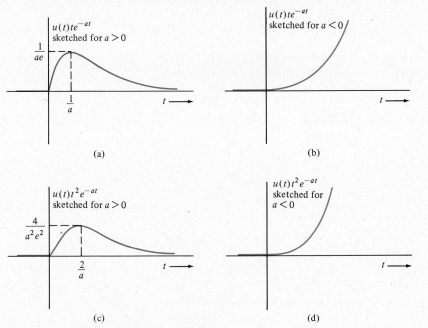

Figure 12-8   Time-weighted exponentials.

as follows:

$$\mathscr{L}[t^2 e^{-at}] = -\frac{d}{ds}\left[\frac{1}{(s+a)^2}\right] = \frac{2}{(s+a)^3}$$

$$\mathscr{L}[t^3 e^{-at}] = -\frac{d}{ds}\left[\frac{2}{(s+a)^3}\right] = \frac{6}{(s+a)^4}$$

$$\mathscr{L}[t^n e^{-at}] = \frac{1 \cdot 2 \cdot 3 \ldots n}{(s+a)^{n+1}} = \frac{n!}{(s+a)^{n+1}} \qquad n = 0, 1, 2, \ldots$$

Time-weighted sinusoids, Figure 12-9(a), have these transforms:

$$\mathscr{L}[t \cos bt] = -\frac{d}{ds}\left[\frac{s}{s^2+b^2}\right] = \frac{s^2-b^2}{(s^2+b^2)^2}$$

$$\mathscr{L}[t \sin bt] = -\frac{d}{ds}\left[\frac{b}{s^2+b^2}\right] = \frac{2bs}{(s^2+b^2)^2}$$

Weightings by higher powers of t may be transformed by repeated differentiation with respect to s. Sinusoids that are both time- and exponentially weighted, Figure 12-9(b), have the following transforms:

$$\mathscr{L}[te^{-at} \cos bt] = -\frac{d}{ds}\left[\frac{s+a}{(s+a)^2+b^2}\right] = \frac{(s+a)^2-b^2}{[(s+a)^2+b^2]^2}$$

$$\mathscr{L}[te^{-at} \sin bt] = -\frac{d}{ds}\left[\frac{b}{(s+a)^2+b^2}\right] = \frac{2b(s+a)}{[(s+a)^2+b^2]^2}$$

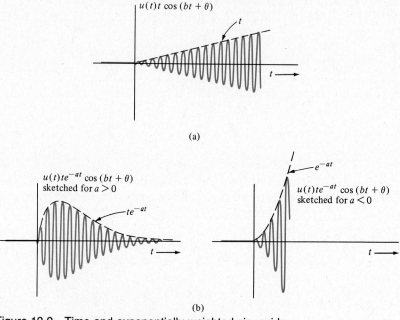

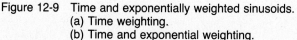

Figure 12-9 Time and exponentially weighted sinusoids.
(a) Time weighting.
(b) Time and exponential weighting.

These results may also be obtained by applying the exponential weighting property to the time-weighted transforms.

12.3.4 Time Scaling

When the time scale of a function is changed, as in the examples of Figure 12-10, the Laplace transform of the scaled function has a simple relationship to the transform of the original function. For a positive constant a,

$$\mathscr{L}[f(at)] = \int_{0^-}^{\infty} f(at)e^{-st}\, dt$$

Letting $t' = at$, $dt' = a\, dt$,

$$\mathscr{L}[f(at)] = \frac{1}{a}\int_{t'=0^-}^{t'=\infty} f(t')e^{-\left(\frac{s}{a}\right)t'}\, dt' = \frac{1}{a}F\left(\frac{s}{a}\right)$$

Consider the specific function

$$f(t) = t^2 e^{-4t}$$

which has Laplace transform

$$F(s) = \frac{2}{(s+4)^3}$$

The time-scaled function

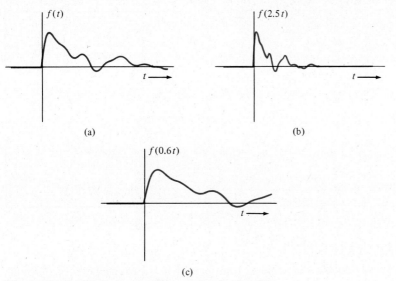

(c)

Figure 12-10   Changing the time scale of a function.

$$g(t) = f(10t) = (10t)^2 e^{-4(10t)} = 100t^2 e^{-40t}$$

has Laplace transform

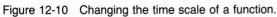

$$F(s) = \frac{1}{10} F\left(\frac{s}{10}\right) = \frac{1}{10} \frac{2}{\left(\dfrac{s}{10} + 4\right)^3} = \frac{200}{(s + 40)^3}$$

D12-2

Find the following Laplace transforms:

(a) $\mathscr{L}[-8t^6]$

 ans. $-5760/s^7$

(b) $\mathscr{L}[6t^3 e^{-2t}]$

 ans. $\dfrac{36}{(s + 2)^4}$

(c) $\mathscr{L}[-3 \cos 8t + 7 \sin 8t]$

 ans. $\dfrac{-3s + 56}{s^2 + 64}$

(d) $\mathscr{L}\{e^{-6t}[4 \cos 2t + 3 \sin 2t]\}$

 ans. $\dfrac{4s + 30}{(s + 6)^2 + 4}$

(e) $\mathscr{L}[6t^3 \sin 7t]$

 ans. $\dfrac{1008s(s^2 - 49)}{(s^2 + 49)^4}$

(f) $\mathscr{L}[10t^2 e^{3t} \cos 4t]$

 ans. $\dfrac{20[(s - 3)^3 - 48](s - 3)}{[(s - 3)^2 + 16]^3}$

(g) $\mathscr{L}[7 \cos (3t + 45°)]$

 ans. $\dfrac{4.95(s - 3)}{s^2 + 9}$

(h) $\mathscr{L}[6 \sin (2t - 30°)]$

 ans. $\dfrac{10.4 - 3s}{s^2 + 4}$

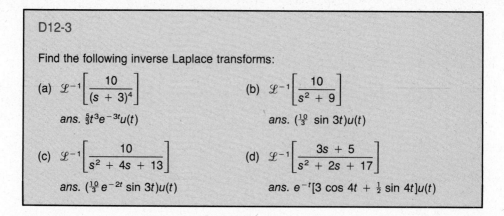

D12-3

Find the following inverse Laplace transforms:

(a) $\mathcal{L}^{-1}\left[\dfrac{10}{(s + 3)^4}\right]$

ans. $\frac{5}{3}t^3e^{-3t}u(t)$

(b) $\mathcal{L}^{-1}\left[\dfrac{10}{s^2 + 9}\right]$

ans. $(\frac{10}{3}\sin 3t)u(t)$

(c) $\mathcal{L}^{-1}\left[\dfrac{10}{s^2 + 4s + 13}\right]$

ans. $(\frac{10}{3}e^{-2t}\sin 3t)u(t)$

(d) $\mathcal{L}^{-1}\left[\dfrac{3s + 5}{s^2 + 2s + 17}\right]$

ans. $e^{-t}[3\cos 4t + \frac{1}{2}\sin 4t]u(t)$

12.3.5 The Time-Shift Relation

A time-shifted function has the Laplace transform that is easily related to the non-shifted functions' transform only under special circumstances. As indicated in Figure 12-11, a function $f(t)$ that is nonzero either side of $t = 0$ includes a portion of the function not involved in $F(s)$ when it is shifted to the right. When $f(t)$ is shifted to the left, part of the function included in $F(s)$ is not included in the transform of the shifted function. It is only when both the original function and the shifted function give zero contribution to the transform prior to $t = 0$ that there is a simple relation between the two transforms.

Consider a function $f(t)$ with Laplace transform $F(s)$. The related time-shifted function $f(t - \tau)$, where τ is a constant, has Laplace transform given by

$$\mathcal{L}[f(t - \tau)] = \int_{0-}^{\infty} f(t - \tau)e^{-st}\, dt$$

Letting $t' = t - \tau$, $t = t' + \tau$, $dt = dt'$,

$$\mathcal{L}[f(t - \tau)] = \int_{0- -\tau}^{\infty} f(t')e^{-s(t' + \tau)}\, dt' = e^{-s\tau}\int_{0- -\tau}^{\infty} f(t')e^{-st'}\, dt'$$

For a positive value of τ, the function $f(t - \tau)$ is a right shift of the function $f(t)$ by amount τ. Provided that $f(t)$ is zero prior to $t = 0$,

$$\mathcal{L}[f(t - \tau)] = e^{-s\tau}\underbrace{\int_{0- -\tau}^{\infty} f(t')e^{-st'}\, dt'}_{\text{negative lower limit}}$$

$$= e^{-s\tau}\left[\underbrace{\int_{0- -\tau}^{0-} f(t')e^{-st'}\, dt'}_{\substack{f(t) \text{ is zero in this} \\ \text{integration interval}}} + \int_{0-}^{\infty} f(t')e^{-st'}\, dt'\right]$$

$$= e^{-s\tau}\int_{0-}^{\infty} f(t')e^{-st'}\, dt' = e^{-s\tau}F(s)$$

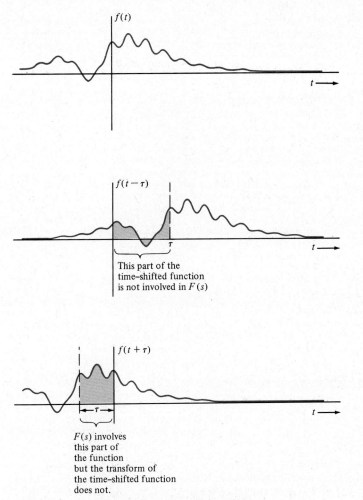

This part of the
time–shifted function
is not involved in $F(s)$

$F(s)$ involves
this part of
the function
but the transform of
the time–shifted function
does not.

Figure 12-11 Time-shifting a function in general.

For a negative value of τ, the function $f(t - \tau)$ is a left shift of the function $f(t)$ by amount $-\tau$. Then, if $f(t - \tau)$ is zero prior to $t = 0$,

$$\mathcal{L}[f(t - \tau)] = e^{-s\tau} \underbrace{\int_{0^- - \tau}^{\infty}}_{\text{positive lower limit}} f(t')e^{-st'}\, dt'$$

$$= e^{-s\tau}\left[\underbrace{\int_{0^-}^{0^- - \tau} f(t')e^{-st'}\, dt'}_{\substack{f(t) \text{ is zero in this} \\ \text{integration interval}}} + \int_{0^- - \tau}^{\infty} f(t')e^{-st'}\, dt'\right]$$

$$= e^{-s\tau}\int_{0^-}^{\infty} f(t')e^{-st'}\, dt' = e^{-s\tau}F(s)$$

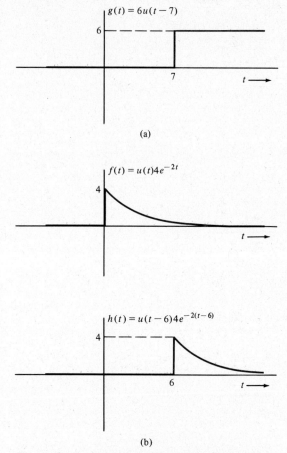

(a)

(b)

Figure 12-12   Examples of time-shifted functions.
(a) Time-shifted step function.
(b) Time-shifted switched-on exponential.

If both the original function and the shifted function are zero prior to $t = 0$, the transforms are related by

$$\mathcal{L}[f(t - \tau)] = e^{-s\tau}F(s)$$

Otherwise, this simple relationship does not hold.

As an example, consider the time-shifted step function of Figure 12-12(a). Its Laplace transform is given by

$$\mathcal{L}[6u(t - 7)] = \frac{6e^{-7s}}{s}$$

Both $6u(t)$ and $6u(t - 7)$ are zero prior to $t = 0$. The time-shifted switched-on exponential $h(t)$ of Figure 12-12(b) is the function $f(t)$ delayed by 6 seconds:

$$H(s) = e^{-6s}F(s) = \frac{4e^{-6s}}{s + 2}$$

D12-4

Sketch the following functions. The unit step function is $u(t)$, and $r(t)$ is the unit ramp function.

(a) $f_1(t) = 3u(t - 3) + u(t - 4) - u(t - 5) - 3u(t - 7)$

(b) $f_2(t) = 5u(t) - 5r(t) - 5u(t - 1)$

(c) $f_3(t) = 3r(t) - 6r(t - 2) + 3r(t - 4)$

(d) $f_4(t) = -5 + r(t) - r(t - 10) - 2r(t - 30) + 2r(t - 35)$

D12-5

Using time-shifted steps and ramps, construct the Laplace transforms of the following functions:

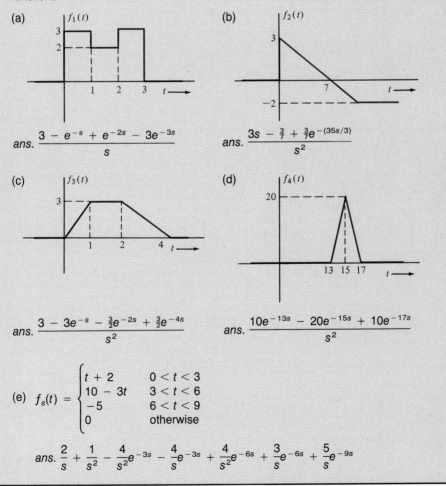

(a) $f_1(t)$

ans. $\dfrac{3 - e^{-s} + e^{-2s} - 3e^{-3s}}{s}$

(b) $f_2(t)$

ans. $\dfrac{3s - \frac{3}{7} + \frac{3}{7}e^{-(35s/3)}}{s^2}$

(c) $f_3(t)$

ans. $\dfrac{3 - 3e^{-s} - \frac{3}{2}e^{-2s} + \frac{3}{2}e^{-4s}}{s^2}$

(d) $f_4(t)$

ans. $\dfrac{10e^{-13s} - 20e^{-15s} + 10e^{-17s}}{s^2}$

(e) $f_s(t) = \begin{cases} t + 2 & 0 < t < 3 \\ 10 - 3t & 3 < t < 6 \\ -5 & 6 < t < 9 \\ 0 & \text{otherwise} \end{cases}$

ans. $\dfrac{2}{s} + \dfrac{1}{s^2} - \dfrac{4}{s^2}e^{-3s} - \dfrac{4}{s}e^{-3s} + \dfrac{4}{s^2}e^{-6s} + \dfrac{3}{s}e^{-6s} + \dfrac{5}{s}e^{-9s}$

12.3.6 Periodic Functions

Using the time delay property, the Laplace transform of a periodic function after $t = 0$ may be easily expressed in terms of the transform of one period of the function. Let the function have period T and let $f_T(t)$ be one cycle of the function in the interval $[0, T)$, as in Figure 12-13(a). Then

$$f(t) = f_T(t) + f_T(t - T) + f_T(t - 2T) + \cdots$$

and

$$F(s) = F_T(s) + e^{-sT}F_T(s) + e^{-2sT}F_T(s) + \cdots$$

$$= F_T(s)[1 + e^{-sT} + (e^{-sT})^2 + (e^{-sT})^3 + \cdots]$$

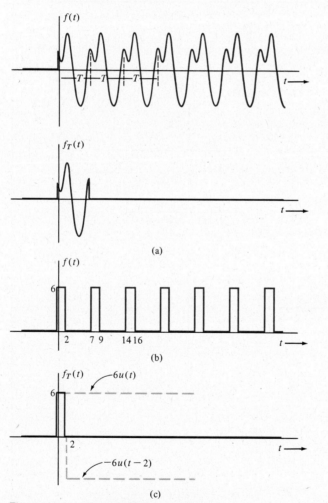

(a)

(b)

(c)

Figure 12-13   Periodic functions after $t = 0$.
  (a) A periodic function after $t = 0$ and its first cycle.
  (b) A periodic pulse train.
  (c) First pulse of the train expressed in terms of step functions.

Using the geometric series formula,

$$1 + \alpha + \alpha^2 + \alpha^3 + \cdots = \frac{1}{1 - \alpha} \qquad |\alpha| < 1$$

(12-1)

$$F(s) = \frac{F_T(s)}{1 - e^{-sT}}$$

Convergence of the series is assured, since

$$\left|e^{-sT}\right| = \left|e^{-\sigma T}e^{-j\omega T}\right| < 1$$

for a sufficiently large convergence factor σ.

Consider the example function $f(t)$, Figure 12-13(b), which is periodic after $t = 0$. As indicated in Figure 12-13(c), the first period of $f(t)$ may be expressed as

$$f_T(t) = 6u(t) - 6u(t - 2) \qquad F_T(s) = \frac{6}{s} - \frac{6e^{-2s}}{s}$$

giving

$$F(s) = \frac{F_T(s)}{1 - e^{-sT}} = \frac{6 - 6e^{-2s}}{s(1 - e^{-7s})}$$

It is not necessary that the repeated function be zero outside the interval $[0, T)$ for relation (12-1) to hold. Figure 12-14 shows a function $f_T(t)$, which overlaps itself when it is summed according to

$$f(t) = f_T(t) + f_T(t - T) + f_T(t - 2T) + \cdots$$

for which relation (12-1) is also correct, even though $f_T(t)$ is not a single, disjoint cycle of $f(t)$.

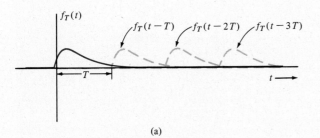

(a)

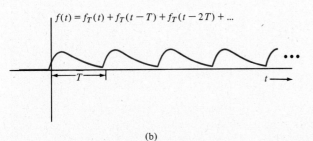

(b)

Figure 12-14   Periodic time shift and addition of an overlapping function.

D12-6

Find the following inverse Laplace transforms:

(a) $\mathcal{L}^{-1}\left[\dfrac{1 - e^{-2s}}{s(1 - e^{-6s})}\right]$

(b) $\mathcal{L}^{-1}\left[\dfrac{4e^{-3s}}{1 - e^{-5s}}\right]$

ans. $\displaystyle\sum_{n=0}^{\infty} [u(t - 6n) - u(t - 2 - 6n)]$

ans. $\displaystyle\sum_{n=0}^{\infty} 4\delta(t - 3 - 5n)$

LAPLACE TRANSFORM PROPERTIES

The following are important basic properties of the Laplace transform:

(a) $\mathcal{L}[Af(t)] = AF(s)$, where A is a constant

(b) $\mathcal{L}[f_1(t) + f_2(t)] = F_1(s) + F_2(s)$

(c) $\mathcal{L}[e^{-at}f(t)] = F(s + a)$

(d) $\mathcal{L}[tf(t)] = -\dfrac{d}{ds} F(s)$

(e) $\mathcal{L}[f(at)] = \dfrac{1}{a} F\left(\dfrac{s}{a}\right)$, where a is a positive constant

Using these properties, the following Laplace transform pairs may be derived:

	$f(t) \quad t \le 0$	$F(s)$
(4)	t	$\dfrac{1}{s^2}$
(5)	t^2	$\dfrac{2}{s^3}$
(6)	t^n	$\dfrac{n!}{s^{n+1}} \quad n = 0, 1, 2, \ldots$
(7)	te^{-at}	$\dfrac{1}{(s + a)^2}$
(8)	t^2e^{-at}	$\dfrac{2}{(s + a)^3}$
(9)	t^ne^{-at}	$\dfrac{n!}{(s + a)^{n+1}} \quad n = 0, 1, 2, \ldots$
(10)	$\cos bt$	$\dfrac{s}{s^2 + b^2}$
(11)	$\sin bt$	$\dfrac{b}{s^2 + b^2}$
(12)	$e^{-at} \cos bt$	$\dfrac{s + a}{(s + a)^2 + b^2}$

(13)	$e^{-at} \sin bt$	$\dfrac{b}{(s+a)^2 + b^2}$
(14)	$t \cos bt$	$\dfrac{s^2 - b^2}{(s^2 + b^2)^2}$
(15)	$t \sin bt$	$\dfrac{2bs}{(s^2 + b^2)^2}$
(16)	$te^{-at} \cos bt$	$\dfrac{(s+a)^2 - b^2}{[(s+a)^2 + b^2]^2}$
(17)	$te^{-at} \sin bt$	$\dfrac{2b(s+a)}{[(s+a)^2 + b^2]^2}$

The following are additional properties of the Laplace transformation:

(f) $\mathcal{L}[f(t - \tau)] = e^{-s\tau}F(s)$, where τ is a constant, providing both $f(t)$ and $f(t - \tau)$ are zero prior to $t = 0$.

(g) If $f_T(t)$ is periodically repeated with period T after $t = 0$, the periodic function has transform

$$F(s) = F_T(s) + e^{-sT}F_T(s) + e^{-2sT}F_T(s) + \cdots$$

$$= \frac{F_T(s)}{1 - e^{-sT}}$$

12.4 DERIVATIVES AND INTEGRALS OF FUNCTIONS

12.4.1 Derivatives

The Laplace transform of the derivative of a function is related to the function itself in a manner that will now be derived:

$$\mathcal{L}\left[\frac{df}{dt}\right] = \int_{0^-}^{\infty} \frac{df}{dt} e^{-st}\, dt$$

Integrating by parts, with

$$u = e^{-st} \qquad du = -se^{-st}\, dt$$
$$dv = df \qquad v = f(t)$$

gives

$$\mathcal{L}\left[\frac{df}{dt}\right] = uv\Big|_{0^-}^{\infty} - \int_{0^-}^{\infty} v\, du = f(t)e^{-st}\Big|_{0^-}^{\infty} + \int_{0^-}^{\infty} f(t)se^{-st}\, dt$$

The upper limit on the evaluation is

$$\lim_{t\to\infty} f(t)e^{-st} = \lim_{t\to\infty} f(t)e^{-\sigma t}e^{-j\omega t} = 0$$

The product $f(t)e^{-\sigma t}$ goes to zero because that is exactly what the convergence factor is to do. Since $e^{-j\omega t}$ is bounded in magnitude, the entire product approaches zero for large t. Thus

$$\mathcal{L}\left[\frac{df}{dt}\right] = -f(0^-) + s \int_{0^-}^{\infty} f(t)e^{-st}\, dt = sF(s) - f(0^-)$$

The transform of the derivative is s times the original transform minus the initial value of the time function.

Repeated application of the derivative relation gives results for higher derivatives:

$$\mathcal{L}\left[\frac{d^2f}{dt^2}\right] = \left[\frac{d}{dt}\left(\frac{df}{dt}\right)\right] = s[sF(s) - f(0^-)] - \left.\frac{df}{dt}\right|_{t=0^-}$$

$$= s^2F(s) - sf(0^-) - f'(0^-)$$

Similarly,

$$\mathcal{L}\left[\frac{d^3f}{dt^3}\right] = \left[\frac{d}{dt}\left(\frac{d^2f}{dt^2}\right)\right]$$

$$= s[s^2F(s) - sf(0^-) - f'(0^-)] - f''(0^-)$$

$$= s^3F(s) - s^2f(0^-) - sf'(0^-) - f''(0^-)$$

In general,

$$\mathcal{L}\left[\frac{d^nf}{dt^n}\right] = s^nF(s) - s^{n-1}f(0^-) - s^{n-2}f'(0^-)$$

$$- \cdots - sf^{[n-2]}(0^-) - f^{[n-1]}(0^-)$$

As an example, the function

$$f(t) = t^2 \sin t$$

has Laplace transform

$$F(s) = \frac{6s^2 - 2}{(s^2 + 1)^3}$$

The time derivative

$$y(t) = \frac{df}{dt} = 2t \sin t + t^2 \cos t$$

has Laplace transform

$$Y(s) = sF(s) - f(0^-) = sF(s) = \frac{6s^3 - 2s}{(s^2 + 1)^3}$$

12.4.2 Derivatives of Discontinuous Functions

The derivatives of functions with discontinuities involve impulses. For example, the step function of Figure 12-15(a),

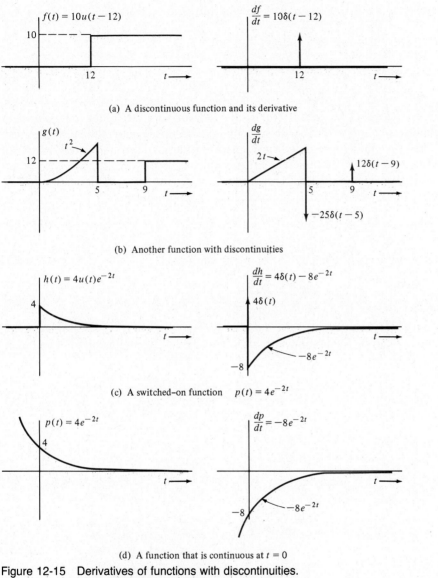

(a) A discontinuous function and its derivative

(b) Another function with discontinuities

(c) A switched-on function $p(t) = 4e^{-2t}$

(d) A function that is continuous at $t = 0$

Figure 12-15   Derivatives of functions with discontinuities.
(a) A discontinuous function and its derivative
(b) Another function with discontinuities
(c) A switched-on function
(d) A function that is continuous at $t = 0$

$$f(t) = 10u(t - 12)$$

has a discontinuity at $t = 12$. The derivative is as indicated, with an impulse at $t = 12$ of strength 10, the amount of the discontinuity. For this function,

$$\frac{df}{dt} = 10\delta(t - 12)$$

This function may be recovered from its derivative by integration:

$$f(t) = \int_{-\infty}^{t} \frac{df}{dt}\, dt = \int_{-\infty}^{t} 10\delta(t - 12)\, dt$$

$$= \begin{cases} 0 & t < 12 \\ 10 & t > 12 \end{cases}$$

Similarly, the $g(t)$ in Figure 12-15(b) has the indicated derivative. Each discontinuity contributes an impulse term, with strength equal to the amount of the discontinuity.

The functions represented by Laplace transforms have derivatives involving impulses at $t = 0$ if the function is discontinuous at the origin. The switched-on exponential of Figure 12-15(c) is discontinuous at $t = 0$, so its derivative has an impulse there, as indicated. The function's Laplace transform is

$$H(s) = \frac{4}{s + 2}$$

Using the derivative relation,

$$\mathscr{L}\left[\frac{dh}{dt}\right] = sH(s) - h(0^-) = \frac{4s}{s + 2}$$

The initial condition term is zero, since $h(t)$ is zero prior to $t = 0$. The presence of the impulse may be seen from the transform by dividing denominator into numerator for one step:

$$\begin{array}{r} 4 \\ s + 2 \overline{\smash{)}\, 4s } \\ \underline{4s + 8} \\ -8 \end{array}$$

$$\mathscr{L}\left[\frac{dh}{dt}\right] = 4 + \frac{-8}{s + 2}$$

$$\frac{dh}{dt} = 4\delta(t) - 8u(t)e^{-2t}$$

For the function of Figure 12-15(d) that is continuous at $t = 0$,

$$H(s) = \frac{4}{s + 2}$$

$$\mathscr{L}\left[\frac{dh}{dt}\right] = sH(s) - h(0^-) = \frac{4s}{s + 2} - 4 = \frac{-8}{s + 2}$$

Even though the process of Laplace transformation "forgets" the function *prior* to $t = 0$, the function's behavior *at* $t = 0$ is included and affects the calculation of derivatives.

12.4.3 Derivatives of Impulses

Laplace transforms that are positive integer powers of the variable s represent derivatives of the impulse:

$$\mathscr{L}^{-1}[s] = \mathscr{L}^{-1}[s \cdot 1] = \frac{d}{dt}\,\delta(t) = \delta'(t)$$

$$\mathscr{L}^{-1}[s^2] = \mathscr{L}^{-1}[s \cdot s] = \frac{d}{dt}\,\delta'(t) = \delta''(t)$$

$$\mathscr{L}^{-1}[s^n] = \delta^{[n]}(t)$$

The derivative of the impulse is defined in terms of the limiting behavior of the derivative of an impulsive function. A triangular impulsive function, Figure 12-16, is especially easy to deal with and has a simple derivative. In the limit as the impulsive function becomes more and more concentrated in the vicinity of the origin, its derivative approaches two sharp narrow pulses, close together and with opposite algebraic signs. The limiting result, which is characteristic of all impulsive functions that have derivatives, is termed a *doublet*.

Higher derivatives of impulses are similarly defined in terms of the limiting behavior of higher derivatives of impulsive functions. Impulse derivatives have the following general sifting properties for functions $f(t)$ that are finite and continuous at the point where the impulse is concentrated:

$$\int_{-\epsilon}^{\epsilon} \delta'(t)\,f(t)\,dt = \left.\frac{df}{dt}\right|_{t=0} = f'(0)$$

$$\int_{a}^{b} \delta'(t-\tau)\,f(t)\,dt = \begin{cases} f'(\tau) & a < \tau < b \\ 0 & \text{otherwise} \end{cases}$$

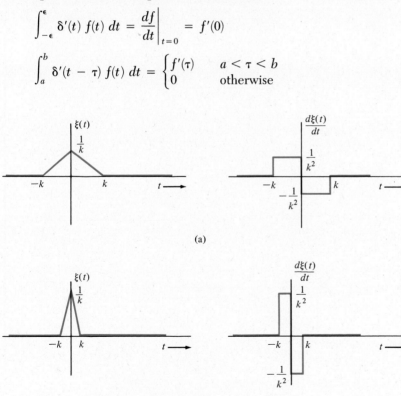

(a)

(b)

Figure 12-16   An impulsive function and its derivative.
(a) Triangular impulsive function.
(b) Triangular impulsive function as its area becomes more concentrated near the origin.

$$\int_{-\epsilon}^{\epsilon} \delta^{[n]}(t)\, f(t)\, dt = f^{[n]}(0)$$

$$\int_{a}^{b} \delta^{[n]}(t - \tau)\, f(t)\, dt = \begin{cases} f^{[n]}(\tau) & a < \tau < b \\ 0 & \text{otherwise} \end{cases}$$

Such expressions, involving impulses as they do, actually refer to limiting behavior. For example, the first of these relations means

$$\lim_{k \to 0} \int_{-\epsilon}^{\epsilon} \frac{d\xi(k,\, t)}{dt}\, f(t)\, dt = f'(0)$$

where $\xi(k,\, t)$ is a family of impulsive functions that approach an impulse as k approaches zero.

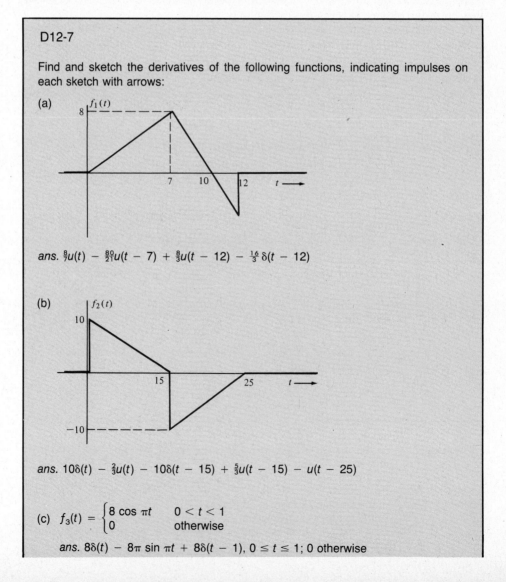

D12-7

Find and sketch the derivatives of the following functions, indicating impulses on each sketch with arrows:

(a)

$f_1(t)$

ans. $\frac{8}{7}u(t) - \frac{80}{21}u(t - 7) + \frac{8}{3}u(t - 12) - \frac{16}{3}\delta(t - 12)$

(b)

$f_2(t)$

ans. $10\delta(t) - \frac{2}{3}u(t) - 10\delta(t - 15) + \frac{5}{3}u(t - 15) - u(t - 25)$

(c) $f_3(t) = \begin{cases} 8 \cos \pi t & 0 < t < 1 \\ 0 & \text{otherwise} \end{cases}$

    ans. $8\delta(t) - 8\pi \sin \pi t + 8\delta(t - 1),\ 0 \le t \le 1;\ 0$ otherwise

(d) $f_4(t) = \begin{cases} 2 - 3t & 0 < t < 6 \\ 4t & 6 < t < 9 \\ 5 & 9 < t < 12 \\ 0 & \text{otherwise} \end{cases}$

ans. $2\delta(t)$, $t = 0$; -3, $0 < t < 6$; $40\delta(t - 6)$, $t = 6$; 4, $6 < t < 9$; $-31\delta(t - 9)$, $t = 9$; $-5\delta(t - 12)$, $t = 12$; 0, otherwise

12.4.4 Integrals

The Laplace transform of the running integral of a function, from time just before $t = 0$ on, is now found in terms of the function itself. Using

$$\mathcal{L}\left[\frac{dg}{dt}\right] = sG(s) - g(0^-)$$

if the function $g(t)$ is taken to be

$$g(t) = \int_{0^-}^{t} f(t)\, dt$$

then

$$\frac{dg}{dt} = f(t)$$

for $t \geq 0$ and

$$g(0^-) = \int_{0^-}^{0^-} f(t)\, dt = 0$$

It follows that

$$\mathcal{L}\left[\frac{dg}{dt}\right] = \mathcal{L}[f(t)] = F(s) = sG(s)$$

$$G(s) = \mathcal{L}\left[\int_{0^-}^{t} f(t)\, dt\right] = \frac{F(s)}{s}$$

Repeated application of this result, for multiple integrals, gives

$$\mathcal{L}\left[\int_{0^-}^{t}\left(\int_{0^-}^{t} f(t)\, dt\right) dt\right] = \frac{F(s)}{s^2}$$

$$\mathcal{L}\left[\underbrace{\int_{0^-}^{t} \cdots \int_{0^-}^{t} f(t)\, d^m t}_{m \text{ integrals}}\right] = \frac{F(s)}{s^m}$$

If the integral extends from before $t = 0^-$,

$$\int_{-\infty}^{t} f(t)\, dt = \int_{-\infty}^{0} f(t)\, dt + \int_{0^-}^{t} f(t)\, dt$$

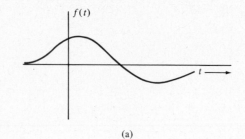

(a)

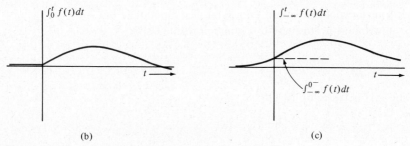

(b)   (c)

Figure 12-17   A function and related running integrals.

The first integral term, between fixed limits $-\infty$ and 0^-, is just a constant. It is the area under the $f(t)$ curve up to time $t = 0^-$, as illustrated in Figure 12-17. Hence

$$\mathcal{L}\left[\int_{-\infty}^{t} f(t)\ dt\right] = \mathcal{L}\left[\int_{-\infty}^{0^-} f(t)\ dt\right] + \mathcal{L}\left[\int_{0^-}^{t} f(t)\ dt\right]$$

$$= \frac{\displaystyle\int_{-\infty}^{0^-} f(t)\ dt}{s} + \frac{F(s)}{s}$$

As a numerical example, consider the function

$$f(t) = e^{2t}$$

which has Laplace transform

$$F(s) = \frac{1}{s - 2}$$

The transform of the running integral, from $t = 0$ (or $t = 0^-$) on, is

$$\mathcal{L}\left[\int_{0}^{t} e^{2t}\ dt\right] = \frac{F(s)}{s} = \frac{1}{s(s - 2)}$$

If the integration starts prior to $t = 0$, say at $t = -1$, the transform of the integral is as follows:

$$\mathcal{L}\left[\int_{-1}^{t} e^{2t}\ dt\right] = \mathcal{L}\left[\int_{-1}^{0} e^{2t}\ dt\right] + \mathcal{L}\left[\int_{0}^{t} e^{2t}\ dt\right]$$

Since

$$\int_{-1}^{0} e^{2t}\, dt = \left.\frac{e^{2t}}{2}\right|_{-1}^{0} = \frac{1 - e^{-2}}{2} = 0.432$$

$$\mathscr{L}\left[\int_{-1}^{t} e^{2t}\, dt\right] = \mathscr{L}[0.432] + \mathscr{L}\left[\int_{0}^{t} e^{2t}\, dt\right] = \frac{0.432}{s} + \frac{1}{s(s-2)}$$

For a lower limit on the integral of $-\infty$, as occurs commonly in the description of physical systems,

$$\mathscr{L}\left[\int_{-\infty}^{t} e^{2t}\, dt\right] = \mathscr{L}\left[\int_{-\infty}^{0} e^{2t}\, dt\right] + \mathscr{L}\left[\int_{0}^{t} e^{2t}\, dt\right]$$

$$= \mathscr{L}\left[\frac{1}{2}\right] + \mathscr{L}\left[\int_{0}^{t} e^{2t}\, dt\right] = \frac{1}{2s} + \frac{1}{s(s-2)}$$

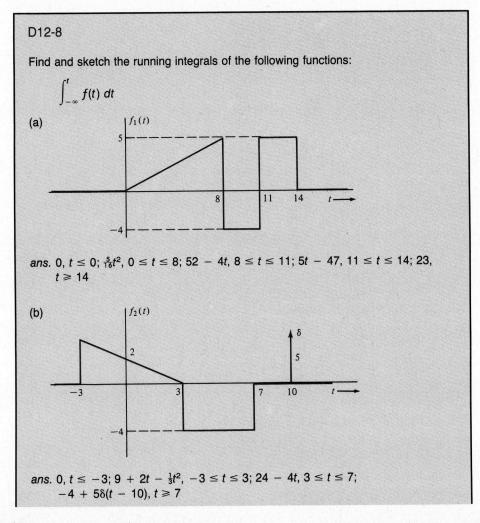

D12-8

Find and sketch the running integrals of the following functions:

$$\int_{-\infty}^{t} f(t)\, dt$$

(a)

ans. $0, t \le 0$; $\frac{5}{16}t^2$, $0 \le t \le 8$; $52 - 4t$, $8 \le t \le 11$; $5t - 47$, $11 \le t \le 14$; 23, $t \ge 14$

(b)

ans. $0, t \le -3$; $9 + 2t - \frac{1}{3}t^2$, $-3 \le t \le 3$; $24 - 4t$, $3 \le t \le 7$; $-4 + 5\delta(t - 10)$, $t \ge 7$

(c) $f_3(t) = 8e^{-3t}[u(t - 2) - u(t - 6)]$

   ans. $0, t \le 2; \frac{8}{3}(e^{-6} - e^{-3t}), 2 \le t \le 6; \frac{8}{3}(e^{-6} - e^{-18}), t \ge 6$

(d) $f_4(t) = \begin{cases} 4e^{3t} & t \le 0 \\ 4e^{-3t} & 0 \le t \end{cases}$

   ans. $\frac{4}{3}e^{3t}, t \le 0; \frac{8}{3} - \frac{4}{3}e^{-3t}, t \ge 0$

DERIVATIVES AND INTEGRALS OF FUNCTIONS

The following are additional properties of the Laplace transformation:

(h) $\mathcal{L}\left[\dfrac{df}{dt}\right] = sF(s) - f(0^-)$

(i) $\mathcal{L}\left[\dfrac{d^2f}{dt^2}\right] = s^2F(s) - sf(0^-) - f'(0^-)$

$\vdots$

(j) $\mathcal{L}\left[\dfrac{d^nf}{dt^n}\right] = s^nF(s) - s^{n-1}f(0^-) - s^{n-2}f'(0^-) - \cdots - f^{n-1}(0^-)$

(k) $\mathcal{L}\left[\displaystyle\int_{0^-}^{t} f(t)\, dt\right] = \dfrac{F(s)}{s}$

(l) $\mathcal{L}\left[\displaystyle\int_{-\infty}^{t} f(t)\, dt\right] = \mathcal{L}\left[\displaystyle\int_{-\infty}^{0^-} f(t)\, dt + \displaystyle\int_{0^-}^{t} f(t)\, dt\right] = \dfrac{\displaystyle\int_{-\infty}^{0^-} f(t)\, dt}{s} + \dfrac{F(s)}{s}$

(m) $\mathcal{L}\left[\underbrace{\displaystyle\int_{0^-}^{t} \cdots \displaystyle\int_{0^-}^{t}}_{m \text{ times}} f(t)\, d^m t\right] = \dfrac{F(s)}{s^m}$

Derivatives of the unit impulse function have transforms as follows:

	$f(t) \qquad t \ge 0$	$F(s)$
(18)	$\dfrac{d}{dt}\,\delta(t) = \delta'(t)$	s
(19)	$\dfrac{d^2}{dt^2}\,\delta(t) = \delta''(t)$	s^2
(20)	$\dfrac{d^n}{dt^n}\,\delta(t) = \delta^{[n]}(t)$	s^n

12.5 LAPLACE TRANSFORM EXPANSION

12.5.1 Partial Fraction Expansion

In the solution of problems involving networks and the differential equations that describe them, it often happens that we wish to invert a Laplace transformed signal;

that is, we wish to find the inverse transformation of a given function of s. If the transform to be inverted can be found in a table of Laplace transform pairs, it is only required to copy the corresponding time function, much as one may do in looking up integrals in an integral table. For more complicated transforms than are commonly listed in tables, the technique of partial fraction expansion allows the expression of transforms as the sum of simpler terms, each of which may be individually found in a transform table.

A rational function is the ratio of two polynomials. Laplace transforms are very often rational functions of the variable s:

$$F(s) = \frac{b_m s^m + b_{m-1} s^{m-1} + \cdots + b_1 s + b_0}{s^n + a_{n-1} s^{n-1} + \cdots + a_1 s + a_0}$$

If the degree n of the denominator polynomial is greater than the degree m of the numerator polynomial, $F(s)$ is termed a *proper* rational function. A proper rational function may be expanded in partial fractions of the form

$$F(t) = \frac{b_m s^m + b_{m-1} s^{m-1} + \cdots + b_1 s + b_0}{(s - s_1)(s - s_2) \cdots (s - s_n)}$$

$$= \frac{K_1}{s - s_1} + \frac{K_2}{s - s_2} + \cdots + \frac{K_n}{s - s_n}$$

provided that none of its *poles*, the roots of the denominator polynomial, $s_1, s_2, \ldots, s_n$, is repeated.

For example, the following rational function with lower-degree numerator polynomial than denominator polynomial expands as shown for some constants K_1 and K_2:

$$F(s) = \frac{-4s + 6}{(s + 2)(s + 3)} = \frac{K_1}{s + 2} + \frac{K_2}{s + 3}$$

$$= \frac{K_1(s + 3) + K_2(s + 2)}{(s + 2)(s + 3)} = \frac{(K_1 + K_2)s + (3K_1 + 2K_2)}{(s + 2)(s + 3)}$$

Equating coefficients in the numerator polynomial,

$$\begin{cases} K_1 + K_2 = -4 \\ 3K_1 + 2K_2 = 6 \end{cases}$$

$$K_1 = \frac{\begin{vmatrix} -4 & 1 \\ 6 & 2 \end{vmatrix}}{\begin{vmatrix} 1 & 1 \\ 3 & 2 \end{vmatrix}} = \frac{-14}{-1} = 14 \qquad K_2 = \frac{\begin{vmatrix} 1 & -4 \\ 3 & 6 \end{vmatrix}}{-1} = \frac{18}{-1} = -18$$

Thus

$$F(s) = \frac{-4s + 6}{(s + 2)(s + 3)} = \frac{14}{s + 2} + \frac{-18}{s + 3}$$

The inverse Laplace transform of $F(s)$ is now easily found:

$$f(t) = \mathcal{L}^{-1}\left[\frac{14}{s+2} + \frac{-18}{s+3}\right] = \mathcal{L}^{-1}\left[\frac{14}{s+2}\right] + \mathcal{L}^{-1}\left[\frac{-18}{s+3}\right]$$

$$= 14e^{-2t} - 18e^{-3t} \qquad t \geq 0$$

It consists of a sum of exponential terms after $t = 0$, one for each term of the partial fraction expansion.

12.5.2 Improper Fractions

A rational function with a numerator polynomial degree that is not lower than the denominator polynomial degree cannot be expanded in partial fractions. However, by dividing denominator into numerator, a remainder that does have lower-degree numerator is obtained. For example,

$$F(s) = \frac{2s^2 + 8s - 3}{(s+2)(s+5)}$$

has numerator and denominator polynomials of equal degree and thus cannot be expanded into partial fractions. An expansion of the form

$$\frac{K_1}{s+2} + \frac{K_2}{s+5} = \frac{(K_1 + K_2)s + (5K_1 + 2K_2)}{(s+2)(s+5)}$$

can only give a first-order numerator polynomial.

Dividing denominator into numerator of $F(s)$ for one step gives the following result:

$$
\begin{array}{r}
2 \\
s^2 + 7s + 10 \overline{\smash{\big)}\, 2s^2 + 8s - 3} \\
2s^2 + 14s + 20 \\
\hline
- 6s - 23
\end{array}
$$

$$F(s) = 2 + \frac{-6s - 23}{s^2 + 7s + 10} = 2 + \frac{-6s - 23}{(s+2)(s+5)}$$

The remainder term can be expanded in partial fractions:

$$\frac{-6s - 23}{(s+2)(s+5)} = \frac{K_1}{s+2} + \frac{K_2}{s+5} = \frac{(K_1 + K_2)s + (5K_1 + 2K_2)}{(s+2)(s+5)}$$

$$\begin{cases} K_1 + K_2 = -6 \\ 5K_1 + 2K_2 = -23 \end{cases}$$

$$K_1 = \frac{\begin{vmatrix} -6 & 1 \\ -23 & 2 \end{vmatrix}}{\begin{vmatrix} 1 & 1 \\ 5 & 2 \end{vmatrix}} = -\frac{11}{3} \qquad K_2 = \frac{\begin{vmatrix} 1 & -6 \\ 5 & -23 \end{vmatrix}}{-3} = -\frac{7}{3}$$

giving

$$F(s) = 2 + \frac{-\frac{11}{3}}{s+2} + \frac{-\frac{7}{3}}{s+5}$$

$$f(t) = 2\delta(t) - \tfrac{11}{3} e^{-2t} - \tfrac{7}{3}e^{-5t} \qquad t \geq 0$$

For the Laplace transform

$$F(s) = \frac{-s^3}{(s + 0.2)(s - 0.1)}$$

division of denominator into numerator polynomial takes two steps before a remainder polynomial is of lower degree than the divisor results:

$$
\begin{array}{r}
-s + 0.1 \\
\hline
s^2 + 0.1s - 0.02 \,\big|\, -s^3 \\
\underline{s^3 - 0.1s^2 + 0.02s } \\
0.1s^2 - 0.02s \\
\underline{0.1s^2 + 0.01s - 0.002} \\
-0.03s + 0.002
\end{array}
$$

$$F(s) = -s + 0.1 + \frac{-0.03s + 0.002}{s^2 + 0.1s - 0.02} = -s + 0.1 + \frac{K_1}{s + 0.2} + \frac{K_2}{s - 0.1}$$

$$= -s + 0.1 + \frac{(K_1 + K_2)s + (-0.1K_1 + 0.2K_2)}{(s + 0.2)(s - 0.1)}$$

$$\begin{cases} K_1 + K_2 = -0.03 \\ -0.01K_1 + 0.2K_2 = 0.002 \end{cases}$$

$$K_1 = \frac{\begin{vmatrix} -0.03 & 1 \\ 0.002 & 0.2 \end{vmatrix}}{\begin{vmatrix} 1 & 1 \\ -0.01 & 0.2 \end{vmatrix}} = 0.0381 \qquad K_2 = \frac{\begin{vmatrix} 1 & -0.03 \\ -0.01 & 0.002 \end{vmatrix}}{0.21} = 0.0081$$

The solution for $f(t)$ involves a doublet:

$$F(s) = -s + 0.1 + \frac{0.381}{s + 0.2} + \frac{0.0081}{s - 0.1}$$

$$f(t) = \begin{cases} 0 & t < 0 \\ -\delta'(t) + 0.1\delta(t) + 0.381e^{-0.2t} + 0.0081e^{0.1t} & t \geq 0 \end{cases}$$

Dividing a rational function's denominator into its numerator for more than the minimum number of steps to obtain a remainder of lower degree than the divisor gives a fractional remainder that is of no benefit.

D12-9

Find the time functions for $t \geq 0$ that have the following Laplace transforms:

(a) $F(s) = \dfrac{-4s + 2}{(s + 3)(s - 6)}$

ans. $-\tfrac{14}{9} e^{-3t} - \tfrac{22}{9} e^{6t}$

(b) $F(s) = \dfrac{5s + 4}{s^2 + 5s + 6}$

ans. $-\frac{1}{4}e^{-t} + \frac{21}{4}e^{-5t}$

(c) $F(s) = \dfrac{4s^2}{s^2 + 9s + 20}$

ans. $4\delta(t) - 100e^{-5t} + 64e^{-4t}$

(d) $F(s) = \dfrac{3s^2 - 2s + 4}{(4s + 1)(2s - 1)}$

ans. $\frac{3}{8}\delta(t) - \frac{151}{96}e^{-(1/4)t} + \frac{59}{48}e^{(1/2)t}$

12.5.3 Rapid Coefficient Calculation

A faster method of calculation of the coefficients of a partial fraction expansion is to perform operations on each side of the equality that result in evaluation of the coefficients. For example, in

$$F(s) = \frac{6s + 10}{(s - 1)(s + 4)} = \frac{K_1}{s - 1} + \frac{K_2}{s + 4}$$

multiplying both sides by the factor $(s - 1)$ gives

$$\frac{(6s + 10)(s - 1)}{(s - 1)(s + 4)} = \frac{K_1(s - 1)}{s - 1} + \frac{K_2(s - 1)}{s + 4}$$

Evaluating both sides at $s = 1$ gives K_1:

$$[(s - 1)F(s)]\Big|_{s=1} = \frac{6 + 10}{1 + 4} = \frac{16}{5} = K_1$$

Similarly,

$$[(s + 4)F(s)]\Big|_{s=-4} = \left[\frac{6s + 10}{s - 1}\right]_{s=-4} = \frac{14}{5}$$

$$= \left[\frac{K_1(s + 4)}{s - 1} + \frac{K_2(s + 4)}{s + 4}\right]_{s=-4} = K_2$$

Hence

$$F(s) = \frac{\frac{16}{5}}{s - 1} + \frac{\frac{14}{5}}{s + 4}$$

$$f(t) = \tfrac{16}{5}e^t + \tfrac{14}{5}e^{-4t} \qquad t \geq 0$$

For the proper rational function

$$F(s) = \frac{6s^2 - 2}{(s + 3)(s + 4)(s - 5)} = \frac{K_1}{s + 3} + \frac{K_2}{s + 4} + \frac{K_3}{s - 5}$$

$$K_1 = [(s + 3)F(s)] \Big|_{s=-3} = \frac{6s^2 - 2}{(s + 44)(s - 5)} \Big|_{s=-3} = \frac{54 - 2}{(1)(-8)} = -6.5$$

$$K_2 = [(s + 4)F(s)] \Big|_{s=-4} = \frac{6s^2 - 2}{(s + 3)(s - 5)} \Big|_{s=-4} = \frac{96 - 2}{(-1)(-9)} = 10.44$$

$$K_3 = [(s - 5)F(s)] \Big|_{s=5} = \frac{6s^2 - 2}{(s + 3)(s + 4)} \Big|_{s=5} = \frac{150 - 2}{(8)(9)} = 2.06$$

which gives

$$F(s) = \frac{-6.5}{s + 3} + \frac{10.44}{s + 4} + \frac{2.06}{s - 5}$$

$$f(t) = [-6.5e^{-3t} + 10.44e^{-4t} + 2.06e^{5t}]u(t)$$

In general, for proper rational functions with nonrepeated denominator roots,

$$F(s) = \frac{\text{numerator polynomial}}{(s - s_1)(s - s_2) \cdots (s - s_n)} = \frac{K_1}{s - s_1} + \frac{K_2}{s - s_2} + \cdots + \frac{K_n}{s - s_n}$$

$$K_i = [(s - s_i)F(s)] \Big|_{s=s_i}$$

For an improper rational function, one with numerator degree greater than or equal to the denominator degree, the evaluation method for calculating partial fraction coefficients will also work, as in the following example.

$$F(s) = \frac{3s^2 - 4}{(s + 1)(s + 2)}$$

$$
\begin{array}{r}
3 \\
s^2 + 3s + 2 \overline{) 3s^2 - 4} \\
3s^2 + 9s + 6 \\
\hline
-9s - 10
\end{array}
$$

$$F(s) = \frac{3s^2 - 4}{(s + 1)(s + 2)} = 3 + \frac{-9s - 10}{(s + 1)(s + 2)} = 3 + \frac{K_1}{s + 1} + \frac{K_2}{s + 2}$$

$$(s + 1)F(s) = 3(s + 1) + K_1 + \frac{K_2(s + 1)}{s + 2}$$

$$K_1 = [(s + 1)F(s)] \Big|_{s=-1} = \frac{3s^2 - 4}{s + 2} \Big|_{s=-1} = -1$$

$$K_2 = [(s + 2)F(s)] \Big|_{s=-2} = \frac{3s^2 - 4}{s + 1} \Big|_{s=-2} = -8$$

$$F(s) = 3 + \frac{-1}{s + 1} + \frac{-8}{s + 2}$$

$$f(t) = 3\delta(t) - e^{-t} - 8e^{-2t} \qquad t \geq 0$$

Notice that if an incorrect expansion is made, for example,

$$F(s) = \frac{3s^2 - 4}{(s + 1)(s + 2)} = \frac{K_1}{s + 1} + \frac{K_2}{s + 2} \qquad \text{(incorrect)}$$

$$K_1 = [(s + 1)F(s)]\Big|_{s=-1} = \frac{3s^2 - 4}{s + 2}\Big|_{s=-1} = -1$$

$$K_2 = [(s + 2)F(s)]\Big|_{s=-2} = \frac{3s^2 - 4}{s + 1}\Big|_{s=-2} = -8$$

one obtains values for K_1 and K_2, but there is no indication from the evaluation that the expansion is not correct.

D12-10

Find the time functions for $t \geq 0$ that have the following Laplace transforms. Use the evaluation method to calculate partial fraction coefficients.

(a) $F(s) = \dfrac{2s - 1}{s(s + 3)}$

ans. $-\frac{1}{3} + \frac{7}{3}e^{-3t}$

(b) $F(s) = \dfrac{s + 3}{(s + 1)(s + 2)}$

ans. $2e^{-t} - e^{-2t}$

(c) $F(s) = \dfrac{10}{s^2 - 4}$

ans. $\frac{5}{2}e^{2t} - \frac{5}{2}e^{-2t}$

(d) $F(s) = \dfrac{s^2 - s + 3}{s(s + 1)(s + 2)}$

ans. $\frac{3}{2} - 5e^{-t} + \frac{9}{2}e^{-2t}$

(e) $F(s) = \dfrac{-4s + 5}{s^3 - 5s^2 + 6s}$

ans. $\frac{5}{6} + \frac{3}{2}e^{2t} - \frac{7}{3}e^{3t}$

(f) $F(s) = \dfrac{s^2 + 2s + 1}{3s^2 - 2s - 1}$

ans. $\frac{1}{2}e^t - \frac{1}{18}e^{-(1/3)t} + \frac{1}{3}\delta(t)$

(g) $F(s) = \dfrac{s^2}{s^2 - 4}$

ans. $\delta(t) + e^{2t} - e^{-2t}$

(h) $F(s) = \dfrac{3s - 2}{s^2 + 8s + 5}$

ans. $\dfrac{-14 + 3\sqrt{11}}{2\sqrt{11}} e^{(-4 + \sqrt{11})t} + \dfrac{14 + 3\sqrt{11}}{2\sqrt{11}} e^{(-4 - \sqrt{11})t}$

12.5.4 Repeated Roots

When the denominator polynomial of a rational function has repeated roots, s_1, the corresponding partial fraction expansion terms are of the form

$$F(s) = \frac{\text{numerator polynomial}}{(s - s_1)^r (s - s_2) \cdots}$$

$$= \frac{K_1}{s - s_1} + \frac{K_2}{(s - s_1)^2} + \cdots + \frac{K_r}{(s - s_1)^r} + \frac{K_{r+1}}{s - s_2} + \cdots$$

The following are examples:

$$\frac{10s + 6}{(s + 4)^3} = \frac{K_1}{s + 4} + \frac{K_2}{(s + 4)^2} + \frac{K_3}{(s + 4)^3}$$

$$\frac{7s^3 - 4}{(s + 3)^2 (s - 2)} = 7 + \frac{K_1}{(s + 3)} + \frac{K_2}{(s + 3)^2} + \frac{K_3}{s - 2}$$

$$\frac{10}{(s + 1)^2 (s - 6)^3} = \frac{K_1}{s + 1} + \frac{K_2}{(s + 1)^2} + \frac{K_3}{(s - 6)} + \frac{K_4}{(s - 6)^2} + \frac{K_5}{(s - 6)^3}$$

As with nonrepeated roots, the expansion coefficients may be found by cross-multiplying and equating coefficients. For example,

$$F(s) = \frac{3s - 2}{s(s + 1)^2} = \frac{K_1}{s} + \frac{K_2}{s + 1} + \frac{K_3}{(s + 1)^2}$$

$$= \frac{K_1(s + 1)^2 + K_2 s(s + 1) + K_3 s}{s(s + 1)^2}$$

$$= \frac{(K_1 + K_2)s^2 + (2K_1 + K_2 + K_3)s + K_1}{s(s + 1)^2}$$

$$\begin{cases} K_1 + K_2 & = 0 \\ 2K_1 + K_2 + K_3 = 3 \\ K_1 & = -2 \end{cases}$$

$$K_1 = -2 \qquad K_2 = 2 \qquad K_3 = 5$$

$$F(s) = \frac{-2}{s} + \frac{2}{s + 1} + \frac{5}{(s + 1)^2}$$

$$f(t) = (-2 + 2e^{-t} + 5te^{-t})u(t)$$

The partial fraction terms of higher than first power have inverse Laplace transforms that are powers of t times exponentials.

If desired, the evaluation method may be used to find some of the partial fraction coefficients. For the previous function,

$$F(s) = \frac{3s - 2}{s(s + 1)^2} = \frac{K_1}{s} + \frac{K_2}{s + 1} + \frac{K_3}{(s + 1)^2}$$

$$sF(s) = \frac{3s - 2}{(s + 1)^2} = K_1 + \frac{K_2 s}{s + 1} + \frac{K_3 s}{(s + 1)^2}$$

$$[sF(s)]\Big|_{s=0} = -2 = K_1$$

Similarly,

$$(s + 1)^2 F(s) = \frac{3s - 2}{s} = \frac{K_1(s + 1)^2}{s} + K_2(s + 1) + K_3$$

$$[(s + 1)^2 F(s)]\Big|_{s=-1} = 5 = K_3$$

But the K_2 coefficient cannot be found in this manner because the resulting expression cannot be evaluated at $s = -1$:

$$(s + 1)F(s) = \frac{3s - 2}{s(s + 1)} = \frac{K_1(s + 1)}{s} + K_2 + \frac{K_3}{s + 1}$$

A method of obtaining the coefficient K_2 in this repeated root expansion is to multiply $F(s)$ by the entire repeated root factor, then differentiate both sides of the expression with respect to s:

$$(s + 1)^2 F(s) = \frac{3s - 2}{s} = \frac{K_1(s + 1)^2}{s} + K_2(s + 1) + K_3$$

$$\frac{d}{ds}[(s + 1)^2 F(s)] = \frac{d}{ds}\left[\frac{3s - 2}{s}\right] = \frac{2}{s^2}$$

$$= \frac{d}{ds}\left[\frac{K_1(s + 1)^2}{s} + K_2(s + 1) + K_3\right]$$

$$= K_1 \frac{2s(s + 1) - (s + 1)^2}{s^2} + K_2$$

Evaluating at $s = -1$

$$\left\{\frac{d}{ds}[(s + 1)^2 F(s)]\right\}_{s=-1} = \left[\frac{2}{s^2}\right]_{s=-1} = 2 = K_2$$

In general,

$$F(s) = \frac{\text{numerator polynomial}}{(s - s_1)^r (s - s_2) \cdots}$$

$$= \frac{K_1}{s - s_1} + \frac{K_2}{(s - s_1)^2} + \cdots + \frac{K_r}{(s - s_1)^r} + \begin{pmatrix} \text{terms for other} \\ \text{different roots} \end{pmatrix}$$

The repeated root coefficients may be found by evaluation according to

$$K_i = \frac{1}{(r - i)!} \left\{ \frac{d^{r-i}}{ds^{r-i}} [(s - s_1)^r F(s)] \right\}_{s = s_1}$$

As a numerical example, consider the following transform:

$$F(s) = \frac{4s}{(s + 2)^2 (s + 3)^3}$$

$$= \frac{K_1}{s + 2} + \frac{K_2}{(s + 2)^2} + \frac{K_3}{s + 3} + \frac{K_4}{(s + 3)^2} + \frac{K_5}{(s + 3)^3}$$

Cross-multiplying and equating coefficients would be quite lengthy, since it would result in a set of five linear algebraic equations in the Ks. The evaluation method is considerably easier:

$$K_2 = [(s + 2)^2 F(s)]_{s = -2} = \frac{4s}{(s + 3)^3} \bigg|_{s = -2} = -8$$

$$K_1 = \left\{ \frac{d}{ds} [(s + 2)^2 F(s)] \right\}_{s = -2} = \left\{ \frac{d}{ds} \left[\frac{4s}{(s + 3)^3} \right] \right\}_{s = -2}$$

$$= \{4(s + 3)^{-3} - 12s(s + 3)^{-4}\}_{s = -2} = 28$$

$$K_5 = [(s + 3)^3 F(s)]_{s = -3} = \frac{4s}{(s + 2)^2} \bigg|_{s = -3} = -12$$

$$K_4 = \left\{ \frac{d}{ds} [(s + 3)^3 F(s)] \right\}_{s = -3} = \left\{ \frac{d}{ds} \left[\frac{4s}{(s + 2)^2} \right] \right\}_{s = -3}$$

$$= \{4(s + 2)^{-2} - 8s(s + 2)^{-3}\}_{s = -3} = -20$$

$$K_3 = \frac{1}{2} \left\{ \frac{d^2}{ds^2} [(s + 3)^3 F(s)] \right\}_{s = -3} = \frac{1}{2} \left\{ \frac{d}{ds} [4(s + 2)^{-2} - 8s(s + 2)^{3}] \right\}_{s = -3}$$

$$= \frac{1}{2} \{-16(s + 2)^{-3} + 24s(s + 2)^{-4}\}_{s = -3} = -28$$

D12-11

Find the time functions for $t \geq 0$ that have the following Laplace transforms:

(a) $F(s) = \dfrac{3s^2 + 2}{(s + 4)^2}$

(b) $F(s) = \dfrac{4s^2}{(s + 3)^3}$

   ans. $3\delta(t) - 24e^{-4t} + 50te^{-4t}$

   ans. $4e^{-3t} - 24te^{-3t} + 36t^2e^{-3t}$

(c) $F(s) = \dfrac{s^2 - 5}{s(s + 1)^2}$

ans. $-5 + 6e^{-t} + 4te^{-t}$

(d) $F(s) = \dfrac{10}{(s + 2)^2(s - 3)^2}$

ans. $\frac{4}{25}e^{-2t} + \frac{2}{5}te^{-2t} - \frac{4}{25}e^{3t} + \frac{2}{5}te^{3t}$

LAPLACE TRANSFORM EXPANSION

A rational function may be expressed in the form

$$F(s) = \frac{b_m s^m + b_{m-1} s^{m-1} + \cdots + b_1 s + b_0}{s^n + a_{n-1} s^{n-1} + \cdots + a_1 s + a_0}$$

For $m < n$ and no repeated roots of the denominator,

$$F(s) = \frac{K_1}{s - s_1} + \frac{K_2}{s - s_2} + \cdots + \frac{K_n}{s - s_n},$$

$$f(t) = [K_1 e^{s_1 t} + K_2 e^{s_2 t} + \cdots + K_n e^{s_n t}]u(t)$$

The coefficients of this partial fraction expansion may be found by cross-multiplying and comparing polynomials or by the evaluation method:

$$K_i = [F(s)(s - s_i)]_{s=s_i}$$

For repeated roots, the corresponding terms in the partial fraction expansion are of the following form:

$$F(s) = \frac{b_m s^m + b_{m-1} s^{m-1} + \cdots + b_1 s + b_0}{(s - s_1)^r (s - s_2) \cdots}$$

$$= \frac{K_1}{s - s_1} + \frac{K_2}{(s - s_1)^2} + \cdots + \frac{K_r}{(s - s_1)^r} + \frac{K_{r+1}}{s - s_2} + \cdots$$

$$f(t) = \left[K_1 e^{s_1 t} + K_2 t e^{s_1 t} + \cdots + K_r \frac{t^{r-1}}{(r - 1)!} e^{s_1 t} + K_{r+1} e^{s_2 t} + \cdots \right] u(t)$$

The partial fraction coefficients for repeated root terms may be found by comparison or by evaluation:

$$K_{r-1} = \left[\frac{d}{ds} (s - s_1)^r F(s) \right]_{s=s_1}$$

$$K_{r-k} = \frac{1}{k!} \left[\frac{d^k}{ds^k} (s - s_1)^r F(s) \right]_{s=s_1}$$

For improper fractions, $m \geq n$, divide denominator into numerator until the remainder is of lower degree than the denominator. Expand the remainder in partial fractions.

12.6 EXPANSIONS WITH COMPLEX ROOTS

12.6.1 Expansion in Cosine and Sine

For a complex conjugate set of Laplace transform denominator roots, the coefficients of the two partial fraction terms are complex conjugates, as can be seen from the evaluation method of calculating the coefficients. For

$$F(s) = \frac{\text{numerator polynomial}}{(s + a - jb)(s + a + jb)(\text{other factors})}$$

$$= \frac{K_1}{s + a - jb} + \frac{K_2}{s + a + jb} + \text{other terms}$$

$$K_1 = [(s + a - jb)F(s)]\Big|_{s = -a + jb}$$

$$K_2 = [(s + a + jb)F(s)]\Big|_{s = -a - jb} = K_1^*$$

To obtain K_2, every j in the expression for K_1 is replaced by $-j$, which is to form the complex conjugate.

The inversion of a set of complex conjugate partial fraction terms may proceed in several ways. While it is true that

$$\mathcal{L}^{-1}\left[\frac{c + jd}{s + a - jb} + \frac{c - jd}{s + a + jb}\right]$$

$$= (c + jd)e^{(-a + jb)t} + (c - jd)e^{(-a - jb)t} \qquad t \geq 0$$

this form of the time function corresponding to a set of complex conjugate partial fractions is not so convenient as exponential-times-sinusoid forms.

Once one of the complex partial fraction coefficients is found, the other need not be calculated; it is the complex conjugate. When two such terms are combined over a second-order denominator, the numerator coefficients are real numbers, and the expression is easily placed in the form of Laplace transform table entries:

$$F(s) = \frac{c + jd}{s + a - jb} + \frac{c - jd}{s + a + jb}$$

$$= \frac{(s + a + jb)(c + jd) + (s + a - jb)(c - jd)}{(s + a - jb)(s + a + jb)}$$

$$= \frac{2c(s + a) - 2db}{(s + a)^2 + b^2} = 2c\left[\frac{s + a}{(s + a)^2 + b^2}\right] - 2d\left[\frac{b}{(s + a)^2 + b^2}\right] \qquad (12\text{-}2)$$

$$f(t) = 2ce^{-at}\cos bt - 2de^{-at}\sin bt$$

$$= e^{-at}(2c\cos bt - 2d\sin bt) \qquad t \geq 0$$

For example, the transform below is already in the form of two combined complex conjugate partial fraction terms. Expansion into a sum of terms of the form of (12-2) allows easy use of the transform tables for inversion.

$$F(s) = \frac{3s + 7}{s^2 + 2s + 10} = \frac{3s + 7}{(s + 1)^2 + (3)^2}$$

$$= 3\left[\frac{s + 1}{(s + 1)^2 + 3^2}\right] + \frac{4}{3}\left[\frac{3}{(s + 1)^2 + 3^2}\right]$$

$$f(t) = 3e^{-t}\cos 3t + \tfrac{4}{3}e^{-t}\sin 3t$$

$$= e^{-t}(3\cos 3t + \tfrac{4}{3}\sin 3t) \qquad t \geq 0$$

The denominator is put in completed square, rather than factored, form and the expression is written as the sum of an exponential-times-cosine and an exponential-times-sine transform. Of course, if the denominator roots are not complex, the square in the denominator cannot be completed with real numbers, and the expansion should be into real root terms, as in the preceding section.

Consider the following transform.

$$F(s) = \frac{10}{(s + 2)(s^2 + 6s + 10)}$$

$$= \frac{10}{(s + 2)(s + 3 - j)(s + 3 + j)}$$

$$= \frac{K_1}{s + 2} + \frac{K_2}{s + 3 - j} + \frac{K_3}{s + 3 + j}$$

$$K_1 = \frac{10}{s^2 + 6s + 10}\bigg|_{s = -2} = \frac{10}{4 - 12 + 10} = 5$$

$$K_2 = \frac{10}{(s + 2)(s + 3 + j)}\bigg|_{s = -3+j} = \frac{10}{(-1 + j)(2j)}$$

$$= \frac{10(-2 + 2j)}{(-2 - 2j)(-2 + 2j)} = \frac{-20 + j20}{8} = -\frac{5}{2} + j\frac{5}{2}$$

$$K_3 = K_2^* = -\tfrac{5}{2} - j\tfrac{5}{2}$$

Combining the two complex conjugate partial fraction terms, there results

$$F(s) = \frac{5}{s + 2} + \frac{-\tfrac{5}{2} + j\tfrac{5}{2}}{s + 3 - j} + \frac{-\tfrac{5}{2} - j\tfrac{5}{2}}{s + 3 + j}$$

$$= \frac{5}{s + 2} + \frac{-5s - 20}{(s + 3)^2 + (1)^2}$$

$$= \frac{5}{s + 2} - 5\left[\frac{s + 3}{(s + 3)^2 + (1)^2}\right] - 5\left[\frac{1}{(s + 3)^2 + (1)^2}\right]$$

$$f(t) = 5e^{-2t} - 5e^{-t} \cos t - 5e^{-t} \sin t$$

$$= 5e^{-2t} + e^{-t}(-5 \cos t - 5 \sin t) \qquad t \geq 0$$

The sinusoidal portion of the response is expressed in quadrature form.

D12-12

Invert the following, expressing any sinusoidal portions of the results in quadrature form.

(a) $F(s) = \dfrac{3s - 2}{s^2 + 2s + 5}$
ans. $e^{-t}(3 \cos 2t - \frac{5}{2} \sin 2t)u(t)$

(b) $F(s) = \dfrac{s}{s^2 - 2s + 5}$
ans. $e^{t}(\cos 2t + \frac{1}{2} \sin 2t)u(t)$

(c) $F(s) = \dfrac{s^2 + s + 1}{s^2 + 4s + 15}$
ans. $\delta(t) - e^{-2t}[3 \cos \sqrt{11}t + (8/\sqrt{11}) \sin \sqrt{11}t]u(t)$

(d) $F(s) = \dfrac{s + 1}{s(s^2 + 6s + 13)}$
ans. $[\frac{1}{13} + e^{-3t}(\frac{-8}{104} \cos 2t - \frac{40}{104} \sin 2t)]u(t)$

(e) $F(s) = \dfrac{s + 1 + e^{-3s}}{s^2 + 2s + 10}$
ans. $(e^{-t} \cos 3t)u(t) + \frac{1}{3}e^{-(t-3)} \sin [3(t - 3)]u(t - 3)$

12.6.2 Expansion in Cosine and Angle

An alternate expression for the time function corresponding to two complex conjugate partial fraction terms involves placing the sinusoidal portion of the result in cosine and angle, rather than quadrature, form. If the coefficients are expressed in polar form,

$$F(s) = \frac{c + jd}{s + a - jb} + \frac{c - jd}{s + a + jb} = \frac{me^{j\alpha}}{s + a - jb} + \frac{me^{-j\alpha}}{s + a + jb}$$

where

$$m = \sqrt{c^2 + d^2} \qquad \alpha = \arctan \frac{d}{c}$$

Then

$$f(t) = e^{-at}[2c \cos bt - 2d \sin bt]$$

$$= e^{-at}\left[2\sqrt{c^2 + d^2} \cos \left(bt + \arctan \frac{d}{c} \right) \right]$$

$$= 2me^{-at} \cos (bt + \alpha) \qquad t \geq 0$$

For example,

$$F(s) = \frac{10s}{s^2 + 6s + 34} = \frac{10s}{(s+3)^2 + (5)^2} = \frac{K_1}{s + 3 - j5} + \frac{K_2}{s + 3 + j5}$$

$$K_1 = \frac{10s}{s + 3 + j5}\Bigg|_{s=-3+j5} = \frac{-30 + j50}{j10} = 5 + j3 = 5.83e^{j31°}$$

$$f(t) = 11.66e^{-3t} \cos (5t + 31°) \qquad t \geq 0$$

Here is another example:

$$F(s) = \frac{3s^2 + s - 7}{s^3 - 4s^2 + 29s} = \frac{3s^2 + s - 7}{s(s^2 - 4s + 29)}$$

$$= \frac{3s^2 + s - 7}{s[(s-2)^2 + (5)^2]} = \frac{K_1}{s} + \frac{K_2}{s - 2 - j5} + \frac{K_3}{s - 2 + j5}$$

$$K_1 = \frac{3s^2 + s - 7}{s^2 - 4s + 29}\Bigg|_{s=0} = -\frac{7}{9}$$

$$K_2 = \frac{3s^2 + s - 7}{s(s - 2 + j5)}\Bigg|_{s=2+j5} = \frac{3(2 + j5)^2 + 2 + j5 - 7}{(2 + j5)(j10)}$$

$$= \frac{-68 + j35}{(2 + j5)(j10)} = \frac{76.5e^{j305°}}{(5.38e^{j68°})(10e^{j90°})} = 1.42e^{j147°}$$

$$f(t) = -\tfrac{7}{9} + 2.84e^{2t} \cos (5t + 147°) \qquad t \geq 0$$

Note that the polar form partial fraction coefficient $me^{j\alpha}$ is that for the term of the form

$$\frac{K}{s + a - jb}$$

D12-13

Invert the following, expressing any sinusoidal portions of the results in cosine and angle form:

(a) $F(s) = \dfrac{8s - 3}{s^2 + 4s + 13}$ ans. $10.2e^{-2t} \cos (3t + 38.4°)u(t)$

(b) $F(s) = \dfrac{10}{4s^2 + s + 16}$ ans. $1.25e^{-0.125t} \cos (2t - 90°)u(t)$

(c) $F(s) = \dfrac{-7s^2 + 10}{s(s^2 + 8)}$ ans. $\tfrac{5}{4} - \tfrac{33}{4} \cos (2 \sqrt{2}t)u(t)$

(d) $F(s) = \dfrac{3s^2 - 2s + 4}{s(s^2 + 4s + 5)}$ ans. $\tfrac{4}{5} + 9.84e^{-2t} \cos (t + 77.1°)u(t)$

(e) $F(s) = \dfrac{3e^{-s}}{s^2 + 2s + 17}$ ans. $\tfrac{3}{4}e^{-(t-1)} \cos [4(t - 1)]u(t - 1)$

EXPANSIONS WITH COMPLEX ROOTS

Each pair of complex conjugate root partial fraction terms for a rational Laplace transform contributes an exponential-times-sinusoidal response term. The following additional Laplace transform table entries are especially useful in inverting complex root terms:

	$f(t)$ $\quad t \geq 0$	$F(s)$
(21)	$e^{-at}(2c \cos bt - 2d \sin bt)$	$\dfrac{c + jd}{s + a - jb} + \dfrac{c - jd}{s + a + jb}$
		$= \dfrac{2c(s + a) - 2db}{(s + a)^2 + (b)^2}$
(22)	$2me^{-at} \cos (bt + \alpha)$	$\dfrac{me^{j\alpha}}{s + a - jb} + \dfrac{me^{-j\alpha}}{s + a + jb}$

12.7 INITIAL AND FINAL VALUES

12.7.1 Initial Value Theorem

The initial value of a function $f(t)$ is the value of the function at $t = 0$, provided that $f(t)$ is continuous at $t = 0$. In the event that $f(t)$ is discontinuous at $t = 0$, then the initial value is the limit

$$f(0) = \lim_{t \to 0^+} f(t)$$

where t approaches $t = 0$ from the right. A function's initial value may be found from the Laplace transform by the relation

$$f(0) = \lim_{s \to \infty} sF(s)$$

where the variable s is taken to be real.

It should be carefully noted that the initial value $f(0)$ and the value $f(0^-)$ used as an initial condition will differ if the function $f(t)$ is discontinuous at $t = 0$. For example, the unit step function, defined as

$$u(t) = \begin{cases} 0 & t < 0 \\ 1 & t \geq 0 \end{cases}$$

has

$$u(0^-) = 0$$

while

$$u(0) = 1$$

To prove the initial value theorem, we note that for large real s in

$$\lim_{s \to \infty} sF(s) = \lim_{s \to \infty} s \int_{0^-}^{\infty} f(t)e^{-st}\, dt = \lim_{s \to \infty} \int_{0^-}^{\infty} f(t)se^{-st}\, dt$$

the integrand,

$$f(t)se^{-st}$$

contributes significantly to the integral only for values of t very near $t = 0$; otherwise the integrand is very small. Hence

$$sF(s) \cong \int_{0-}^{\infty} f(0)se^{-st} \, dt = f(0) \int_{0-}^{\infty} se^{-st} \, dt$$

for sufficiently large s, where the approximation improves with increasing s. It follows that

$$\lim_{s \to \infty} sF(s) = \lim_{s \to \infty} f(0) \int_{0-}^{\infty} se^{-st} \, dt$$

$$= f(0) \lim_{s \to \infty} \frac{se^{-st}}{s} \bigg|_{0-}^{\infty} = f(0)$$

The initial value theorem is especially easy to apply to rational Laplace transforms, where it is seen that the highest powers of s in the numerator and the denominator determine the initial values. For example, if

$$Y_1(s) = \frac{10s^3 - 6s^2 + 10s - 1}{s^5 + 4s^4 + 3s^2 + s + 6}$$

$$y_1(0) = \lim_{s \to \infty} sY_1(s) = \lim_{s \to \infty} \frac{10s^4 - 6s^3 + 10s^2 - s}{s^5 + 4s^4 + 3s^2 + s + 6} = 0$$

If

$$Y_2(s) = \frac{-2s^3 + 7s^2 - 2s + 9}{3s^4 - 2s^3 + 8s^2 + 10s + 1}$$

$$y_2(0) = \lim_{s \to \infty} sY_2(s) = \lim_{s \to \infty} \frac{-2s^4 + 7s^3 - 2s^2 + 9s}{3s^4 - 2s^3 + 8s^2 + 10s + 1} = -\frac{2}{3}$$

If

$$Y_3(s) = \frac{-2s^4}{s^4 + 8s^2 + 6s + 3}$$

$$y_3(0) = \lim_{s \to \infty} sY_3(s) = \lim_{s \to \infty} \frac{-2s^5}{s^4 + 8s^2 + 6s + 3} = -\infty$$

Time delay factors approach zero for large s, as with

$$Y_4(s) = \frac{e^{-2s}(s^3 + s^2 - 4)}{s^3 + 6s^2 + 5s + 7}$$

for which

$$y_4(0) = \lim_{s \to \infty} sY_4(s) = \lim_{s \to \infty} \frac{e^{-2s}(s^4 + s^3 - 4s)}{s^3 + 6s^2 + 5s + 7}$$

12.7.2 Final Value Theorem

The final value of a function $f(t)$ is

$$\lim_{t \to \infty} f(t)$$

provided the limit exists. The limit does not exist for sinusoidal functions for example, since these functions oscillate with constant amplitude, regardless of the value of t. Provided the final value exists *and is finite*, it is given, in terms of the function's Laplace transform, by

$$\lim_{t \to \infty} f(t) = \lim_{s \to 0} sF(s)$$

This result is known as the *final value theorem*. It will not be proved here, although it is easy to verify for all of the Laplace transforms of functions with final values that are listed in this chapter.

The operation performed in applying the final value theorem is that of computing, by the evaluation method, the partial fraction coefficient K of a term K/s in the function's expansion.

D12-14

Find the initial and final values, if they exist, of the functions with the following Laplace transforms:

(a) $F(s) = \dfrac{6s - 5}{s^2 + 2s + 10}$

    ans. 6, 0

(b) $F(s) = \dfrac{-3s^3 + 2s^2 + 8s + 9}{s^3 + 2s^2 + 6s}$

    ans. $-\infty$, $\frac{3}{2}$

(c) $F(s) = \dfrac{6s - 5}{s^2 - 2s + 10}$

    ans. 6, no final value

(d) $F(s) = \dfrac{9s^4 - 1}{s^3 + 2s^2 + 10s}$

    ans. ∞, $-\frac{1}{10}$

(e) $F(s) = \dfrac{8s^2 + 3}{(s + 1)^2(s + 2)(s^2 + s + 12)}$

    ans. 0, 0

INITIAL AND FINAL VALUE THEOREMS

The initial value of a function and its Laplace transform are related by

$$\lim_{t \to 0+} f(t) = f(0) = \lim_{s \to \infty} sF(s)$$

Provided that the final value of $f(t)$ exists and is finite,

$$\lim_{t \to \infty} f(t) = \lim_{s \to 0} sF(s)$$

This limit is simply the calculation of the coefficient K of a K/s term in the partial fraction expansion of $F(s)$.

12.8 LAPLACE TRANSFORM SOLUTION OF DIFFERENTIAL EQUATIONS

The Laplace transform provides a convenient method of differential equation solution, in which boundary conditions at $t = 0^-$ are automatically incorporated into the calculations at an early stage. Consider the differential equation

$$\frac{d^2y}{dt^2} + 5\frac{dy}{dt} + 6y = 7e^{-t}$$

with initial conditions

$$y(0^-) = 4 \qquad y'(0^-) = -8$$

Laplace transforming the equation, there results

$$s^2Y(s) - sy(0^-) - y'(0^-) + 5[sY(s) - y(0^-)] + 6Y(s) = \frac{7}{s+1}$$

$$s^2Y(s) - 4s + 8 + 5sY(s) - 20 + 6Y(s) = \frac{7}{s+1}$$

$$(s^2 + 5s + 6)Y(s) = \frac{7}{s+1} + 4s + 12 = \frac{4s^2 + 16s + 19}{s+1}$$

$$Y(s) = \frac{4s^2 + 16s + 19}{(s+1)(s^2 + 5s + 6)} = \frac{4s^2 + 16s + 19}{(s+1)(s+2)(s+3)}$$

$$= \frac{K_1}{s+1} + \frac{K_2}{s+2} + \frac{K_3}{s+3}$$

$$K_1 = \left.\frac{4s^2 + 16s + 19}{(s+2)(s+3)}\right|_{s=-1} = \frac{4 - 16 + 19}{(1)(2)} = \frac{7}{2}$$

$$K_2 = \left.\frac{4s^2 + 16s + 19}{(s+1)(s+3)}\right|_{s=-2} = \frac{16 - 32 + 19}{(-1)(1)} = -3$$

$$K_3 = \left.\frac{4s^2 + 16s + 19}{(s+1)(s+2)}\right|_{s=-3} = \frac{36 - 48 + 19}{(-2)(-1)} = \frac{7}{2}$$

The equation solution, with the given initial conditions, is thus

$$y(t) = \tfrac{7}{2}e^{-t} - 3e^{-2t} + \tfrac{7}{2}e^{-3t} \qquad t \geq 0$$

Since the Laplace transform of a function does not depend upon the function's behavior prior to $t = 0$, equation solutions obtained in this manner hold only for $t \geq 0$.

Equations with running integrals, such as those from electrical networks, may also be routinely solved by Laplace transforming, solving for the transform of the solution, then inverting the transform to find the corresponding function of time for $t \geq 0$. For example, the equation

$$\frac{dy}{dt} + 2y + 10 \int_{-\infty}^{t} y \, dt = 6t$$

when Laplace transformed becomes

$$sY(s) - y(0^-) + 2Y(s) + \frac{10}{s} Y(s) + \frac{10}{s} \int_{-\infty}^{0^-} y(t) \, dt = \frac{6}{s^2}$$

If it is known that the initial conditions are

$$y(0^-) = 3 \qquad \int_{-\infty}^{0^-} y(t) \, dt = -\tfrac{6}{10}$$

then

$$sY(s) - 3 + 2Y(s) + \frac{10}{s} Y(s) - \frac{6}{s} = \frac{6}{s^2}$$

$$(s^3 + 2s^2 + 10s)Y(s) = 3s^2 + 6s + 6$$

$$Y(s) = \frac{3s^2 + 6s + 6}{s(s^2 + 2s + 10)} = \frac{3s^2 + 6s + 6}{s[(s + 1)^2 + 3^2]}$$

$$= \frac{K_1}{s} + \frac{K_2}{s + 1 - j3} + \frac{K_3}{s + 1 + j3}$$

$$K_1 = \left. \frac{3s^2 + 6s + 6}{s^2 + 2s + 10} \right|_{s=0} = \frac{3}{5}$$

$$K_2 = \left. \frac{3s^2 + 6s + 6}{s(s + 1 + j3)} \right|_{s = -1+j3} = \frac{3(-1 + j3)^2 + 6(-1 + j3) + 6}{(-1 + j3)(j6)}$$

$$= \frac{3(-8 - j6) - 6 + j18 + 6}{(-1 + j3)(j6)} = \frac{-24}{(-1 + j3)(j6)}$$

$$= \frac{24e^{j180°}}{(3.16e^{j108°})(6e^{j90°})} = 1.3e^{-j18°}$$

so that

$$y(t) = \tfrac{3}{5} + 2.6e^{-t} \cos (3t - 18°) \qquad t \geq 0$$

When simultaneous integrodifferential equations are Laplace transformed, the result is a corresponding set of simultaneous algebraic equations involving the variable s. Laplace transforming the simultaneous equations

$$\begin{cases} \dfrac{dx_1}{dt} - x_2 = 3e^{-4t} \\[2ex] 4\displaystyle\int_{-\infty}^{t} x_1\, dt + x_2 = 0 \end{cases}$$

with zero initial conditions

$$x_1(0^-) = 0 \qquad \int_{-\infty}^{0^-} x_1(t)\, dt = 0$$

gives the simultaneous algebraic equations

$$\begin{cases} sX_1(s) - X_2(s) = \dfrac{3}{s+4} \\[2ex] \dfrac{4}{s} X_1(s) + X_2(s) = 0 \end{cases}$$

Solving, say, for $X_1(s)$

$$X_1(s) = \frac{\begin{vmatrix} \dfrac{3}{s+4} & -1 \\[2ex] 0 & 1 \end{vmatrix}}{\begin{vmatrix} s & -1 \\[2ex] \dfrac{4}{s} & 1 \end{vmatrix}} = \frac{\dfrac{3}{s+4}}{s + \dfrac{4}{s}}$$

$$= \frac{3s}{(s+4)(s^2+4)} = \frac{3s}{(s+4)(s-2j)(s+2j)}$$

$$= \frac{K_1}{s+4} + \frac{K_2}{s-2j} + \frac{K_3}{s+2j}$$

Then

$$K_1 = \left.\frac{3s}{s^2+4}\right|_{s=-4} = \frac{-12}{16+4} = -\frac{3}{5}$$

$$K_2 = \left.\frac{3s}{(s+4)(s+2j)}\right|_{s=2j} = \frac{6j}{(4+2j)(4j)}$$

$$= \frac{6}{(4.47e^{j27°})(4)} = 0.336e^{-j27°}$$

so the solution for $x_1(t)$ after $t = 0$ is

$$x_1(t) = -\tfrac{3}{5}e^{-4t} + 0.672 \cos{(2t - 27°)}$$

D12-15

Use Laplace transform methods to obtain the solutions to the following integrodifferential equations after time $t = 0$, given the indicated boundary conditions:

(a) $\dfrac{dy}{dt} + 3y = \delta(t) \qquad y(0^-) = -5$

ans. $-4e^{-3t}$

(b) $2\dfrac{dy}{dt} - 3y = 4 \cos 6t \qquad y(0^-) = 0$

ans. $\tfrac{12}{51}e^{(3/2)t} - \tfrac{12}{51} \cos 6t - \tfrac{16}{51} \sin 6t$

(c) $\dfrac{dy}{dt} + 4y = 10 \qquad y(0^-) = 6$

ans. $\tfrac{5}{2} + \tfrac{7}{2}e^{-4t}$

(d) $3\dfrac{dy}{dt} + 4\displaystyle\int_{0^-}^{t} y\, dt = 5u(t) \qquad y(0^-) = 7$

ans. $\tfrac{7}{3} \cos \sqrt{\tfrac{4}{3}}\,t + \tfrac{5}{3}\sqrt{\tfrac{3}{4}} \sin \sqrt{\tfrac{4}{3}}\,t$

(e) $\dfrac{d^2y}{dt^2} + 2\dfrac{dy}{dt} + 5y = 0$

$y(0^-) = 0 \qquad \dfrac{dy}{dt}\bigg|_{t=0^-} = 4$

ans. $2e^{-t} \sin 2t$

(f) $\dfrac{d^2y}{dt^2} + 5\dfrac{dy}{dt} + 6y = 4e^{-3t}$

$y(0^-) = 2 \qquad \dfrac{dy}{dt}\bigg|_{t=0^-} = 0$

ans. $-3e^{-3t} - 4te^{-3t} + 5e^{-2t}$

(g) $\dfrac{dy}{dt} + 2y + 5\displaystyle\int_{-\infty}^{t} y\, dt = 0$

$y(0^-) = -3 \qquad \displaystyle\int_{-\infty}^{0^-} y\, dt = 4$

ans. $e^{-t}(-3 \cos 2t - \tfrac{17}{2} \sin 2t)$

(h)

$$\frac{dx_1}{dt} = -3x_1 + x_2 + 4e^{-t}$$

$$\frac{dx_2}{dt} = -2x_1$$

$$y = 3x_1$$

$$x_1(0^-) = 0 \qquad x_2(0^-) = 4$$

ans. $36e^{-t} - 12te^{-t} - 36e^{-2t}$

LAPLACE TRANSFORM SOLUTION OF DIFFERENTIAL EQUATIONS

To solve a differential equation for time $t \geq 0$, Laplace transform it, solve for the transform of the solution, then invert to find the solution. This technique also applies to simultaneous coupled equations and when running integrals are involved.

Initial conditions at $t = 0^-$ are routinely incorporated into the transformed equations.

12.9 LAPLACE TRANSFORM SOLUTION OF SWITCHED NETWORKS WITH ZERO INITIAL CONDITIONS

When the initial condition terms in network equations are all zero, it is simple to write Laplace transformed network equations directly from the network. In terms of the network, zero initial conditions mean that all inductor currents and all capacitor voltages are zero at $t = 0^-$, just prior to time $t = 0$.

Suppose that the network of Figure 12-18(a) is connected at time $t = 0$, with zero initial conditions. Systematic simultaneous mesh equations, in terms of $i_1(t)$ and $i_2(t)$ are:

$$\left(3\frac{d}{dt} + 4\int_{-\infty}^{t} dt\right)i_1(t) - \left(4\int_{-\infty}^{t} dt\right)i_2(t) = 6e^{-2t}$$

$$-\left(4\int_{-\infty}^{t} dt\right)i_1(t) + \left(1 + 4\int_{-\infty}^{t} dt\right)i_2(t) = -5$$

Laplace transforming these equations when the initial conditions are zero involves simply replacing the derivative operations with s, the integrals with $1/s$, and dealing with Laplace transformed signals:

$$\left(3s + \frac{4}{s}\right)I_1(s) - \left(\frac{4}{s}\right)I_2(s) = \frac{6}{s+2}$$

$$-\left(\frac{4}{s}\right)I_1(s) + \left(1 + \frac{4}{s}\right)I_2(s) = -\frac{5}{s}$$

In terms of the network, the transformed equations may be written directly, by replacing all signals by their Laplace transforms and dealing with the element impedances, as functions of s, as shown in Figure 12-18(b).

The Laplace transformed network represents the relationships between the

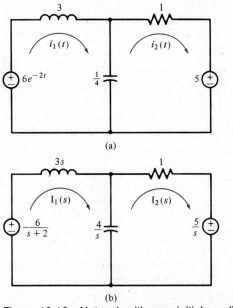

Figure 12-18   Network with zero initial conditions.
(a) Original network.
(b) Laplace transformed network.

Laplace transforms of the network signals, just as the original network represents signal relations as functions of time. All network solution methods, including equivalent circuits, nodal equations and mesh equations apply to the problem of determining signal transforms from the transformed network, just as they do to finding the signals as functions of time from the original network.

The solution for the mesh current $I_1(s)$ is, from the transformed equations,

$$
I_1(s) = \frac{\begin{vmatrix} \dfrac{6}{s+2} & -\dfrac{4}{s} \\[2mm] -\dfrac{5}{s} & \left(1 + \dfrac{4}{s}\right) \end{vmatrix}}{\begin{vmatrix} \left(3s + \dfrac{4}{s}\right) & -\dfrac{4}{s} \\[2mm] -\dfrac{4}{s} & \left(1 + \dfrac{4}{s}\right) \end{vmatrix}}
$$

$$
= \frac{\left(\dfrac{6}{s+2}\right)\left(1 + \dfrac{4}{s}\right) - \dfrac{20}{s^2}}{\left(3s + \dfrac{4}{s}\right)\left(1 + \dfrac{4}{s}\right) - \dfrac{16}{s^2}} = \frac{\dfrac{6s^2 - 4s - 40}{s^2(s+2)}}{\dfrac{3s^2 + 12s + 4}{s}}
$$

$$
= \frac{2s^2 - \tfrac{4}{3}s - \tfrac{40}{3}}{s(s+2)(s^2 + 4s + \tfrac{4}{3})}
$$

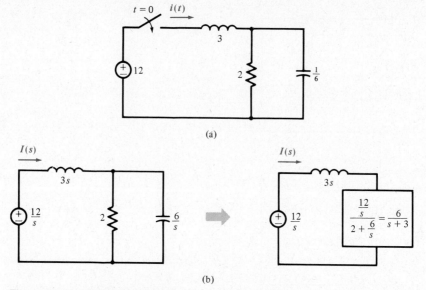

(a)

(b)

Figure 12-19   Switched network with zero initial conditions.
(a) Original network.
(b) Laplace transformed network after $t = 0$.

This rational function expands in partial fractions as

$$I_1(s) = \frac{-5}{s} + \frac{\frac{1}{2}}{s + 2} + \frac{-3.76}{s + 2.633} + \frac{9.26}{s + 0.367}$$

so that

$$i_1(t) = -5 - \tfrac{1}{2}e^{-2t} - 3.76e^{-2.633t} + 9.26e^{-0.367t}$$

after $t = 0$. The other mesh current, or any other network voltage or current after $t = 0$, can be found similarly.

The switched network of Figure 12-19(a) has zero initial conditions, since the inductor current and the capacitor voltage are initially zero. The Laplace transformed network is shown in Figure 12-19(b), where equivalent circuits are used to obtain

$$\left(3s + \frac{6}{s + 3}\right) I(s) = \frac{12}{s}$$

$$I(s) = \frac{12}{s(3s^2 + 9s + 6)} = \frac{4}{s(s + 1)(s + 2)}$$

$$= \frac{K_1}{s} + \frac{K_2}{s + 1} + \frac{K_3}{s + 2}$$

Then

$$K_1 = \frac{4}{(s + 1)(s + 2)}\Bigg|_{s=0} = 2$$

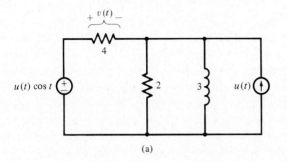

(a)

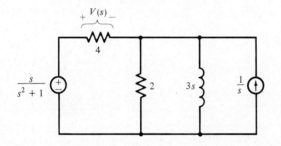

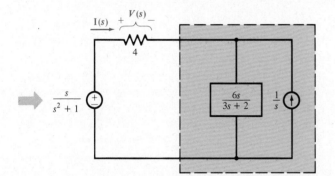

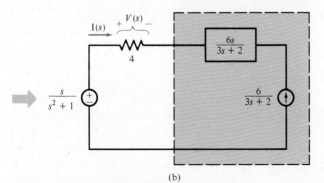

(b)

Figure 12-20   Another switched network with zero initial conditions.
(a) Original network.
(b) Laplace transformed network after $t = 0$.

$$K_2 = \frac{4}{s(s + 2)}\bigg|_{s=-1} = -4$$

$$K_3 = \frac{4}{s(s + 1)}\bigg|_{s=-2} = 2$$

giving

$$i(t) = 2 - 4e^{-t} + 2e^{-2t} \qquad t \geq 0$$

In the switched network of Figure 12-20, a Thevénin equivalent is used to obtain the solution. The network initial conditions are zero, since the inductor current is zero before $t = 0$.

$$\left(4 + \frac{6s}{3s + 2}\right)I(s) = \frac{s}{s^2 + 1} - \frac{6}{3s + 2}$$

$$\frac{18s + 8}{3s + 2} I(s) = \frac{-3s^2 + 2s - 6}{(s^2 + 1)(3s + 2)}$$

$$I(s) = \frac{-3s^2 + 2s - 6}{(18s + 8)(s^2 + 1)} = \frac{-\frac{1}{6}s^2 + \frac{1}{9}s - \frac{1}{3}}{(s + \frac{4}{9})(s^2 + 1)}$$

$$V(s) = 4I(s) = \frac{-\frac{2}{3}s^2 + \frac{4}{9}s - \frac{4}{3}}{(s + \frac{4}{9})(s^2 + 1)} = \frac{K_1}{s + \frac{4}{9}} + \frac{K_2}{s + j} + \frac{K_2^*}{s - j}$$

Solving for the partial fraction coefficients, there results

$$K_1 = \frac{-\frac{2}{3}s^2 + \frac{4}{9}s - \frac{4}{3}}{s^2 + 1}\bigg|_{s=-\frac{4}{9}} = -1.37$$

$$K_2 = \frac{-\frac{2}{3}s^2 + \frac{1}{9}s - \frac{1}{3}}{(s + \frac{4}{9})(s - j)}\bigg|_{s=-j} = \frac{3 + 2j}{9 + 4j} = \frac{3.61e^{j33.7°}}{9.85e^{j24°}} = 0.367e^{j9.7°}$$

and

$$v(t) = -1.37e^{-(4/9)t} + 0.734 \cos (t + 9.7°) \qquad t \geq 0$$

D12-16

Solve the following networks for the indicated signals after time $t = 0$. The network initial conditions are zero.

(a)

(b)

ans. $\frac{1}{2}e^{-(1/3)t} - \frac{1}{2}e^{-t}$

ans. $\frac{35}{8}e^{-6t} - \frac{35}{8}e^{-2t}$

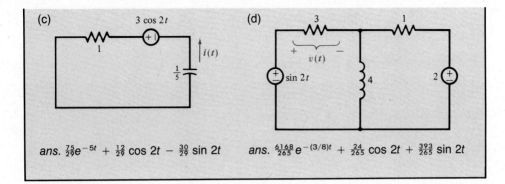

(c)

3 cos 2t

$i(t)$

ans. $\frac{75}{29}e^{-5t} + \frac{12}{29}\cos 2t - \frac{30}{29}\sin 2t$

(d)

ans. $\frac{6168}{265}e^{-(3/8)t} + \frac{24}{265}\cos 2t + \frac{393}{265}\sin 2t$

LAPLACE TRANSFORM SOLUTION OF SWITCHED NETWORKS WITH ZERO INITIAL CONDITIONS

To solve a network with zero initial conditions for $t \geq 0$, solve the transformed network for the Laplace transform of the signal of interest, then invert to find the signal as a function of time. In the transformed network, all signals are replaced by their Laplace transforms, and all elements by their impedances, as functions of s.

12.10 GENERAL SOLUTION OF SWITCHED NETWORKS

Nonzero network initial conditions could be incorporated by writing equations for the network, Laplace transforming those equations, including the initial conditions, then solving. It is far more convenient, however, to represent nonzero initial capacitor voltages and inductor currents by sources in an equivalent network with zero initial conditions.

The sink-reference, voltage-current relation for a capacitor is

$$v_C(t) = \frac{1}{C}\int_{-\infty}^{t} i_C(t)\,dt$$

After time $t = 0^-$, the capacitor voltage is

$$v_C(t) = \frac{1}{C}\int_{-\infty}^{0^-} i_C(t)\,dt + \frac{1}{C}\int_{0^-}^{t} i_C(t)\,dt$$

$$= v_C(0^-) + \frac{1}{C}\int_{0^-}^{t} i_C(t)\,dt$$

which is identical to the voltage-current relation for an initially charged capacitor in series with a constant voltage source equal to the initial capacitor voltage, $V = v_C(0^-)$. This equivalence is illustrated in Figure 12-21(a). It is thus always possible to convert any problem involving an initially charged capacitor to one with an uncharged capacitor in series with a constant voltage source.

In the Laplace transformed network, the capacitor has impedance $1/sC$, and

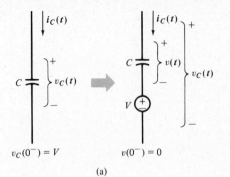

$$v_C(0^-) = V \qquad\qquad v(0^-) = 0$$

(a)

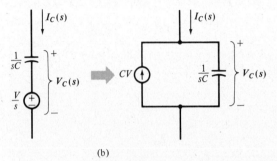

(b)

Figure 12-21 Representing initial capacitor charge with a source.
(a) Equivalence of an initially charged capacitor to an uncharged capacitor in series with a constant voltage source.
(b) Laplace transformed equivalences.

the constant voltage V in the equivalent has transform V/s. If it is more convenient to deal with current sources in the network solution, as when nodal equations are written, the Thévenin-Norton transformation of Figure 12-21(b) may be used to obtain an equivalent involving a current source in parallel with an uncharged capacitor. The current source has a constant Laplace transform and thus represents the time function that is an impulse of strength CV.

The switched network of Figure 12-22(a) has a nonzero initial capacitor voltage. Before $t = 0$, the $10u(t)$ step source is zero, and the forced capacitor voltage is

$$v(0) = -5$$

as shown in Figure 12-22(b). The Laplace transformed network, incorporating the nonzero initial condition, is shown in Figure 12-22(c). For it,

$$\left(3 + \frac{1}{2s}\right)I(s) = \frac{10}{s} - \left(-\frac{5}{s}\right) - \frac{5}{s}$$

$$I(s) = \frac{20}{6s + 1}$$

$$V(s) = \frac{1}{2s}I(s) - \frac{5}{s} = \frac{-5s + \frac{5}{6}}{s(s + \frac{1}{6})} = \frac{K_1}{s} + \frac{K_2}{s + \frac{1}{6}}$$

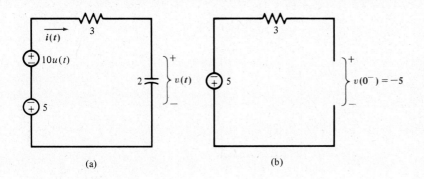

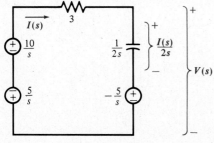

(c)

Figure 12-22   Switched network with nonzero initial condition.
(a) Network with nonzero initial condition.
(b) Finding the initial capacitor voltage.
(c) Laplace transformed network after $t = 0^-$.

Then

$$K_1 = \left. \frac{-5s + \frac{5}{6}}{s + \frac{1}{6}} \right|_{s=0} = 5$$

$$K_2 = \left. \frac{-5s + \frac{5}{6}}{s} \right|_{s=-\frac{1}{6}} = -10$$

giving

$$v(t) = 5 - 10e^{-(1/6)t} \qquad t \geq 0$$

An initial inductor current may also be represented by a source, as indicated in Figure 12-23(a). The sink-reference current-voltage relation for an inductor is

$$i_L(t) = \frac{1}{L} \int_{-\infty}^{t} v_L(t) \, dt = \frac{1}{L} \int_{-\infty}^{0^-} v_L(t) \, dt + \frac{1}{L} \int_{0^-}^{t} v_L(t) \, dt$$

$$= i_L(0^-) + \frac{1}{L} \int_{0^-}^{t} v_L(t) \, dt$$

after $t = 0^-$, and this is equivalent to a constant current source in parallel with an inductor with zero initial current. The Laplace transformed equivalent is shown in

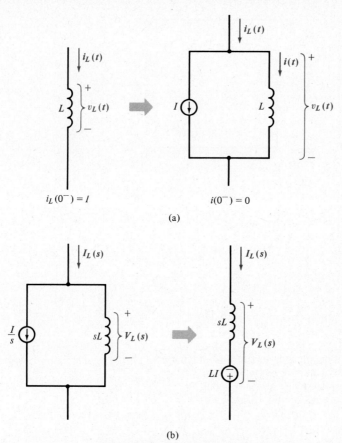

(a)

(b)

Figure 12-23 Representing initial inductor current with a source.
(a) Equivalence of an initial inductor current to a current source in parallel with an inductor with zero initial current.
(b) Laplace transformed equivalences.

Figure 12-23(b), and a Norton-Thévenin transformation gives a series equivalent involving an impulsive voltage source.

In the example network of Figure 12-24(a), the inductor current and the capacitor voltage at $t = 0^-$ are the forced constant values before the switching, as calculated in Figure 12-24(b). Using the current source equivalents for the inductor and the capacitor, the Laplace transformed network of Figure 12-24(c), with all elements in parallel, results. The two-node equation for $V_C(s)$ is then

$$\left(\frac{2}{s} + 5 + 3s\right)V_C(s) = \frac{60}{s} + 36$$

$$V_C(s) = \frac{36s + 60}{3s^2 + 5s + 2} = \frac{12s + 20}{(s + \frac{2}{3})(s + 1)} = \frac{K_1}{s + \frac{2}{3}} + \frac{K_2}{s + 1}$$

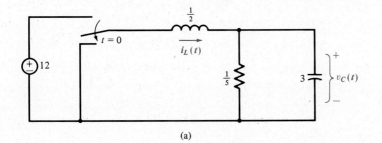

(a)

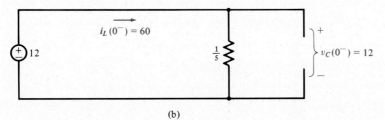

(b)

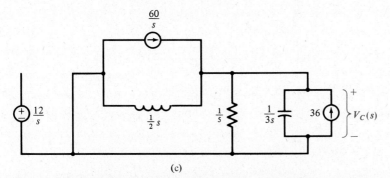

(c)

Figure 12-24   Another switched network with nonzero initial conditions.
(a) Network with nonzero initial conditions.
(b) Finding the initial inductor current and capacitor voltage.
(c) Laplace transformed network after $t = 0^-$.

The coefficients of this partial fraction expansion are

$$K_1 = \left. \frac{12s + 20}{s + 1} \right|_{s = -\frac{2}{3}} = 36$$

$$K_2 = \left. \frac{12s + 20}{s + \frac{2}{3}} \right|_{s = -1} = -24$$

so

$$v_C(t) = 36e^{-(2/3)t} - 24e^{-t} \qquad t \ge 0$$

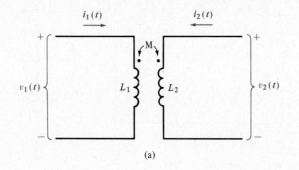

(a)

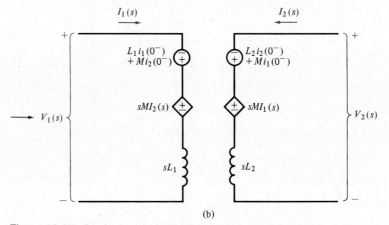

(b)

Figure 12-25   Laplace transformed coupled inductor model.

For coupled inductors, Figure 12-25(a), the voltages are related to the currents by

$$\begin{cases} v_1(t) = L_1 \dfrac{di_1}{dt} + M \dfrac{di_2}{dt} \\[2mm] v_2(t) = M \dfrac{di_1}{dt} + L_2 \dfrac{di_2}{dt} \end{cases}$$

When these equations are Laplace transformed, the following equations result:

$$\begin{cases} V_1(s) = sL_1I_1(s) - L_1i_1(0^-) + sMI_2(s) - Mi_2(0^-) \\ V_2(s) = sMI_1(s) - Mi_1(0^-) + sL_2I_2(s) - L_2i_2(0^-) \end{cases}$$

These are represented by the Laplace transformed network model of Figure 12-25(b), in which sources to account for nonzero initial conditions are added to the usual controlled voltage source model for coupled inductors. Other Laplace transformed coupled inductor models, analogous to those in Section 9.9, may be derived from this model by transformation, if desired.

D12-17

Solve the following networks for the indicated signals for $t \geq 0$, using Laplace transform methods:

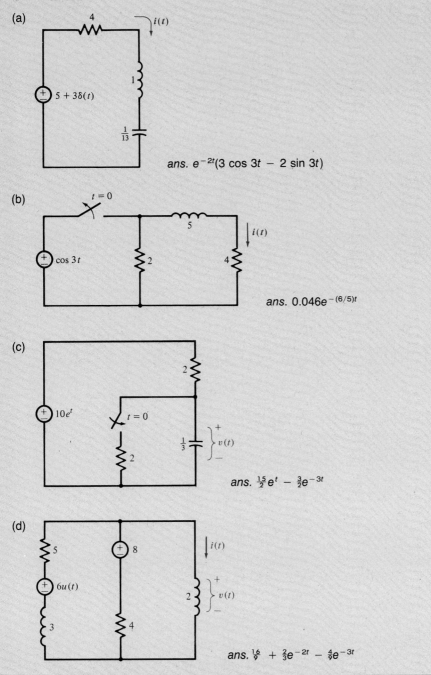

(a)

ans. $e^{-2t}(3 \cos 3t - 2 \sin 3t)$

(b)

ans. $0.046e^{-(6/5)t}$

(c)

ans. $\frac{15}{2}e^t - \frac{3}{2}e^{-3t}$

(d)

ans. $\frac{16}{9} + \frac{2}{3}e^{-2t} - \frac{4}{9}e^{-3t}$

GENERAL SOLUTION OF SWITCHED NETWORKS

If the network initial conditions are not all zero, initial capacitor voltages and inductor currents may be replaced by equivalent sources, and zero initial conditions:

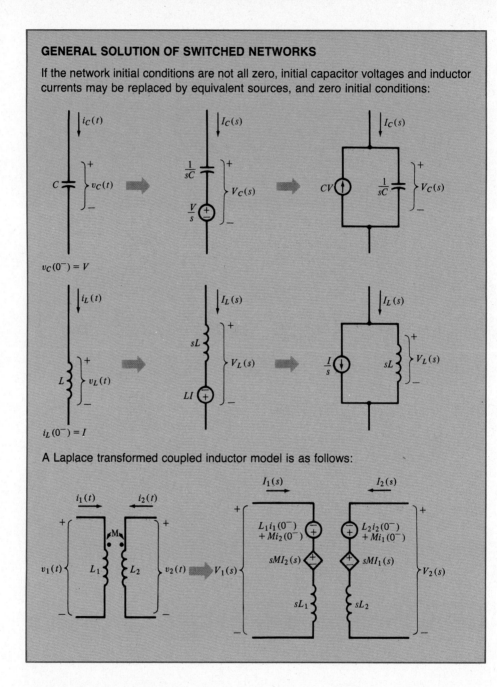

A Laplace transformed coupled inductor model is as follows:

12.11 CONVOLUTION

Occasionally, it is desirable to calculate the time function corresponding to the product of two Laplace transforms directly. Let $f_1(t)$ and $f_2(t)$ be two functions of time, with $F_1(s)$ and $F_2(s)$ their Laplace transforms:

$$g(t) = \mathcal{L}^{-1}[F_1(s)F_2(s)] = \int_{\sigma-j\infty}^{\sigma+j\infty} F_1(s)F_2(s)e^{st}\,ds \qquad (12\text{-}3)$$

To derive the desired relationship,

$$F_2(s) = \int_{0^-}^{\infty} f_2(t)e^{-st}\,dt = \int_{0^-}^{\infty} f_2(\tau)e^{-s\tau}\,d\tau \qquad (12\text{-}4)$$

is to be substituted into (12-3). The variable of integration in (12-4) is changed from t to τ before substitution, since t is already used as a variable in (12-3). Then

$$g(t) = \int_{\sigma-j\infty}^{\sigma+j\infty} F_1(s)\left[\int_{0^-}^{\infty} f_2(\tau)e^{-s\tau}\,d\tau\right]e^{st}\,ds$$

and combining the exponentials and interchanging the order of integration gives

$$g(t) = \int_{0^-}^{\infty} f_2(\tau)\underbrace{\left[\int_{\sigma+j\infty}^{\sigma-j\infty} F_1(s)e^{s(t-\tau)}\,ds\right]}_{f_1(t-\tau)u(t-\tau)\,d\tau}\,d\tau$$

The quantity in brackets above is the inverse Laplace transform of $F_1(s)$, with t replaced by $(t-\tau)$, as indicated, so that

$$g(t) = \int_{0^-}^{\infty} f_2(\tau)f_1(t-\tau)u(t-\tau)\,d\tau$$

Since $u(t-\tau)$ is zero for $\tau > t$, the τ integration may be rewritten as follows:

$$g(t) = \mathcal{L}^{-1}[F_1(s)F_2(s)] = \int_{0^-}^{t} f_1(t-\tau)f_2(\tau)\,d\tau$$

$$= f_1(t) \otimes f_2(t)$$

The similar relation for Fourier transforms does not involve the step function $u(t-\tau)$ and so results in an integral extending from $\tau = -\infty$ to $\tau = +\infty$. The symbol $\otimes$ is a shorthand to denote the convolution integral. Interchanging the order of $f_1(t)$ and $f_2(t)$, it is also true that

$$g(t) = \mathcal{L}^{-1}[F_1(s)F_2(s)] = \int_{0^-}^{t} f_2(t-\tau)f_1(\tau)\,d\tau$$

$$= f_2(t) \otimes f_1(t)$$

that is, the order that two functions are convolved is of no consequence to the result.

The operation of convolution is illustrated in Figure 12-26 for two simple example functions. For each value of the variable t, the convolution is the area under the product

$$f_1(t-\tau)f_2(\tau)$$

from $\tau = 0^-$ to $\tau = t$. Graphically, one can reverse the time axis on $f_1(\tau)$ to form $f_1(-\tau)$, then time-shift the function various amounts t, determining the area under the product, as in the example.

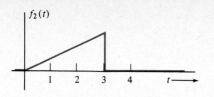

(a)

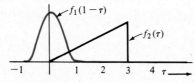

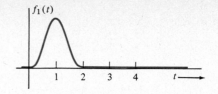

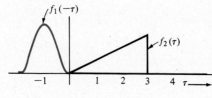

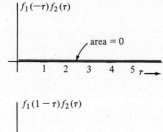

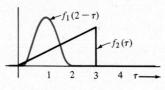

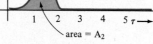

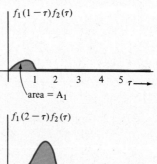

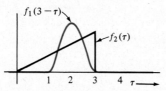

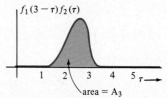

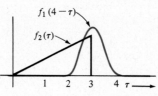

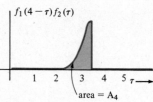

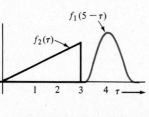

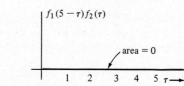

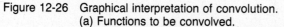

(b)

Figure 12-26   Graphical interpretation of convolution.
          (a) Functions to be convolved.
          (b) Integrals for various values of the variable t.
          (c) Resulting convolution.

Figure 12-26   (*Continued*)

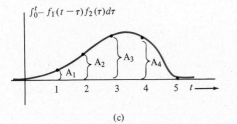

(c)

Besides being a powerful computational tool, convolution is often useful in visualizing network response. For example, in the Laplace transformed network of Figure 12-27(a),

$$V(s) = \frac{10}{s}\left(\frac{\frac{4}{s}}{2 + \frac{4}{s}}\right) = \left(\frac{10}{s}\right)\left(\frac{2}{s + 2}\right)$$

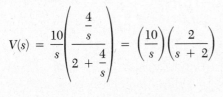

(a)

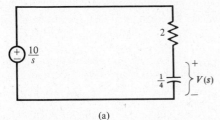

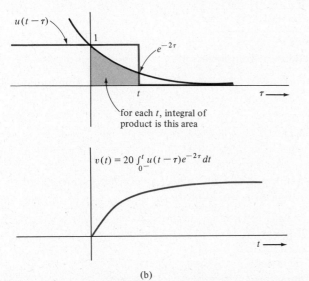

(b)

Figure 12-27   Finding switched network response using convolution.
(a) Laplace transformed network.
(b) Convolution of two time functions.

which involves the product of the source transform

$$F_1(s) = \frac{10}{s}$$

and the transform

$$F_2(s) = \frac{2}{s + 2}$$

The corresponding two time functions are

$$f_1(t) = 10u(t) \qquad f_2 = 2u(t)e^{-2t}$$

Their convolution,

$$v(t) = \mathcal{L}^{-1}\left[\left(\frac{10}{s}\right)\left(\frac{2}{s + 2}\right)\right] = \int_{0^-}^{t} 20u(t - \tau)e^{-2\tau}\, d\tau$$

is constructed graphically in Figure 12-27(b).

When a network has a single source, the Laplace transform of any network voltage or current consists of the source transform multiplied by a function of s, depending on network elements, that is the transfer function relating that voltage or current to the source. For example, in the single-source network of Figure 12-28(a),

$$I(s) = \left(\frac{\dfrac{1}{2s}}{\dfrac{1}{2s} + 3 + 4s}\right)\frac{6s}{s^2 + 4} = \left(\frac{1}{8s^2 + 6s + 1}\right)\left(\frac{6s}{s^2 + 4}\right)$$

$$V(s) = 4sI(s) = \left(\underbrace{\frac{4s}{8s^2 + 6s + 1}}_{\substack{\text{Transfer} \\ \text{function}}}\right)\left(\underbrace{\frac{6s}{s^2 + 4}}_{\substack{\text{Source} \\ \text{transform}}}\right)$$

As a function of time, the network signal $v(t)$ is the convolution of the time function having Laplace transform

$$H(s) = \frac{4s}{8s^2 + 6s + 1}$$

with the source function.

Were the source instead a unit impulse $\delta(t)$, with Laplace transform 1, Figure 12-28(b), the network response transform would be $H(s)$:

$$V_{\text{impulse}}(s) = H(s) = \left(\underbrace{\frac{4s}{8s^2 + 6s + 1}}_{\substack{\text{Transfer} \\ \text{function}}}\right)\underbrace{}_{\substack{\text{Source} \\ \text{function}}} \qquad (1)$$

Hence the transfer function involved is the Laplace transform of the network response to a unit impulse source. The response to any source is thus the convolution

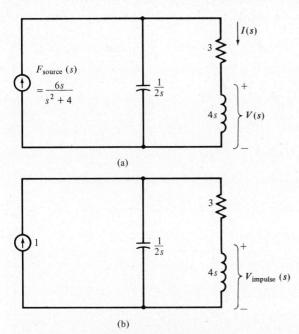

Figure 12-28 Convolving impulse response and the source function.
(a) Original source.
(b) Unit impulse source.

of impulse response with the source function:

$$v(t) = v_{impulse}(t) \otimes f_{source}(t)$$

A convolution relation also may be developed for the Laplace transform of the product of two functions of time in terms of the individual time function transforms. Beginning with

$$E(s) = \mathcal{L}[f_1(t)f_2(t)] = \int_{0^-}^{\infty} f_1(t)f_2(t)e^{-st} \, dt$$

and substituting

$$f_2(t) = \int_{\sigma-j\infty}^{\sigma+j\infty} F_2(s)e^{st} \, ds = \int_{\sigma-j}^{\sigma+j} F_2(x)e^{xt} \, dx$$

there results

$$E(s) = \int_{0^-}^{\infty} f_1(t)\left[\int_{\sigma-j\infty}^{\sigma+j\infty} F_2(x)e^{xt} \, dx\right]e^{-st} \, dt$$

$$= \int_{\sigma-j\infty}^{\sigma+j\infty} F_2(x)\underbrace{\left[\int_{\sigma-j\infty}^{\sigma+j\infty} f_1(t)e^{-(s-x)t} \, dt\right]}_{F_1(s-x)} \, dx$$

$$= \int_{\sigma-j\infty}^{\sigma+j\infty} F_1(s-x)F_2(x) \, dx$$

D12-18

Graphically convolve the pairs of functions given, obtaining a sketch of

$$y(t) = h(t) \otimes f(t)$$

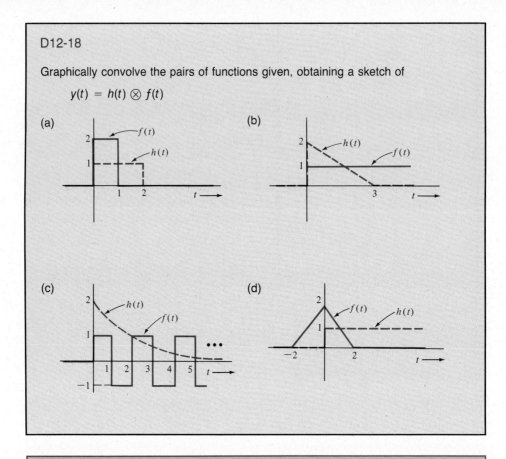

(a)

(b)

(c)

(d)

CONVOLUTION

The inverse Laplace transform of the product of the transforms of two functions is the convolution of the time functions:

$$\mathcal{L}^{-1}[F_1(s)F_2(s)] = \int_{0_-}^{t} f_1(t - \tau)f_2(\tau)\, d\tau = \int_{0_-}^{t} f_2(t - \tau)f_1(\tau)\, d\tau$$

$$= f_1(t) \otimes f_2(t)$$

The response of a network to a single source is the convolution of the network's unit impulse response with the source function. The Laplace transform of the unit impulse response is the transfer function relating the response signal to the source.

The Laplace transform of the product of two functions is a convolution of the individual transforms:

$$\mathcal{L}[f_1(t)f_2(t)] = \int_{\sigma - j\infty}^{\sigma + j\infty} F_1(s - x)F_2(x)\, dx$$

$$= \int_{\sigma - j\infty}^{\sigma + j\infty} F_2(s - x)F_1(x)\, dx$$

CHAPTER TWELVE PROBLEMS

Basic Problems

Laplace Transform Calculation
1. Calculate (setting up and evaluating the integrals involved) the Laplace transforms of the following functions:

(a)

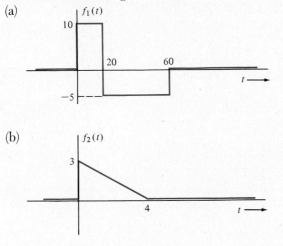

(b)

(c) $f_3(t) = \begin{cases} e^t & 2 \le t < 3 \\ 0 & \text{otherwise} \end{cases}$

(d) $f_4(t) = \begin{cases} 0 & t < 5 \\ 6 & t \ge 5 \end{cases}$

2. Compare $\mathcal{L}[e^{7t}]$ and $\mathcal{L}[u(t)e^{7t}]$

Laplace Transform Properties
3. Find the following Laplace transforms:
   (a) $\mathcal{L}[-5t^3e^{4t}]$
   (b) $\mathcal{L}[7t^4e^{-5t} \sin 6t]$
   (c) $\mathcal{L}[t^2e^{-t}(\cos 2t + 2 \sin 2t)]$
   (d) $\mathcal{L}[5t \cos 3t - 8t \sin 3t]$
   (e) $\mathcal{L}[10e^{-3t} \cos (5t + 50°)]$
   (f) $\mathcal{L}[t^2 \sin (3t - 120°)]$

4. Verify that

$$\mathcal{L}[e^{-3t}] = \mathcal{L}[e^{-t}e^{-2t}] = \frac{1}{s + 3}$$

using the exponential weighting formula.

5. Find the following inverse Laplace transforms:

(a) $\mathcal{L}^{-1}\left[\dfrac{1}{(s - 5)^3}\right]$

Adjustments are made to filter networks used to enhance the quality of the image of an industrial television camera. The camera is focused on a test pattern, then the high frequencies present in the video signal voltage are boosted slightly to increase the sharpness of edges of objects in the picture. (*Photo courtesy of General Electric Co.*)

(b) $\mathcal{L}^{-1}\left[\dfrac{1}{3s^2 + 6s + 15}\right]$

(c) $\mathcal{L}^{-1}\left[\dfrac{s}{s^2 + 2s + 10}\right]$

(d) $\mathcal{L}^{-1}\left[\dfrac{2s + 7}{s^2 + 16}\right]$

(e) $\mathcal{L}^{-1}\left[\dfrac{10}{(s^2 + 4s + 5)^2}\right]$

(f) $\mathcal{L}^{-1}\left[\dfrac{-8s + 13}{s^2 + 10s + 26}\right]$

6. Using time-shifted steps and ramps, construct the Laplace transforms of the following functions:

(a)

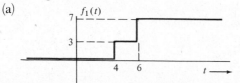

(b)

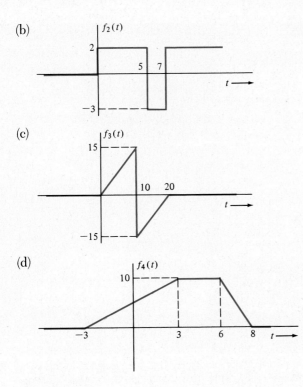

(c)

(d)

Periodic Functions

7. Find the Laplace transforms of the following functions that are periodic after $t = 0$.

(a)

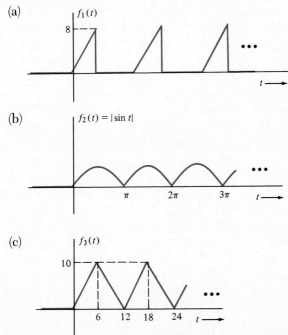

(b) $f_2(t) = |\sin t|$

(c)

Signal voltage waveforms are stored digitally for display and manipulation in this advanced oscilloscope. The keyboard allows the user to position the waves on the screen, to change scales and to perform calculations of signal properties. (*Photo courtesy of Tektronix, Inc.*)

8. A certain function $f(t)$ consists of

$$g(t) = \begin{cases} 3e^t & 0 \le t < 2 \\ 0 & \text{otherwise} \end{cases}$$

repeated periodically at 5-second intervals:

$$f(t) = g(t) + g(t - 5) + g(t - 10) + \cdots$$

Find $F(s)$. Do not calculate $G(s)$; construct it from known transforms.

Derivatives
9. Find and sketch the derivatives of the following functions, indicating impulses on each sketch with arrows:

(a)

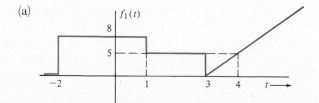

(b)

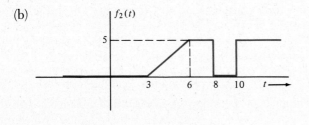

(c) $f_3(t) = \begin{cases} 5 \cos 3t & 0 < t < \pi \\ 0 & \text{otherwise} \end{cases}$

10. Find the Laplace transforms of the derivatives of each of the following functions:
   (a) $f_1(t) = 4 + 3t$
   (b) $f_3(t) = (4 + 3t)u(t)$

(c)

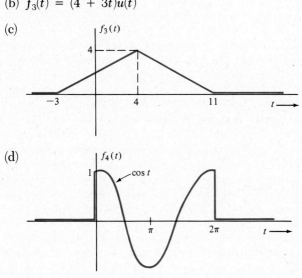

(d)

Integrals
11. Find and sketch the running integrals of the following functions:

(a)

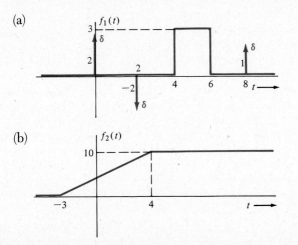

(b)

(c) $f_3(t) = \begin{cases} \cos t & 0 \leq t < \dfrac{3\pi}{2} \\ 0 & \text{otherwise} \end{cases}$

Partial Fraction Expansion

12. The denominator polynomials of the following rational functions have real, distinct roots. Find the time functions for $t \geq 0$ that have these Laplace transforms:

(a) $F(s) = \dfrac{4}{(s + 3)(s + 7)}$

(b) $F(s) = \dfrac{3s + 5}{7(s + 1)(s - 4)}$

(c) $F(s) = \dfrac{3s^2 + 2s + 4}{5(s^2 + 4s + 3)}$

(d) $F(s) = \dfrac{-2(s + \frac{1}{4})}{4(s + \frac{1}{2}s + \frac{1}{3})}$

(e) $F(s) = 3 + \dfrac{4}{6s^2 + 8s + 2}$

(f) $F(s) = \dfrac{s^2 - 9}{(s + 1)(s + 2)(s + 3)}$

(g) $F(s) = \dfrac{3s}{s^2 + 3s + 1}$

(h) $F(s) = \dfrac{s^3 + 2}{6s^2 + 4s - 2}$

13. The denominator polynomials of the following rational functions have real, repeated roots. Find the time functions for $t \geq 0$ that have these Laplace transforms.

(a) $F(s) = \dfrac{3s + 7}{s^2 + 10s + 25}$

(b) $F(s) = \dfrac{s(s^2 - 9)}{(s + 2)^3}$

(c) $F(s) = \dfrac{4s - 1}{s^3 + 6s^2 + 9s}$

(d) $F(s) = \dfrac{-5s^3 + 2s^2 + 1}{(s + 2)(s + 4)^2}$

(e) $F(s) = \dfrac{27}{(s + 3)(s^2 - 9)}$

14. Find the time functions for $t \geq 0$ that have the following Laplace transforms:

(a) $F(s) = \dfrac{4 + e^{-s}}{(s + 2)(s + 3)}$

(b) $F(s) = \dfrac{e^{-2s} - e^{-3s}}{2s^2 + 8s + 6}$

(c) $F(s) = \dfrac{3s^2 - se^{-s}}{s^2 + 5s + 6}$

(d) $F(s) = \dfrac{s^3 - (1/s)e^{-4s}}{(s + 3)^3}$

(e) $F(s) = \dfrac{1 + e^{-s}}{1 - e^{-s}}\left[\dfrac{s}{(s + 1)(s + 2)}\right]$

15. The denominator polynomials of the following rational functions have distinct roots, some of which are complex. Find the time functions for $t \geq 0$ that have these Laplace transforms:

(a) $F(s) = \dfrac{s}{s^2 - 2s + 5}$

(b) $F(s) = \dfrac{s^2 + 3s + 2}{s^2 + 7s + 20}$

(c) $F(s) = \dfrac{-7s}{s(s^2 + 9)}$

(d) $F(s) = \dfrac{s + 1}{s(s^2 + 6s + 13)}$

(e) $F(s) = \dfrac{10s(s + 2)}{(s + 1)(s + 1 + j2)(s + 1 - j2)}$

(f) $F(s) = \dfrac{s^2 - 4}{(s + 3)(s^2 + 2s + 17)}$

(g) $F(s) = 2 + \dfrac{3s - 4}{s^2 + 9}$

16. Invert the following, expressing any sinusoidal portions of the results in the cosine and angle form:

(a) $F(s) = \dfrac{s + 8}{s^2 + 9}$

(b) $F(s) = \dfrac{3s}{s^2 + 6s + 25}$

(c) $F(s) = \dfrac{\frac{1}{2}(s + 3)}{s^2 - 4s + 5}$

(d) $F(s) = \dfrac{4s + 1}{s(s^2 + s + 8)}$

(e) $F(s) = \dfrac{s^2}{(s + 2)(s^2 + 4s + 29)}$

Initial and Final Values

17. Find the initial and final values, if they exist, of the functions with the following Laplace transforms:

(a) $F(s) = \dfrac{6s + 7}{s^2 + 2s + 10}$

(b) $F(s) = \dfrac{4s - 5}{s^3 + 3s^2 + 2s}$

(c) $F(s) = \dfrac{10}{s(s + 3)^2(s^2 + 2s + 8)}$

(d) $F(s) = \dfrac{3s^2 + 2s}{(s + 2)^2(s^2 + s + 1)}$

(e) $F(s) = \dfrac{e^{-3s}}{4s^2 + 3s + 2}$

Differential Equation Solution

18. Use Laplace transform methods to obtain the solutions $y(t)$ to the following integrodifferential equations after time $t = 0$, given the indicated boundary conditions:

(a) $\dfrac{dy}{dt} - 3y = 2e^{-4t}$

$y(0^-) = 5$

(b) $\dfrac{d^2y}{dt^2} + 4\dfrac{dy}{dt} + 5y = 0$

$y(0^-) = 6$

$y'(0^-) = -2$

(c) $6\dfrac{dy}{dt} + 5y + \displaystyle\int_{0^-}^{\infty} y\,dt = \delta(t)$

$y(0^-) = 3$

(d) $8y + 3\displaystyle\int_{-\infty}^{t} y(t)\,dt = 4e^{-2t}$

$\displaystyle\int_{-\infty}^{0^-} y(t)\,dt = 10$

(e) $\begin{cases} \dfrac{dx_1}{dt} = -5x_1 + x_2 \\[2mm] \dfrac{dx_2}{dt} = -4x_2 + e^{3t} \\[2mm] y = x_1 - 2x_2 \\[1mm] x_1(0^-) = 6 \\ x_2(0^-) = 8 \end{cases}$

Switched Networks

19. Solve the following networks for the indicated signals after time $t = 0$. The network initial conditions are zero.

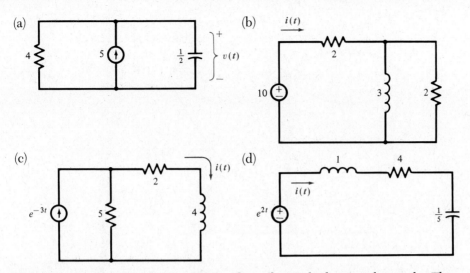

20. Find the Laplace *transforms* of the indicated switched network signals. The network initial conditions are zero.

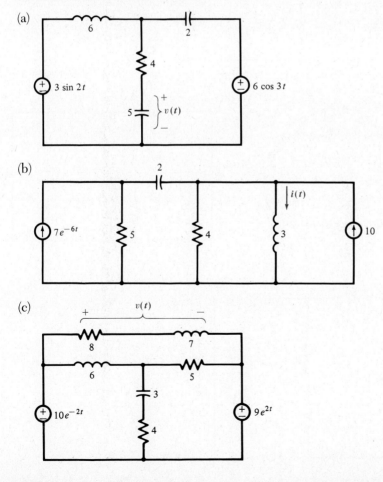

(d)

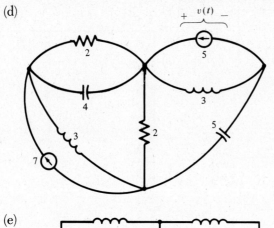

(e)

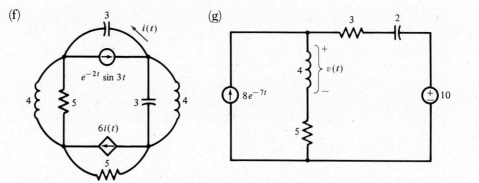

(f)

(g)

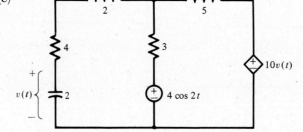

(h)

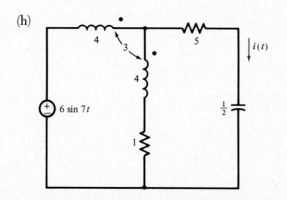

21. Solve the following networks for the indicated signals, using Laplace transform methods. Network initial conditions are given.

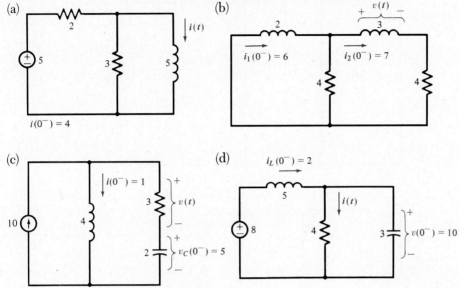

(a)

(b)

(c)

(d)

22. In each of the following switched networks, find all initial inductor currents and all initial capacitor voltages:

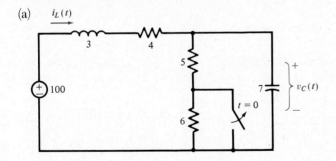

(a)

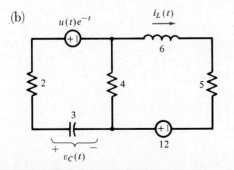

(b)

(c)

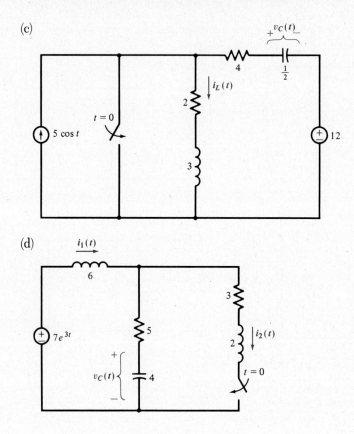

(d)

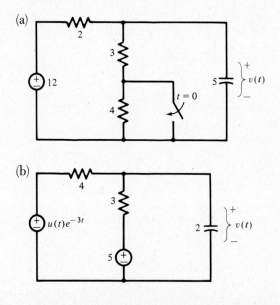

23. Find the indicated network signals after time $t = 0$:

(c)

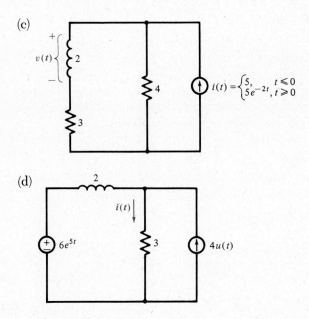

$$i(t) = \begin{cases} 5, & t \leqslant 0 \\ 5e^{-2t}, & t \geqslant 0 \end{cases}$$

(d)

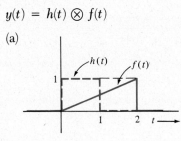

$6e^{5t}$ 3 $4u(t)$

Convolution

24. Graphically convolve the pairs of functions given, obtaining a sketch of

$$y(t) = h(t) \otimes f(t)$$

(a)

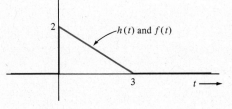

(b)

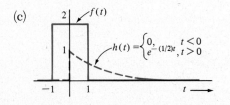

(c)

$$h(t) = \begin{cases} 0, & t < 0 \\ e^{-(1/2)t}, & t > 0 \end{cases}$$

Practical Problems

Step Approximation of Functions

A general function may be easily approximated after time $t = 0$ by a succession of evenly spaced steps, as shown. This is an especially useful technique for calculating approximate Laplace transforms for complicated functions and for fitting a piecewise constant function to numerical data. For the example function,

$$h_{approx}(t) = 1.5u(t) + (2.9 - 1.5)u(t - 0.1) + (3.2 - 2.9)u(t - 0.2) +$$

$$(2.5 - 3.2)u(t - 0.3) + (2.0 - 2.5)u(t - 0.4) + \cdots$$

$$= 1.5u(t) + 1.4u(t - 0.1) + 0.3u(t - 0.2) - 0.7u(t - 0.3)$$

$$- 0.5u(t - 0.4) + \cdots$$

$$H_{approx}(s) = \frac{1.5 + 1.4e^{-0.1s} + 0.3e^{-0.2s} - 0.7e^{-0.3s} - 0.5e^{-0.4s} + \cdots}{s}$$

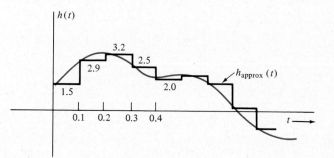

Notice that each successive step is in the amount of the change in the function from the last step. In general, a function $f(t)$ may be approximated by steps evenly spaced at intervals Δt in time according to

$$f_{approx}(t) = f(0)u(t) + [f(\Delta t) - f(0)]u(t - \Delta t)$$

$$+ [f(2\Delta t) - f(\Delta t)]u(t - 2\Delta t) + \cdots$$

25. Find a delayed step approximation of the Laplace transform of the following function. Use steps at 0.1-second intervals.

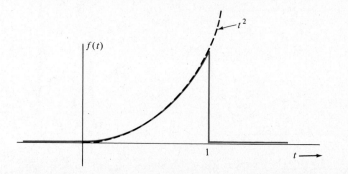

26. Using a delayed step approximation, find an approximate Laplace transform of the function

$$f(t) = \begin{cases} 10 \sin t & 0 \le t \le \pi \\ 0 & \text{otherwise} \end{cases}$$

Choose an appropriate step size, Δt.

27. Samples of a function $f(t)$ within the interval $[0, 3]$ are given in the accompanying table. If $f(t) = 0$ for $t > 3$, find

(a) $F(s)$ approximately, using a delayed step approximation

(b) $F(s)$ approximately, using a sum of delayed ramps

SAMPLES OF $f(t)$

t	$f(t)$
0	5.5
0.5	4.1
0.7	2.2
0.8	0.0
1.2	-1.0
1.5	0.0
2.0	2.2
2.5	4.4
2.8	-1.5

Piecewise Construction from Impulses

Integration of impulses and derivatives of impulses may be used to advantage in the calculation of Laplace transforms of functions that are, piecewise, powers of time. The function of (a) is piecewise linear. Its derivative is depicted in (b). The discontinuities at $t = 0$ and $t = 3$ contribute impulses to the derivative with strengths equal to the amount of the discontinuity, as shown. The derivative and the function are related by

$$f(t) = \int_{0^-}^{\infty} \frac{df}{dt} \, dt \qquad t \ge 0$$

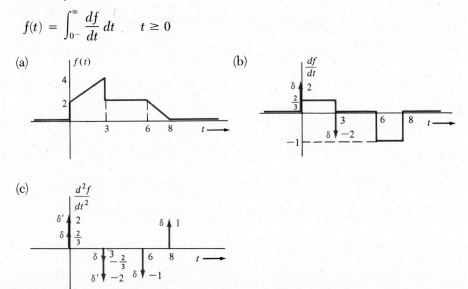

The second derivative of the function $f(t)$ in (c) involves only impulses and derivatives of impulses. Its Laplace transform is especially simple:

$$\mathscr{L}\left[\frac{d^2f}{dt^2}\right] = \mathscr{L}[2\delta'(t) + \tfrac{1}{3}\delta(t) - 2\delta'(t - 3) - \tfrac{1}{3}\delta(t - 3)$$

$$- \delta(t - 6) + \delta(t - 8)]$$

$$= 2s + \tfrac{2}{3} - 2se^{-3s} - \tfrac{2}{3}e^{-3s} - e^{-6s} + e^{-8s}$$

Then since

$$f(t) = \int_{0-}^{t}\left(\int_{0-}^{t}\frac{d^2f}{dt^2}\,dt\right)dt \qquad t \geq 0$$

$$F(s) = \frac{2s + \tfrac{2}{3} - 2se^{-3s} - \tfrac{2}{3}e^{-3s} - e^{-6s} + e^{-8s}}{s^2}$$

28. Express the functions as integrals of impulses and impulse derivatives to obtain their Laplace transforms:

(a)   (b)

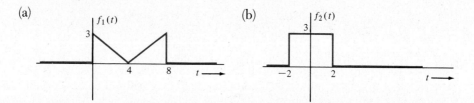

(c)

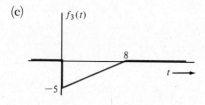

Advanced Problems

Derivatives

29. Find the Laplace transform of

$$f(t) = \frac{dg}{dt} = \frac{d}{dt}(10 - 3e^{2t})$$

by first performing the differentiation, then Laplace transforming. Then use the derivative relation instead to obtain the transform of the derivative from the transform of $g(t)$. Both methods should, of course, give the same result.

30. Using the sifting properties of $\delta^{[n]}(t)$, (the nth derivative of the impulse), show that

$$\int_{-\epsilon}^{\epsilon} \delta^{[n]}(t) \, dt = 0 \qquad n = 1, 2, 3, \ldots$$

Integrals
31. Find the Laplace transform of

$$f(t) = \int_{-\infty}^{t} e^{6t} \, dt$$

by first performing the integration, then Laplace transforming. Then use the integral relation instead to obtain the transform from the integrand. Both methods should, of course, give the same result.

32. Develop a relation for

$$\mathscr{L}\left[\int_{-\infty}^{t} \int_{-\infty}^{t} f(t) \, d^2t\right]$$

in terms of various constants and $F(s)$, analogous to that given for

$$\mathscr{L}\left[\int_{-\infty}^{t} f(t) \, dt\right]$$

Periodic Functions
33. Show that the basic square wave has Laplace transform

$$F(s) = \frac{1}{s} \tanh\left(\frac{\pi s}{2}\right)$$

where the hyperbolic tangent is

$$\tanh x = \frac{e^x - e^{-x}}{e^x + e^{-x}}$$

Differential Equation Solution
34. Use Laplace transform methods to obtain the *general* solution of

$$\frac{d^2y}{dt^2} + 3\frac{dy}{dt} = 4e^{2t}$$

Chapter Thirteen
Active Network Design

Engineering design is the process of devising a system, component, or process to meet desired needs. It is a decision-making process (often iterative) in which the basic sciences, mathematics, and engineering sciences are applied to convert resources optimally to meet a stated objective. Among the fundamental elements of the design process are the establishment of objectives and criteria, synthesis, analysis, construction, testing, and evaluation. Central to the process are the essential and complementary roles of synthesis and analysis.

From *Criteria for Accrediting Programs
in Engineering in the United States,*
50th Annual Report of the Accreditation Board
for Engineering and Technology

13.1 PREVIEW

This final chapter brings together the basic tools of network analysis and applies them to the design of active filters using operational amplifiers. Active filter design is both of great importance in practice and a fine tool for teaching design.

The concepts and interrelationships of transfer functions, pole-zero plots, and frequency response are reviewed and connected with the descriptions of frequency-selective filters. Low-pass, high-pass, bandpass, bandstop, and all-pass filters are covered. The realization of these filters via cascade, tandem, or mixed compositions of first and second order sections is emphasized.

The beginning student is at the point where the advantages of Bode plots can be appreciated while learning their mechanics. A careful explanation of Bode plot fundamentals is given. Then the classical Butterworth, Chebyshev, and Bessel filters are developed, and transformations in frequency and from low-pass to the other filter types are shown.

The summing amplifier design with operational amplifiers introduced in Chapter 3 is now reviewed. Simple methods for the design of adjustable-gain amplifiers, using potentiometer control, are developed. Single-amplifier filters involving buffered RC networks and RC networks with feedforward are used, and the second order Sallen and Key circuits are presented.

Filters of the state variable type are constructed by interconnecting summing integrators and summing amplifiers. When the amplifiers involved have adjustable gains, adjustable filters result. Special emphasis is given to the biquadratic structure and to the several important ways to arrange second order filter adjustments.

When you complete this chapter, you should know—

1. how the frequency response of networks or systems is related to their transfer functions;
2. how pole-zero plots and frequency response are related;
3. about important classes of filters, including low-pass, high-pass, bandpass, bandstop, and all-pass filters;
4. how to depict frequency response in the form of Bode plots;
5. about classical filters of the Butterworth, Chebyshev, and Bessel types;
6. how to use potentiometers in operational amplifier circuits to obtain amplifiers with adjustable gains;
7. how to design first, second, and higher order filters using operational amplifiers;
8. how to design adjustable active filters.

13.2 NETWORK AND SYSTEM DESCRIPTIONS

13.2.1 Transfer Functions

The signals in most electrical networks and other, similar systems are related by linear, time-invariant differential equations. When a network has a single applied signal (the input) and a single signal of interest (the output), variables may be eliminated between the equations to obtain a single differential equation relating the input and the output. Such input-output equations are of the form

$$\begin{pmatrix} \text{linear combination} \\ \text{of derivatives of} \\ \text{the output signal} \end{pmatrix} = \begin{pmatrix} \text{linear combination} \\ \text{of derivatives of} \\ \text{the input signal} \end{pmatrix}$$

An alternative, algebraic description of the relation between input and output in a network or system is the *transfer function*. There are three equivalent definitions of the transfer function:

1. In terms of exponential signals
2. In terms of Laplace transforms
3. In terms of impulse response

Each of these will now be considered in turn.

Except in special cases, if the input signal is exponential, varying as e^{st}, then the output will have a forced component that also varies as e^{st}. The ratio of forced exponential output to exponential input will then be constant, depending only on the specific value of s involved. The transfer function relating a network or system output and input is the ratio of forced exponential output to exponential input, as a function of the exponential constant s. Transfer functions are a generalization of the impedance concept; here, the ratio need not necessarily be of sink reference voltage to current in a network element. Figure 13-1(a) illustrates the idea.

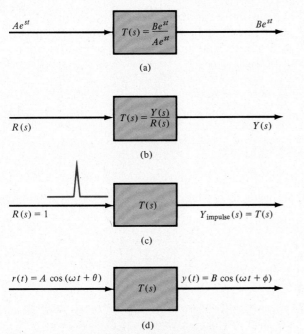

(a)

(b)

(c)

$r(t) = A \cos(\omega t + \theta)$ $y(t) = B \cos(\omega t + \phi)$

$T(s)$

(d)

Figure 13-1   Transfer function properties.
(a) Exponential signal ratio.
(b) Laplace transform ratio.
(c) Laplace transform of the impulse response.
(d) Forced sinusoidal response from the transfer function.

Consider, for example, a network or system with the input-output equation

$$\frac{d^2y}{dt^2} + 3\frac{dy}{dt} + 2y = -4\frac{dr}{dt} + 5r$$

If the input and output signals are each exponential, varying as e^{st},

$$r(t) = Ae^{st} \qquad y(t) = Be^{st}$$

they are related by

$$\left(\frac{d^2}{dt^2} + 3\frac{d}{dt} + 2\right)Be^{st} = \left(-4\frac{d}{dt} + 5\right)Ae^{st}$$

$$(s^2 + 3s + 2)Be^{st} = (-4s + 5)Ae^{st}$$

and their ratio is

$$T(s) = \frac{y(t)}{r(t)}\bigg|_{\substack{\text{when each} \\ \text{varies as } e^{st}}} = \frac{Be^{st}}{Ae^{st}} = \frac{-4s + 5}{s^2 + 3s + 2}$$

The transfer function is the ratio of exponential output to exponential input; when the ratio is formed, the terms associated with the output derivatives appear in the denominator of $T(s)$, while the input derivative terms are in the numerator.

A transfer function is also the ratio of the Laplace transform of the output signal to the Laplace transform of the input signal, when all initial conditions are zero, as indicated in Figure 13-1(b). Laplace transforming the same input-output equation,

$$\frac{d^2y}{dt^2} + 3\frac{dy}{dt} + 2y = -4\frac{dr}{dt} + 5r$$

with zero initial conditions, gives the same result as with the exponential signal definition:

$$(s^2 + 3s + 2)Y(s) = (-4s + 5)R(s)$$

$$T(s) = \frac{Y(s)}{R(s)} = \frac{-4s + 5}{s^2 + 3s + 2}$$

If the input signal is a unit impulse, Figure 13-1(c),

$$r(t) = \delta(t) \qquad R(s) = 1$$

the output signal, the response to an impulse input, is given by

$$Y_{impulse}(s) = T(s) \cdot 1$$

$$y_{impulse}(t) = \mathcal{L}^{-1}[T(s)]$$

The transfer function is thus also the Laplace transform of the network or system impulse response.

13.2.2 Frequency Response

When the input signal is sinusoidal, the forced component of the output is also sinusoidal, with the same frequency as the input. Although the sinusoidal output can be found by using Laplace transform methods, the transform is more complicated than is necessary because it also includes the natural component of the output. Instead, the exponential ratio property of transfer functions may be used to advantage.

The sinor corresponding to the input

$$r(t) = A \cos(\omega t + \theta)$$

is

$$\underline{r}(t) = Ae^{j(\omega t + \theta)} = A \cos(\omega t + \theta) + jA \sin(\omega t + \theta)$$

The sinor is exponential, and the real part of the input sinor is the input signal of interest. The output sinor is given by

$$\underline{y}(t) = T(s = j\omega)\underline{r}(t) = T(s = j\omega)Ae^{j(\omega t + \theta)}$$

$$= Be^{j(\omega t + \phi)} = B \cos(\omega t + \phi) + jB \sin(\omega t + \phi)$$

and the real part of the output sinor is the output signal of interest:

$$y(t) = \text{Re}[\underline{y}(t)] = B \cos(\omega t + \phi)$$

The amplitude of the forced sinusoidal output is the input amplitude multiplied by the magnitude of the transfer function for $s = j\omega$:

$$B = |T(s = j\omega)| \cdot A$$

The phase angle of the forced sinusoidal output is the angle of the input plus the angle of the transfer function for $s = j\omega$:

$$\phi = \theta + \underline{/T(s = j\omega)}$$

The magnitude of a transfer function for $s = j\omega$ is thus the ratio of sinusoidal output amplitude to the input sinusoidal amplitude:

$$\frac{B}{A} = |T(s = j\omega)|$$

The angle of the transfer function is the *phase shift* between the output and input sinusoids:

$$\phi - \theta = \underline{/T(s = j\omega)}$$

As a numerical example, suppose a network or system with transfer function

$$T(s) = \frac{10}{s + 4}$$

has input

$$r(t) = 3 \cos (4t + 20°)$$

When evaluated at $s = j4$, the transfer function is

$$T(s = j4) = \frac{10}{j4 + 4} = \frac{10}{4\sqrt{2}e^{j45°}}$$

It has magnitude

$$|T(s = j4)| = \frac{5}{2\sqrt{2}}$$

and angle

$$\underline{/T(s = j4)} = -45°$$

The forced sinusoidal output is then

$$y(t) = 3\left(\frac{5}{2\sqrt{2}}\right) \cos (4t + 20° - 45°) = \frac{15}{2\sqrt{2}} \cos (4t - 25°)$$

Knowing the magnitude and angle of a transfer function, as functions of frequency ω, allows calculation of forced sinusoidal response to any sinusoidal input. The *amplitude ratio* and *phase shift* functions,

$$M(\omega) = |T(s = j\omega)| \qquad \Phi(\omega) = \underline{/T(s = j\omega)}$$

specify how any sinusoidal signal is changed in amplitude and phase in passing through a network or system. $M(\omega)$ and $\Phi(\omega)$ are collectively termed the system's *frequency response*.

For a system with transfer function

$$T(s) = \frac{4s}{s + 3}$$

the frequency response is given by

$$T(s = j\omega) = \frac{4j}{j\omega + 3}$$

The amplitude ratio is

$$M(\omega) = |T(s = j\omega)| = \frac{4}{\sqrt{9 + \omega^2}}$$

and the phase shift is

$$\Phi(\omega) = \underline{/T(s = j\omega)} = 90° - \arctan(\omega/3)$$

These are plotted in Figure 13-2. From these frequency response plots, it is simple to find the sinusoidal output due to any sinusoidal input. For example, using the plots, an input

$$r(t) = 10 \cos(1.6t + 50°)$$

produces an output 1.9 times as large in amplitude and with 62° greater phase shift:

$$y(t) = (1.9)(10) \cos(1.6t + 50° + 62°)$$

$$= 19 \cos(1.6t + 112°)$$

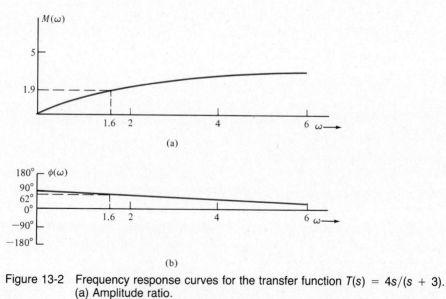

Figure 13-2　Frequency response curves for the transfer function $T(s) = 4s/(s + 3)$.
(a) Amplitude ratio.
(b) Phase shift.

D13-1

Find the frequency response amplitude ratio and phase shift functions for each of the following transfer functions:

(a) $T(s) = \dfrac{-3}{s + 7}$

   ans. $3/\sqrt{\omega^2 + 49}$ $180° - \arctan(\omega/7)$

(b) $T(s) = \dfrac{3s + 2}{s}$

   ans. $\sqrt{9\omega^2 + 4}/\omega$ $\arctan(3\omega/2) - 90°$

(c) $T(s) = \dfrac{1}{s^2 + 2s + 1}$

   ans. $1/\sqrt{4\omega^2 + (1 - \omega^2)^2}$ $-\arctan[2\omega/(1 - \omega^2)]$

(d) $T(s) = \dfrac{10s}{s^2 + 2s + 10}$

   ans. $10\omega/\sqrt{4\omega^2 + (10 - \omega^2)^2}$ $90° - \arctan[2\omega/(10 - \omega^2)]$

13.2.3 Pole-Zero Plots

The transfer functions of electrical networks and similar systems are rational functions of the complex frequency variable s; that is, they can be expressed as the ratio of two polynomials in s:

$$T(s) = \frac{b_m s^m + b_{m-1} s^{m-1} + \cdots + b_0}{a_n s^n + a_{n-1} s^{n-1} + \cdots + a_0}$$

$$= \left(\frac{b_m}{a_n}\right) \frac{(s - s_a)(s - s_b) \cdots (s - s_m)}{(s - s_1)(s - s_2) \cdots (s - s_n)}$$

As introduced in Section 6.8, a pole-zero plot of a rational function is constructed by indicating the locations of the zeros of $T(s)$ with 0s and the poles of $T(s)$ with Xs on a drawing of the complex plane. It is useful, too, to include the multiplying constant (b_m/a_n) with the plot so that the plot completely specifies $T(s)$. As an example, the rational function

$$T(s) = \frac{s - 1}{s^2 + 4s + 13} = \frac{(1)(s - 1)}{(s + 2 - 3j)(s + 2 + 3j)}$$

has the pole-zero plot of Figure 13-3(a).

Recall that a rational function may be graphically evaluated at a specific value of the variable $s = p$ as follows:

$$|T(s = p)| = \left(\frac{b_m}{a_n}\right) \frac{\text{product of distances of zeros to } p}{\text{product of distances of poles to } p}$$

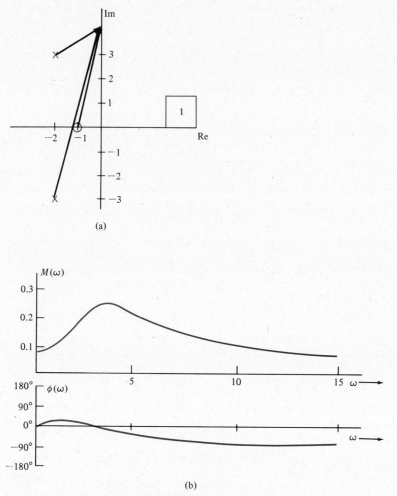

(a)

(b)

Figure 13-3 Frequency response from a pole-zero plot.
(a) Graphical evaluation for points along the imaginary axis.
(b) Amplitude ratio and phase shift curves for the example function.
(c) The rubber sheet analogy.

and

$$\underline{/T(s = p)} = (\text{sum of zero angles to } p) - (\text{sum of pole angles to } p)$$

$$+ (180° \text{ if the multiplying constant is negative})$$

Frequency response involves evaluating a transfer function at various values of $s = j\omega$; that is, at points along the positive imaginary axis on the complex plane. By imagining graphical evaluation of $T(s)$ along the imaginary axis, it is relatively easy to sketch

$$M(\omega) = |T(s = j\omega)| = \left(\frac{b_m}{a_n}\right) \frac{\text{product of distances of zeros to } s = j\omega}{\text{product of distances of poles to } s = j\omega}$$

Figure 13-3   (*Continued*)

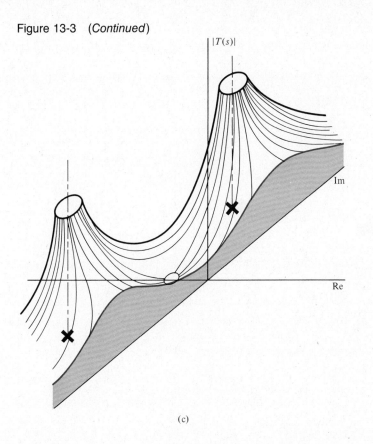

(c)

and

$$\Phi(\omega) = \underline{/T(s = j\omega)} = \text{(sum of zero angles to } s = j\omega)$$

$$- \text{(sum of pole angles to } s = j\omega)$$

$$+ (180° \text{ if the multiplying constant is negative)}$$

as has been done for the example in Figure 13-3(b).

Another way of visualizing the amplitude ratio $M(\omega)$ from a pole-zero plot is to imagine a very flexible rubber sheet stretched over the complex plane. The sheet is poked up by high, thin rods at each of the pole locations and is tacked down to the plane at each zero, as in Figure 13-3(c). The height of the sheet as a function of location s on the complex plane represents the magnitude of $T(s)$. The height of the sheet along the imaginary axis is the shape of

$$M(\omega) = |T(s = j\omega)|$$

13.2.4 Stability

Systems are said to be *stable* if their impulse response decays to zero with time. If the impulse response of a system "blows up," it means that the system's natural response is composed of one or more terms that expand with time. As the transfer function is the Laplace transform of the system's unit impulse response, the impulse response as a function of time is

$$y_{impulse}(t) = \mathcal{L}^{-1}[T(s)]$$

Expanding $T(s)$ in partial fractions,

$$T(s) = \frac{K_1}{s - s_1} + \frac{K_2}{s - s_2} + \cdots + \frac{K_n}{s - s_n}$$

and

$$y_{impulse}(t) = K_1 e^{s_1 t} + K_2 e^{s_2 t} + \cdots + K_n e^{s_n t} \qquad t \geq 0$$

For the impulse response to approach zero with time, each of these terms must decay to zero, which is the case only if $s_1, s_2, \ldots, s_n$ all are negative or (in the case of conjugate poles) have negative real parts. The system represented by the transfer function $T(s)$ is thus stable if and only if all of the poles of $T(s)$ are in the left half of the complex plane. That is, for stability, all poles of $T(s)$ must be to the left of the imaginary axis.

D13-2

Sketch approximate frequency response plots (both amplitude ratio and phase shift versus frequency) for transfer functions with the following pole-zero plots:

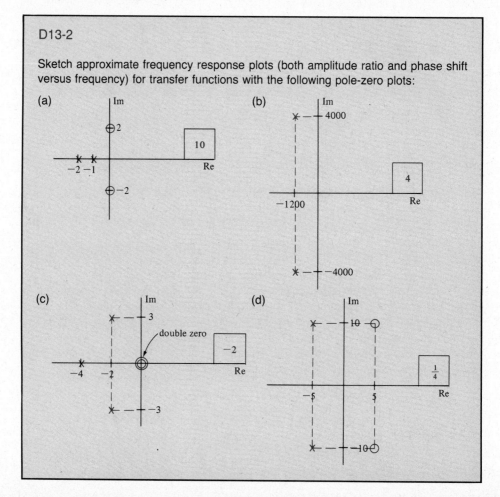

NETWORK AND SYSTEM DESCRIPTIONS

Networks and systems are described by their transfer functions, which are—

1. the ratio of output to input, as a function of the exponential constant s, when each varies as e^{st};
2. the ratio of the Laplace transform of the output to the Laplace transform of the input, when the initial conditions are zero;
3. the Laplace transform of the output when the input is a unit impulse.

A pole-zero plot of a rational transfer function is a drawing, on the complex plane, showing the locations of the roots of the numerator polynomial (the function's zeros) and the roots of the denominator polynomial (the poles). It is convenient also to incorporate the multiplying constant of the function in the drawing.

The frequency response of a network or system with transfer function $T(s)$ consists of the amplitude ratio

$$M(\omega) = |T(s = j\omega)|$$

and the phase shift

$$\Phi(\omega) = \underline{/T(s = j\omega)}$$

as functions of frequency. The forced sinusoidal output due to a sinusoidal input of radian frequency ω has amplitude modified by $M(\omega)$ and added phase shift $\Phi(\omega)$.

The frequency response of a transfer function can be constructed graphically from the transfer function's pole-zero plot. The amplitude ratio is also easily visualized, using the rubber sheet analogy.

A system is stable if and only if its impulse response decays to zero with time. The transfer functions of stable systems have all poles in the left half of the complex plane.

13.3 FILTERS

13.3.1 Filter Classification

A filter is an electrical network or other system that is frequency-selective. Input signal frequencies are changed in amplitude and/or in phase by a filter, in some desirable way, to form the output. Usually, only stable filters are of interest.

Filters are classified broadly according to whether their purpose is primarily to affect the amplitudes of the incoming signal frequencies, their phases, or both. In audio applications, since the human ear is insensitive to phase of a sinusoidal signal, it is the amplitude ratio of a filter that is of primary importance. In digital signal transmission, although the amplitude ratio is of some concern, phase shift is crucial if incoming pulses are to emerge still looking like pulses. One can easily understand this situation by imagining a squarewave signal. Moderate changes in harmonic amplitudes cause only moderate distortion in the shape of the wave. Changes in the relative phases of squarewave harmonics, however, have major effects on the resulting waveshape. In other applications, such as those involving video signals, both the amplitude ratio and the phase shift of a filter must be carefully controlled.

Filters for which the amplitude ratio $M(\omega)$ is of primary importance and the phase shift is of little concern are first considered here. These are classified according to the shape of $A(\omega)$: low-pass, high-pass, bandpass, and bandstop. Then all-pass filters, which have constant amplitude ratio and otherwise affect only phase, are discussed. When needed, these two filter classes in combination give filters with prescribed amplitude ratio and phase shift properties.

A low-pass filter approximates the amplitude ratio indicated in Figure 13-4(a). Ideally, frequencies up to some *cutoff* frequency ω_C are passed with equal amplitude ratio. Frequencies above ω_C are not passed by the ideal filter. Frequencies between 0 and ω_C are said to be in the *passband* of the filter, while frequencies above ω_C are in the filter's *stopband*. In practical low-pass filters, low-order examples of which are shown in Figure 13-4(b), (c), and (d), the boundary between passband and stopband is not so distinct. Then, the cutoff frequency, alternatively called the *rolloff frequency*, is defined conveniently.

In general, a low-pass filter has a frequency response amplitude ratio that is nonzero for $\omega = 0$ and approaches zero for large ω. The transfer function for $\omega = 0$, termed the *dc gain* of the filter, is the ratio of the constant terms in the transfer function's numerator and denominator polynomials. The amplitude ratio at $\omega = 0$ is the magnitude of the dc gain. For example, the low-pass filter

$$T_3(s) = \frac{s - 3}{s^3 + 2s^2 + 2s + 1}$$

has

$$\text{dc gain} = T_3(s = j0) = \frac{-3}{1} = -3$$

$$M(0) = |\text{dc gain}| = |T_3(s = j0)| = 3$$

At high frequencies, the amplitude ratio of a transfer function is dominated by the highest powers of s in the numerator and denominator polynomials. For the amplitude ratio to approach zero at high frequencies, there must be a higher power of s in the denominator than in the numerator. For $T_3(s)$,

$$\lim_{\omega \to \infty} M(\omega) = \lim_{\omega \to \infty} |T_3(s = j\omega)| \approx \lim_{\omega \to \infty} \left| \frac{j\omega}{(j\omega)^3} \right| = 0$$

as required for a low-pass filter.

A high-pass filter has an amplitude ratio that approximates the ideal curve of Figure 13-5(a), where frequencies below ω_C are not passed and those above ω_C are passed with equal amplitude ratio. The passband is at high frequencies, and the stopband is at low frequencies. For practical high-pass filters such as those in Figure 13-5(b), (c), and (d), it is more appropriate to call the boundary between stopband and passband the rolloff frequency, as with low-pass filters.

In general, a high-pass filter has a zero amplitude ratio at $\omega = 0$ and a ratio that approaches a nonzero constant at high frequencies. The transfer function of a high-pass filter must then have one or more factors of s in its numerator (but not its denominator) polynomial so that the amplitude ratio is zero for $s = j0$. It must have equal orders of numerator and denominator polynomials so that at high frequencies,

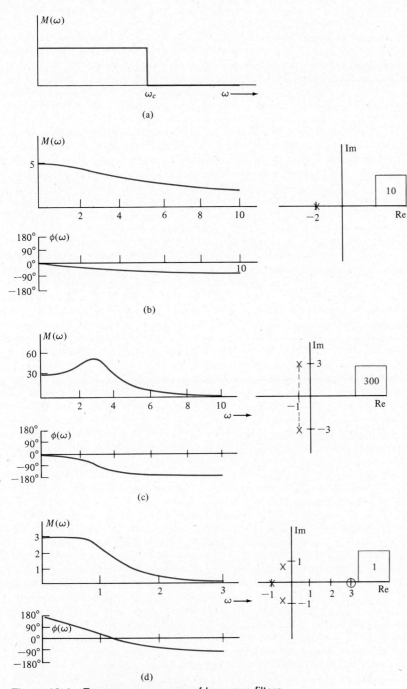

Figure 13-4 Frequency response of low-pass filters.
    (a) Ideal amplitude ratio.
    (b) $T_1(s) = 10/(s + 2)$.
    (c) $T_2(s) = 300/(s^2 + 2s + 10)$.
    (d) $T_3(s) = (s - 3)/(s^3 + 2s^2 + 1)$.

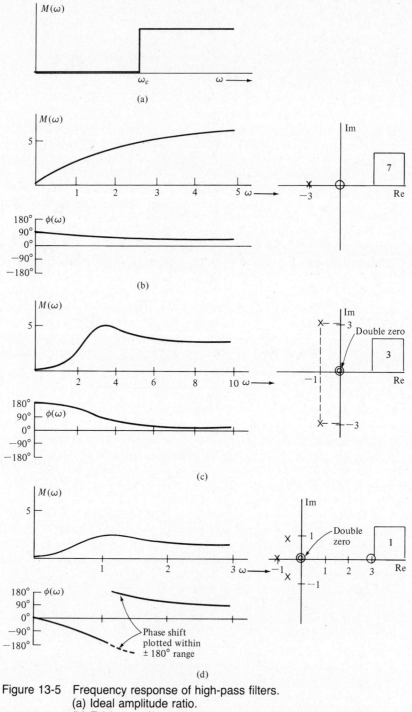

Figure 13-5 Frequency response of high-pass filters.
(a) Ideal amplitude ratio.
(b) $T_4(s) = 7s/(s + 3)$.
(c) $T_4(s) = 3s^2/(s^2 + 2s + 10)$.
(d) $T_6(s) = (s^3 - 3s^2)/(s^3 + 2s^2 + 2s + 1)$.

where the highest power of s terms dominate, the transfer function approaches a nonzero constant. The limiting value of a transfer function for large ω is termed its *high frequency gain*. For the example high-pass transfer function

$$T_6(s) = \frac{s^2(s-3)}{s^3 + 2s^2 + 2s + 1}$$

the dc gain is

$$T_6(s = j0) = 0$$

At high frequencies,

$$\text{high frequency gain} = \lim_{\omega \to \infty} T(s = j\omega) \approx \lim_{\omega \to \infty} \frac{(j\omega)^3}{(j\omega)^3} = 1$$

$$\lim_{\omega \to \infty} M(\omega) = |\text{high frequency gain}| = 1$$

The phase shift curve for this filter has been plotted within a strict $\pm 180°$ range in Figure 13-5(d).

Bandpass filters are used to select a frequency interval, as is indicated by the ideal amplitude ratio of Figure 13-6(a). A bandpass filter has nearly uniform amplitude ratio in a range of frequency about the *center frequency* ω_C of the filter. The width of this passband is termed the *bandwidth* of the filter. At low and high frequencies, in the filter's stopbands, the amplitude ratio approaches zero.

The frequency responses of some practical example bandpass filters of low order are given in Figure 13-6(b), (c), and (d). In general, the transfer function of a bandpass filter must have one or more factors of s in the numerator polynomial so that its amplitude ratio goes to zero for $\omega = 0$. It must have a higher order denominator polynomial than numerator polynomial so that the amplitude ratio approaches zero at high frequencies. The filter

$$T_7(s) = \frac{s}{s^2 + 2s + 101}$$

for example, has

$$M(0) = |T_7(s = j0)| = 0$$

and

$$\lim_{\omega \to \infty} M(\omega) = \lim_{\omega \to \infty} |T_7(s = j\omega)| \approx \lim_{\omega \to \infty} \left| \frac{j\omega}{(j\omega)^2} \right| = 0$$

A bandpass filter must necessarily be of second or higher order.

Bandstop (or *notch*) filters have the same amplitude ratio at dc and at high frequency, with the amplitude ratio going to zero or near zero at an intermediate frequency. The filter of Figure 13-7(a),

$$T_9(s) = \frac{6(s^2 + 10)}{s^2 + 2s + 10}$$

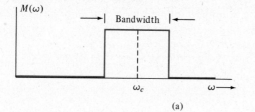

(a)

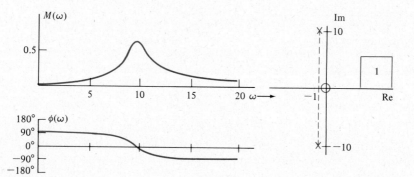

(b)

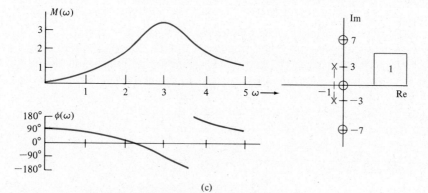

(c)

Figure 13-6   Frequency response of bandpass filters.
(a) Ideal amplitude ratio.
(b) $T_7(s) = s/(s^2 + 2s + 101)$.
(c) $T_8(s) = s(s^2 + 49)/(s^2 + 2s + 10)^2$.

has

$$M(0) = |T_9(s = j0)| = 6$$

$$\lim_{\omega \to \infty} M(\omega) = \lim_{\omega \to \infty} |T_9(s = j\omega)| \approx \left| \frac{6(j\omega)^2}{(j\omega)^2} \right| = 6$$

The amplitude ratio is zero at $\omega = \sqrt{10}$,

$$M(\sqrt{10}) = |T_9(s = j\sqrt{10})| = \left| \frac{6(-10 + 10)}{-10 + 2j\sqrt{10} + 10} \right| = 0$$

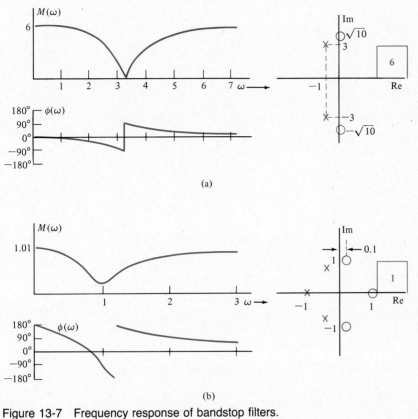

Figure 13-7 Frequency response of bandstop filters.
(a) $T_9(s) = (6s^2 + 60)/(s^2 + 2s + 10)$.
(b) $T_{10}(s) = (s^2 - 0.2s + 1.01)(s - 1)/(s^3 + 2s^2 + 2s + 1)$.

and so this is a bandstop filter with critical frequency or *notch frequency*

$$\omega = \sqrt{10} \text{ rad/s}$$

In the bandstop filter of Figure 13-7(b), the amplitude ratio of the notch does not reach zero, possibly because of slight inaccuracies in the values of the elements with which the filter is constructed.

The purpose of an all-pass filter is to modify the relative phases of the sinusoidal components of a signal without affecting their relative amplitudes. For such a filter, the frequency response amplitude ratio is constant, while the phase shift varies with frequency. A constant amplitude ratio is obtained by having mirror-image transfer function poles and zeros, as in the examples of Figure 13-8. For the filter to be stable, all of its poles must be in the left half of the complex plane. If, for each left half-plane pole there is a corresponding right half-plane zero in the transfer function, the frequency response amplitude ratio

$$M(\omega) = \frac{|\text{multiplying constant}|(\text{product of pole distances to the point } s = j\omega)}{\text{product of zero distances to the point } s = j\omega}$$

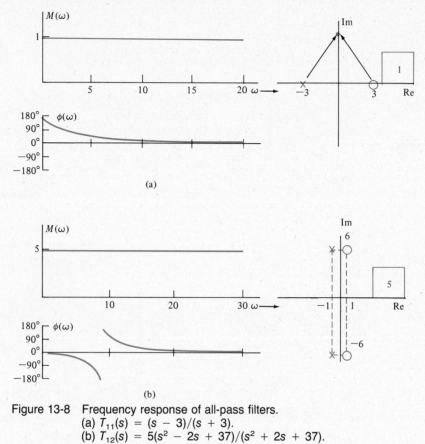

Figure 13-8 Frequency response of all-pass filters.
(a) $T_{11}(s) = (s - 3)/(s + 3)$.
(b) $T_{12}(s) = 5(s^2 - 2s + 37)/(s^2 + 2s + 37)$.

will be constant. The phase shift will vary with ω:

$$\Phi(\omega) = \text{(sum of zero angles to the point } s = j\omega)$$

$$- \text{(sum of pole angles to the point } s = j\omega)$$

$$\pm \text{(180° if the multiplying constant is negative)}$$

All-pass filters are used to improve the phase shift properties of a system. The phase shift

$$\Phi(\omega) = 0°$$

cannot be obtained in a stable, nontrivial system. The next best phase shift for many applications is a phase shift proportional to frequency,

$$\Phi(\omega) = -\tau\omega$$

where τ is a constant. This is the phase shift of a constant time delay of the signal, and thus all its sinusoidal signal components passing through the system. A sinusoidal signal of radian frequency ω,

$$f(t) = A \cos(\omega t + \theta)$$

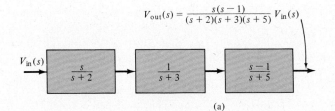

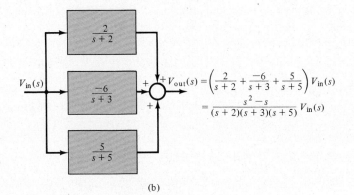

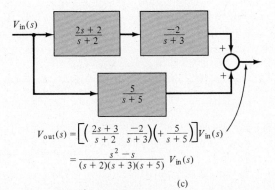

Figure 13-9   Filters composed of first- and second-order sections.
(a) Cascade connection.
(b) Tandem connection.
(c) A mixed composition.

when delayed τ seconds is

$$f(t - \tau) = A \cos (\omega t + \theta - \omega \tau)$$

and is shifted in phase by $\omega\tau$ radians, an amount proportional to the frequency, ω.
The delay $\tau(\omega)$ of a filter is the negative slope of its phase shift curve,

$$\tau(\omega) = -\frac{d\Phi(\omega)}{d\omega}$$

If $\Phi(\omega)$ is in radians and ω is in radians per second, the delay τ is in seconds. The ideal situation is that of a constant time delay for all frequencies. In practice, the amount of delay in a system varies with frequency, and the purpose of most all-pass filters is to add additional delay so that the sum of system and filter delay is approximately constant.

13.3.2 Filter Structures

Although methods exist for directly synthesizing filters of high order, most filter design is done by interconnecting first and second order filter sections. This approach generally results in designs that are easier to adjust, less sensitive to component changes with time and temperature, and simpler to modify and service.

There are two fundamental ways in which a filter transfer function can be decomposed into first and second order sections. In a cascade connection, the individual sections are connected in a chain, the output of one section being the input to the next section, and so on, as in the example of Figure 13-9(a). For real filter poles and zeros, these sections can each be of first order. Complex conjugate sets of poles and/or zeros must be realized as a unit, in a second order section.

The other basic type of transfer function decomposition is the tandem connection, as in the example of Figure 13-9(b). Each of the filter sections is driven by the filter input, and the section outputs are summed to form the overall filter output. The tandem connection involves a partial fraction expansion of the transfer function, with any complex conjugate pairs of pole terms combined into a single component section.

Combinations of cascade and tandem connections of first and second order filter sections, as in the example of Figure 13-9(c), are also commonly used in practice.

D13-3

Determine the types of the filters with the following transfer functions:

(a) $T(s) = \dfrac{3s^2 - 2s}{(s + 5)^2}$

ans. high-pass

(b) $T(s) = \dfrac{4}{3s^2 + 2s + 1}$

ans. low-pass

(c) $T(s) = \dfrac{-3s}{s^2 + 2s + 10}$

ans. bandpass

(d) $T(s) = \dfrac{6(s^2 - 2s + 17)(s - 3)}{(s^2 + 2s + 17)(s + 3)}$

ans. all-pass

(e) $T(s) = \dfrac{s^2 + 100}{10s^2 + 3s + 1000}$

ans. bandstop

(f) $T(s) = \dfrac{s^3}{(s^2 + s + 100)^2}$

ans. bandpass

(g) $T(s) = \dfrac{-3s^2 + 2s + 4}{s^2 + s + 100}$

ans. none of the types discussed

(h) $T(s) = \dfrac{(s - 3)^2}{(s^2 + s + 100)^2}$

ans. low-pass

FILTERS

A filter is a frequency-selective network or system. Common types of filters are low-pass, high-pass, bandpass, bandstop, and all-pass.

The delay of a filter is the negative slope of the filter's phase shift:

$$\tau(\omega) = -\frac{d\Phi(\omega)}{d\omega}$$

A pure time delay of a signal by a filter requires constant amplitude ratio and a phase shift with negative slope, proportional to frequency.

Higher order filters are usually constructed by interconnecting first order and second order filter sections in cascade, in tandem, or in a combination of cascade and tandem.

13.4 BODE PLOTS

13.4.1 Amplitude Ratio in Decibels

When a transfer function is decomposed into the product of simpler terms,

$$T(s) = T_1(s)T_2(s)T_3(s) \cdots$$

the overall amplitude ratio is the product of the individual ratios,

$$|T(s = j\omega)| = |T_1(s = j\omega)| \, |T_2(s = j\omega)| \, |T_3(s = j\omega)| \cdots$$

and the overall phase shift is the sum of the individual phase shifts:

$$\underline{/T(s = j\omega)} = \underline{/T_1(s = j\omega)} + \underline{/T_2(s = j\omega)} + \underline{/T_3(s = j\omega)} + \cdots$$

By dealing with the logarithm of the amplitude ratio rather than the ratio itself, the amplitude measure may be made to add for a product of terms, since the log of a product is the sum of the individual logs.

For historical reasons dating back to studies of hearing by Alexander Bell [1847–1922], the logarithmic measure of amplitude ratio most commonly used is the decibel function

$$dB(\omega) = 20 \log_{10} M(\omega)$$

Table 13-1 lists the decibels corresponding to various amplitude ratios. A unity ratio is 0 dB, a ratio of 10 is 20 dB, of 100 is 40 dB, and so forth.

Table 13-1 DECIBELS CORRESPONDING TO VARIOUS AMPLITUDE RATIOS

AMPLITUDE RATIO	DECIBELS	AMPLITUDE RATIO	DECIBELS
10^{-10}	-200	1	0
10^{-9}	-180	2	6.02
10^{-8}	-160	3	9.54
10^{-7}	-140	4	12.04
10^{-6}	-120	5	13.97
10^{-5}	-100	6	15.56
10^{-4}	-80	7	16.9
10^{-3}	-60	8	18.06
		9	19.08
0.002	-53.98		
0.003	-50.46	10	20
0.004	-47.96	20	26.02
0.005	-46.03	30	29.54
0.006	-44.44	40	32.04
0.007	-43.10	50	33.97
0.008	-41.94	60	35.56
0.009	-40.92	70	36.9
		80	38.06
0.01	-40	90	39.08
0.02	-33.98		
0.03	-30.46	100	40
0.04	-27.96	200	46.02
0.05	-26.03	300	49.54
0.06	-24.44	400	52.04
0.07	-23.10	500	53.97
0.08	-21.94	600	55.56
		700	56.9
0.1	-20	800	58.06
0.2	-13.98	900	59.08
0.3	-10.46		
0.4	-7.96	10^3	60
0.5	-6.03	10^4	80
0.6	-4.44	10^5	100
0.7	-3.10	10^6	120
0.8	-1.94	10^7	140
		10^8	160
		10^9	180
		10^{10}	200

It is very useful to plot frequency response dB and phase shift on a logarithmic scale of frequency. When this is done, good straight line approximations are easily constructed for simple transfer functions, and transfer function models of experimental data are apparent. Frequency response plots in this form are known as Bode plots, after Hendrik W. Bode [1905–1982], a researcher at Bell Telephone Laboratories and professor at Harvard University.

13.4.2 Real Root Terms

A transfer function factor

$$T_1(s) = s$$

has dB function

$$dB(\omega) = 20 \log_{10} |j\omega| = 20 \log_{10} \omega$$

which plots, versus $\log_{10} \omega$, as a straight line with slope 20 dB per decade of ω, as shown in Figure 13-10(a). The phase shift of this transfer function is

$$\Phi = \underline{/j\omega} = 90°$$

as shown. The transfer function term

$$T_2(s) = \frac{1}{s}$$

has dB function

$$dB(\omega) = 20 \log_{10} \left| \frac{1}{j\omega} \right| = -20 \log_{10} \omega$$

which, on a logarithmic scale for ω, is a straight line with slope -20 dB per decade of ω. The phase shift is

$$\Phi = \underline{/\frac{1}{j\omega}} = -90°$$

as shown in Figure 13-10(b).

A transfer function factor of the form

$$T_3(s) = \frac{s + a}{a}$$

with positive constant a, has dB function

$$dB(\omega) = 20 \log_{10} \left| \frac{j\omega + a}{a} \right| = 20 \log_{10} \left| j\frac{\omega}{a} + 1 \right|$$

which is sketched in Figure 13-10(c). This dB curve is approximated by two straight line segments, through $\omega = a$, the *break frequency*, as shown. For ω less than the break frequency a, the dB curve is approximately

$$dB = 20 \log_{10} (1) = 0$$

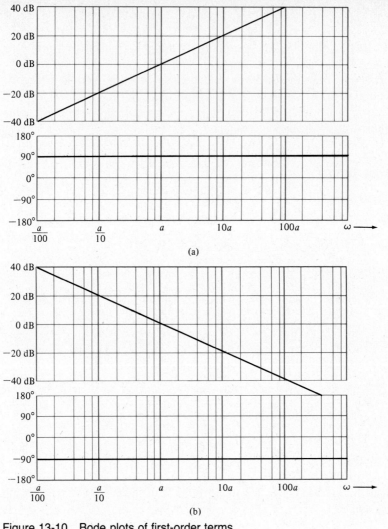

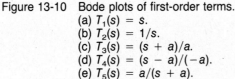

Figure 13-10 Bode plots of first-order terms.
   (a) $T_1(s) = s$.
   (b) $T_2(s) = 1/s$.
   (c) $T_3(s) = (s + a)/a$.
   (d) $T_4(s) = (s - a)/(-a)$.
   (e) $T_5(s) = a/(s + a)$.

For ω larger than the break frequency, the dB curve is approximately

$$dB(\omega) = 20 \log_{10}\left(\frac{\omega}{a}\right)$$

which is a straight line with slope 20 dB per decade of ω. The error between the approximation and the actual dB curve is greatest at $\omega = a$, where

$$dB = 20 \log_{10} |j + 1| = 20 \log_{10} \sqrt{2} = 3.01$$

In practice, one sketches the straight line approximation, then rounds the corner through the 3-dB point to give an accurate dB sketch.

Figure 13-10 (*Continued*)

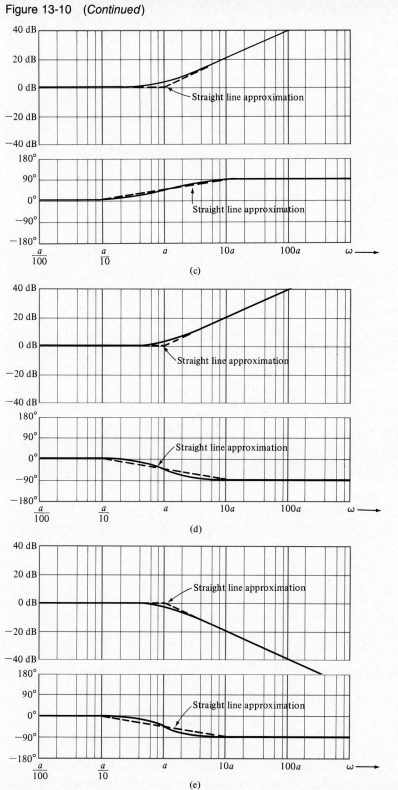

The phase shift curve for a term of the form of $T_3(s)$ is also shown in Figure 13-10(c). It is approximated well by three straight line segments. The phase shift is nearly 0° up to one-tenth the break frequency a. It is very nearly 90° at frequencies ten times the break frequency upward. Between $\omega = a/10$ and $\omega = 10a$, the phase shift is approximated well by a straight line passing through 45° at $\omega = a$, as shown. The maximum error of this three line segment approximation is only about 6°.

Bode plots and straight line approximations for terms of the form

$$T_4(s) = \frac{s - a}{-a}$$

and

$$T_5(s) = \frac{a}{s + a}$$

with positive a, are shown in Figure 13-10(d) and (e). $T_4(s)$ has the same dB curve but the negative of the phase shift curve of $T_3(s)$. The dB curve for $T_5(s)$ breaks downward above $\omega = a$, and its phase shift is the negative of that for $T_3(s)$. A term of the form

$$T_6(s) = \frac{-a}{s - a}$$

with positive a, would represent an unstable system.

Bode plots for more complicated transfer functions are constructed by expressing the transfer function as a product of simple terms. The overall dB curve is the sum of the individual dB curves, and the overall phase shift is the sum of the phase shifts contributed by the individual terms.

For example, consider the transfer function

$$T(s) = \frac{s^2 + 10s}{s^2 + 2s + 1} = 10(s)\left(\frac{1}{s + 1}\right)^2\left(\frac{s + 10}{10}\right)$$

The factor of 10 contributes

$$dB = 20 \log_{10} = 20$$

to the dB curve. The factor s contributes a line with slope 20 dB per decade, passing through 0 dB at $\omega = 1$. Below $\omega = 1$, the remaining terms are at 0 dB, and the straight line approximation to the sum of the dB curves is a line with slope 20 dB passing through 20 dB at $\omega = 1$. Above $\omega = 1$, the two factors

$$\left(\frac{1}{s + 1}\right)$$

each contribute -20 dB per decade slope. Above $\omega = 10$, the

$$\left(\frac{s + 10}{10}\right)$$

factor contributes an additional $+20$ dB per decade slope. The overall straight line approximation is shown in Figure 13-11. Rounding the corners through a point

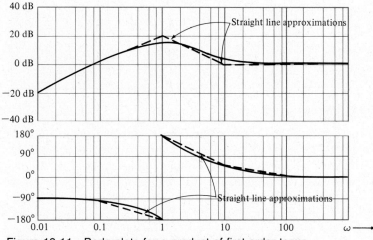

Figure 13-11   Bode plots for a product of first-order terms.

6 dB below the lines at $\omega = 1$ (because there are two poles at $s = -1$) and through a point 3 dB above the lines at $\omega = 10$ gives a good approximation to the actual dB curve.

The phase shift curve for this example is similarly constructed. The 90° contribution of the s factor, the linear contributions to the phase shift between $\omega = 0.1$ and $\omega = 10$ of each of the two $1/(s + 1)$ factors, and the contribution of the $(s + 10)/10$ factor between $\omega = 1$ and $\omega = 100$ are all summed to form a piecewise linear approximation to the phase shift curve. The result, in Figure 13-11, is sketched within a $\pm 180°$ range.

D13-4

Sketch Bode plots (both amplitude ratio in dB and phase shift) for filters with the following transfer functions, all of which have real poles and zeros:

(a) $T(s) = \dfrac{10}{s + 1}$

(b) $T(s) = \dfrac{10s}{s + 2}$

(c) $T(s) = \dfrac{s - 2}{s + 8}$

(d) $T(s) = \dfrac{3s + 1}{2s + 1}$

(e) $T(s) = \dfrac{s}{s^2 + 4s + 4}$

(f) $T(s) = \dfrac{4s - 1}{s^2 + 10s}$

(g) $T(s) = \dfrac{10s^2}{s^2 + 5s + 6}$

(h) $T(s) = \dfrac{s^2 - 9}{3s^2 + 7s + 2}$

13.4.3 Complex Root Terms

For Bode plots, the real axis poles and zeros of a transfer function are accommodated easily and well by taking the terms one at a time and approximating their frequency response with the straight line segments. For complex poles and zeros, simple straight line approximations are seldom sufficiently accurate to use in practice. When a Bode plot is to be constructed for a transfer function with complex roots, it is most convenient to group each pair of conjugate terms into a single second-order term with real coefficients, of the form

$$\frac{a_0}{s^2 + a_1 s + a_0}$$

for pole pairs, or

$$\frac{s^2 + b_1 s + b_0}{b_0}$$

for complex pairs of zeros.

Bode plots for several representative sets of conjugate pole pairs

$$\frac{\omega_r^2}{s^2 + \left(\dfrac{1}{Q}\right)\omega_r + \omega_r^2}$$

in terms of the resonant (or critical or center) frequency ω_r and the quality factor Q are given in Figure 13-12. The phase shift passes through $90°$ at the resonant fre-

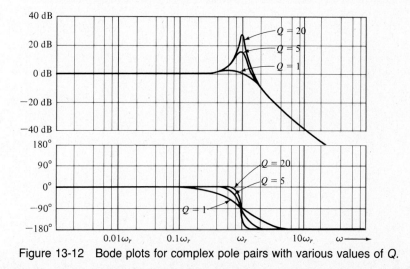

Figure 13-12　Bode plots for complex pole pairs with various values of Q.

quency, and the amplitude ratio has a peak in the vicinity of ω_r. The larger the Q, the more pronounced the resonance peak and the more sharply the phase shift changes near resonance.

Bode plots for the specific three transfer functions

$$T_1(s) = \frac{101}{s^2 + 2s + 101}$$

$$T_2(s) = \frac{s^2 + 2s + 101}{101}$$

$$T_3(s) = \frac{s^2 - 2s + 101}{101}$$

are illustrated in Figure 13-13. Complex left half-plane zero pairs, of the form

$$\frac{s^2 + \left(\frac{1}{Q}\right)\omega_r s + \omega_r^2}{\omega_r^2}$$

have a dB curve and phase shift that are each negatives of those for the poles. Complex right half-plane zero pairs, of the form

$$\frac{s^2 - \left(\frac{1}{Q}\right)\omega_r s + \omega_r^2}{\omega_r^2}$$

have a dB curve that is the negative of that for the pole term but phase shift that is the same as that for the poles. A term of the form

$$\frac{\omega_r^2}{s^2 - \left(\frac{1}{Q}\right)\omega_r s + \omega_r^2}$$

represents poles in the right half of the complex plane and so would belong to an unstable system.

The construction of a Bode plot for the transfer function

$$T(s) = \frac{s - 10}{s(s^2 + 2s + 10)} = (-1)\left(\frac{1}{s}\right)\left(\frac{s - 10}{-10}\right)\left(\frac{10}{s^2 + 2s + 10}\right)$$

is shown in Figure 13-14. In Figure 13-14(a), Bode plots are sketched for the second order term that involves complex poles,

$$\frac{10}{(s + 1 + j3)(s + 1 - j3)} = \frac{10}{s^2 + 2s + 10} = \frac{\omega_r^2}{s^2 + \left(\frac{1}{Q}\right)\omega_r s + \omega_r^2}$$

for which

$$\omega_r = \sqrt{10} = 3.16$$

$$\frac{\omega_r}{Q} = 2 \qquad Q = 1.58$$

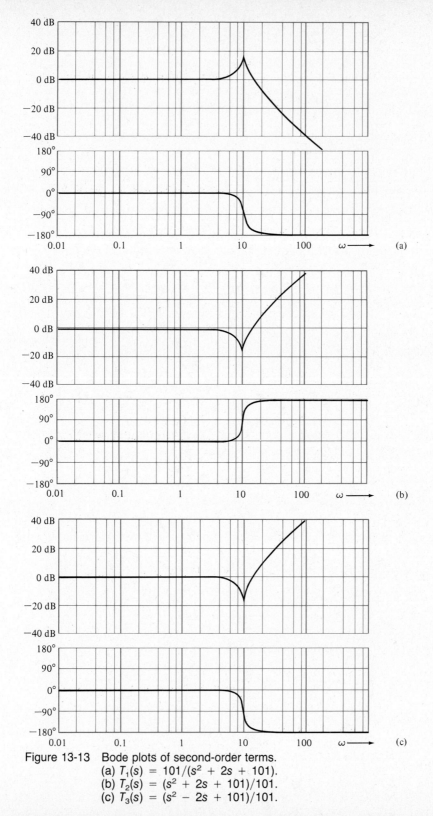

Figure 13-13 Bode plots of second-order terms.
(a) $T_1(s) = 101/(s^2 + 2s + 101)$.
(b) $T_2(s) = (s^2 + 2s + 101)/101$.
(c) $T_3(s) = (s^2 - 2s + 101)/101$.

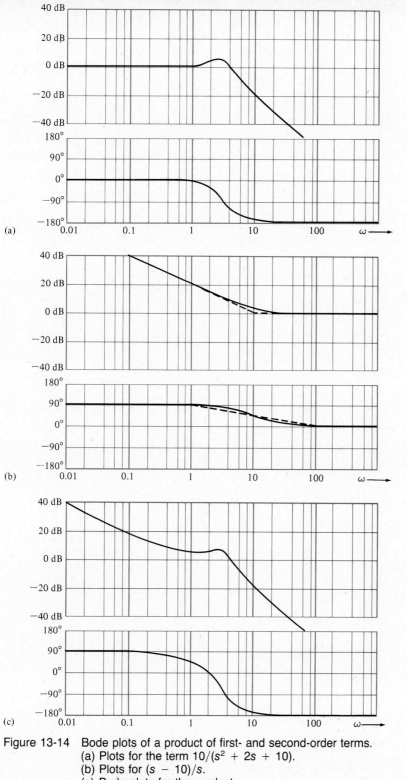

(a)

(b)

(c)

Figure 13-14   Bode plots of a product of first- and second-order terms.
(a) Plots for the term $10/(s^2 + 2s + 10)$.
(b) Plots for $(s - 10)/s$.
(c) Bode plots for the product.

These are sketched by interpolating between the representative curves in Figure 13-12.

In Figure 13-14(b), a Bode plot for the product of terms

$$(-1)\left(\frac{1}{s}\right)\left(\frac{s-10}{-10}\right)$$

is constructed. The Bode plots for the entire function $T(s)$, shown in Figure 13-14(c), consists of the sums of the corresponding dB and phase shift curves in Figure 13-14(a) and (b).

D13-5

Sketch Bode plots (both amplitude ratio in dB and phase shift) for filters with the following transfer functions:

(a) $T(s) = \dfrac{13}{s^2 + 4s + 13}$

(b) $T(s) = \dfrac{s}{s^2 + 6s + 13}$

(c) $T(s) = \dfrac{s^2 + 6s + 10}{s^2 + 3s + 2}$

(d) $T(s) = \dfrac{s^2 - 6s + 10}{s^2 + 20s + 200}$

BODE PLOTS

Bode plots are frequency response curves for which the frequency axis has a logarithmic scale and where the amplitude ratio is plotted in terms of

$$dB(\omega) = 20 \log_{10} M(\omega)$$

In this arrangement, the individual dB and phase shift curves of transfer function terms each add to form the curves for products of terms. And first order transfer function terms may be quickly sketched from simple straight line approximations.

A second order transfer function term for a pair of complex conjugate poles, in terms of resonant frequency ω_r and Q, is expressed in the form

$$T(s) = \frac{a_0}{s^2 + a_1 s + a_0} = \frac{\omega_r^2}{s^2 + (1/Q)\omega_r s + \omega_r^2}$$

and may be sketched from standard curves. Complex conjugate zero terms are handed similarly.

13.5 CLASSICAL FILTERS

13.5.1 Butterworth Low-Pass Filters

Low-pass filters are fundamental to filter design because high-pass, bandpass, and other filters can be derived from a low-pass filter through transformations. For convenience, standard low-pass filters are normalized for an $\omega = 1$ rolloff frequency and a dc gain of unity (or nearly so). By scaling the complex frequency variable s, the rolloff frequency is moved to any desired value.

Classical low-pass filters are of the "all pole" type. Their transfer functions have a number of poles equal to their order and no zeros:

$$T(s) = \frac{b_m}{s^n + a_{n-1}s^{n-1} + \cdots + a_1 s + a_0}$$

At high frequencies, the filter's amplitude ratio then rolls off at the maximum rate for a filter of order n:

$$\lim_{\omega \to \infty} M(\omega) = \lim_{\omega \to \infty} |T(s = j\omega)| \approx \lim_{\omega \to \infty} \left| \frac{b_m}{(j\omega)^n} \right|$$

The amplitude ratio of a transfer function $T(s)$ is

$$M(\omega) = |T(s = j\omega)|$$

and the square of the amplitude ratio may be written conveniently as

$$M^2(\omega) = T(s = j\omega)T^*(s = j\omega) = T(s = j\omega)T(s = -j\omega)$$

where * denotes the complex conjugate. For example, for the transfer function

$$T(s) = \frac{3s}{s + 4}$$

$$M^2(\omega) = \left| \frac{3j\omega}{j\omega + 4} \right|^2 = \left(\frac{3j\omega}{j\omega + 4} \right)\left(\frac{-3j\omega}{-j\omega + 4} \right) = \frac{9\omega^2}{\omega^2 + 16}$$

For the normalized Butterworth low-pass filter of order n, the square of the amplitude ratio is chosen to be

$$M^2(\omega) = \frac{1}{1 + \omega^{2n}} = T(s = j\omega)T(s = -j\omega)$$

The product $T(s)T(-s)$ is obtained from $T(j\omega)T(-j\omega)$ by substituting s/j for each ω in $M^2(\omega)$:

$$T(s)T(-s) = [T(s = j\omega)T(s = -j\omega)]_{\omega = s/j} = M^2(s/j) = \frac{1}{1 + \dfrac{s^{2n}}{j^{2n}}}$$

Since $j^2 = -1$,

$$T(s)T(-s) = \frac{1}{1 + (-1)^n s^{2n}}$$

The poles of $T(s)T(-s)$ are thus given by

$$s^{2n} = (-1)^{n+1}$$

and are evenly spaced on a circle of unit radius about the origin of the complex plane, as indicated in Figure 13-15(a). For a stable filter, the n of these poles in the

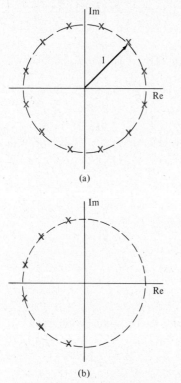

(a)

(b)

Figure 13-15   Pole locations for Butterworth low-pass filters.
(a) Roots of $s^{2n} = (-1)^{n+1}$.
(b) Left half-plane roots.

left half-plane must be the poles of $T(s)$, Figure 13-15(b). The mirror-image n right half-plane poles are those of $T(-s)$.

Butterworth filters, then, have frequency response amplitude ratios given by

$$M(\omega) = \frac{1}{\sqrt{1 + \omega^{2n}}}$$

where n is the order of the filter. The filters each have unity dc gain and, at $\omega = 1$,

$$M(1) = \frac{1}{\sqrt{1 + 1}} = \frac{1}{\sqrt{2}}$$

regardless of filter order. The normalized Butterworth low-pass filters thus have amplitude ratios that each pass through

$$20 \log_{10} M(1) = 20 \log_{10} \frac{1}{\sqrt{2}} = -3.01 \text{ dB}$$

at $\omega = 1$.

Table 13-2 TRANSFER FUNCTIONS OF BUTTERWORTH LOW-PASS FILTERS

$$\frac{1}{s + 1}$$

$$\frac{1}{s^2 + 1.414s + 1}$$

$$\frac{1}{s^3 + 2s^2 + 2s + 1}$$

$$\frac{1}{s^4 + 2.61s^3 + 3.41s^2 + 2.61s + 1}$$

$$\frac{1}{s^5 + 3.24s^4 + 5.24s^3 + 5.24s^2 + 3.24s + 1}$$

$$\frac{1}{s^6 + 3.86s^5 + 7.46s^4 + 9.14s^3 + 7.46s^2 + 3.86s + 1}$$

Table 13-2 lists the transfer functions of normalized Butterworth low-pass filters of orders one through six. Bode plots for the third order filter are given in Figure 13-16. As with Butterworth filters of all orders, the amplitude ratio decreases monotonically with frequency and is 0.707 times the dc gain (down -3.01 dB) at the rolloff frequency. The ratio decreases at the rate of $-20n$ dB per decade at high frequencies, where n is the filter order. The third order filter amplitude ratio eventually rolls off at the rate of -60 dB with each decade of frequency. The amplitude ratios and delays of Butterworth filters of various orders are plotted in Figure 13-17.

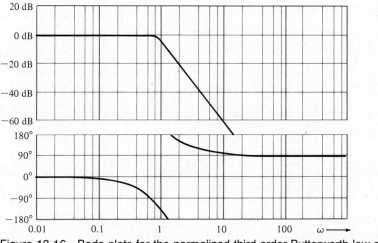

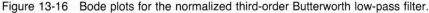

Figure 13-16   Bode plots for the normalized third-order Butterworth low-pass filter.

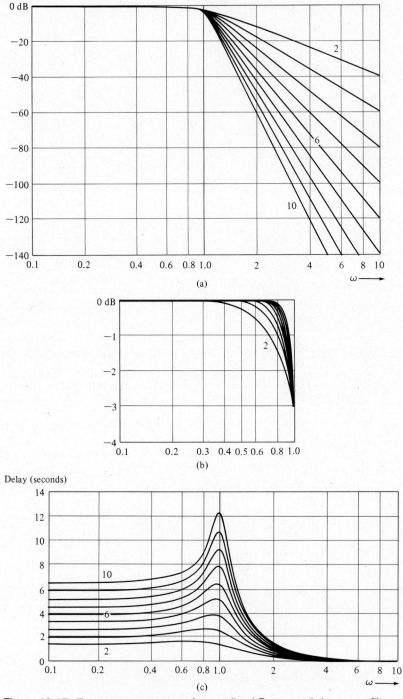

Figure 13-17 Frequency response of normalized Butterworth low-pass filters of various orders.
(a) Amplitude ratio.
(b) Closeup of the amplitude ratio near the rolloff frequency.
(c) Delay.

13.5.2 Chebyshev Low-Pass Filters

Chebyshev filters are based on the Chebyshev polynomials $C_n(x)$, which are defined by the recursive relations

$$C_0 = 1$$

$$C_1(x) = x$$

$$C_{n+1}(x) = 2xC_n(x) - C_{n-1}(x)$$

Table 13-3 lists Chebyshev polynomials through order six. Several of the Chebyshev polynomials are plotted in Figure 13-18(a), where it is seen that each oscillates between ± 1 for values of its argument in the range ± 1. For arguments outside ± 1, the Chebyshev polynomials (except C_0) grow rapidly in magnitude.

Normalized Chebyshev low-pass filters have amplitude ratios with square

$$M^2(\omega) = \frac{1}{1 + \epsilon^2 C_n^2(\omega)} = T(s = j\omega)T^*(s = j\omega)$$

$$= T(s = j\omega)T(s = -j\omega)$$

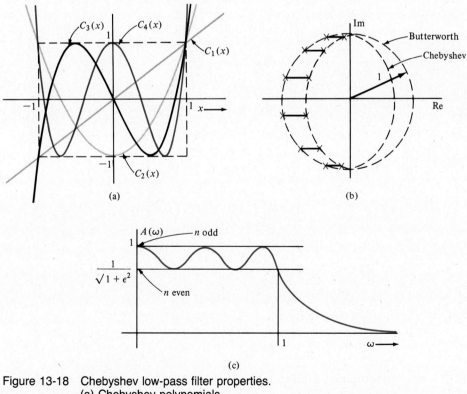

(a)

(b)

(c)

Figure 13-18   Chebyshev low-pass filter properties.
(a) Chebyshev polynomials.
(b) Pole locations.
(c) Character of amplitude ratio.

Table 13-3 CHEBYSHEV POLYNOMIALS

$C_0 = 1$

$C_1(x) = x$

$C_{n+1}(x) = 2xC_n(x) - C_{n-1}(x)$

$C_0 = 1$

$C_1(x) = x$

$C_2(x) = 2x^2 - 1$

$C_3(x) = 4x^3 - 3x$

$C_4(x) = 8x^4 - 8x^2 + 1$

$C_5(x) = 16x^5 - 20x^3 + 5x$

$C_6(x) = 32x^6 - 48x^4 + 18x^2 - 1$

where ϵ is a constant factor between 0 and 1 and n is the filter order. The product $T(s)T(-s)$ is obtained from $T(j\omega)T(-j\omega)$, as for the Butterworth filters, by substituting s/j for each ω in $M^2(\omega)$. This product has poles located on an ellipse of eccentricity ϵ on the complex plane, as indicated in Figure 13-18(b). The left half-plane poles are those of $T(s)$, while the symmetrical right half-plane poles on the ellipse are those of $T(-s)$.

In the passband, for ω between 0 and 1, a Chebyshev filter's amplitude ratio

$$M(\omega) = \frac{1}{\sqrt{1 + \epsilon^2 C_n^2(\omega)}}$$

oscillates between

$$\frac{1}{\sqrt{1 + 0}} = 1$$

when $C_n(\omega) = 0$, and

$$\frac{1}{\sqrt{1 + \epsilon^2}}$$

when $C_n(\omega) = \pm 1$. This character is shown in Figure 13-18(c). For n odd, $C_n(0) = 0$, so $M(0) = 1$. For n even, $C_n(0) = \pm 1$, so

$$M(0) = \frac{1}{\sqrt{1 + \epsilon^2}}$$

For $\omega > 1$, $C_n^2(\omega)$ increases rapidly, so $M(\omega)$ decreases.

Consider, for example, a second order Chebyshev low-pass filter with

$$\epsilon = 0.5$$

so that the ripple amount is given by

$$\frac{1}{\sqrt{1 + \epsilon^2}} = 0.894$$

For this filter

$$M^2(\omega) = \frac{1}{1 + (0.5)^2 C_3^2(\omega)} = \frac{1}{1 + 0.25(2\omega^2 - 1)^2} = \frac{1}{\omega^4 - \omega^2 + 0.75}$$

and

$$T(s)T(-s) = M^2(\omega)\bigg|_{\omega = s/j} = \frac{1}{s^4 + s^2 + 0.75}$$

$$= \frac{1}{(s + 0.428 + j0.826)(s + 0.428 - j0.826)(s - 0.428 + j0.826)(s - 0.428 - j0.8)}$$

so

$$T(s) = \frac{1}{(s + 0.428 + j0.826)(s + 0.428 - j0.826)} = \frac{1}{s^2 + 0.855s + 0.866}$$

By selecting the value of ϵ, a designer controls the filter's *passband ripple*, the amount of variation in $M(\omega)$ in the passband. Small ϵ gives small passband ripple, but also results in a smaller initial rate of decrease in $M(\omega)$ in the stopband. In practice, one chooses ϵ to achieve the required limit on passband ripple, and the filter order n to obtain the necessary rolloff characteristics.

Transfer functions of several normalized Chebyshev low-pass filters are listed in Table 13-4. For $\epsilon = 0.509$, the passband ripple of a Chebyshev filter is

$$\frac{1}{\sqrt{1 + \epsilon^2}} = 0.89$$

Table 13-4 TRANSFER FUNCTIONS OF CHEBYSHEV LOW-PASS FILTERS

1-dB PASSBAND RIPPLE ($\epsilon = 0.509$)

$$\frac{1.97}{s + 1.97}$$

$$\frac{0.983}{s^2 + 1.098s + 1.103}$$

$$\frac{0.491}{s^3 + 0.988s^2 + 1.238s + 0.491}$$

$$\frac{0.246}{s^4 + 0.953s^3 + 1.454s^2 + 0.743s + 0.276}$$

0.1-dB PASSBAND RIPPLE ($\epsilon = 0.153$)

$$\frac{6.552}{s + 6.552}$$

$$\frac{2.345}{s^2 + 3.313s + 2.372}$$

$$\frac{1.939}{s^3 + 1.638s^2 + 2.630s + 1.939}$$

$$\frac{1.783}{s^4 + 0.829s^3 + 2.026s^2 + 2.627s + 1.804}$$

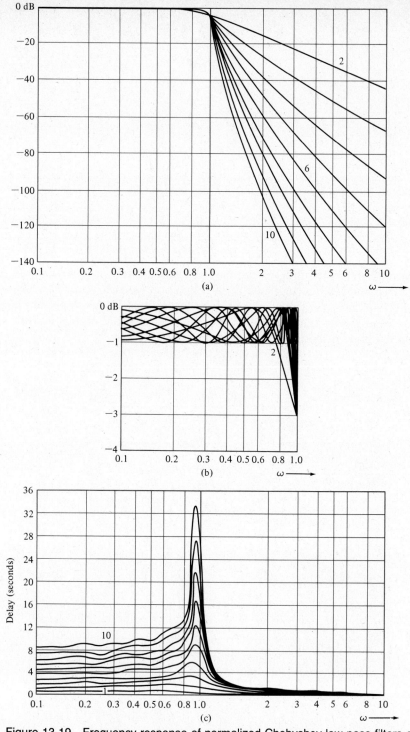

Figure 13-19 Frequency response of normalized Chebyshev low-pass filters of various orders, with $\epsilon = 0.509$, which gives 1-dB passband ripple.
(a) Amplitude ratio.
(b) Closeup of the amplitude ratio in the passband.
(c) Delay.

which is 1 dB, since

$$20 \log_{10} (0.89) = 1$$

Also listed are transfer functions for Chebyshev filters with 0.1-dB passband ripple. For these, the rolloff in the vicinity of $\omega = 1$ is not so steep as for filters with larger ripple, although the amplitude ratios of all such filters eventually approach a rolloff rate of $-20n$ dB per decade, where n is the filter order.

The amplitude ratios and delays of 1-dB ripple Chebyshev filters of various orders are plotted in Figure 13-19. For these, the frequency has been scaled so that each filter has amplitude ratio -3 dB at $\omega = 1$, as is commonly done in filter design tables. The delays of Chebyshev filters are much farther than Butterworth filters from the desired constant curves in the passband.

13.5.3 Bessel Low-Pass Filters

The Bessel filters are derived from Bessel polynomials according to

$$T(s) = \frac{P_n(0)}{P_n(s)}$$

where $P_n(s)$ is the degree-n Bessel polynomial with coefficients reversed. These are defined by the recursive relations

$$P_0 = 1$$

$$P_1(s) = s + 1$$

$$P_n(s) = (2n - 1)P_{n-1}(s) + s^2 P_{n-2}(s)$$

The transfer functions of normalized Bessel low-pass filters through the sixth order are listed in Table 13-5. The poles of these filters lie on a curve outside the semicircle of Butterworth poles, as shown in Figure 13-20.

Table 13-5 TRANSFER FUNCTIONS OF BESSEL LOW-PASS FILTERS

$$\frac{1}{s + 1}$$

$$\frac{3}{s^2 + 3s + 3}$$

$$\frac{15}{s^3 + 6s^2 + 15s + 15}$$

$$\frac{105}{s^4 + 10s^3 + 45s^2 + 105s + 105}$$

$$\frac{945}{s^5 + 15s^4 + 105s^3 + 420s^2 + 945s + 945}$$

$$\frac{10,395}{s^6 + 21s^5 + 210s^4 + 1260s^3 + 4725s^2 + 10,395s + 10,395}$$

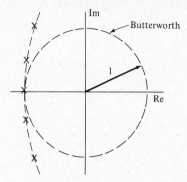

Figure 13-20 Bessel low-pass filter pole locations.

Frequency response amplitude ratio and delay curves for the Bessel filters are given in Figure 13-21. It can be shown that each of these filters has the most nearly constant delay in the passband that is possible with an all-pole filter of that order. The tradeoff, of course, is in the sharpness of the rolloff.

13.5.4 Transformations

Changing the frequency scale of a filter involves scaling the s variable in the transfer function. To move a filter's behavior at $\omega = 1$ to $\omega = a$, each s in the filter's original transfer function is replaced by (s/a). For example, the normalized third order Butterworth low-pass filter, which has rolloff frequency $\omega = 1$,

$$T_1(s) = \frac{1}{s^3 + 2s^2 + 2s + 1}$$

is converted to a third order Butterworth low-pass filter with 1000-Hz rolloff frequency by the substitution

$$s \rightarrow \frac{s}{2\pi \cdot 1000}$$

giving

$$T_2(s) = \frac{1}{\left(\dfrac{s}{2\pi \cdot 1000}\right)^3 + 2\left(\dfrac{s}{2\pi \cdot 1000}\right)^2 + 2\left(\dfrac{s}{2\pi \cdot 1000}\right) + 1}$$

For a dc gain of 10, rather than unity, the transfer function is

$$T_3(s) = 10T_2(s) = \frac{10(2000\pi)^3}{s^3 + 2(2000\pi)s^2 + 2(2000\pi)^2 s + (2000\pi)^3}$$

Bode plots for the scaled filter, $T_3(s)$, are given in Figure 13-22(a). The substitution

$$s \rightarrow \frac{1}{s}$$

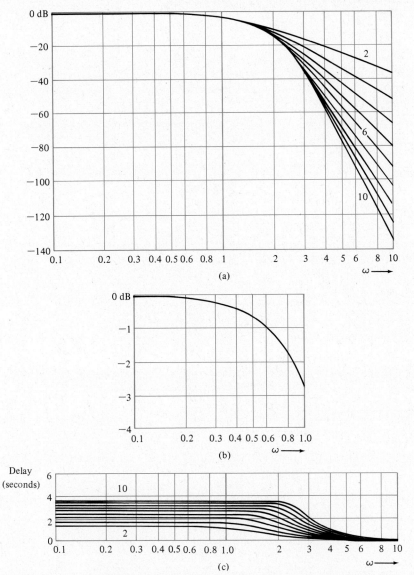

Figure 13-21   Frequency response of normalized Bessel low-pass filters of various orders.

(a) Amplitude ratio.
(b) Closeup of the amplitude ratio in the passband.
(c) Delay.

converts a normalized low-pass filter to a high-pass filter with rolloff frequency at $\omega = 1$. For example, substituting $(1/s)$ for each s in the normalized 1-dB passband ripple Chebyshev low-pass filter,

$$T_4(s) = \frac{0.491}{s^3 + 0.988s^2 + 1.238s + 0.491}$$

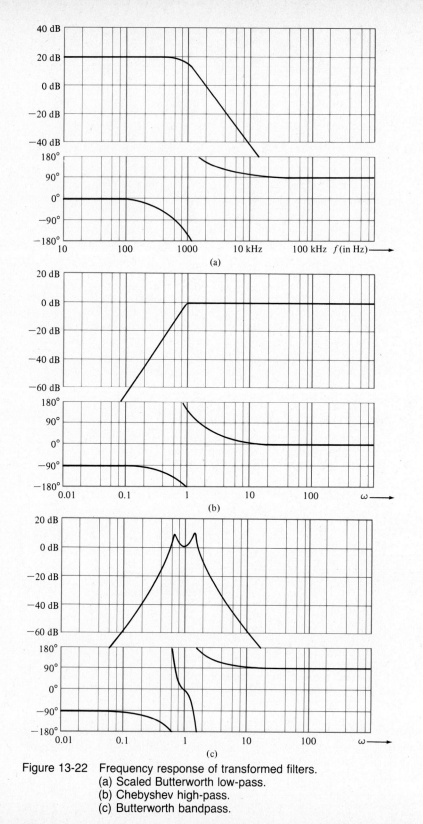

Figure 13-22   Frequency response of transformed filters.
(a) Scaled Butterworth low-pass.
(b) Chebyshev high-pass.
(c) Butterworth bandpass.

gives $\quad T(s) = \dfrac{s^3}{s^3 + 2.521s^2 + 2.012s + 2.037}$

The frequency response of this filter is plotted in Figure 13-22(b). The new high-pass filter has 1-dB ripple in its passband above $\omega = 1$. If it were desired to obtain a high-pass filter with 20-Hz rolloff frequency and a high-frequency gain of 12, the scaling

$$T_6(s) = 12T_5\left(\frac{s}{2\pi \cdot 20}\right)$$

would then be used.

A change of variables that converts a normalized low-pass filter to a bandpass filter is the substitution

$$s \rightarrow \frac{s^2 + 1}{s}$$

When applied to the third order normalized Butterworth low-pass filter,

$$T_1(s) = \frac{1}{s^3 + 2s^2 + 2s + 1}$$

this substitution gives

$$T_7(s) = \frac{1}{\left(\dfrac{s^2 + 1}{s}\right)^3 + 2\left(\dfrac{s^2 + 1}{s}\right)^2 + 2\left(\dfrac{s^2 + 1}{s}\right) + 1}$$

$$= \frac{s^3}{s^6 + 2s^5 + 4s^4 + 5s^3 + 4s^2 + 2s + 1}$$

which results in a transfer function of twice the original low-pass filter order. Bode plots for $T_7(s)$ are given in Figure 13-22(c).

D13-6

Using transformations of normalized low-pass filter transfer functions, find the transfer functions of the following:

(a) Second order Butterworth high-pass, with a 50-Hz rolloff frequency and a high-frequency gain of 10.

   ans. $10s^2/(s^2 + 443s + 98,596)$

(b) Second order Butterworth bandpass (derived from a first order low-pass filter) with a 1-kHz center frequency.

   ans. $(\text{constant})s^2/[s^2 + 6280s + (6280)^2]$

(c) Third order Bessel high-pass with a 0.1-Hz rolloff frequency.

   ans. $(\text{constant})s^3/[s^3 + \frac{1}{10}s + \frac{1}{250}s + \frac{1}{15,000}]$

(d) Fourth order Bessel bandpass (derived from a second order low-pass filter) with a 10-rad/s center frequency and a gain of 20 at 10 rad/s.

ans. $\pm 6000s^2/(s^4 + 30s^3 + 500s^2 + 3000s + 10{,}000)$

CLASSICAL FILTERS

A choice between Butterworth, Chebyshev, and Bessel filters is generally made by the designer, according to the relative importance of the shapes of the amplitude ratio and the phase shift curves. Chebyshev filters give uniformity of passband amplitude ratio at the expense of constancy in the passband filter delay. Bessel filters have nearly constant passband delay, but lack sharpness of rolloff. Butterworth filters have intermediate passband amplitude ratio and delay characteristics.

The substitution

$$s \to \frac{1}{s}$$

in a transfer function converts a normalized low-pass filter to a normalized high-pass filter, while the substitution

$$s \to \frac{s^2 + 1}{s}$$

converts a low-pass filter of order n to a bandpass filter of order $2n$.

Scaling the complex frequency by the substitution

$$s \to \left(\frac{s}{a}\right)$$

in a transfer function expands or contracts the frequency scale, so that the behavior of the original filter at $\omega = 1$ occurs at $\omega = a$ in the scaled filter.

13.6 ADJUSTABLE-GAIN AMPLIFIERS

Summing amplifier design using operational amplifiers was developed in Chapter 3. It was shown that, within broad ranges of resistor values, the circuit of Figure 13-23(a) has output voltage

$$v_{\text{out}} = -\frac{R_F}{R_a} v_a - \frac{R_F}{R_b} v_b - \frac{R_F}{R_c} v_c - \cdots$$

$$+ \left(1 + \frac{R_F}{R_A}\right)(R_1\|R_2\|R_3\|\cdots)\left(\frac{v_1}{R_1} + \frac{v_2}{R_2} + \frac{v_3}{R_3} + \cdots\right)$$

where

$$R_A = R_a\|R_b\|R_c\|\cdots$$

The notation $\|$ denotes parallel combination of the resistances:

$$R_1\|R_2\|R_3\|\cdots = \frac{1}{(1/R_1) + (1/R_2) + (1/R_3) + \cdots}$$

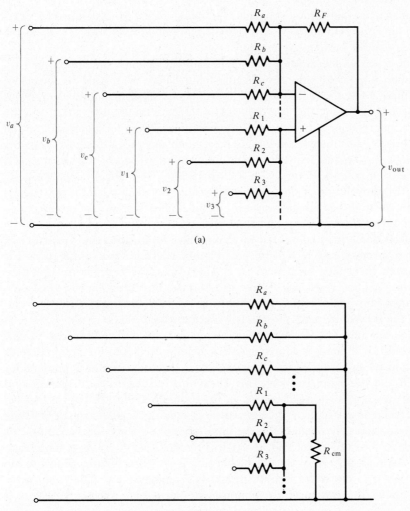

Figure 13-23   General summing amplifier.
(a) Structure of the amplifier with both inverting and noninverting inputs.
(b) Resistances seen by input signal sources.

To cause cancellation of the effects on the output of the average input bias currents when it is important to do so, the resistors should be chosen so that

$$R_1 \| R_2 \| R_3 \| \cdots = R_F \| R_a \| R_b \| R_c \| \cdots$$

The details of designing amplifiers to achieve output voltages consisting of any de-sired linear combination of a set of input voltages should be carefully reviewed by the reader. Now, because of the complexity of some of the circuits to be studied, the ground symbol will always be used to denote the power supply common.

   In Chapter 3, it was shown that, to the input signal sources, the inverting inputs have input resistance very nearly equal to the series summing resistors R_a,

$R_b, \ldots$. For moderate gain coefficients, the noninverting inputs appear to signal sources, as in Figure 13-23(b), where the common-mode resistance R_{cm} is on the order of 200 MΩ for inexpensive op amps.

To construct amplifiers with adjustable gain coefficients, the appropriate resistors in an op amp circuit can be made adjustable. There are several disadvantages to this method, however. These include the difficulty of maintaining average bias current effect cancellation, the interdependence of gains on resistor values, and the nonproportional relation between gain and resistance, except for adjustment of the feedback resistor, R_F. Instead, designers most often use potentiometers, resistors with sliding taps, to obtain adjustable gains. When the voltage v_1 of a signal source is connected to a potentiometer, as in Figure 13-24(a), the tapped voltage v_2 may be varied, by adjusting the slider, from zero (when the slider is all the way down) to v_1 (when the slider is all the way up). If the slider fraction is denoted by α, $v_2 = 0$ when $\alpha = 0$, and $v_2 = v_1$ when $\alpha = 1$.

If a potentiometer is connected to an amplifier input, the input resistance of the amplifier forms a load R_L on the potentiometer so that the voltage v_2 is not proportional to the slider fraction α, as is shown in the set of curves in Figure 13-24(a). For a loading resistance R_L five or ten times the potentiometer resistance R, the relation between v_2 and v_1 is very nearly proportional to the slider fraction; otherwise, the relation departs substantially from a straight line.

Viewed another way, the potentiometer circuit has a Thévenin equivalent, Figure 13-24(b), that consists of a voltage that is proportional to the potentiometer slider fraction α, in series with a resistance that depends on the potentiometer resistance R and the slider fraction. This Thévenin resistance adds to the op amp circuit input resistors such as R_a and R_1 in the example circuits of Figure 13-24(c) and (d); thus the amplifier gains change with a change in the slider position. However, if R is substantially smaller than the input resistor to which the potentiometer slider is connected, this effect will be small. Provided that R_a or R_1 is more than five or ten times as large as R, the adjustable gain is very nearly proportional to the slider position. The departure from a straight line relationship is then comparable to the unevenness in resistance distribution of an inexpensive potentiometer.

Suppose that it is desired to design an adjustable gain amplifier with input-output voltage relation

$$v_{out} = -A_1 v_1 - 4v_2 + A_2 v_3$$

where A_1 is adjustable from 0 to 2, and A_2 is adjustable from 0.5 to 1. An ordinary op amp circuit is first designed for the maximum gain magnitudes:

$$v_{out} = -2v_1 - 4v_2 + v_3$$

The result of this design, easily developed with the methods of Section 3.8, is shown in Figure 13-25(a).

Then potentiometers (called "pots" for short) are placed at the adjustable inputs, as shown in Figure 13-25(b). For nearly proportional adjustment and relatively small effect on other amplifier gains, each pot's resistance should be no more than one-fifth the value of the series input resistor connected to its slider arm. If the input voltage connected to a pot is from another op amp output or similar signal source, the potentiometer resistance should be at least 2 kΩ, so that excessive signal

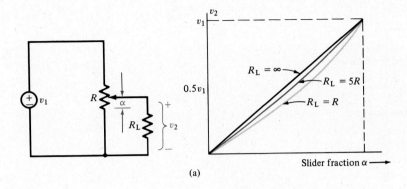

(a)

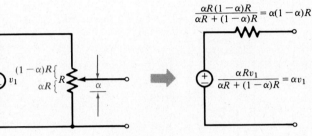

(b)

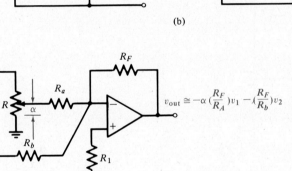

(c)

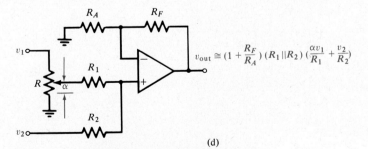

(d)

Figure 13-24   Using potentiometers for gain control.
(a) Effect of potentiometer loading.
(b) Thévenin equivalent of potentiometer circuit.
(c) Potentiometer for inverting input.
(d) Potentiometer for noninverting input.

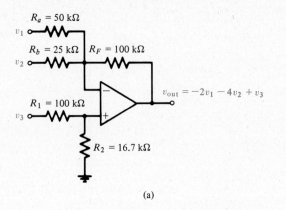

(a)

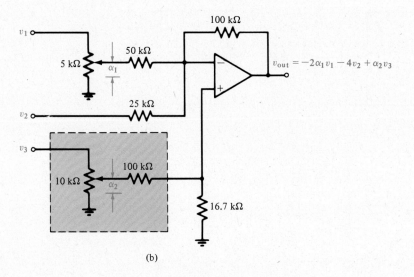

(b)

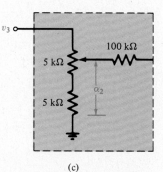

(c)

Figure 13-25   Design of an adjustable-gain amplifier.
(a) Amplifier with maximum required gains.
(b) Full adjustment of the v_1 and v_3 gains.
(c) Using a buildout resistor to limit the v_1 gain adjustment range.

source current does not flow at normal signal voltages. Rescaling of the amplifier's external resistors may be necessary in order to meet these constraints. With the potentiometers in place for the example

$$v_{out} = -2\alpha_1 v_1 - 4v_2 + \alpha_2 v_3$$

where the coefficient of v_1 is adjustable from 0 to -2 and the coefficient of v_3 is adjustable from 0 to 1.

To obtain a v_3 gain adjustment from 0.5 to 1, rather than from 0 to 1, the user could just not turn the 10-kΩ potentiometer down below the halfway position. For that matter, a mechanical stop could be placed on this potentiometer's slider travel. This would be wasteful of the available precision of adjustment and would be quite impractical if the adjustment were to be limited to a much smaller range, say a gain adjustable from 0.98 to 1.0. A simple and effective solution is to add a potentiometer *buildout resistor*, as shown in Figure 13-25(c). The full-range potentiometer and buildout resistor form a potentiometer with limited adjustment range.

D13-7

Design adjustable-gain amplifiers with the following specifications. More than one op amp may be used for each solution, if desired.

(a) $v_{out} = A_1 v_1 + A_2 v_2,$ where $0 \le A_1 \le 10$ and $0 \le A_2 \le \frac{1}{10}$.

(b) $v_{out} = A(v_1 - v_2),$ where $0 \le A \le 2$.

(c) $v_{out} = v_1 - v_2 - A v_3,$ where $10 \le A \le 20$.

(d) $v_{out} = A_1(v_1 + 3v_2) + A_2 v_3,$ where $0 \le A_1 \le 4$ and $-1 \le A_2 \le 1$

ADJUSTABLE-GAIN AMPLIFIERS

Fixed-gain summing amplifier design is developed in detail in Chapter 3.

Adjustable gains are best obtained with potentiometers. For potentiometer settings to have small effects on other amplifier gains and the bias current cancellation, the resistor connected to a potentiometer slider at the op amp input should be at least five times the potentiometer resistance.

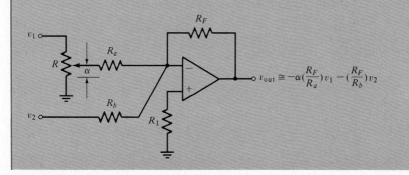

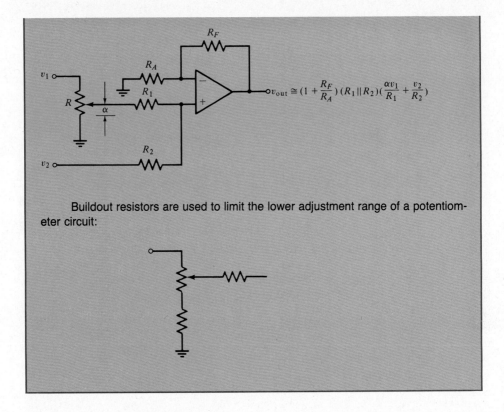

Buildout resistors are used to limit the lower adjustment range of a potentiometer circuit:

13.7 SINGLE-AMPLIFIER FILTERS

13.7.1 Buffered *RC* Networks

First order filters have transfer functions with a single pole and, at most, one zero. A first order low-pass filter has a transfer function of the form

$$T_{\text{LP}}(s) = \frac{b_0}{s + a_0}$$

where b_0 and a_0 are constants. Normally, a_0 is positive, so that the transfer function pole is in the left half of the complex plane, and the filter is stable. The constant a_0 is the *rolloff* (or *break* or *cutoff*) frequency of the filter, in rad/s. The dc gain of the filter is

$$T_{\text{LP}}(0) = b_0/a_0$$

In the circuit of Figure 13-26(a), the capacitor voltage in an *RC* network is amplified by an operational amplifier. The amplifier input is at the noninverting terminal, which to the *RC* network is virtually an open circuit. The function of the amplifier here is to present a high input resistance to the *RC* network, a low output resistance at the output terminals, and possibly some gain. In this application the amplifier is termed a *buffer amplifier*. The capacitor voltage in this circuit is given by the voltage divider rule:

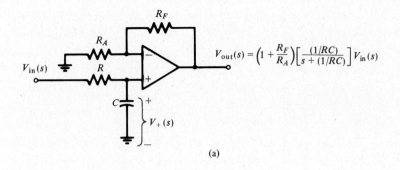

(a)

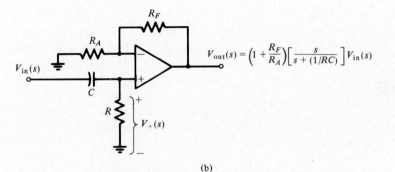

(b)

Figure 13-26 First-order filters consisting of buffered RC networks.
(a) Low-pass.
(b) High-pass.

$$V_+(s) = \frac{\dfrac{1}{sC}}{R + \dfrac{1}{sC}} V_{in}(s) = \frac{\dfrac{1}{RC}}{s + \dfrac{1}{RC}} V_{in}(s)$$

The amplifier output voltage is then

$$V_{out}(s) = \left(1 + \frac{R_F}{R_A}\right)V_+(s) = \left(1 + \frac{R_F}{R_A}\right)\left(\frac{\dfrac{1}{RC}}{s + \dfrac{1}{RC}}\right)V_{in}(s)$$

If a first order low-pass filter with a dc gain of 10 and a 1 kHz = 6280 rad/s rolloff frequency is desired, the transfer function needed is

$$T(s) = \frac{10(6280)}{s + 6280} = \left(1 + \frac{R_F}{R_A}\right)\left(\frac{\dfrac{1}{RC}}{s + \dfrac{1}{RC}}\right)$$

Choosing

$$C = 0.01 \; \mu\text{F} = 10^{-8} \; \text{F} \qquad R = \frac{10^8}{6280} = 16 \; \text{k}\Omega$$

gives

$$T(s) = \left(1 + \frac{R_F}{R_A}\right)\left(\frac{6280}{s + 6280}\right)$$

from which

$$\frac{R_F}{R_A} = 9$$

For average bias current balance

$$R_F \| R_A = R_F \| \tfrac{1}{9} R_F = \tfrac{1}{10} R_F = 16 \; \text{k}\Omega$$

For constant signals, as the bias currents are, the capacitor behaves as an open circuit. Then

$$R_F = 160 \; \text{k}\Omega \qquad R_A = 17.8 \; \text{k}\Omega$$

There are, of course, many other reasonable choices of element values.

A first order high-pass filter has a transfer function of the form

$$T_{\text{HP}}(s) = \frac{b_1 s}{s + a_0}$$

The coefficient a_0 is positive for a stable filter and is the rolloff frequency (in rad/s) of the filter. The filter's high-frequency gain is the coefficient b_1. The buffered RC network with series input capacitor in Figure 13-26(b) is a first order high-pass filter. The voltage across the resistor R, in view of the high input resistance at the op amp's noninverting input terminal, is

$$V_+(s) = \frac{R}{R + \dfrac{1}{sC}} V_{\text{in}}(s) = \frac{s}{s + \dfrac{1}{RC}} V_{\text{in}}(s)$$

This voltage is amplified to form the operational amplifier output voltage,

$$V_{\text{out}}(s) = \left(1 + \frac{R_F}{R_A}\right)\left(\frac{s}{s + \dfrac{1}{RC}}\right) V_{\text{in}}(s)$$

Suppose a first order high-pass filter is to be designed to have a high frequency gain of 10 and a 1-kHz rolloff frequency. The desired transfer function is

$$T(s) = \frac{10s}{s + 6280} = \left(1 + \frac{R_F}{R_A}\right)\left(\frac{s}{s + \dfrac{1}{RC}}\right)$$

Choosing

$$C = 0.005 \; \mu\text{F} = 5 \times 10^{-9} \; \text{F}$$

results in a realistic value for R:

$$R = \frac{10^9}{5(6280)} = 31.8 \text{ k}\Omega$$

Then

$$1 + \frac{R_F}{R_A} = 10 \qquad \frac{R_F}{R_A} = 9$$

If it is desired to balance the average bias currents,

$$R_A \| R_F = R$$

$$R_A \| 9R_A = \frac{9}{10} R_A = 31.8 \text{ k}\Omega$$

$$R_A = 35.3 \text{ k}\Omega$$

$$R_F = 9R_A = 318 \text{ k}\Omega$$

13.7.2 The Feedforward Circuit

A first-order filter circuit that does not use the op amp simply as a buffer amplifier is shown in Figure 13-27(a). This arrangement is said to have *feedforward* of the input signal because a portion of the input signal is applied directly to the op amp's noninverting input. The transfer function relating $V_{out}(s)$ and $V_{in}(s)$ may be found from the usual op amp gain formula with resistances replaced by impedances. The parallel combination of the impedances R_F and $1/sC$ is

$$Z_F = \frac{R_F\left(\dfrac{1}{sC}\right)}{R_F + \dfrac{1}{sC}} = \frac{R_F}{R_F Cs + 1}$$

The op amp gain formula with impedances gives

$$V_{out}(s) = -\frac{Z_F}{R_A} V_{in}(s) + \left(1 + \frac{Z_F}{R_A}\right)\left(\frac{R_2}{R_1 + R_2}\right) V_{in}(s)$$

$$= \left[-\frac{\dfrac{R_F}{R_A}}{R_F Cs + 1} + \left(1 + \frac{\dfrac{R_F}{R_A}}{R_F Cs + 1}\right)\left(\frac{R_2}{R_1 + R_2}\right) \right] V_{in}(s)$$

$$= \left(\frac{R_2}{R_1 + R_2}\right)\left[\frac{s + \left(\dfrac{1}{R_A C}\right)\left(\dfrac{R_A}{R_F} - \dfrac{R_1}{R_2}\right)}{s + \dfrac{1}{R_F C}} \right] V_{in}(s)$$

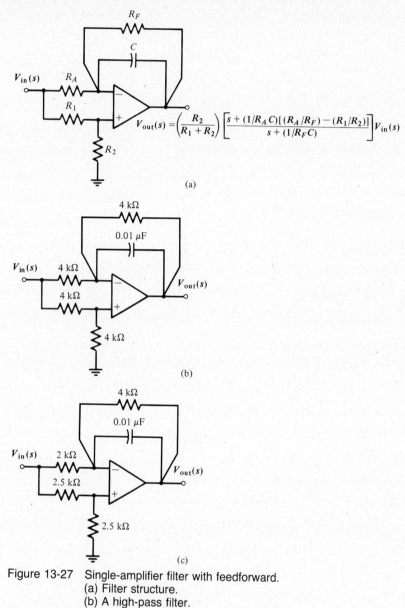

Figure 13-27   Single-amplifier filter with feedforward.
(a) Filter structure.
(b) A high-pass filter.
(c) An all-pass filter.

A high-pass filter of this type has

$$R_A R_2 = R_F R_1$$

so that

$$T(s) = \left(\frac{R_2}{R_1 + R_2}\right)\left(\frac{s}{s + \dfrac{1}{R_F C}}\right)$$

The filter example in Figure 13-27(b) has a high frequency gain of $\frac{1}{2}$ and a rolloff frequency of 4 kHz. Its transfer function is

$$T(s) = \frac{\frac{1}{2}s}{s + 25,120}$$

For an all-pass filter, choosing

$$R_F R_1 = 2 R_A R_2$$

gives a transfer function of the form

$$T(s) = \left(\frac{R_2}{R_1 + R_2}\right)\left(\frac{s - \dfrac{1}{R_F C}}{s + \dfrac{1}{R_F C}}\right)$$

The specific example of an all-pass filter in Figure 13-27(c) has transfer function

$$T(s) = \frac{\frac{1}{2}\left(s - \frac{1}{25120}\right)}{s + \frac{1}{25120}}$$

13.7.3 Second Order Filters

More complicated arrangements, involving a number of capacitors equal to the filter order, may be used to construct higher order filters using a single op amp. Some of the better second-order circuits are shown in Figure 13-28. The circuits shown are known as Sallen and Key circuits, after their inventors. The transfer functions of the filters shown are low-pass, bandpass, and high-pass, respectively and have the following transfer functions:

$$T_{\text{LP}}(s) = \frac{\dfrac{1}{R_1 R_2 C_1 C_2}}{s^2 + \dfrac{R_1 + R_2}{R_1 R_2 C_1} s + \dfrac{1}{R_1 R_2 C_1 C_2}}$$

$$T_{\text{BP}}(s) = \frac{-\dfrac{1}{R_1 C_1} s}{s^2 + \left(\dfrac{1}{R_3 C_1} + \dfrac{1}{R_3 C_2}\right)s + \dfrac{R_1 + R_2}{R_1 R_2 R_3 C_1 C_2}}$$

$$T_{\text{HP}}(s) = \frac{s^2}{s^2 + \left(\dfrac{1}{R_2 C_1} + \dfrac{1}{R_2 C_2}\right)s + \dfrac{1}{R_1 R_2 C_1 C_2}}$$

Third order Sallen and Key filter circuits, which are more complicated yet, are also occasionally used in practice.

The design of one of these filters involves choosing combinations of resistor

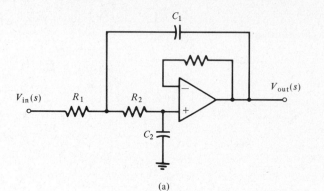

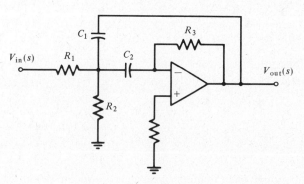

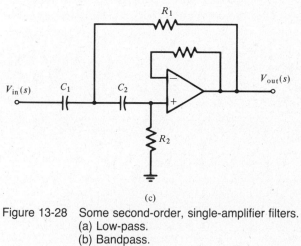

Figure 13-28 Some second-order, single-amplifier filters.
(a) Low-pass.
(b) Bandpass.
(c) High-pass.

and capacitor values to give the desired transfer function coefficients, with resistor values such that the simple op amp model is accurate. Because of the complicated dependence of the transfer function coefficients on component values, it is difficult to adjust (or *tune*) these filters.

D13-8

Design single-amplifier filters with the following transfer functions:

(a) $T(s) = \dfrac{10}{s + 2}$

  ans. A buffered *RC* network solution has $R = 50$ kΩ, $C = 10$ μF, and a non-inverting amplifier gain of 5.

(b) $T(s) = \dfrac{10s}{s + 1000}$

  ans. A buffered *RC* network solution has $R = 10$ kΩ, $C = 0.1$ μF, and a noninverting amplifier gain of 10.

(c) $T(s) = \dfrac{50s}{10s + 1}$

  ans. A buffered *RC* network solution has $R = 100$ kΩ, $C = 1$ μF, and a non-inverting amplifier gain of 5.

(d) $T(s) = \dfrac{10^6}{s + 10^6}$

  ans. A buffered *RC* network solution has $R = 10$ kΩ, $C = 100$ pF, and a noninverting amplifier gain of 10.

SINGLE-AMPLIFIER FILTERS

Simple first order low-pass and high-pass filters may be constructed with buffered *RC* networks.

  The feedforward circuit, in which the input signal is applied to both the inverting and the noninverting op amp inputs, can be designed to produce high-pass, all-pass, and more general first order transfer functions.

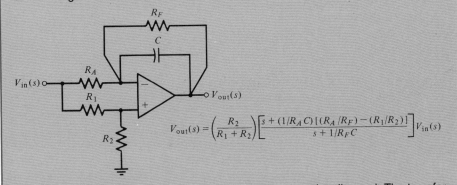

$$V_{out}(s) = \left(\frac{R_2}{R_1 + R_2}\right)\left[\frac{s + (1/R_A C)\,[(R_A/R_F) - (R_1/R_2)]}{s + 1/R_F C}\right]V_{in}(s)$$

  Higher-order filters using a single op amp are occasionally used. The transfer functions of single op amp filters generally depend in a complicated way on the individual component values.

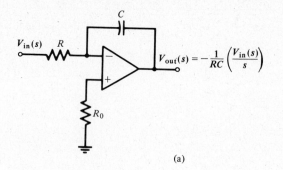

(a)

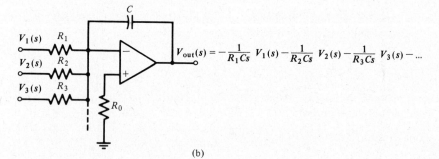

(b)

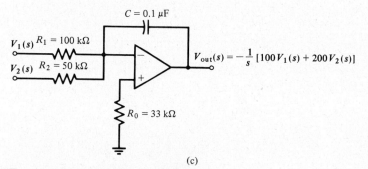

(c)

Figure 13-29 Operational integrators.
(a) Inverting integrator.
(c) Summing inverting integrator.
(c) Numerical example.

13.8 FIRST ORDER STATE VARIABLE FILTERS

13.8.1 The Operational Integrator

When the feedback resistor of an operational amplifier is replaced by a capacitor, Figure 13-29(a), an integrator results. Using impedance relations, where the capacitor impedance replaces R_F in the amplifier gain relation, the amplifier input and output voltage Laplace transforms are related by

$$V_{out}(s) = -\frac{Z_F}{R} V_{in}(s) = -\frac{1}{RsC} V_{in}(s) = -\frac{1}{RC}\left[\frac{V_{in}(s)}{s}\right]$$

The integrator has a negative gain of $-1/RC$ and produces an output voltage that is proportional to the integral of the input voltage.

A summing, inverting integrator is shown in Figure 13-29(b). For it,

$$V_{out}(s) = -\frac{Z_F}{R_1} V_1(s) - \frac{Z_F}{R_2} V_2(s) - \frac{Z_F}{R_3} V_3(s) - \cdots$$

$$= -\frac{1}{R_1 Cs} V_1(s) - \frac{1}{R_2 Cs} V_2(s) - \frac{1}{R_3 Cs} V_3(s) - \cdots$$

The specific design of Figure 13-29(c) is for a summing, inverting amplifier with Laplace transform input-output voltage relation

$$V_{out}(s) = -\frac{1}{s}[100V_1(s) + 200V_2(s)] = -\frac{1}{R_1 Cs} V_1(s) - \frac{1}{R_2 Cs} V_2(s)$$

Choosing

$$C = 0.1 \ \mu\text{F} = 10^{-9} \ \text{F}$$

then

$$\frac{1}{R_1 C} = 100 \qquad R_1 = \frac{1}{100C} = 100 \ \text{k}\Omega$$

$$\frac{1}{R_2 C} = 200 \qquad R_2 = \frac{1}{200C} = 50 \ \text{k}\Omega$$

To the constant average input bias currents, the capacitor behaves as an open circuit. For average bias current balance,

$$R_0 = R_1 \| R_2 = 100 \ \text{k} \| 50\text{k} = 33 \ \text{k}\Omega$$

Bias current balance is especially important in an integrator because a constant signal, when integrated, becomes an ever-expanding ramp signal.

Filters of the state variable type consist of interconnections of summing amplifiers and integrators.

D13-9

Design operational integrators with the following relations between the Laplace transform of the output voltage $V_{out}(s)$ and the input voltage transforms $V_1(s)$, $V_2(s)$, ... Arrange for cancellation of average input bias current effects.

(a) $V_{out}(s) = \dfrac{-V_1(s)}{100s}$

  ans. One solution has $R = 1 \ \text{M}\Omega$, $C = 100 \ \mu\text{F}$.

(b) $V_{out}(s) = -\dfrac{1}{s}\left[V_1(s) + 10V_2(s) + \left(\dfrac{1}{2}\right)V_3(s)\right]$

  ans. One solution has $R_a = 100 \ \text{k}\Omega$, $R_b = 10 \ \text{k}\Omega$, $R_c = 200 \ \text{k}\Omega$, $C = 10 \ \mu\text{F}$.

13.8.2 Low-Pass Structure

The arrangement of Figure 13-30(a) forms a first order low-pass filter of the state variable type. It may be viewed as a summing, inverting integrator with inputs through the R_1 and R_2 resistors. $V_{in}(s)$ enters at the R_1 input, and $V_{out}(s)$ is fed back through the R_2 resistor:

$$V_{out}(s) = -\frac{1}{R_1 C s} V_{in}(s) - \frac{1}{R_2 C s} V_{out}(s)$$

$$\left(1 + \frac{1}{R_2 C s}\right) V_{out}(s) = -\frac{1}{R_1 C s} V_{in}(s)$$

$$V_{out}(s) = \left(\frac{-\dfrac{1}{R_1 C}}{s + \dfrac{1}{R_2 C}}\right) V_{in}(s)$$

Alternatively, the operational amplifier gain formula may be used to derive this filter's transfer function, using impedances in place of resistances. The parallel

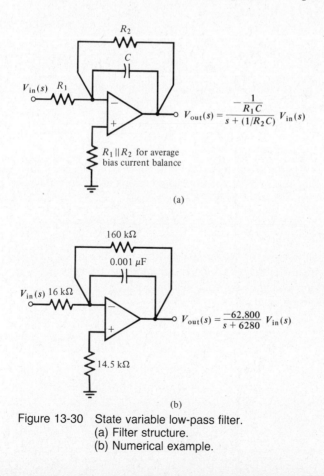

$$V_{out}(s) = \frac{-\dfrac{1}{R_1 C}}{s + (1/R_2 C)} V_{in}(s)$$

$R_1 \| R_2$ for average
bias current balance

(a)

$$V_{out}(s) = \frac{-62,800}{s + 6280} V_{in}(s)$$

(b)

Figure 13-30   State variable low-pass filter.
(a) Filter structure.
(b) Numerical example.

combination of R_2 and C has impedance

$$Z_F = \frac{R_2/sC}{R_2 + (1/sC)} = \frac{R_2}{R_2sC + 1}$$

so

$$V_{out}(s) = -\frac{Z_F}{R_1} V_{in}(s) = \frac{-(R_2/R_1)}{R_2sC + 1} V_{in}(s) = \frac{-(1/R_1C)}{s + (1/R_2C)} V_{in}(s)$$

A first order low-pass filter with a dc gain of -10 and 1-kHz rolloff frequency has transfer function

$$T(s) = \frac{-10(6280)}{s + 6280} = \frac{-\dfrac{1}{R_1C}}{s + \dfrac{1}{R_2C}}$$

Choosing

$$C = 0.001 \ \mu F$$

gives

$$R_1 = \frac{10^9}{62,800} = 16 \ k\Omega \qquad R_2 = \frac{10^9}{6280} = 160 \ k\Omega$$

which are of acceptable size. For average bias current balance,

$$R_3 = R_1 \| R_2 = 14.5 \ k\Omega$$

This design is shown in Figure 13-30(b).

D13-10

Design first order low-pass filters with the state variable structure to meet the following specifications. As there are many different design solutions, the answers given here are only the transfer functions to be realized.

(a) Rolloff frequency 15 radians per second and dc gain $-\frac{1}{2}$.

ans. $T(s) = -7.5/(s + 15)$

(b) Rolloff frequency 500 radians per second and dc gain with magnitude 40.

ans. $T(s) = \pm 20,000/(s + 500)$

(c) Rolloff frequency 50 Hz and dc gain -10.

ans. $T(s) = -3140/(s + 314)$

(d) Rolloff frequency 7 kHz.

ans. $T(s) = constant/(s + 43,960)$

13.8.3 Other First Order Filters

First order state variable filters having transfer functions with first order numerator polynomials are constructed by summing a linear combination of the low-pass filtered input signal and the input signal itself, as in Figure 13-31(a). When there is a following cascade filter stage, the summing amplifier may often be eliminated by performing the summation in the following filter stage.

As an example, consider a high-pass filter with transfer function

$$T(s) = \frac{10s}{s + 2500} = 10 + 10\left(\frac{-2500}{s + 2500}\right)$$

This transfer function can be considered to be the summation of 10 times the input plus 10 times the signal produced by passing the input through the low-pass filter with transfer function

$$T_{LP}(s) = \frac{-2500}{s + 2500} = \frac{-(1/R_1 C)}{s + (1/R_1 C)}$$

The component low-pass filter is designed to have a unity dc gain magnitude so that small input signals will produce voltages of comparable amplitude at the low-pass filter's op amp output.

In the design example of Figure 13-31(b), the choices

$$C = 0.01 \ \mu\text{F} \qquad R_1 = 40 \ \text{k}\Omega$$

are made. Then for average bias current balance in the low-pass filter op amp,

$$R_0 = R_1 \| R_1 = 20 \ \text{k}\Omega$$

The summing amplifier is to have two gain-of-ten inputs:

$$V_{out}(s) = 10V_{in}(s) + 10V_{LP}(s) = 10\left(1 + \frac{-2500}{s + 2500}\right)V_{in}(s)$$

For this, the choice

$$R_3 = R_4 = 10 \ \text{k}\Omega$$

gives

$$V_{out}(s) = \left(1 + \frac{R_F}{R_A}\right)(10\text{k}\|10\text{k})\left[\frac{V_{in}(s)}{10\text{k}} + \frac{V_{LP}(s)}{10\text{k}}\right]$$

$$= \left(1 + \frac{R_F}{R_A}\right)\left[\frac{1}{2}V_{in}(s) + \frac{1}{2}V_{LP}(s)\right]$$

from which

$$\frac{R_F}{R} = 19$$

and

$$R_F = 95 \ \text{k}\Omega \qquad R_A = 5 \ \text{k}\Omega$$

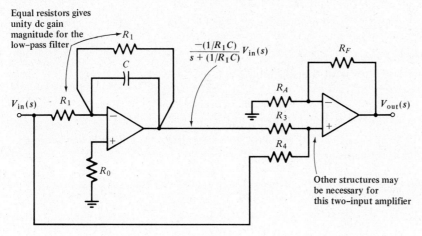

(a)

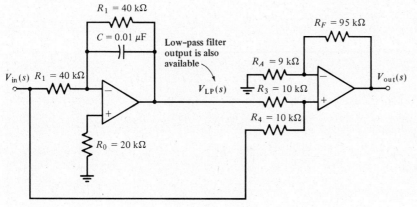

(b)

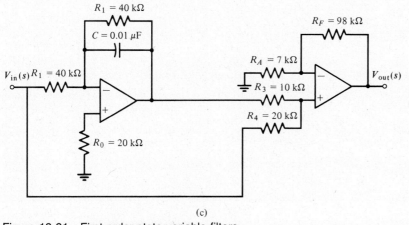

(c)

Figure 13-31   First order state variable filters.
(a) General structure.
(b) High-pass filter.
(c) All-pass filter.

gives

$$R_F \| R_A \; = \; R_3 \| R_4$$

average bias current balance.

An all-pass filter with transfer function

$$T(s) \; = \; \frac{5(s \; - \; 2500)}{s \; + \; 2500} \; = \; 5 \; + \; 10\!\left(\frac{-\,2500}{s \; + \; 2500}\right)$$

is shown in Figure 13-31(c). The previous low-pass filter component is used, but the summing amplifier gains are changed so that

$$V_{\text{out}}(s) \; = \; 5V_{\text{in}}(s) \; + \; 10V_{\text{LP}}(s) \; = \; \left(1 \; + \; \frac{R_F}{R_A}\right)(R_3 \| R_4)\!\left[\frac{V_{\text{LP}}(s)}{R_3} \; + \; \frac{V_{\text{in}}(s)}{R_4}\right]$$

$$= \; (15)\!\left[\frac{2}{3}\,V_{\text{LP}}(s) \; + \; \frac{1}{3}\,V_{\text{in}}(s)\right] \; = \; 10\!\left[-\frac{2500}{s \; + \; 2500}\,V_{\text{in}}(s)\right] \; + \; 5V_{\text{in}}(s)$$

$$= \; \frac{5(s \; - \; 2500)}{s \; + \; 2500}\,V_{\text{in}}(s)$$

D13-11

Design first order state variable filters to meet the following specifications. As there are many different design solutions, the answers given here are only the transfer functions to be realized.

(a) High-pass, with rolloff frequency 400 radians per second and high frequency gain $\frac{1}{4}$.

ans. $T(s) \; = \; \dfrac{0.25\,s}{s \; + \; 400}$

(b) All-pass, with critical frequency 20 rad/s and dc gain -5.

ans. $T(s) \; = \; \dfrac{5s \; - \; 100}{s \; + \; 20}$

(c) High-pass, with rolloff frequency 35 Hz and high frequency gain magnitude 2.

ans. $T(s) \; = \; \dfrac{\pm 2s}{s \; + \; 219.8}$

(d) All-pass, with critical frequency 2.5 kHz and unity dc gain.

ans. $T(s) \; = \; \dfrac{15{,}700 \; - \; s}{s \; + \; 15{,}700}$

13.8.4 Adjustable First Order Filters

A major advantage of the state variable filter structure is the ease with which adjustable filters may be designed and constructed. Figure 13-32 illustrates the use of

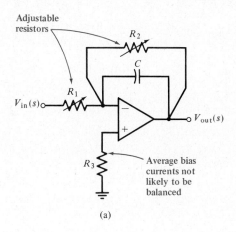

(a)

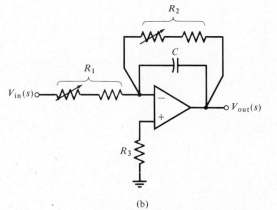

(b)

Figure 13-32   Adjustable resistor values for filter tuning.
(a) Full-range resistor adjustments.
(b) Limited resistor adjustment.

adjustable resistors, in whole or in part, for R_1 and R_2 in a state variable low-pass filter. Limited adjustment of these resistances is often provided in fixed coefficient filter circuits to compensate for inaccuracies in components. When it is desired to construct a user-adjustable filter, potentiometers are usually used because, with them, one may obtain proportional adjustments with negligible interaction with other parameters.

Figure 13-33(a) shows how a single potentiometer is used to vary the dc gain of a state variable low-pass filter without significantly affecting the filter's rolloff frequency. For small potentiometer loading effect, the series input resistor R_1 should be at least five times the potentiometer resistance R.

The arrangement of Figure 13-33(b) results in an adjustable denominator coefficient in the transfer function:

$$T(s) = \frac{-(1/R_1C)}{s + (\alpha/R_2C)} = \frac{b_0}{s + a_0}$$

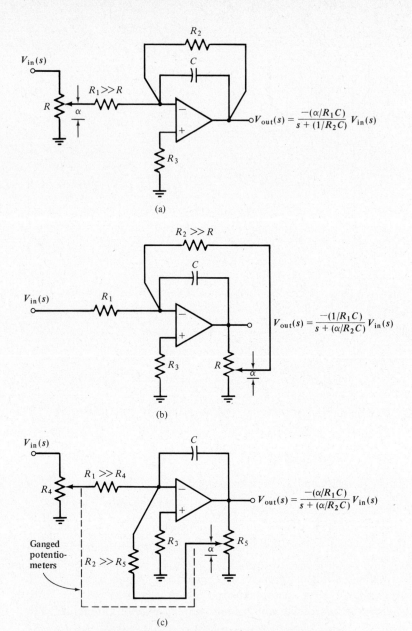

Figure 13-33   Adjustable low-pass filters using potentiometers.
(a) DC gain adjustment.
(b) Rolloff frequency adjustment with effect on dc gain.
(c) Constant dc gain and adjustable rolloff frequency.
(d) Design example.

The rolloff frequency is thereby adjustable, but as the filter's dc gain is the ratio

$$b_0/a_0 = -\frac{R_2}{R_1 \alpha}$$

Figure 13-33   (*Continued*)

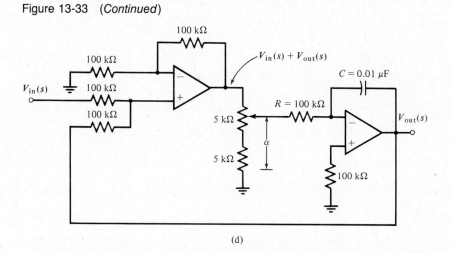

(d)

it, too, varies with this adjustment. Adjusting both the integrator input signals proportionally, using a *ganged* set of two potentiometers sharing the same mechanical shaft as indicated in Figure 13-33(c), results in a design where the rolloff frequency is adjustable, while the dc gain remains constant.

Another, related arrangement is given in Figure 13-33(d), where the two signals that are summed and integrated in the single op amp state variable filter are summed with a separate amplifier, passed through a single potentiometer, then integrated. For this circuit,

$$V_{out}(s) = -\frac{\alpha}{RCs}[V_{in}(s) + V_{out}(s)] = \frac{-100\alpha}{s}[V_{in}(s) + V_{out}(s)]$$

$$V_{out}(s) = \frac{-100\alpha}{s + 100\alpha} V_{in}(s)$$

A limited range of adjustment,

$$0.5 \leq \alpha \leq 1$$

is used, giving a rolloff frequency that is adjustable between 50 and 100 rad/s, while maintaining constant dc gain.

An adjustable high-pass filter that uses this adjustable low-pass filter as a component is shown in Figure 13-34. The high-pass filter's transfer function is given by

$$V_{out}(s) = \left(1 + \frac{R_F}{R_A}\right)(R_1\|R_2)\left[\frac{1}{R_1}\left(\frac{-100\alpha}{s + 100\alpha}\right)V_{in}(s) + \frac{1}{R_1} V_{in}(s)\right]$$

$$= (5)\left(\frac{1}{2}\right)\left(\frac{-100\alpha}{s + 100\alpha} + 1\right)V_{in}(s) = \frac{(5/2)s}{s + 100\alpha} V_{in}(s)$$

where

$$0.5 \leq \alpha \leq 1$$

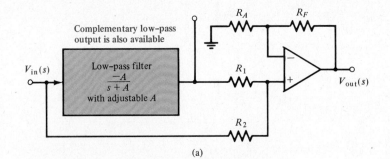

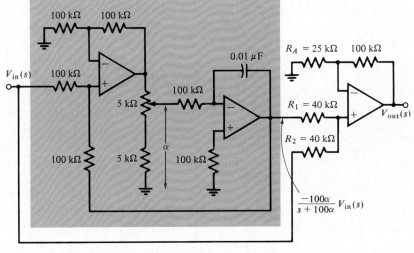

(b)

Figure 13-34 An adjustable high-pass filter.
(a) Filter structure.
(b) Final design.

D13-12

Design adjustable filters with the following transfer functions:

(a) $T(s) = \dfrac{3As}{s + 10}$ where $0 \le A \le 1$

(b) $T(s) = \dfrac{-15A}{s + 50A}$ where $0.5 \le A \le 1$

(c) $T(s) = \dfrac{10s - 100A}{s + 10A}$ where $0.8 \le A \le 1$

(d) $T(s) = \dfrac{1000A_1 A_2}{s + 100A_2}$ where $0 \le A_1 \le 1$ and $0.4 \le A_2 \le 1$

FIRST ORDER STATE VARIABLE FILTERS

The summing operational integrator produces an output voltage that is (ideally) the integral of the sum, with negative gains, of the applied input voltages.

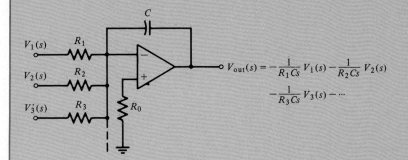

$$V_{out}(s) = -\frac{1}{R_1 Cs} V_1(s) - \frac{1}{R_2 Cs} V_2(s)$$
$$-\frac{1}{R_3 Cs} V_3(s) - \cdots$$

State variable filters consist of interconnected summing amplifiers and operational integrators. The first order state variable low-pass filter consists of a summing integrator with feedback. Other first order state variable filters are constructed by summing the input signal with the low-pass filter output. When filter stages are cascaded, this summation can often be performed in the following stage rather than with a separate amplifier.

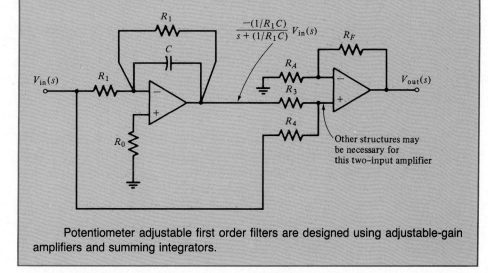

Potentiometer adjustable first order filters are designed using adjustable-gain amplifiers and summing integrators.

13.9 SECOND ORDER STATE VARIABLE FILTERS

13.9.1 Biquadratic Filter Structure and Scaling

Second order filters have transfer functions with two poles and up to two zeros:

$$T(s) = \frac{b_2 s^2 + b_1 s + b_0}{s^2 + a_1 s + a_0}$$

When filter poles are complex, they must be realized together, so second order filters are important components of higher order filters.

For a set of complex poles, it is helpful to express a second order transfer function denominator polynomial in the form

$$T(s) = \frac{\text{numerator}}{s^2 + \left(\dfrac{1}{Q}\right)\omega_r s + \omega_r^2}$$

where ω_r is the resonant frequency of the filter (in rad/s), and Q is the filter's quality factor. These terms are consistent with the definitions of resonant frequency and Q of series and parallel RLC networks. The admittance of a series RLC network, for instance, is

$$Y(s) = \frac{1}{Z(s)} = \frac{1}{sL + R + \dfrac{1}{sC}}$$

$$= \frac{\dfrac{1}{L}s}{s^2 + \dfrac{R}{L}s + \dfrac{1}{LC}}$$

The resonant frequency (in rad/s) is

$$\omega_r = \sqrt{\frac{1}{LC}}$$

and

$$Q = \frac{\omega_r L}{R} = \frac{L}{R\sqrt{LC}}$$

The filter of Figure 13-35(a) involves two integrators and a summing amplifier in a feedback arrangement called a *biquadratic* structure. Three different op amp output voltages of this second order state variable filter are indicated. One provides high-pass, another bandpass, and the other low-pass filtering. Representative frequency response amplitude ratios are shown in Figure 13-35(b).

The filter voltages are related as follows:

$$\begin{cases} V_{\text{LP}}(s) = -\dfrac{1}{R_2 C_2 s} V_{\text{BP}}(s) = -\dfrac{A_1}{s} V_{\text{BP}}(s) \\[3mm] V_{\text{BP}}(s) = -\dfrac{1}{R_1 C_1 s} V_{\text{HP}}(s) = -\dfrac{A_2}{s} V_{\text{HP}}(s) \\[3mm] V_{\text{HP}}(s) = -\dfrac{R_F}{R_a} V_{\text{in}}(s) - \dfrac{R_F}{R_b} V_{\text{LP}}(s) + \left(1 + \dfrac{R_F}{R_a \| R_b}\right)\left(\dfrac{R_4}{R_3 + R_4}\right) V_{\text{BP}}(s) \end{cases}$$

$$= -A_3 V_{\text{in}}(s) - A_4 V_{\text{LP}}(s) + A_5 V_{\text{BP}}(s)$$

Eliminating the variables V_{BP} and V_{LP} among the three equations gives

$$V_{HP}(s) = -A_3 V_{in}(s) - A_4\left(-\frac{A_1}{s}\right)\left[-\frac{A_2}{s}V_{HP}(s)\right] + A_5\left[-\frac{A_2}{s}V_{HP}(s)\right]$$

$$\left(1 + \frac{A_1 A_2 A_4}{s^2} + \frac{A_2 A_5}{s}\right)V_{HP}(s) = -A_3 V_{in}(s)$$

$$V_{HP}(s) = \frac{-A_3 s^2}{s^2 + A_2 A_5 s + A_1 A_2 A_4}V_{in}(s) = T_{HP}(s)V_{in}(s)$$

Solving for the bandpass output voltage,

$$V_{BP}(s) = -\frac{A_2}{s}V_{HP}(s) = \frac{A_2 A_3 s}{s^2 + A_2 A_5 s + A_1 A_2 A_4}V_{in}(s) = T_{BP}(s)V_{in}(s)$$

and similarly

$$V_{LP}(s) = -\frac{A_1}{s}V_{BP}(s) = \frac{-A_1 A_2 A_3}{s^2 + A_2 A_5 s + A_1 A_2 A_4}V_{in}(s) = T_{LP}(s)V_{in}(s)$$

It is important to keep each op amp's output voltage properly scaled, so that all of the op amps have approximately the same maximum output amplitude for a given input voltage amplitude. For a biquadratic filter with relatively high Q, the largest filter amplitude ratios occur approximately at the resonant frequency

$$\omega_r = \sqrt{A_1 A_2 A_4}$$

At this frequency, the transfer functions from the input to the op amp outputs have magnitudes

$$|T_{HP}(s = j\omega_r)| = \frac{A_3 \sqrt{A_1 A_2 A_4}}{A_2 A_5}$$

$$|T_{BP}(s = j\omega_r)| = \frac{A_3}{A_5}$$

$$|T_{LP}(s = j\omega_r)| = \frac{A_1 A_3}{A_5 \sqrt{A_1 A_2 A_4}}$$

Equating these three quantities so that, approximately, the maximum signal amplitudes at the three op amp outputs will be equal gives

$$A_1 = A_2 \qquad A_4 = 1$$

The scaled biquadratic filter is shown in Figure 13-35(c). For it, the two integrators are made to be identical, so that $A_1 = A_2$, and the summing amplifier gain of $V_{LP}(s)$ is made -1. The transfer functions of the scaled biquadratic filter are

$$T_{HP}(s) = \frac{-A_3 s^2}{s^2 + A_1 A_5 s + A_1^2}$$

$$T_{BP}(s) = \frac{A_1 A_3 s}{s^2 + A_1 A_5 s + A_1^2}$$

$$T_{LP}(s) = \frac{-A_1^2 A_3}{s^2 + A_1 A_5 s + A_1^2}$$

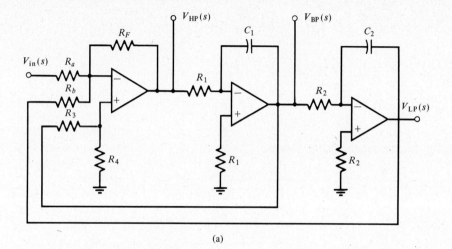

(a)

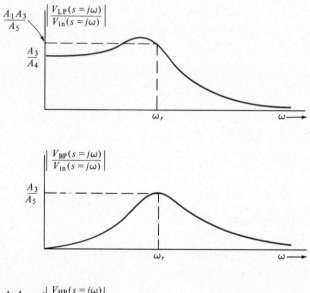

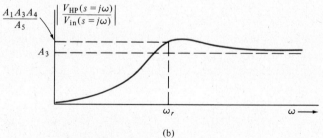

(b)

Figure 13-35 The biquadratic filter.
    (a) Biquadratic structure in general.
    (b) Typical frequency response amplitude ratios.
    (c) Scaled structure.

Figure 13-35 *(Continued)*

$$V_{LP}(s) = -\frac{1}{R_1Cs} V_{BP}(s) \quad V_{BP}(s) = -\frac{A_1}{s} V_{BP}(s)$$

$$V_{HP}(s) = -A_3 V_{in}(s) - V_{LP}(s) + A_5 V_{BP}(s) \quad V_{BP}(s) = -\frac{1}{R_1Cs} V_{HP}(s) = -\frac{A_1}{s} V_{HP}(s)$$

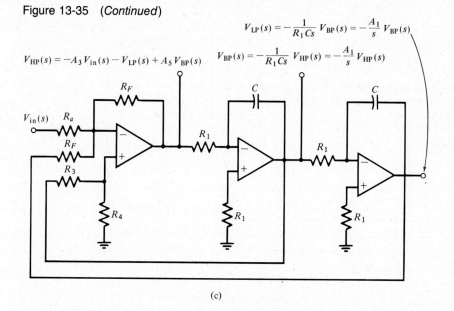

(c)

where the inverting integrators each have gain

$$-A_1 = -\frac{1}{R_1C}$$

and where the summing amplifier has gains given by

$$V_{HP}(s) = -A_3V_{in}(s) - V_{LP}(s) + A_5V_{BP}(s)$$

Suppose that it is desired to obtain a second order bandpass filter with the specific transfer function

$$T_{BP}(s) = \frac{s}{s^2 + s + 100}$$

Using the scaled biquadratic state variable structure, a bandpass filter has transfer function of the form

$$T_{BP}(s) = \frac{A_1A_3}{s^2 + A_1A_5s + A_1^2}$$

from which

$$A_1^2 = 100 \qquad\qquad A_1 = 10$$

$$A_1A_5 = 10A_5 = 1 \qquad A_5 = 0.1$$

$$A_1A_3 = 10A_3 = 1 \qquad A_3 = 0.1$$

Each of the integrators for the completed design in Figure 13-36 has negative gain

$$-\frac{1}{R_1C} = -10 = -A_1$$

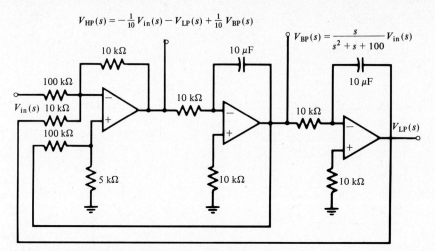

$$V_{\mathrm{HP}}(s) = -\frac{1}{10}V_{\mathrm{in}}(s) - V_{\mathrm{LP}}(s) + \frac{1}{10}V_{\mathrm{BP}}(s)$$

$$V_{\mathrm{BP}}(s) = \frac{s}{s^2 + s + 100}V_{\mathrm{in}}(s)$$

Figure 13-36 Bandpass filter design example.

and the summing amplifier has input-output relation

$$
\begin{aligned}
V_{\mathrm{HP}}(s) &= -\frac{10k}{100k}V_{\mathrm{in}}(s) - \frac{10k}{10k}V_{\mathrm{LP}}(s) + \left(1 + \frac{10k}{100k\|10k}\right)\left(\frac{5k}{100k + 5k}\right)V_{\mathrm{BP}}(s) \\
&= -0.1V_{\mathrm{in}}(s) - V_{\mathrm{LP}}(s) + 0.1V_{\mathrm{BP}}(s) \\
&= -A_3V_{\mathrm{in}}(s) - V_{\mathrm{LP}}(s) + A_5V_{\mathrm{BP}}(s)
\end{aligned}
$$

as required.

D13-13

Design biquadratic filters with the following transfer functions:

(a) $T(s) = \dfrac{-17}{s^2 + 2s + 17}$

(b) $T(s) = \dfrac{-s}{(s + 10)^2}$

(c) $T(s) = \dfrac{100s}{4s^2 + 2s + 2}$

(d) $T_1(s) = \dfrac{-s^2}{s^2 + 4s + 13}$

and

$T_2(s) = \dfrac{-13}{s^2 + 4s + 13}$

13.9.2 Adjustable Designs

When the gains of a biquadratic state variable filter are made adjustable, the resulting filters have adjustable transfer functions. There are a variety of important ways that adjustments can be arranged.

One type of adjustable filter is shown in Figure 13-37(a). For it, the gains A_1, A_3, and A_5 are each separately adjustable with potentiometers. As A_1 is the magnitude of the gain of each of the two integrators in the scaled filter, two potentiometers, mechanically ganged on the same shaft, are used for the A_1 adjustment. The bandpass output of this filter, for example, is related to the input by

$$V_{\mathrm{BP}}(s) = \frac{A_1 A_3 s}{s^2 + A_1 A_5 s + A_1^2} V_{\mathrm{in}}(s) = \frac{A_3 \omega_r s}{s^2 + \left(\dfrac{1}{Q}\right)\omega_r s + \omega_r^2} V_{\mathrm{in}}(s)$$

where

$$A_1 = \omega_r \qquad A_5 = 1/Q$$

In this arrangement, the ganged potentiometers that control A_1 directly adjust the resonant frequency ω_r. The A_5 gain pot directly adjusts the inverse of the filter Q. The filter's gain at resonance is

$$|T_{\mathrm{BP}}(s = j\omega_r)| = \frac{|A_3|\omega_r^2}{\left(\dfrac{1}{Q}\right)\omega_r^2} = Q|A_3|$$

which for a fixed Q can be set as desired by adjusting A_3. Q adjustment does not affect ω_r, but it does change the gain at resonance.

In some applications, different adjustments are desirable. For example, in the filter of Figure 13-37(b), ganged potentiometers are used to make adjustments of the gains A_3 and A_5 proportional to one another. Letting

$$A_3 = \alpha A_3' \qquad A_5 = \alpha A_5'$$

where α is the common potentiometer ratio, the resulting bandpass filter transfer function is

$$T_{\mathrm{BP}}(s) = \frac{\alpha A_1 A_3' s}{s^2 + \alpha A_1 A_5' s + A_1^2}$$

As before, the adjustable gain A_5 controls $1/Q$ without affecting ω_r. The filter's gain at resonance is now

$$|T_{\mathrm{BP}}(s = j\omega_r)| = \left|\frac{A_3'}{A_5'}\right|$$

which is now independent of the Q adjustment.

The bandwidth of a second order bandpass filter is

$$BW = \frac{\omega_r}{Q}$$

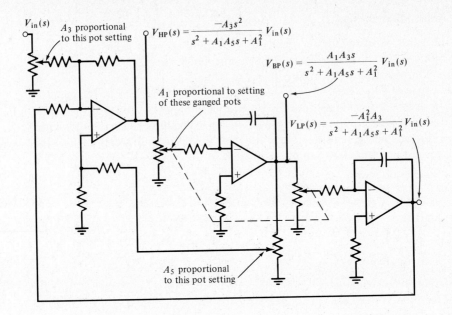

$$V_{HP}(s) = \frac{-A_3 s^2}{s^2 + A_1 A_5 s + A_1^2} V_{in}(s)$$

$$V_{BP}(s) = \frac{A_1 A_3 s}{s^2 + A_1 A_5 s + A_1^2} V_{in}(s)$$

$$V_{LP}(s) = \frac{-A_1^2 A_3}{s^2 + A_1 A_5 s + A_1^2} V_{in}(s)$$

A_3 proportional to this pot setting

A_1 proportional to setting of these ganged pots

A_5 proportional to this pot setting

(a)

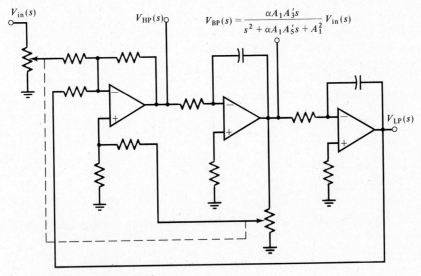

$$V_{BP}(s) = \frac{\alpha A_1 A_3' s}{s^2 + \alpha A_1 A_5' s + A_1^2} V_{in}(s)$$

A_3 and A_5 each proportional to ganged pot setting

(b)

Figure 13-37   Adjustable biquadratic filters.
(a) Adjustable gain coefficients.
(b) Adjustable Q, fixed gain and frequency.
(c) Adjustable frequency, fixed gain and bandwidth.

Figure 13-37   *(Continued)*

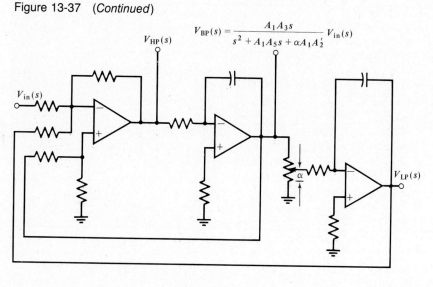

$$V_{BP}(s) = \frac{A_1 A_3 s}{s^2 + A_1 A_5 s + \alpha A_1 A_2'} V_{in}(s)$$

(c)

a definition that is consistent with the bandwidth definition for series and parallel *RLC* networks. The potentiometer in the circuit of Figure 13-37(c) controls the filter's resonant frequency, while filter bandwidth is held constant. Because both of the integrators do not have identical gains, $A_1 \neq A_2$, this biquadratic filter is no longer scaled. Instead, the bandpass transfer function is of the form

$$T(s) = \frac{A_1 A_3 s}{s^2 + A_1 A_5 s + A_1 A_2} = \frac{A_1 A_3 s}{s^2 + A_1 A_5 s + \alpha A_1 A_2'}$$

where $A_2 = \alpha A_2'$.

The resonant frequency of this filter is

$$\omega_r = \sqrt{\alpha A_1 A_2'}$$

so the potentiometer adjustment of ω_r is proportional to the square root of the fraction α. The filter's gain at resonance,

$$|T(s = j\omega_r)| = \left| \frac{A_3}{A_5} \right|$$

is not affected by α, but

$$Q = \frac{\omega_r}{A_1 A_5} = \frac{\sqrt{\alpha A_1 A_2'}}{A_1 A_5}$$

varies in proportion to ω_r, so that the filter bandwidth remains constant:

$$BW = \frac{\omega_r}{Q} = A_1 A_5$$

D13-14

Design adjustable biquadratic filters with the following transfer functions:

(a) $T(s) = \dfrac{5As}{s^2 + 2As + 17}$ where $0 \le A \le 1$

(b) $T(s) = \dfrac{-100A}{s^2 + 2s + 100A}$ where $0.5 \le A \le 1$

(c) $T(s) = \dfrac{s^2}{s^2 + As + 10A^2}$ where $0.9 \le A \le 1$

(d) $T(s) = \dfrac{5A_1A_2s}{s^2 + 0.2A_2s + 1}$ where $0 \le A_1 \le 1$ and $0.3 \le A_2 \le 1$

13.9.3 All-Pass, Bandstop, and Other Filters

Summing the three outputs of a biquadratic filter in various amounts, Figure 13-38(a), will result in a filter with any desired second order transfer function. As an example, the transfer function

$$T(s) = \frac{s^2 + 9}{s^2 + 2s + 9}$$

that of a bandstop filter with notch frequency at $\omega = 3$, will be designed.

A scaled biquadratic filter with transfer functions

$$T_{\mathrm{HP}}(s) = \frac{-A_3 s^2}{s^2 + A_1 A_5 s + A_1^2} = \frac{-s^2}{s^2 + 2s + 9}$$

$$T_{\mathrm{BP}}(s) = \frac{A_1 A_3 s}{s^2 + A_1 A_5 s + A_1^2} = \frac{3s}{s^2 + 2s + 9}$$

$$T_{\mathrm{LP}}(s) = \frac{-A_1^2 A_3}{s^2 + A_1 A_5 s + A_1^2} = \frac{-9}{s^2 + 2s + 9}$$

is first designed. For this biquadratic filter, Figure 13-38(b),

$$A_1 = \frac{1}{R_1 C} = 3$$

$$A_3 = \frac{R_F}{R_a} = 1$$

$$A_5 = \left(1 + \frac{R_F}{R_a \| R_F}\right)\left(\frac{R_4}{R_3 + R_4}\right) = \frac{2}{3}$$

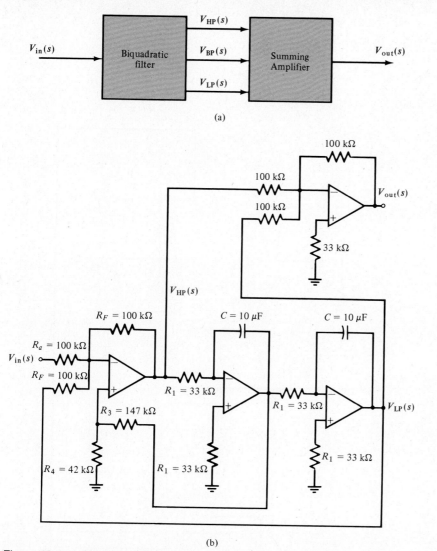

(b)

Figure 13-38 Obtaining other second-order filters, using a biquadratic section.
(a) A general structure.
(b) Numerical bandstop filter example.

The choices

$$R_1 = 33 \text{ k}\Omega$$

$$C = 10 \text{ μF}$$

$$R_F = R_a = 100 \text{ k}\Omega$$

$$R_3 = 147 \text{ k}\Omega$$

$$R_4 = 42 \text{ k}\Omega$$

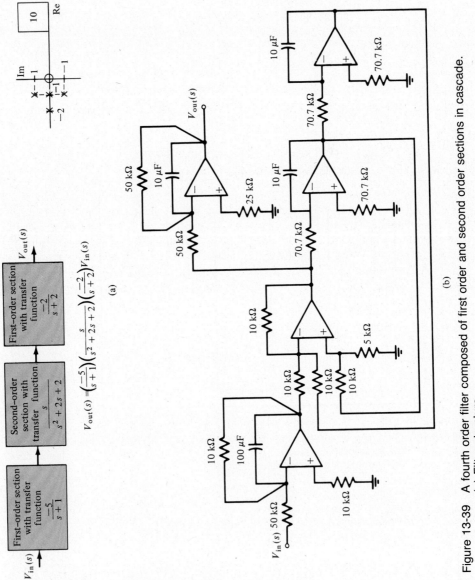

Figure 13-39 A fourth order filter composed of first order and second order sections in cascade.
(a) Filter structure.
(b) Complete design.

are made, giving

$$V_{HP}(s) = \frac{-s^2}{s^2 + 2s + 9} V_{in}(s)$$

$$V_{BP}(s) = \frac{3s}{s^2 + 2s + 9} V_{in}(s)$$

$$V_{LP}(s) = \frac{-9}{s^2 + 2s + 9} V_{in}(s)$$

To obtain

$$V_{out}(s) = \frac{s^2 + 9}{s^2 + 2s + 9} V_{in}(s) = -V_{HP}(s) + 0 \cdot V_{BP}(s) - V_{LP}(s)$$

an additional summing amplifier, with two gains of -1, is used.

The design of a fourth order filter with transfer function

$$T(s) = \frac{10s}{(s + 1)(s + 2)(s^2 + 2s + 2)}$$

is shown in Figure 13-39. A cascade connection of first and second order filter sections is used. There are, of course, many ways to decompose most transfer functions into products of terms. Complex root terms must be realized together, however, and it is helpful to distribute the transfer function zeros and the algebraic signs of the terms to give the simplest individual section designs. The choice made here is the decomposition

$$T(s) = \left(\frac{-5}{s + 1}\right)\left(\frac{s}{s^2 + 2s + 2}\right)\left(\frac{-2}{s + 2}\right)$$

which involves two first order low-pass filter sections and a second order bandpass filter with complex conjugate poles. The specific order of these sections shown in Figure 13-39(a) is chosen.

The individual filter sections, shown connected in cascade in Figure 13-39(b), are designed by using the methods of the previous sections.

D13-15

Design biquadratic filters to meet the following specifications. As there are many different design solutions, the answers given here are only the transfer functions to be realized.

(a) All-pass, with gain -0.5, $Q = 5$, and critical frequency 20 radians per second.

$$\text{ans. } T(s) = \frac{-0.5(s^2 - 4s + 400)}{s^2 + 4s + 400}$$

(b) Bandstop, with notch frequency 60 Hz, unity dc gain, and $Q = 20$.

$$\text{ans. } T(s) = \frac{s^2 + 142,130}{s^2 + 18.85s + 142,130}$$

(c) Adjustable all-pass, with gain $= 10$, $Q = 10$, and critical frequency adjustable from 1 kHz to 2.5 kHz.

$$\text{ans. } T(s) = \frac{10(s^2 - 1570ks + 2.47 \times 10^8 k^2)}{s^2 + 1570ks + 2.47 \times 10^8 k^2} \quad \text{with } 0.4 \leq k \leq 1$$

SECOND ORDER STATE VARIABLE FILTERS

The biquadratic state variable filter arrangement provides second order high-pass, bandpass, and low-pass filters. It is important to scale the internal gains in a biquadratic filter so that the operational amplifier output voltages have comparable maximum amplitudes.

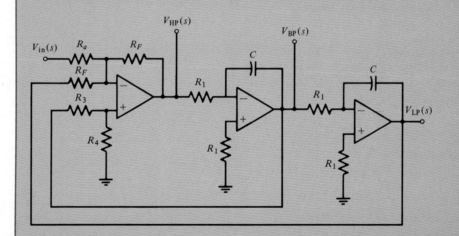

For the scaled filter,

$$V_{HP}(s) = \frac{-A_3 s^2}{s^2 + A_1 A_5 s + A_1^2} V_{in}(s)$$

$$V_{BP}(s) = \frac{A_1 A_3 s}{s^2 + A_1 A_5 s + A_1^2} V_{in}(s)$$

$$V_{LP}(s) = \frac{-A_1^2 A_3}{s^2 + A_1 A_5 s + A_1^2} V_{in}(s)$$

Adjustable filters are designed with potentiometers to vary the gains A_1, A_3, and A_5.

Summing the high-pass, bandpass, and low-pass outputs of a biquadratic filter in proper amounts gives a second order filter with any desired numerator polynomial.

CHAPTER THIRTEEN PROBLEMS

Basic Problems

Frequency Response

1. Find the frequency response amplitude ratio and phase shift functions for each of the following transfer functions:

(a) $T(s) = \dfrac{5}{s(s + 2)}$

(b) $T(s) = \dfrac{s}{s^2 + 4}$

(c) $T(s) = \dfrac{-s}{s^2 + 4s + 17}$

(d) $T(s) = \dfrac{1 + e^{-s}}{s + 3}$

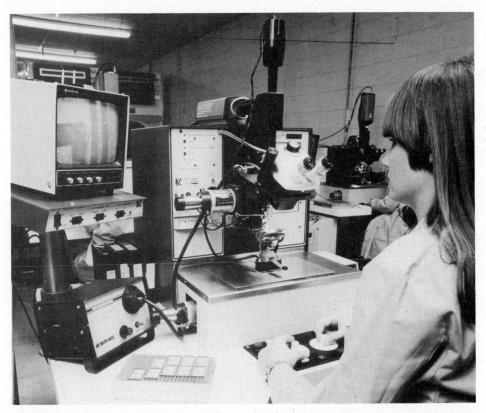

The interconnection of integrated circuits in a single assembly is done by bonding very fine gold wires from chip to chip. (*Photo courtesy of Datel–Intersil, Inc.*)

Poles and Zeros

2. Sketch pole-zero plots for each of the following transfer functions:

   (a) $T(s) = \dfrac{s - 2}{4s(s + 3)}$

   (b) $T(s) = \dfrac{2s^2 + 6}{(s + 4)^2}$

   (c) $T(s) = \dfrac{4s^2 - 1}{s^2 + 2s + 17}$

   (d) $T(s) = \dfrac{s^2 - 4s + 5}{s(s^2 + 4s + 5)}$

3. Sketch approximate frequency response plots (both amplitude ratio and phase shift versus frequency) on linear scales for the transfer functions with the following pole-zero plots:

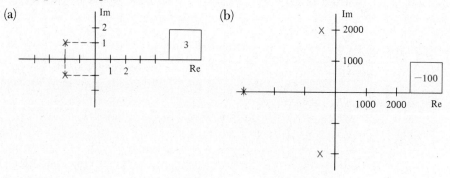

Bode Plots

4. Sketch Bode plots (both amplitude ratio in dB and phase shift) for filters with each of the following transfer functions:

   (a) $T(s) = \dfrac{s}{10s + 1}$

   (b) $T(s) = \dfrac{5s - 1}{s + 5}$

   (c) $T(s) = \dfrac{s^2 - 8s + 16}{s^2 + 8s + 16}$

   (d) $T(s) = \dfrac{s + 3}{s^2 + 9s + 18}$

   (e) $T(s) = \dfrac{1000s}{s^2 + 2s + 100}$

   (f) $T(s) = \dfrac{16}{s(s^2 + 4s + 8)^2}$

(g) $T(s) = \dfrac{s^2 + 2s + 10}{s^2 + 7s + 10}$

(h) $T(s) = \dfrac{100s^2}{(s + 1)(s + 2 + j)(s + 2 - j)}$

Filter Transformations

5. Using transformations of normalized low-pass filter transfer functions, find the transfer functions of the indicated filters:

   (a) Third order Chebyshev high-pass with 100-Hz rolloff frequency, 1-dB pass-band ripple, and 0-dB high frequency gain.

   (b) Fifth order Butterworth low-pass with dc gain 10 and rolloff frequency 5 Hz.

   (c) Fourth order Chebyshev bandpass (derived from a second order low-pass filter) with 0.1-dB passband ripple and 2 rad/s center frequency.

   (d) Fourth order Bessel high-pass with 2-kHz rolloff frequency and unity high frequency gain.

Adjustable-Gain Amplifiers

6. Design adjustable-gain amplifier circuits, using potentiometer gain adjustment, to meet the following specifications. More than one op amp may be used for each solution, if desired.

   (a) $v_{out} = A_1 v_1 + A_2 v_2$, where $0 \le A_1 \le 5$ and $2 \le A_2 \le 4$.

   (b) $v_{out} = A(3v_1 - 4v_2)$, where $0.4 \le A \le 1$.

   (c) $v_{out} = -A_1 v_1 - A_2 v_2$, where $0 \le A_1 \le 5$ and $2 \le A_2 \le 4$.

   (d) $v_{out} = A_1 v_1 - A_2(2v_1 + 3v_2)$, where $-1 \le A_1 \le 1$ and $0 \le A_2 \le 1$.

First-Order Filter Design

7. Design first order filters, of the buffered RC network type, with the following specifications:

   (a) Low-pass with a dc gain of 5 and a rolloff frequency of 20 radians per second.

   (b) High-pass with a 2-kHz rolloff frequency and a 10-dB high frequency gain.

   (c) Low-pass with 6-dB dc gain and 500-Hz rolloff frequency.

   (d) High-pass with a 10 radian per second rolloff frequency and a gain of 2 at 10 rad/s.

8. Design first order filters having the feedforward structure, with the following transfer functions:

   (a) $T(s) = \dfrac{s}{s + 3}$

   (b) $T(s) = \dfrac{10s - 1000}{s + 100}$

   (c) $T(s) = \dfrac{3s + 10^4}{s + 10^4}$

   (d) $T(s) = \dfrac{10s - 4}{5s + 1}$

9. Design first order filters of the state variable type with the following specifications:
   (a) Low-pass, with a dc gain of magnitude $\frac{1}{2}$ and a rolloff frequency of 10 kHz.
   (b) Low-pass, with a dc gain of -100 and a rolloff frequency of 0.4 Hz.
   (c) High-pass, with a high frequency gain of magnitude $\frac{1}{2}$ and a rolloff frequency of 20 Hz.
   (d) Having both a low-pass and a high-pass output, with 200-Hz rolloff frequency, and with dc gain of the low-pass filter and high frequency gain of the high-pass filter each -10.

Adjustable First Order Filters

10. Design first order filters that are adjustable with a single (not ganged) potentiometer as specified:
   (a) Low-pass, with constant dc gain of 200 and a rolloff frequency adjustable between 10 Hz and 100 Hz.
   (b) High-pass, with rolloff frequency 1 kHz and high frequency gain adjustable between -10 dB and $+10$ dB.
   (c) All-pass, with unity gain and critical frequency adjustable in the range 5–10 Hz.
   (d) Transfer function

$$T(s) = \frac{3s + 2 + A}{s + 3A}$$

   where $1 \le A \le 10$.

Second Order Biquadratic Filters

11. Design biquadratic filters with the following transfer functions:
   (a) $T(s) = \dfrac{-200}{s^2 + 10s + 1000}$

   (b) $T(s) = \dfrac{40s}{s^2 + 4s + 500}$

   (c) $T(s) = \dfrac{(1/2)s}{3s^2 + s + 40}$

   (d) $T(s) = \dfrac{-10s^2}{s^2 + 3s + 100}$

12. Design biquadratic filters to meet the following specifications.
   (a) Low-pass, with poles at $s = -1 \pm j5$ and a dc gain of 2.
   (b) Bandpass, with a resonant frequency of 100 Hz, a damping ratio of 0.3, and a gain at resonance of 4.
   (c) High-pass, with resonant frequency 5 radians per second, unity high frequency gain, and a Q of 10.
   (d) Bandpass, with resonant frequency 1.5 kHz, and bandwidth 200 Hz.

Responses of a high order filter circuit to a squarewave input are displayed on the screen of an oscilloscope as fine adjustments are made to the filter parameters. (*Photo courtesy of Tektronix, Inc.*)

Adjustable Second Order Filters

13. Design adjustable biquadratic filters with the following transfer functions:

(a) $T(s) = \dfrac{A}{s^2 + 0.4s + A}$ with $0.1 \le A \le 5$

(b) $T(s) = \dfrac{As}{s^2 + As + 20}$ with $0.1 \le A \le 10$

(c) $T(s) = \dfrac{-10A^2}{s^2 + 2As + 100A^2}$ with $5 \le A \le 20$

(d) $T(s) = \dfrac{A_1 s^2}{s^2 + A_2 s + A_2 A_3}$

   with $-5 \le A_1 \le 5$, $1 \le A_2 \le 2$, and $1 \le A_3 \le 10$

Practical Problems

Frequency Response

14. Frequency response amplitude measurements on an unknown filter have resulted in the data in Table 13-6. If all of a filter's poles and zeros are in the left half of the complex plane, and if the transfer function multiplying constant is

Table 13-6 EXPERIMENTAL FREQUENCY RESPONSE DATA

FREQUENCY (IN kHz)	INPUT AMPLITUDE (IN mV)	OUTPUT AMPLITUDE (IN mV)
0.3	5.0	10.2
0.6	5.5	8.37
1.0	5.0	5.81
1.5	4.8	3.75
3.0	4.9	2.26
4.0	4.9	1.91
5.5	6.5	2.09
8.0	12.1	3.15
10.0	10.5	2.41
12.3	10.3	2.05

positive, the filter is said to be *minimum phase*. Find a second order minimum phase transfer function model for this system.

Single-Amplifier Filters

15. The buffered *RC* network shown has a transfer function that is high-pass, with a high frequency gain that can be less than unity. Find the transfer function in terms of the element values (assuming the simple op amp model), then design a high-pass filter with transfer function

$$T(s) = \frac{0.1s}{s + 200}$$

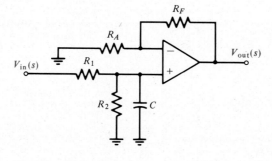

16. Design a filter of the first order feedforward type with transfer function

$$T(s) = \frac{10s - 5}{3s + 40}$$

17. Design single-amplifier filters with the following transfer functions:

(a) $T(s) = \dfrac{1000}{s^2 + 10s + 1000}$

(b) $T(s) = \dfrac{-3s}{s^2 + s + 200}$

(c) $T(s) = \dfrac{s^2}{s^2 + 300s + 4000}$

State Variable Filters

18. Find the transfer function in terms of the potentiometer fraction α:

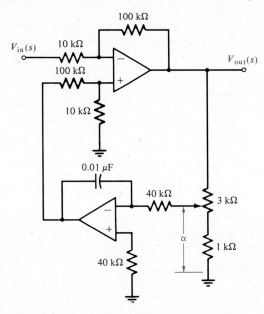

19. The first order filter shown may be adjusted with the potentiometer from low-pass to high-pass. Show that

$$V_2(s) = \alpha V_1(s) + (1 - \alpha)V_{in}(s)$$

and, in terms of the element values and the potentiometer fraction α, find the transfer function relating $V_{out}(s)$ and $V_{in}(s)$.

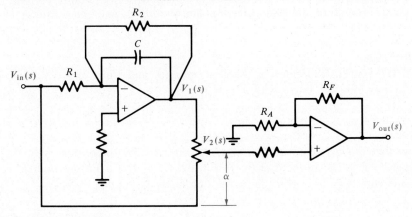

Advanced Problems

Sinusoidal Response

20. Consider a sinusoidal input

$$r(t) = A \cos (\omega t + \theta)$$

$$R(s) = \frac{(A/2)e^{j\theta}}{s - j\omega} + \frac{(A/2)e^{-j\theta}}{s + j\omega}$$

to a network or system with transfer function $T(s)$ and zero initial conditions. The Laplace transform of the output consists of the forced sinusoidal component plus the natural response:

$$Y(s) = \frac{(B/2)e^{j\phi}}{s - j\omega} + \frac{(B/2)e^{-j\phi}}{s + j\omega} + \text{other terms}$$

Using Laplace transform methods, show that

$$\frac{B}{A} = |T(s = j\omega)|$$

and that

$$\phi - \theta = \underline{/T(s = j\omega)}$$

which is the result obtained in the text with sinors.

Frequency Response
21. For the transfer function

$$T(s) = \frac{-3(s^2 + 5s + 6)}{(s + 1)(s^2 + s + 5)}$$

(a) Find three other stable transfer functions with the same amplitude ratio;
(b) Find three other stable transfer functions with the same phase shift.

Chebyshev Filters
22. Find the transfer functions of the first, second, and third order normalized Chebyshev low-pass filters with $\epsilon = 0.3$.

Active Filters
23. Show that the given op amp circuit is a *differencing integrator*, with $v_{\text{out}}(t)$ proportional to the integral of $(v_1 - v_2)$ if $R_1 C_1 = R_2 C_2$. Proper operation of this network depends upon the two RC products' being nearly equal.
(a) Design a differencing integrator for which

$$V_{\text{out}} = \frac{50}{s}[V_1(s) - V_2(s)]$$

(b) For arbitrary R_1, C_1, R_2, and C_2, find the relationship between $V_{\text{out}}(s)$, $V_1(s)$, and $V_2(s)$.

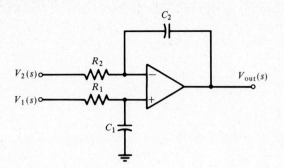

24. Design an adjustable bandstop filter with transfer function

$$T(s) = \frac{s^2 + As + 100}{s^2 + s + 100}$$

where A is adjustable (with a potentiometer) in the range between -1 and $+1$. The potentiometer is said to control the "notch depth."

Index

84 85 86 7 6 5 4 3 2 1